Image Sequence Processing and Dynamic Scene Analysis

NATO ASI Series

(Advanced Science Institutes Series)

A series presenting the results of activities sponsored by the NATO Science Committee, which aims at the dessemination of advanced scientific and technological knowledge, with a view to strengthening links between scientific communities.

The Series is published by an international board of publishers in conjunction with the NATO Scientific Affairs Division

A Life Sciences **B Physics**	Plenum Publishing Corporation London and New York
C Mathematical and Physical Sciences	D. Reidel Publishing Company Dordrecht and Boston
D Behavioural and Social Sciences **E Applied Sciences**	Martinus Nijhoff Publishers The Hague, The Netherlands
F Computer and Systems Sciences **G Ecological Sciences**	Springer-Verlag Berlin Heidelberg New York

Series F: Computer and System Sciences No. 2

Image Sequence Processing and Dynamic Scene Analysis

Edited by

T. S. Huang

Springer-Verlag Berlin Heidelberg New York Tokyo 1983

Proceedings of the NATO Advanced Study Institute on Image Sequence Processing and Dynamic Scene Analysis

Professor Dr. Thomas S. Huang
Coordinated Science Laboratory, University of Illinois, 1101 West Springfield Avenue, Urbana, IL 61801, USA

Published in corporation with NATO Scientific Affaires Division

ISBN 3-540-11997-3 Springer-Verlag Berlin Heidelberg New York Tokyo
ISBN 0-387-11997-3 Springer-Verlag New York Heidelberg Berlin Tokyo

Printed in Germany

Printing: Beltz Offsetdruck, Hemsbach; Bookbinding: J. Schäffer OHG, Grünstadt
2145/3140-543210

PREFACE

This volume contains the proceedings of the NATO Advanced Study Institute on "Image Sequence Processing and Dynamic Scene Analysis" held 21 June - 2 July, 1982 in Hotel Maritim, Braunlage/Harz, Federal Republic of Germany. The organizing committee of the institute consists of T.S. Huang (Director), H.G. Musmann (Co-Director), H.H. Nagel (Consultant), and C.E. Liedtke and W. Geuen (Local arrangement).

This Institute was devoted to the rapidly emerging field of image sequence processing and dynamic scene analysis which has many important applications including target tracking, television bandwidth compression, highway traffic monitoring, and analysis of heart wall motion for medical diagnosis.

The lectures and discussions in this Institute fell into three overlapping categories: Motion estimation; pattern recognition and artificial intelligence techniques in dynamic scene analysis; and applications.

1) Motion estimation - One of the most important problems in image sequence analysis and dynamic scene analysis is displacement and motion estimation. For example, in interframe coding using temporal DPCM, displacement estimation and compensation can improve efficiency significantly. Also, estimated motion parameters can be powerful cues in target segmentation, detection, and classification. In this Institute, a number of recently developed techniques for displacement and motion estimation were discussed.

2) Pattern Recognition and Artificial Intelligence Techniques - Many image sequence processing tasks involve segmentation, feature extraction, and object detection and classification. These fall into the realm of pattern recognition. Since dynamic scenes are usually very complicated, the use of heuristics is often necessary in doing the analysis. Another important aspect is the representation of knowledge. These considerations fall into the realm of artificial intelligence. In this Institute, several lectures were devoted to discussing the state-of-the-art of PR and AI as related to image sequence analysis.

3) Applications - In the main lectures and short contributory presentations, the following applications among others were discussed: Bandwidth compression for broadcast TV, TV conferenceing, and RPV; target tracking using video data; image sequence analysis in radiology and nuclear medicine; tracking biological cells; and cloud pattern analysis.

During the Institute, main invited lectures were given by H.H. Nagel, J.K. Aggarwal, C. Cafforio, A. Netravali, F. Kretz, S. Sabri, H.G. Musmann, T.S. Huang, T.D. Williams, M.D. Levine, K.H. Höhne, C. Gaudeau, and A. Oosterlinck. In addition, more than twenty-five short contributory papers describing recent research results were presented. Also, two panel discussion sessions were held on "present state and trends in image coding" and "Future research directions in image sequence analysis". Written versions of almost all of the invited and contributory papers are included in this volume.

Many people contributed to the success of the Institute. The organizing committee would like to express their deep gratitude and appreciation to Dr. C. Sinclair and Dr. M. DiLullo, NATO Scientific Affairs Division, for their support and encouragement; Dr. Tilo Kester and Mrs. Barbara Kester, International Transfer of Science and Technology, for their invaluable help in the logistics; Mr. A. Künzli and his staff at the Hotel Miratim, for providing an ideal environment for meetings such as this. Last, but certainly the most, the organizing committee would like to thank all the lecturers and other participants, upon whose enthusiasm and technical expertise the reputation of the Institute must be built.

Urbana, Illinois
Christmas 1982

Thomas S. Huang

TABLE OF CONTENTS

PREFACE

PART I OVERVIEW

Hans-Hellmut Nagel
Overview on Image Sequence Analysis 2

J.K. Aggarwal and W.N. Martin
Dynamic Scene Analysis 40

PART II IMAGE SEQUENCE CODING

J.D. Robbins and A.N. Netravali
Recursive Motion Compensation: A Review 75

C. Cafforio and F. Rocca
The Differential Method for Image Motion Estimation 104

Francis Kretz
Edges in Visual Scenes and Sequences: Application to Filtering, Sampling and Adaptive DPCM Coding 125

Shaker Sabri
Movement-Compensated Interframe Prediction for NTSC Color TV Signals 156

H.G. Musmann
Coding of Colour TV Signals with 34 MBit/s Transmission Rate 200

Hans C. Bergmann
Analysis of Different Displacement Estimation Algorithms for Digital Television Signals 215

Richard A. Jones and Carl D. Bowling
An Adaptive Gradient Approach to Displacement Estimation 235

Z. Houkes
Motion Parameter Estimation in TV-Pictures 249

R. Lenz and A. Gerhard
Image Sequence Coding Using Scene Analysis and Spatio-Temporal Interpolation 264

Siegfried Beyer
Two Motion Adaptive Interframe Coding Techniques for Air to Ground Video Signals 275

C. Labit and A. Benveniste
Motion Estimation in a Sequence of Television Pictures 292

P. Robert and C. Labit
Comparative Study Between Intra- and Inter-Frame Prediction Schemes 307

J.F. Abramatic, Ph. Letellier, and M. Nadler
A Narrow-Band Video Communication System for the Transmission of Sign Language Over Ordinary Telephone Lines 314

Giovanni Zarone
Classification and Block Coding of the Frame Difference Signal 337

A.W. Lohmann and Ch. Thum
Histograms of Image Sequence Spectra 350

PART III SCENE ANALYSIS AND INDUSTRIAL APPLICATIONS

B.L. Yen and T.S. Huang
Determining 3-D Motion and Structure of a Rigid Body Using Straight Line Correspondences 365

Thomas D. Williams and Frank Glazer
Comparison of Feature Operators for use in Matching Image Pairs 395

Peter Spoer
Displacement Estimation for Objects on Moving Backgroud 424

Ulrich Boes
Linear Filtering in Image Sequences 437

H. Westphal
Photometric Stereo For Moving Objects 448

L.S. Dreschler and H.-H. Nagel
On the Selection of Critical Points and Local Curvature Extrema of Region Boundaries for Interframe Matching 457

Claus -E. Liedtke
Image Segmentation Considering Properties of the Human Visual System 471

W. Geuen
A Fast Edge Detection Algorithm Matching Visual Contour Perception 483

K.H. Bers, M. Bohner, and P. Fritsche
Image Sequence Analysis for Target Tracking 493

W.E. Snyder and S.A. Rajala
Track Acquisition of Sub-Pixel Targets 502

Volker Graefe
A Pre-Processor for the Real-Time Interpretation of Dynamic Scenes 519

H.G. Meissner and E.D. Dickmanns
Control of an Unstable Plant by Computer Vision 532

S. Grinaker
Real-time Processing of Rasterscan Images 549

D. Cano and M. Bénard
3-D Kalman Filtering of Image Sequences 563

P. Bouthemy and A. Benveniste
Atmospheric Disturbances Tracking in Satellite Images 580

Guenter Warnecke
Aspects of Dynamic Scene Analysis in Meteorology 594

PART IV BIOMEDICAL APPLICATIONS

K.H. Höhne and M. Böhm
Processing and Analysis of Radiographic Image Sequences 602

C. Gaudeau, G. Routier, D. Dzeidzina, and C. Longevialle
Image Sequence Processing and Pattern Recognition of Biomedical Pictures 624

Martin D. Levine, Peter B. Noble, and Youssry M. Youssef
A Rule-Based System for Characterizing Blood Cell Motion 663

H. Oswald
Three Dimensional Imaging from Computed Tomograms 710

H. Bunke, G. Sagerer, and H. Niemann
Model Based Analysis of Scintigraphic Image Sequences of the Human Heart 725

Subject Index 741

PART I

OVERVIEW

OVERVIEW ON IMAGE SEQUENCE ANALYSIS

Hans-Hellmut Nagel
Fachbereich Informatik
Universität Hamburg
Schlüterstraße 70
2000 Hamburg 13
FR Germany

Abstract

Attempts to concisely describe dynamic phenomena recorded by image sequences tend to model the depicted scene as a configuration of objects which exhibit well defined (sequences of) state transitions. Current research in image sequence analysis is concerned with the design of computer-internal representations for objects and associated dynamic phenomena.

One goal is the ability to automatically extract an adequate description for a specific image sequence by generally applicable concepts, representational tools, and procedures. A framework is suggested which decomposes this task for motion-related phenomena into subproblems. The recent literature is presented within this framework to emphasize the common aspects of and the relations between various published approaches.

1. Introduction

Machine analysis of image sequences requires the development of computer-internal representations for (quasi-)invariant entities in a depicted scene as well as for discernible changes in appearance and configuration of such entities. Many frame-to-frame changes can be attributed to quite general phenomena such as relative motion between observer, scene components, and light sources. Although image sequence analysis emerged out of a variety of attempts to solve application problems, current research tends to shift towards modeling those more general phenomena.

NATO ASI Series, Vol. F2
Image Sequence Processing and Dynamic Scene Analysis
Edited by T.S. Huang

Often such more fundamental approaches result from efforts to improve application-oriented solutions. Illustrative examples are mentioned in the following section which provides a concise outline of different application areas concerned with image sequence analysis. The third section concentrates on approaches which attempt to interprete frame-to-frame changes as motion-related phenomena. A hierarchy of abstractions is suggested which provides a framework for sketching the relation between recent contributions to the analysis of images and image sequences.

The literature relating to three topics in this hierarchy of abstractions is treated in separate sections in order to prevent that the numerous references to be quoted obscure the structure of this framework. Section four discusses various aggregation techniques which have been investigated to construct descriptors for characteristic grayvalue distributions and their frame-to-frame variation. The fifth section outlines various approaches to determine displacement vector fields by interframe match of functional approximations to local grayvalue distributions. The derivation of 3-D object descriptions from image sequences and the related problem of space trajectory determination for objects or observers is discussed in the sixth section. Finally, the seventh section sketches approaches which are based on the detection and analysis of interframe differences and relates these approaches to the methods treated in preceding sections.

2. Application of Image Sequence Analysis

The following outline of this topic is based on a comprehensive literature survey by NAGEL 81b. Additional material can be found in special issues edited by AGGARWAL and BADLER 80 and more recently by SNYDER 81. In this section, only selected references will be quoted for illustration.

Interframe coding - treated extensively elsewhere in this volume - may serve as a prime example for continuous improvements towards modeling dynamic phenomena in real-world scenes. Efforts to reduce the amount of image data which have to be transmitted for adequate rendition at the receiver site were originally restricted to purely local picture processing operations. Among these, MOUNTS 69 studied the 'conditional replenishment' technique which may be looked at as a precursor to more recent attempts to interprete difference pictures. Interpretation of difference pictures have been developed quite independently for automatic evaluation rather than mere transmission of image sequences. Analogously, 'movement compensation' discussed by ROCCA 69 may be looked at as a precursor for recent efforts to determine displacement vector or optical flow fields in order to derive 3-D descriptions for objects and their space trajectories. The techniques for estimation of displacement vector fields in both research areas turn out to be closely related if not already identical. Moreover, ideas from more global, semantically oriented approaches towards image sequence evaluation begin to be studied for image coding purposes - see, e.g., MAY and WOLF 82.

Image sequences obtained from airborne or satellite sensors are routinely analysed in order to detect and monitor changes. Evaluation of image series recorded throughout the growth and harvest period can result in more reliable cover type mapping as well as improved estimates of crop yield - see LANDGREBE 81. Very important is the determination of cloud displacement vector fields. These are used to estimate wind velocity distributions which in turn are employed for weather prediction and meteorological modeling. In the latter context, several attempts have been reported recently to describe non-local configurations of cloud displacement vectors in a quantitative manner - see, e.g., HAASS 81, BOUTHEMY and BENVENISTE 82 as well as the contribution by WARNECKE 82 in this volume.

Angiocardiography, blood circulation, and studies of metabolism - the latter via the accumulation and wash-out of radioactively marked pharmaceuticals - are the primary areas of medical interest for the evaluation of temporal image sequences. These topics will be treated in a separate contribution to this volume by HÖHNE. In addition, several groups study temporal variations of body surface potential maps, for example TACCARDI et al. 76 or SCHOFFA et al. 80, see also SCHOFFA and MAYER 81.

Biomedical applications are concerned with the study of growth, transformation, and transport phenomena. Other groups like Levine and collaborators investigate the automatic evaluation of image sequences obtained by time-lapse cinemicrophotography in order to study reaction phenomena by quantitative cell tracking - see FERRIE et al. 82 as well as the contribution by LEVINE and YOUSSEF in this volume.

Similar questions - although on a macroscopic scale - are of interest in behavioral studies. Examples from the recent literature pertain to the training of athletes (DONKOH 80), the behavior of cows and pigs in modern stable and feeding arrangements (BOXBERGER et al. 81) or research on artificial limbs.

Architects who have to design pedestrian circulation-areas would appreciate quantitative data about how pedestrians move about in halls and corridors. Efforts to extract such data from TV-frame sequences (HARTLEY et al. 78) could be considered behavioral studies. They might as well be assigned to a separate topic like object tracking which is of special concern in the case of traffic monitoring, target tracking and visual feedback for automated navigation. The few successful systems for visual tracking of moving objects reported so far exploit a priori knowledge about the scene domain in a judiciously balanced selection of heuristics, for example GILBERT et al. 80 for rockets, HIRZINGER et al. 80 for aeroplanes and cars, BERS et al. 80 for road scenes, and YACHIDA et al. 81 for fishes swimming in a vat. Visual feedback becomes increasingly important, too, for another area - robotics. It is not surprising, therefore, to find researchers from robotics interested in the progress of image sequence analysis - see BIRK and KELLEY 81.

The last topic to be mentioned here concerns spatial rather than temporal image sequences. Spatial image sequences are obtained, e.g., from series of real microtome slices as well as 'virtual slices' such as the focus plane of a microscope or an arbitrarily selected planar section from a 3-D stack of tomographic data. The methods for tracking an elongated nerve cell in a series of microtome slices or for extracting automatically an organ from a 3-D stack of tomographic data (SHANI 80) are closely related to those used in the evaluation of temporal image sequences - see, e.g., MORGENTHALER and ROSENFELD 81, SLOAN and HRECHANYK 81, or UDUPA et al. 82.

Such an observation can be generalized - the same technical problems are recognized in different application areas. The registration problem is a prominent example. It arises in the course of cloud displacement measurements and during the construction of mosaics from satellite or aerial image series, during the evaluation of film recorded in behavioral studies, X-ray films of the beating heart or image series from microtome slices.

Technical solutions to the registration problem as well as other problems may be transferred from one application area to another. This applies especially if the solution method is based on a quantitative approach with a well understood mathematical background. As an example, various methods for the determination of displacement vector fields should be evaluated for image coding, movement of the heart wall in X-ray image sequences, for cloud motion as well as the derivation of 3-D descriptions from monocular image sequences. Similarly, the spectral/temporal clustering techniques developed for the study of metabolism with radionuclid imagery may be explored for crop classification and crop yield estimation - and vice versa.

An in-depth evaluation of the literature about the various applications reveals an increasingly discernible tendency to move beyond approaches which merely classify individual pixels or image areas. The genuine challenge associated with the analysis of image sequences consists in the design, activation, and updating of dynamic descriptions as well as the required data structures. This tendency can be exposed most clearly if the gamut of potential explanations for observable changes in an image sequence is restricted to a specific subset - namely the hypothesis that changes must be attributed to relative motion between an image sensor and one or more macroscopic objects in the observed scene. Results will nevertheless be relevant for a broad range of applications. Moreover, techniques are available to describe the geometry and motion of solid objects in sufficiently compact form, i.e. the scene domain should not pose problems in itself.

3. A Hierarchy of Abstractions for the Description of Changes Related to Motion

The interpretation of an image sequence will be presented as an abstraction process - see figure 1 - which searches for the smallest number of parameters required for a consistent description of the recorded data. The absolute number of parameters may still be rather large, but it will in general be small in comparison to the amount of input data, for example 10 MByte per second in the case of B/W broadcast-TV.

Figure 1

RAW DATA - sequences of digitized images

Aggregation of pixels into LOW-LEVEL DESCRIPTORS

AGGREGATION of low-level DESCRIPTORS

* 2-D spatial aggregation within image plane
* temporal aggregation across image frames
* combined spatio-temporal aggregation without recourse to 3-D concepts

3-D GEOMETRIC DESCRIPTION of a SINGLE OBJECT

* rigid 3-D configuration of points
* rigid 3-D wire-frame model
* rigid 3-D surface model
* configuration of rigid components linked at joints
* object representation with deformable surfaces

Elementary NATURAL LANGUAGE DESCRIPTION

* OBJECT ==> NOUN
* MOTION ==> VERB

Configuration of independently movable objects ==> SCENE

* SCENE ==> NOUN
* scene CHANGE ==> VERB

Aggregation of spatio-temporal changes ==> EVENT

Aggregation of events ==> HISTORY

Abstraction from individual histories ==> SCHEMA (script)

Concepts for the description of schemas

* ACTOR
* CAUSAL RELATION
* RULE of behavior
* LAW of nature

Natural language provides the means to describe a complex development at various levels of abstraction. A hierarchy of abstractions from image sequence data towards natural language notions is employed to organize the references. In the following we assume that only gradual changes occur between consecutive frames of a sequence. Otherwise, just the kind of scene-specific knowledge which we desire to extract from the observations would already be required to bridge the gaps in the input data.

3.1 Low-level descriptors

A first level of abstraction is the localization of grayvalue variations characteristic enough to be found again in a subsequent frame - see, e.g., HALL et al. 80. Examples are edge segments (KORIES 81, STOCKMAN et al. 82). How to define an 'edge' in general is still an open question. MARR and HILDRETH 80 studied 'zero-crossings' of the Laplacian applied to the convolution of a rotationally symmetric Gaussian filter function with the grayvalue distribution. The zero-crossings of this so-called 'Mexican hat' operator correspond to locations of large grayvalue gradients.

Heuristic "points of interest" have been defined by MORAVEC 79+80 and by HANNAH 80. Difficulties with these interest operators have been discussed by DRESCHLER 81. Considering the grayvalue as a function of image plane coordinates led to the application of concepts from differential geometry to the grayvalue 'surface'. KITCHEN and ROSENFELD 80 as well as DRESCHLER and NAGEL 81 developed a characterization of a 'grayvalue corner': it is the location of sharpest turn in the curve of maximum grayvalue gradient. YAM and DAVIS 81 selected locations of high planar curvature in contour lines of binarized images. This approach can be considered to be equivalent to the aforementioned characterization if the contour lines of binarized images coincide with maximum gradient lines of the original grayvalue images. The relation between these approaches have been investigated by DRESCHLER and NAGEL 82.

Another possibility is the selection of primitive regions as descriptors (PRICE and REDDY 79). The subimages selected as reference grayvalue templates for cross correlation search can be considered, too, as elementary region descriptors.

3.2 Aggregation of low-level descriptors

It is useful to distinguish between three categories of aggregation:

(i) spatial aggregation in the 2-D image plane;
(ii) temporal aggregation across image frames for local or more complex non-local descriptors;
(iii) combined spatio-temporal aggregation.

Aggregation or grouping of descriptors has been discussed, e.g., by MARR 76 and STEVENS 78. Segmentation of a grayvalue image, too, belongs to category (i) of purely 2-D spatial aggregation. However, approaches which exploit "motion for segmentation" should be assigned to category (iii). Since each of these three categories should be illustrated by a variety of references, they are treated separately in section four.

At this level of abstraction we exclude from consideration those approaches which mediate aggregation through processes relying on 3-D models or computations which are considered to be characteristic for the next higher level of abstraction.

3.3 3-D geometric description of a single object

The next abstraction level explains the multitude of observations as different projections of a 3-D object moving in space - taking into account the associated phenomena such as variation in the illumination of moving object surfaces or self-occlusion in the case of a non-convex opaque object.

The previously discussed 2-D level of abstraction may be skipped in favor of an immediate spatio-temporal aggregation in 3-D space provided the associated search can be tolerated - see, e.g., NEUMANN 80 - or controlled through exploitation of a priori knowledge. O'ROURKE 81 employed a Hough-transform and cluster analysis in a dynamically quantized parameter space to determine the structure and motion parameters for random dot patterns on the surfaces of two translucent rotating cylinders. The knowledge about differently oriented planar surfaces in a road scene has been exploited by WILLIAMS 80 to determine the observer motion from descriptor displacements in an image sequence recorded by a movie camera.

A priori knowledge about specific structures and motions of objects or sensors can be used in model-based approaches. WALLACE and MITCHELL 80 provide an example for the case of a single rigid object. They track an airplane by relying on a database of projected contours derived from a systematic sample of aspects of 3-D CAD-type airplane models. O'ROURKE and BADLER 80 investigated the movement of a jointed object by analyzing simulated image sequences derived from a human body model based on spheres.

In the general case where no scene-specific a priori knowledge is available, the task of deriving a 3-D description may be organized - at least conceptually - into the following steps.

A <u>single rigid object</u> may be roughly approximated as a fixed 3-D configuration of points - see the separate section six. Such point configurations can be extended to wire-frame models provided it can be determined which vertices are connected by edges. The articles of KRAASCH et al. 79, RADIG et al. 80, ASADA et al. 80, and TSUJI et al. 80 offer examples. Techniques to interpolate surfaces can be adapted from related work in stereopsis as it has been described by GRIMSON 81a. WESTPHAL and NAGEL 81, WESTPHAL 82, and WEBB 81 study frame-to-frame variations of grayvalues in order to infer the surface structure. Surface properties such as grayvalue, texture, color, reflectivity, could be estimated from the image sequence in order to obtain a more complete description consistent with the observed time varying object image. An early attempt has been described by NAGEL 78a. The determination of translatory (CLOCKSIN 80) or curvilinear motion (PRAZDNY 80+81a, BRUSS and HORN 81) for an image sensor is a variant of the single rigid object problem.

BAUMGART 74+75 derived a volume description for various isolated objects. He extracted the contour of the object image in orthographic projections taken from different angles. Each contour was extended along the line of sight to form a cylinder. The object description was obtained as the intersection of such cylinders. This approach requires knowledge of the relative angle between the different lines of sight for the orthographic projections - usually obtained by rotating the object on a lazy Susan in front of the recording camera. The resulting polyhedral object description has been criticized by MARR and NISHIHARA 76 as difficult to manipulate:

* no real sense of uniqueness for its derivation; it was judged, therefore, to be virtually useless for later recognition.
* lack of any natural representation for the articulation of parts of an object.
* difficulty to answer overall questions like where the object is pointing, given only a set of polyhedra each of which describes a small part.
* complex way in which joins between polyhedra have to be specified.

Some of these objections may become less troublesome provided the growing experience with the CAD technique of 'solid modeling' (REQUICHA and VOELCKER 82) is taken into account. MARTIN and AGGARWAL 81a+82 reported results from renewed activities in the direction explored by BAUMGART 74.

A <u>single jointed object</u> can be described as an assembly of rigid component objects. Aggregation of observations into component objects can be attempted in the image plane - RASHID 80, FLINCHBAUGH and CHANDRASEKARAN 81 - or in 3-D space. Examples for the latter approach have been provided by ASADA et al. 80+82, WEBB 81, WEBB and AGGARWAL 81.

3.4 Natural language descriptions of isolated objects and their movements

The next level of abstraction establishes the connection with natural language concepts. Objects may be associated with nouns and adjectives which provide an access path to a priori knowledge about these objects or observations at other times and scenes. AGIN 81 provides an example for the static situation.

BADLER 75 studied already the association between certain motion verbs and knowledge about the object location in space as a function of time. This problem area has been further investigated by TSOTSOS et al. 80 who studied knowledge bases to connect specific motion verbs to observations extracted from medical image sequences. Neumann and collaborators reported first results to describe car movements extracted from TV-frame sequences of traffic scenes by verbs and problems associated with these approaches (MARBURGER et al. 81, NEUMANN 82).

3.5 Configurations of indepently moving objects

A scene may be described as a composition of single objects. But additional questions arise, first of all the possibility of occlusion. If object representations approximating those used for computer graphic methods have been derived such as by DRESCHLER and NAGEL 81, regular hidden surface algorithms can be applied to predict object configurations and compare these with actual images. MARTIN and AGGARWAL 81a as well as WARD and CHIEN 81 investigated various heuristics which provide 2-D image plane cues to occlusion. Shadows can be conceived as the effects of lightsource occlusion by an opaque object and may, therefore, be handled by an appropriate modification of techniques employed for hidden surfaces (SHAFER and KANADE 82). The complexity of natural language descriptions has to be extended to describe scenes as configurations of objects and to introduce verbs in order to describe temporal changes in a scene - see, e.g., OKADA 80.

3.6 Possible further steps

These may be indicated by the following more complex abstractions.

- Temporal scene changes can be aggregated into events. Such events may mark, too, the transition between two different kinds of motion - a complicated problem explored by MARR and VAINA 80 and more recently by TSOTSOS 81.
- Tsuji and collaborators study approaches to aggregate events extracted from line drawing sequences into short histories - see TSUJI et al. 77+80, ABE et al. 81. A different - more linguistically oriented - approach towards this goal has been pursued intensively by OKADA 79+80.
- Finally, one might attempt to abstract from individual histories by describing them as function of laws of nature, as consequences of the intention of an actor, or of behavioral rules adhered to by an actor. Such approaches complete the link to well established investigations about story understanding - see, e.g., SCHANK and ABELSON 77.

4. Descriptor Construction by Aggregation within and between Images

If attribute values provide sufficient discrimination to evaluate tentative matches between a pair of descriptors from different frames, a search for optimum descriptor pairings is performed, often based on cross-correlation (MORAVEC 79, HANNAH 80) or variants thereof (JAYARAMAMURTHY and JAIN 81). If the descriptors are line segments, the matching score may evaluate length and orientation of tentatively paired descriptors in addition to their position. In most cases, however, the continuity hypothesis is exploited to select descriptor pairings associated with an interframe displacement which is similar to displacements found in the neighborhood. This can be achieved by iterative removal of lower scoring candidate displacements which are inconsistent with a higher scoring displacement (KORIES 81). Sometimes the descriptor attributes are only evaluated to determine an initial score for each tentative match. A subsequent relaxation approach modifies these scores using heuristic compatibility functions in order to obtain a locally smooth displacement vector field (PRAGER 79, BARNARD and THOMPSON 80, DRESCHLER 81, DRESCHLER and NAGEL 81).

If the 2-D spatial configuration of descriptors is expected to remain essentially the same from one image frame to the next one, the entire configuration may be represented as a graph. Standard graph processing methods can then be applied to match such graphs from different frames. Descriptor attributes can be exploited to exclude tentative matches between incompatible nodes (KRAASCH et al. 79, RADIG et al. 80, JACOBUS et al. 80). If the configuration to be matched covers a sizable fraction of a frame, the number of descriptors and therefore the cost associated with the required graph search tends to become very large. One remedy consists in organizing neighboring elementary descriptors into relational structures and to employ such structures as complex descriptors, thus reducing the number of nodes to be considered in the graph match (RADIG 81a+b, CHENG and HUANG 81).

A priori knowledge about objects in the scene domain or about the relative motion between objects and sensor may be exploited to reduce the search required for spatial or temporal aggregation. In blocksworld scenes, one may use specific line junctions - in combination with a hierarchical match approach (ROACH and AGGARWAL 79) - and transition tables containing the legal frame-to-frame descriptor transformations (ASADA et al. 80). Likewise, knowledge about sensor motion facilitates the determination of associated displacement vector fields (LAWTON 81). An analogous situation exists for the determination of binocular disparity in stereopsis where the relative sensor positions are assumed to be known (GRIMSON 81a, MAYHEW and FRISBY 81).

In case the only descriptor attributes are its position coordinates with respect to the image frame, interframe aggregation will be evaluated based on a non-local function of coordinate differences. A global translation of a descriptor configuration may be determined by a relaxation method (RANADE and ROSENFELD 80) or by a hierarchical Hough transform (YAM and DAVIS 81). Both approaches reportedly tolerate some deviation from the assumed global translation. ULLMAN 79 investigated the "minimal mapping" between two descriptor configurations from different frames. It determines the minimal - i.e. no superfluous link - cover between the descriptor configurations which simultaneously minimizes a cost function. Besides admitting split as well as merge competition it has other attractive properties, but may not be applicable in all situations - especially if feature points may appear or disappear due to object movements in the scene (DRESCHLER 81, AGGARWAL et al. 81).

So far, spatial and temporal aggregation were considered to be separate processes. Heuristic search procedures have been investigated for the simultaneous aggregation of point descriptors within a frame - i.e. in 2-D space - and between consecutive frames, i.e. in time (FLINCHBAUGH and CHANDRASEKARAN 81). RASHID 80 employed temporal aggregation across more than two frames. 2-D spatial aggregation into separate objects was then performed through a cluster analysis based on a minimum spanning tree comprising descriptor positions and displacements from many frames. A similar approach was subsequently employed to aggregate descriptors from a single object into its constituent rigid components. Another example for this category is represented by the work of TSUJI et al. 80 which, however, uses more descriptors such as regions and line segments in addition to points - see also DRESCHLER and NAGEL 78.

The tree-translation approach developed by FAN and FU 81 should be assigned to this category, too. An image plane configuration of descriptors for moving cars is represented by a tree. A syntax-directed tree-translation procedure has been designed to describe the frame-to-frame changes in descriptor positions. A possible application could be the determination of projected car trajectories for traffic monitoring.

5. Determination of Displacement Vector Fields

The abstraction of characteristic spatial grayvalue variations into descriptors and their interframe match is only one way to determine a displacement vector field which represents the geometric - as opposed to radiometric - aspects of a mapping from one frame to the next one. Functional approximation of the spatial grayvalue variation facilitates the use of analytic methods to determine such a mapping.

5.1 Local displacement estimates based on first-order approximations

It is assumed that the grayvalue image can be modeled as a mosaic of areas within which the grayvalue varies at most linearly with the pixel coordinates, i.e. (I_x = $\delta I/\delta x$, etc.)

$$I(x,y) = I(x_o,y_o) + I_x(x-x_o) + I_y(y-y_o) \tag{1}$$

Neglecting boundary effects, the temporal change of grayvalue inside the i-th area is solely attributed to a shift $\vec{U} = (u,v)^T$ of the sloping grayvalue plane, i.e.

$$I_{ti} = \partial I_i/\partial t = -(\nabla I_i)^T * \vec{U} \tag{2}$$

Since each area i is characterized by a fixed value for the grayvalue gradient ∇I_i, we obtain one equation per area for the two unknown components of the displacement vector $\vec{U}$.

This problem can be circumvented if it is assumed that one displacement component is known - especially to have the value zero as, for example, in the case of binocular disparity provided the coordinate system is aligned with the sensor baseline.

If the displacement is assumed to be constant across at least two different areas, we obtain two independent equations and thus may determine both components of $\vec{U}$. Extending this assumption to N > 2 areas, the resulting overdetermined matrix equation

$$\vec{I}_t = G\,\vec{U} \tag{3}$$

with $I_{ti} = \frac{\delta Ii}{\delta t}$, $G_{ix} = \frac{\delta Ii}{\delta x}$, $G_{iy} = \frac{\delta Ii}{\delta y}$ for i = 1,2,...N

can be solved - provided $\det(G^TG) \neq 0$ - in the form

$$\vec{U} = (G^TG)^{-1}\,G^T\vec{I}_t. \tag{4}$$

SCHALKOFF 79, SCHALKOFF and MCVEY 79 extended this approach to include planar rotation and scale change. Recent investigations by HUANG et al. 82 indicate that this extension does not seem to be necessary for windows of limited size. A review of approaches based on first-order approximations to the grayvalue distribution including their use for motion-compensated interframe coding can be found in NAGEL 81b - see also NETRAVALI and ROBBINS 80.

Cluster analysis in the (u,v)-space after a Hough transformation from the image data is another possibility to determine $\vec{U}$. In favorable situations, more than one displacement vector can be identified (CAFFORIO and ROCCA 76, FENNEMA and THOMPSON 79). In essence, this approach represents a particular implementation to obtain a maximum likelihood estimate for $\vec{U}$. Let

$$p(\vec{U}, \vec{X}, I(\vec{X},t_1), I(\vec{X},t_2))$$

represent the probability that a displacement $\vec{U}$ at the image position $\vec{X}$ explains the grayvalue transition from $I(\vec{X},t_1)$ at frame time t_1 to $I(\vec{X},t_2)$ at frame time t_2. The image has to be segmented into regions of maximum size such that for each region the integral

$$\int_{\vec{X}\,\varepsilon\,\text{region}} p(\vec{U}, \vec{X}, I(\vec{X},t_1), I(\vec{X},t_2))\, d\vec{X} = p(\vec{U}, \text{region}) \implies \text{Maximum} \tag{5}$$

is maximized, i.e. the most likely displacement vector for this region is selected. Since this problem is rather complex, one usually attempts to solve the simpler problem of finding the most likely displacement vector for a predefined region. Estimates of the image region and associated displacement vector could be improved by iterative approaches, especially through adaptation from frame to frame. Related techniques have been studied recently by DARMON 82.

5.2 <u>First-order approximations with supplementary assumptions</u>

It is unknown in general which image areas experience the same displacement $\vec{U}$ - especially if object rotation in 3-D space cannot be excluded or the boundary between two areas comprises an occluding edge segment separating object images with different displacements. HORN and SCHUNCK 81 replaced the assumption of constant displacement by postulating that the displacement should vary smoothly with the image plane coordinates. This should cope approximately with 3-D object rotations. They formulated this idea by requiring that the squared gradient of the displacement components should be minimized. Since noise influences the estimates of I_t as well as I_x and I_y, equation (2) cannot be expected to be obeyed exactly by the estimated values. These authors therefore minimize the following integral

$$\iint dxdy \left\{ \left[I_t + I_x u + I_y v \right]^2 + \alpha^2 \bullet \left[\left(\frac{\delta u}{\delta x}\right)^2 + \left(\frac{\delta u}{\delta y}\right)^2 + \left(\frac{\delta v}{\delta x}\right)^2 + \left(\frac{\delta v}{\delta y}\right)^2 \right] \right\} \implies \text{Minimum} \tag{6}$$

Here α^2 represents the relative weight between the two contributions to the error integral. Using the calculus of variation, they obtain

$$u = \bar{u} - I_x \bullet (I_x \bar{u} + I_y \bar{v} + I_t) \,/\, (\alpha^2 + I_x{}^2 + I_y{}^2) \tag{7a}$$

$$v = \bar{v} - I_y \bullet (I_x \bar{u} + I_y \bar{v} + I_t) \,/\, (\alpha^2 + I_x{}^2 + I_y{}^2) \tag{7b}$$

where $\bar{u}$ and $\bar{v}$ represent the local spatial average of the displacement components resulting from an approximate expression for their Laplacian. These equations estimate the displacement at each image location and may be solved by iteration.

YACHIDA 81 extended this approach by postulating that the displacement should vary smoothly both in space and time. He computed the average values $\bar{u}$ and $\bar{v}$ not only from a 2-D spatial neighborhood around the image location in the current frame, but included the same neighborhoods from the preceding and the following frame. Moreover, he suggested to initialize the displacement components u and v with values derived from interframe descriptor matches whereever available.

There remains, however, the conceptual difficulty that the displacement may change abruptly across occlusion or shadow edges. Neither the assumption of a spatially constant nor of a smoothly varying displacement is appropriate in such a situation to supplement equation (2) in order to determine both components u and v. SCHUNCK and HORN 81 advanced probabilistic considerations to exclude at least the major failures such as those investigated by DINSE et al. 81.

5.3 Higher-order functional approximations for displacement estimation

It appears necessary to look for a more local constraint in order to obtain the additional information required to estimate both components of $\vec{U}$. The edge - i.e. the grayvalue transition between two areas with different gradients - could be taken into account by extending the Taylor expansion up to second order terms rather than terminating it after the first order terms as in equation (2). At first glance this does not appear to help (SNYDER et al. 80) - unless the context restricts again the uncertainty to one component as in the case pointed out by PRAZDNY 81b. WEBB 81 employed second order expansions of the grayvalue function to estimate the coefficients of an affine mapping between local neighborhoods around image locations in different frames - provided the proper displacement between the corresponding locations was known.

NAGEL 82a studied a minimization approach analogous to equation (5), but replacing the assumption of smoothly varying displacements by a second order model

$$\iint dxdy \; [\; I(x,y,t_2) - I(x-u,y-v,t_1) \;]^2 \tag{8a}$$

$$\iint dxdy \; [\; I(x,y,t_2) - I_o - I_x(x-u) - I_y(y-v) - \frac{1}{2}I_{xx}(x-u)^2 - I_{xy}(x-u)(y-v) - \frac{1}{2}I_{yy}(y-v)^2 \;]^2 \tag{8b}$$

where $I_{xx} = \delta^2 I/\delta x^2$ at time t_1, etc. The term I_{xy} has been made to zero by aligning the coordinate system with the principal curvature directions of $I(\vec{X},t_1)$ at the center of the integration window. This expression has to be minimized by appropriate choice of u and v. One obtains a system of two coupled equations with up to third powers in u and v. NAGEL 82a could show that a solution for u and v without recourse to iterative methods is possible in the special situation of 'grayvalue corners' which can be characterized quantitatively (see section 3.1) by the requirements

$$\partial I(x,y,t_1)/\partial x = I1_x = \text{maximum} \; ; \qquad \partial I(x,y,t_1)/\partial y = I1_y = 0 \tag{9a}$$

$$\partial^2 I(x,y,t_1)/\partial x^2 = I1_{xx} = 0 \; ; \qquad \partial^2 I(x,y,t_1)/\partial y^2 = I1_{yy} = \text{maximum} \neq 0 \tag{9b}$$

Exploitation of these conditions results in the following solutions of equation (8b):

$$\hat{u} = - (\overline{I2} - \overline{I1} - \tfrac{1}{2} I1_{yy} \hat{v}^2) \; / \; I1_x \tag{10a}$$

$$\hat{v} = - I2_y \; / \; I1_{yy} \tag{10b}$$

BERGMANN 82 has derived an expression formally similar to equation (10b) for a correction term $\Delta\mu$ in an iterative estimation procedure for the displacement μ. He used cross correlation as a similarity function and simplified his final results by taking the cross correlation function to be circular symmetric. For the purpose of comparison we may, therefore, restrict the discussion to the one-dimensional case. Let the cross correlation $C(\mu)$ be defined as

$$C(\mu) = \int I2(y) \bullet I1(y-\mu) \, dy \tag{11}$$

BERGMANN 82 obtained the result

$$\hat{\mu} = \mu_{k+1} = \mu_k + \Delta\mu \tag{12a}$$

with

$$\Delta\mu = - \{ \partial C(\mu)/\partial\mu \} \; / \; \{ \partial^2 C(\mu)/\partial\mu^2 \} = - C_\mu / C_{\mu\mu} \tag{12b}$$

In order to illustrate the difference between equations (10b) and (12b), we give an explicit expression for $C(\mu)$ based on the assumptions underlying the approach of NAGEL 82a, namely approximation of the grayvalue distribution $I(y,t_1) = I1(y)$ by a second order polynomial with the characteristics corresponding to equations (9). Moreover, the grayvalue distribution at time t_2 is assumed to be a version of $I(y,t_1)$ shifted by Δy, i.e.

$$I(y,t_2) = I2(y) = I1(y-\Delta y) \tag{13}$$

We therefore obtain

$$\partial^2 I1(y)/\partial y^2 = \partial^2/\partial y^2 \, [\, I1_0 + \tfrac{1}{2} I1_{yy} y^2 \,] = I1_{yy} \tag{14a}$$

$$I2_y = \frac{\partial}{\partial y} I2(y) = \frac{\partial}{\partial y} [\, I1_0 + \tfrac{1}{2} I1_{yy} (y-\Delta y)^2 \,] = I1_{yy}(y-\Delta y) \tag{14b}$$

where we have to average the derivatives over the integration window or - in this case equivalent - take them at the center, i.e. for y=0. This results in

$$I2_y = -I1_{yy}\Delta y \tag{14c}$$

Equation (10b) then yields the estimate

$$\hat{v} = - I2_y / I1_{yy} = \Delta y \tag{15}$$

in agreement with our expectations.

If we introduce the same expressions for I1(y) and I2(y) into the formulas of BERGMANN 82, equation (11) yields

$$C(\mu) = \int [I1_0 + \frac{1}{2}I1_{yy}(y-\Delta y)^2] \bullet [I1_0 + \frac{1}{2}I1_{yy}(y-\mu)^2] \, dy \tag{16}$$

The derivatives with respect to μ can be computed under the integral. If we subsequently replace the integration by a summation over all raster points y_i in the summation window, we obtain

$$C_\mu(\mu) = - \sum_y [I1_0 + \frac{1}{2}I1_{yy}(y-\Delta y)^2] \; I1_{yy}(y-\mu) \tag{17a}$$

$$C_{\mu\mu}(\mu) = \sum_y [I1_0 + \frac{1}{2}I1_{yy}(y-\Delta y)^2] \; I1_{yy} \tag{17b}$$

These sums can be evaluated exploiting the symmetry of the summation window whereby odd powers of y vanish. Introduction of the results for equations (17) into equations (12) then yields

$$\mu_{k+1} = \mu_k + \{ -\mu_k - I1_{yy}\bullet\overline{y^2}\bullet\Delta y / [I1_0 + \frac{1}{2}I1_{yy}(\overline{y^2} + \Delta y^2)] \} \tag{18a}$$

$$\hat{\mu} = \mu_{k+1} = - \Delta y \{ I1_{yy}\bullet\overline{y^2} / [I1_0 + \frac{1}{2}I1_{yy}(\overline{y^2} + \Delta y^2)] \} \tag{18b}$$

This result for the displacement estimate is contrary to our expectations. The reason for the discrepancy between the results in equation (15) and (18b) can be traced back to the fact that BERGMANN 82 maximized the cross correlation in equation (11) between the subimages from frame time t_1 and t_2 whereas NAGEL 82a minimized the squared difference given by equation (8a). These two approaches are only equivalent if the 'picture energy' in the search window is virtually independent of the displacement μ (see, e.g., DUDA and HART 73, pp.279/280). This implicit assumption for the use of cross correlation is definitely violated around a grayvalue corner specified as in equations (9). In case the image sequence does not contain too many such grayvalue corners, this particular difficulty might not be too detrimental for the application of BERGMANN's technique to image coding.

Preliminary application of equations (10) by NAGEL and ENKELMANN 82 to an image sequence of a street scene yielded encouraging results. It is important that no starting values for an iterative solution search have to be provided in this case as opposed to other approaches. However, NAGEL 82a has developed an iterative procedure to refine displacement estimates obtained in this manner even for grayvalue distributions which do not conform exactly to the conditions (9) for a 'grayvalue corner'.

The mathematical definition of a 'grayvalue corner' characterizes a visually prominent feature point well localized in both image plane dimensions. It is satisfying that both components of the displacement vector for such a feature point can be determined by a mathematically well founded procedure, not only by a heuristic interframe match.

6. 3-D Rigid Point Configurations from Image Sequences

This section outlines the relation between various approaches to determine the relative position of points in a rigid 3-D configuration based on a series of 2-D frame coordinates for the corresponding image points. Frame-to-frame variations in the relative position of image points are in this case attributed exclusively to translation and rotation of the fixed 3-D point configuration relative to the camera system and to the effects of orthographic or perspective projection into the image plane.

The "structure-from-motion" theorem of ULLMAN 79 states: Given three distinct orthographic views of four non-coplanar points in a rigid configuration, the structure and motion compatible with the three views are uniquely determined. This formulation implies that the inherent ambiguities for orthographic projection - reflection about the frontal plane, absolute distance to the object and its translation in depth - represent equivalent solutions.

6.1 Two-frame approach based on non-linear equations

In the case of perspective projection, ULLMAN 79 studied only the situation of an object rotation around a vertical axis parallel to the image plane, followed by a translation - yielding his so-called "polar equation". NAGEL 81a derived a general fourth order equation in the three parameters describing the object motion between two frames. This equation combined measurements of three points in two views comprising ULLMAN's polar equation as a special case. A total of three such equations based on measurements from five points in two views are required to determine the rotation parameters and hence the translation as well as the relative 3-D point positions up to a common scale factor. The same technique used to derive this equation has also been employed to rederive the optical flow equations of PRAZDNY 80 based on purely algebraic manipulations. A simpler derivation of the equations in NAGEL 81a together with an intuitive geometric interpretation has been given by NAGEL and NEUMANN 81. Independently, similar equations and an iterative procedure to solve them for the purpose of calibrating a stereo camera setup have been described by CLARKSON 81.

If the free parameters fixing the initial coordinate system and the global scale factor in a (5 point, 2 frame)-problem are chosen, but none of the unknowns are eliminated, one obtains a nonlinear system of equations with 18 unknowns. ROACH and AGGARWAL 80 used a modified Levenberg-Marquardt algorithm to minimize the squared deviation from equality for the resulting expressions by descent search in this 18-dimensional parameter space.

6.2 Multi-frame minimization approach

BONDE and NAGEL 79 - see also NAGEL 80 - have studied already a minimization approach for a multi-frame problem. In order to reduce the dimensionality of the search space, unknowns have been eliminated by algebraic manipulations. Moreover, the object motion has been restricted to a plane. This approach provided useful results for actual measurements obtained from image sequences of blocksworld scenes (KRAASCH et al. 79, RADIG et al. 80) and street scenes (DRESCHLER and NAGEL 81).

6.3 The problem of uniqueness

It has recently been shown by LONGUET-HIGGINS 81 and independently by TSAI and HUANG 81 that a solution without the necessity to solve non-linear equations can be obtained if the image coordinates of at least eight points in two frames have been measured. TSAI and HUANG 81 proved that a unique solution - up to a global scale factor - can be found provided the object points do not reside on two planes with one of the two planes containing the origin nor do they lie on a cone containing the origin. This, then, represents the equivalent of Ullman's "structure from motion" theorem for the perspective projection case. Similar questions have been investigated for a long time in photogrammetry. Especially relevant are the results about so-called 'dangerous surfaces or volumes', i.e. special 3-D point configurations for which no unambiguous solutions can be obtained from measurements in planar images - see SCHWIDEFSKY and ACKERMANN 76 (page 165, especially the work by Hofmann 1953).

The basic equation of TSAI and HUANG 81 is derived here in a manner related to the derivations by NAGEL and NEUMANN 81 - see also LONGUET-HIGGINS 81. Assume that the image plane coordinate system $\vec{X} = (x,y)^T$ is embedded into a 3-D coordinate system $\vec{\Xi} = (\xi,\zeta,\eta)^T$ such that the image plane has the coordinate $\zeta = f$ - see figure 2. The notation for the 3-D coordinate system deliberately deviates from the alphabetical order so that the axes represent a right-handed coordinate system despite adherence to the usual choice of the (x,y)-system for image plane coordinates.

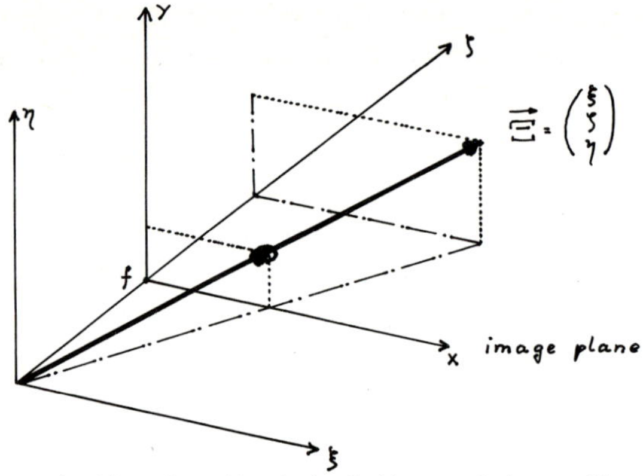

Figure 2

The center of projection is situated at the origin. The image plane coordinates (x,y) are related to the 3-D coordinates by

$$x = \frac{\xi}{\zeta} \bullet f \qquad \text{and} \qquad y = \frac{\eta}{\zeta} \bullet f \tag{19a+b}$$

Let the coordinates of a point at frame times t_1 and t_2 be indicated by concatenation of 1 and 2, respectively. Relative motion between the camera and a 3-D point configuration from time t_1 to time t_2 can be described by a rotation $\boldsymbol{R}$ around an axis through the origin and a subsequent translation $\Delta\vec{\Xi}$.

$$\vec{\Xi}2 = \boldsymbol{R} \bullet \vec{\Xi}1 + \Delta\vec{\Xi} \tag{20}$$

A vector product between $\Delta\vec{\Xi}$ and both sides of equation (20) and a subsequent scalar product with $\boldsymbol{R} \bullet \vec{\Xi}1$ yields

$$[\ \vec{\Xi}2 \times \Delta\vec{\Xi}\ ] * \boldsymbol{R} \bullet \vec{\Xi}1 = 0 \tag{21}$$

Since the vector product between $\vec{\Xi}2$ and $\Delta\vec{\Xi}$ can be written as a multiplication between the vector $\vec{\Xi}2$ and an antisymmetric matrix containing the components of $\Delta\vec{\Xi}$, equation (21) can be written in the form

$$(\vec{\Xi}2)^T \bullet \begin{pmatrix} 0 & -\Delta\zeta & \Delta\eta \\ \Delta\zeta & 0 & -\Delta\xi \\ -\Delta\eta & \Delta\xi & 0 \end{pmatrix} \boldsymbol{R} \bullet \vec{\Xi}1 = 0 \tag{21c}$$

or

$$(\vec{\Xi}2)^T \bullet \boldsymbol{E} \bullet \vec{\Xi}1 = 0 \tag{21d}$$

Equation (21d) can be multiplied by $f^2/(\zeta_1 \zeta_2)$ in order to obtain

$$(\vec{X}2)^T \bullet \boldsymbol{E} \bullet \vec{X}1 = 0 \tag{21e}$$

TSAI and HUANG 81 discuss in detail how the parameters specifying $\Delta\vec{\Xi}$ and $\boldsymbol{R}$ can be determined from a set of equations (21e) for eight pairs of corresponding image plane measurements $\{ \vec{X}1_i , \vec{X}2_i \}$, $i = 1,2,..,8$.

Once these parameters are known, the 3-D coordinates for each point can be inferred up to a scale factor by solving a linear equation for $\zeta 1$:

$$x2 = \frac{\xi 2}{\zeta 2} \bullet f = f \bullet \frac{R_{11}\xi 1 + R_{12}\zeta 1 + R_{13}\eta 1 + \Delta\xi}{R_{21}\xi 1 + R_{22}\zeta 1 + R_{23}\eta 1 + \Delta\zeta} \tag{22a}$$

$$= f \bullet \frac{(R_{11}x1 + R_{12}f + R_{13}y1)\zeta 1/f + \Delta\xi}{(R_{21}x1 + R_{22}f + R_{23}y1)\zeta 1/f + \Delta\zeta} \tag{22b}$$

as well as an analogous equation for y2. Once $\zeta 1$ has been obtained from the measured values of $\vec{X}1$ and $\vec{X}2$, the other coordinates $\xi 1$ and $\eta 1$ can be determined using equations (19).

Equation (22b) shows explicitly that in case of pure rotation around an axis through the origin - i.e. $\Delta\xi = \Delta\eta = \Delta\zeta = 0$ - it is impossible to determine $\zeta 1$ and thus $\xi 1$ or $\eta 1$ because in this case $\zeta 1$ drops out from equation (22b).

6.4 Relation to the focus of expansion

The 'focus of expansion (FoE)' is defined as the location where a line from the origin in the direction of translatory motion $\Delta\vec{\Xi}$ intersects the image plane, i.e. the FoE has the image plane coordinates

$$x_{FoE} = \frac{\Delta\xi}{\Delta\zeta} \bullet f \quad \text{and} \quad y_{FoE} = \frac{\Delta\eta}{\Delta\zeta} \bullet f \tag{23}$$

According to the definition of the displacement vector $\vec{U} = (u,v)^T$ we may write

$$\vec{U} = \vec{X}2 - \vec{X}1 \tag{24}$$

Introducing this into equation (21e) specialized for the case of pure translation - i.e. $\boldsymbol{R}$ given by the 3x3 identity matrix - yields

$$(\vec{X}1 + \vec{U})^T \bullet \boldsymbol{E} \bullet \vec{X}1 = [\ (\vec{X}1 + \vec{U}) \times \Delta\vec{\Xi}\ ] * \vec{X}1 \qquad = 0 \tag{25a}$$

$$= [\qquad \vec{U} \times \Delta\vec{\Xi}\ ] * \vec{X}1 \qquad = 0 \tag{25b}$$

If we subtract from equation (25b) the following expression which vanishes identically,

$$[\ \vec{U} \times \Delta\vec{\Xi}\ ] * (\ - \frac{f}{\Delta\zeta} \bullet \Delta\vec{\Xi}) \quad = \quad 0 \tag{26}$$

we obtain

$$[\ \vec{U} \times \Delta\vec{\Xi}\ ]\ *\ (\vec{X}1\ -\ \overrightarrow{FoE}) \quad = \quad 0 \tag{27a}$$

or

$$[\ \vec{U} \times (\vec{X}1\ -\ \overrightarrow{FoE})\ ]\ *\ \Delta\vec{\Xi} \quad = \quad 0 \tag{27b}$$

Since both $\vec{U}$ and $\vec{X}1 - \overrightarrow{FoE}$ are vectors with ζ-component equal to zero, only the ζ-component of the vectorproduct can be different from zero, i.e. equation (27b) is equivalent to

$$[\ \vec{U} \times (\vec{X}1\ -\ \overrightarrow{FoE})\ ]_{\zeta} \bullet \Delta\zeta \quad = \quad 0 \tag{28}$$

If there is a non-zero translation component Δζ along the line of sight, it follows from equation (28) that

$$[\ \vec{U} \times (\vec{X}1\ -\ \overrightarrow{FoE})\ ] \quad = \quad 0 \tag{29a}$$

or

$$\vec{U} \sim (\vec{X}1\ -\ \overrightarrow{FoE}) \tag{29b}$$

i.e. in the case of pure translation with a component along the line of sight, the displacement vector $\vec{U}$ at image plane position $\vec{X}1$ is oriented along the direction from the FoE to the point $\vec{X}1$ and is proportional to the length of this vector.

6.5 Decomposition of displacement vector fields

The discussion in the preceding two sections has illustrated two statements:

- Only the contribution of translatory motion allows the extraction of quantitative information about the truly three-dimensional arrangement of a rigid point configuration in space.
- Translation results in a particular pattern of displacement vectors, namely radial from the focus of expansion.

PRAZDNY 81a employed some geometric arguments to suggest a technique by which an optical flow field could be decomposed into contributions from translatory and rotatory motion. Optical flow is understood to denote the displacement vector field in the limit of vanishing time interval between two consecutive frames.

This decomposition will be derived here by purely algebraic manipulations. We start from equation (20) which can be written in the case of incremental rotation by the angle $\omega\Delta t$ around an axis through the origin given by a unit vector $\vec{r}$:

$$\vec{\Xi}2 = \vec{\Xi}1 + [\vec{r} \times \vec{\Xi}1]\omega\Delta t + \Delta\vec{\Xi} \tag{30}$$

From this we obtain

$$\Delta x = x2 - x1 = \left(\frac{\xi 2}{\zeta 2} - \frac{\xi 1}{\zeta 1}\right) = \frac{\xi 1 + \omega\Delta t(r_\zeta \eta 1 - r_\eta \zeta 1) + \Delta\xi}{\zeta 1 + \omega\Delta t(r_\eta \xi 1 - r_\xi \eta 1) + \Delta\zeta} - \frac{\xi 1}{\zeta 1} \tag{31a}$$

$$\Delta y = y2 - y1 = \left(\frac{\eta 2}{\zeta 2} - \frac{\eta 1}{\zeta 1}\right) = \frac{\eta 1 + \omega\Delta t(r_\xi \zeta 1 - r_\zeta \xi 1) + \Delta\eta}{\zeta 1 + \omega\Delta t(r_\eta \xi 1 - r_\xi \eta 1) + \Delta\zeta} - \frac{\eta 1}{\zeta 1} \tag{31b}$$

For sufficiently small time intervals Δt, both the angle $\omega\Delta t$ of rotation as well as the incremental translation $\Delta\zeta$ will be small compared to $\zeta 1$. We therefore may approximate the first fraction in equation (31) according to the formula $1/(1+\varepsilon)\approx 1-\varepsilon$ and neglect products with higher than linear powers of Δt or components of $\Delta\vec{\Xi}$. These operations yield

$$\Delta x = \frac{\xi 1}{\zeta 1} + \omega\Delta t\bullet\left(r_\zeta\bullet\frac{\eta 1}{\zeta 1} - r_\eta\right) + \frac{\Delta\xi}{\zeta 1} - \frac{\xi 1}{\zeta 1}\bullet\omega\Delta t\bullet\left(r_\eta\bullet\frac{\xi 1}{\zeta 1} - r_\xi\bullet\frac{\eta 1}{\zeta 1}\right) - \frac{\xi 1}{\zeta 1}\bullet\frac{\Delta\zeta}{\zeta 1} - \frac{\xi 1}{\zeta 1} \tag{32a}$$

$$\Delta y = \frac{\eta 1}{\zeta 1} + \omega\Delta t\bullet\left(r_\xi - r_\zeta\bullet\frac{\xi 1}{\zeta 1}\right) + \frac{\Delta\eta}{\zeta 1} - \frac{\eta 1}{\zeta 1}\bullet\omega\Delta t\bullet\left(r_\eta\bullet\frac{\xi 1}{\zeta 1} - r_\xi\bullet\frac{\eta 1}{\zeta 1}\right) - \frac{\eta 1}{\zeta 1}\bullet\frac{\Delta\zeta}{\zeta 1} - \frac{\eta 1}{\zeta 1} \tag{32b}$$

In the limit of vanishing Δt equations (32) yield

$$u = \lim_{\Delta t\to 0}\frac{\Delta x}{\Delta t} = \omega\bullet\{\, x1\bullet y1\bullet r_\xi + y1\bullet r_\zeta - (x1^2+1)\bullet r_\eta \,\} - \frac{\Delta\zeta}{\zeta 1}\bullet(x1 - x_{FoE}) \tag{33a}$$

$$v = \lim_{\Delta t\to 0}\frac{\Delta y}{\Delta t} = \omega\bullet\{\, (y1^2+1)\bullet r_\xi - x1\bullet r_\zeta - x1\bullet y1\bullet r_\eta \,\} - \frac{\Delta\zeta}{\zeta 1}\bullet(y1 - y_{FoE}) \tag{33b}$$

Equations (33) represent the decomposition of the optical flow field into the rotatory contribution with the factor ω and the translatory contribution with the factor $-\Delta\zeta/\zeta 1$. In the case of pure translation with $\Delta\zeta\neq 0$, equations (33) simplify to an expression encountered already in equation (29b) - only that here the factor of proportionality is given explicitly. Note that the sign of this factor is consistent with our expectations: for receding motion - away from the center of projection - the component $\Delta\zeta$ will be positive and hence the optical flow will be aligned opposite to the vector from the FoE to the point $\vec{X}1$, i.e. the optical flow will occur towards the FoE.

6.6 Remarks

The preceding sections outline mathematical relations between various recent approaches to interprete the displacement vector field. All these approaches assume that a single set of rotation and translation parameters is sufficient to describe the displacement vector field. This is strictly applicable only in the case of an observer moving through a completely stationary environment. Any moving object in the scene requires an additional set of these parameters in order to describe the displacement vector field associated with its motion. This raises the problem to segment the image into regions each of which can be described by an associated set of motion parameters. Since such regions need not be contiguous - for example in the situation of a moving car partially occluded by swaying branches of a tree or by walking pedestrians - the analysis of displacement vector fields represents a genuine problem regarding both its determination as well as its interpretation.

7. Change Detection and Analysis

7.1 Detection of changes

Change detection may be exploited to control detailed processing of input data provided the sensor does not move relative to the major part of the scene and if the illumination remains constant. For a sensor with a signal-to-noise ratio better than 45 db, grayvalue differences between the same image locations from two frames should be significant if they exceed 2-3 levels out of 256 (8 bit grayvalue quantization). If noise is spatially uncorrelated, one may suppress isolated changes or use small image areas to test for significant differences. A likelihood test originally developed by YAKIMOVSKY 76 for edge detection has been adapted by NAGEL 78a for temporal change detection. In this test, grayvalues in each test area are assumed to be drawn independently from a univariate normal distribution with constant central value. NAGEL and REKERS 82 used instead a normal distribution accommodating a linear variation of the grayvalue with the pixel position in the test area. A systematic grayvalue slope in a transition area between image regions may therefore be distinguished from "random fluctuations", for example in highly textured areas. This allowed a tightening of change detection criteria which enabled the derivation of a surprisingly good basis for a mask of a moving object in frame n by disjunctive combination of change locations detected in the comparison of frame n with frames (n-1) and (n+1).

This experience encouraged a theoretical investigation with the goal to establish a quantitative relation between change detection and the determination of displacement vector fields. If the model of grayvalue distribution in image sequences is restricted to a mosaic with constant grayvalues - possibly with additional normally distributed noise -, interframe changes can only be taken care of by stochastic approaches or by recourse to model-based interpretation of segmented image frames. If, however, the model of grayvalue distribution accommodates at least an additional linear variation of grayvalues with pixel position, interframe grayvalue changes at an image location can be attributed to a displacement of the grayvalue distribution - see NAGEL 82b.

Let V_i (i=1,2) denote the variance of grayvalues within the corresponding test windows to be compared. Suppose the first test window contains m grayvalue samples and the other one n samples. Let V_o denote the variance if the grayvalue samples from both test windows are combined into a single sample set. Maximum likelihood estimates are derived for V_j (j=0,1,2) based on the assumption that the grayvalues in a sample set are independent samples from a normal distribution characterizing each set.

The likelihood test decides in favor of compatibility between the two test windows if the ratio

$$V_o^{m+n} \;/\; (V_1^{m} * V_2^{n}) \tag{34}$$

turns out to be smaller than a threshold.

In accordance with the approach of NAGEL 82a, the grayvalue distribution in test window i is approximated by a second order polynomial

$$I(x,y,i) \;=\; Ii(\vec{X}) \;=\; Ii_0 \;+\; (\nabla Ii)^T * \vec{X} \;+\; \frac{1}{2}\vec{X}^T \bullet \mathcal{F}i \bullet \vec{X} \tag{35}$$

where $\mathcal{F}i$ is a 2x2 matrix formed by the second partial derivatives of Ii with respect to the image plane coordinates (x,y). The variance V_0 for the combined test windows can be written in the form

$$V_o \;=\; \frac{1}{2}(V_1 + V_2) \;+\; \frac{1}{n}\,\Sigma\,\{\; \Delta I_o \;+\; \Delta(\vec{\nabla} I)^T * \vec{X} \;+\; \frac{1}{2}\vec{X}^T \bullet \Delta\mathcal{F} \bullet \vec{X} \;\}^2 \tag{36}$$

where the symbol Δ denotes the difference between the corresponding terms from the two test windows 1 and 2 which are assumed to be of equal size n=m.

The displacement vector $\vec{U}$ is introduced by a requirement according to equation (8a), only in a form symmetric with respect to frame 1 and 2. The integral is replaced by the sum over all pixels within the test window.

$$V_U \;=\; \frac{1}{n}\,\Sigma\,\{\; I2(\vec{X}+\vec{U}) \;-\; I1(\vec{X}-\vec{U}) \;\}^2 \tag{37}$$

If the polynomial approximations for $I1(\vec{X})$ and $I2(\vec{X})$ are introduced in this expression for V_U, some lengthy manipulations eventually allow to combine the expression for V_U given by equation (37) with that given for V_0 in the form (see NAGEL 82b)

$$V_o = \frac{1}{2}(V_1 + V_2) + \frac{1}{4}\{ V_u + (\vec{U}^T \bullet \Delta F \bullet \vec{U})^2 \} - (\Delta\vec{\nabla} I \bullet F0 \bullet \vec{U}) \bullet \frac{1}{n}\Sigma x^2$$

$$+ \frac{1}{2}\{ (\vec{\nabla} I0)^T * \vec{U} \} \bullet \{ (\vec{U}^T \bullet \Delta F \bullet \vec{U}) - \overline{\Delta I} \} \qquad (38)$$

The variance V_0 of the combined sample set can therefore be described as the arithmetic average of the variances in the two test windows, the variance V_U due to the fit of interframe differences as given by equation (37), various terms depending on the interframe displacement vector $\vec{U}$ and the average grayvalue difference $\overline{\Delta I}$. $\vec{\nabla} I0$ and $F0$ represent the arithmetic averages of the corresponding terms from test windows 1 and 2. To illustrate the significance of equation (38), neglect for a moment the second order terms ΔF and $F0$. The resulting expression can be written in the form

$$V_o = \frac{1}{2}(V_1 + V_2) + \{ (\Delta\frac{\partial I}{\partial x})^2 + (\Delta\frac{\partial I}{\partial y})^2 \} \bullet \frac{1}{n}\Sigma x^2 + \{ (\vec{\nabla} I0)^T * \vec{U} \}^2 \qquad (39)$$

In this special case it becomes obvious how the parameters obtained from the fits of a polynomial to the picture functions in both test windows are combined with the displacement vector $\vec{U}$ to determine the variance V_o of the combined greyvalue sample set.

The dissimilarity grading approach - see Nagel 78b - based on equation (34) has thus been related quantitatively to the displacement vector estimation approach. Since the latter one is applicable even in the case of non-constant camera parameters - for example with a panning or zooming camera -, the analysis of image sequences can concentrate on the determination and interpretation of displacement vector fields.

7.2 Interpretation of changes

The displacement of an homogeneous object between the previous and the current frame generates change areas with different interpretations - see fig. 3. The horizontally shaded area corresponds to the object in the current frame covering previously visible background whereas the vertically shaded area corresponds to background uncovered since the previous frame due to the object displacement. The cross-hatched area will not exhibit changes - resulting in disconnected change areas if the displacement is parallel to object boundaries as in figure 3a. If the change area - or the unchanged object area "surrounded" by change areas as in figure 3b - can be correctly interpreted, one may deduce the grayvalue characteristics of the moving object and exploit this knowledge to determine the entire object image - a special kind of spatio-temporal aggregation. JAIN and NAGEL 79 investigated edge coincidences between change area contour and object boundary to find the proper interpretation. JAIN et al. 79 exploited this observation in an attempt to extent an image area considered to be part of the object image by region growing into the complete object image. JAIN 81 presented a decision tree classifier in order to interprete change areas, based on a case analysis which comprised scale changes in the object image - for example due to approaching or receding motion. His approach attempts to exploit in qualitative terms the same information which has been analyzed quantitatively in section 6.4 and 6.5. YALAMANCHILI et al. 80 investigated geometric properties such as the collinearity of contour segments from different change areas - see figure 3a - or the "surrounding" indicated in figure 3b to select a seed region within the object image for region growing.

Figure 3

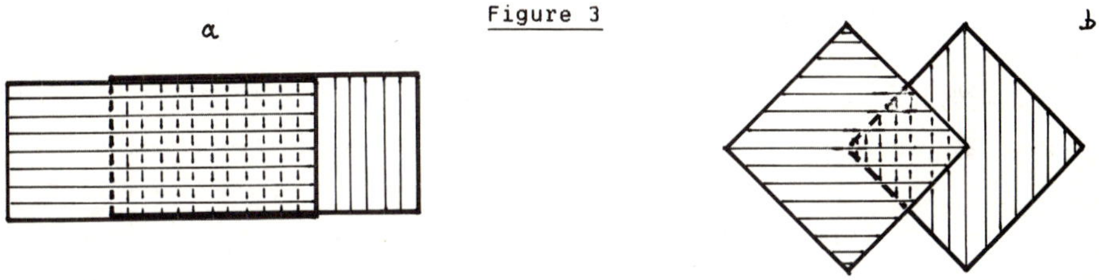

Another variant of spatio-temporal aggregation has been investigated by THOMPSON 80. He decomposed an image frame into elementary regions with homogeneous grayvalue and estimated the displacement in grayvalue transition areas - i.e. around region boundaries - based on techniques discussed in section five (FENNEMA and THOMPSON 79). The displacement estimates were then used as region attributes in a region merging process - a task related to the questions discussed in section 6.6.

Experience has shown that these approaches tend to break down if they are applied to digitized TV-frames of real-world scenes. According to our investigations, some of the reasons are:

- the decision whether a contour element of a change area corresponds to an object boundary element in the previous or the current frame can be distorted by noise;
- noisy boundaries introduce considerable uncertainty into geometric inference processes;
- coincidental alignment of object and background edge segments, especially in textured background;
- the well-known combinatorial problems of region growing approaches for image segmentation.

These problems are basically caused by the necessity to economize processing power and storage space for datastructures. This often causes an overemphasis on a few selected aspects, combined with hidden or very specialized assumptions which results in brittle systems.

8. Concluding Remarks

Image sequence analysis as a field of activity is in a transition phase. The continuing evolution of storage and processing technologies facilitates expanding work on applications. Underneath this striving for - often ad hoc - solutions to pressing application problems, one can discern a growing body of knowledge about how to tackle more fundamental questions.

This is especially evident for the determination of 3-D object descriptions from monocular image sequences. Contacts to neighboring disciplines like photogrammetry have been established. The importance of methods from computer-aided design and graphical representation is evident.

The reliable derivation and analysis of displacement vector fields is likely to widen the common ground with physiological and psychological investigations of visual perception regarding dynamic phenomena. Another contact to watch is the interaction with computational linguistics in a broad sense. Computer-internal representations for high-level concepts about dynamic phenomena are required in both areas.

The preceding sections sketched a number of approaches towards various questions with the intention to illustrate the steady maturation of image sequence analysis as a discipline.

9. Acknowledgements

I thank F. Leberl / Graz for drawing my attention to certain publications in photogrammetry and W. Hofmann / Karlsruhe for making available a copy of his thesis about dangerous surfaces. H.C. Bergmann, C. Cafforio, T.S. Huang, B. Neumann, and S. Ullman helped me by providing copies of preprints and other literature. I thank the members of our group for numerous discussions. B. Radig and Mrs. R. Jancke supported directly or indirectly the editing work for this contribution for which I want to thank them.

10. References

Abe et al. 81
A Plot Understanding System on Reference to Both Image and Language
N. Abe, I. Soga, and S. Tsuji, IJCAI-81, pp. 77-84

Aggarwal and Badler 80
Special Issue on Motion and Time-Varying Imagery (November 1980)
J.K. Aggarwal and N.I. Badler (eds.)
IEEE Trans. Pat. Anal. Machine Intel. PAMI-2, No. 6

Aggarwal et al. 81a
Correspondence Processes in Dynamic Scene Analysis
J.K. Aggarwal, L.S. Davis, and W.N. Martin
Proc. IEEE 69 (1981) 562-572

Aggarwal et al. 81b
Survey: Representation Methods for Three-Dimensional Objects
J.K. Aggarwal, L.S. Davis, W.N. Martin, and J.W. Roach
in L.N. Kanal and A. Rosenfeld (eds.)
Progress in Pattern Recognition vol. 1, pp. 377-391
North-Holland Publ. Co., Amsterdam New York Oxford 1981

Agin 81
Hierarchical Representation of Three-Dimensional
Objects Using Verbal Models
G.J. Agin
IEEE Trans. Pat. Anal. Machine Intel. PAMI-3 (1981) 197-204

Asada et al. 80
Three Dimensional Motion Interpretation for the Sequence
of Line Drawings
M. Asada, M. Yachida, and S. Tsuji, ICPR-80, pp. 1266-1273

Asada et al. 82
Understanding of Three-Dimensional Motions in Blocks World
M. Asada, M. Yachida, and S. Tsuji
IEEE Trans. Pat. Anal. Machine Intel. PAMI-4 (1982) (to appear)

Badler 75
Temporal Scene Analysis: Conceptual Descriptions of
Object Movements
N.I. Badler, Technical Report No. 80, February 1975
Dept. Computer Science, University of Toronto

Barnard and Thompson 80
Disparity Analysis of Images
S.T. Barnard and W.B. Thompson
IEEE Trans. Pat. Anal. Machine Intel. PAMI-2 (1980) 333-340

Baumgart 74
Geometric Modeling for Computer Vision
B.G. Baumgart, STAN-CS-74-463 (October 1974)
Computer Science Dept., Stanford University, Stanford/CA

Baumgart 75
A Polyhedron Representation for Computer Vision
B.L.Baumgart, Proc. NCC, AFIPS 44 (1975) 589-596

Bergmann 82
Displacement Estimation Based on the Correlation of Image Segments
H.C. Bergmann, Proc. IEE International Conference on Electronic
Image Processing, University of York/UK, July 26-28, 1982
see also this volume

Bers et al. 80
Object Detection in Image Sequences
K.H. Bers, M. Bohner, and H. Gerlach, ICPR-80, pp. 1317-1319

Birk and Kelley 81
An Overview of the Basic Research Needed to Advance
the State of Knowledge in Robotics
J.R. Birk and R.B. Kelley
IEEE Trans. Systems, Man, and Cybernetics SMC-11 (1981) 574-579

Bonde and Nagel 79
Deriving a 3-D Description of a Moving Rigid Object from Monocular TV-Frame Sequences
T. Bonde and H.-H. Nagel, WCATVI-79, pp. 44-45

Bouthemy and Benveniste 82
Atmospheric Disturbances Tracking in Satellite Images
P. Bouthemy and A. Benveniste, preprint (July 1982)
IRISA, Campus de Beaulieu, 35042 Rennes Cedex, France

Boxberger et al. 81
Einsatz der Nahbereichsphotogrammetrie in der Tierbeobachtung
J. Boxberger, T. Jungbluth, M. Stephani, H. Auernhammer, und A. Zips
Schriftenreihe der Landtechnik Weihenstephan
Institut fuer Landtechnik, Bayer. Landesanstalt fuer Landtechnik
Freising-Weihenstephan 1981

Bruss and Horn 81
Passive Navigation
A.R. Bruss and B.K.P. Horn
AI-Memo 645 (August 1981), MIT, Cambridge/MA

Cafforio 82
Remarks on the Differential Method for the Estimation of Movement in Television Images
C. Cafforio, Signal Processing 4 (1982) 45-52

Cafforio and Rocca 76
Methods for Measuring Small Displacements of Television Images
C. Cafforio and F. Rocca
IEEE Trans. Information Theory IT-22 (1976) 573-579

Cafforio and Rocca 79
Tracking Moving Objects in Television Images
C. Cafforio and F. Rocca, Signal Processing 1 (1979) 133-140

Cheng and Huang 81
Image Recognition by Matching Relational Structures
J.K. Cheng and T.S. Huang, IEEE PRIP-81, pp. 542-547

Clarkson 81
A Procedure for Camera Calibration with Image Sequences
K.L. Clarkson
Proc. Image Understanding Workshop, L.S. Baumann (ed.), April 1981, pp. 175-177, Science Applications, Inc., Arlington/VA

Clocksin 80
Perception of Surface Slant and Edge Labels from Optical Flow: A Computational Approach
W.F. Clocksin, Perception 9 (1980) 253-269

Darmon 82
A Recursive Method to Apply the Hough Transform to a Set of Moving Objects, C.A. Darmon
Proc. Int. Conf. Acoustics, Speech, and Signal Processing
Paris, May 3-5, 1982, pp. 825-829

Davis and Yam 80
A Generalized Hough-like Transformation for Shape Recognition
L.S. Davis and S. Yam
TR-134 (February 1980), Department of Computer Science
University of Texas at Austin, Austin/TX

Dinse et al. 81
Untersuchung von Verschiebungsvektorfeldern in Bildfolgen
T. Dinse, W. Enkelmann, und H.-H. Nagel
in B. Radig (Hrsgb.) Modelle und Strukturen
4. DAGM Symposium Hamburg 6.-8. Oktober 1981
Informatik-Fachberichte 49 pp. 69-75
Springer Verlag Berlin-Heidelberg-New York 1981

Donkoh 80
Computer Analysis Helps Train Athletes
S. Donkoh, CIPS Review (July/August 1980) 14-15

Dreschler 81
Ermittlung markanter Punkte auf den Bildern bewegter Objekte und Berechnung einer 3D-Beschreibung auf dieser Grundlage
L. Dreschler, Dissertation (Juni 1981)
Fachbereich Informatik der Universitaet Hamburg

Dreschler and Nagel 78
Using "Affinity" for Extracting Images of Moving Objects from TV-Frame Sequences, L. Dreschler and H.-H. Nagel
IfI-HH-B-44/78 (February 1978)
Fachbereich Informatik, Universitaet Hamburg

Dreschler and Nagel 81
Volumetric Model and 3D-Trajectory of a Moving Car Derived from Monocular TV-Frame Sequences of a Street Scene
L. Dreschler and H.-H. Nagel, IJCAI-81, pp. 692-697
see also Computer Graphics and Image Processing Fall 1982 (in press)

Dreschler and Nagel 82
On the Selection of Critical Points and Local Curvature Extrema of Region Boundaries for Interframe Matching
L. Dreschler and H.-H. Nagel
ICPR-82 (in press); see also this volume.

Duda and Hart 73
Pattern Classification and Scene Analysis
R.O. Duda and P.E. Hart, John Wiley _Sons, New York, 1973

Fan and Fu 81
Tree Translation and its Application to Traffic Image Sequence Analysis
T.I. Fan and K.S. Fu, IEEE PRIP-81, pp. 184-188

Fennema and Thompson 79
Velocity Determination in Scenes Containing Several Moving Objects, C.L. Fennema and W.B. Thompson
Computer Graphics and Image Processing 9 (1979) 301-315

Ferrie et al. 82
Cell Tracking: A Modeling and Minimization Approach
F.P. Ferrie, M.D. Levine, and S.W. Zucker
IEEE Trans. Pat. Anal. Machine Intel. PAMI-4 (1982) 277-291

Flinchbaugh and Chandrasekaran 81
A Theory of Spatio-Temporal Aggregation for Vision
B.E. Flinchbaugh and B. Chandrasekaran
Artificial Intelligence 17 (1981) 387-407

Gilbert et al. 80
A Real-Time Video Tracking System
A.L. Gilbert, M.K. Giles, G.M. Flachs, R.B. Rogers, and Y. Hsun U
IEEE Trans. Pat. Anal. Machine Intel. PAMI-2 (1980) 47-56

Grimson 81a
A Computer Implementation of a Theory of Human Stereo Vision
W.E.L. Grimson
Phil. Trans. Royal Society of London B 292 (1981) 217-253

Grimson 81b
A Computational Theory of Visual Surface Interpolation
W.E.L. Grimson, AI-Memo 613 (June 1981)
Artificial Intelligence Laboratory, MIT, Cambridge/MA

Haass 81
Cloud Tracking from Satellite Pictures
U.L. Haass, Ph.D. Dissertation (Spring 1981)
Electrical Engineering Department
Colorado State University, Fort Collins/CO

Hall et al. 80
The Selection of Critical Subsets for Signal, Image, and Scene Matching
E.L. Hall, D.L. Davies, and M.E. Casey
IEEE Trans. Pat. Anal. Machine Intel. PAMI-2 (1980) 313-322

Hannah 80
Bootstrap Stereo
M.J. Hannah
Proc. Image Understanding Workshop, L.S. Baumann (ed.)
April 1980, pp. 201-208, Science Applications, Inc., Arlington/VA

Hartley et al. 78
Preliminary Studies into the CAD of Pedestrian
Circulation Areas
M.G. Hartley, R.C. Waterfall, and A.J. Fisher
Universities Transport Study Group,
Ninth Annual Conference, City University, London
January 1978, pp. 693-700

Hirzinger et al. 80
Automated TV-Tracking of Moving Objects -
the DFVLR-Tracker and Related Approaches
G. Hirzinger, K. Landzettel, and W. Snyder, ICPR-80, pp. 1255-1261

Horn and Schunck 81
Determining Optical Flow
B.K.P. Horn and B.G. Schunck
Artificial Intelligence 17 (1981) 185-203

Huang et al. 82
Interframe Coding with General Two-Dimensional Motion Compensation
T.S. Huang, Y.P. Hsu, and R.Y. Tsai, Proc. Int. Conf. Acoustics,
Speech, and Signal Processing, Paris, May 3-5, 1982, pp. 464-466

Inigo and McVey 81
CCD Implementation of a Three-Dimensional Video-Tracking Algorithm
R.M. Inigo and E.S. McVey
IEEE Trans. Pat. Anal. Machine Intel. PAMI-3 (1981) 230-240

Jacobus et al. 80
Motion Detection and Analysis by Matching Graphs of
Intermediate-Levels Primitives
C.J. Jacobus, R.T. Chien, and J.M. Selander
IEEE Trans. Pat. Anal. Machine Intel. PAMI-2 (1980) 495-510

Jain 81
Extraction of Motion Information from Peripheral Processes
R. Jain
IEEE Trans. Pat. Anal. Machine Intel. PAMI-3 (1981) 489-503

Jain and Nagel 79
On the Analysis of Accumulative Difference Pictures
from Image Sequences of Real World Scenes
R. Jain and H.-H. Nagel
IEEE Trans. Pat. Anal. Machine Intel. PAMI-1 (1979) 206-214

Jain et al. 79
Segmentation through the Detection of Changes due to Motion
R. Jain, W.N. Martin, and J.K. Aggarwal
Computer Graphics and Image Processing 11 (1979) 13-34

Jayaramamurthy and Jain 81
Segmentation of Textured Dynamic Scenes
S.N. Jayaramamurthy and R. Jain, IEEE PRIP-81, pp. 91-93

Kitchen and Rosenfeld 80
Gray-Level Corner Detection
L. Kitchen and A. Rosenfeld, TR-887 (April 1980)
Computer Science Center, University of Maryland, College Park/MD

Kories 81
Determination of Displacement Vector Fields for General
Camera Motions
R. Kories, IEEE PRIP-81, pp. 115-117

Kraasch et al. 79
Automatische Dreidimensionale Beschreibung bewegter Gegenstaende
R. Kraasch, B. Radig, W. Zach
in: Angewandte Szenenanalyse, J.P. Foith (ed.)
Informatik Fachberichte 20, pp. 208-215

Springer Verlag, Berlin-Heidelberg-New York 1979

Landgrebe 81
Analysis Technology for Land Remote Sensing
D.A. Landgrebe, Proc. IEEE 69 (1981) 628-642, especially p. 638

Lawton 81
Optic Flow Field Structure and Processing Image Motion
D.T. Lawton, IJCAI-81, pp. 700-703

Longuet-Higgins 81
A Computer Algorithm for Reconstructing a Scene from Two Projections
H.C. Longuet-Higgins, Nature 293 (September 1981) 133-135

Lucas and Kanade 81
An Iterative Image Registration Technique with an Application to Stereo Vision
B.D. Lucas and T. Kanade, IJCAI-81, pp. 674-679

Marburger et al. 81
Natural Language Dialogue about Moving Objects in an Automatically Analyzed Traffic Scene
H. Marburger, B. Neumann, and H.-J. Novak, IJCAI-81, pp. 49-51

Marr 76
Early Processing of Visual Information
D. Marr, Phil. Trans. Royal Society of London B 275 (1976) 483-524

Marr 78
Representing Visual Information - a Computational Approach
D. Marr, in Computer Vision Systems, pp. 61-80
A.R. Hanson and E.M. Riseman (eds.), Academic Press New York 1978

Marr and Hildreth 80
Theory of Edge Detection
D. Marr and E. Hildreth
Proc. Royal Society of London B 207 (1980) 187-217

Marr and Nishihara 76
Representation and Recognition of the Spatial Organization of Three-Dimensional Shapes
D. Marr and H.K. Nishihara, A.I.-Memo 377 (August 1976)
Artificial Intelligence Laboratory, MIT, Cambridge/MA

Marr and Nishihara 78
Representation and Recognition of the Spatial Organization of Three Dimensional Shapes
D. Marr and H.K. Nishihara, Proc. Royal Society B 200 (1978) 269-294

Marr and Vaina 80
Representation and Recognition of the Movements of Shapes
D. Marr and L. Vaina, AI-Memo 597 (October 1980)
Artificial Intelligence Laboratory, MIT, Cambridge/MA

Martin and Aggarwal 81a
Occlusion in Dynamic Scene Analysis, W.N. Martin and J.K. Aggarwal in Digital Image Processing, J.C. Simon and R.M. Haralick (eds.)
D. Reidel Publ. Co., Dordrecht/Holland and Boston/MA 1981
pp. 579-590

Martin and Aggarwal 81b
Occluding Contours in Dynamic Scenes
W.N. Martin and J.K. Aggarwal, IEEE PRIP-81, pp. 189-192

Martin and Aggarwal 82
Dynamic Scenes and Object Descriptions
W.N. Martin and J.K. Aggarwal, Proc. Int. Conf. Acoustics, Speech, and Signal Processing, Paris, May 3-5, 1982, pp. 859-862

May and Wolf 82
Picture Coding with Motion Analysis for Low Rate Transmission
F. May and W. Wolf
Proc. ICC '82 vol. 1, Philadelphia/PA, June 1982

Mayhew and Frisby 81
Psychophysical and Computational Studies towards a Theory of Human Stereopsis
J.E.W. Mayhew and J.P. Frisby
Artificial Intelligence 17 (1981) 349-385

Moravec 79
Visual Mapping by a Robot Rover
H.P. Moravec, IJCAI-79, pp. 598-600

Moravec 80
Obstacle Avoidance and Navigation in the Real World by a Seeing Robot Rover
H.P. Moravec, Ph.D. Thesis, Department of Computer Science STAN-CS-80-813, Stanford University
available, too, as CMU-RI-TR-3 (September 1980)
Robotics Institute, Carnegie-Mellon University, Pittsburgh/PA

Morgenthaler and Rosenfeld 81
Multidimensional Edge Detection by Hypersurface Fitting
D.G. Morgenthaler and A. Rosenfeld
IEEE Trans. Pat. Anal. Machine Intel. PAMI-3 (1981) 482-486

Mounts 69
A Video Encoding System with Conditional Picture-Element Replenishment
F.W. Mounts, Bell System Techn. J. 48 (1969) 2545-2554

Nagel 78a
Formation of an Object Concept by Analysis of Systematic Time Variations in the Optically Perceptible Environment
H.-H. Nagel
Computer Graphics and Image Processing 7 (1978) 149-194

Nagel 78b
Analysis Techniques for Image Sequences
H.-H. Nagel, IJCPR-78, pp. 186-211

Nagel 80
From Digital Picture Processing to Image Analysis
H.-H. Nagel
Proc. International Conference on Image Analysis and Processing
V. Cantoni (ed.), Pavia/Italy, October 22-24, 1980, pp. 27-40

Nagel 81a
On the Derivation of 3D Rigid Point Configurations from Image Sequences
H.-H. Nagel, IEEE PRIP-81, pp. 103-108

Nagel 81b
Image Sequence Analysis: What Can We Learn from Applications?
H.-H. Nagel, in Image Sequence Analysis, pp. 19-228
T.S. Huang (ed.), Springer Verlag Berlin-Heidelberg-New York 1981

Nagel 82a
Displacement Vectors Derived from Second Order Intensity Variations in Image Sequences
H.-H. Nagel
Computer Graphics and Image Processing 1982 (in press)

Nagel 82b
On Change Detection and Displacement Vector Estimation in Image Sequences, H.-H. Nagel
Pattern Recognition Letters, October 1982 (in press)

Nagel and Enkelmann 82
Investigation of Second Order Greyvalue Variations to Estimate Corner Point Displacements
H.-H. Nagel and W. Enkelmann, ICPR-82 (in press)

Nagel and Neumann 81
On 3D Reconstruction from Two Perspective Views
H.-H. Nagel and B. Neumann, IJCAI-81, pp. 661-663

Nagel and Rekers 82
Moving Object Masks Based on an Improved Likelihood Test

H.-H. Nagel and G. Rekers, ICPR-82 (in press)
Netravali and Robbins 80
Motion-Compensated Coding: Some New Results
A.N. Netravali and J.D. Robbins
Bell System Technical J. 59 (1980) 1735-1745
Neumann 80
Motion Analysis of Image Sequences for Object Grouping and Reconstruction
B. Neumann, ICPR-80, pp. 1262-1265
Neumann 81
3D-Information aus mehrfachen Ansichten
B. Neumann, in B. Radig (Hrsgb.) Modelle und Strukturen
4. DAGM Symposium Hamburg 6.-8. Oktober 1981
Informatik-Fachberichte 49 pp. 93-111
Springer Verlag Berlin-Heidelberg-New York 1981
Neumann 82
Towards Natural Language Description of Real-World Image Sequences
B. Neumann, 12. Jahrestagung der Gesellschaft fuer Informatik 1982
Informatik-Fachberichte, (in press)
Springer-Verlag Berlin-Heidelberg-New York 1982
O'Rourke 81
Motion Detection Using Hough Technique
J. O'Rourke, IEEE PRIP-81, pp. 82-87
O'Rourke and Badler 80
Model-Based Image Analysis of Human Motion Using Constraint Propagation, J. O'Rourke and N.I. Badler
IEEE Trans. Pat. Anal. Machine Intel. PAMI-2 (1980) 522-536
Okada 79
SUPP: Understanding Moving Picture Patterns Based on Linguistic Knowledge
N. Okada, IJCAI-79, pp. 690-692
Okada 80
Conceptual Taxonomy of Japanese Verbs for Understanding Natural Language and Picture Patterns
N. Okada, Proc. COLING-80, pp. 127-135
see, too, Conceptual Taxonomy of Japanese Verbs and Sentence Production from Picture Pattern Sequences
Technical Report, Information Science and Systems Engineering
Oita University, Oita 870-11/Japan (December 1980)
Prager 79
Segmentation of Static and Dynamic Scenes
J.M. Prager, COINS Technical Report 79-7 (May 1979)
Computer and Information Science
University of Massachusetts, Amherst/MA
Prazdny 80
Egomotion and Relative Depth Map from Optical Flow
K. Prazdny, Biological Cybernetics 36 (1980) 87-102
Prazdny 81a
Determining the Instantaneous Direction of Motion from Optical Flow Generated by a Curvilinearly Moving Observer
K. Prazdny
Computer Graphics and Image Processing 17 (1981) 238-248
Prazdny 81b
A Simple Method for Recovering Relative Depth Map in the Case of a Translating Sensor
K. Prazdny, IJCAI-81, pp. 698-699
Price and Reddy 79
Matching Segments of Images, K. Price and R. Reddy
IEEE Trans. Pat. Anal. Machine Intel. PAMI-1 (1979) 110-116
Radig 81a
Inferential Region Extraction in TV-Sequences
B. Radig, IJCAI-81, pp. 719-721

Radig 81b
Image Region Extraction of Moving Objects
B.M. Radig, in Image Sequence Analysis, pp. 311-354
T.S. Huang (ed.), Springer Verlag Berlin-Heidelberg-New York 1981
Radig et al. 80
Matching Symbolic Descriptions for 3-D Reconstruction of Simple Moving Objects
B. Radig, R. Kraasch, and W. Zach, ICPR-80, pp. 1081-1084
Ranade and Rosenfeld 80
Point Pattern Matching by Relaxation
S. Ranade and A. Rosenfeld, Pattern Recognition 12 (1980) 269-275
Rashid 80
Towards a System for the Interpretation of Moving Light Displays
R.F. Rashid
IEEE Trans. Pat. Anal. Machine Intel. PAMI-2 (1980) 574-581
Requicha and Voelcker 82
Solid Modeling: A Historical Summary and Contemporary Assessment
A.A.G. Requicha and H.B. Voelcker
IEEE Computer Graphics and Applications 2, No. 2 (March 1982) 9-24
Roach and Aggarwal 79
Computer Tracking of Objects Moving in Space
J. Roach and J.K. Aggarwal
IEEE Trans. Pat. Anal. Machine Intel. PAMI-1 (1979) 127-135
Roach and Aggarwal 80
Determining the Movement of Objects from a Sequence of Images
J.W. Roach and J.K. Aggarwal
IEEE Trans. Pat. Anal. Machine Intel. PAMI-2 (1980) 554-562
Rocca 69
Television Bandwidth Compression Utilizing Frame-to-Frame Correlation and Movement Compensation
F. Rocca
Symposium on Picture Bandwidth Compression
MIT, Cambridge/MA, April 1969, T.S. Huang and O.J. Tretiak (eds.)
Gordon and Breach, New York 1972, pp. 675-693
Schalkoff 79
Algorithms for a Real-Time Automatic Video Tracking System
R.J. Schalkoff, Ph.D. Thesis (May 1979)
Dept. Electrical Engineering, Univ. of Virginia, Charlottesville/VA
Schalkoff and McVey 79
Algorithm Development for Real-Time Automatic Video Tracking Systems
R.J. Schalkoff and E.S. McVey
Proc. 3rd International Computer Software and Applications Conference, Chicago/IL, November 1979, pp. 504-511
see also IEEE Trans. Pat. Anal. Machine Intel. PAMI-4 (1982) 2-10
Schank and Abelson 77
Scripts, Plans, Goals, and Understanding
R.C. Schank and R.P. Abelson
Lawrence Erlbaum Associates, Publishers, Hillsdale/NJ 1977
Schoffa and Mayer 81
Merkmalsreduzierung mit Fourier-Deskriptoren bei elektrischen Potentialkarten des Herzens
G. Schoffa und R. Mayer
in B. Radig (Hrsgb.) Modelle und Strukturen
4. DAGM Symposium Hamburg 6.-8. Oktober 1981
Informatik-Fachberichte 49 pp. 248-254
Springer Verlag Berlin-Heidelberg-New York 1981
Schoffa et al. 80
Image Processing in Body Surface Mapping
G. Schoffa, R. Mayer, and W. Postweiler
Suppl. Mikroskopie (Wien) 37 (1980) 444-447
Schwidefsky and Ackermann 76

Photogrammetrie , K. Schwidefsky und F. Ackermann
B.G. Teubner, Stuttgart/FR Germany 1976

Schunck and Horn 81
Constraints on Optical Flow Computation
B.G. Schunck and B.K.P. Horn, IEEE PRIP-81, pp. 205-210

Shafer and Kanade 82
Using Shadows in Finding Surface Orientation
S.A. Shafer and T. Kanade
CMU-CS-81-100 (January 1982), Computer Science Department
Carnegie-Mellon University, Pittsburgh/PA

Shani 80
A 3-D Model-Driven System for the Recognition of
Abdominal Anatomy from CT Scans
U. Shani, ICPR-80, pp. 585-591; see also TR-82 (August 1981)
Computer Science Department, University of Rochester, Rochester/NY

Sloan and Hrechanyk 81
Surface Reconstruction from Sparse Data
K.R. Sloan and L.M. Hrechanyk, IEEE PRIP-81, pp. 45-48

Snyder 81
Computer Analysis of Time-Varying Images, W.E. Snyder (ed.)
Special issue IEEE Computer 14, No. 8 (August 1981)

Snyder et al. 80
Image Modeling, the Continuity Assumption and Tracking
W.E. Snyder, S.A. Rajala, and G. Hirzinger
ICPR-80, pp. 1111-1114

Stevens 78
Computation of Locally Parallel Structure
K.A. Stevens
Proc. Image Understanding Workshop, L.S. Baumann (ed.)
Pittsburgh/PA, November 14-15, 1978, pp. 92-102
Science Applications, Inc., Arlington/VA 22209
see also: Biological Cybernetics 29 (1978) 19-28

Stockman et al. 82
Matching Images to Models for Registration and Object
Detection via Clustering
G. Stockman, S. Kopstein, and S. Benett
IEEE Trans. Pat. Anal. Machine Intel. PAMI-4 (1982) 229-241

Taccardi et al. 76
Body-Surface Mapping of Heart Potentials
B. Taccardi, L. De Ambroggi, and C. Viganotti
in: The Theoretical Basis of Electrocardiology
C.V. Nelson and D.B. Geselowitz (eds.)
Clarendon Press, Oxford 1976, pp. 436-466

Thompson 80
Combining Motion and Contrast for Segmentation
W.B. Thompson
IEEE Trans. Pat. Anal. Machine Intel. PAMI-2 (1980) 543-549

Thompson et al. 81
Relaxation Labeling Using Staged Updating
W.B. Thompson, K.M. Mutch, J.K. Kearney, and R.L. Madarasz
IEEE PRIP-81, pp. 449-451

Tsai and Huang 81
Uniqueness and Estimation of Three-Dimensional Motion
Parameters of Rigid Objects with Curved Surfaces
R.Y. Tsai and T.S. Huang
Report R-921 (October 1981), Coordinated Science Laboratory
University of Illinois at Urbana-Champaign
see also T.S. Huang, this volume

Tsotsos 81
Temporal Event Recognition: An Application to Left
Ventricular Performance
J.K. Tsotsos, IJCAI-81, pp. 900-907

Tsotsos et al. 80
A Framework for Visual Motion Understanding
J.K. Tsotsos, J. Mylopoulos, H.D. Covvey, and S.W. Zucker
IEEE Trans. Pat. Anal. Machine Intel. PAMI-2 (1980) 563-573

Tsuji et al. 77
Understanding a Simple Cartoon Film by a Computer Vision System
S. Tsuji, A. Morizono, and S. Kuroda, IJCAI-77, pp. 609-610

Tsuji et al. 80
Tracking and Segmentation of Moving Objects in Dynamic Line Images
S. Tsuji, M. Osada, and M. Yachida
IEEE Trans. Pat. Anal. Machine Intel. PAMI-2 (1980) 516-522

Udupa et al. 82
Boundary Detection in Multidimensions
J.K. Udupa, S.N. Srihari, and G.T. Herman
IEEE Trans. Pat. Anal. Machine Intel. PAMI-4 (1982) 41-50

Ullman 79
The Interpretation of Visual Motion
S. Ullman, The MIT Press, Cambridge/Mass., 1979

Wallace and Mitchell 80
Analysis of Three-Dimensional Movement Using Fourier Descriptors
T.P. Wallace and O.R. Mitchell
IEEE Trans. Pat. Anal. Machine Intel. PAMI-2 (1980) 583-588

Ward and Chien 80
Analysis of Time-Varying Imagery Through the Representation of Position and Shape Changes
M.O. Ward and Y.T. Chien, ICPR-80, pp. 1236-1238

Ward and Chien 81
Occlusion Analysis in Time-Varying Imagery
M.O. Ward and Y.T. Chien, IEEE PRIP-81, pp. 504-507

Webb 81
Shape and Structure from Motion of Objects
J.A. Webb, Ph.D. Dissertation (December 1981)
University of Texas at Austin, Austin/TX

Webb and Aggarwal 80
Observing Jointed Objects
J.A. Webb and J.K. Aggarwal, ICPR-80, pp. 1246-1250

Webb and Aggarwal 81
Structure from Motion of Rigid and Jointed Objects
J.A. Webb and J.K. Aggarwal, IJCAI-81, pp. 686-691

Westphal 82
Photometric Stereo for Moving Objects
H. Westphal, this volume; see also ICPR-82 (in press)

Westphal and Nagel 81
Zur Ermittlung gekruemmter Oberflaechen bewegter Objekte aus Bildfolgen
H. Westphal und H.-H. Nagel
in B. Radig (Hrsgb.) Modelle und Strukturen
4. DAGM Symposium Hamburg 6.-8. Oktober 1981
Informatik-Fachberichte 49, pp. 123-129
Springer Verlag Berlin-Heidelberg-New York 1981

Williams 80
Depth from Camera Motion in a Real World Scene
T.D. Williams
IEEE Trans. Pat. Anal. Machine Intel. PAMI-2 (1980) 511-516

Yachida 81
Determining Velocity Map by 3-D Iterative Estimation
M. Yachida, IJCAI-81, pp. 716-718

Yachida et al. 81
Automatic Analysis of Moving Images
M. Yachida, M. Asada, and S. Tsuji
IEEE Trans. Pat. Anal. Machine Intel. PAMI-3 (1981) 12-20

Yakimovsky 76
Boundary and Object Detection in Real World Images
Y. Yakimovsky, J. ACM 23 (1976) 599-618

Yalamanchili et al. 80
Differencing Operations for the Segmentation of Moving Objects in Dynamic Scenes
S. Yalamanchili, W.N. Martin, and J.K. Aggarwal
ICPR-80, pp. 1239-1242; see also IEEE PRIP-81, pp. 211-216 and Computer Graphics and Image Processing 18 (1982) 188-201

Yam and Davis 80
Image Registration Using Generalized Hough Transforms
S. Yam and L.S. Davis
TR 80-7 (December 1980), Computer Sciences Department University of Texas at Austin, Austin/TX

Yam and Davis 81
Image Registration Using Generalized Hough Transforms
S. Yam and L.S. Davis, IEEE PRIP-81, pp. 526-533

Zucker and Hummel 81
A Three-Dimensional Edge Operator
S.W. Zucker and R.A. Hummel
IEEE Trans. Pat. Anal. Machine Intel. PAMI-3 (1981) 324-331

DYNAMIC SCENE ANALYSIS

J. K. Aggarwal

Laboratory for Image and Signal Analysis
and
Department of Electrical Engineering
The University of Texas at Austin
Austin, Texas, 78712 / U. S. A.

and

W. N. Martin

Department of Applied Mathematics
and Computer Sciences
University of Virginia
Charlottesville, VA 22901

ABSTRACT: The three major components of dynamic scene analysis, namely segmentation, occlusion and the computation of three-dimensional information from images are discussed in depth. Segmentation refers to the process of determining features of interest, occlusion analysis includes the deriving of changes due to projection perspective, and computation of three-dimensional information entails the constructing of structural models and describing motions from image information. The purpose of the review is to give the reader a coherent view of the issues and the manner in which researchers are currently addressing these issues. Detailed descriptions of the systems developed at The University of Texas are presented.

NATO ASI Series, Vol. F2
Image Sequence Processing and Dynamic Scene Analysis
Edited by T.S. Huang

1.0 Introduction

Dynamic Scene Analysis, also referred to as the Analysis of Time-Varying Imagery, is concerned with the processing of a sequence or a collection of images. The ultimate goal of the analysis is to assimilate information from the sequence as a whole that cannot be obtained from any one image by itself. The sequence of images usually represents a scene as sampled by a sensor at instants close in time and may arise from a variety of scenarios. Examples include motions of objects in a scene where the sensor is fixed, motion of the sensor relative to the scene or a combination of the two motions. A variety of applications have motivated the present research in Dynamic Scene Analysis. These include industrial automation and inspection, robotics, navigation, automatic surveillance and biomedical engineering. The research area of Dynamic Scene Analysis is rather new, however, it is receiving considerable attention as is evident from the Advanced Study Institute at Braunlage, other conferences and the recent literature [1-8].

The present paper addresses the important issues and ingredients of Dynamic Scene Analysis. Specifically, the three issues discussed in this review paper are segmentation, occlusion and the computation of three-dimensional information from images. Here segmentation refers to the process of determining features of interest together with distinguishing interesting changes from uninteresting changes and relating the features and components in one image to those of the succeeding images. Occlusion analysis includes deriving structural changes due to projection perspective and the appearance or disappearance of objects. Finally, the computation of three-dimensional information entails constructing structural models and describing three-dimensional motions through the analysis of two-dimensional image information. The next three sections discuss Dynamic Scene Segmentation, Occlusion in Image Sequences and Three-Dimensional Information from Images.

2.0 Dynamic Scene Segmentation

In almost every static scene analysis task, the first step is segmentation, i.e., to locate the significant scene components, to extract features from the image, or to separate the image into meaningful regions. Dynamic scene analysis is no different but must also consider dividing the images into parts that are changing and parts that are constant, or finding the the moving parts in each element of the sequence of images. In order to account for change and movement, information must be combined from consecutive frames or subsequences of images.

There are two distinct approaches to segmentation in dynamic scenes: Feature Based Segmentation and Pixel Based Segmentation. Feature based segmentation consists of finding edges, corners, boundaries, or surfaces in each of two images and then establishing a correspondence between various features in the two images. The process of establishing correspondence is at times difficult, especially if one has noisy images. Thus, the analysis proceeds with the static scene segmentation of each of the two images, and then establishes a feature correspondence between consecutive images to determine the changes in the images. Pixel based segmentation compares the two images at the pixel level by methods such as differencing, correlation or temporal-spatial gradient. In each case, pixel level comparisons are made and velocity estimates are assigned to various pixel positions. The velocity estimates become the basis for segmentation.

Neither type of process yields unique answers but generally the end product is a description of the moving parts of a scene. Each of the approaches makes the assumption that image components which move together are parts of the same underlying object in the scene. The above procedures are illustrated in the following discussion.

2.1 Feature Based Analysis

A variety of features have been used in segmenting each of the images and in establishing correspondence for the moving parts in a sequence of images. The list of features includes corners, straight edges, curvilinear edge segments, centroid, area, major and minor axes for moment of inertia, and others. The choice of features depends upon the problem domain and the assumptions that may be made about the moving parts. Two systems are briefly discussed to illustrate feature based analysis.

Corners and Straight Edges: Aggarwal and Duda [9] consider the motion of polygonal figures which are arbitrarily complex in shape, and possibly contain holes. In this case the polygons are software generated to appear as planar objects moving in planes parallel to the image plane. The parallel projection essentially creates the silhouette of the objects.

The overlapping of the actual polygons creates new vertices while removing occluded vertices and edges. The new vertices are referred to as "false" vertices and the visible vertices of the actual polygons are called "real" vertices. One of the main functions of the system is to classify the vertices of the input image into the appropriate one of these two categories. This classification process is facilitated by two characteristics of the input domain. First, no "false" vertex can have an interior angle which measures less than 180 degrees. Second, any vertex which changes its angular measure between two frames must be a "false" vertex. The first characteristic is due to the polygonal nature of the objects, while the restriction to rigid polygons assures the second. However, these two characteristics do not provide enough information to directly classify every vertex. There are vertices with obtuse interior angles which are not "false" vertices and there are "false" vertices which do not change their angular measure. One further restriction is necessary, and it is that no more than one "real" vertex can appear or become occluded between any two consecutive frames. The importance of this restriction is that it allows the system to determine the type of change that has occurred between two consecutive frames. This determination is based on the difference in the number of vertices having acute interior angles along with the difference in the number of vertices having obtuse interior angles.

The correspondence is established based on the nature of the vertex (i.e., acute or obtuse), the lengths of the polygons' sides, etc. In this fashion, the moving parts are isolated for further processing. Figure 2.1 shows an example of a sequence of images to which this process has been applied.

<u>Curvilinear Boundaries</u>: The system of Martin and Aggarwal [10] analyzes scenes containing figures with curvilinear boundaries in a manner similar to the system just described. The input is again restricted so that the objects move independently in planes parallel to the image plane. However, instead of software-generated images, homogeneously shaded, opaque, planar figures are moved in front of a TV camera to produce a sequence of images. The camera approximates an orthogonal projection into the digital images which are preprocessed to extract the boundaries of the figures [11]. The figure shading and the camera setup give rise to images in which overlapping figures are merged into single apparent objects. The task of the system is thus to derive descriptions of the

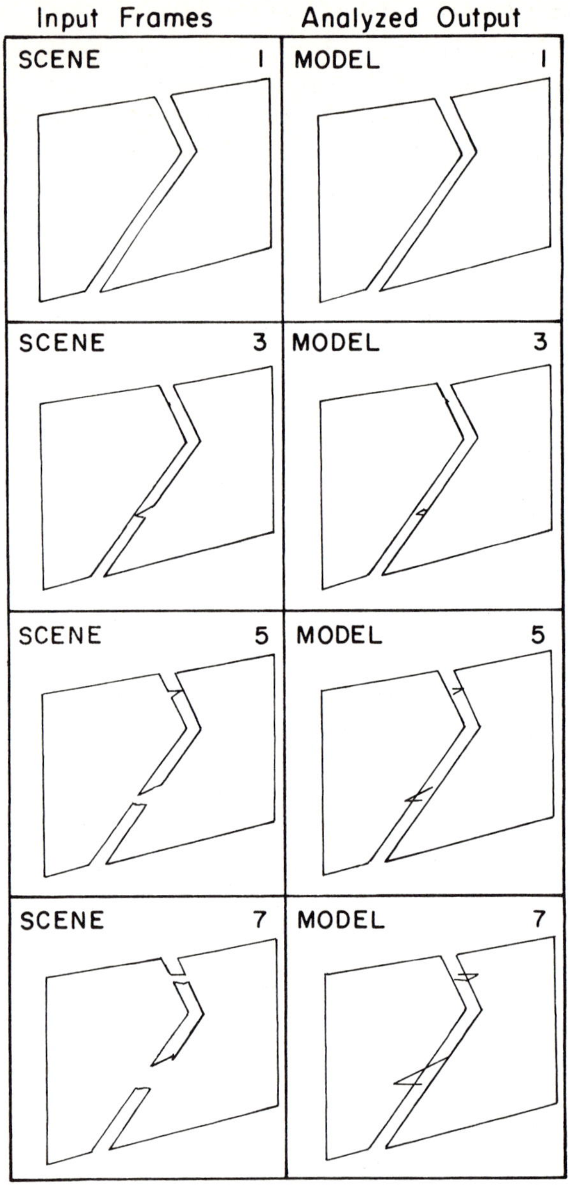

Figure 2.1. A sequence of polygonal objects illustrating feature based segmentation.

constituent actual figures and their motions by analyzing the apparent objects of the sequence of images. The analysis of the sequence is performed on pairs of consecutive images from the sequence and is based upon identifying shapes which are common to both images of any given pair. The matched shapes are interpreted as two views of the same object. In this way the moving objects can be tracked throughout the sequence while motion measurements are made from the displacements between the matched views.

The tokens used by this system are circular arcs approximated by portions of the object boundaries. The arcs are derived by analyzing the subtended angle versus arc length, ψ-s, a function of the boundary as measured from an arbitrary starting point on that boundary. This function is useful because intervals of constant slope in the ψ-s function correspond to boundary sections of constant curvature, i.e., circular arcs. The appropriate intervals are determined by forming a piecewise straight line approximation of the pictorial graph of the ψ-s function. The set of straight lines in the ψ-s function approximation effectively decomposes the object boundary into a set of arcs. Figure 2.2 shows an object before and after being segmented into arcs by this process.

The shape representation, as entered in the data base which contains all the relevant information derived from the sequence of images, includes the coordinate list of the object boundary, the straight-line description of the ψ-s function, and pointers relating specific boundary sections to the appropriate elements of the straight-line set. This representation separates clearly the information needed for the shape matching from the information required in the movement measurement process. In fact the ψ-s function is invariant to translation and rotation (see [10] for minor qualifications) and is processed to eliminate the effects of arbitrarily choosing its starting point. This separation is in accordance with the system's use of the constancy in shape of the actual figures in order to interpret the movement of the apparent objects.

The initial correspondence is based on matching the tokens through their shape attributes, but is again aided by the higher level constraint imposed by the token ordering along object boundaries. Contiguous arcs from one image which match, in the same order, contiguous arcs from the second image are grouped into edge segments. This matching is performed by first choosing two arcs, one from each image of a consecutive pair, whose ψ-s function lines

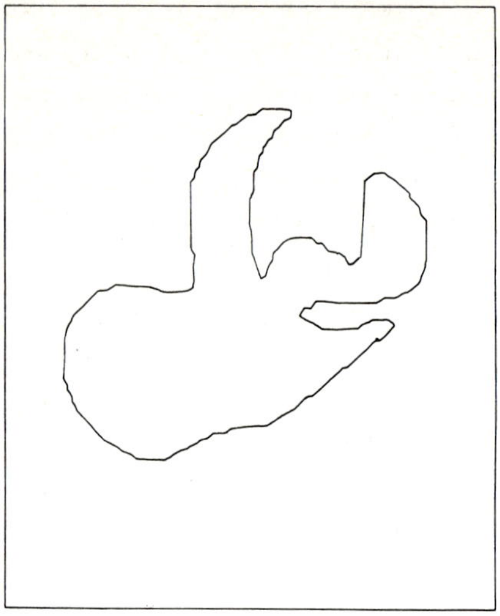

(a)

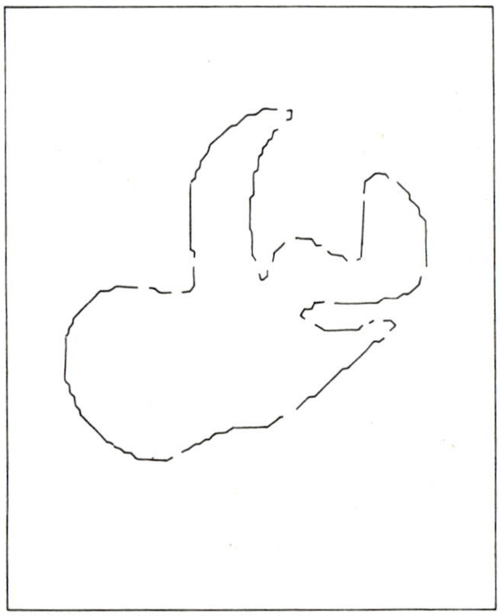

(b)

Figure 2.2. (a) A curvilinear object.
(b) Feature based segmentation into circular arcs and straight lines.

have similar slopes and lengths. From these "seed" arcs an edge segment can be "grown" by adding continuous arcs to either end of the already matched segments until a dissimilarity in the curves is found. The dissimilarity of two curves is measured by the area between the normalized pictorial graphs of their ψ-s functions. Two arcs are declared dissimilar when the measured value exceeds a preset threshold.

Edge segments grown in this way represent the portions of the object boundaries which have retained their shape through the sequence. Thus an edge segment relates two views of some part of an actual figure. The displacement between two such views provides motion measurements for the given edge segment. These measurements are then used to group the edge segments into object models under the assumption that edge segments which exhibit a common motion belong to the same object. Figure 2.3 shows two input frames and the derived models.

In contrast to the above research where detailed analysis is carried out on sections of the object boundaries, Chow and Aggarwal [12] compute measures over the complete boundaries, e.g., the centroid, major and minor axes, etc. These measures are used in conjunction with a predictive scheme to perform the dynamic scene segmentation for blob-like figures.

2.2. Pixel Based Analysis

The techniques of differencing and cross-correlation have been used extensively for comparing a pair of given images in several applications. In addition, in time-varying imagery, techniques based on temporal-spatial gradients have been developed. These techniques will be discussed in the following sections.

Differencing: A comparison of two images of a scene will indicate that the images differ in areas affected by the motions of objects. One method of comparison is to "subtract" or "difference" two images and record the results in another image referred to as the difference image. If the absolute value of the difference is above a preset threshold, the corresponding pixel is set to 1; otherwise it is set to 0. On analyzing a pair of images of a scene, it is assumed that all points in the two images and those of the difference image are referenced to a common grid, i.e. the three images are assumed to be registered.

Several researchers have used the difference images to characterize objects and their motions. For example, references [13-16]

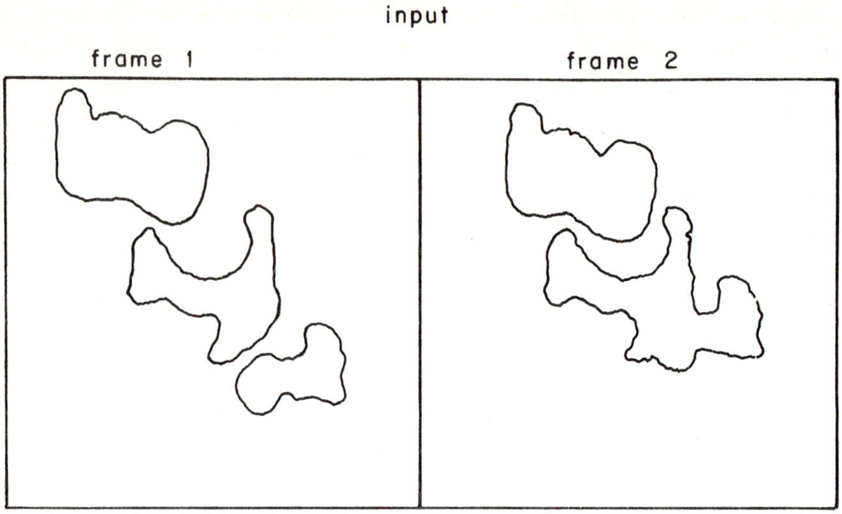

(a)

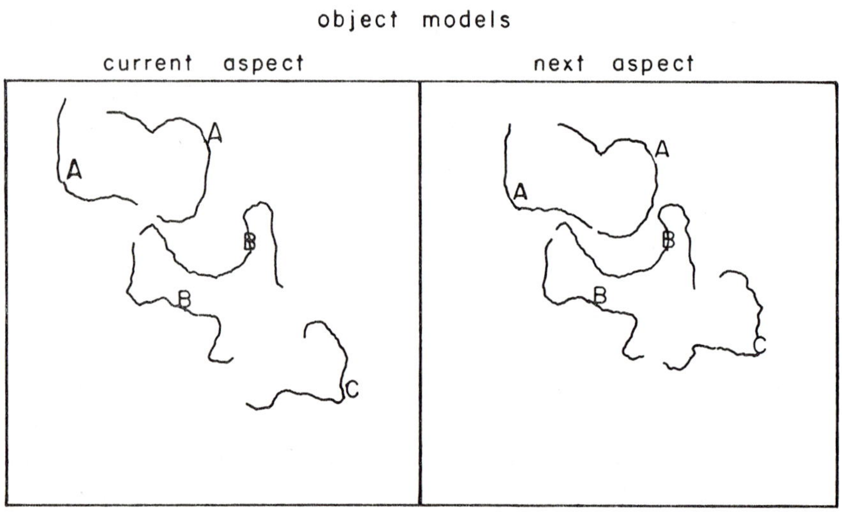

(b)

Figure 2.3. (a) Input images.
(b) Object models based on feature based segmentation and velocity measurements.

describe a variety of results obtained. The differencing technique has been applied to synthetic scenes, laboratory scenes, and real world scenes. In general the technique is applicable to both polygonal and curvilinear objects. However, theoretical analysis is available only for the case of polygonal objects. For best results, the objects are assumed to be of a homogeneous gray level. If an object of interest were to comprise areas having distinct but uniform gray levels, each homogeneous portion could be treated as an individual object. It would then remain for some higher level process, possibly using a common motion constraint to determine that all the individually identified segments were parts of the same object.

By examining the difference image, object samples can be determined and then expanded, e.g. "grown" [13,14], to effectively include all image points that correspond to the object. An example illustrating the use of differencing is given in Figure 2.4.

Cross-Correlation Analysis: Given two images and a small window in the first image, the purpose of cross-correlation is to find the region in the second image that matches the windowed region of the first image. This situation is illustrated in the Figure 2.5a,b with the matching accomplished as in template matching.

Let the second image be denoted by $f(x,y)$, the window of the first image by $w(x,y)$, and the shifted version of the window by $w(x-m,y-n)$, then their cross correlation function may be defined as

$$R(m,n) = \sum_x \sum_y f(x,y)w(x-m,y-n)$$

where the summation is taken over the region where $w(x,y)$ is defined, and (m,n) vary over the entire image $f(x,y)$. The range of the summation and the positions of the window and figure are illustrated in Figure 2.5c. As (m,n) vary, $R(m,n)$ changes and reaches a maximum at the place where $w(x,y)$ best matches $f(x,y)$. A more complicated correlation function is given by

$$R'(m,n) = R(m,n)/D(f)$$

$$\text{with } D(f) = \sqrt{\sum_x \sum_y f^2(x,y)}$$

where both summations are defined over the region where $w(x-m,y-n)$ is non-zero. The denominator varies with the position of

(a)

(b)

<u>Figure 2.4.</u> (a),(b) Two images of a laboratory scene.

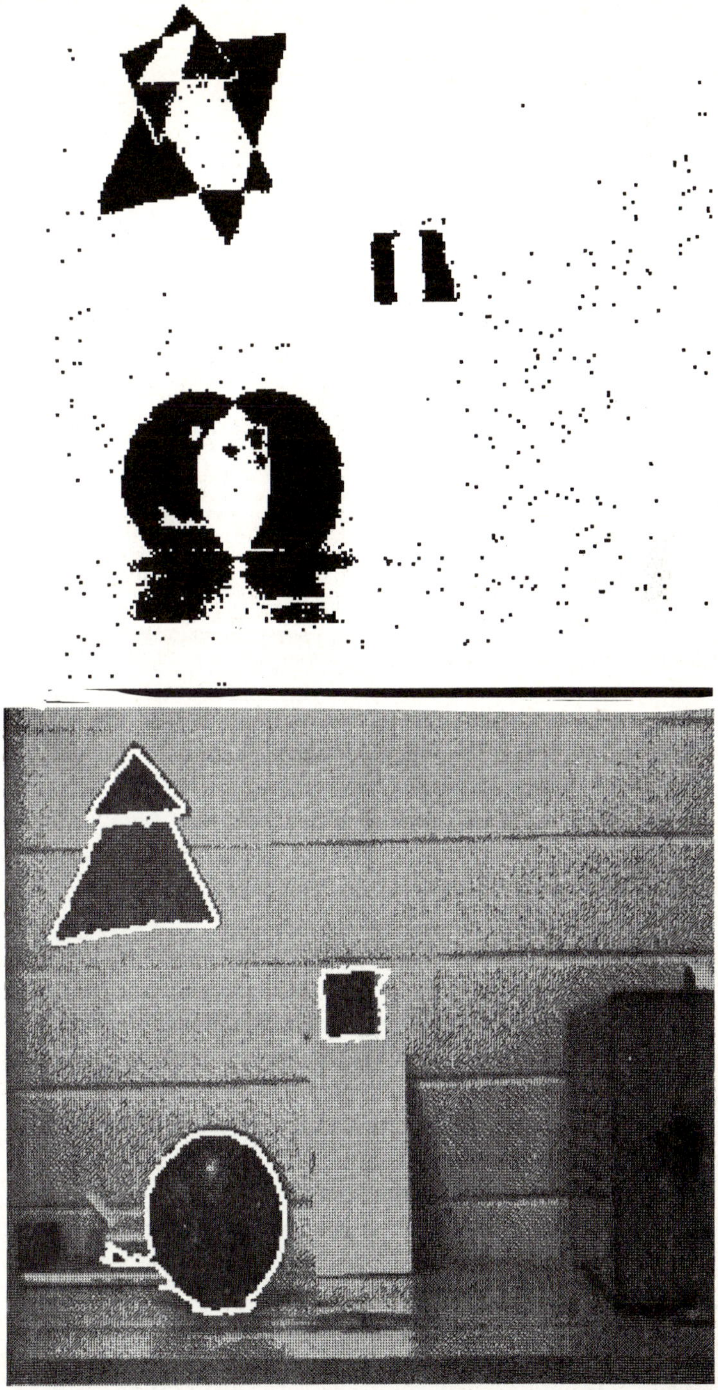

(c)

(d)

Figure 2.4. cont. (c) The difference picture of the two images.
(d) The results of the region growing process using the derived object samples.

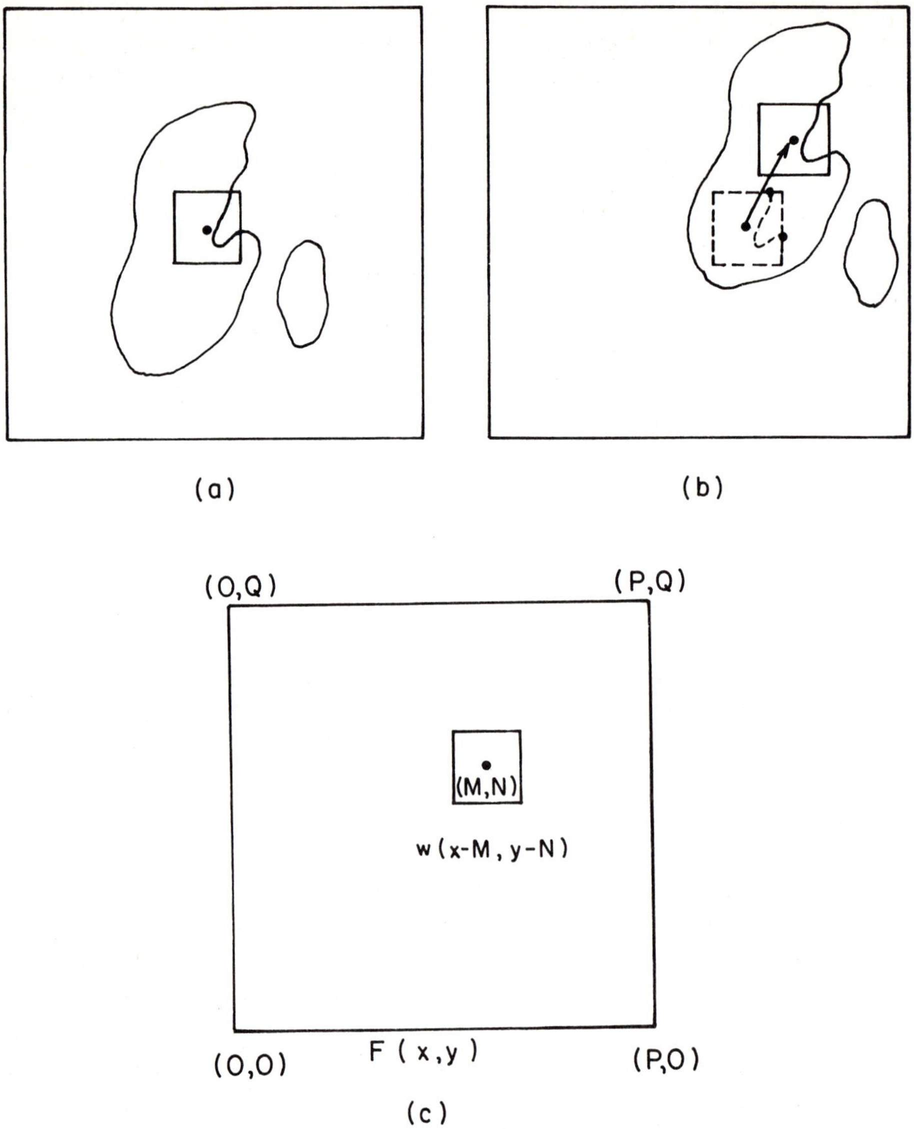

Figure 2.5. (a),(b) Two images of curvilinear objects with the windows indicating the matched regions, and the arrow denoting the movement of the object.
(c) The range of summation and the position of the window.

$w(x-m,y-n)$ and it tends to sharpen the peaks of $R(m,n)$. Details on cross-correlation are found in the book [17], whereas the application of cross-correlation to cloud motion analysis is found in [18,19].

Temporal-spatial Gradient: Let $f_1(x,y)$ and $f_2(x,y)$ denote the image intensities at the two instants, t_1 and t_2. During the intervening time interval, the image has moved by the amount Δx and Δy in the x- and y- directions, respectively. Now

$$\Delta f(x,y) = f_2(x,y)-f_1(x,y)$$

also $$f_1(x,y) = f_2(x+\Delta x,y+\Delta y)$$

therefore $$\Delta f(x,y) = f_2(x,y)-f_2(x+\Delta x,y+\Delta y)$$

Thus, at each point one can calculate Δf, $\partial f/\partial x$ and $\partial f/\partial y$; and obtain a linear equation for Δx and Δy. If one has two points, one can obtain two linear equations, and their solution would lead to the determination of Δx and Δy provided the equations were independent. In practice, one uses considerable redundancy in the number of points and obtains a least mean square solution. An excellent set of examples illustrating segmentation using temporal-spatial gradient and contrast is given by Thompson [20].

2.3 General Observations on Dynamic Scene Segmentation

Two broad approaches toward dynamic scene segmentation based on pixel and feature have been briefly presented and illustrated by example systems. The end product of the two approaches are similar, i.e., the location and the motion of moving parts in the sequence of scenes. It may be observed that the feature based methods must be supplemented by a correspondence procedure, whereas correspondence is a by-product of the pixel based analysis. This apparent superiority of pixel based techniques is quickly defeated by the simple observation that in the presence of occlusion, pixel based methods often break down. The same is true in case of structural changes in the objects. A graphical comparison of two types of methods is presented in Figure 2.6.

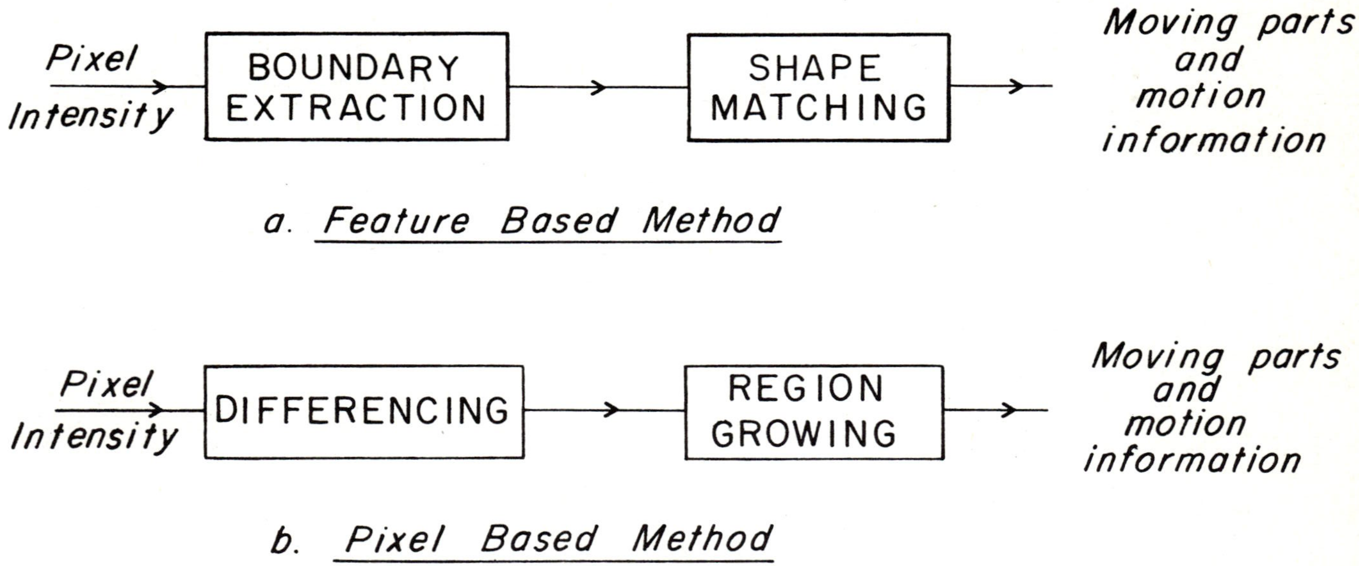

Figure 2.6. Examples of methods for the extraction of moving parts.

3.0 Occlusion

Occlusion occurs whenever the image to be analyzed is a projection of some three-dimensional scene onto a two-dimensional plane. In this general case there is always a background obscured by the objects which are considered to be the foreground. For objects widely spaced over a homogeneous background, e.g., paintings on a museum wall or a pair of birds flying in a clear sky, there is no problem in understanding the image. The background is understood to be homogeneous so that the characteristics of the obscured portions are indicated by the visible sections. The foreground objects are assumed to have image characteristics which are distinct from the background making the foreground objects readily detectable in the image. In addition, the spacing of the objects assures that the presence of the features on one object will not interfere with the analysis of the remaining foreground objects. However, if the background has a complex structure, e.g., the museum wall has a highly patterned covering, or if the foreground objects are closely arranged in some structure, e.g., a flock of birds flying in the same direction, then the classic "figure-ground" problem arises, for example see [21]. In the figure-ground problem, the spatial relationships between disjoint elements of the viewed scene combine to interfere with the perception of the individual elements. In its full generality, this is a psychophysical problem where the preconceptions and expectations of the viewer play an important part in perception. For a fuller discussion of this topic, the reader is referred to [22].

3.1 Scene Domain Imposed Constraints

In abstract geometrical patterns, both the quantity and subtlety of the inter-element relationships are greater than those occurring in typical natural scenes. Similarly, the constraints imposed by the three-dimensional structure and distribution of the objects appearing in typical scenes are greater than those in abstract patterns. For instance, the boundary edges of noncontiguous objects are rarely collinear in natural scenes. This consideration makes reasonable the assumption that if the image of a scene contains disjoint edges which are collinear, then those edges correspond to a single boundary in the scene and the discontinuity is caused by the boundary being partially obscured in the given view. Barrow and Tenenbaum [23] argue that certain psychological phenomena, such as subjective contour, are the result of the human visual system attempting to use such evidence of occlusion as a cue to apparent depth.

An elegant example of how scene domain constraints can be used in understanding occlusion is the system developed by Waltz [24]. In this case, the domain is that of scenes having a single light source illuminating a set of planar-faced objects whose vertices are trihedral. The strong constraints imposed by this scene domain are primarily embedded in a junction classification and line labeling scheme generalized from the system first discussed by Huffman [25] and Clowes [26]. Junctions are the line drawing representations of the vertices in the scene, thus the trihedral restriction of the object vertices provides extensive constraints on both the types of junctions possible and the allowable labelings of the lines forming those junctions. In particular, certain of the labeled junction types can only arise through cases of occlusion, and thus, when found in the drawings, provide a reliable indication of occlusion.

Many pictures are inherently ambiguous, and no information derived from the image can resolve the uncertainties. Frequently, it is not that the image has no consistent interpretation, but rather that there are several mutually exclusive interpretations which are each independently consistent. The choice among such alternatives must be based on the expectations or goals of the viewer, not simply on features actually exhibited in the image.

Several factors are fundamental to the understanding of scenes containing occluding objects. First, the concept of occlusion is used at a very early stage in the human visual system in order to provide interpretations in terms of apparent depth. Second, effective cues to occlusion can be derived from scene-domain constraints. And third, occlusion necessarily results in the loss of information available about the obscured object, thus causing uncertainties in the interpretation of the image. Finally, it may be observed that the use of occlusion cues may involve the complex integration of information taken from areas which are widely separated in the image and that the resolution of some occlusion ambiguities depends on external expectations and goals.

3.2 Occlusion in Image Sequences

The discussion up to this point has dealt with the implications of occlusion on the analysis of simple images. For the remainder of this section the focus will be on time-varying images. The question addressed is: How is the complexity of the occlusion analysis problem affected by the addition of time variation? The broad answer to this question is that time variation simplifies some aspects of the problem, complicates other aspects, and introduces several new problems. These points are discussed on a general level in the following.

The time variation can simplify the initial feature extraction phase of processing through both the redundancy inherent in the dynamic scenes and the opportunities provided for acquiring new information. Typically the sampling rate along the time axis is such that the majority of the scene does not change through short sequences of images. This property has been exploited for data reduction by frame-to-frame encoding of video signals [27], but can also be used to attenuate noise and produce more reliable feature values. The new information can be obtained from the changing views of the objects in the scene. For instance, if one of the occluding objects in the foreground is moving, then additional portions of the objects that it is obscuring will become visible in each successive image. Similarly, since any three-dimensional object is self-obscuring, the object's motion will usually bring into view previously unseen portions of the object. This concept has been used in a system [28] which forms a description of a planar-faced object from a sequence of views taken while the object rotates. The description is in terms of the object faces and their interconnections, which are "learned" as previously hidden faces become visible. New views also result from changes in the orientation of the image plane caused by eye (camera) movement. In these situations, areas of ambiguity in a given image may be clarified by the additional information contained in the subsequent images.

The continual change in the information content of the images, which is an advantage when the change adds information, can be a disadvantage when the change results in a reduction of available information. In each of the information-adding cases discussed above, there can be a complementary aspect in which information is lost. For instance, the moving foreground object is probably proceeding to obscure some other objects or even other portions of the same object that it is elsewhere uncovering. This aspect raises the question as to what can be said about previously visible features once they are no longer visible. If a recently obscured feature is part of an object which is still partially visible, then the relationship of the feature to the currently visible portion, as determined in preceding images, can be used to infer the location and orientation of the feature in the present scene. This type of implication is based on the assumption that the object is rigid and thus that the spatial relationships of the various features of an object will remain constant through time. This is an extremely important scene-domain constraint.

The information flux in time-varying images also creates new problems at the image segmentation and object identification levels. The problems encountered here involve the additional "semantic noise" [29] exhibited in time-varying images. Typical systems for static image analysis must be capable of interfacing with preprocessors which occasionally fail to detect, erroneously produce, or incorrectly locate image feature descriptions. Systems for time-varying images will have similar preprocessing problems but must furthermore be prepared to interpret features which, through time, may take on different values yet signify the same scene component semantically. For example, the effects of shadows on a textured outdoor surface, e.g., a gravel road bed, will vary as the sun angle changes throughout the day.

This problem of identifying "apparently different but semantically identical objects" [30] indicates a fundamental concept in the analysis of time-varying images: in order to understand the changes that a given aspect of an entity in a scene may be undergoing, there must be some form of constancy in other aspects of that same entity to serve as the identifying features of the entity. This is particularly important when there are several objects moving about the scene, because the simple detection of change cannot attribute that change to the proper object.

As an illustration, consider the illusion depicted in Fig. 3.1. Here four identical disks are attached pairwise to the ends of two cross members which are slightly offset in depth and spin in opposite directions about the center point. They exhibit constancy in both shape and color. These features make it easy to track the disks while they are moving through positions such as that of Fig. 3.1a. However, when the position shown in Fig. 3.1b is reached, the constancies no longer serve as identifying features and thus admit an ambiguity to the interpretation of the position displayed in Fig. 3.1c.

Is the pair of velocities labeled A in Fig. 3.1c the correct interpretation or is the pair labeled B correct? An assumption of

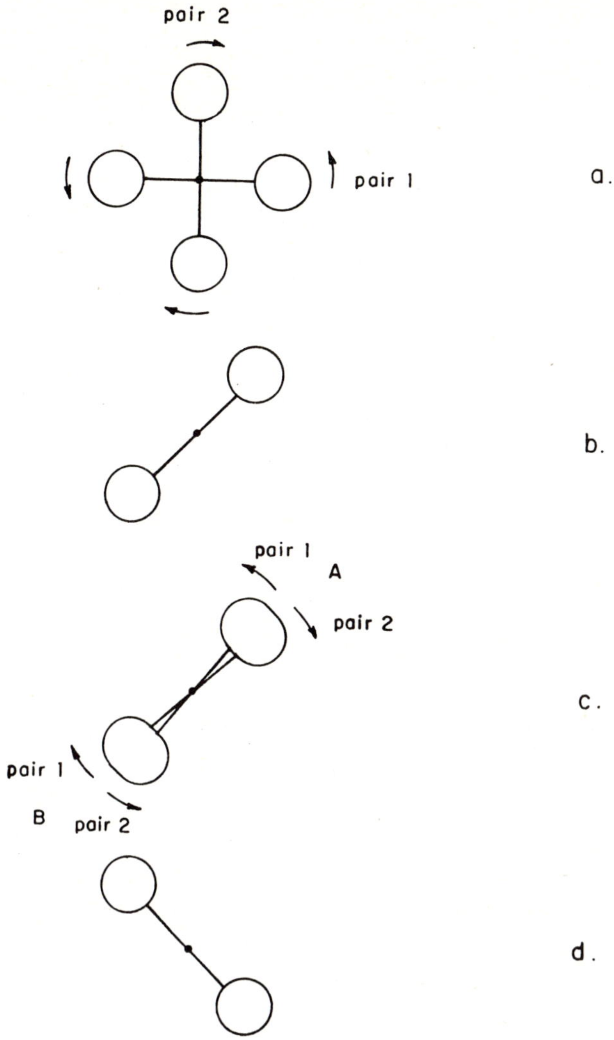

Figure 3.1. Four spinning disks forming a motion illusion.

minimal velocity change for each object results in a perception according to the velocities labeled A. In such a perception the disks appear to have circular paths and pass completely through one another at positions such as those shown in Fig. 3.1b and d. A rather more complicated proximity criterion, which holds that the disk last viewed completely in a given quadrant will return immediately to that quadrant, yields a perception according to the velocities labeled B. In this latter case, each disk sweeps both back and forth through a given quadrant as bounded by the positions shown in Fig. 3.1b and d. At these positions the disks appear to "bounce" off each other thus exactly reversing their velocities.

The two cases discussed above can, however, be understood in terms of two different types of constancy, one involving velocity, and the other, occupancy. These constancies can only be used to resolve the ambiguity in an indirect way because the ambiguity occurs when one is trying to understand the progression from an image of a position such as that of Fig. 3.1b to the immediately succeeding image. In the constant velocity case, for example, the instantaneous velocity is measured as the displacement in disk location between two successive images. But the location of the disk in the image after that of Fig. 3.1b is precisely what is in question. Thus the analysis of these two images in isolation cannot resolve the ambiguity. Instead the velocity information must be derived from the preceding images in which the constancies of shape and color can be used to locate each disk, thereby allowing the calculation of its velocity. The velocity information can then be applied to the given pair of images as part of a predictive analysis or as the criterion for a hypothesis and test procedure.

4.0 Three-Dimensional Information from Images

Early studies of sequences of images were motivated by the desire to analyze two-dimensional motion, for example, the satellite imagery of clouds. Several researchers also considered abstract models of two-dimensional motion using polygonal as well as curvilinear figures. The use of planar figures and parallel projection allowed these systems to ignore considerations of the third dimension. In contrast to the above purely two-dimensional works (reviewed in [1]), certain researchers have considered scenes containing objects undergoing three-dimensional motion. The initial research, however, analyzed only the image plane motions taking the two-dimensional approximation to be adequate. For example, in the work of Jain and Nagel [31] and Yachida et al [32], there was no attempt to recover the three-dimensional structure of the objects or their three-dimensional motion. This emphasis on two-dimensional motion was a natural outgrowth of the research. Recovery of three-dimensional structure of objects and the parameters of motion is certainly more complex.

In the present discussion it is assumed that the low-level processing problems have already been solved, i.e., the feature points on the surface of the objects have already been extracted in each of the images and the correspondence between the feature points in various images have been established. As mentioned earlier, this is a non-trivial task. Later in this section, we shall consider another scenario where this establishing of correspondence is not necessary. The correspondence problem is further complicated by the disappearance of points on an object due to occlusion from other objects, self occlusion as points rotate out of view, and shadows. Also, the assumption of rigidity plays an important role.

Ullman [33] considers the problem of determining the three-dimensional structure of an object from its two-dimensional projections. Under the assumptions of object rigidity and parallel projection, Ullman proved that three distinct views of four non-coplanar points in a rigid configuration, enable one to determine uniquely the motion and structure compatible with the given three views. Roach and Aggarwal [34] give an alternate solution for the case of central projection. They showed that two views of five points leads to 18 nonlinear equations whose solution yields the three-dimensional structure of points under consideration. Bonde and Nagel [35] consider a restricted case of the above general three-dimensional motion. Badler [36] uses a spherical projection model and is able to predict the point positions in succeeding images of moving objects. More recently Tsai and Huang [Chapter 1 in ref. 7, 37] reformulate the problem in terms of five unknown motion parameters and show certain existence results. Also Nagel [38] has derived a compact vector equation to determine three-dimensional points from two-dimensional image points. In view of the several formulations and different results, it may be emphasized that the results obtained depend on the following assumptions: (i) the nature of projection: parallel, central or spherical; (ii) the number of points and the existence of any relationships among these points, e.g., object rigidity; and (iii) the number of available views. In all of the above works, rigidity of the object under consideration and pre-establishment of the correspondence of points are assumed.

Our group at The University of Texas at Austin has considered three distinct problems: the recovery of three-dimensional structure under the assumption of central projection [34]; the motion of articulated objects under parallel projection [39]; and the derivation of volumetric descriptions from occluding contours with viewpoint specifications [40]. These projects will be briefly discussed.

Feature Points from Rigid Objects: The image of a point under central projection is a function of the point's three-dimensional position, the focal length of the camera, and the location and orientation of the camera's lens relative to the global coordinate system. Information about the camera's position is needed to relate the position of points given in two-dimensional focal plane coordinates to the global three-dimensional coordinate system. The necessary camera information is the camera focal length F, the orientation angles Θ,ϕ, and κ of the camera to the global coordinate system, and the three-dimensional coordinates of the lens center (X_0,Y_0,Z_0). The three angles orient the camera to the global coordinate system as follows: (assume for simplicity that the camera lens center has been translated to (0,0,0) of the global coordinate system) Θ is a rotation about the X-axis that brings the optical axis into the X-Z plane, ϕ is a rotation about the Y-axis so that the optical axis is aligned with the Z-axis, and κ is a rotation about the Z-axis so that the x', y' axes of the focal plane are aligned with the global X,Y axes. The use of primes in this section in general denotes the focal plane coordinate system. It is, of course, impossible to determine the original (x,y,z) position of a point from a single image. The best we can do is to determine a line in space on which the point falls. Further explanations of the equations in this section may be found in [34], and [41].

However, we want to know how much of the original three-dimensional information can be recovered given only a sequence of images of a moving object. It is possible to show that any sequence of images is inherently ambiguous. That is, there are an infinite number of objects that produce the same sequence of images. The objects are all similar in structure and movement.

In the following we discuss how to find the movement and three-dimensional model of points on an object's surface from a sequence of noise-free images up to a scaling factor; that is, by setting the scaling factor to an arbitrary value we can find a particular movement and model for the points on the object.

In the description above it was assumed that the camera is stationary and the object is moving. It is convenient to reformulate the problem such that the object is stationary and the camera moves. Under this formulation the three-dimensional structure and motion of an object can be derived from two views each containing five feature points from the object surface such that not all five of the surface points are in the same plane.

The solution can be obtained from a system of non-linear equations specified in the following manner. The global coordinates of each point are variable, so five points produce 15 variables. The global coordinates and the Θ,ϕ,κ orientation angles for each camera position are also variable producing 12 more variables. Thus, there are a total of 27 variables in the problem. Each point produces two projection equations per camera position for a total of 20 nonlinear equations. To make the number of equations and unknowns come out the same, seven variables must be known including one variable that will determine the scaling factor.

Six of the variables are specified by assuming the first camera position is coincident with the global axis system, that is, set the X_0, Y_0, Z_0 position and Θ, ϕ, κ orientation angles of the first camera to zero. In addition, the z-component of any one of the five points is set to an arbitrary positive constant. We mentioned earlier that the best result possible in locating the three-dimensional position of a point on an object is to find (sx, sy, sz) where s is an arbitrary scaling factor. By setting the z-component of the position of a point to an arbitrary constant, we are fixing the scaling factor. Once the z-component of a point is known, the x and y components can also be found using the inverse of the projection equations (see [34] that is, two of the 20 equations can be solved directly using the given z-component). There are now 18 projection equations in 18 unknowns; the equations of projection, however, are nonlinear. The situation is shown in Figure 4.1.

The system of nonlinear projection equations explained above can be solved by using a modified finite difference Levenberg-Marquardt algorithm due to Brown [42-44] without strict descent

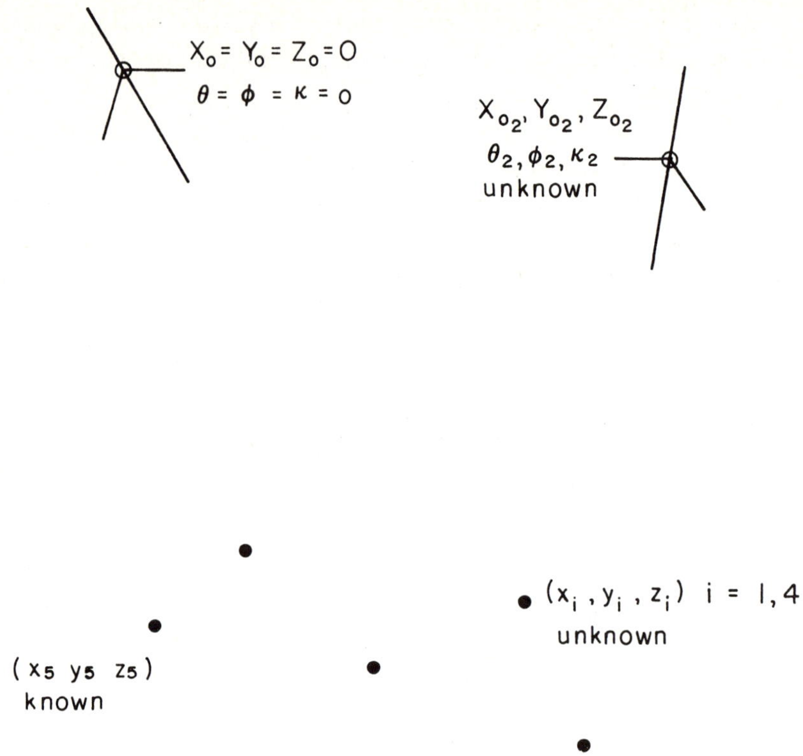

Figure 4.1. Camera configuration and point coordinates showing known and unknown parameters.

that minimizes the least-squared error of the 18 equations. The method employed is iterative and requires an initial guess for each unknown parameter.

This work is somewhat like the camera calibration systems of Sobel [45] and Yakimovsky and Cunningham [46]. In their work multiple images of points together with a central projection model and numerical methods are used to determine camera parameters such as focal length, position, and orientation. These studies, however, have considerably more information about the three-dimensional positions of points than we are assuming. Thus, the problems being solved and the information given for the calibration systems are different from the work described in this section.

Implicit in this work are two very important assumptions: that the objects being observed are rigid and that the images of the object are noise free and thus completely accurate. To test the effect of the second assumption on the numerical method described above, from one to four pixels were randomly added to or subtracted from the exact photocoordinate data for a moving object. This perturbation of the data causes extreme instability in the numerical solutions. However, one of the main reasons for using a least-squared error technique to solve a problem is to make adjustments to observations that contain error (noise). Adjustment is only possible, however, when there are more equations than unknowns. Two views of five points are therefore inadequate for noisy data since there are the same number of equations as unknowns. Two views of six points or three views of four points produce 22 equations in 21 unknowns using the same problem model discussed above. Examination of experimental runs using overdetermined systems of equations shows that minimal overdetermination is not very accurate. It is only with considerable overdetermination (two views of 12 or even 15 points; three views of seven or eight points) that the results become accurate.

<u>A Motion Constraint to Recover Structure</u>: In this section we will consider a system [39] that uses a motion constraint to derive three-dimensional structures from sequences of feature point images. The constraint is that the motion of each rigid object in the scene consists of a translation and a rotation about an axis that is <u>fixed in direction</u>. Of course, given two distinct frames from a dynamic scene containing a moving rigid object, a transformation taking the first three-dimensional position of the object to the second three-dimensional position can always be defined that

consists of only one translation and one rotation [47]. However, the restriction is that the direction of the axis of rotation is fixed throughout the sequence, or at least over subsequences of frames.

To understand the importance of this restriction consider first, a dynamic scene containing an object that is simply rotating about a stationary axis. The surface of the object is assumed to produce identifiable feature points in the images under an orthogonal projection without occlusion. Clearly, as the object rotates, each surface point will trace out a circle about the axis of rotation. The parallel projection assures that the feature point, i.e., image position, of each surface point will trace out an ellipse in the image plane. Now, the axes of the ellipse can be used to specify the orientation, in three-dimensions, of the plane containing the original circle and thereby specify the direction of the axis of rotation. Deriving this information for the given feature points of an object allows one to build a model of the three-dimensional structure of the object from the sequence of images. This simple case indicates the basic steps in the process.

In general, since the motion constraint does not demand zero translation nor a completely fixed axis of rotation, the surface points will not actually be tracing circles and their projections will not be ellipses. If, however, an arbitrary surface point is chosen as the reference point and the object motion is considered in a coordinate axis system having the reference point as origin, then the motion in that system follows the case described above. In other words, if the motion could be normalized to a coordinate axis system with a surface point as origin then the remaining surface points would trace circles about an axis through the reference point, the projected positions would trace ellipses and the three-dimensional structure could be derived.

Clearly, this normalization is not directly possible, but it has been shown [39] that the given image sequence can be normalized to yield the same effect. In particular, the feature point images are translated so that the image plane origin coincides in each case with the projected position of an arbitrarily selected surface point. The processing proceeds as follows: the image sequence is acquired; feature points are detected for various surface points of the object; the feature point images are normalized to have a selected feature point as the origin of each image; the parameters of the traced ellipses are determined; and finally, the three-dimensional structure is derived.

There are two major advantages of this approach. First, it can accomodate dynamic scenes containing several independently moving objects by assuming that a set of feature points mutually satisfying the motion constraint, i.e., each element of the set traces an ellipse relative to the same reference point, constitutes the feature points for a single object. In detail, a feature point is selected as the reference point, the images are normalized to this reference point, and all ellipse tracing points are included in an object description. These points are deleted from further consideration and the process is iterated until every feature point is part of some object model. Ullman [33] used his rigidity constraint in a similar manner.

The second advantage is that analytically only two feature points need be detectable in three consecutive frames to derive the three-dimensional model. In this way articulated objects with only two feature points on each part can be analyzed. By "articulated object" is meant an object comprising several rigid parts connected through various joints that allow distinct but not independent movements for each part. The experiments of Johansson [48] provide a good example of articulated objects with minimal numbers of feature points. In those experiments people with reflectors attached to their major joints moved about in front of a camera that was calibrated to record only the positions of the reflectors. The system discussed in this section adequately analyzed an example taken from a Johansson experiment as reported in [49].

<u>Occluding Contours in Dynamic Scenes</u>: In this section we will describe the results of work [40] done in the pursuit of two major goals. The first goal is the development of a dynamic scene analysis system that does not depend completely on feature point measurements. The second goal is the development of a scheme for representing three-dimensional objects that is descriptive of surface detail, yet remains functional in the context of structure from motion in dynamic scenes.

To lessen the dependency on feature point detection, occluding contours with viewpoint specifications are used. The term "occluding contour" means the boundary in the image plane of the silhouette generated by an orthogonal projection. Silhouettes can most often be formed by a simple thresholding of the intensity values. A connected component analysis [50, pp. 336-347] of the resulting binary valued image yields the boundary of the object silhouette. An ordered list of the image plane coordinates of the resulting boundary constitutes the initial representation of the occluding contour. Throughout the analysis of the dynamic image, however, another representation, referred to as the rasterized area [51], of the contour will also be used. For its use here, the most significant attribute of this representation is that given an area so represented and an arbitrary segment on a line parallel to the "raster direction" it is a simple process to determine what portions of the given segment intersect the area, i.e., to clip that segment to the represented area.

The three-dimensional structure to be derived from the sequence of occluding contours is a bounding volume approximation to the actual object. For this reason the representation incorporated in this system is based on volume specification through a "volume segment" data structure. The volume segment representation is a generalization to three-dimensions of the rasterized area description. For the rasterized area, each of the segments denoted a rectangular area. The generalization to three dimensions is to have each segment represent a volume, i.e., a rectilinear parallelepiped with edges parallel to the coordinate axes. In addition to grouping collinear segments into lists, the set of segment lists is partitioned so that the subsets contain lists having coplanar segments. The primary dimension of the parallelepiped specified by a segment is the length of the segment. The second dimension is given by the inter-segment spacing within the plane of the segment, while the third dimension is the inter-plane distance. The latter two dimensions are specified to be uniform throughout the volume segment representation.

The primary advantage of this structure in general situations is that the process of determining whether an arbitrary point is within the surface boundary consists of a simple search of three ordered lists: select a "plane" by z-coordinate; select a "line" by x-coordinate; and finally, check for inclusion of the y-coordinate in a segment.

This volume segment representation is created from a dynamic image by two processes. The first process combines information from frames 1 and 2 of the dynamic image to form an initial volume segment representation. The second process then accepts each succeeding frame in order to refine the approximation represented by the volume segment structure. Thus these processes analyze the occluding contours with their view orientations to initially construct and to continually refine the volume segment representation of the object generating the contours. Algorithm summaries of the two processes are given in more detail in [52,53].

5.0 Conclusions

The three major ingredients of Dynamic Scene Analysis, namely segmentation, occlusion and computation of three-dimensional information from images have been discussed in depth. The approaches to dynamic scene segmentation may be broadly divided into two classes - pixel and feature based techniques. Each has its advantages and disadvantages. It appears that a combination of the two approaches may be necessary to accomplish segmentation in a difficult dynamic scene segmentation task. Occlusion introduces some really difficult problems in dynamic scene analysis. Effective cues to occlusion may be derived from scene-domain constraints. The use of these cues depends upon the ability to integrate information derived from widely separated spatial areas in an image or widely separated temporal events in a sequence. The techniques for computation of three-dimensional information from images are still in their early stages of evolution. The solution of a large number of nonlinear equations and the sensitivity of solutions to noise pose serious hurdles in a satisfactory computation of three-dimensional information from images. The absence of an ideal description of three-dimensional structure of objects compounds the difficulties. Further, in each of the three areas discussed above, the correspondence problem plays an important role.

In view of the active participation at the Braunlage Advanced Study Institute, it appears that the next decade will be an exciting era in Dynamic Scene Analysis. In addition to the development of new techniques for the solution of problems discussed above, one will witness the application of dynamic scene analysis to many new areas.

References

1. W. N. Martin and J. K. Aggarwal, "Dynamic Scene Analysis: A Survey," Computer Graphics and Image Processing, Vol. 7, No. 3, pp. 356-374, June 1978.

2. H.-H. Nagel, "Analysis Techniques for Image Sequences," Proc. 4th International Joint Conference on Pattern Recognition, Kyoto, Japan, November 1978, pp. 186-211.

3. J. K. Aggarwal and N. I. Badler, eds., Abstracts for the Workshop on Computer Analysis of Time-Varying Imagery, Philadelphia, PA, April 1979.

4. J. K. Aggarwal and N. I. Badler, guest eds., Special Issue on Motion and Time-Varying Imagery, IEEE Trans. on Pattern Analysis and Machine Intelligence Vol. PAMI-2, No. 6, November 1980.

5. W. E. Snyder, guest ed., Computer Analysis of Time-Varying Images, Computer, Vol. 14, No. 8, August 1981.

6. J. K. Aggarwal, guest ed., Special Issue of Computer Graphics and Image Processing, (to appear).

7. T. S. Huang, Image Sequence Analysis, Springer-Verlag, New York, 1981.

8. NATO Advanced Study Institute, on Image Sequence Processing and Dynamic Scene Analysis, Advance Abstracts of Invited and Contributory Papers, June 21-July 2, 1982, Braunlage, W. Germany.

9. J. K. Aggarwal and R. O. Duda, "Computer Analysis of Moving Polygonal Images," IEEE Trans. on Computers, Vol. C-24, pp. 966-976, Oct. 1975.

10. W. N. Martin and J. K. Aggarwal, "Computer Analysis of Dynamic Scenes Containing Curvilinear Figures," Pattern Recognition, Vol. 11, pp. 169-178, 1979.

11. J. W. McKee and J. K. Aggarwal, "Finding the Edges of the Surfaces of Three-Dimensional Curved Objects by Computer," Pattern Recognition, Vol. 7, pp. 25-52, 1975.

12. C. K. Chow and J. K. Aggarwal, "Computer Analysis of Planar Curvilinear Moving Images," IEEE Trans. on Computers, Vol. C-26, pp. 179-185, 1977.

13. R. Jain, W. N. Martin and J. K. Aggarwal, "Segmentation Through the Detection of Changes Due to Motion," Computer Graphics and Image Processing, Vol. 11, pp. 13-34, 1979.

14. S. Yalamanchili, W. N. Martin and J. K. Aggarwal, "Extraction of Moving Object Descriptions Via Differencing," Computer Graphics and Image Processing, Vol. 18, pp. 188-201, 1982.

15. R. Jain and H.-H. Nagel, "On the Analysis of Accumulative Difference Pictures from Image Sequences of Real World Scenes," IEEE Trans. on Pattern Analysis and Machine Intelligence Vol. PAMI-1, No. 2, pp. 206-214, 1979.

16. H.-H. Nagel, "Formation of an Object Concept by Analysis of Systematic Time Variations in the Optically Perceptible Environment," Computer Graphics and Image Processing, Vol. 7, No. 2, pp. 149-194, 1978.

17. R. C. Gonzalez and P. Wintz, Digital Image Processing, Addison Wesley Publ. Co., Inc., Reading, MA, 1977, pp. 383-386.

18. J. A. Leese, C. S. Novak and V. R. Taylor, "The Determination of Cloud Pattern Motions from Geosynchronous Satellite Image Data," Pattern Recognition, Vol. 2, pp. 279-292, Dec. 1970.

19. J. A. Leese, C. S. Novak and B. B. Clark, "An Automated Technique for Obtaining Cloud Motion from Geosynchronous Satellite Data Using Cross-Correlation," J. Applied Meteorology, Vol. 10, pp. 118-132, Feb. 1971.

20. W. B. Thompson, "Combining Motion and Contrast for Segmentation," IEEE Trans. on Pattern Analysis and Machine Intelligence, Vol. PAMI-2, No. 6, pp. 543-549, 1980.

21. D. R. Hofstadter, Godel, Escher, Bach: An Eternal Golden Braid, Vintage Books, New York, 1980.

22. J. K. Aggarwal and W. N. Martin, "Analyzing Dynamic Scenes Containing Multiple Moving Objects," in Image Sequence Analysis, T. S. Huang, ed., Springer-Verlag, New York, 1981, pp. 355-380.

23. H. G. Barrow and J. M. Tenenbaum, "Recovering Intrinsic Scene Characteristics from Images," in Computer Vision Systems, A. R. Hanson and E. M. Riseman, eds., Academic Press, New York, 1978.

24. D. Waltz, "Understanding Line Drawings of Scenes with Shadows," in Psychology of Computer Vision, P. H. Winston, ed., McGraw-Hill, New York, 1975, pp. 19-92.

25. D. Huffman, "Impossible Objects as Nonsense Sentences," in Machine Intelligence 6, B. Meltzer and D. Michie, eds., Edinburgh University Press, Edinburgh, Scotland, 1971.

26. M. Clowes, "On Seeing Things," Artificial Intelligence, Vol. 2, No. 1, pp. 79-116, 1971.

27. F. W. Mounts, "A Video Encoding System With Conditional Picture-Element Replenishment," Bell Syst. Tech. J., 48, pp. 2545-2554, 1969.

28. S. A. Underwood and C. L. Coates, "Visual Learning from Multiple Views," IEEE Trans. on Computers, Vol. C-24, pp. 651-661, 1975.

29. A. Guzman, "Decomposition of a Visual Scene into Three-Dimensional Bodies," in Computer Methods in Image Analysis, J. K. Aggarwal, R. O. Duda and A. Rosenfeld, eds., IEEE Press, New York, 1977, pp. 324-337.

30. R. P. Futrelle and M. J. Potel, "The System Design for GALATEA, an Interactive Real-Time Computer Graphics System for Movie and Video Analysis," Computer Graphics, Vol. 1, pp. 115-121, 1975.

31. R. Jain and H.-H. Nagel, "On a Motion Analysis Process for Image Sequences from Real World Scenes," IFI-HH-B-48/78, Institut fuer Informatik der Universitat Hamburg, Hamburg, Germany, April 1978.

32. M. Yachida, M. Asada and S. Tsuji, "Automatic Motion Analysis of Moving Objects from the Record of Natural Process," Proc. 4th International Joint Conference on Pattern Recognition, Kyoto, Japan, November 1978, pp. 726-730.

33. S. Ullman, The Interpretation of Visual Motion, MIT Press, Cambridge, MA, 1979.

34. J. W. Roach and J. K. Aggarwal, "Determining the Movement of Objects from a Sequence of Images," IEEE Trans. on Pattern Analysis and Machine Intelligence, Vol. PAMI-2, No. 6, pp. 554-562, 1980.

35. T. Bonde and H.-H. Nagel, "Deriving a 3-D Description of a Moving Rigid Object from Monocular TV-Frame Sequences," WCATVI-79, Philadelphia, PA, pp. 44-45.

36. N. Badler, "Temporal Scene Analysis: Conceptual Descriptions of Object Movements," Ph.D. dissertation, University of Toronto, Toronto, Ont., Canada, 1975.

37. R. Y. Tsai and T. S. Huang, "Estimating Three-Dimensional Motion Parameters of a Rigid Planar Patch," Proc. of the IEEE Conf. on Pattern Recognition and Image Processing, Dallas, TX, August 1981, pp. 94-97.

38. H.-H. Nagel, "On the Derivation of 3-D Rigid Point Configurations from Image Sequences," Proc. of the IEEE Conf. on Pattern Recognition and Image Processing, Dallas, TX, August 1979, pp. 103-108.

39. J. A. Webb and J. K. Aggarwal, "Structure from Motion of Rigid and Jointed Objects," to appear in Artificial Intelligence.

40. W. N. Martin and J. K. Aggarwal, "Occluding Contours in Dynamic Scenes," Proc. of the IEEE Conf. on Pattern Recognition and Image Processing, Dallas, TX, August 1981, pp. 189-192.

41. J. W. Roach, "Determining the Three-Dimensional Motion and Model of Objects from a Sequence of Images," Ph.D. dissertation, Department of Computer Sciences, The University of Texas, Austin, TX, May 1980.

42. K. M. Brown and J. E. Dennis, "Derivative Free Analogues of the Levenberg-Marquardt and Gauss Algorithms for Nonlinear Least Squares," Numer. Math., Vol. 18, pp. 289-297, 1972.

43. K. Levenberg, "A Method for the Solution of Certain Non-Linear Problems in Least Squares," Quart. Appl. Math., Vol. 2, pp. 164-168, 1944.

44. D. W. Marquardt, "An Algorithm for Least-Squares Estimation of Nonlinear Parameters," J. SIAM, Vol. 11, No. 2, 1963.

45. I. Sobel, "On Calibrating Computer Controlled Cameras for Perceiving 3-D Scenes," Artificial Intelligence, Vol. 5, No. 2, pp. 185-198, 1974.

46. Y. Yakimovsky and R. Cunningham, "A System for Extracting Three-Dimensional Measurements from a Stereo Pair of TV Cameras," Computer Graphics and Image Processing, Vol. 7, No. 2, pp. 195-210, 1978.

47. J. Coffin, Vector Analysis, second edition, John Wiley & Sons, New York, 1911.

48. G. Johansson, "Visual Motion Perception,"Scientific American, Vol. 232, No. 6, pp. 76-89, June 1975.

49. J. A. Webb and J. K. Aggarwal, "Visually Interpreting the Motion of Objects in Space," Computer, Vol. 14, No. 8, pp. 40-46, August 1981.

50. A. Rosenfeld and A. C. Kak, Digital Picture Processing, Academic Press, New York, 1976.

51. W. M. Newman and R. F. Sproull, Principles of Interactive Computer Graphics, second edition, McGraw-Hill, New York, 1979.

52. W. N. Martin and J. K. Aggarwal, "Analyzing Dynamic Scenes," Laboratory for Image and Signal Analysis, The University of Texas at Austin, TR-81-5, December 1981.

53. W. N. Martin and J. K. Aggarwal, "Volumetric Descriptions of Objects from Multiple Views," to appear.

Acknowledgements

This research was supported by the Air Force Office of Scientific Research under grant AFOSR 82-0064.

The authors thank Chiun-Hong Chien for proofreading, Sudha Yalamanchili for the generation of the pictures and Linda Melnick for inputting the manuscript.

PART II

IMAGE SEQUENCE CODING

RECURSIVE MOTION COMPENSATION: A REVIEW

by

J. D. Robbins
and
A. N. Netravali

Bell Laboratories
Holmdel, NJ 07733

ABSTRACT

We present algorithms for estimating frame-to-frame changes in intensity of a television scene. The changes can be a result of object motion in a TV scene or variation of illumination. By a gradient search technique, which seeks to minimize a functional of the interframe intensity prediction error, we estimate two parameters associated with these intensity changes - translational displacement of objects and a scaling factor associated with illumination changes. Algorithms for the parameter estimation are given in both the picture element and transform domain. Using the estimates of these parameters, we also develop, for motion compensated interframe coding, adaptive predictors and a segmenter to determine which pels need to be transmitted. We describe several coder variations and compare them by computer simulations using several scene sequences. Simulations indicate that motion compensation reduces the bit rates by 30 to 70 percent compared to conditional replenishment.

NATO ASI Series, Vol. F2
Image Sequence Processing and Dynamic Scene Analysis
Edited by T.S. Huang

RECURSIVE MOTION COMPENSATION: A REVIEW

by

J. D. Robbins
and
A. N. Netravali

Bell Laboratories
Holmdel, NJ 07733

1. INTRODUCTION

Television signals contain a significant amount of frame-to-frame redundancy. Some of this redundancy is removed by techniques of conditional replenishment, in which only picture elements that have changed from the previous frame by at least a threshold amount are transmitted.[1-5] Conditional replenishment has recently been improved by displacement compensation.[6-14] The displacement compensation schemes have been developed primarily to compensate for translational displacement of objects in uniform illumination. This is done by estimating the translation of the object in the scene and using it for predictive coding by taking differences of elements with respect to appropriately displaced elements in the previous frame. Such schemes have been developed in both the picture element (pel) domain and the transform domain. Extensions have been made to models of intensity changes that incorporate not only the translational displacement but also variation of illumination.[14] Interframe coders based on these principles work by:

(i) Segmenting each television frame into two parts, one part which is predictable from the previous data, and the unpredictable part.

(ii) Transmitting two types of information: (a) addresses specifying the location of the picture elements in the unpredictable area, and (b) information (usually quantized prediction error) by which the intensities of the unpredictable area can be updated.

(iii) Matching the coder bit rate to the channel rate. Since the motion in a real television scene occurs randomly and in bursts, the amount of information about the unpredictable area will change as a function of time. It is transmitted over a constant bit rate channel by, (a) storing it in a buffer prior to transmission to smooth out the transmitted information rate, and (b) using the buffer fullness to regulate the encoded bit rate by varying the amplitude, spatial, and temporal resolution of the television signal. Intensities of the picture elements (pels) in the unpredictable areas are transmitted by predictive coding.

In this paper we review recursive methods of estimating displacement and other parameters in picture element or transform domain and give simulation results for monochrome sequences. Many extensions of our techniques have been made but have not been considered in this review. For example, motion compensation for sequences of color frames in composite and component domain have not being considered.[15-17] Also, we do not consider use of motion estimation in the process of interpolation that is required in interframe coding.[18-19] The paper is organized as follows. In the next section we describe several algorithms for estimation of displacement and other parameters. The third section describes the use of these parameters for coding and finally the results of simulations using these algorithms are presented.

2. *MOTION ESTIMATION*

The algorithms to be described in this section are recursive. They start with an initial estimate of the motion parameters and revise these estimates at predetermined rates (e.g., at every pel along a scanning line or, at every nth pel along the scanning line).

Let $I(\underline{x}_k,t)$ denote the intensity of a scene at the kth sample point $\underline{x}_k$ of a scan line, and let $I(\underline{x}_k,t-\tau)$ denote the intensity at the same spatial location in the previous frame. If the scene consists of an object that is undergoing pure translation, then, neglecting background,

$$I(\underline{x}_k,t) = I(\underline{x}_k-\underline{\mathbf{D}},t-\tau), \tag{1}$$

where $\underline{D}$ is the displacement of the object in one frame interval τ. Pel-recursive displacement estimation attempts to estimate $\underline{D}$ by minimizing the squared value of the displaced frame difference.

$$DFD(\underline{x}_k, \hat{\mathbf{D}}) = I(\underline{x}_k, t) - I(\underline{x}_k - \hat{\mathbf{D}}, t-\tau), \tag{2}$$

recursively with k using a steepest descent algorithm of the form:

$$\hat{\mathbf{D}}_{k+1} = \hat{\mathbf{D}}_k - \tfrac{1}{2}\, \epsilon\, \nabla_{\hat{\underline{D}}_k} \left[DFD(\underline{x}_k, \hat{\underline{D}}_k)\right]^2, \tag{3a}$$

where $\nabla_{\hat{\underline{D}}_k}$ is the two-dimensional gradient operator with respect to $\hat{\underline{D}}_k$. Carrying out this operation in (3a) and using (2) yields

$$\hat{\mathbf{D}}_{k+1} = \hat{\mathbf{D}}_k - \epsilon\, DFD(\underline{x}_k, \hat{\mathbf{D}}_k)\ \nabla I(x_k - \hat{\mathbf{D}}_k, t - \tau), \tag{3b}$$

where $\nabla = \nabla_{\underline{x}}$ is the two-dimensional spatial gradient operator with respect to horizontal and vertical coordinates x_1 and x_2 in $\underline{x} = (x_1, x_2)^T$:

$$\nabla I(\underline{x}_k - \hat{\mathbf{D}}_k \cdot t - \tau) = \begin{bmatrix} \frac{\partial}{\partial x_1} \\ \frac{\partial}{\partial x_2} \end{bmatrix} I(\underline{x}, t - \tau)\}_{\underline{x}=\underline{x}_k - \hat{\mathbf{D}}_k}. \tag{4}$$

Superscript T denotes transpose of a vector or matrix.

Algorithm 1

The above displacement estimator requires multiplication at each iteration which is undesirable for hardware implementation, and is therefore simplified to:

$$\hat{\underline{\mathbf{D}}}_{k+1} = \hat{\underline{\mathbf{D}}}_k - \epsilon\, sgn\left[DFD\,(\underline{x}_{k+1},\hat{\underline{\mathbf{D}}}_k)\right]\cdot sgn\left[\nabla I\,(\underline{x}_{k+1}-\hat{\underline{\mathbf{D}}}_k, t-\tau)\right] \tag{5}$$

where

$$sgn\,(z) = \begin{cases} -1 & ,\ \text{if } Z \leftarrow T \\ 0 & ,\ \text{if } |Z| \leq T \\ +1 & ,\ \textit{otherwise.} \end{cases} \tag{6}$$

The above recursion to update $\hat{\underline{D}}_k$ is carried out only in the moving areas of the current frame, i.e., for those pels where

$$\sum_{j=-p}^{+P} |I\ (\underline{x}_{k+j}, t) - I\ (x_{k+j,t-\tau})| \geq \textit{Threshold}$$

2.1 *MULTIPLE PEL MOTION ESTIMATION (ALGORITHMS 2 AND 3)*

The above algorithm can be extended by considering the displaced frame differences at many picture elements to estimate $\underline{D}$. Two algorithms exist. In the first (Algorithm 2), $\hat{\underline{D}}$ can be updated from sample to sample by using a steepest-descent algorithm to minimize a weighted sum of the squared displaced frame differences at some previously transmitted neighboring picture elements. Thus,

$$\hat{\underline{\mathbf{D}}}_{i+1} = \hat{\underline{\mathbf{D}}}_i - \epsilon \cdot ½ \cdot \nabla_{\hat{\underline{\mathbf{D}}}_i}\left[\sum_{j=0}^{p} W_j\,[DFD\,(\underline{x}_{k-j},\hat{\underline{\mathbf{D}}}_i)]^2\right], \tag{7}$$

where $W_j \geq 0$ and $\sum_{j=0}^{p} W_j = 1.$

This can be expanded as:

$$\hat{\underline{\mathbf{D}}}_{i+1} = \hat{\underline{\mathbf{D}}}_i - \epsilon \left\{ \sum_{j=0}^{p} W_j DFD(\underline{x}_{k-j}, \hat{\underline{\mathbf{D}}}_i) \cdot \nabla [I(\underline{x}_{k-j} - \hat{\underline{\mathbf{D}}}_i, t - \tau)] \right\}. \tag{8}$$

This can be generalized further using a different function of the errors (i.e., DFD(. , .)) instead of the square function. The difference between eqs. (8) and (5) is that, to estimate a new value $\hat{\underline{D}}_{i+1}$ of $\underline{D}$ from the old value $\hat{\underline{D}}_i$, the displaced frame difference is evaluated at several neighboring picture elements rather than just one picture element. This has the effect of smoothing the update term [that is, the second term on the right-hand side of eq. (5)].

The second algorithm using multiple pels (ALGORITHM 3) is a least-mean-square estimator based on the intensity in the previous frame at a location displaced by the old estimate of displacement. Thus, assuming eq. (1), the displaced frame difference of eq. (2) can be written as

$$\begin{aligned} DFD\ (\underline{x}_k, \hat{\underline{\mathbf{D}}}_i) &= I(\underline{x}_k - \underline{\mathbf{D}}, t - \tau) - I(\underline{x}_k - \hat{\underline{\mathbf{D}}}_i, t - \tau) \\ &= -(\underline{\mathbf{D}} - \hat{\underline{\mathbf{D}}}_i)^T \nabla I(\underline{x}_k - \hat{\underline{\mathbf{D}}}_i, t - \tau) + \textit{other terms}, \end{aligned} \tag{9}$$

Quantity $(\underline{\mathbf{D}} - \underline{\mathbf{D}}_i)$ can be estimated by standard techniques of linear least-mean square by using DFD(.,.) on the left-hand side as an observation and treating the higher order terms on the right-hand side as noise. This gives us an algorithm of the form

$$\hat{\underline{\mathbf{D}}}_{i+1} = \hat{\underline{\mathbf{D}}}_i - \epsilon \left[\sum_{j=0}^{p} \nabla I(\underline{x}_{k-j} - \hat{\underline{\mathbf{D}}}_i, t - \tau) \nabla I^T(\underline{x}_{k-j} - \hat{\underline{\mathbf{D}}}_i, t - \tau) \right]^{-1} \tag{10}$$

$$\cdot \left[\sum_{j=0}^{p} DFD(\underline{x}_{k-j}, \hat{\underline{\mathbf{D}}}_i) \nabla I(\underline{x}_{k-j} - \hat{\underline{\mathbf{D}}}_i, t - \tau) \right].$$

The matrix inverse in the above equation can be approximated by

$$M_i^{-1} = \frac{1}{\Delta_i} \begin{bmatrix} \sum_{j=0}^{p} LDIF_j^2 & -\sum_{j=0}^{p} LDIF_j EDIF_j \\ -\sum_{j=0}^{p} LDIF_j EDIF_j & \sum_{j=0}^{p} EDIF_j^2 \end{bmatrix}$$

with

$$\Delta_i = \sum_{j=0}^{p} EDIF_j^2 \sum_{j=0}^{p} LDIF_j^2 - \left\{ \sum_{j=0}^{p} EDIF_j \; LDIF_j \right\}^2 , \tag{11}$$

where

$EDIF_j$ = Element difference at $(\underline{x}_{k-j} - \hat{\underline{\mathbf{D}}}_i)$ in the previous frame (approximating the horizontal component of ∇I).

$LDIF_j$ = Line difference at $(\underline{x}_{k-j} - \hat{\underline{\mathbf{D}}}_i)$ in the previous frame (approximating the vertical component of ∇I).

In evaluating the matrix inverse, one must be careful that the matrix is not singular. Singularity can be a result of not averaging enough samples. If, on the other hand, a large number of samples are averaged, then displacement averaging may result. To avoid this, whenever M_i came close to being singular (as indicated by its determinant), no update was performed. The second term of Eq. (10) is given by

$$\psi_i = \begin{bmatrix} \sum_{j=0}^{p} DFD(\underline{x}_{k-j} - \hat{\mathbf{D}}_i, t-\tau) \ EDIF_j \\ \\ \sum_{j=0}^{p} DFD(\underline{x}_{k-j} - \hat{\mathbf{D}}_i, t-\tau) \ LDIF_j \end{bmatrix} \tag{12}$$

2.2 GAIN COMPENSATION. (ALGORITHM 4)

In several situations the model of image intensity of Equation (1) is not adequate. In these situations, a multiplicative factor (called Gain) which may vary temporally and spatially describes the intensity changes. Thus, the model of eq. (1) can be generalized to

$$I(\underline{x}, t) = \rho_1 I(\underline{x}, t-\tau) \tag{13a}$$

and

$$I(\underline{x}, t) = \rho_2 I(\underline{x} - \mathbf{D}, t-\tau). \tag{13b}$$

Equations (13a) and (13b) are alternative models of frame-to-frame intensity variation.

The compensation for the above variation of intensity can be accomplished by estimating ρ_1, ρ_2, and $\underline{\mathbf{D}}$ using gradient-type algorithms as before. The algorithm for compensation of (13a) is:

$$\hat{\rho}_1^{i+1} = \hat{\rho}_1^i + \epsilon_1 \Big[I(\underline{x}, t) - \hat{\rho}_1^i I(\underline{x}, t-\tau) \Big] I(\underline{x}, t-\tau). \tag{14}$$

For compensation of (13b), both ρ_2 and $\underline{\mathbf{D}}$ need to be estimated. This is accomplished by

$$\hat{\rho}_2^{i+1} = \hat{\rho}_2^i + \epsilon_1 DFD(\underline{x},\hat{\rho}_2^i, \hat{\underline{\mathbf{D}}}^i) I(\underline{x}-\hat{\underline{\mathbf{D}}}^i, t-\tau) \tag{15}$$

$$\hat{\underline{\mathbf{D}}}^{i+1} = \hat{\underline{\mathbf{D}}}^i - \epsilon_2 \hat{\rho}_2^i DFD(\underline{x}, \hat{\rho}_2^i, \hat{\underline{\mathbf{D}}}^i) \nabla I(\underline{x} - \hat{\underline{\mathbf{D}}}^i, t - \tau), \tag{16}$$

where DFD(.,.,.) is defined by

$$DFD(\underline{x}, \hat{\rho}_2, \hat{\underline{\mathbf{D}}}) = I(\underline{x},t) - \hat{\rho}_2 I(\underline{x}-\hat{\underline{\mathbf{D}}}, t-\tau) \tag{17}$$

and ϵ_1 and ϵ_2 are positive scalar constants. The iterations may proceed from sample to sample along a scan line. Use of algorithms (14) to (17) assumes that ρ_1, ρ_2, and $\underline{\mathbf{D}}$ vary sufficiently slowly spatially - an assumption that appears to be often valid in practice.

We note in passing that eq. (13b) is the appropriate model for intensity variation due to translational motion of objects. For uniform illumination, ρ_2 will be unity and $\underline{\mathbf{D}}$ will be equal to the displacement. However, we can see that eq. (13a) also provides a description of object motion. Because of this, eq. (13a), with $\hat{\rho}_i$ obtained from (14), in some cases approximates variations of intensity due to object translation. This occurs if the parameter ρ_1 varies sufficiently slowly to be learned by (14). An example of this is shown in Fig. 1, which shows a vertical edge in intensity that is displaced from one frame to the next in the horizontal direction by two picture elements. Recurrence relationship (14) is carried out in Fig. 1, where a plot of $\hat{\rho}_1$ along the scan line is shown. Also shown is the frame difference and the gain-compensated frame difference GFDIF:

$$GFDIF(\underline{x},\hat{\rho}_1,t) = I(\underline{x},t) - \hat{\rho}_1 I(\underline{x},t-\tau). \tag{18}$$

It is seen then that the gain-compensated frame difference is able to provide a certain degree of displacement compensation and, therefore, results in errors that are less

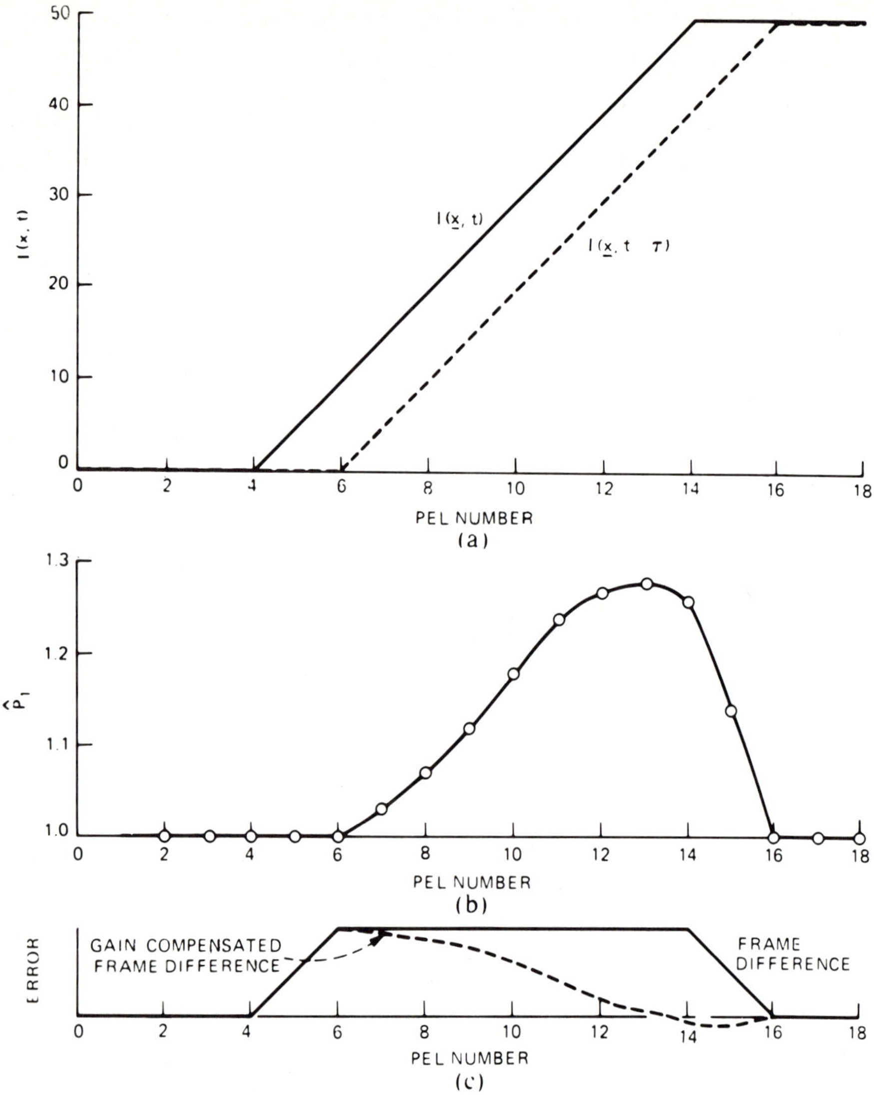

Figure 1. (a) A synthetic vertical edge whose intensity for any scan line is plotted as a function of horizontal position for two consecutive frames. The horizontal shift in the edge is 2 pels per frame. (b) Plot of the value of ρ generated recursively from eq. (14), starting with a value of 1.0. Recursion is assumed to proceed pel by pel in the direction of scanning. (c) Plot of frame difference (or the prediction error in the conditional replenishment coders) and the gain-compensated frame difference (or the prediction error in the gain-compensated coders) as a function of pel position.

than the frame differences traditionally used in conditional replenishment.

2.3 MOTION ESTIMATION IN TRANSFORM DOMAIN. (ALGORITHM 5)

The recursive algorithm for displacement estimation can also be carried out in transform domain. In this case, one obtains coefficients by linear transformation and operates on these coefficients much like on the picture elements.

Let a field of video be partitioned into rectangular blocks of pels each having dimension N_r rows and N_c columns ($N_r \times N_c$). Let $\underline{x}_q = (x_{1q}, x_{2q})^T$ denote the coordinates of the upper left-hand pel of the qth block, where the blocks in each row of blocks are numbered from left to right with $q = 0,1,2,\cdots$. We number the $N = N_r N_c$ pel intensities of block q in a column-scanning fashion and denote them by a column vector $I(\underline{x}_q, t)$. Let the N component vector $\underline{\phi}_n$ be the nth basis vector of a nonzero but otherwise arbitrary real linear transform, and denote the nth coefficient of the qth block of this transform in the present frame by $c_n(q)$, where

$$c_n(q) = I^T(\underline{x}_q, t)\underline{\phi}_n \tag{19}$$

and n is numbered from 0 to N - 1. The displaced previous frame value of this coefficient is

$$\hat{c}_n(q, \hat{\underline{\mathbf{D}}}) = I^T(\underline{x}_q - \hat{\underline{\mathbf{D}}}, t-\tau)\underline{\phi}_n, \tag{20}$$

where $I(\underline{x}_q - \hat{\underline{\mathbf{D}}}, t-\tau)$ is the column vector of intensities of the displaced qth block of the previous frame and $\hat{\underline{\mathbf{D}}}$ is the estimated displacement of the moving object. Computation of the elements in $I(\underline{x}_q - \hat{\underline{\mathbf{D}}}, t-\tau)$ generally requires an interpolation among the given previous-frame pel intensities. Prediction of $c_n(q)$ of (5) by $\hat{c}_n(q, \hat{\underline{\mathbf{D}}})$ of (20) results in coefficient prediction error

$$e_n(q,\hat{\underline{\mathbf{D}}}) = [I(\underline{x}_q,t) - I(\underline{x}_q - \hat{\underline{\mathbf{D}}},t-\tau)]^T \underline{\phi}_n. \tag{21}$$

The algorithm defined in this section attempts to decrease the squared-prediction errors $e(q,\hat{\underline{\mathbf{D}}})^2$ in a coefficient-recursive manner by steepest descent iteration of the form

$$\hat{\underline{\mathbf{D}}}_{n+1}(q) = \hat{\underline{\mathbf{D}}}_n(q) - \frac{\epsilon}{2} \nabla_{\hat{\underline{\mathbf{D}}}_n}(q) e_n^2(q,\hat{\underline{\mathbf{D}}}_n(q))$$

$$= \hat{\underline{\mathbf{D}}}_n(q) - \epsilon e_n(q,\hat{\underline{\mathbf{D}}}_n(q)) G_n(q) \tag{22a}$$

for $n = 0,1, \cdots$, M-2, M $\leq$ N, and $q = 0,1,2, \cdots$, with

$$\hat{\underline{\mathbf{D}}}_o(q) = \hat{\underline{\mathbf{D}}}_{M-i}(q-1)$$

$$- \epsilon e_{M-1}(q-1,\hat{\underline{\mathbf{D}}}_{M-1}(q-1)) G_{M-1}(q-1). \tag{22b}$$

In (22), $G_n(q)$ is the coefficient gradient vector

$$G_n(q) = \nabla I^T(\underline{x}_q - \hat{\underline{\mathbf{D}}}_n(q),t-\tau)\underline{\phi}_n. \tag{23}$$

Note that (22) operates upon coefficients 0 through $M-1$, where for generality we assume $M \leq N$. Iteration in (22) progresses as follows. The initial displacement estimate of the qth block ($q > 1$), $\hat{\underline{\mathbf{D}}}_o(q)$ is formed by updating the final displacement estimate $\hat{\underline{\mathbf{D}}}_{M-1}(q-1)$ of the previous block as in (22b). The next displacement estimate of the qth block $\hat{\underline{\mathbf{D}}}_1(q)$ is formed from (22a) with $n = 0$. Iteration progresses in the qth block by (22a) with $n = 1,2, \cdots ,M - 2$, resulting finally in displacement estimate $\hat{\underline{\mathbf{D}}}_{M-1}(q)$ which, when updated in (22b) (with $q \rightarrow q = 1$), forms the initial displacement estimate $\hat{\underline{\mathbf{D}}}_o(q+1)$ of block $q+1$. This iteration procedure continues along all horizontal blocks of raster. The procedure is started in the $q = 0$ block with an arbitrarily chosen initial displacement estimate $\hat{\underline{\mathbf{D}}}_0(0)$ followed by iterations of (22a) for $n = 0,1, \cdots ,M - 2$ and $q = 0$. We assume the $\hat{\underline{\mathbf{D}}}_0(0)$ is zero.

3. CODING

The above section describes various algorithms for estimation of parameters from frame sequences containing motion. In this section we describe their use for interframe coding.

ALGORITHMS 1,2,3

The displacement is calculated as in Equation 5 for algorithm 1 and as in Section 2.2 for algorithms 2 and 3. The motion compensated coder then predicts intensity $I(\underline{x}_k,t)$, using either the previous frame intensity $I(\underline{x}_k,t-\tau)$ or displaced previous frame intensity $I(\underline{x}_k-\hat{\underline{\mathbf{D}}}_{k-1},t-\tau)$, based on which results in less prediction error for certain already transmitted neighbors. We note, if a point $\underline{x} - \underline{D}$ does not lie on the grid formed by the pels, then interpolation is required to evaluate $I(\underline{x}_k-\hat{\mathbf{D}}_{k-1},t-\tau)$. The displacement at either the previous pel or the previous line element is used to form the displaced previous frame prediction of the present pel. This allows the receiver to compute displaced previous frame predictions without explicit transmission of the displacement. If the magnitude of the prediction error exceeds a predetermined threshold, the coder transmits a quantized version of the prediction error, and the necessary addressing information to the receiver.

ALGORITHMS 4

Two cases of the gain compensation algorithm exist. In the first case (ALGORITHM 4a), only gain compensation is attempted, whereas in the second case (ALGORITHM 4b), both displacement and gain are compensated.

Let the frame difference, FDIF, and the gain-compensated frame difference, GFDIF, be defined as

$$FDIF(\underline{x},t) = I(\underline{x},t) - I(\underline{x},t-\tau). \tag{24}$$

$$GFDIF(\underline{x},\rho_1,t) = I(\underline{x},t) - \rho_1 I(\underline{x},t-\tau). \tag{25}$$

Then the intensity at pel X (Fig. 2) is predicted by:

$$P_z = \begin{cases} I(\bar{X},t-\tau), \ \ if \sum_{\underline{x}\in\{J,K,L\}} |FDIF(\underline{x},t)| \\ \qquad \leq \sum_{\underline{x}\in\{J,K,L\}} |GFDIF(\underline{x},t,\hat{\rho}_1^C)|, \\ \hat{\rho}_1^K I(\bar{X},t-\tau), \ \ otherwise \end{cases} \tag{26}$$

where $\hat{\rho}_1^K$ is the previously estimated $\hat{\rho}_1$ at location K. The iteration for ρ_1 proceeds along the scan line according to eq. (14). However, to simplify multiplications that are necessary for implementing eq. (14), the algorithm was modified as

$$\hat{\rho}_1^{i+1} = \hat{\rho}_1^i + \epsilon_1 sgn[I(\underline{x},t) - \hat{\rho}_1^i I(\underline{x},t-\tau)], \tag{27}$$

In our simulations, $\hat{\rho}_1$ was constrained to be in the interval [15/16,17/16], and ϵ_1 was taken to be 1/128. Iterations were carried out pel by pel in the direction of the scanning with the last value of $\hat{\rho}_1$ on a line being used as an initial estimate for the next line. This coder requires a simple modification of the frame difference conditional replenishment coder. A block diagram of the coder is shown in Fig. 3.

The ALGORITHM 4b consists of gain and translation compensation used simultaneously. In this case, the prediction for the present pel X of Fig. 2 is made by switching between the three predictors given below, where τ_1 and τ_2 are frame and field intervals, respectively.

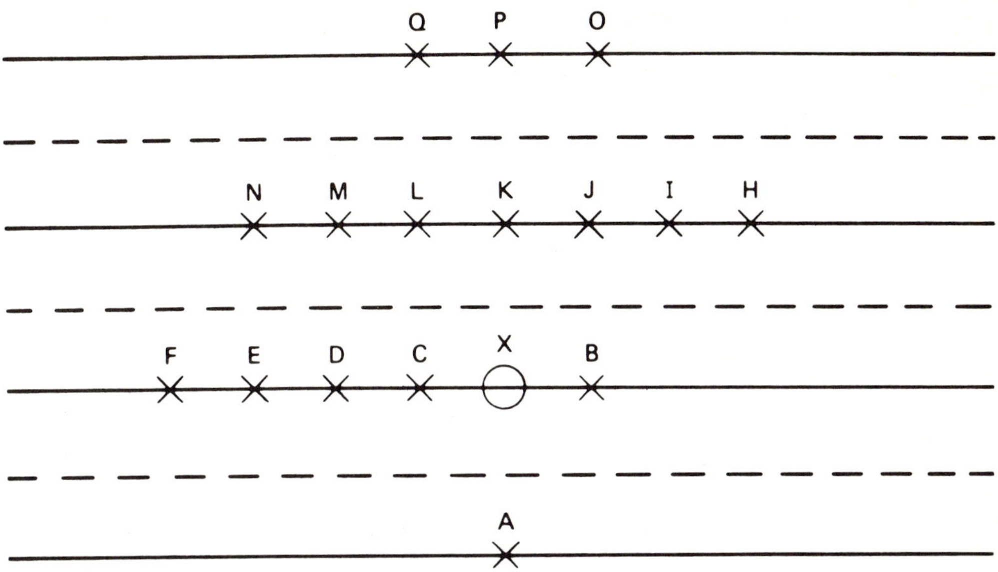

Figure 2. Configuration of picture elements for weighted error calculation. Pel X is being predicted. Dotted lines denote scan lines from previous field.

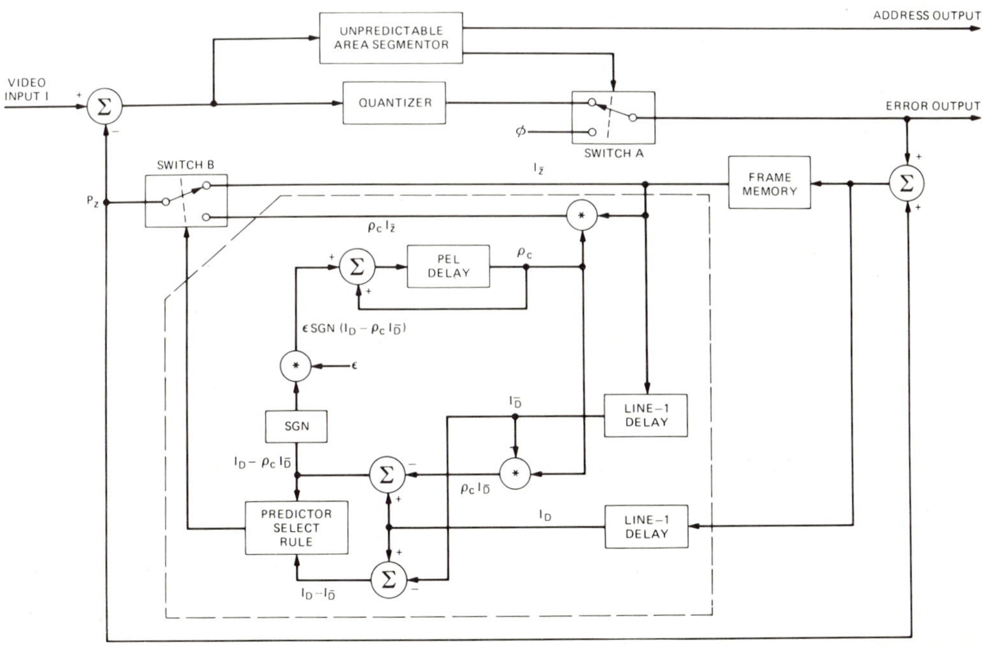

Figure 3. Block diagram of a gain-compensated coder. Without the dotted section, the DPCM loop of a frame-difference conditional replenishment encoder is implemented. Thus, gain compensation requires a simple modification of the conditional replenishment coder.

$$P_1 = I(\bar{X}, t-\tau_1)$$

$$P_2 = \hat{\rho}_1^C I(\bar{X}, t-\tau_1)$$

$$P_3 = \hat{\rho}_2^C I(\bar{X} - \hat{\underline{\mathbf{D}}}^C, t - \tau_2), \tag{28}$$

where $\hat{\underline{\mathbf{D}}}^K, \hat{\rho}_2^K$ and $\hat{\rho}_1^K$ are the estimated translation and gain for pel K from the previous line. $\hat{\rho}_1^K$ is estimated recursively according to eq. (27) with $\tau = \tau_1$. $\hat{\rho}_2^K, \hat{\underline{\mathbf{D}}}^K$ are estimated using a simplification of eq. (15) and (16):

$$\hat{\rho}_2^{i+1} = \hat{\rho}_2^i + \epsilon_1 sgn[DFD(\underline{x}, \hat{\rho}_2^i, \hat{\underline{\mathbf{D}}}^i)]$$

$$\hat{\underline{\mathbf{D}}}^{i+1} = \hat{\underline{\mathbf{D}}}^i - \epsilon_2 sgn[DFD(\underline{x}, \hat{\rho}_2^i, \hat{\underline{\mathbf{D}}}^i)] \cdot sgn[\nabla I(\underline{x} - \hat{\underline{\mathbf{D}}}^i, t-\tau)], \tag{29}$$

where $\tau = \tau_2$. As in the gain-compensated coder, initial estimates of $\hat{\rho}_1, \hat{\rho}_2$, and $\hat{\underline{\mathbf{D}}}$ of the beginning of a line were taken to be the corresponding estimates at the last pel of the previous line. ϵ_1 and ϵ_2 are taken to be 1/128, and 1/16, respectively. Iteration proceeds pel by pel in the direction of scanning. The selection of the predictor for a pel from the above three predictors is made by choosing the one that resulted in the least error for the adjacent previous line elements. Referring to Fig. 2, the selection rule is as follows:

$$P^X = \begin{cases} P_1 \ ; \ if \ E_1 = MIN(E_1, E_2, E_3), \\ P_2 \ ; \ if \ E_2 = MIN(E_1, E_2, E_3) \ and \ (E_1 \neq E_2), \\ P_3 \ ; \ otherwise, \end{cases} \tag{30}$$

where

$$E_1 = \sum_{\underline{x} \epsilon \{J,K,L\}} |FDIF(\underline{x},t)|$$

$$E_2 = \sum_{\underline{x} \epsilon \{J,K,L\}} |GFDIF(\underline{x},t,\hat{\rho}_1^K)|$$

$$E_3 = \sum_{\underline{x} \epsilon \{J,K,L\}} |DFD(\underline{x},\hat{\rho}_2^K,\underline{\mathbf{D}}^K)|$$

and $MIN(\cdot,\cdot,\cdot)$ is the minimum value of its three variables.

ALGORITHM 5

In the motion compensated hybrid transform/DPCM coder shown in Fig. 4, the nth coefficient of the qth present field block is predicted by the nth coefficient of either the displaced or the nondisplaced block of the previous frame depending on which was better for the $(n\text{-}1)^{th}$ coefficient, where the displacement is an estimate of the frame-to-frame translation of a moving object.

Such a motion compensated transform encoder transmits a quantized version of the coefficient prediction error $e_n(q,\hat{\underline{\mathbf{D}}}(q))$ to the receiver whenever its magnitude $|e_n(q,\hat{\underline{\mathbf{D}}}(q)|$ exceeds a given threshold, thereby enabling the decoder to update its displacement estimate $\hat{\underline{\mathbf{D}}}_n(q)$ as in equations (3) and (4), as well as correcting its prediction of coefficient $c_n(q)$. Both the encoder and decoder use the updated displacement estimate in predicting the next coefficient, and the process continues. We note that since only previously transmitted information is used for displacement updating, no separate transmission of displacement is necessary.

4. SIMULATION RESULTS

The coder algorithms described in the previous section were simulated by a computer. Several representative scenes were used.

The first scene, called Judy, consists of 64 frames (2:1 interlaced fields) of 256×256 samples each, obtained at 30 times a second and sampled at Nyquist rate from a video signal of 1 MHz bandwidth, and contains head-and-shoulders view of a person engaged in a rather active conversation. The portion of a frame classified as moving area varies from 15 to 51 percent. Also, the motion is not translational, and different parts of the scene move differently (such as lips, eyes, and head). Four frames of this scene are shown in Fig. 5.

The second scene, called Mike and Nadine, consists of 64 frames with the same resolution as the scene Judy and contains a panned full body view of two people briskly walking around each other on a set with severe nonuniform illumination. The percentage of a frame classified as moving area varied from 92 to 96 percent. Four frames from this sequence are shown in Fig. 6.

The interframe coders (both the conditional replenishment coder simulated for the comparison purposes and various versions of the motion compensated coders) used a quantizer that had 35 levels as shown in Figure 7.

Prediction error was sent in a quantized form only if it was higher than a threshold of 3 (out of 255 corresponding to 8-bit signals). This gave a picture quality which appeared reasonable; that is, the coding degradations were visible but not annoying. The total bits per frame were calculated by adding bit requirements for quantized prediction error (whenever it was sent) and addressing. Error bits were computed by evaluating the entropy of the quantized prediction error, and the address bits were computed by one-dimensional run length coding of the unpredictable picture elements along a scan line.

4.1 ALGORITHMS 1, 2, and 3

The performance of the coder with Algorithm 1 for the scenes Judy, and Mike and Nadine are shown in Figures 8 and 9 respectively. It is seen that improvement over conditional replenishment is substantial for the scene Judy (about 70 percent) compared to Mike and Nadine (about 30 percent).

Results of simulation of Algorithms 2 and 3 for the scene Judy are shown in Figure 10.

To reduce the computational burden, only a part (240×240 array) of the frame was coded. Figure 6 shows the plots of bits per field as a function of the field number for the 120 fields of the scene Judy. Results of Algorithm 1 are plotted for comparison. It is seen from Fig. 10 that both algorithms 2 and 3 are about 10 percent better than algorithm 1. There is a slight preference for algorithm 3 over algorithm 2. Configuration 3, which uses some picture elements not yet available to the receiver, does about 20 percent better than algorithm 1. The use of a larger number of picture elements for evaluation of the error term has the effect of averaging displacements which might not be so useful if these displacements are nonuniform.

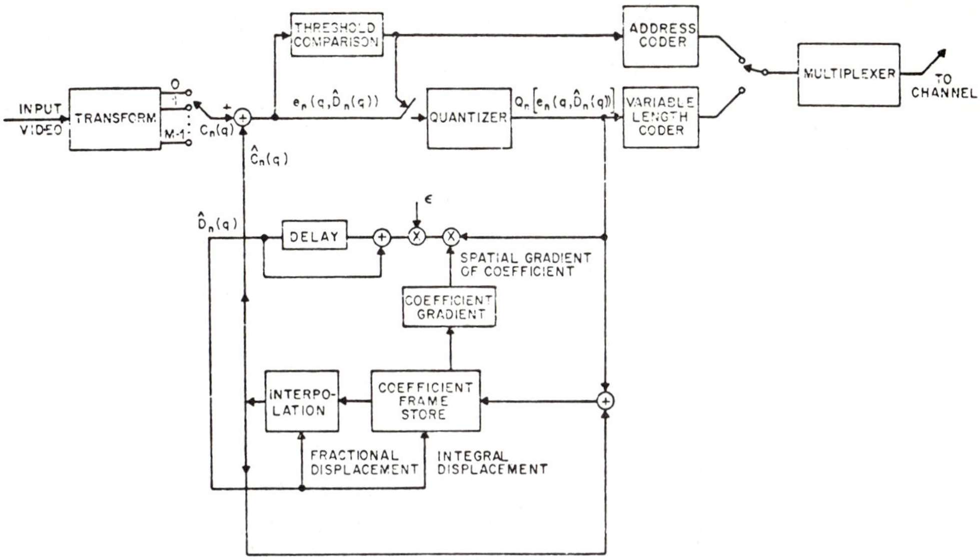

Figure 4 A Block Diagram of Motion-Compensated Transform/DPCM Coder.

Figure 5. Four frames of the scene Judy.

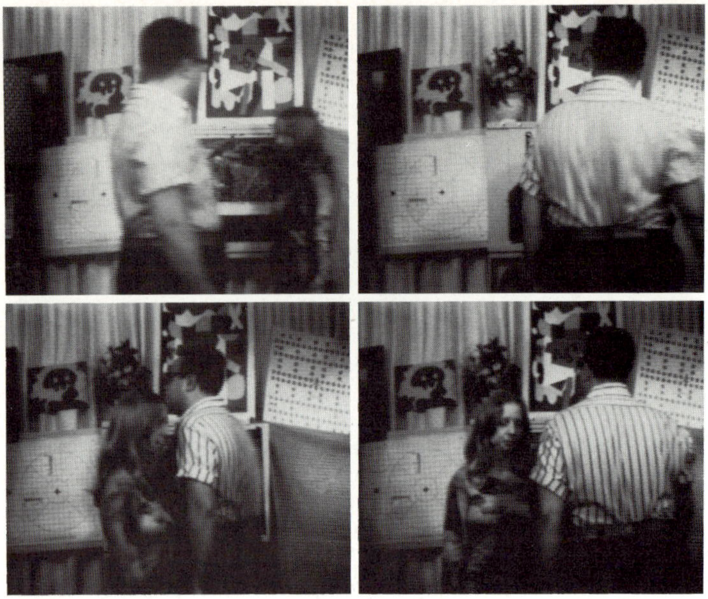

Figure 6. Four frames of the scene Mike and Nadine.

0, 5, 12, 19, 28, 37, 46, 57, 68, 79, 90, 103, 116, 129, 142, 155, 168, 181

Figure 7. Positive representative levels of the Quantizer (on an 8-bit, 0-255 scale).

4.2 ALGORITHMS 4

Figure 11 shows the simulation results using algorithm 4 for both the scenes. For the (head and shoulders) scene Judy, where the illumination is close to uniform, the gain-compensated coder results in average bit rates that are about 50 percent below those of the frame difference conditional replenishment coder. The displacement-compensated coder, on the other hand, results in about 61 percent decrease. Thus, gain compensation reduces the coder bit rate by a significant amount without the increase in the complexity associated with the displacement-compensated encoder. The gain- and displacement-compensated encoder reduce the bit rate by 63.4 percent for Judy. This additional decrease due to gain compensation is small, perhaps, as a result of relatively uniform illumination and lack of shadows in the scene. For the scene Mike and Nadine, again, most of the decrease in bit rate is provided by displacement-compensation and gain-compensation results in bit rates that are between the conditional replenishment and displacement compensation. Also, combining gain and displacement compensation provides an additional decrease of 7.3 percent.

4.3 ALGORITHM 5

Figures 12, 13 and 14 give results of interframe coding in transform domain for the scene Judy. Results of conditional replenishment in transform domain is given in Figure 12. It is seen from this figure that a one-dimensional, four-element cosine transform is about 5 to 10 percent worse than a 2×2 cosine transform. As the transform size is increased, the bit rate dropped further; the 2×8 block using the cosine transform did about 15 percent better than the 2×2 block with cosine transform.

The results of motion compensated coding in transform domain for the scene JUDY are given in Fig. 13. It is seen from this figure that motion compensation in the transform domain is about 20 to 40 percent better than conditional replenishment in the transform domain. In Fig. 14 results of motion compensation in the transform domain are given, which use different types of transforms and block sizes. This figure shows that the cosine transform with 2×4 block does the best. Increasing the block size increasers the bit rate, perhaps as a result of uncompensable area being in small isolated fragments. This result is in contrast with the result obtained with larger block sizes in conditional replenishment, where a larger block size, such as 2×8, gave better results than a smaller block size, such as 2×4.

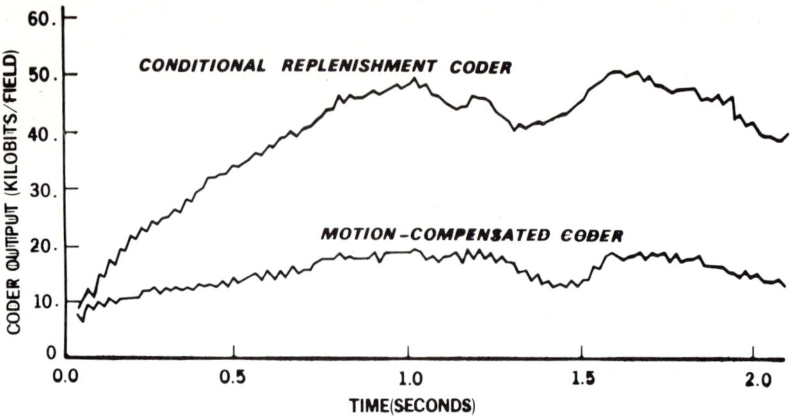

Figure 8. Performance of Conditional Replenishment and Algorithm 1 for Sequence 'JUDY'.

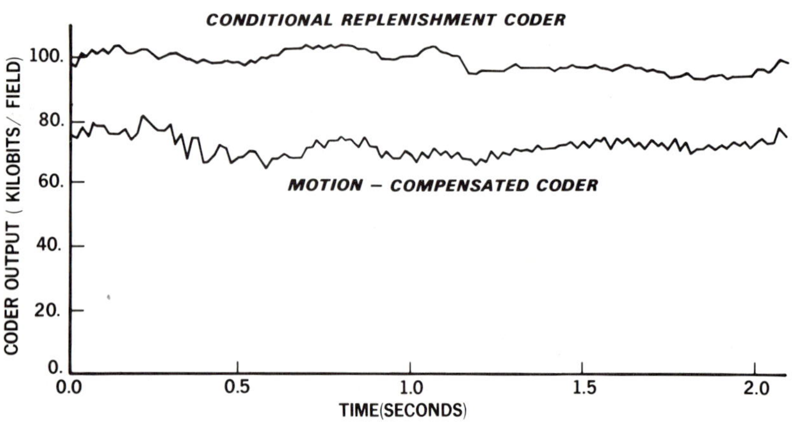

Figure 9. Performance of Conditional Replenishment and Algorithm 1 for Sequence 'MIKE and NADINE'. This sequence was chosen as a severe test for the coders. It encompasses camera panning, complex motion and nonuniform illumination.

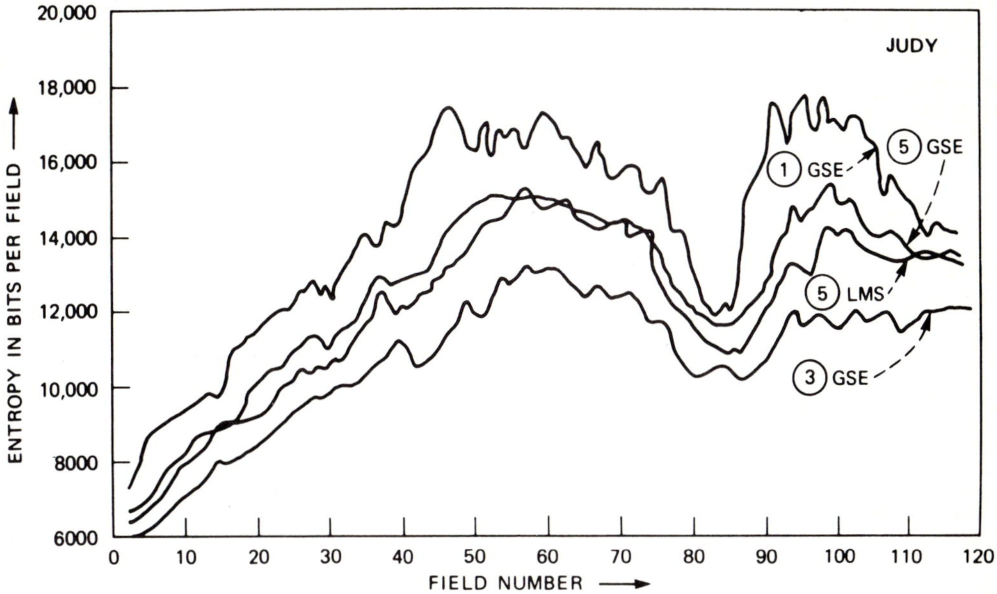

Figure 10. Plots of bits per field are given for the scene Judy. Configuration 1 corresponds to the motion-compensation Algorithm 1. Configuration 5 is shown for the Gradient of Summed Error (GSE) and Least Mean Square (LMS) algorithms. Configuration 3 uses some pels for displacement estimation that are not yet available to the receiver.

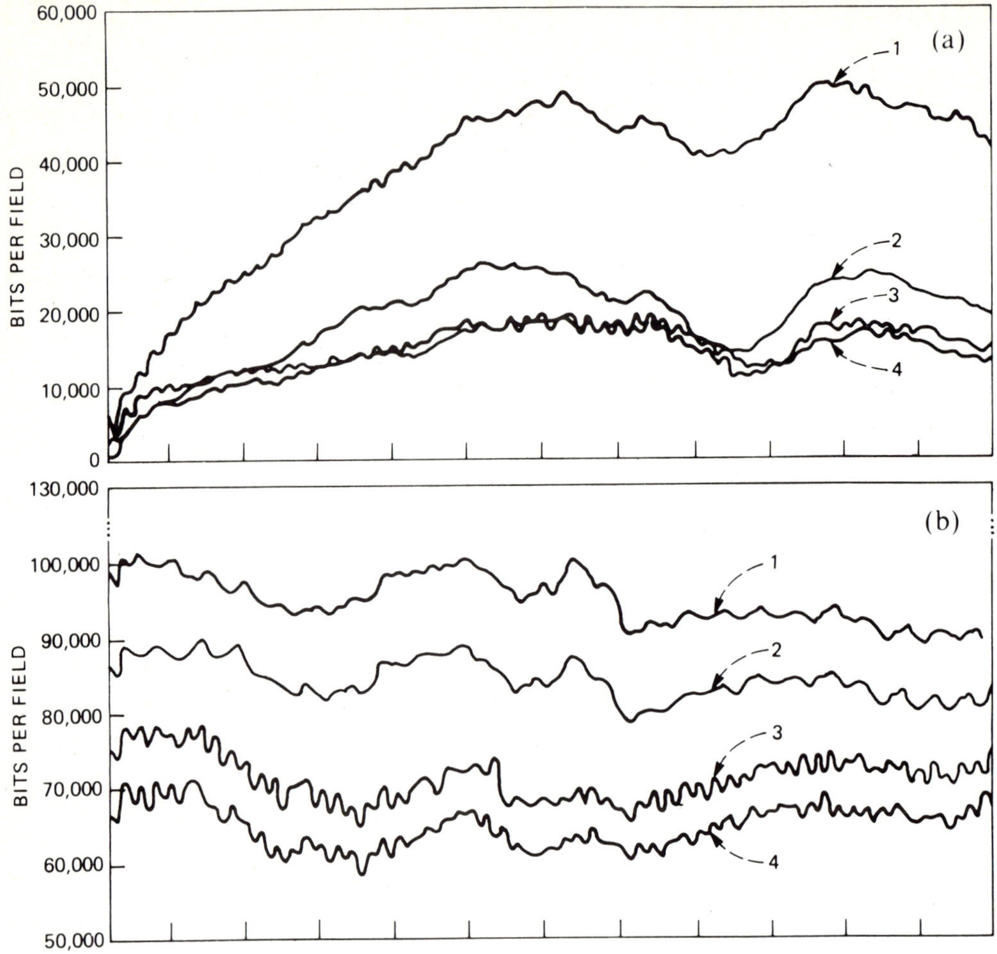

Figure 11. Plots of bits/field versus the field number using the four coders for the scenes (a) Judy, and (b) Mike and Nadine. The four coders are: (1) frame difference conditional replenishment; (2) gain compensation; (3) displacement compensation; (4) gain and displacement compensation.

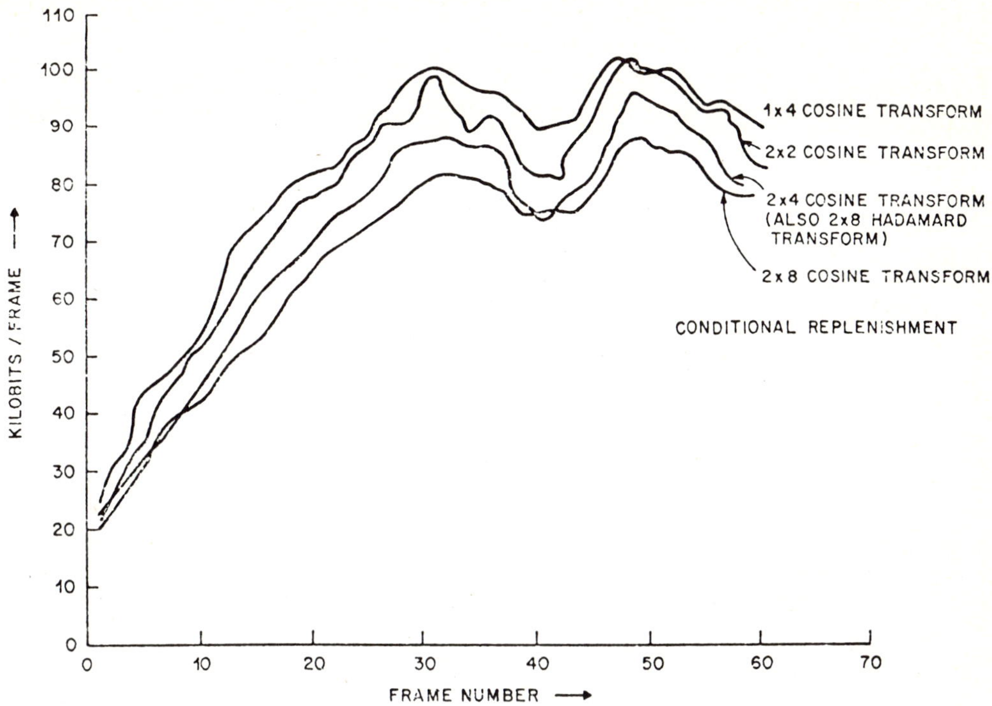

Figure 12. Performance of conditional replenishment and motion-compensated transform coders. Kilobits/frame are plotted as a function of the frame number for a typical sequence containing active motion of a head-and-shoulders view.

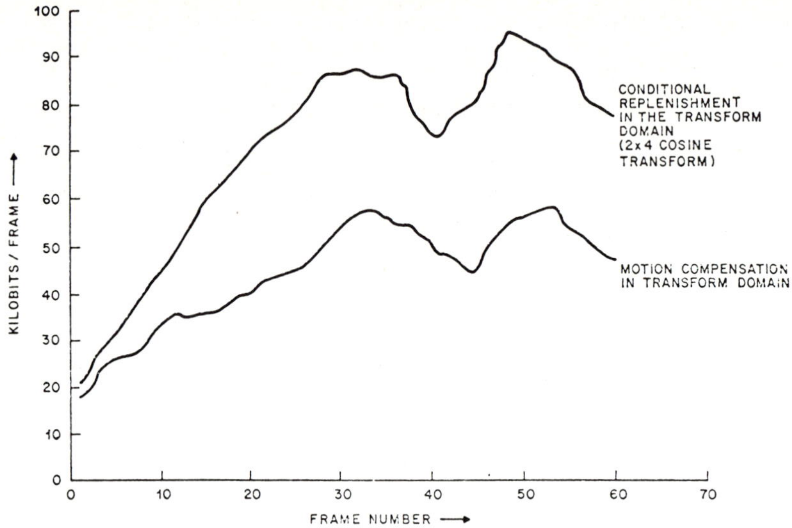

Figure 13. Performance of Conditional Replenishment and Motion-Compensated Transform Coders.

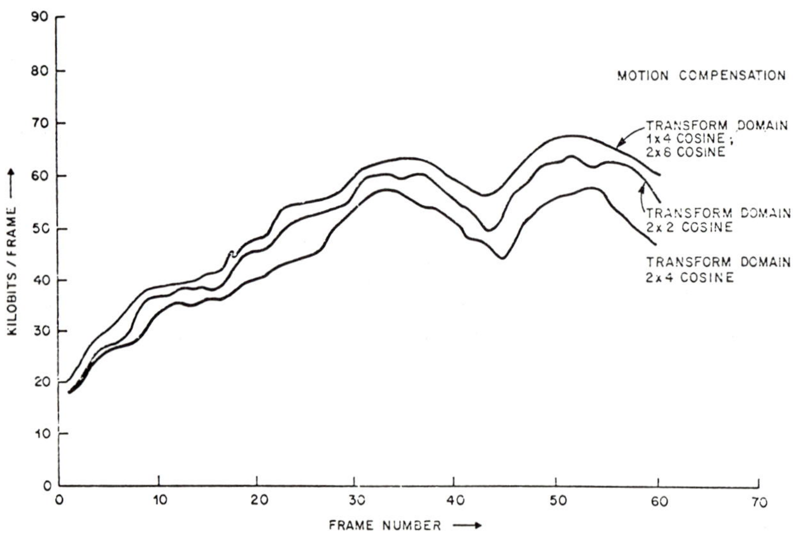

Figure 14. Performance of Motion-Compensated Coder with Different Transforms.

We found that more coefficients could be dropped from transmission altogether in the motion compensated transform coder than in the conditional replenishment transform coder. For example, with a 2×4 block, and cosine transform, only five coefficients were needed in motion compensation, compared to the seven coefficients that were necessary for conditional replenishment. Unfortunately, however, the effect of dropping a larger number of coefficients did not result in a large bit rate reduction since the number of bits that were required for these coefficients was very small.

Coarser quantization of the higher order coefficients was possible in motion compensation, compared to conditional replenishment, without significantly degrading picture quality. An increase of the predictability thresholds of the first coefficient resulted in rapid degradation of the picture quality. As the predictability threshold was increased, the block structure of the transform became clearly evident, and the frame-to-frame effects of the visibility of the block structure was found rather annoying. For higher order coefficients, however, the picture quality was not a sensitive function of the predictability thresholds.

5. SUMMARY AND CONCLUSIONS

We have presented five algorithms for compensation of frame-to-frame changes of intensity in a television scene. These changes are a result of motion of objects and changes in illumination (e.g., shadows). Both pel and transform domain representation of intensity are considered. Algorithms of varying complexity are considered: some are implementable with the current electronic circuits, whereas some are significantly complex, requiring fast matrix inversions. Computer simulations on several frame sequences are used to evaluate the efficiency of algorithms. Our general conclusion is that the motion compensation reduces the coder bit rates by 30 to 70 percent compared to conditional replenishment.

REFERENCES

[1] F. W. Mounts, "A Video Encoding System Using Conditional Picture-Element Replenishment," B.S.T.J., *48*, No. 7 (September 1969), pp. 2545-2554.

[2] B. G. Haskell, F. W. Mounts, and J. C. Candy, "Interframe Coding of Videotelephone Pictures," Proc. IEEE, *60*, No. 7 (July 1972), pp. 792-800.

[3] J. O. Limb, R. F. W. Pease, and K. A. Walsh, "Combining Intraframe and Frame-to-Frame Coding for Television," B.S.T.J., *53*, No. 6 (August 1974), pp. 1137-1173.

[4] T. Ishiguro, K. Iinuma, Y. Iijima, T. Koga, S. Azami, and T. Mune, "Composite Interframe Coding of NTSC Television Signals," 1976 National Telecommunications Conference Record, Vol. 1, Dallas, Texas, (November 1976), pp. 6.4-1 to 6.4-5.

[5] B. G. Haskell, P. L. Gordon, R. L. Schmidt, and J. V. Scattaglia, "Interframe Coding of 525-Line Monochrome Television at 1.5 Mbs," IEEE Trans. Commun. COM-25 (October 1977), pp. 1461-1467.

[6] F. Rocca, "Television Bandwidth Compression Utilizing Frame-to-Frame Correlation and Movement Compensation," Symposium on Picture Bandwidth Compression, (M.I.T., Cambridge, Mass., 1969), Gordon and Breach, 1972.

[7] B. G. Haskell and J. O. Limb, "Predictive Video Encoding Using Measured Subject Velocity," January 1972, U.S. Patent 3,632,865.

[8] C. Cafforio and F. Rocca, "Methods for Measuring Small Displacements of Television Images," IEEE Trans. Inform. Theory, IT-22, No. 5 (September 1976), pp. 573-579.

[9] B. G. Haskell, "Entropy Measurements for Nonadaptive and Adaptive, Frame-to-Frame, Linear Predictive Coding of Video Telephone Signals," B.S.T.J., *54*, No. 6 (August 1975), pp. 1155-1174.

[10] A. N. Netravali and J. D. Robbins, "Motion-Compensated Television Coding, Part-I," B.S.T.J., *58*, No. 3 (March 1979), pp. 629-668.

[11] J. D. Robbins and A. N. Netravali, "Interframe Television Coding Using Movement Compensation," International Conference on Communications, Boston, Mass., June 1979.

[12] J. A. Stuller and A. N. Netravali, "Transform Domain Motion Estimation," B.S.T.J., *58*, No. 7 (September 1979), pp. 1673-1702.

[13] A. N. Netravali and J. A. Stuller, "Motion-Compensated Transform Coding," B.S.T.J., *58*, No. 7 (September 1979), pp. 1703-1718.

[14] J. A. Stuller, A. N. Netravali and J. D. Robbins, "Interframe Television Coding Using Gain and Displacement Compensation," B.S.T.J., *59*, No. 7, September 1980, pp. 1227-1240.

[15] K. A. Prabhu and A. N. Netravali, "Pel-Recursive Motion Compensated Color Coding," International Conference on Communications, Philadelphia, PA, June 1982.

[16] K. A. Prabhu and A. N. Netravali, "Motion Compensated Component Color Coding," IEEE Transactions on Communications (to appear).

[17] K. A. Prabhu and A. N. Netravali, "Motion Compensated Composite Color Coding," IEEE Transactions on Communications (to appear).

[18] A. N. Netravali and J. D. Robbins, "Spatial Subsampling in Motion Compensated Television," B.S.T.J., (to appear).

[19] A. N. Netravali and J. D. Robbins, "Motion Compensated Temporal Interpolation," Picture Coding Symposium, Montreal, 1981.

THE DIFFERENTIAL METHOD FOR IMAGE MOTION ESTIMATION

C.Cafforio, F. Rocca
Dipartimento di Elettronica,Politecnico di Milano
Piazza Leonardo da Vinci 32, 20133 MILANO (Italy)

Abstract

Small variations of the apparent velocity of objects imaged by television cameras can be efficiently measured comparing the frame differences with the spatial variations of the luminance. The main lobe of the autocovariance of a block of picture elements of an image is first modeled as a paraboloid; afterwards, the comparison with the mutual covariance between two successive video frames, allows to determine their relative displacement. The curvatures of the autocovariance taken at the origin are proportional to the m.s. slopes of the image along the coordinate axes; it can be shown that the slopes should be calculated using a centered differentiation, rather than a non centered one, to avoid bias in the estimation of the displacement. To be able to precisely measure the displacement along a given direction, it is necessary that the m.s. slope in that direction be high enough; in fact, the measure of temporal and spatial image derivatives only constrains the projection of the actual displacement along the image gradient. More complex, but continuous, motion fields can be estimated with simple extensions of the original formulation of the differential method. A stochastic model for the motion field is therefore introduced, in order to derive a recursive estimation technique. The component of the motion parallel to the local image gradient is updated, on a pel by pel basis, while the other one is approximately kept constant. The practical results are sufficient for image coding, but not good enough to get a nicely smooth velocity field in real situations.

1. Introduction

The estimation of the apparent motion of television images has several applications like coding for redundancy reduction, image understanding, robotics, telemetry, remote guidance and so on. The motion of the imaged object can be estimated either using the correlation function, at high costs, or using image differentials, i.e. comparing luminance variations in time and space. This last method was introduced in /1/, /2/ and then fully analyzed in /3/. Applications to real time tracking /4/, missile guidance /5/, reconstruction of the three dimensional motion of an object /6/ have also been considered.

The authors have also remarked that, in order to avoid estimation bias, the image differentials should be centered and that the maximal incertitude in the displacement cannot exceed the radius of the main lobe of the image autocovariance function /7/.

NATO ASI Series, Vol. F2
Image Sequence Processing and Dynamic Scene Analysis
Edited by T.S. Huang

Since different objects can move on the screen in different directions, it is natural to look for variations of the method in which a complete velocity field is obtained. This can be got by dividing the image in blocks and finding a vector for each block /8/, /9/. Another approach, introduced in /10/ analyzes a pel-recursive technique for motion estimation, to be used in image coding. A discussion of these techniques is given in the book /6/.

In paper /11/, the effects of the dimensions of the block on the precision of the estimate have been considered. It is shown that if the block is small (say 4 by 4 picture elements or smaller) then, practically only the component of the motion along the prevalent image gradient can be determined. Therefore the unobservable component has to be estimated on the basis of some prior measurement, introducing a recursion based on the stochastic structure of the motion field.

In the same paper formulas are derived for local velocity estimation, optimally combining all the available informations. When the block dimensions are reduced to the single pel, the 2 by 2 matrix P that has to be inverted to determine both components of the motion is singular and pseudoinversion techniques should be used. Finally, in /11/ this recursive method is compared with that proposed by Netravali /10/.

The aim of this paper is to summarize these results, identifying possibilities and limits of the differential method.

In section 2 the basic idea will be presented. A statistical interpretation, valid for large block sizes, will be given in section 3; it will explain the need of centered approximations of image differentials and the limits of the method when the displacement is not smaller that the width of the main lobe of the autocovariance function of the image. An example of hardware realization together with some experimental results will then be shown in section 4 and 5.

The analysis of the effects of smaller block sizes is carried out in section 6 where the distribution of the eigenvalues of matrix P is considered.

In section 7 , a model for the velocity field is introduced and formulas for the optimal combination of measurements in the actual block and prior information are given. These formulas, that lead to a pel recursive technique for motion estimation are compared with the similar ones proposed by Netravali in an earlier paper.

Some recent experimental results are discussed in section 8 and the conclusions are drawn in section 9.

2. The algorithm: a short review

An object imaged by a TV camera and moving by pure translation generates frame differences. Their values $\Delta_T(x,y)$ may be expressed as a function of the vectorial components of the object displacement in a frame period:

$$\Delta_T(x,y) = i(x+\alpha,y+\beta)-i(x,y)+n(x,y) = \frac{\partial i}{\partial x}\alpha+\frac{\partial i}{\partial y}\beta+\varepsilon(x,y) \tag{1}$$

where $i(x,y)$ is the image luminance function, α and β are the horizontal (x) and vertical (y) displacement components, $n(x,y)$ is a term due to the noise added to luminance; $\varepsilon(x,y)$ accounts for both noise and Taylor series truncation.
Of course in a digital environment derivatives must be approximated by finite differences:

$$\partial i(n\Delta,m\Delta)/\partial x \cong \Delta_x(n,m) = [i(n\Delta+\delta^+,m\Delta)-i(n\Delta-\delta^-,m\Delta)]/(\delta^++\delta^-)$$

where δ^-,δ^+ are integer multiples of the sampling step Δ. Later it will be shown that the values for δ^-, δ^+ must be carefully chosen. To be consistent with (1) temporal differences Δ_T are always taken with $\delta^+ = 1$, $\delta^- = 0$.
Relation (1) holds for all points of the imaged object in translational motion. By means of linear regression an estimate for α, β is therefore obtainable (in pixels/frame).

$$\begin{vmatrix} \hat{\alpha} \\ \\ \hat{\beta} \end{vmatrix} = \begin{vmatrix} \Sigma_i \Delta X_i^2 & \Sigma_i \Delta X_i \Delta Y_i \\ \\ \Sigma_i \Delta X_i \Delta Y_i & \Sigma_i \Delta Y_i^2 \end{vmatrix}^{-1} \cdot \begin{vmatrix} \Sigma_i \Delta T_i \Delta X_i \\ \\ \Sigma_i \Delta T_i \Delta Y_i \end{vmatrix} = P^{-1}b = Qb \tag{2}$$

$$\hat{\alpha} = \frac{\sigma_{\Delta T}}{\sigma_{\Delta X}} \frac{r_{\Delta T\Delta X} - r_{\Delta X\Delta Y} r_{\Delta T\Delta Y}}{1-r^2_{\Delta X\Delta Y}} \quad , \quad \hat{\beta} = \frac{\sigma_{\Delta T}}{\sigma_{\Delta Y}} \frac{r_{\Delta T\Delta Y} - r_{\Delta X\Delta Y} r_{\Delta T\Delta X}}{1-r^2_{\Delta X\Delta Y}} \tag{3}$$

$\sigma_{\Delta X}$ and $r_{\Delta X \Delta Y}$ are defined as follows:

$$\sigma^2_{\Delta X} = \frac{1}{N} \sum_{i,j \in A_{obj}} \Delta X^2(i,j) \quad , \quad r_{\Delta X\Delta Y}\sigma_{\Delta X}\sigma_{\Delta Y} = \frac{1}{N} \sum_{i,j \in A_{obj}} \Delta X(i,j)\Delta Y(i,j)$$

where N is the total number of pixels belonging to the moving area (A_{obj}).
If the mean values of ΔX, ΔY, ΔZ are zero or very small, as indeed they are when A_{obj} covers the entire moving object, then $\sigma_{\Delta X}$, $r_{\Delta X\Delta Y}$ are estimates for standard devia - tions and cross-correlation factors of the various luminance differences. However the validity of (3) is not based on this correspondence. The procedure, here derived for pure translation, could be generalized /5,6/ to other movements like rotation, change of scale, etc.
In the monodimensional case one would get

$$\hat{\alpha} = \Sigma_i \, \Delta T_i \, \Delta X_i \, / \, \Sigma_i \, \Delta X_i^2 \tag{4}$$

It is important to notice that when the image gradient ΔX is large the estimate of the displacement will be small. On the other hand it is likely that this happens where some image edge is present and then the Taylor series is not accurate. In the opposite case of small gradient , there are also problems due to image quantization. In fact in these zones the outcome of the prediction is often some integer multiple of a fraction of the sampling interval.
This fact indicates that some precautions should be taken to avoid too small a

denominator in (4), for instance by adding a positive constant.

3. Large block sizes: an interpretation in terms of the autocorrelation function

A more complete understanding of the algorithm and its limits may be gained by using a different and more general approach. In this way the results in the previous section appear as a particular case.

Should the image autocorrelation function and the cross-correlation function of two consecutive frames be known for all points of the plane, to get a measure of relative displacement would be an easy task. To measure such functions is, however, a very long procedure. In a digital environment (the only viable one) it is only possible to determine their values at integer multiples of the sampling interval. In order to detect displacements with an accuracy greater than this interval one has to revert to interpolation, based on a continuous model of the object.

The operation of an algorithm for motion estimation may then be traced as follows:

(a) A model is assumed for the image-autocorrelation function;

(b) A local identification of the parameters of the model is carried out by measuring standard deviations and cross-correlation factor of horizontal and vertical luminance differences (not affected by displacements); the same model, but for a spatial translation is assumed for the cross-correlation function of two successive images;

(c) Measuring the cross-correlation of temporal and spatial differences, the cross-correlation function of two following images can be estimated and thus their relative displacement can be determined.

Based on this, we can now interpret the algorithm described in Section 2. In (1) the Taylor series expansion is truncated and only first order derivatives are taken into account. This may be seen equivalent to modeling the image autocorrelation function

$$R(x,y) = \frac{1}{N} \sum_{k,j \in A_{obj}} i(k\Delta, j\Delta)\, i(k\Delta + x, j\Delta + y)$$

with a quadratic function. In fact,

$$(-1)^{m+n} \frac{\partial^{2(m+n)} R(0,0)}{\partial X^{2m} \partial Y^{2n}} = \frac{1}{N} \sum_{k,j \in A_{obj}} \left| \frac{\partial^{m+n} i(k\Delta, j\Delta)}{\partial x^{m}\, \partial y^{n}} \right|^{2} .$$

Now, if luminance derivatives of order greater than 1 are neglected, then the autocorrelation function derivatives of order greater than 2 can be neglected too. As a consequence, at least near the origin, R(x,y) can be modeled as:

$$R_{M(x,y)} = Ax^2 + By^2 + Cxy + D.$$

Throughout all this section R(.,.) will be used to denote the true image autocorrelation function and $R_M(.,.)$ to denote the model used to approximate it.

The four coefficients A, B, C, D have to be determined from the image to get the best fit to the data. In effect just three coefficients are enough since the algorithm is differential and the constant D is unessential.

It is straightforward to show that:

$$\sigma^2_{\Delta X} = 2[R(0,0)-R(\delta^-+\delta^+,0)]/(\delta^++\delta^-)^2,$$

$$\sigma^2_{\Delta Y} = 2[R(0,0)-R(0,\delta^-+\delta^+)]/(\delta^++\delta^-)^2,$$

$$r_{\Delta X\Delta Y}\sigma_{\Delta X}\sigma_{\Delta Y} = [R(-\delta^+,\delta^+)-R(\delta^-,\delta^+)-R(-\delta^+,-\delta^-)+R(\delta^-,-\delta^-)]/(\delta^++\delta^-)^2,$$

$$r_{\Delta X\Delta T}\sigma_{\Delta X}\sigma_{\Delta T} = [R(\delta^-,0)-R(\alpha+\delta^-,\beta)-R(-\delta^+,0)+R(\alpha-\delta^+,\beta)]/(\delta^++\delta^-) ,$$

$$r_{\Delta Y\Delta T}\sigma_{\Delta Y}\sigma_{\Delta T} = [R(0,\delta^-)-R(\alpha,\beta+\delta^-)-R(0,-\delta^+)+R(\alpha,\beta-\delta^+)]/(\delta^++\delta^-)$$

where δ^-,δ^+ are measures of how far away (in the two directions) from the current picture element are the points used in the finite-difference approximations to image luminance derivatives.

The measurements of $\sigma^2_{\Delta X}$, $\sigma^2_{\Delta Y}$, $r_{\Delta X\Delta Y}$ allow the identification of the autocorrelation function $R_{M(x,y)}$ but for the constant D.

From the measurements of $r_{\Delta T\Delta X}$, $r_{\Delta T\Delta Y}$ it is possible to determine the differences in the values of the mutual correlation function at the points $(\alpha-\delta^+,\beta)$ $(\alpha+\delta^-,\beta)$ and at points $(\alpha,\beta-\delta^+)$ $(\alpha,\beta+\delta^-)$.

Using now the hypothesis that the mutual correlation function is identical to the autocorrelation function but for the displacement α, β, it is possible to find $\hat{\alpha}$, $\hat{\beta}$ estimates of the components of the displacement vector, minimizing the squared difference between $[R_M(\hat{\alpha}-\delta^+,\hat{\beta})-R_M(\hat{\alpha}+\delta^-,\hat{\beta})]$, $[R_M(\hat{\alpha},\hat{\beta}-\delta^+)-R_M(\hat{\alpha},\hat{\beta}+\delta^-)]$ and the corresponding measured values.

Of course, the method relies very much on the model assumed for the autocorrelation function. The model used throughout this paper is simple but still effective. Higher order polynomials or other functions could be used instead. The process of identification would, however, become more involved and more displacement-independent measures should be gathered from the data.

It is now easily understood why only a limited range of accurate estimation can be attained.

The quadratic function modeling allows a good description of the main lobe of the image-autocorrelation-function only. It is then obvious that a precise estimation of the displacement is possible only if its value is not greater than the mean width of this lobe. Outside of this range a poor description is available for the image autocorrelation function and a good estimate is therefore impossible.

4. Centered vs. non-centered luminance differences

The need for centered-difference approximations of luminance derivatives is more easily accounted for in a simplified framework.

In a 1-dimensional case (such as raster scanning of a scene in which only perfectly horizontal movement is present) the quadratic model reduces the shape of the auto - correlation function to a parabola.

If non-centered luminance differences are used $\delta^- = 0$, $\delta^+ = \Delta$, where Δ is the sampling interval.

The parabola is so determined as to fit $[R(0)-R(\Delta)]$. Therefore the model for the true autocorrelation function is accurate up to 1 sampling step approximately. For greater values it becomes less and less accurate.

The estimation of displacement is obtained looking for the value $\hat{\alpha}$ for which

$$[R_M(\hat{\alpha}-\Delta)-R_M(\hat{\alpha})] = [R(\alpha-\Delta)-R(\alpha)].$$

It is easy to understand from fig. 1 that a good estimation is possible only if the displacement is small. Underestimation results otherwise. On the other hand the sign of displacement plays a role on the accuracy of estimation. A fit is always made to

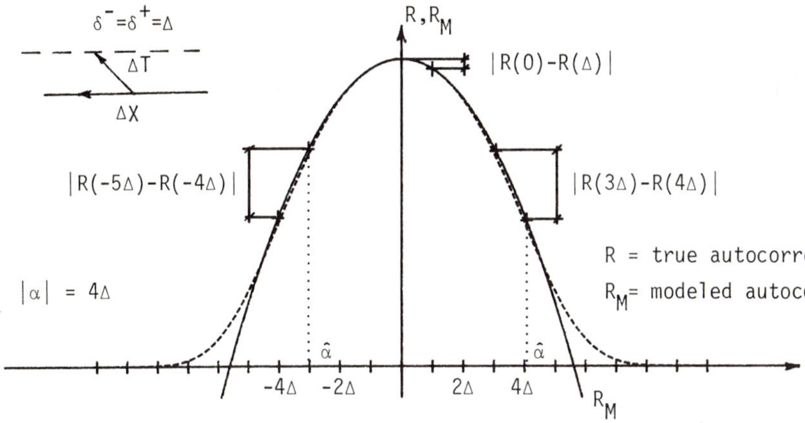

Fig. 1. 1-dimensional estimation: non centered-differences.

$[R(\alpha-\Delta)-R(\alpha)]$.For a fixed absolute value $|\alpha|$ an arc of the parabola nearer the central lobe is used when the displacement is positive. As a consequence the range of accurate estimation is not symmetric and is wider for positive displacements, narrower for negative ones (subject to the assumed values of δ^-, δ^+) /5/.

The use of non-centered luminance differences has another drawback: noise superimposed to the signal results in biased estimation for both variance and cross-correlation factors. A non-zero displacement measure results even if the scene is completely still but noisy; in general the estimate is heavily disturbed, as already pointed out in /2/.

Much better results are obtained if centered-difference approximations to luminance derivatives are used instead.

Now ($\delta^- = \delta^+ = \Delta$) and the parabola is determined by measuring $[R(0)-R(2\Delta)]$. The range of better accuracy is therefore shifted around two pixels displacement. However $\hat{\alpha}$ is now defined as the value for which

$[(R_M(\hat{\alpha}-\Delta)-R_M(\hat{\alpha}+\Delta)] = [R(\alpha-\Delta)-R(\alpha+\Delta)]$.

The two sample-values of R(.) are now symmetric around the α point. A perfectly symmetric behaviour in the performance of the algorithm ensues for both positive and negative displacements (see fig. 2). Also noise has less objectionable effects.

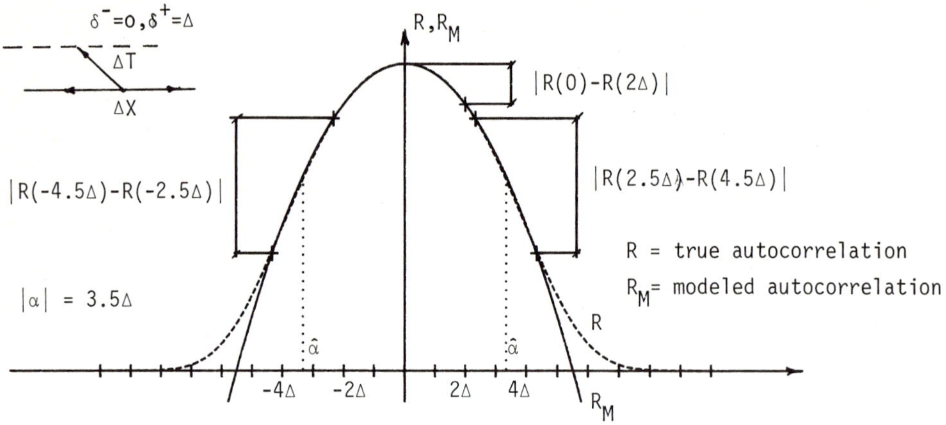

Fig. 2. 1-dimensional estimation: centered-differences.

Current pixel is not used to measure horizontal and vertical differences ($\Delta X,\Delta Y$). As a consequence cross-correlation factors are no longer biased, if, as supposed, noise samples are uncorrelated. No improvement results in the estimation of variances. Taking into account noise effects, formulas (3) become

$$\hat{\alpha} = \frac{\alpha_{\Delta T}}{\alpha_{\Delta X}} \; \frac{r_{\Delta T\Delta X}-r_{\Delta X\Delta Y}r_{\Delta T\Delta Y}+2r_{\Delta T\Delta X}\sigma_n^2/\sigma_{\Delta Y}^2}{1-r_{\Delta X\Delta Y}^2+\dfrac{4\sigma_n^4}{\sigma_{\Delta X}^2\sigma_{\Delta Y}^2}+2\sigma_n^2\left(\dfrac{1}{\sigma_{\Delta X}^2}+\dfrac{1}{\sigma_{\Delta Y}^2}\right)} ,$$

$$\hat{\beta} = \frac{\sigma_{\Delta T}}{\sigma_{\Delta Y}} \; \frac{r_{\Delta T\Delta Y}-r_{\Delta X\Delta Y}r_{\Delta T\Delta X}+2r_{\Delta T\Delta Y}\sigma_n^2/\sigma_{\Delta X}^2}{1-r_{\Delta X\Delta Y}^2+\dfrac{4\sigma_n^4}{\sigma_{\Delta X}^2\sigma_{\Delta Y}^2}+2\sigma_n^2\left(\dfrac{1}{\sigma_{\Delta X}^2}+\dfrac{1}{\sigma_{\Delta Y}^2}\right)} .$$

Here a zero-displacement measure results for still scenes ($r_{\Delta T\Delta X} = r_{\Delta T\Delta Y} = 0$) notwithstanding noise added to signal. Furthermore noise has always and only the effect of underestimating displacement components.

Noise power adds to signal power and consequently the measure of $[R(0)-R(2\Delta)]$ is augmented by noise variance. A "stretching" of the parabola and successive underestimation ensues.

Of course filtering may be used. Low-pass filtering has a two-fold beneficial effect: it reduces noise power and widens the central lobe of image autocorrelation function. Both effects improve the accuracy of measures and, provided that a linear phase filter is used, low-pass-filtering is to be considered as a useful signal conditioning even when noise is small.

5. Hardware implementation

Eqs. (3) can be rewritten , with a little algebra, as

$$\hat{\alpha} = \frac{1}{2}\frac{\sigma_{\Delta T}}{\sigma_{\Delta X}}\left(\frac{r_{\Delta T\Delta X}+r_{\Delta T\Delta Y}}{1+r_{\Delta X\Delta Y}} + \frac{r_{\Delta T\Delta X}-r_{\Delta T\Delta Y}}{1-r_{\Delta X\Delta Y}}\right), \quad \hat{\beta} = \frac{1}{2}\frac{\sigma_{\Delta T}}{\sigma_{\Delta Y}}\left(\frac{r_{\Delta T\Delta X}+r_{\Delta T\Delta Y}}{1+r_{\Delta X\Delta Y}} - \frac{r_{\Delta T\Delta X}-r_{\Delta T\Delta Y}}{1-r_{\Delta X\Delta Y}}\right)$$

If we now make use of the approximate relation

$$r_{ab}\sigma_a = \frac{k}{N}\sum_{i,j} a_{ij}\,\mathrm{sgn}(b_{ij})$$

where a,b denote any two variables and k a constant whose value is related to their joint probability density function (with a value of $\sqrt{2/\pi}$ if the variables are jointly gaussian), eqs. (4) may be rewritten as

$$\hat{\alpha} = \frac{1}{2}\left|\frac{\sum_{i,j}|\Delta T_{ij}|(-1)^{\alpha_{ij}}\beta_{ij}}{\sum_{i,j}|\Delta X_{ij}|\,\beta_{ij}} + \frac{\sum_{i,j}|\Delta T_{ij}|(-1)^{\alpha_{ij}}\bar{\beta}_{ij}}{\sum_{i,j}|\Delta X_{ij}|\,\bar{\beta}_{ij}}\right|,$$

$$\hat{\beta} = \frac{1}{2}\left|\frac{\sum_{i,j}|\Delta T_{ij}|(-1)^{\alpha_{ij}}\beta_{ij}}{\sum_{i,j}|\Delta Y_{ij}|\,\beta_{ij}} - \frac{\sum_{i,j}|\Delta T_{ij}|(-1)^{\alpha_{ij}}\bar{\beta}_{ij}}{\sum_{i,j}|\Delta Y_{ij}|\,\bar{\beta}_{ij}}\right| \qquad (5)$$

with α_{ij}, β_{ij}, $\bar{\beta}_{ij}$ defined by

$\alpha_{ij}=0$, if $\mathrm{sgn}\Delta T_{ij}=\mathrm{sgn}\Delta X_{ij}$; $\beta_{ij}=0$, if $\mathrm{sgn}\Delta X_{ij}\neq\mathrm{sgn}\Delta Y_{ij}$,

$\alpha_{ij}=1$, if $\mathrm{sgn}\Delta T_{ij}\neq\mathrm{sgn}\Delta X_{ij}$; $\beta_{ij}=1$, if $\mathrm{sgn}\Delta X_{ij}=\mathrm{sgn}\Delta Y_{ij}$,

$\bar{\beta}_{ij}=1-\beta_{ij}$.

In eqs. (5) six different summations can be seen. A further simplification is easily obtained. Let us consider the summations in the numerators. For all the values of the indexes i,j the term to be added to the first summation is $|\Delta T_{ij}|(-1)^{\alpha_{ij}}\beta_{ij}$ and the one to be added to the other is $|\Delta T_{ij}|(-1)^{\alpha_{ij}}\bar{\beta}_{ij}$. As either $\beta_{ij} = 0$ and $\bar{\beta}_{ij} = 1$ or the opposite, we can consider the single term $|\Delta T_{ij}|(-1)^{\alpha_{ij}}$ as being added to one summation or the other depending on the value of β_{ij}.
Similar results hold for the remaining four summations on the denominators.

It follows that only three arithmetic units will suffice, each connected to two banks of registers whose selection is controlled by the signal β_{ij}. Fig. 3 represents the general configuration of the hardware estimator. A close match to the structure of eqs. (5) is evident. A little explanation is needed only for the "window and marker generator" and the signal M/S. In a television image, different subjects may be present, each with a different movement. We have then to define the part of the image where we should apply our algorithm: this is the function of the "window generator".

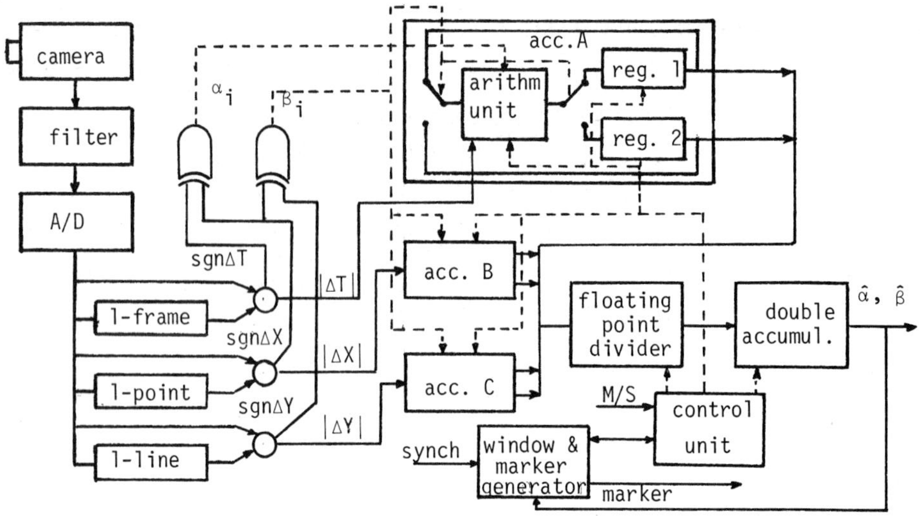

Fig. 3. Block diagram of the estimator.

The window is square, and encompasses pixels of both fields; its size and position can be chosen by the operator.

Moreover, as it was pointed out in Section 2, not all the windowed pixels, but only the "moving" ones have to be considered as input to the algorithm. In fact, the signal M/S is the output of a segmenting machine which determines if a pixel belongs to a moving or fixed area. This signal acts in logic AND with the output of the window generator.

It is not our concern to deal here with segmentation algorithms, but the interested reader may refer to /3/.

To experiment on the precision of the estimate, we have used the outputs of the estimator to update the position of the window at the end of every frame. The center of the window is visualized on the TV image by means of a marker. In this way, a good feeling of the algorithm accuracy is obtained by observing how exactly the marker follows the subject in its movements. Since the precision of the algorithm allows the detection of movements of a small fraction of pixel per frame, the minimum resolvable displacement was chosen to be 1/16 of a pixel. The window could be moved, in each frame, only by an integer numer of pixels. The residual fractional displacement not applied was taken into account in the next frame.

6. Experimental results for large block sizes.

All the results presented heretofore have been experimentally tested for their validity. The "image" is 512 by 512 pixels from the central area of a standard CCIR 625 lines, 50 field/second TV signal sampled at 10.7 MHz.

A rectangular window (from 128 by 128 down to 32 by 32 pixels) is positioned on the image. The algorithm is then applied to all moving pixels within this window. As a result the average displacement of the objects present in it is measured. Good performance has been observed even when the smaller window (see fig. 4) is used.

Fig. 4. Moving mannequin head used as test object.

Frame-differences have been displayed in order to visually judge how much their amplitude could be reduced through the compensation of movement.
Fig. 5 shows the displayed frame-differences before (fig. 5(a)) and after the compensation of the movement (fig. 5(b)), measured in the area evidenced in fig.4

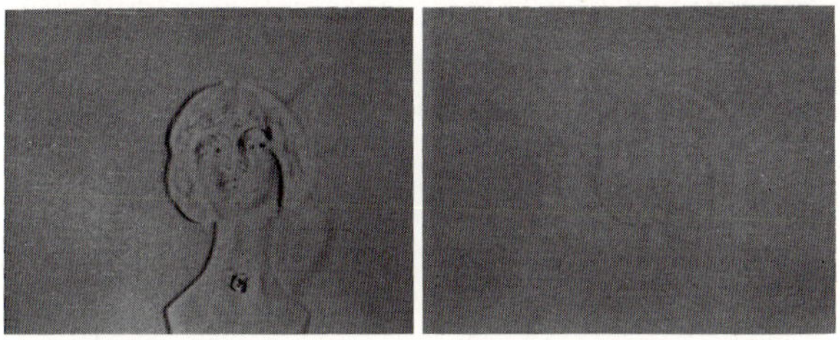

Fig. 5. Temporal luminance differences obtained without (a) and with (b) movement compensation.

Visual judgement of course can give only a rough idea. To obtain objective results the probability-density of frame-differences is real-time measured within the moving head area and over a presettable number of frames. Fig. 6 shows two sample

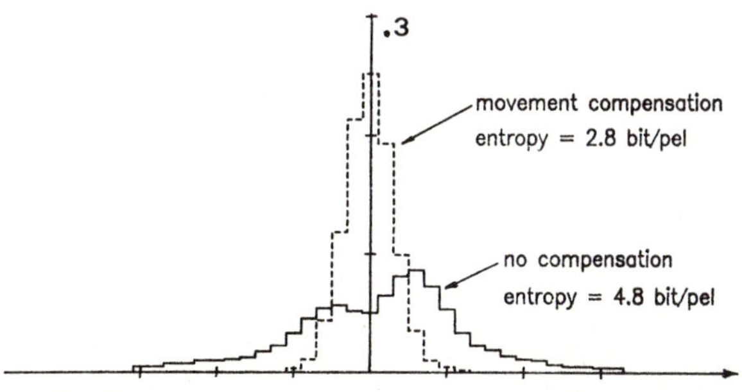

Fig. 6. Histograms of frame-differences in the moving area with and without movement compensation.

histograms obtained in the same situation with and without movement compensation together with the frame-differences entropies calculated through them.

Fig. 7 is a diagram of the same two entropies plotted versus estimated displacement and gives an accurate idea of the savings in bandwidth that could be obtained through the use of the proposed algorithm in a conditional replenishment movement compensating video encoder.

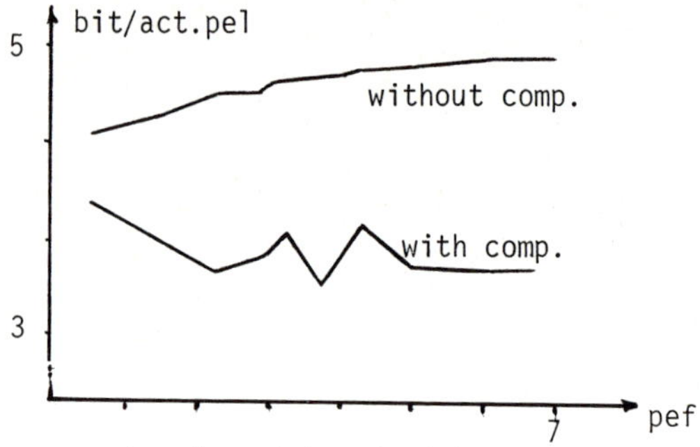

Fig. 7. Frame-differences entropies with and without movement compensation versus measured displacement.

7. Smaller block sizes and singularity of matrix P.

Up to now we have not considered the possibility of singularity of the matrix P in formula (2). In fact this is not relevant until the block size is large. We will now consider smaller block sizes and therefore we will calculate the two eigenvalues of P:

$$\lambda_{1,2} = \frac{\Sigma\Delta X_i^2 + \Sigma\Delta Y_i^2 \pm \sqrt{(\Sigma\Delta X_i^2)^2 + (\Sigma\Delta Y_i^2)^2 - 2\Sigma\Delta X_i^2\Sigma\Delta Y_i^2 + 4(\Sigma\Delta X_i\Delta Y_i)^2}}{2}$$

$\lambda_1 > \lambda_2$.

The correspondent eigenvectors $u_{1,2}$ have directions $\theta_{1,2} = \mathrm{arctg}(\lambda_{1,2} - \Sigma_i\Delta X_i^2)/\Sigma_i\Delta X_i\Delta Y_i$.

If $\Sigma_i\Delta X_i\Delta Y_i = 0$, then $\theta_1 = 0$, $\theta_2 = \pi/2$. If $\lambda_2 = 0$, then the eigenvector u_1 is directed along the image gradient and it is impossible to retrieve the other component of the motion. For big blocks and isotropic images it is easy to see that P tends to the identity matrix but for a non essential constant. However, it is important to determine how big the block should be in order to get λ_2 substantially non-zero. This can be easily carried out with a Gaussian model for the image. Experimental results are more telling, though.

In fig. 8 a-d we plot the histograms of the eigenvalues of P(normalized to the number of picture elements within the block),for different block sizes for the part of the image "Judy" shown in fig. 9. It is evident that the estimation of the motion in the direction orthogonal to the prevalent gradient can be carried out only when the block size is 8 by 8 or larger, coherently with previous experimental results /4/.

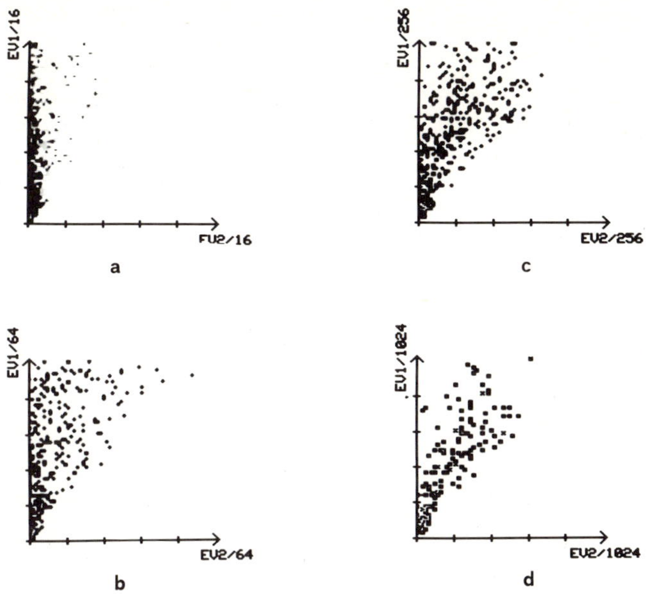

Fig. 8. Histograms of the eigenvalues of matrix P. (a) block size 4 by 4; (b) block size 8 by 8; (c) block size 16 by 16; (d) block size 32 by 32.

Fig. 9. Image analyzed (part of "Judy", courtesy of Bell Labs.).

The fact that only one motion component can be found may be quite irrelevant for image coding, since it would not change the prediction by much. However, it is important when the whole motion field has to be consistently determined to avoid instabilities.

To obtain that, a model for the motion field has to be introduced, in order to provide an estimate of the unobservable component.

8. Combination of estimates from adjacent blocks

The previous section has shown that the displacement measure can be obtained with accuracy only if the block size is large. This may be a serious drawback. Greater block sizes will yield more accurate displacement measures if a rigid object is translating. It is not difficult to conceive situations where this is far from true. In this case larger block sizes mean only better estimates of the average displacement within the block.

A possible solution could be to refine the model in (1) in order to take into account rotational movement /6/. In this paper this hypothesis will not be investigated. Smaller block sizes are appealing in that they seem to be able to cope easily with multiple displacements or even with a continuous displacement field. Unfortunately we know that their use gives measures with too high a spread.

It is, however, highly improbable that displacements in adjacent blocks be independent. In effect they are correlated and this correlation is higher when the blocks are small in size. This is the argument in support of large block size but the same concept may be used to increase the accuracy of displacement measures obtained with small blocks. In fact each estimate may be made more precise using the estimate already obtained for an adjacent block.

It is possible to proceed more formally now. Let v_i be the estimate of the vector (α,β) obtained within the i-th block and v_{i-1} the one for the (i-1)-th one.

$$\hat{v}_{i-1} = v(i-1)+n_1 \quad , \quad v_i = v(i) + Q\, n_2 \tag{6}$$

where v(i) denotes the true velocity of the object in the i-th block and n_1, n_2 are noise terms which account for the estimation inaccuracies. Note that n_1 has the dimension of a velocity while n_2 has the same dimension of the product $\Delta T \Delta X$; the matrix Q has been defined in section 2.

A linear combination $\hat{v}_i$ of the two estimates could be an estimate for the velocity in the i-th block better than v_i

$$\hat{v}_i = A\, \hat{v}_{i-1} + B\, v_i \tag{7}$$

A and B are 2 by 2 matrices and may be determined with the condition that $E[(v(i)-\hat{v}_i)^2]$ be a minimum.

Noise terms n_1 and n_2 may be considered as independent both with respect to one another and to v(i), with zero mean and autocovariance matrices:

$$R_{n_1n_1} = \begin{vmatrix} \sigma^2_{n_{1x}} & r_{n_1}\,\sigma_{n_{1x}}\,\sigma_{n_{1y}} \\ r_{n_1}\,\sigma_{n_{1x}}\,\sigma_{n_{1y}} & \sigma^2_{n_{1y}} \end{vmatrix} \quad , \quad R_{n_2n_2} = \sigma_n^2\, I$$

with I unit matrix. Then (see Appendix for details):

$$A = \{I-R_{vv}(0)[R_{vv}(0)+QR_{n_2n_2}Q^*]^{-1}\}\{[R_{vv}(0)+R_{n_1n_1}]R^{-1}{}_{vv}(d)- -R_{vv}(d)[R_{vv}(0)+QR_{n_2n_2}Q^*]^{-1}\}^{-1} \tag{8}$$

$$B = \{R_{vv}(0)R^{-1}{}_{vv}(d)-R_{vv}(d)[R_{vv}(0)+R_{n_1n_1}]^{-1}\} \cdot$$

$$\{[R_{vv}(0)+QR_{n_2n_2}Q^{*}]R^{-1}{}_{vv}(d)-R_{vv}(d)[R_{vv}(0)+R_{n_1n_1}]^{-1}\}^{-1}$$

where all terms are 2 by 2 matrices. $R_{vv}(d)$ is the autocorrelation matrix of the true displacement and is the only a priori knowledge required about the velocity field.

In order now to use formulas (8) we need a model for the velocity field to calculate $R_{vv}(d)$.

A general television scene may be made up of different objects, each moving in a different direction with no relation to the others.

Within a single frame, different areas may be located in which the displacement is almost constant but independent of displacements in contiguous areas. These areas roughly correspond to the objects. The intersections of their boundaries with any straight line may be considered as a realization of a Poisson process /11/ whose intensity denotes the average number of intersections per unit length.

The displacement value within each area may be considered another r.v. Reasonable hypotheses are that its amplitude has zero mean and a probability density function uniformly decreasing for large arguments. Its direction has a uniform probability density in the range $-\pi/2$ $\pi/2$.

With these hypotheses the autocorrelation matrix $R_{vv}(d) = \sigma_v^2 \exp(-\lambda d)\ I$, where σ_v^2 is the variance of velocity within a given area and $\exp(-\lambda d)$ is the probability that points at a distance d lie within a same constant velocity zone.

If the block size is greater than 1, then $R_{vv}(d)$ must be calculated for velocities averaged over the whole block:

$$R_{vv}(d) = \frac{\sigma_v^2}{N^2} \sum_{\text{all } N^2 \text{ pairs}} p\ (\text{pair within the same area})\ I$$

$$= \sigma_v^2 \exp(-f(\bar{d}))\ I$$

where $\bar{d}$ is the distance between centers of continuous blocks. Therefore $R_{vv}(d) = \sigma_v^2 \exp(-\mu)I$. If, for the sake of simplicity, we pose $R_{n_1n_1} = \sigma_{n_1}^2\ I$, then

$$A = U\,\mathcal{A}\,U^{*} \qquad , \qquad B = U\,\mathcal{B}\,U^{*}$$

where U is a matrix whose columns are the eigenvectors of the matrix $P = Q^{-1}$ with eigenvalues λ_i, and $\mathcal{A}$, $\mathcal{B}$ are diagonal matrices with elements

$$a_{ii} = \frac{\exp(-\mu)}{1+\frac{\sigma_{n_1}^2}{\sigma_v^2}+\frac{\sigma_{n_1}^2\lambda_i^2}{\sigma_{n_2}^2}+\frac{\sigma_v^2\lambda_i^2}{\sigma_{n_2}^2}(1-\exp(-2\mu))} \quad ; \quad b_{ii} = \frac{1+(\sigma_v^2/\sigma_{n_1}^2)(1-\exp(-\mu))}{1+\frac{\sigma_v^2}{\sigma_{n_1}^2}(1-\exp(-2\mu))+(\frac{\sigma_{n_2}^2}{\sigma_{n_1}^2}+\frac{\sigma_{n_2}^2}{\sigma_v^2})\frac{1}{\lambda_i^2}} \tag{9}$$

These formulas show that, should one of the two eigenvalues be very small, the new estimate would contribute only to the velocity component in the direction of the dominant eigenvector.

Of course, if the noise term n_1 is not white, matrixes A and B are expressed by different, more involved formulas (see appendix). The basic idea remains altogether the same.

Eigenvalues depend on the block size. If this size is smaller than 8 by 8 picture elements, the smallest eigenvalue is practically zero. This has been shown in section 7. Then, the best thing to do in this situation is to update the velocity estimate on a pel by pel basis.

9. Pel-recursive algorithm

When the velocity estimate is carried out pel by pel, the matrix P is singular. The non-zero eigenvalue has the value $\lambda_1 = \Delta X^2+\Delta Y^2$ and the related eigenvector is a unity vector in the direction of the image luminance gradient $\nabla f = (\Delta x,\Delta y)$. The velocity estimate can therefore be updated only in that direction.

If $\sigma_{n1}^2 \ll \sigma_v^2$ and $\mu \ll 1$, then

$$a_{11} = \frac{\varepsilon^2}{\varepsilon^2+\lambda_1} \quad , \qquad a_{22} = 1$$

$$b_{11} = k\,\frac{\lambda_1}{\lambda_1+\varepsilon^2} \quad , \qquad b_{22} = 0$$

where

$$k = (\sigma_{n1}^2/\sigma_{n2}^2+\mu\sigma_v^2/\sigma_{n2}^2)/(\sigma_{n1}^2/\sigma_{n2}^2+2\mu\;\sigma_v^2/\sigma_{n2}^2\;)$$

and

$$\varepsilon^2 = (\sigma_{n1}^2/\sigma_{n2}^2+2\mu\sigma_v^2/\sigma_{n2}^2)/(\Delta X^2+\Delta Y^2)$$

Dimensionality in the previous formulas should be carefully considered, owing to the different physical meaning of the two noise contributions n_1 and n_2.

Taking into account the results of section 2, we get:

$$\hat{v}_i = \begin{vmatrix} \frac{\varepsilon^2}{\varepsilon^2+|\nabla|^2}\,\frac{\Delta X^2}{|\nabla|^2}+\frac{\Delta Y^2}{|\nabla|^2} & -\frac{\Delta X\Delta Y}{\varepsilon^2+|\nabla|^2} \\ -\frac{|\nabla|^2}{\varepsilon^2+|\nabla|^2}\,\frac{\Delta X\Delta Y}{|\nabla|^2} & \frac{\Delta X^2}{|\nabla|^2}+\frac{\varepsilon^2}{\varepsilon^2+|\nabla|^2}\,\frac{\Delta Y^2}{|\nabla|^2} \end{vmatrix} \hat{v}_{i-1}+k\,\frac{\Delta T}{\varepsilon^2+|\nabla|^2}\nabla f \qquad (10)$$

This equation implies that at each step a new estimate $\hat{v}_i$ is obtained. Apart from weighting factors, this new estimate changes only the component of the previous estimate in the direction of the local image gradient. v_i is obtained using the differential algorithm, which is very accurate when small displacements have to be measured; underestimation would result otherwise.

The same procedure may, however, be implemented in another way, using displaced frame differences /9/.

Once the estimate $\hat{v}_{i-1}$ has been obtained, one of the two frames used to measure temporal differences may be displaced according to this quantity.

In this case, instead of (6) we get:

$$\hat{v}_{i-1} = v(i-1)+n_1 \quad ; \quad v_i = v(i)-\hat{v}_{i-1} + Qn_2$$

and v_i results smaller than in (6). The optimal estimate for velocity at location i may be obtained as

$$\hat{v}_i = (A+B)\ \hat{v}_{i-1} + Bv_i \tag{11}$$

where the matrices A and B are the ones already calculated.
This pel recursive algorithm closely resembles the one in /10/. It is easily amenable to a meaningful physical interpretation.
A local measure of displacement may be obtained only by relating movement induced temporal luminance differences to the spatial luminance differences within the same image area. In fact a moving imaged object can generate temporal differences only if its luminance is not constant. In frame areas where luminance changes occur only along one spatial direction, a displacement component may be defined along that direction only. An orthogonal component of any value would not affect temporal differences and therefore cannot be defined.
When the area in which the displacement is estimated shrinks to become a single picture element, the spatial luminance change is expressed in value and direction by the luminance gradient. Therefore the only displacement component that can be measured lies in that direction. However physical objects are compact and therefore the missing information may be gathered by measures obtained from neighbouring elements.
In /10/ the displacement component update is $C\Delta T\nabla f$, here it is $k\Delta T\nabla f/(\varepsilon^2+|\nabla f|^2)$.
The only difference lies in the normalizing factor.
In either case the updating velocity component lies on the gradient direction and is proportional to temporal luminance differences. These, however, may be high even for small displacements, where high spatial luminance variations occur due to contours. A correction term is needed to account for this effect. This correction must include the ε^2 term to avoid the problems which would arise when ∇f is very small or zero, due to the noise superimposed on the image.
Experimental results, to be shown in the next section, confirm this observation.
In fact the algorithm (a) deduced from /10/ has more stability problems than the one proposed in this paper (b). In fact the former requires a very small value for the constant C (1/1000 or less), while k = 1 and ε^2 = 100 are satisfactory values for the constants in the proposed procedure (the r.m.s. value for the gradient is about 140).
Moreover the two methods give different results when used in a movement compensation scheme.

10. Experimental results for small block sizes

All the theoretical results presented heretofore have been experimentally tested.

The various parameters present in the formulas have been assigned several values in order to test the algorithms' sensitivity and stability. Typical values, which ensure stability have been already mentioned.

On the other hand a bidimensional recursion has proved to be better than a monodimensional one. Therefore results shown refer to pel-recursive algorithms in which v_{i-1} has been taken as the mean between estimates $\hat{v}$ obtained for the previous pel along the line and for the homologous pel in the previous line.

Only one field from each frame has been used, in order to avoid possible entanglements with field interleaving. This accounts for the unusual aspect ratio of all the figures which represent a sample frame, temporal differences and velocity fields.

Fig. 10 shows one sample frame of the TV sequence used to test the usefulness of the algorithms within a movement compensating video coder.

Fig. 10. Sample field of the TV sequence used to test the various algorithms.

Three algorithms have been tested and the entropy of the residual frame to frame luminance differences within the active area of the field has been used as a test measure.

The first one, used as a reference, is movement compensation with a single displacement estimate for the whole field, obtained applying the differential algorithm over the moving area (the same as the one used for real time experiments).

The second one is algorithm (a), the third is the one of section 9 (b).

Fig. 11 a shows temporal frame differences without movement compensation. Figs. 11 b-c-d show similar results obtained after movement compensation with the aforesaid algorithms.

In translating a field, in order to compensate for movement, bilinear interpolation of luminance values has been used.

Fig. 12 summarizes these same results showing the histograms of the frame differences in figs. 11, cumulated in the moving area alone. The entropy of each histogram is shown too.

The entropy values are very similar, notwithstanding more marked differences noticeable in figs. 11.

The simulations with the pel-recursive algorithms used the velocity measure obtained at location i along a line to compensate for movement at location i+1. This has been done in order to simulate a coding scheme in which velocity fields are not transmitted, but are recalculated at the receiving end. Entropy values are, however, only slightly affected by this approximation.

A careful examination of the frame differences shows that the main contribution

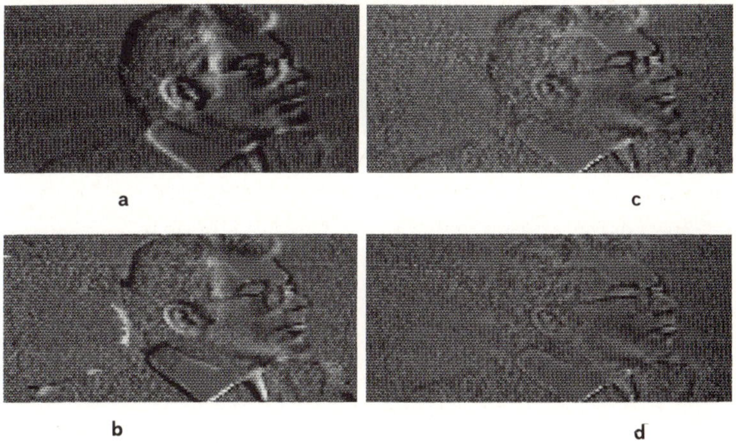

Fig. 11. Frame-to-frame differences obtained with: a) no compensation; b) movement compensation with a single displacement measure for the whole moving area; c) movement compensation with algorithm a); d) movement compensation with algorithm b).

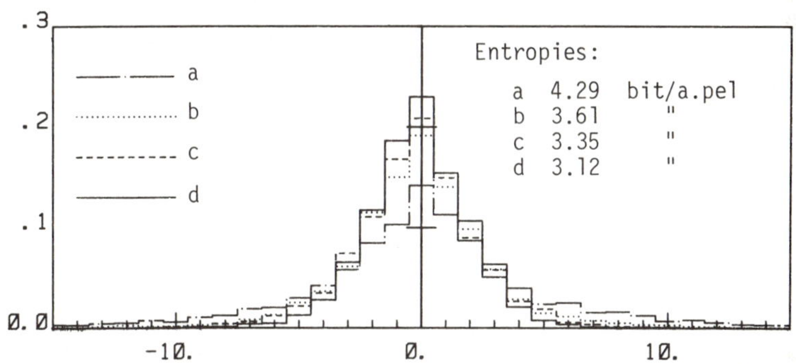

Fig. 12. Histograms of the frame differences shown in fig. 4, calculated over the moving area only (a,b,c and d same as for fig. 4).

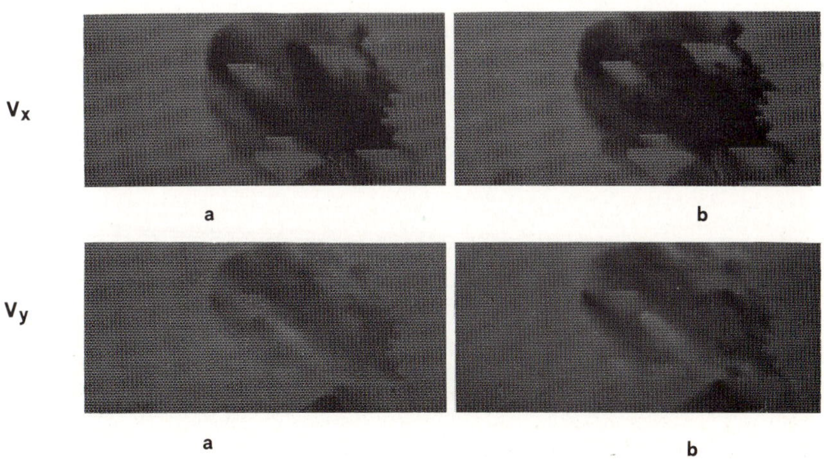

Fig. 13. Velocity fields obtained with algorithms (a) and (b).

to entropy comes from noise which, in principle, is not necessary to the description of the image. Its suppression, however, could impair the performance of a hypothetical receiver. A careful dose of dummy luminance information could possibly be an elegant solution to this problem.

Figs.13 show the horizontal and vertical velocity fields produced by the algorithms (a) and (b).

The outline of the moving area of the image is clearly visible.

Figs. 14 show the histograms of both velocity components accumulated within the moving area.

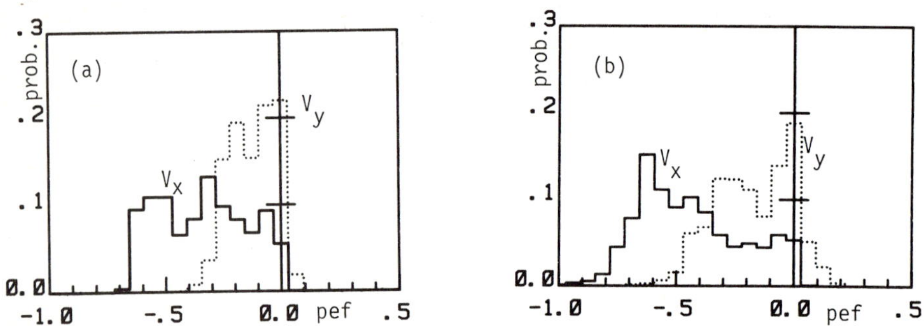

Fig. 14. Histograms of the displacement components shown in fig. 6. As a reference the output of the differential method applied to the whole moving area was Vx = -.417, Vy = -.160.

11. Conclusions

This paper has shown that it is possible to use the differential algorithm even with very small block sizes. In that case, the obtained measure is only the component of the displacement along the local luminance gradient. If this is not enough, or in order to get a more homogeneous velocity field, a recursion scheme proves quite valuable. In either case of large and small block size, the obtained results are very good for coding purposes, but not quite so if a physical velocity field is the goal. Looking at figs. 13, for example, in fact, one would like to see more smooth variations.

In the authors' opinion a better model of the moving object could greatly improve the results.

Appendix

Given

$$\hat{v}_{i-1} = v(i-1)+n_1 \quad , \quad v_i = v(i)+Q\ n_2 \qquad \text{(A1)}$$

their linear combination

$$\hat{v}_i = A\ \hat{v}_{i-1} + B\ v_i \qquad \text{(A2)}$$

which minimizes the mean square error $E[(v(i)-\hat{v}_i)^2]$ can be obtained easily. The

2 by 2 matrixes A and B result to be:

$$A = \{E[v(i)v^*_{i-1}]E[v_i v^*_{i-1}]^{-1} - E[v(i)v^*_i]E[v_i v^*_i]^{-1}\}\cdot$$
$$\cdot\{E[v_{i-1}v^*_{i-1}]E[v_i v^*_{i-1}]^{-1} - E[v_{i-1}v^*_i]E[v_i v^*_i]^{-1}\}^{-1}$$
$$B = \{E[v(i)v^*_i]E[v_{i-1}v^*_i]^{-1} - E[v(i)v^*_{i-1}]E[v_{i-1}v^*_{i-1}]^{-1}\}\cdot \tag{A3}$$
$$\cdot\{E[v_i v^*_i]E[v_{i-1}v^*_i]^{-1} - E[v_i v^*_{i-1}]E[v_{i-1}v^*_{i-1}]^{-1}\}^{-1}$$

where the various v's are column vectors and * denotes transposition. Therefore $E[vv^*]$ results a 2 by 2 matrix .
After combining (A3) with (A1) and (A2)

$$A = \{R_{vv}(d)[R_{vv}(d)+QR^*_{n_1n_2}]^{-1} - R_{vv}(0)[R_{vv}(0)+QR_{n_2n_2}Q^*]^{-1}\}\cdot$$
$$\cdot\{[R_{vv}(0)+R_{n_1n_1}][R_{vv}(d)+QR^*_{n_1n_2}]^{-1} - [R_{vv}(d)+R_{n_1n_2}Q^*]$$
$$[R_{vv}(0)+QR_{n_2n_2}Q^*]^{-1}\}^{-1}$$
$$B = \{R_{vv}(0)[R_{vv}(d)+R_{n_1n_2}Q^*]^{-1} - R_{vv}(d)[R_{vv}(0)+R_{n_1n_1}]^{-1}\}\cdot \tag{A4}$$
$$\cdot\{[R_{vv}(0)+QR_{n_2n_2}Q^*][R_{vv}(d)+R_{n_1n_2}Q^*]^{-1} - [R_{vv}(d)+QR^*_{n_1n_2}]$$
$$[R_{vv}(0)+R_{n_1n_1}]^{-1}\}^{-1}$$

where $R_{vv}(x) = E[v(i)v^*(i-x)]$ and $R_{ab} = E[ab^*]$.
As already pointed out in section 2, matrix Q is

$$Q = \begin{vmatrix} \Sigma_k\ \Delta x^2_k & \Sigma_k \Delta X_k \Delta Y_k \\ \Sigma_k \Delta X_k \Delta Y_k & \Sigma_k \Delta Y^2_k \end{vmatrix}^{-1} = P^{-1}$$

where summations are carried out over i-th block. If Λ is the diagonal matrix of eigenvalues and U is a matrix whose columns are the eigenvectors of matrix P,

$$P = U\,\Lambda\,U^* \qquad \text{and} \qquad Q = U\,\Lambda^{-1}\,U^* \tag{A5}$$

A model for $R_{vv}(x)$ has already been given in Section 8.
The two noise terms n_1 and n_2 can be reasonably supposed independent and consequently $R_{n_1n_2} = 0$. Moreover we may consider n_2 as white: $R_{n_2n_2} = \sigma^2_{n_2}\, I$ but this same hypothesis would result not appropriate for n_1. This term must account for the spread of estimate $\hat{v}_{i-1}$ and it is unlikely that this could be modeled by means of two independent and equal variance components along the two coordinate axes. Therefore we should pose

$$R_{n_1 n_1} = \sigma_{n_1}^2 \, N \, N^* \qquad N = M \begin{vmatrix} \sigma_1/\sigma & 0 \\ 0 & \sigma_2/\sigma \end{vmatrix}$$

with M an orthogonal matrix.

In this case formulas (A4) may be modified according to the aforesaid assumptions but they are not amenable to an easily understandable format.

However, if N = I, then all matrices in A4 are diagonal but for Q. In this case, using (A5) and $U^* U = I$, (A4) modify into

$$A = U \mathcal{A} U^* \quad , \qquad B = U \mathcal{B} B^*$$

Here the elements of the diagonal matrices $\mathcal{A}$ and $\mathcal{B}$ whose values are given in (9), can be understood as weighting factors for the components of $\hat{v}_{i-1}$ and v_i in a new coordinate system identified by the two orthogonal eigenvectors of matrix Q.

Bibliography

/1/ J.O. Limb and J.A. Murphy "Measuring the speed of moving objects from television images" IEEE Trans. on Communications, vol. COM-23, pp. 474-478, Apr. 1975

/2/ J.O. Limb and J.A. Murphy "Estimating the velocity of moving images from television signals" Computer Graphics and Image Processing, 4 (1975) pp. 311-327

/3/ C. Cafforio and F. Rocca "Methods for measuring small displacements of television images" IEEE Trans. on Inf. Theory, vol. IT-22, Sept. 1976, pp. 573-579.

/4/ C. Cafforio, F. Rocca "Tracking moving objects in television images" Signal Processing 1 (1979) pp. 133-140.

/5/ S.R.J. Axelsson "On optimum algorithms for imaging tracking systems" Proc. EUSIPCO 80 Lausanne, Switzerland, Sept. 16-18 1980, North-Holland, Amsterdam,80

/6/ Image sequence analysis - Editor T.S. Huang, Springer-Verlag, 1981

/7/ C. Cafforio "Remarks on the differential method for the estimation of movement in television images" Signal Processing 4 (1982) pp. 45-52

/8/ F. Rocca "Television bandwidth compression utilizing frame-to-frame correlation and movement compensation" Symposium on Picture Bandwidth Compression (MIT Cambridge, Mass., 1969) Gordon and Breach 1972

/9/ Y. Ninomiya "Motion correction for interframe coding systems", Tech. Group Image Engineering (IE) IECE Japan, Paper IE78-6, May 1978

/10/ A.N. Netravali and J.D. Robbins "Motion compensated television coding - Part I" Bell Syst. Tech. J., vol. 58, pp. 631-670, March 1979

/11/ F. Rocca and S. Zanoletti "Bandwidth reduction via movement compensation on a model of the random video process" IEEE Trans. on Comm., vol. COM-20, Oct. 1972 pp. 960-965.

EDGES IN VISUAL SCENES AND SEQUENCES : APPLICATION TO FILTERING, SAMPLING AND ADAPTIVE DPCM CODING.

Francis KRETZ
C.C.E.T.T., Rennes, FRANCE

I – INTRODUCTION : NON STATIONARITY AND ADAPTIVE CODING

Although quite natural in scene analysis, the consideration of local edge occurrence and of picture structure implying a certain kind of nonstationarity has not been used much within the framework of picture coding. Some years ago, mainly in order to analyse coding impairments on pictures and their perception by the human visual system, we introduced the notion of «local critical configurations» and this led us to consider pictures as a more or less random concatenation of these configurations each with their own statistical properties [1]. Extension to moving sequences is straightforward, the concatenation being now dynamical and each configuration bearing its own motion parameters. This approach departs clearly from the usual stationary approach the merit of which was the design of the two main coding families : DPCM coding using a local prediction and the quantization of the error signal ; transform coding using quantization of the transform coefficients on a block-by-block basis. But, more efficient coding schemes for bit rate reduction needed to be defined by taking advantage as much as possible of the specific properties of pictures or sequences and of the human visual system. We refer here to [2] for a recent general review on picture coding and to [3] more specifically for a review on sequence coding.

On the one hand, an optimal coding algorithm will thus be based on vision properties such as perception of noise, blur,... and in general, of the impairments produced by filtering, sampling, segmenting or quantizing. Masking properties are, here, to be considered, that is the local loss of perception accuracy due to severe changes in pictures or sequences : masking of edges (spatial masking) is one example ; masking of flashes or sudden scene switches (temporal masking) and masking of moving edges (spatio-temporal masking) are other examples. Adaption of quantization schemes is generally involved in order to take advantage of these properties.

On the other hand, an optimal coding algorithm will make use of source properties and we will focalise this paper on this matter. An efficient way would be to adapt coding algorithms locally to the specific properties of the local picture configurations, «locally» being considered in space and time. Basically our idea is to segment pictures or sequences into configurations with low activity (generally carrying low visual information) and configurations with high activity, mainly spatial and spatiotemporal edges carrying

NATO ASI Series, Vol. F2
Image Sequence Processing and Dynamic Scene Analysis
Edited by T.S. Huang

most of the visual information (structure). Local estimation of edges parameters (orientation and motion) is thus implied.

Our approach specifically stresses edges : the suggested segmentation of pictures or sequences considers edges as stationarity breaks or «ruptures», i.e. they are the locii where one (locally) stationary model —associated with one uniform or textured quasi-stationary area— is replaced by another one. These stationarity breaks are also very likely the locii where large contrasts occur and thus where any coding scheme is put in difficulties as they are usually based on stationarity assumptions. One structural properties of edges is that they lie spatially along one dimension (longitudinally). In this sense, they can be considered as «lineal» whereas quasi-stationary configurations are «areal», lying spatially in two dimensions [4]. Although the a priori probability of a pixel to lie on an edge is small (say 5 %), edges are fundamental for visual perception (a picture without edges would not be not much informative) and also for processing because they correspond to sharp signal discontinuities.

Various other nonstationary models of pictures have been proposed. Hunt [5] suggested a space-variant local mean and space-variant local autocorrelation function with constant shape, allowing each picture to be transformed into a stationary one using polynomial coordinate warp. This model is limited to second order statistics and to smooth variations of mean and autocorrelation, a solution we believe unsufficient for taking into account edges in pictures. Maxemchuk [6] considered pictures to be composed of areas of texture, each with different 2-D autoregressive stationary models, a model we think unappropriate for coding because it doesnot introduce high local activity configurations as such. These are the frontiers of stationary areas and they are the most difficult to encode as compared with texture configurations. Our approach better corresponds to the one from Zschunke [7] where edge pixels are predicted according to an estimation of the local edge orientation whereas uniform areas pixels are predicted using previous pixel prediction. Another solution is to decompose pictures into two additive components [8], one carrying edges and plateaus, the other carrying textures and being assumed stationary. This implies twice encoding edge pixels but it allows the second component to be encoded by transform coding. Considering the specific constraints of TV signals coding we will recall below, this last solution seems less promising ; it is better fitted to the coding of still (frozen) pictures.

We will describe an algorithm estimating local orientation recursively along spatial edges which is used for intrafield adaptive prediction. The longitudinal recursion allows to filter along edges the noise affecting of local orientation estimates and makes the algorithm robust to source noise as well as to channel noise. For edge pixels and from the filtered orientation estimate, an intrafield adaptive prediction is derived .

In the case, now, of sequence coding using motion estimation and motion compensated prediction (see [3]), block-structured algorithms have been designed. They need homogeneity of the motion within a block, a situation which fails on edges which separate areas with different motions. The pel-recursive algorithms are more attractive because estimation can be made very local although Netravali's algorithms can be looked at as involving only «soft» adaptation : the motion estimate need several pixels after a break in motion stationarity (i.e. after a moving edge) to reach the proper new value. Oscillation or even divergence can occur with sharp and fast moving edges. We will extend here the recursive estimation of edge orientation to edge motion estimation, this allowing precise estimation of motion on edges and enabling a pel-recursive algorithm to be then properly initialized to estimate motion on moving areas.This (re)initialization by the motion estimate on edges can be seen as a «hard» adaption process.

Before getting into more details, it is useful to recall here that our approach is directed towards coding of television signals mainly for broadcast applications. Four constraints have to be followed :

source constraint : we are dealing with sharp contrasted pictures usually with a large number of edges ; in moving sequences, motions can be of all kinds (see typology in Part 2), possibly affecting the whole frame ; even for still scenes, coding TV signals significantly differs from still (frozen) picture coding because of temporal source noise.

quality constraint : perception thresholds for all impairments in presence must not be exceeded under standardized normal viewing conditions.

hardware constraint : high speed real time implementation is needed (50 or 60 fields per second, sampling rates higher than say, 9 MHz) ; cost of equipment may need processings to be easily integrated but coding boards are only a small part of a complete codec which need analogue interfaces, synchronization, multiplexing video channel with sound channels, data channels, error protection, etc...

channel constraint : channels of standard bit rates must be used taken from the telephony PCM hierarchy (140, 34, 8 or 2 Mbits/s in Europe) ; quality must be maintained for channel error rate of, say, 10^{-6} and kept acceptable up to, say, 10^{-3}. Of course, a certain amount of protection against errors can be inserted.

Part 2 of this paper will be devoted to our nonstationary approach of TV scenes and sequences. Part 3 will consider linear processings –filtering and sampling– of spatiotemporal edges. Part 4 will describe the recursive edge orientation estimation algorithm from C. Richard and will summarize our comparison study of adaptive predictions. Part 5 will introduce the recursive edge motion estimation algorithm from C. Labit. These two researchers worked at IRISA in Rennes for their Doctor-Engineer theses under close cooperation with CCETT and the active participation of A. Benveniste (IRISA) is to be deeply acknowledged.

II – AN ESSENTIALLY NONSTATIONARY APPROACH OF TV SCENES AND SEQUENCES

1 – Fourier spectrums of TV pictures

It is interesting to study the limits of the usual stationary stochastic models of pictures. For this purpose, we computed 512 x 512 Fourier transforms of center parts of eleven TV pictures (12 MHz, 8-bit digital pictures of 624 pixels by 576 lines corresponding to the 625 line standard). Fourier amplitude was twice logarithmically compressed before visualisation. Figure 1, taken from [9], shows three of these pictures, the mean of all eleven pictures and the associated Fourier spectrums.

The spectrums of pictures «Boats and lighthouse» and «Girl with toys» show at high frequencies accumulations of spectral energy corresponding to the strongest features of the pictures : long, contrasted and sharp edges for the first picture ; + 45° and – 45° periodic structures from the wall paper of the second picture. The third picture «Fruits» has a spectrum which better suits usual autocorrelation models because this picture contains mainly textures (energy on vertical or horizontal frequency axis is due to false edges produced by the periodic nature of discrete Fourier transform). The mean spectrum of the set of eleven typical TV pictures still show some features from the component spectrums.

Thus, the usual models would correspond to quite a large ensemble average, they are in general inadequate to describe individual pictures [9]. Also, Fourier analysis has severe drawbacks : information in the spectral space is delocalised, spectrums of isolated objets being superposed and only the most contrasted or repetitive features become apparent in the Fourier domain. Thus, a space (or space-time) domain approach is necessary and furthermore only a nonstationary and local approach can put the stress on the main features of scenes.

2 – Picture configurations : a «mosaic» model for pictures and sequences

As mentionned in the Introduction, in relation to the main processings used in picture coding, it is meaningful to decompose pictures or sequences into local configurations of different types [1] :

quasi-uniform areas with spatial gradient lower than, say, 2 % of dynamic range per pixel. Their spatial extension must be higher than some tens of pixels in every directions. Many types of impairments can occur : granular noise, false contours due to quantizing or segmenting (block structure, moving areas segmentation, ...), various kind of interferences such as sampling patterns or effects of transmission errors and also, loss of details. Objective parameters of these areas are their average luminance or color and their geometry. The visibility of impairments is generally high.

areas of low contrasted details with local contrast lower than 5 % of dynamic range per pixel, each detail being small (few pixels). Apart from just mentionned impairments, the most classical one is the loss of details (blur) due to filtering, sampling or coding. Objective parameters are the same as above plus parameters describing the texture. Visibility of impairments is also high but the visual system can still be linearized around the average color of the area and described by three spatiotemporal inhomogeneous filters followed by a statistical detector.

areas of contrasted details with contrasts and gradients more than say 15 % of the dynamic range and the dynamic range per pixel respectively, each detail being small. In addition to impairments like noise, effects of transmission errors, ..., two impairments mainly occur : one is the loss of contrast of

Figure 1. Three typical TV pictures and their Fourier spectrums (three top rows). Mean of eleven TV pictures and its Fourier spectrum (bottom row). From reference [9].

details (more precisely here than the loss of details) due to filtering or coding ; the other is «moiré» due to sampling (aliasing) of periodic textures. Visibility starts to be decreased and is quite complex for modelisation due to the large number of parameters that can describe a texture.

edges with constrasts higher than 15 % of dynamic range, transition width of one or two pixels, lying longitudinally on some tens of pixels. Classical impairments are loss of sharpness (another kind of blur), staircase rendition due to subsampling or to segmentation (more generally periodic perturcations on the transition either still, moving or flickering [9]), noise (edge busyness) and in some cases, overshoot or ringing due to filtering, effects of error concealment techniques, etc... Objective parameters of edges are the luminances or colors on both sides of the transition, the width of the transition, the edge orientation and length. In comparison with uniform areas, the visibility of impairments is reduced due to visual masking but, in the case of linear processings (sampling, filtering), the impairments amplitudes increase linearly with edge contrast and thus faster than masking, so they become more and more visible as contrast increases.

It must be clear that these four categories of picture configurations are introduced here merely for analysis. They are no sharp boundaries between them : as local contrast increase, we move from uniform areas to contrasted textures ; as the viewing distance or sampling density increases, we start considering an area of texture as a mosaic of small areas separated by edges.

Each type of configurations have its own statistical properties. The global statistical properties of pictures that have been classically established reflect those of the three first types of configurations because edge pixels are quite rare on pictures. We will in the next section give a summary of statistical properties established specifically for edges and edge pixels.

In *picture sequences*, scenes can be segmented into objects. Each object has a certain motion (or possibly no motion) and is itself decomposed into different connex picture configurations of the kind described above. The extension of the «mosaic» model to a «dynamic mosaic» model is easy : we only need to add to the objective parameters of each configuration parameters of its motion. Section II.5 will describe these parameters along with a typology of motions.

3 – Statistical properties of edges and edge pixels in TV pictures

We refer here to a study [4] of edge properties done on the same eleven pictures typical of television scenes we mentionned in section II.1. Various kinds of statistical measurements were intended : edge orientation and length histograms, orientation correlation along edges, histogram of runlengths between edges, statistics of transition width and histograms of gradients for edge pixels and for isolated pixels which were defined as pixels with a significant local gradient but not lying on an edge.

In order to be able to derive these statistics, we used Hueckel's algorithm which is based on a local least squares fit of the original picture within a small circular window with a model of an edge crossing that window. A slight modification of the algorithm were made to speed it up. When an edge segment has been validated at a given location (using a threshold on the fit distance), it is extended in order to locate the next window for a new fit. The algorithm is a bit sophisticated but it provides edge detection and edge following along with an excellent estimation of the local orientation of edges as well as other parameters such as the elementary length of the edge segment crossing the window. For pixels corresponding to a validated edge segment, other local statistical measurements were done, for example concerning transition width and horizontal as well as vertical gradients.

Figure 2 shows two orientation histograms. The general tendancy (Fig. 2b) is that horizontal and vertical

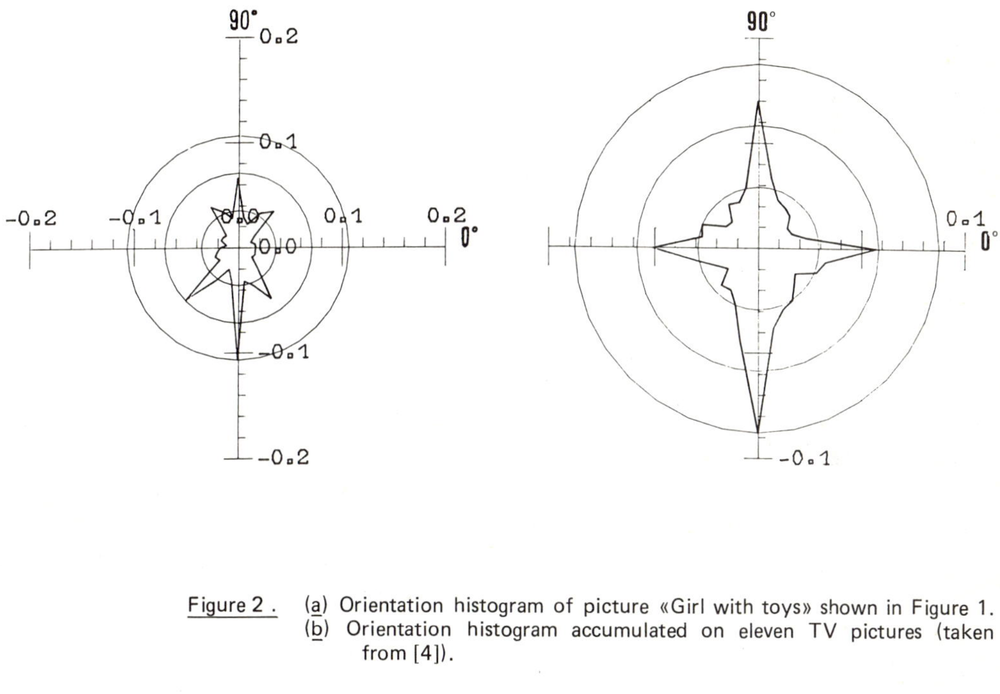

Figure 2 . (a) Orientation histogram of picture «Girl with toys» shown in Figure 1.
(b) Orientation histogram accumulated on eleven TV pictures (taken from [4]).

orientations are the most probable, about 3 times more than ± 45° orientations. But, due to the large number of edges in pictures, oblique edges are still present. The observed anisotropy can be explained by the frequent occurence of horizontal and vertical edges in common visual scenes (buildings, fourniture, trees, horizon, ...). But it is instructive to see that a particular picture (see Fig 2a) can largely depart from the general trend. Picture «Girl and toys» displays a large number of ± 45° edges as already mentionned in section II.1.

The comparison between these orientation histograms and the Fourier spectrums is not usely easy. For example, the spectrum of the average of eleven pictures would give the impression of a certain isotropy but the involved representation which necessitate heavy compression of amplitudes may well not show factors of 3 in probabilities. Edge orientation estimation as was done is of course a much more sensitive method.

Figure 3 shows some vertical gradient (difference between two horizontally adjacent pixels) histograms. Small gradients were skipped here to better show the interesting range of gradients above 10 % of dynamic range. The top curse is obtained from all pixels. The middle one corresponds to edge pixels (pixels

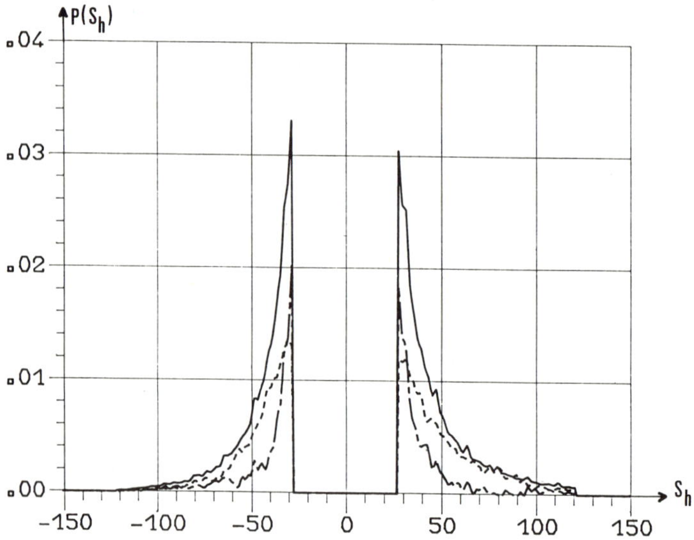

Figure 3 . Vertical gradient (horizontal difference) histogram accumulated from top to botton over all pixels, over edge pixels and over isolated pixels (taken from [4]).

lying on an edge as detected by the modified Hueckel's algorithm) : for gradients higher than 50 to 70, the «all pixels» histogram comes from edge pixels. The bottom curve corresponds to isolated pixels (pixels where horizontal and vertical gradients are higher than 10 % of dynamic range but not belonging to a detected edge). It shows that above a horizontal difference of say 60, isolated edge pixels become unprobable. Thus, if it were possible to correctly predict edge pixels from orientation estimation, a one-to-two reduction could be obtained for the dynamic range of the associated adaptive prediction error. Such an edge-prediction scheme will be considered in part IV.

The other statistical measurements of [4] show a strong correlation of successive local orientations along edges, this correlation being isotropic. This implies a predominance of *straight line edges*. Edge length has a probability distribution close to an exponential function whereas edge runlength statistics (distance between two edges in a given direction) departs from an exponential law but are very similar for horizontal and vertical directions.

Given these statistics, a quantitative nonstationary model could be constructed. This was tried in the case of one-dimensional signals [10] : from runlength distribution law, edges can be randomly positionned ; between edges, the signal is modelized by a Markov process with gaussian transition law and parameters independant from one run to the next. Amplitude, variance, gradient histograms can fit quite well real picture statistics. A two-dimensional model with edges of random orientation would much complicate the situation. But it is interesting to mention here the work by Modestino [11] based on random rectangular or oblique tesselation (partition) of the plane with uniform level in each area and given correlation of levels between adjacent parts. One sample picture looks like a mosaic with edges parallel to one of the two chosen directions. The model is considered globally stationary and the 2-D autocorrelation function is derived. The generalization to a more realistic polygonal partitionning is proposed but it becomes too complicated to handle for deriving second-order statistics. This kind of model is a first step towards a more realistic mosaic model taking into account statistical properties of each of the different types of configurations, although our approach would be to keep the point of view local, that is, to consider the model as essentially non-stationary and not reduce it at the end to second-order statistics.

4 – Some properties of eye movements in scanning visual scenes or sequences

A section on eye movements may seem at first look strange here, but in order to design a relevant typology of motions in visual sequences, it is useful to have some information about the way the final users will perceive them.

The visual scan of a still picture is largely unconscious (apart from voluntary fixations). Its structure is quite well known [12] : it is composed of *fixations* and *saccades*. During a fixation, the visual axes of both eyes are still during 70 to 350 ms for the visual system to perform a local analysis of the area that is fixated and to decide for the location of the next fixation. Then, a saccade is activated which brings the visual axes very quikly (15 to 50 ms) to a new location. The indicated durations are taken from [13], a study in which a series of eye movements measurements were obtained (25 subjects have viewed 25 slides 40 s each). The perception of the scene is thus achieved mainly from the fixations, vision being much degraded during saccades. The first fixations are usually directed towards the main spots of interest in the scene, then the visual scan is concerned by more detailed information.

In the case of the scan of moving sequences, apart from fixations and saccades on non moving areas, other types of eye activity can occur : *pursuit eye movements* when the visual system tracks or tries to track a moving object, *correcting saccades* when tracking gets imperfect enough. P. Viviani [14] designed with us an original experiment to measure the eye tracking capability in situations as realistic as possible. The eye system needed to be motivated by the tracking mode ; reading of texts and tables of digits moving sinusoidally along horizontal, vertical and oblique directions were choosen. Figure 4 shows the results obtained on moving texts relative to results on non moving texts. Amplitude and frequency have both an effect

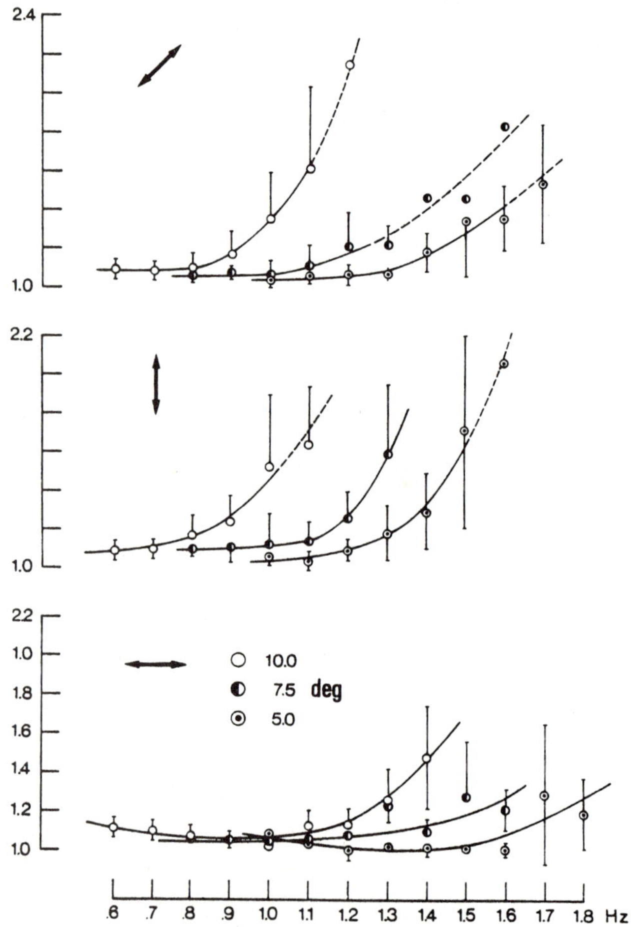

Figure 4 . Relative reading time of moving texts (average of 10 subjects). Abscissa is the frequency of the sinusoidal motion ; ordinate is the ratio of reading time in moving situations over reading time in non moving situation ; parameter is the amplitude of the motion (half peak-to-peak). Upper graphic corresponds to oblique (45^0) motions, middle one to vertical motions, lower one to horizontal motions.

which cannot be strictly summarized by using the maximum speed of the motion (the product of both with 2π), but using this parameter and for an increase of 20 % of reading time, we get the following (arbitrary) limits of eye tracking capability : 56.5 to 78.5 deg./s for horizontal, 41 to 53.5 deg./s for vertical and 44 to 56.5 deg./s for oblique motions.

Horizontal eye tracking capability is thus seen to be better than vertical or oblique ones which are close to eachother. One could argue that reading a text favours horizontal tracking but this is not quite true as the second experiment with tables of digits has shown us [14] : reading time of columns with vertical motion is 27 % higher than reading time of rows with horizontal motion. Also we showed that eye tracking is worse if motion is perpendicular to the task (reading rows with vertical motion for example) rather than parallel to it (+ 30 % for reading columns, + 14 % for reading rows).

These values give hints for the limits of regular motion speeds above which the eye is not able to track correctly the moving objects. Below these limits, the motion of the object can be assumed to be correctly nullified by the eye movements, as in the case of fixations of still objects. Foveas (centers of retinas) thus play a fundamental role allowing precise perception, but peripheries of retinas are also important because they are used to detect events of potential interest which may induce a saccade. Of course, memory is active all the time during the observation (short-term or long term memory, unconsciously or not).

5 – Typology of motions in television sequences

Our analysis [15] is that a dynamic visual scene is composed of different objects. Some of them may be still (relative to the floor on which the camera is set), they usually correspond to the background ; some may be moving. The camera which is used to analyse the scene can be still with no change of its parameters or the parameters of the camera analysis can be continuously modified in a number of ways as we will see below. Thus, relatively to a still referential, the floor on which the camera is set, two types of movements can occur simultaneously : the *object movement* and the *camera movement*. The movement which is «seen» by the video processing systems, i.e. the *video signal movement*, is the vectorial combination of both movements. It is also the movement which is displayed on the TV screen (more precisely there is proportionality between them, the units being different). But the movement which is seen by the observer, i.e. the *retinal movement*, is the combination of both object and camera movements, plus the *eyes movements*.

It is clear that the *critical situations* for the visual discrimination of impairments corresponds *to small and slow retinal movements* (in general, psychovisual sensitivity decreases with increasing speeds on retina). This type of movements occur when the movement displayed on the screen is itself small and slow (the visual system being in this case in the fixation mode) ; but, it can also occurs when the visual system is able to track an object moving on the screen. So, it is usefull to distinguish between *trackable* and *non-trackable* movements, the trackability being considered here from the «point of view» of the human oculomotor system.

Different parameters are needed to describe a *moving object* area ; some values will be given below to qualify some ranges of parameters but they are to be considered as indications more than precise values :

The (local) motion vector (or vectors) which describes the speed and orientation of the movement (pixel by pixel or globally for the area). It can be expressed in visual angle per second (degree/s) or in number of pixels per frame (viewed at 6 times picture height, the TV screen width is 12.7^{o} ; for the 625-line system sampled with 520 samples per active line, or 10 MHz sampling frequency, 1 pixel/frame = 0.53 degree/s). A *slow* movement is lower than about 1 degree/s ; a *fast* one is higher than 10 degree/s. Visual tracking capability is higher than that : 40 to 80 degree/s as we described it above. Acuity starts to decrease when *retinal* motion is higher than 2.5 degree/s [16].

the visual field concerned by the motion, i.e. the spatial extent of the moving area (height, width, diameter ... in degrees or number of pixels). A *local* movement (small area) has an extension lower than 1° : it can be completely viewed within the fovea. An *extended* movement (large area) would correspond to more than 2 to 3°.

the total amplitude of the motion, i.e. its total displacement (in degrees or number of pixels). A *small* movement would be beyond 1° (within fovea). A *large* movement would correspond to a displacement of more than 2 to 3°.

the geometry of the movement which is the spatial distribution of the local motion vectors within the moving area. Elementary movements are the *translation* and the *rotation*. A translation perpendicular to camera optical axis produces a uniform distribution of motion vectors. A translation parallel to the camera optical axis produces a radial distribution. A rotation produces a tangential distribution of vectors around the rotation axis. At a given instant, these different types of distributions can be combined and also the speeds can vary with time.

the trackability of the movement which is related to its temporal continuity (as opposed to its suddenness or randonness). The cut-off frequency of the oculomotor system is around 1 to 2 Hz (see Figure 4 and a measured transfer function in [12], p. 348).

Camera motions can be described by some of the parameters given above, but they are movements of a large extent (affecting the whole picture). Different geometrical distribution of motion vectors are produced by the different situations. The main types of camera movements are :

– *fixed spot* : no camera movement (!).

– *travelling* : the camera is moved on a chariot, usually «parallel» to the scene. Travellings towards or from the scene –perpendicularly to it– have similarities with zoomings ;

– *panning* : the camera is rotated around its vertical axis ; it is similar to «parallel» travelling ;

– *tilting* : the camera is rotated or moved up or down ;

– *zooming* : in this case, the camera is fixed but action is made on its optics (zooming in or out).

Of course, all these movements can vary in speed and they can be combined in a large number of ways (zooming while travelling, ...).

Apart from object and camera motions, other spatiotemporal events can occur on a TV sequence, for example, *temporal changes* and *special effects*. Temporal changes can occur either *locally* (appearance of the background becoming visible due to a move of an object which was hiding it), or simultaneously *on all the screen* (in case of camera switching). Special effects are very often used in TV production. They can introduce *movements* (movement of the wipe in case of an overlay, ...) or *changes* (change of focus, for example).

Another approach for a typology is to consider the criticalness of motion situations with respect to the various kind of interframe processings :

For the *movement detector*, one of the problems is to be able to discriminate *low-contrasted moving objects* from noise, not to filter them out in noise reduction processing or ignore them in conditional replenishment coding.

In *conditional replenishment coding of TV signals*, it is necessary to handle multi-moving objects situations and also situations where *movements of large extent* occur (camera movements or scene change).

For *movement estimation* and motion compensated prediction, *sharp contrasted moving edges* may disturb the estimator and also unpredictable movements such as the eyes or lips movements must not be impaired.

6 – Statistical modelling of moving sequences

Very few studies have been devoted to these aspects, apart from studies of frame to frame autocorrelation or difference signal statistics (see references and some further measurements in [17]). These authors [17] showed that the image dynamic activity can be separated into two distinct classes : one, called «continuous activity», corresponds to scene and camera motions ; the other is the class of «scene switches» events that were shown on four different TV programs to behave close to a poissonian interval model. The mean interval between switches varies from 5 to 23 s with the type of program.

Scene switches can thus be interpreted as fundamental (temporal) continuity breaks and between two such breaks, a sequence can be described as a sample of a dynamic mosaic model with continuously varying objects and camera motions, and where moving and still edges are the spatial and spatio-temporal stationarity breaks between still or moving, textured or not, areas. Of course, this is a qualitative structural model and many statistical measurements for each type of configurations would be needed to make it more quantitative.

III – SPATIOTEMPORAL EDGES AND LINEAR PROCESSINGS : FILTERING AND SAMPLING

In the case of linear processings, mainly filtering and sampling of pictures or sequences, Fourier analysis is very helpful although it involves usually the only consideration of sinusoidal patterns. We have shown [9] that a 3D Fourier analysis can also be applied to spatiotemporal edges. For this, we define an *idealised edge* as being straight and animated by a uniform motion. This idealisation can be seen as a local approximation in space and time of a real moving edge. Let u and v describe the transversal and longitudinal coordinates, the edge –either line or border– is easily represented using its transversal transition function c (arbitrary) and speed $V_\perp$. Its 3D Fourier transform can then be derived [9] :

$$e\,(x, y, t) = c\,(u - V_\perp . t) \qquad \text{and} \qquad E\,(f_x, f_y, f_t) = C\,(f_u) \,.\, \delta^{(2)}(f_v, f_t + V_\perp f_u),$$

where C is the 1D Fourier transform of c and $\delta^{(2)}$ is a 2D Dirac function. Figure 5 shows the spatiotemporal and Fourier representations of that edge. A localisation window s could limit its extension : at first approximation [9], it will replace the 2D Dirac by a function concentrating energy along the axis defined by $f_v = 0$ and $f_t + V_\perp f_u = 0$.

An ideal line-edge corresponds to c being a Dirac function, so C = 1 ; an ideal border-edge corresponds to c being a step function, so $C\,(f_u) = 1 / (2i\pi f_u)$.

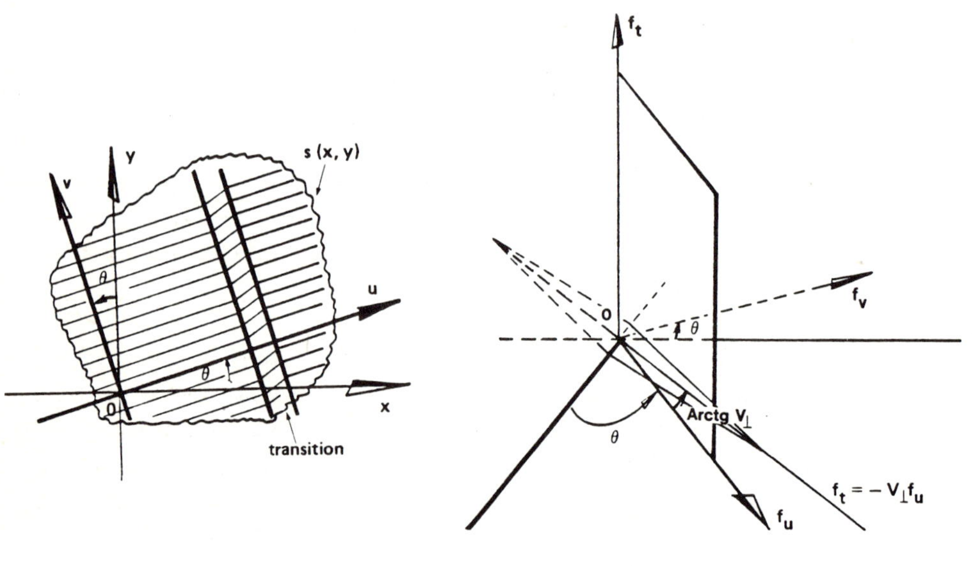

Figure 5. Spatial representation of an edge assumed to move with transversal speed $V_\perp$ and its 3D Fourier spectrum.

Let now $F(f_x, f_y, f_t)$ be an arbitrary *spatiotemporal filter*. It is easy to see that its effect on a moving edge is only on the transversal dimension, and we have shown [9] that the filtered edge Fourier transform is :

$$FE(f_x, f_y, f_t) = F_\perp(f_u) \cdot C(f_u) \cdot \delta^{(2)}(f_v, f_t + V_\perp f_u)$$

with $F_\perp(f_u) = F(f_u \cos\theta, f_u \sin\theta, -V_\perp f_u)$.

For example, for a purely temporal filter $F(f_t)$, the equivalent transversal filter is $F_\perp(f_u) = F(-V_\perp f_u)$, its bandwidth being inversely proportional to $V_\perp$.

Another type of impairments affecting edges can be studied in the Fourier domain and will be proved to be useful for interpreting general spatiotemporal sampling impairments (aliased edges). We consider here an edge which is still in mean but the transition phase (position) of which is varying longitudinally and temporally with respective periods S and P : $e(x, y, t) = c[u - \beta(v, t)]$. Its Fourier transform is [9] :

$$E(f_x, f_y, f_t) = C(f_u) \sum_{n,p} \beta_{n,p}\, \delta^{(2)}(f_v - n/S, f_t - p/P),$$

where the $\beta_{n,p}$ coefficients can be derived from $\beta(v, t)$. The spectrum of this phase-impaired edge is thus composed of replications of the original spectrum at multiples of 1/S and 1/P. For example, in case of a still impairment, $\beta_{n,p} = \alpha_n$ and the replication is only on f_v axis. Extension to an edge uniformly moving in average can be done by changing u into $u - V_\perp t$ and thus f_t into $f_t + V_\perp f_u$.

General *spatiotemporal sampling* is performed by multiplying a picture sequence $e(x, y, t)$ by a sampling pattern $\psi(x, y, t)$ which is usually a periodic sum of 3D Dirac functions. Its Fourier transform Ψ is then also a periodic sum of 3D Dirac functions now in the Fourier domain. The Fourier transform of the sampled sequence is thus :

$$E_S(f_x, f_y, f_t) = E * \Psi = \sum_{m,n,p} K_{m,n,p}\, E(f_x - X_{m,n,p}, f_y - Y_{m,n,p}, f_t - T_{m,n,p}),$$

where $(X_{m,n,p}, Y_{m,n,p}, T_{m,n,p})$ is one Dirac function of Ψ. The spectrum of the sequence is thus composed of replications of the original spectrum over a periodic 3D grid.

For simplification, Figure 6 is restricted to the spatial domain with still edges. It shows first in the spatial domain and then in the Fourier domain the effect of sampling. Two of the basic sampling structure [9] have been used here : the «orthogonal» one with pixel positions lying on columns and the «line-quincunx» one (QL) with half-pixel position shifts each line of the same field. (In the Fourier planes, QT corresponds to another structure with half-pixel shifts each field). The lower figures only consider Dirac functions of the sampling pattern which are the closest to the origin.

With good approximation, we can only retain the first harmonics of the impaired edge spectrums. The comparison of sampled edge spectrums with periodic phase edge spectrums allows to interprete directly in

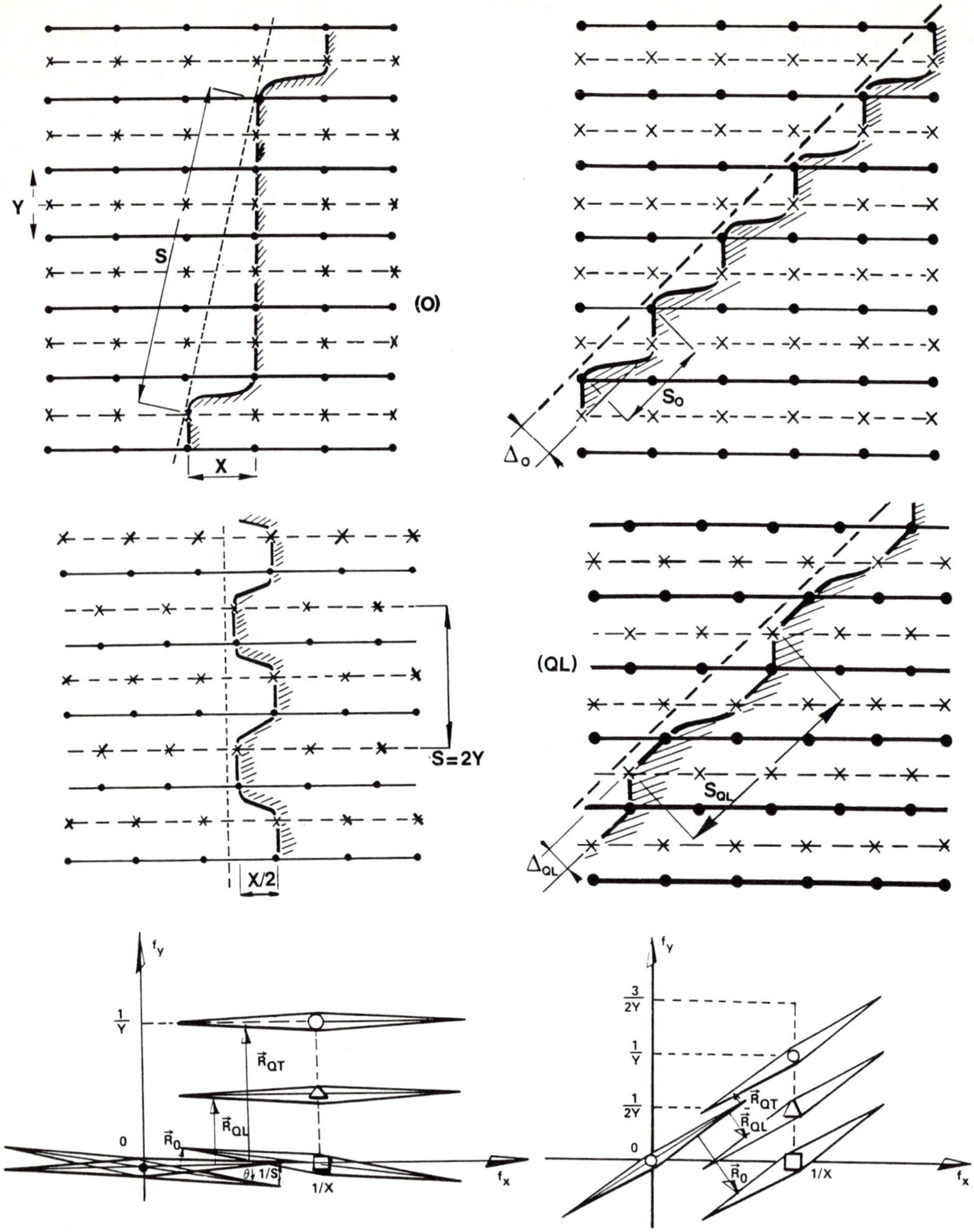

Figure 6. Upper figures : sampling of a quasi vertical edge and of an oblique edge with the orthogonal sampling structure. Middle figures : sampling of a vertical and an oblique edge with line-quincunx sampling structure. Lower figures : Fourier representation of the sampling of a vertical or quasi-vertical edge (left) and of an oblique edge (right).

the Fourier domain the main parameter of the alias impairment : the distance between original and alias spectrums is directly the inverse of the longitudinal period of the sampling impairment along the edge : for example, $R_{QL} = 1/S_{QL}$. This result can be extended to the general spatiotemporal case except that for the temporal parameter the distance between original and alias spectrum has to be taken parallel to f_t axis [9].

The study [9] that is briefly summarized here has extensively considered either filtering or sampling of sinusoidal patterns and of edges with arbitrary transition function. The optimum situation is likely to be when filtering and sampling impairments are equally visible : this calls for a compromise between filters and sampling structures, the solution of which necessitates a quantitative description of the interaction of both processings on edges and a precise visibility criterion. Nethertheless, we have laid down the first steps for the general analysis of *spatiotemporal resolution* with the essential consideration of spatiotemporal edges.

IV – SPATIAL EDGES, ORIENTATION ESTIMATION AND INTRAFIELD ADAPTIVE PREDICTION

1 – Graham and Zschunke algorithms

From Figure 3, one conclusion was that if a predictor could be adapted to the local orientation and if edges could be predicted, the dynamic range of the DPCM quantizer (difference between outer levels) could be significantly reduced.

A DPCM coding scheme [2] is based on a prediction P evaluated at every current pixel X. A difference signal is derived (prediction error), $\delta = X - P$, which is quantized and coded. The local and distant decodings give $\widehat{\delta} = Q(\delta)$ and finally $\widehat{X} = P + \widehat{\delta}$ is the DPCM decoded signal. For stability and causality, the predictor at the coder and at the decoder is based on *previously decoded pixels* $\widehat{X}_v$ from a causal neighbourhood of X. Usually linear predictors are used : $P = \Sigma\, a_v \widehat{X}_v$. In the case of an adaptive predictor, the prediction coefficients are adapted, for example to the local orientation estimate θ. Figure 7 gives the notation of pixels in a causal neighbourhood of the current pixel and shows the discretisation of orientation angles we used.

The first adaptive prediction scheme is due to Graham [18] : P is either A or C whether $|B-A| - |B-C| \geqslant 0$ or not. Pixels can thus have only two states and this doesnot take into account edges specifically. Zschunke's algorithm [7] separates non-edge pixels and edge pixels. The first ones are simply predicted using previous pixel predictor, P = A. The other ones are adaptively predicted using for example the following set of predictors : A, AB, B, (B+ C)/2, C, (C+ D)/2, D and E. The adaption process can be analysed in two steps [19]. First, the optimal «orientation» is obtained at every previously decoded pixel $\widehat{X}$ which is an edge pixel. This is done by comparing $\widehat{X}$ with the set of predictors which could have been used to predict it. The best one is retained to define the best orientation ; if no one is satisfactory, horizontal orientation is retained. Then, now at the current pixel X , the algorithm looks at neighbouring pixels A, AB,

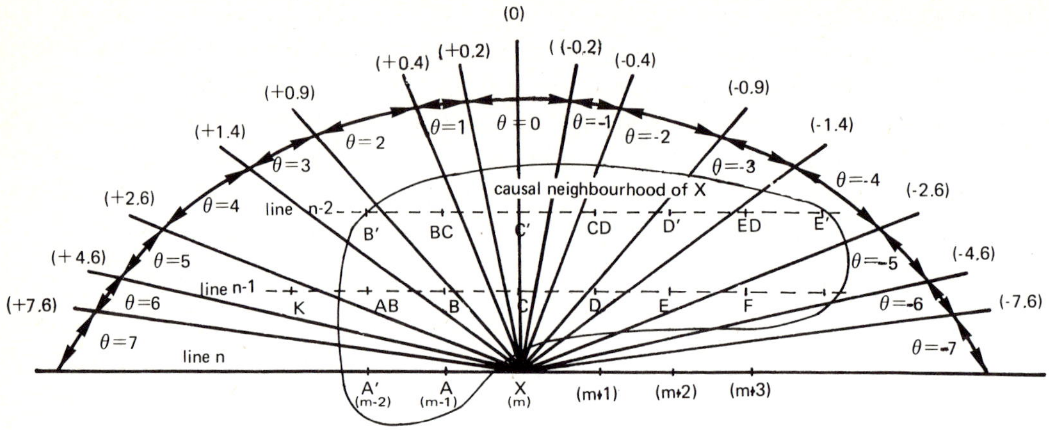

Figure 7. Causal neighbourhood of current pixel X for DPCM prediction. Discretisation of orientation angles.

B, C, D and E. If all are labelled with an horizontal orientation, prediction A is used for X. If one or more are labelled with a given orientation (say, best predictor being C' for pixel B), that orientation is predicted to be the best one for X (in the example, X will thus be predicted by D). In conflicting situations, some decision has to be taken [19].

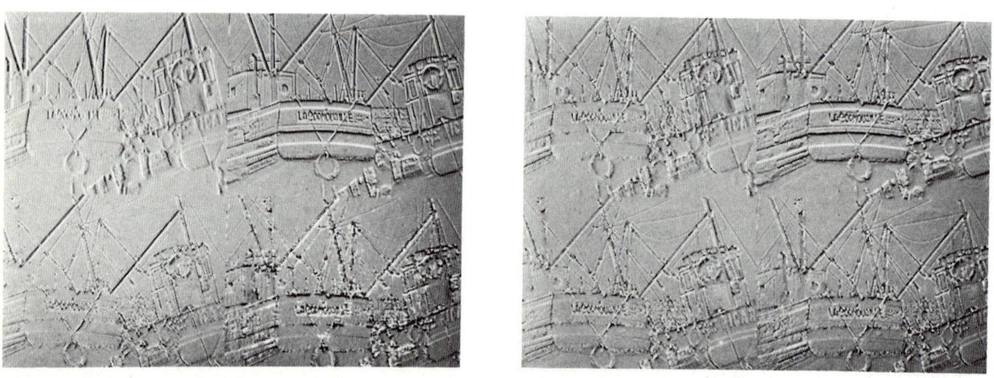

Figure 8. Prediction error pictures displaying Y = 127 + X − P. Photographs are taken from a TV monitor screen (12 MHz x 8 bit digital pictures). (a) from left to right and top to bottom : prediction A, (A+ D)/2, Graham, Zschunke algorithms ; (b) : algorithms B, B*, C and C*. The original picture is «Boats and lighthouse» shown in Figure 1 or 9.

Figure 8a shows pictures of error prediction ($Y = 127 + X - P$) : Zschunke algorithm is seen to lead to quite a noisy prediction on edges. In fact, some improvements can be made as mentionned in [19]. The reduction of the predictor set to A, B, C and D is a first improvement, likely to be explained by the orientation estimation imprecision. Further, a different strategy can be used for labelling the current line pixels : from previous line edge pixels $\hat{X}$, observed orientations can be used for extrapolation to current line (in the above example, X will be predicted by D if D is an edge pixel itself best predicted by D'). Thus, we now determine predicted edge pixels at current line not by a neighbourhood process but by an orientation extrapolation process. But even so the modified Zschunke algorithm is not enoughly efficient and it is still much sensitive to transmission errors. This seems to be due to the fact that only simple and very local tests are used for orientation estimation.

2 – The edge-recursive orientation estimator from C. Richard

The starting point of this algorithm [20] was to get an estimation of the local edge orientation that was *robust* to picture noise (DPCM noise on $\hat{X}$ and effects of the spatial discretisation due to sampling) as well as to channel noise (effects of transmission errors). The algorithm is based on the geometrical structure of edges : they can be followed from line to line and thus local estimates can be recursively filtered along edges to reduce orientation estimation noise.

We will only summarise here the main steps of the algorithm. At pixel $\hat{X}_{m,n}$, a *vertical edge element* VE is observed simply if $|\hat{X}_{m,n} - \hat{X}_{m-1,n}| \geqslant 25$ and a *horizontal edge element* HE if $|\hat{X}_{m,n} - \hat{X}_{m,n-1}| \geqslant 25$. Then, *connected edge elements* are looked for : a representation of edge elements with bars between pixels such as in Figure 9a is useful, connexity of edge elements corresponding to its geometrical meaning.

Now, a *connection* is a set of connected edge element for pixels at a given line with at least one vertical edge element (indeed if no VE occurs, no edge following will be possible but it is not necessary because prediction A is likely to be the best in this case). *Edges* can now be defined as sets of connections connected from line to line through VEs. An edge thus resembles a staircase and by analogy it will be segmented into *stairs* associated to each VE except for purely vertical edges with only VEs. The precise definition of a stair is : (i) at the beginning of an edge, the first connection is the first stair ; (ii) then, a new stair is observed at line n if there is a connection with at least one VE and one HE or if connections with only VEs are observed at lines n–2, n–1 and n.

At each stair k, three integers are used to summarize the *local observations* : $\bar{H}_k$ and $\bar{V}_k$, respectively the number of HEs and VEs connected in this stair ; $\bar{S}_k$, either + 1 if the first HE on the left of the stair is connected to a VE at previous line or – 1 if not. Each stair is a step in the recursive filtering process which leads to the filtered variables H_k, V_k and F_k :

- at first stair (k = 1) : $H_1 = \bar{H}_1$, $V_1 = \bar{V}_1$, $F_1 = \bar{S}_1$

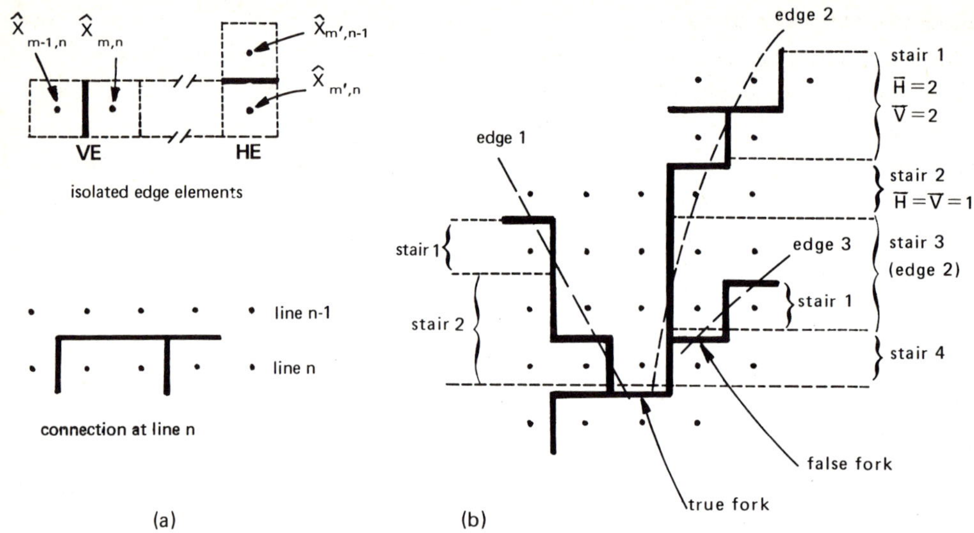

Figure 9. (a) isolated vertical and horizontal edge elements ; connected edge elements at a line ; (b) edges and stairs ; false and true forks.

then $(k \geqslant 2)$:

$$\begin{aligned} H_k &= \left| H_{k-1}\, S_{k-1} + \gamma_k\, (\overline{H}_k\, \overline{S}_k - H_{k-1}\, S_{k-1}) \right| \\ V_k &= V_{k-1} + \gamma_k\, (\overline{V}_k - V_{k-1}) \\ F_k &= F_{k-1} + \overline{S}_k \;, \quad S_k = \operatorname{sgn} F_k \,. \end{aligned}$$

At edge beginnings, γ_k is lowered such that $\gamma_2 = 1/2$ and $\gamma_k = 3/8$ when $k \geqslant 3$. The *edge state vector* gathers all necessary informations for recursivity : the stair number k (saturated to 3 or 4) which carry some information on edge length and allows to determine γ_k ; the filtered variables H_k , V_k , F_k which allows local (filtered)orientation estimation

$$|\theta_k| = \operatorname{Arctan}\,(V_k/H_k) \;, \operatorname{sgn} \theta_k = \operatorname{sgn} F_k \;;$$

plus another variable used for the edge sign break test [20].

The above equations correspond to two steps of the algorithm, namely *initialization* and *filter recursion*, the latter being based on the most likely edge model (see section II.3), θ = constant + noise, which describes straight edges. The recursive process allows to follow curvilinear edges but in the case of various edge accidents like *elbows* (abrupt changes in edge direction), *forks* (fusion of few edges into one) or *crossings* (intersection of edges), it is compulsary to stop the inertia of recursion and restart from local observations (otherwise prediction will not be correctly adapted at these locations). Two *edge model break detection tests* are used : one looks to a significant difference between the local $\overline{\theta}_k$ derived from $\overline{H}_k$, $\overline{V}_k$ and $\overline{S}_k$ and the filtered one θ_{k-1} ; the other detects significant changes in mean of F_k [20]. In case of a significant break, *reinitilization* is done. In case of a fork or a crossing, before applying the break tests an *equivalent edge state vector* has to be used. It is based on a kind of weighted average of the state vectors of the merging edges [20]. This is done in such a way that false crossings (small edge with only one stair merging in a longer one) donot lead to reinitialization.

3 – Adaptive prediction algorithms B*, D and E

The previous section has described orientation estimation at previously decoded pixels $\hat{X}$. The orientation estimate is first discretized into 16 classes (see Figure 7) and is used first to label pixels $\hat{X}$ which are edge pixels, that is, in a stair at the right of any VE (non edge pixels would have been thus correctly predicted by prediction A). Adaptive prediction using orientation estimation is to be applied to pixels X estimated to be edge pixels at current line, namely *predicted edge pixels* or «dangerous» pixels [20]. Their positions are predicted using an *orientation extrapolation process* (see end of section IV.1 and Figure 10) : if $\hat{X}_{m,n-1}$ is an edge pixel with $\theta = 0$, $X_{m,n}$ is the corresponding predicted edge pixel at current line ; if $\theta = 1$, 2 or 3, predicted edge pixels are $X_{m,n}$ and $X_{m+1,n}$; if $\theta = 4$, they are $X_{m+1,n}$ and $X_{m+2,n}$ (symmetrically for negative θs). Edges pixels with $|\theta| \geqslant 5$ donot project at current line ; it is the same for edge pixels of first stairs (texture pixels or edge beginnings). Thus, at current line, predicted edge pixels are labelled with the θ from previous line ($|\theta| \leqslant 4$),the other are *(predicted) non edge pixels.*

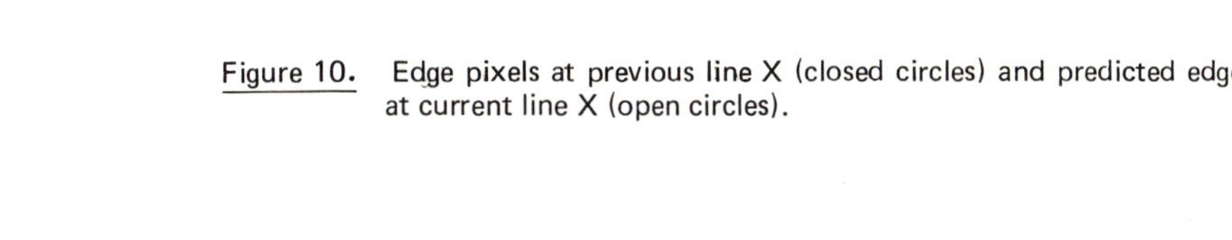

Figure 10. Edge pixels at previous line X (closed circles) and predicted edge pixels at current line X (open circles).

Two types of prediction algorithms have been designed. The first one implies a given predictor for each class of orientations θ , its prototype is algorithm B [20] and we modified and reoptimized it into algorithm

B* and better into algorithm D [19] (see Table I) :

predictor \ θ	-4	-3	-2	-1	0	1	2	3	4	non edge pixels
algorithm B*	$\frac{D+E}{2}$	$\frac{3D+A}{4}$	$\frac{C+D}{2}$	$\frac{3C+D}{4}$	C	C	$\frac{B+C}{2}$	$\frac{A+C}{2}$	$\frac{A+B}{2}$	$\frac{A+C}{2}$
algorithm D	$\frac{A+D}{2}$ if $\theta < 0$				$\frac{A+C}{2}$ if $\theta \geqslant 0$ or non edge pixels					

Table I — Definition of algorithms B* and D.

The second type of algorithms adds to the recursive orientation adaption an adaption based on local observations on a causal neighbourhood of current pixels : binary variables are used associated to couples of adjacent pixels. For example $i_1 = 1$ if $|A - B| > 25$, $i_1 = 0$ otherwise. The prototype is algorithm C [20] which uses five of such binary local gradient variables. For each orientation class and for non edge pixels, 32 states can occur, the algorithm assigns to them a set of predictors (2 for edge pixels, 4 for non edge pixels) . We modified it into algorithm C* [19] and we introduce here an algorithm E which combines algorithm D for (predicted) edge pixels (Table I) and algorithm C* for (predicted) non edge pixels because algorithm D is better than C* on edge pixels, the contrary being true for non edge pixels :

$$\text{prediction of non edge pixels by algorithms C* or E} \begin{cases} = (A+C)/2 & \text{if } (0, i_2, i_3, i_4) \text{ or } (1, 1, i_3, 0) \\ = (A+D)/2 & \text{if } (1, 1, i_3, 1) \\ = A & \text{if } (1, 0, i_3, i_4), \end{cases}$$

i_1 to i_4 being the binary variables associated to gradients $|A - B|$, $|C - B|$, $|C - D|$, $|D - E|$ respectively and $(1, 1, i_3, 0)$ meaning that variable i_3 can be 0 or 1 in this case.

All these algorithms can be called «hard adaption» processes : first, on a given line, the algorithm switches between edge pixels and non-edge pixels and also along a followed edge, the smooth recursion process is reinitialized immediately when local tests working in parallel get positive.

4 — Results and discussion

Figure 8b (see above) shows prediction error pictures for algorithms B, B*, C and C* which compare well with Graham or Zschunke algorithms (Figure 8a) as regards to prediction noisyness and error prediction dynamic range. Algorithms D and E would be even better. Figure 11 shows processed pictures where the

Figure 11. Pictures processed by DPCM loop with prediction error truncation to + and − 50. From left to right and top to bottom : the original (part of the picture «Boats and lighthouse» of Figure 1) ; prediction (A+C)/2 ; algorithm B* ; algorithm D.

DPCM quantizer has been replaced in the DPCM loop by a saturation scheme ($\hat{\delta} = \delta$ if $|\delta| < 50$, $\hat{\delta} = 50$ if $\delta \geqslant 50$, $\hat{\delta} = -50$ if $\delta \leqslant -50$). The fixed predictor (A + C)/2 is not satisfactory (blur of diagonal edges), algorithms B* and D very well behave with this reduced quantizer dynamic range. Table II gives the values of probabilities that the prediction error exceeds 50 in absolute :

P(|X−P|>50|1 or 2)	pred. A	pred. $\frac{A+C}{2}$	$A+\frac{C-B}{2}$	Zschunke	alg. B*	alg. D	alg. E
1 = predicted edge pixels (5,4% of pixels)	0,278	0,130	0,172	0,168	0,120	0,105	0,105
2 = pred. non edge pixels (94,6% of pixels)	0,012	0,014	0,009	0,023	0,014	0,014	0,008

Table II — Probabilities that the absolute prediction error exceeds 50 for a set of fixed predictors and adaptive prediction algorithms (statistics done on the whole picture «Boats and lighthouse» digitized at 12 MHz shown in Figure 1 or 11).

It can be seen that a gain is achieved from a fixed predictor like (A + C)/2. This is a good fixed predictor which is even better than Zschunke algorithm, certainly due to the criticalness of the picture we used displaying sharp and contracted edges. But we have the feeling that the proposed adaptive algorithm are not yet optimized in all its parameters, the task being quite complex. Two aspects would have to be deepened [19] : one is the *orientation selectivity and phase resilience of predictors* used in the adaptive prediction scheme. Indeed, the efficiency of algorithm D on edges may be interpreted by its use of poorly orientation selective predictors. So, the discretization of orientation θ and the orientation selectivity of predictors used for each θ must be matched together. Also, the sensitivity of predictor to phase (position) errors on edges would have to be studied. The second aspect concerns the prediction of edge pixels positions at current line which may need to be studied for optimization. We are here getting to a complex question : what are the limits for the local predictibility of (real) edges ?

Another aspects of adaption is the sensitivity to transmission errors. We have found [19] that Graham and Zschunke algorithm were quite sensitive to them. Algorithm B is less sensitive but certainly still unacceptable. Algorithms B*, C, C* are significantly less sensitive and algorithms D and E would be close to them. The important difference between algorithms B and B* is the predictor used for non-edge pixels, respectively prediction A and (A + C)/2. Prediction A, as prediction A + (C–B)/2, propagates the initial error to the end of a line. If the initial error is in a non-edge area, it propagates and will perturbate next edge. If it occurs on an edge, it usually propagates and fades along the edge but it may produce a horizontal propagation from the edge over the adjacent non-edge area. The result is disastrous and there are thus two reasons for which on non-edges areas the prediction scheme (adaptive or not) must not propagate errors without quick fading. Algorithms C, C* and E are adaptive on non-edge areas but they satisfy to this constraint because they use for much of the pixels (A + C)/2 or (A+ D)/2 predictors.

The results [19] show that it is possible to get a bit rate reduction down to 3.5 bits/pixel at 12 MHz with very good quality and error resilience. An objective such 3 bits/pixel may be obtained using adaptive quantization which is much complementary to adaptive prediction. One of the main problem here is the sensitivity of combined quantization and prediction adaption schemes, but J. Dewitte [21] has found satisfying solutions to this. The complexity of algorithms such D or E is not that much and a hardware implementation has been realized [22].

IV – SPATIOTEMPORAL EDGES AND MOTION ESTIMATION

We will consider in this part the case of dynamic scenes : the most sophisticated techniques involve motion estimation [3] in order to compensate prediction using displacement estimates. Improving prediction enables to reduce the spread of non predictable areas and thus the amount of information to be transmitted. These techniques seem well adapted to situations like slow panning, travelling, tilting or zooming (see section II.5) which occur in television programs and imply extended moving areas. Usual conditional replenishment techniques cannot cope with these situations but interframe coding of television programs using motion compensation will still need to handle contrasted and sharp moving edges (because of good quality cameras), multiple moving objects one hiding the other, etc..., without any a priori knowledge of the scene. For reasons given in the introduction, we have chosen pel-recursive types of algorithms. But, we felt that under these severe conditions, Netravali's algorithms [23-24] were insufficient as involving only soft adaption and thus long convergence after a motion break. Figure 12 illustrates this. Classical model

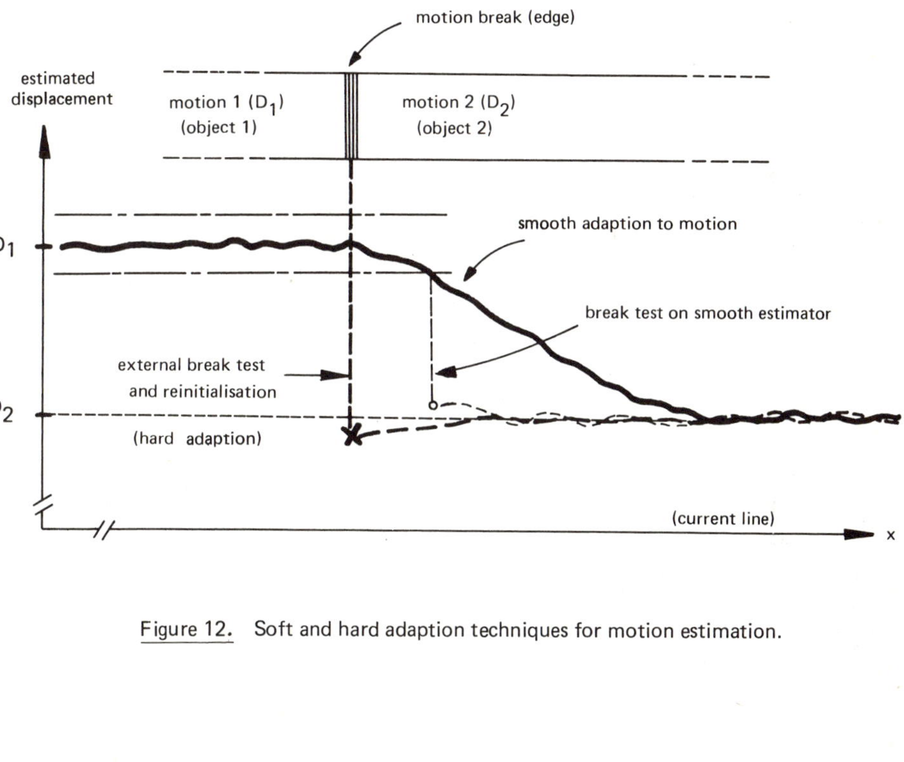

Figure 12. Soft and hard adaption techniques for motion estimation.

breaks tests could be used but they introduce a delay. Thus a proper «hard adaption» scheme needs an external model breaks test and a local estimation process to detect breaks and reinitialize the pel-recursive algorithm which will be used in areas where motion doesnot change abruptly. From our analysis (section II), it is clear that motion breaks are likely to occur on boundaries of objects and thus on edges. The idea is then to be able to estimate motion parameters of edges [25-26] and use it for motion compensation coding [27]. With this approach, edges donot cause troubles ; to the contrary, as they are contrasted, precise motion estimation can be performed. We will now describe this approach which can be designed as *edge-adaptive pel-recursive motion compensation*. We will restrict ourselves here to the edge motion estimator. Its use for motion estimation at every pel and the presentation of preliminary results are given in [25-28].

The basic idea for this estimator is that, if we consider an ideal spatiotemporal edge, $e(x, y, t) = c(x\cos\theta + y\sin\theta - V_\perp t)$, as in Part III, for each frame t_o, each line y_o or each column x_o, we get 2D representations of that spatiotemporal edge. It is easy to see that in each of these planes, either (x, y), (x, t) or (y, t), this representation is an edge and furthermore its «orientation» in these planes directly indicates respectively its spatial orientation θ , its horizontal displacement along x axis per time unit, Δx, and its vertical displacement along y axis, Δy. These quantities are not independant : θ and Δx are sufficient to determine the edge (transversal) displacement vector, $D_\perp$, with components $D_{\perp x} = \Delta x \cos^2\theta$ and $D_{\perp y} = \Delta y \sin\theta\cos\theta$ which fundamentally differ from Δx and Δy (see Figure 13).

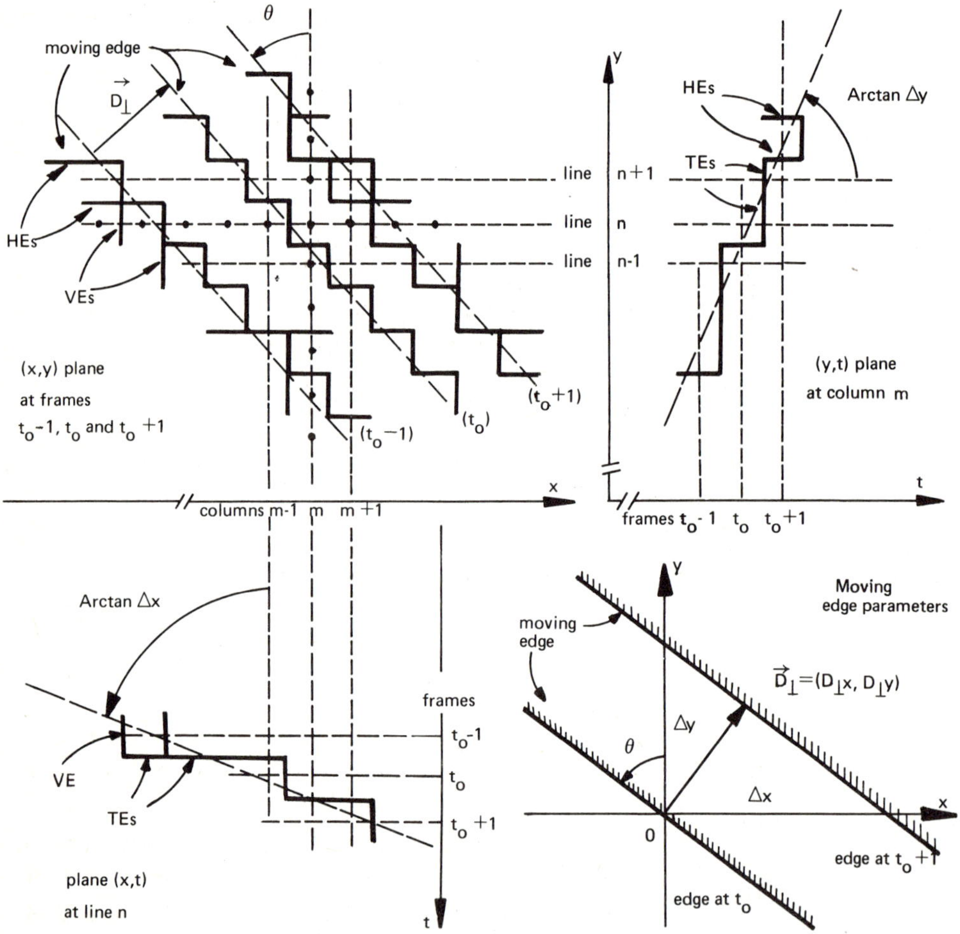

Figure 13. Discretized representation of a spatiotemporal edge on the three basic spatiotemporal planes showing VEs, HEs and TEs. In this example, pixel values are 40 at the left of the edge, 120 at its right and 80 at pixels within connections. Thresholds for VE, HE and TE are 25. Right bottom corner : spatiotemporal edge parameters.

These are the spatiotemporal edge parameters for the ideal case. Our spatiotemporal edge model is thus : θ = constant + noise, Δx = constant + noise. As can be seen in Figure 13, we have added to the already defined (Part IV) vertical and horizontal edge elements, VEs and HEs, *temporal edge elements* or TEs with the same thereshold, 25, now on the interframe difference.

The estimation algorithm for horizontal displacement Δx can directly proceed from the (spatial) orientation estimation algorithm described in section IV.2 (we choosed to estimate Δx because it is more suited to the line-by-line scanning than Δy estimation which would involve column memories). Let now Δ be the (horizontal) displacement and consider a given spatiotemporal stair with connected HEs, VEs and TEs. Let k be its stair number for the (x, y) representation and l for (x, t) case. $\bar{\Delta}$ will be the local observations (number of TEs at this stair) and Δ the filtered value. C. Labit designed three types of algorithms with three different types of recursion [25] :

Algorithm MC1 uses a temporal recursion in the (x, t) plane and thus needs one automata per edge for each (x, t) plane, i.e. for each line. The filtering process is :

$$\Delta_l = \gamma_l \bar{\Delta}_l + (1 - \gamma_l) \Delta_{l-1} .$$

Algorithm MC1bis uses a double spatial-temporal recursion in both (x, t) and (x, y) planes. For each edge and each line, it needs a double automata which performs :

$$\Delta_{k,l} = \gamma_{k,l} \bar{\Delta}_{k,l} + (1 - \gamma_{k,l}) [\alpha_{k,l} \Delta_{k,l-1} + (1 - \alpha_{k,l}) \Delta_{k-1,l}] .$$

In fact, these algorithms give similar results and need a significant realization complexity. On long enough moving edges, the motion parameters are seen to be much correlated from line to line and thus a simpler algorithm were designed.

Algorithm MC2 uses only a spatial recursion, i.e. in the (x, y) plane. It thus only needs to increase the dimension of the edge state vector already used for orientation estimation to incorporate now horizontal displacement information :

$$\Delta_k = \alpha_k \bar{\Delta}_k + (1 - \alpha_k) \Delta_{k-1} , \qquad \text{with } \bar{\Delta}_k = \frac{LED_k + RED_k}{2} .$$

In fact, here the local observation is slightly modified to be the mean of left end displacement LED (displacement from previous frame on the left end of the stair) and right end displacement RED. This eliminates taking into account TEs at endings of a stair which often occur and would bias the observations. Some other modifications from the spatial algorithm were introduced ; the main one is *the validation process*. Long enough edges are a posteriori validated (6 lines) in order edge motion estimation to be performed only on valid motion breaks : short edges (textures) and thick edges are thus excluded.

Algorithm MC2 performs as well as the two others and from the spatial orientation estimation algorithm realization [22], it only requires a frame memory plus only very little increase of hardware. Figure 14 presents the result of displacing previous frame using the MC2 estimates. Part of a complex moving sequen-

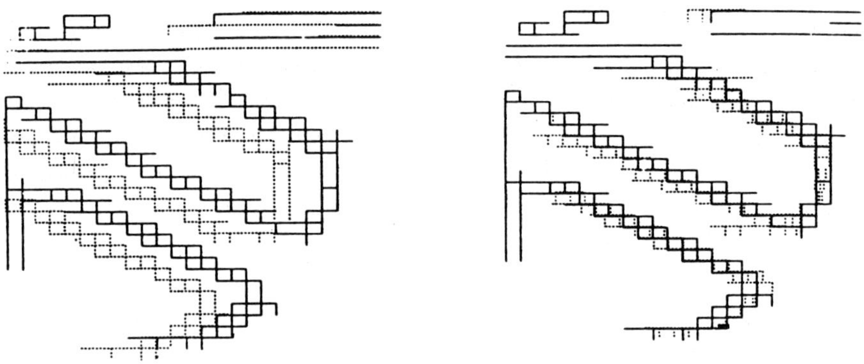

Figure 14. Left : Spatial edges on current picture (full lines) and previous picture (dotted lines). Right : same with displaced previous picture (dotted lines) : the TV sequence «Castle» involves zoom and panning plus a rotating castle. It was digitized (12 MHz) at FTZ, Darmstadt, Germany with courtesy of L. Stenger.

ce «Castle» is shown. Motions are here quite fast (3 to 5 pixels/frame). Results [25-27] show the significant advantage of the edge motion estimation first in reducing displaced frame differences on moving edges and secondly, in avoiding the pel-recursive algorithm to diverge or oscillate on quasi-uniform areas (the way in which the area pel-recursive algorithm is initialized using the edge motion estimator is described in [25-27]). This algorithm is well fitted to complex television scenes with sharp contrasted moving edges and involving a large variety of motion types. Sequence «Castle» is quite critical for that because it contains mainly edges and plateaus. We are at the stage of preliminary results [28] and an extensive comparison of the motion compensation coding schemes has to be done on a large variety of pictures sequences.

VI – CONCLUSION

This paper has been devoted to spatial edges and to spatio-temporal edges (moving edges). We have shown that they are fundamental configurations of pictures and sequences. Indeed, they correspond to stationarity breaks which are the locii where a 2D or 3D stationary model is replaced by another one. They are thus most likely associated with high local constrats. Ideal edges (straight edges animated with uniform motion) were used for studying impairments produced by linear processes such as filtering or sampling : this approach usefully complements the usual analysis of spatiotemporal resolution which restricts itself to sinusoidal spatiotemporal patterns.

In two other situations, we have described adaptive prediction schemes for ADPCM coding of TV pictures or sequences. They take specifically advantage of edges : one scheme is an adaptive intrafield prediction using a recursive edge orientation estimator, the other is its extension to motion compensated interframe prediction with a recursive edge motion estimator. Both are based on local observations of spatial or spatiotemporal edge parameters which are then filtered recursively along the edges. The ideal edge model is used locally to define the proper recursion and model break tests are used to stop the filtering process where necessary in order to allow fast reinitialization. These algorithms can be considered as being based on edge recursive hard adaption. A continuing question would be to investigate to what extent we have approached real edges predictibility in TV pictures or sequences.

This stress on edges is not to forget other types of configurations and parameters (color, texture,...) in the structure of pictures and sequences. Our approach was based on a basically nonstationary modelling and the general situation of TV sequences was summarized by considering them as samples of a dynamic mosaic model continuously evolving between temporal breaks which are usually refered to as scene switches. Unfortunately, we are far from a quantitative statistical description of this model but it was used to draw the structure of the adaptive coding schemes we have described.

REFERENCES :

[1] F. KRETZ, J-L. BOUDEVILLE : «Dégradation des signaux d'images et qualité subjective en codage numérique : visibilité du flottement de contour», Annales des Télécommunications, 31, No 9-10, sept-oct. 1976, pp. 333-356.

[2] A.N. NETRAVALI, J.O. LIMB, «Picture coding : a review», Pr. of the IEEE, 68, No 3, March 1980, pp. 366-406.

[3] E. DUBOIS, B. PRASADA, M. S. SABRI, «Image sequence coding», Chap. 3 in 'Image sequence analysis', edited by T.S. HUANG, Springer-Verlag, 1981, pp. 229-287.

[4] N. KESKES, F. KRETZ, H. MAITRE, «Statistical study of edges in TV pictures», IEEE Tr. on Comm., 27, No 8, August 1979, pp. 1239-1247.

[5] B.R. HUNT, «Nonstationary statistical image models and their application to image data compression», Computer graphics and image processing, 12, No 2, 1980, pp. 173-186.

[6] N.F. MAXEMCHUK, J.A. STULLER, «An adaptive interframe DPCM codec based upon nonstationary image model», Bell Syst. Techn. J., 58, No 6, July-August 1979, pp. 1395-1412.

[7] W. ZSCHUNKE, «DPCM picture coding with adaptive prediction», IEEE Tr. on Comm., 25, No 11, nov. 1977, pp. 1295-1302.

[8] J.K. YAN, D.J. SAKRISON, «Encoding of images based on a two-component source model», IEEE Tr. on Comm., 25, No 11, nov. 1977, pp. 1315-1322.

[9] F. KRETZ, J. SABATIER, «Echantillonnage des images de télévision : analyse dans le domaine spatio-temporel et dans le domaine de Fourier», Annales des Télécommunications, 36, No 3-4, mars-avril 1981, pp. 231-273.

[10] B. FAUST, «Modélisation du signal d'image à l'aide de processus aléatoires non stationnaires», rapport No 78007, Ecole Nationale Supérieure des Télécommunications, Laboratoire Image, juin 1978, Paris, 35 p.

[11] J.W. MODESTINO, R.W. FRIES, V.L. VICKERS, «Stochastic image models generated by random tessellations of the plane», Computer graphics and image processing, 12, 1980, pp. 74-97.

[12] R.W. DITCHBURN, «Eyes movement and visual perception», Clarendon Press, 1973, 421 p.

[13] P. VIVIANI, A. MONOT, P. SALLIO, F. KRETZ, «Caractéristiques générales des mouvements exploratoires oculaires d'images fixes», Revue Radiodiffusion Télévision, No 71, janv.-mars 1982, pp. 30-39.

[14] P. VIVIANI, «L'exploration visuelle de textes alphanumériques en mouvement», Final report of a contract CCETT/Laboratoire de Physiologie Neurosensorielle du CNRS, Paris, déc. 1981, 145 p.

[15] F. KRETZ, P. SALLIO, «Typology of movements in television scenes, critical situations», International Workshop on image processing : real-time edge and motion detection/estimation, Rennes, France, 17-21 sept. 1979, pp. 13.1-13.11 (Technical document CCETT-CTN/T/1/80).

[16] G. WESTHEIMER, S-P. McKEE, «Visual acuity in the presence of retinal-image motion», J. Opt. Soc. Am., 65, No 7, July 1975, pp. 847-850.

[17] D.C. COLL, G.K. CHOMA, «Image activity characteristics of broadcast television», IEEE Tr. on Comm., october 1976, pp. 1201-1206.

[18] R.E. GRAHAM, «Predictive quantizing of television pictures», IRE Wescon Conv. Record, USA (1978), 2, Pt 4, pp. 147-157.

[19] F. KRETZ, «Codage MIC Différentiel à prédiction adaptative en télévision : qualité visuelle, réduction de débit et susceptibilité aux erreurs de transmission», Annales des Télécommunications, 37, No 5-6, mai-juin 1982.

[20] C. RICHARD, «Estimation récursive de l'état local des contours d'images et application à la prédiction adaptative en codage différentiel», Thèse 3ème cycle, Rennes, IRISA, nov. 1979, 195 p. also : C. RICHARD, A. BENVENISTE, F. KRETZ, «Recursive estimation of local characteristics of edges in TV pictures» (submitted).

[21] J. DEWITTE, «Un système de codage MIC différentiel adaptatif avec critère psychovisuel de qualité», Thèse Docteur-Ingénieur, Rennes, INSA, dec. 1981, 135 p.

[22] F.B. JORGENSEN, G. MICHEL, C. WAGNER, «PREDITE : a real-time processor for bandwidth compression in TV transmission», Proc. of ICASSP, Paris, May 1982, pp. 1195-1198.

[23] A.N. NETRAVALI, J.D. ROBBINS, «Motion-compensated television coding : Part I», Bell Syst. Techn. J., 58, No 3, March 1979, pp. 631-670.

[24] J.A. STULLER, A.N. NETRAVALI, J.D. ROBBINS, «Interframe television coding using gain and displacement compensation», Bell Syst. Techn. J., 59, No 7, Sept. 1980, pp. 1227-1240.

[25] C. LABIT, «Estimation du mouvement dans une séquence d'images de télévision», Thèse de Docteur-Ingénieur, Rennes, IRISA, février 1982, 119 p.

[26] C. LABIT, A. BENVENISTE, «Motion of edges and motion estimation in a sequence of TV pictures», Pr. of the ICASSP, Paris, May 1982, pp. 460-463.

[27] C. LABIT, A. BENVENISTE, «Motion estimation in a sequence of TV pictures», ASI—Conference on Image Sequence processing and dynamic scene analysis, 21 June-2 July, 1982, Braunlage, RFA, (this book).

[28] P. ROBERT, C. LABIT, «Comparative study between intra- and interframe prediction schemes», idem (this book).

MOVEMENT-COMPENSATED INTERFRAME PREDICTION FOR NTSC COLOUR TV SIGNALS

by

SHAKER SABRI
Bell-Northern Research Ltd.
3, Place du Commerce
Nuns' Island, Montreal
Quebec, Canada
H3E 1H6

ABSTRACT

This paper presents techniques for interframe prediction of colour NTSC signals using motion estimation and compensation techniques. A key element in movement-compensated prediction is the displacement estimation algorithm. A number of techniques for performing this estimation have been presented in the literature [1-13]. Some new techniques giving improved performance are presented in this paper. Also a movement-compensated composite prediction approach for coding of the NTSC colour signals is presented.

The performance of different displacement estimation algorithms is evaluated and compared for both component and composite approaches. A database of several colour NTSC video sequences is used in the evaluation.

Work presented in this paper has been funded by The International Telecommunications Satellite Organization INTELSAT under contract INTEL-114,1980.

NATO ASI Series, Vol. F2
Image Sequence Processing and Dynamic Scene Analysis
Edited by T.S. Huang

1. INTRODUCTION

Interframe video coders generally use some form of predictive coding [9]. In this technique, previously coded and transmitted picture elements from current and previous frames are used to form a prediction of the current frame picture elements. The prediction error is quantized, statistically coded and transmitted. The coding efficiency, i.e., the amount of bit rate reduction, depends on the accuracy of the prediction process. For standard previous frame predictors, this efficiency largery depends on the percentage of the picture area classified as being stationary. Situations such as a camera pan or zoom will result in a large unpredictable picture area. This will lead to a higher transmission bit rate to avoid visible impairments in the picture. Movement-compensated prediction [1-10] offers a promising approach to improve the performance of interframe coders. In the majority of cases, TV pictures contain the same objects but displaced (moving) from one frame to the next. In movement-compensated prediction, the frame to frame displacements of different objects are estimated, and the prediction is formed by the displaced previous frame elements. In this way the prediction performance can be improved substantially, resulting in a reduction of the bit rate required to transmit the TV signals as compared to standard previous frame prediction.

The NTSC colour TV signal can be processed in either composite or component form. The choice between the two approaches is generally influenced by several issues such as performance, implementation complexity and interface requirements. In this paper, techniques for composite interframe prediction that operate directly on the composite signal are presented for different sampling frequencies. This is carried out for both movement-compensated and previous frame predictors, and the efficiency of the composite and component approaches are compared.

A key element in movement-compensated prediction is the displacement estimation algorithm. A number of techniques for performing this estimation have been presented in the literature [1-13]. Some new techniques giving improved performance are presented in this paper. Also, the composite coding approach described above is adapted to all these algorithms. The performance of these displacement estimation algorithms is evaluated and compared for both component and composite prediction.

2. COMPOSITE INTERFRAME PREDICTION

The composite prediction approach has been of interest for some time [15,16,17]. In this approach the prediction is formed directly from the composite signal. Generally, most of the results that have been reported in the literature indicate that the composite predictors are not as efficient as the component predictors. While this is largely true for intrafield composite predictors, composite interframe predictors will be shown to give comparable performance to that of component predictors.

In the component prediction approach, the composite NTSC signal is demodulated into its three main components using two-dimensional (2-D) digital filtering techniques. This process is carried out outside the feedback loop and each signal is coded separately. The composite prediction process can be viewed as a two step process. In the first step, a demodulation process takes place to separate the components, then a reconstruction process is carried out to form a composite prediction having the same properties as the pel to be predicted in terms of chrominance signal phase. This process can be carried out either by implicit or explicit demodulation. The difference between the component and the composite approaches is in the place where the demodulation / reconstruction process takes place. In component coding, these two processes must occur outside the DPCM feedback loop and the demodulation must be explicit. In the composite coding approach, the two processes are carried out inside the DPCM feedback loop, and may be realized explicitly or implicitly.

In intraframe DPCM, the prediction is formed using previously transmitted (processed) elements. In order to carry out an effective demodulation process, knowledge of future picture elements is essential. Since these are not available at both receiver and transmitter, proper demodulation and reconstruction cannot be carried out. Therefore, most of the intraframe composite predictors realize a pseudo demodulation process and as a result their performance is generally lower than that of the component approach.

In the interframe case the situation is different. Predictors which utilize elements from the previous frame (or field) can realize an effective demodulation (explicit or implicit) since all previous frame elements are available at both the receiver and transmitter.

In this section an approach for handling the composite signal directly is presented. First this approach is presented for the standard interframe prediction case where no movement compensation is utilized. This approach is then extended for movement-compensated prediction. Several issues related to this approach will be discussed.

2.1 Previous Frame Composite Prediction

The composite NTSC colour signal could be expressed as:

$$[U] = [Y] + [C] \tag{1}$$

where $[Y]$ is the luminance signal and $[C]$ is the chrominance signal (I and Q) modulated by the colour subcarrier whose frequency is $f_{sc} = 3.58\ MHz$.

Because of the relation between the line, field, and colour subcarrier frequencies, the chrominance phase changes by 180° every frame.

The composite signal may be separated into its luminance and chrominance components as follows:

$$\begin{aligned}[Y] &= [U]*[H_L] \\ [C] &= [U]*[H_C]\end{aligned} \tag{2}$$

where,

$$[H_L] + [H_C] = [I].$$

$[H_L]$ and $[H_C]$ are the impulse responses of luminance and chrominance filtering operators respectively, "*" denotes the convolution process (in this case 2-D convolution), and $[I]$ is the unity operator.

In order to take into consideration the chrominance phase relation between successive frames, the previous frame composite prediction $[P]$ is expressed as:

$$\begin{aligned}[P] &= [Y] - [C] \\ &= [[I] - 2[H_C]]*[U] \\ &= [H_p]*[U].\end{aligned} \tag{3}$$

$[H_p]$ is therefore the impulse response of the composite previous frame predictor. Relation (3) is valid for any sampling frequency. However, the form and complexity of $[H_p]$ will vary depending on the sampling frequency. The composite previous frame prediction process could be viewed as an implicit separation and reconstruction (with proper phase relation) of the video signal. This process could be implemented explicitly or implicitly.

Because of the fact that the composite prediction is placed in the feedback loop of the DPCM system, there are some practical limitations on the predictor's order and the representation accuracy of the coefficients. These are mainly due to timing and speed constraints of high speed digital processing technology. Careful choice of the sampling frequency of the composite signal will alleviate these problems. For $2*f_{sc}, 3*f_{sc}$, and $4*f_{sc}$ sampling frequencies, composite predictors are easily realizable. In the following sections some of these predictors are presented.

(i) $4*f_{sc}$ sampling frequency:

Using the maximally flat design criterion, composite predictors are obtained as follows:

$$[H_C] = (1/16)\begin{bmatrix} 1 & 0 & -2 & 0 & 1 \\ -2 & 0 & 4 & 0 & -2 \\ 1 & 0 & -2 & 0 & 1 \end{bmatrix} \tag{4}$$

$[H_C]$ is the 2-D impulse response of a bandpass digital filter that extracts the chrominance signal. Using relation (3) the corresponding interframe composite predictor $[H_p]$ is:

$$[H_p] = (1/8)\begin{bmatrix} -1 & 0 & 2 & 0 & -1 \\ 2 & 0 & 4 & 0 & 2 \\ -1 & 0 & 2 & 0 & -1 \end{bmatrix} \tag{5}$$

Relation (5) gives the lowest predictor order at this sampling frequency. The coefficients are expressed in the form of 2^{-n} and, therefore, the multiplications are realized by simple shift operations. Only 4 bit accuracy is required to represent these coefficients. The next higher order predictor at this sampling frequency is given as:

$$[H_C] = (1/64)\begin{bmatrix} -1 & 0 & 4 & 0 & -6 & 0 & 4 & 0 & -1 \\ 2 & 0 & -8 & 0 & 12 & 0 & -8 & 0 & 2 \\ -1 & 0 & 4 & 0 & -6 & 0 & 4 & 0 & -1 \end{bmatrix} \tag{6}$$

$$[H_p] = (1/32)\begin{bmatrix} 1 & 0 & -4 & 0 & 6 & 0 & -4 & 0 & 1 \\ -2 & 0 & 8 & 0 & 20 & 0 & 8 & 0 & -2 \\ 1 & 0 & -4 & 0 & 6 & 0 & -4 & 0 & 1 \end{bmatrix} \tag{7}$$

In this case multipliers (or lookup tables) are required for implementation and 6 bit accuracy is needed for representation of coefficients. Higher order predictors than the one given by (7) are also possible [20].

ii) 2^*f_{sc} **sampling frequency:**

The same composite predictors given by (5) and (7) can be used for this sampling frequency, except that the zeroes in the impulse response are removed.

(iii) 3^*f_{sc} **sampling frequency:**

Using the same procedure outlined in the previous section, composite predictors at the 3^*f_{sc} sampling frequency are obtained as follows:

$$[H_C] = (1/36)\begin{bmatrix} 1 & 2 & -6 & 2 & 1 \\ -2 & -4 & 12 & -4 & -2 \\ 1 & 2 & -6 & 2 & 1 \end{bmatrix} \tag{8}$$

and the corresponding predictor $[H_p]$ is given by

$$[H_p] = (1/18)\begin{bmatrix} -1 & -2 & 6 & -2 & -1 \\ 2 & 4 & 6 & 4 & 2 \\ -1 & -2 & 6 & -2 & -1 \end{bmatrix} \tag{9}$$

Relation (9) represents the minimum predictor order required to give satisfactory performance at this sampling frequency. The next higher order predictor is given by:

$$[H_p] = (1/54)\begin{bmatrix} 2 & -3 & -6 & 14 & -6 & -3 & 2 \\ -4 & 6 & 12 & 26 & 12 & 6 & -4 \\ 2 & -3 & -6 & 14 & -6 & -3 & 2 \end{bmatrix} \tag{10}$$

Based on the above discussions, it is clear that the 4^*f_{sc} and 2^*f_{sc} sampling frequencies offer significant advantages in terms of predictor complexity. Fig. 1 illustrates a composite predictor for 4^*f_{sc} sampling.

2.2 Movement-Compensated Interframe Composite Prediction

The composite previous frame prediction technique presented in the previous section will have a comparable performance to the previous frame prediction of the component prediction approach, as will be shown in the evaluation sections. In this section, a movement-compensated prediction approach that operates directly on the composite colour signal is presented. Therefore, the advantages of both movement compensation and direct processing of the composite signal are maintained. In this case, both the displacement estimation and the prediction have to be obtained from the composite colour signal. In this section we will discuss how the movement-compensated prediction, and relevant quantities required to calculate the displacement estimates, are obtained. Algorithms for calculation of the displacement estimates will be discussed in section 3. In the following discussions we will concentrate on the 4^*f_{sc} sampling frequency case, though this approach is also applicable to other sampling frequencies.

At any given pel $I(x, \tau)$ and a given displacement estimate $\underline{d}$, it is required to calculate the displaced frame pel value $I(\underline{x}-\underline{d}, \tau-1)$ and the spatial gradients. The displaced frame is used as prediction and also in the calculation of the displacement estimates. The spatial gradients are necessary for calculation of the displacement estimates and could also be used for interpolation.

The composite prediction consists of a luminance and a chrominance part. Therefore an explicit or implicit demodulation process is required. A realization of this approach using an explicit demodulation process is shown in Fig. 2. The integral part of the displacement estimate is used to access a window of L lines by P pels in the previous frame. The minimum size of this window is equal to the dimensions of the 2-D filter impulse response required for separation of luminance and chrominance components. However, the actual size of this window is influenced by the specific memory hardware realization structure. In Fig. 3 a possible memory organization structure is shown. In Fig. 4, an illustration of the different pels used to form the prediction is shown.

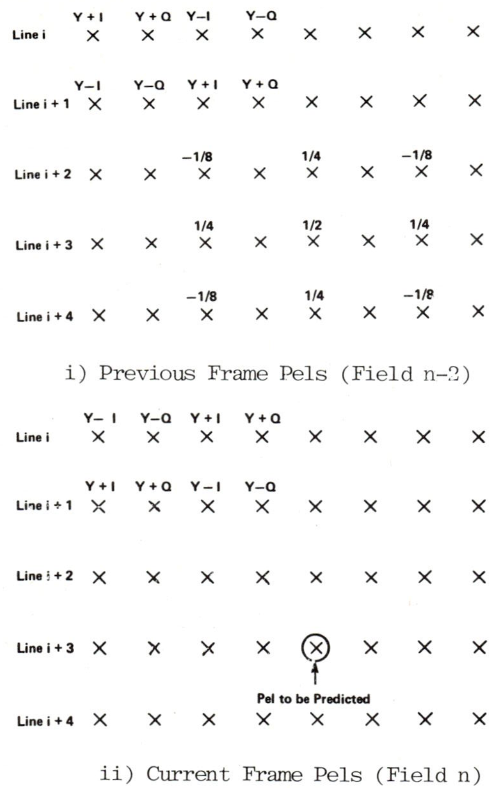

Fig. 1. Composite Interframe Predictor at $4*f_{sc}$ Sampling Frequency

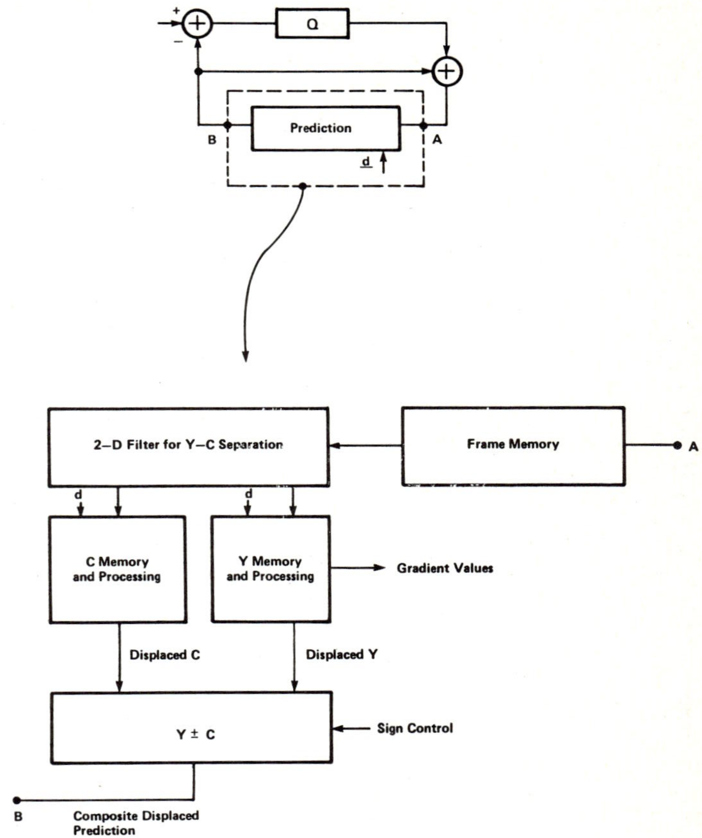

Fig. 2. Movement Compensated Composite Prediction

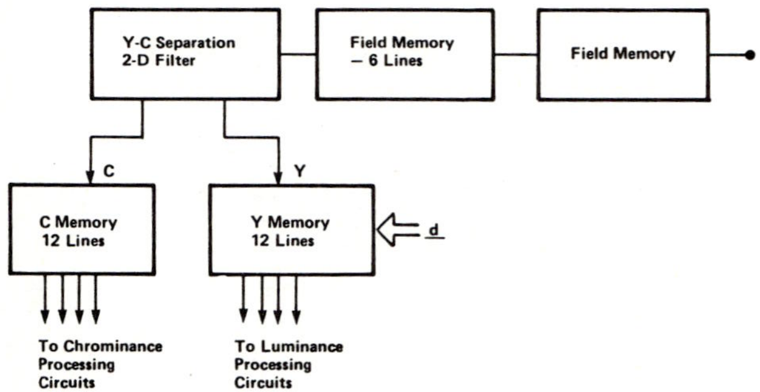

Fig. 3. Possible Memory Organization for the Composite Prediction Approach

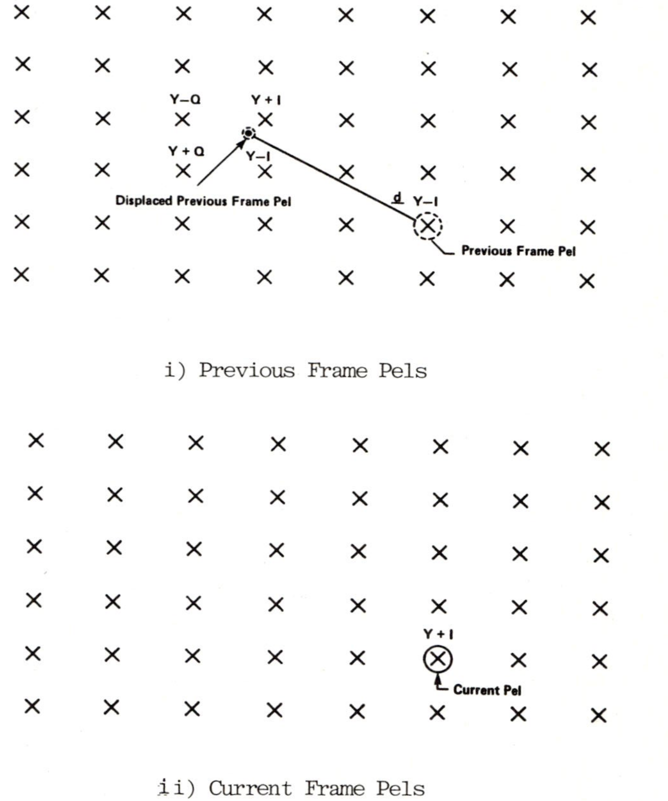

Fig. 4a. Different Picture Elements (Pels) Used in Forming the Displaced Composite Prediction

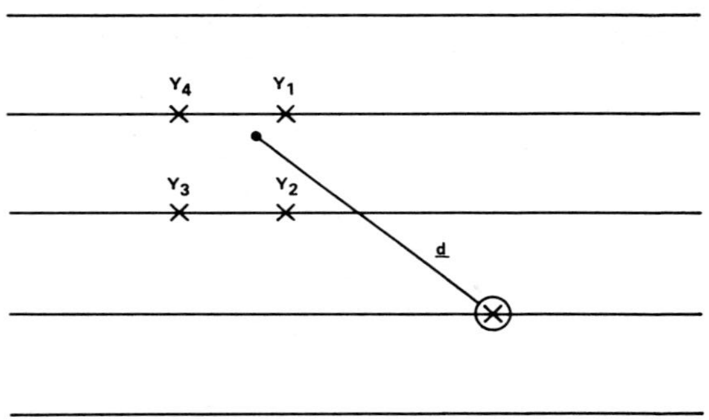

i) Previous Frame Pels Used to Form the Luminance Part of the Composite Prediction

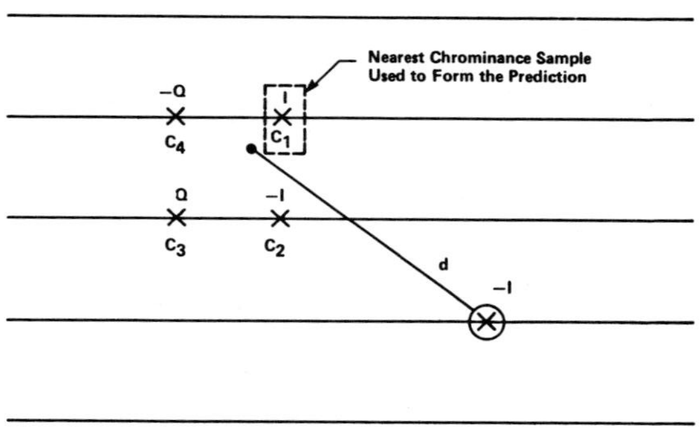

ii) Previous Frame Chrominance Pels

Fig. 4.b. Luminance and Chrominance Pels Used to Form the Composite Prediction at $4*f_{sc}$ Sampling

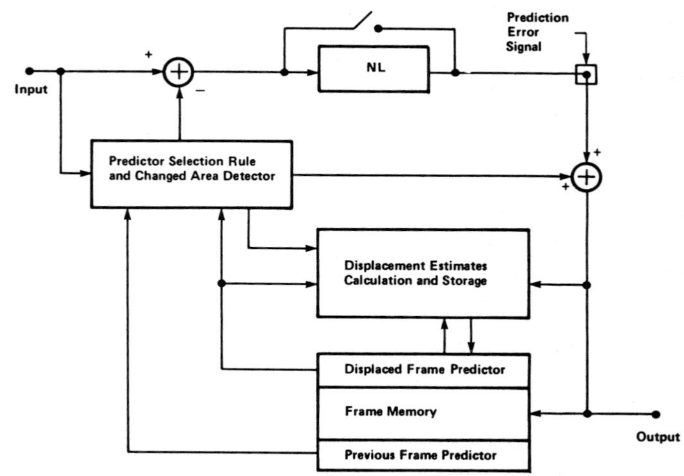

Fig. 5. Basic Configuration Used to Evaluate the Displacement Estimation Algorithms

The 2-D chrominance filter $[H_C]$ described in the previous section is used to separate the luminance and the chrominance signals. The luminance part of the prediction is obtained by interpolation using the four surrounding luminance pels as shown in Fig. 4.b and the fractional part of the displacement estimate. Spatial gradients ,i.e., vertical and horizontal gradients are derived also from the four luminance pels. The chrominance part of the composite prediction is selected to be the nearest chrominance sample with the proper chrominance (I or Q) component. No interpolation is needed in this case as the chrominance signals, I and Q, have a much lower bandwidth than the luminance signal. The displaced luminance part of the prediction is added to the displaced chrominance (after proper sign change, if necessary) to form the displaced composite prediction. The gradient information and the composite displaced prediction are subsequently used in calculating the displacement estimates, as will be described in the following sections.

3. MOVEMENT-COMPENSATED PREDICTION

In television imagery the actual displacement field is not known, and hence the accuracy of the displacement estimates obtained using any of the algorithms under consideration could not be directly determined. In order to assess the performance of each of the algorithms, several indirect measures are used: entropy of prediction error, root mean square value of prediction error (RMSPE), and percentage of picture area that is predictable, i.e., prediction error equals zero.

The entropy of the prediction error gives an indication of the number of bits required to transmit the video signal if an optimum variable- word-length code is used. The RMSPE gives an indication of the magnitude of the prediction error to be transmitted, and consequently affects the bit rate required for transmission. Finally, in interframe predictive coding systems, areas of the picture which are predictable normally are not transmitted. Therefore, the larger the predictable area the lower the bit rate needed to transmit the signal.

The issues to be examined can be classified into two categories. The first relates to performance and includes investigation of different parameters such as block size, sensitivity to noise, threshold values, type of signals (composite or component), etc. The second category relates to implementation complexity. The choice between algorithms is influenced by the specific application and normally involves a compromise between performance and complexity.

The evaluation of displacement estimation techniques could be carried out in several contexts, depending on the intended application. For the purpose of this paper, these are carried out within an interframe recursive type structure ,e.g., interframe DPCM or a recursive temporal filter. The structure shown in Fig. 5 is used in both the coding and noise reduction applications [18,19].

In Fig. 5 two predictions are formed, the previous frame prediction and the displaced previous frame prediction. Both predictors share the same field or frame memory. The choice of which predictor to use is based on the absolute value of the prediction error, i.e. the predictor which gives lower prediction error is selected. It should be noted that this rule is considered an ideal selection rule which could be directly used in the noise reduction application [18] as it requires the knowledge of the input signal. However, if this selection rule is used for the coding application, information on which predictor is used must be transmitted to the receiver. This will add to the transmitted information overhead. Since the main objective of this section is to evaluate the performance of the displacement estimation techniques, it is desirable to isolate imperfections caused by non-ideal selection rules. The implications of non-ideal switching rules are discussed in [19].

The changed area segmentor is a thresholding circuit whose function is to detect if the frame difference signal exceeds a predetermined threshold. The corresponding picture element is then classified as part of a changing or moving area. This information is used in the displacement estimation algorithms. The nonlinear element (NL) is used to suppress noise and is included here as part of the study to investigate the effects of noise on the performance of the displacement estimators. The nonlinearity is defined by the three parameters (P_s, P_e, α_s) and normally α_e is equal to 1. The displacement estimator contains necessary elements to implement the algorithms under investigation and perform other functions such as interpolation, storage of previous displacement estimate if needed, and calculation of quantities such as gradients and partial sums. The configuration of this element will vary from one algorithm to another.

3.1 DISPLACEMENT ESTIMATION ALGORITHMS

In the following sections the algorithms that have been investigated will be described and their performance evaluated.

3.1.1 ALGORITHM I:

This algorithm is based on the pel recursive method [1]. The displacement estimate is given by

$$\underline{d}^{i} = \underline{d}^{i-1} - \epsilon . DFD(\underline{x}, \underline{d}^{i-1}) . \nabla I(\underline{x} - \underline{d}^{i-1}, t - \tau). \tag{11}$$

where $\underline{d}^{i}$ is the displacement estimation at the i th iteration, $\underline{d}^{i-1}$ is the previous displacement estimate, $DFD(\underline{x}, \underline{d}^{i-1})$ is the displaced frame difference,

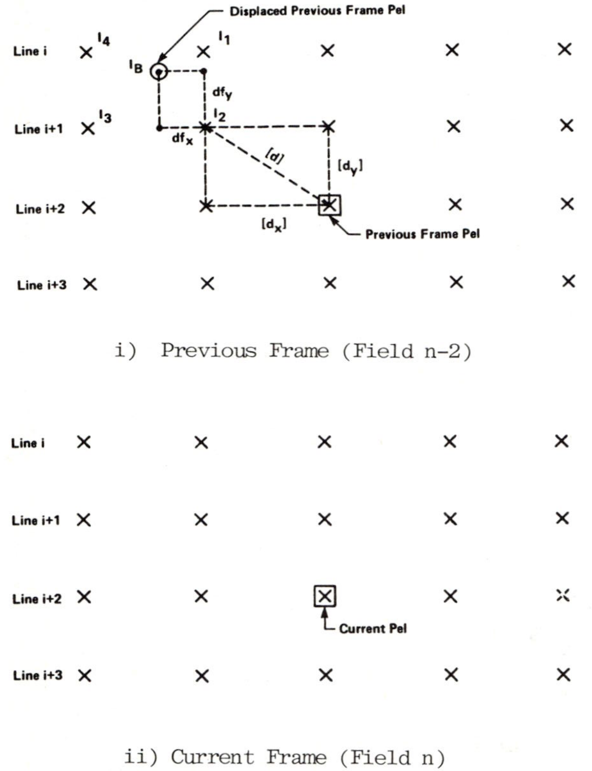

Fig. 6. Illustration of Different Pels Used in Calculation of the Displacement Estimates

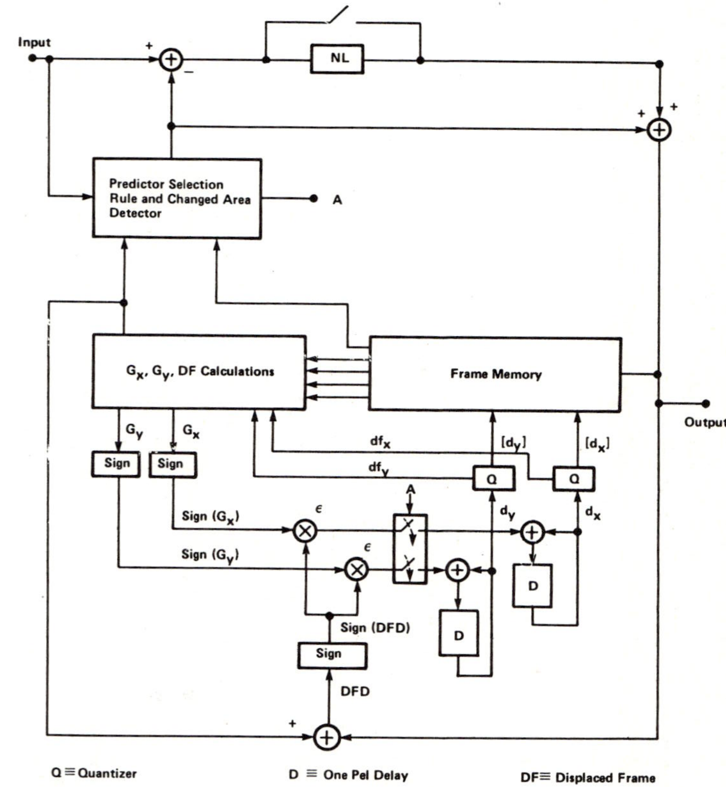

Fig. 7. Implementation of Displacement Estimation Algorithm I for the Component Approach

∇I is the spatial gradient, and ϵ is the convergence control parameter. Implementation of the pel recursive method using (11) requires multiplications, and interpolation to evaluate DFD and ∇I. It could be simplified without seriously affecting the performance as follows [1]

$$d_x^i = d_x^{i-1} - \epsilon .sign(DFD(\underline{x}, \underline{d}^{i-1})).sign(G_x(\underline{x} - [\underline{d}^{i-1}], t - \tau))$$

$$d_y^i = d_y^{i-1} - \epsilon .\ sign(DFD(\underline{x}, \underline{d}^{i-1})).sign(G_y(\underline{x} - [\underline{d}^{i-1}], t - \tau)). \tag{12}$$

where $\underline{d}^i = (d_x^i, d_y^i)$, d_x^i and d_y^i are the horizontal and vertical components of the displacement estimate, $[d^i]$ is the integral value of the displacement estimate. G_x and G_y are the horizontal and vertical components of the spatial gradient. The sign function is defined as:

$$sign(z) = \begin{cases} \frac{z}{|z|} & z \neq 0 \\ 0 & z = 0 \end{cases} \tag{13}$$

In (12), the multiplications have been eliminated, and the interpolation process is required only to calculate the displaced frame difference signal.

Update of the displacement estimate is carried out in the scan direction and is disabled in the stationary area of the picture, i.e., where the frame difference signal is less than or equal to a certain threshold value K_m. The vertical and horizontal gradient calculations and interpolation are illustrated in Fig. 6. The spatial gradients G_x and G_y are given as:

$$G_x = \frac{I_4 - I_1 + I_3 - I_2}{2}$$

$$G_y = \frac{I_1 - I_2 + I_4 - I_3}{2} \tag{14}$$

The corresponding displaced frame element $\hat{I}_B$ is obtained by interpolation as:

$$\hat{I}_B = I_2 + df_x(I_3 - I_2) + df_y(I_1 - I_2) + df_x df_y[I_4 - I_1 + I_2 - I_3] \tag{15}$$

where df_x and df_y are the fractional part of the displacement estimates in the horizontal and vertical directions respectively. The displaced frame difference is given by

$$DFD(\underline{x}, \underline{d}^{i-1)} = I_B - \hat{I}_B \tag{16}$$

Fig. 7 shows a possible implementation of Algorithm I for component (luminance) signal. In this case the previous frame prediction is a one frame delay, and the displaced frame difference is obtained by interpolation. Note that in this configuration only one interpolator is required as the same displaced frame value is used for both prediction and update of displacement estimate. In some cases more than one interpolation process may be needed [3]. The displacement estimation is carried out using previously processed picture elements which, in the coding application, are available at both the receiver and transmitter. In this case, the displacement estimates need not be transmitted.

For the composite signal, a possible implementation is shown in Fig. 10. As discussed before, the different quantities of relation (12) have be to obtained from the composite signal. An explicit demodulation process is performed to separate the luminance and the chrominance. Gradient calculations G_x and G_y are obtained from the luminance only. This is followed by interpolation to form the luminance part of the displaced frame signal DFY. For the chrominance signals (I and Q) no interpolation is needed because of their low bandwidth. The chrominance part of the composite displaced frame is chosen as the nearest pel (among 4 pels) with the proper chrominance signal (I or Q). The composite displaced frame signal is formed by adding the displaced luminance signal to the displaced chrominance signal with proper sign change. The displaced frame difference is formed by subtracting the current composite output signal from the displaced composite frame signal. Note that the displaced frame difference could not be obtained from the luminance part only, as this requires the knowledge of future output picture elements to perform a proper separation process.

Variations of the configurations shown in Fig. 8 are also possible. However, the current configuration is efficient in terms of storage requirements, since the video signal is maintained in composite form. Otherwise two separate memories have to be used to store the luminance and chrominance components. In Fig. 8, the block containing the displacement estimates calculations and update is identical to the corresponding part in Fig. 7.

The parameter ϵ controls the convergence of the algorithm and normally is chosen as large as possible subject to a stability constraint. For relation (11) ϵ is normally chosen in order of 10^{-2} to 10^{-3}. However, since the algorithm in the form of (11) requires multiplications and is more complex to implement, the simplified version as shown by relation (12) is the one considered for implementation. In this case ϵ is chosen to be 1/16.

In relation (12) the parameter ϵ influences several factors related to the performance of the estimator:

1) Rate of convergence: The displacement estimate update term in (12) has a value of ϵ. Therefore, for a given displacement of d_1 pels (or lines), the algorithm will take d_1/ϵ iterations for the displacement estimate d^i to converge to the actual displacement d_1. For example, if the actual displacement is 2 pels per frame and $\epsilon = 1/16$, the algorithm will need at

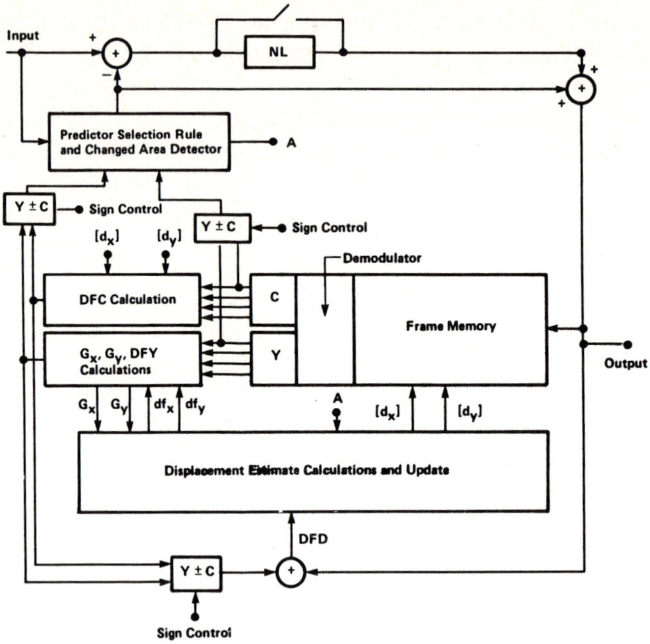

Fig. 8. Implementation of Displacement Estimation Algorithm I for the Composite Approach

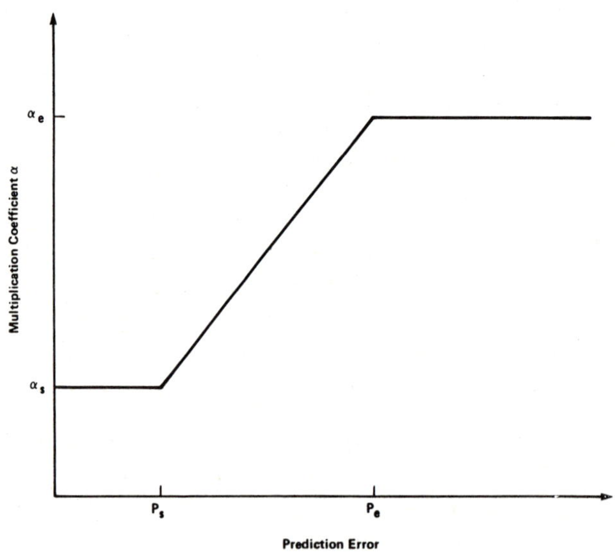

Fig. 9. Nonlinear Characteristic (NL)

least 32 iterations to converge under noise-free conditions. In practice, this noise-free condition rarely exists and hence the algorithm will require a larger number of iterations to converge because of noise. Choosing a larger value for ϵ will speed up the convergence, however it causes problems regarding stability of the algorithm and serious oscillations will result. In addition increasing ϵ will influence the accuracy of the displacement as will be discussed next.

2) Accuracy of the displacement estimate: In television imagery, very large displacements occur infrequently. Even when they do occur, in situations such as fast camera pan or a very fast moving object (e.g., racing car), the spatio-temporal details are degraded because of the limited spatio-temporal response of the television cameras or imaging devices. In these situations the spatial and temporal gradients are inaccurate, and as a result the displacement estimate will be inaccurate for very large displacements. In view of this, the integral part of the displacement estimate $[\underline{d}]$ is limited to $\pm$ 7 pels/frame horizontally and $\pm$ 4 lines/frame vertically. This implies that at most 4 bits (3 bits for magnitude and one for sign) are needed to represent the integral part. The accuracy of the fractional part of the displacement estimate $(df_x \quad or \quad df_y)$ is influenced by the value of ϵ, as the minimum displacement estimate update magnitude is equal to ϵ. For example if $\epsilon = 1/16$, the fractional part $(df_x \quad or \quad df_y)$ needs 4 bits representation. As a result, the accuracy of interpolation and/or prediction is limited to 1/16 pel or 1/16 line.

The performance of the Algorithm I, is illustrated In Fig. 10, Fig. 11 and Fig. 12 under different noise conditions. Point A is when the prediction error signal is processed through the nonlinearity shown in Fig. 9 which is intended to remove part of the noise in the original signal. For points B, C and D this nonlinearity is bypassed. Point B represents the original signal. Points C and D represent the original signal with added computer generated noise. The signal-to-added-noise ratio (SNR) at points C and D are 43 db and 36 db respectively.

As can be seen from Fig. 10, for video sequence "Exercise Girl", movement-compensated prediction using Algorithm I results in 50 percent reduction in RMSPE as compared to standard previous frame prediction. The RMSPE for Algorithm I and previous frame prediction does increase proportionally to noise level in the signal. The percentage of predictable area increases when movement compensation is used as shown in Fig. 12.

In order to reduce the displacement estimation algorithm sensitivity to noise, proper segmentation into stationary and moving areas is needed. Complex segmentation algorithms are possible; however, in many instances

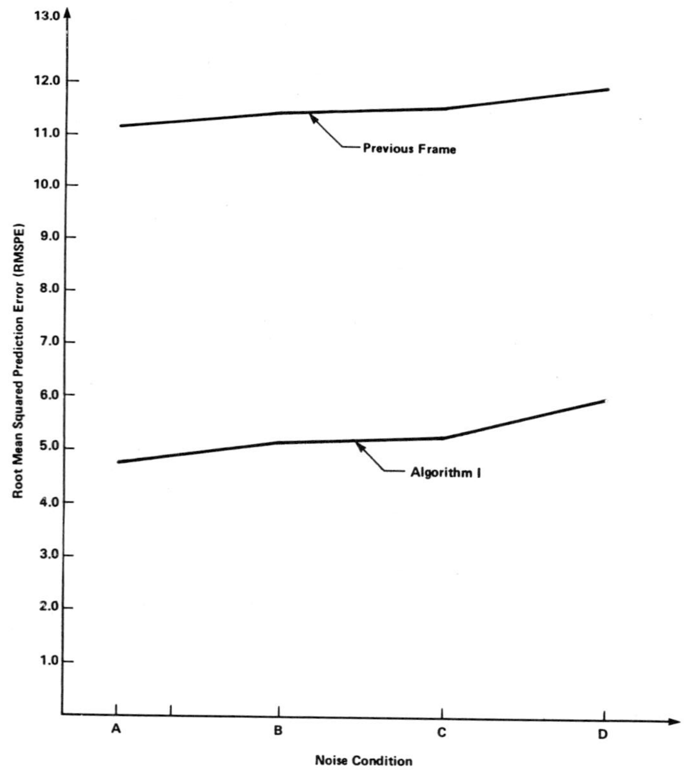

Fig. 10. Improvement in RMSPE for Algorithm I Under Different Noise Conditions. Video Sequence "Exercise Girl", Luminance Signal, $k_m = 4$, $f_s = 14.3$ MHz

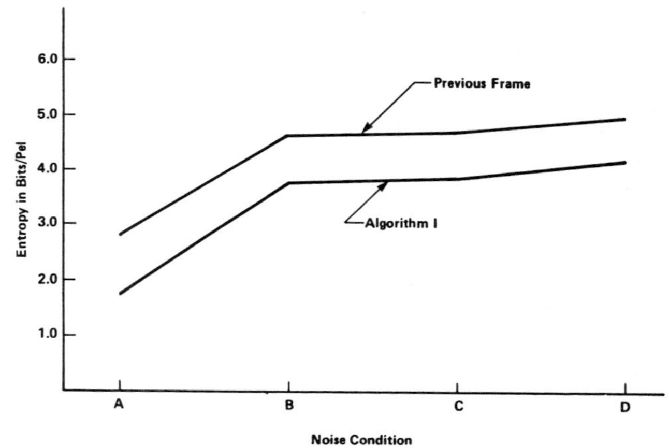

Fig. 11. Entropy Under Different Noise Conditions for Algorithm I. Video Sequence "Exercise Girl", Luminance Signal, $k_m = 4$, $f_s = 14.3$ MHz

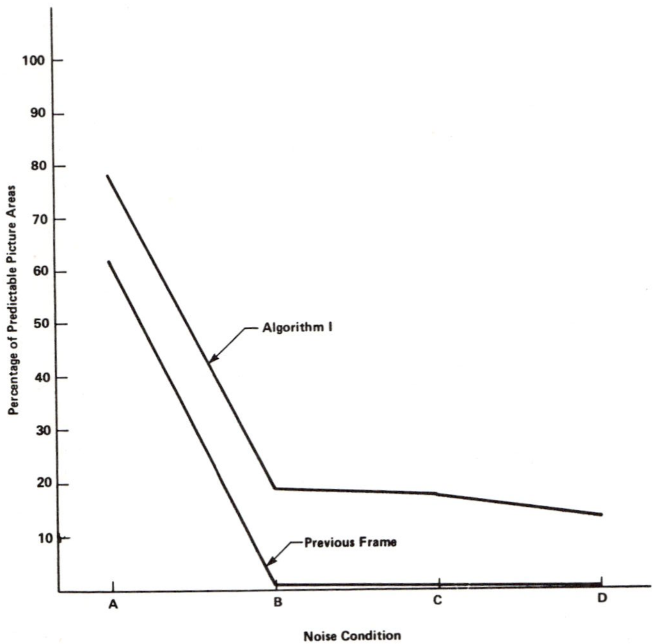

Fig. 12. Percentage of Predictable Areas Under Different Noise Conditions. Video Sequences "Exercise Girl", Luminance Signal, $k_m = 4$, $f_s = 14.3$ MHz

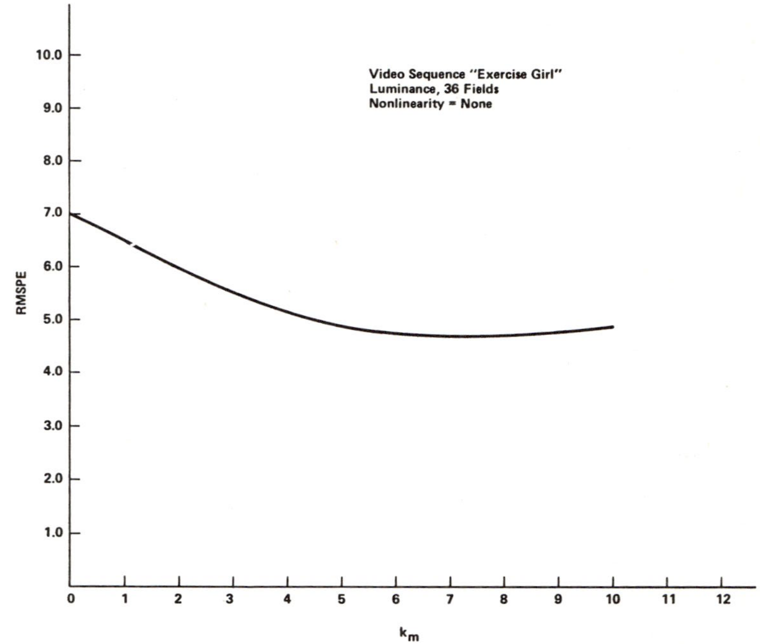

Fig. 13. Relation Between RMSPE and Threshold Value k_m for Algorithm I

distinction between noise and low contrast fine details in the picture is extremely difficult due to the lack of an accurate noise model for TV imagery. For practical purposes, the segmentation used is based on comparing the previous frame prediction error (frame difference) signal to a threshold K_m. Updating of the displacement estimation is disabled whenever the frame difference is less than or equal to K_m. If K_m is set too low this will cause the noise to be classified as a changing area. Raising this threshold too high would result in moving areas being classified as stationary and consequently the displacement estimate update process is unnecessarily disabled. Ideally K_m should be varied according to the noise level in the signal if it could be measured. Fig. 13 shows the performance of Algorithm I for different values of K_m. We notice that as the value of K_m is reduced, the performance of the algorithm deteriorates very rapidly. This is due to the fact that noise causes the stationary area of the picture to be classified as a changing area. Therefore the displacement update process is enabled and as a result the algorithm starts to diverge. In addition, in moving areas, as the displaced frame difference DFD value is reduced because the algorithm starts to converge, noise value becomes comparable to the value of DFD. This will cause the convergence process to halt and again the algorithm starts to diverge.

3.1.2 ALGORITHM II:

This algorithm was presented in [10]. Each field is subdivided into virtual blocks of N pels by M lines each. For each block a displacement estimate is obtained using the following simplified relation:

$$d_x^i = d_x^{i-1} - \frac{\sum_{MA} DFD(\underline{x}, \underline{d}^{i-1}).sign(G_x)}{\sum_{MA} |G_x|}$$

$$d_y^i = d_y^{i-1} - \frac{\sum_{MA} DFD(\underline{x}, \underline{d}^{i-1}).sign(G_y)}{\sum_{MA} |G_y|} \tag{17}$$

where $\underline{d}^i = (d_x^i, d_y^i)$, d_x^i and d_y^i are the displacement estimates in the horizontal and vertical directions, G_x and G_y are the horizontal and vertical gradients respectively, and DFD is the displaced frame difference. All these quantities are defined in section 3.1.1, and will not be redefined here. The summations in the (17) are carried out over the moving area within the block, i.e., for those pels whose frame differences exceed the threshold value K_m.

For prediction, the previous field displacement estimate for the same block, $\underline{d}^{i-1}$, is used to calculate the movement-compensated prediction.

Fig. 14 shows a possible implementation of Algorithm II as defined by relation (17) for the component (luminance) signal. Notice that the

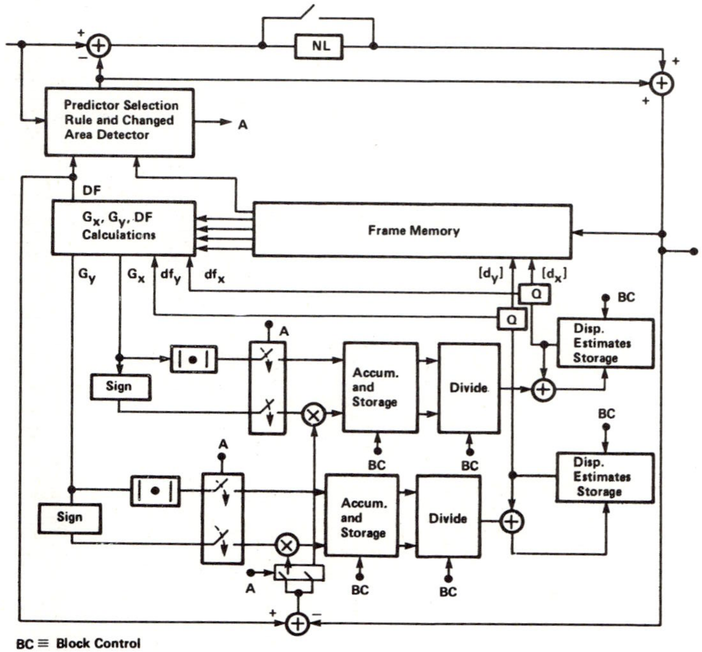

Fig. 14. Implementation of Displacement Estimation Algorithm II for the Component Approach

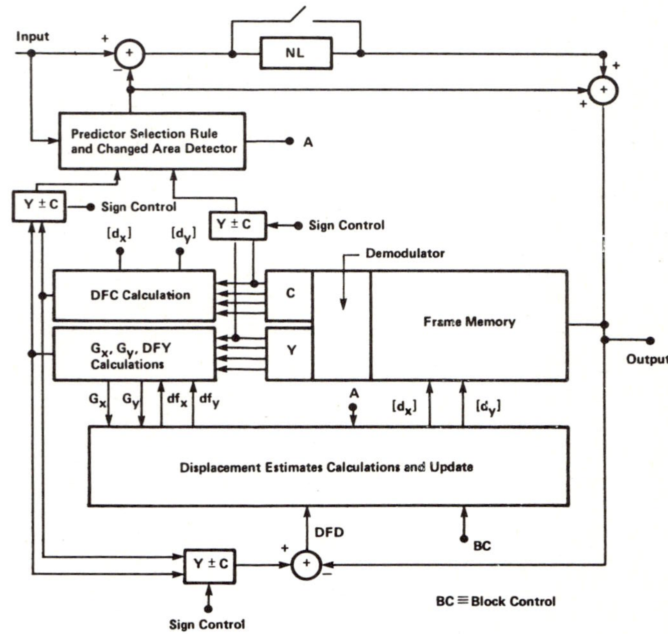

Fig. 15. Implementation of Displacement Estimation Algorithm II for the Composite Approach

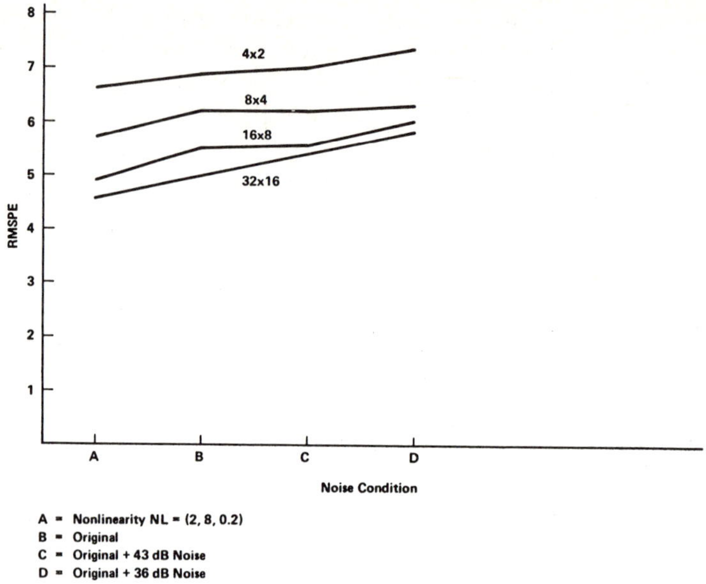

Fig. 16. Performance of Algorithm II (RMSPE) Under Different Noise Conditions and for Different Block Sizes. Video Sequence "Exercise Girl", 36 Fields, $k_m = 4$, Luminance Only

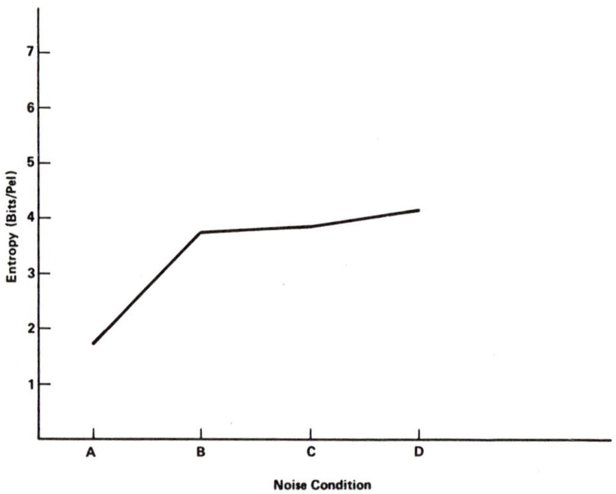

Fig. 17. Performance of Algorithm II (Entropy) Under Different Noise Conditions. Video Sequence "Exercise Girl", Luminance Only 36 Fields, $k_m = 4$, Block Size = 32 x 16

displacement estimates need not be transmitted to the receiver in the coding application as all calculations use previously processed data. Four sets of accumulators are used to perform the summations in relation (17). At the end of each block the division is performed. Previous field displacement estimates are stored as shown in Fig. 14 and are used to form the displaced prediction DF. The multipliers shown are used only to change the sign of the displaced frame difference.

Fig. 15 shows a possible implementation of Algorithm II for the composite approach. The displacement estimates calculations and update are identical to those of Fig. 14. Gradient and displaced frame calculations are performed in the same way given in section 3.1.1.

The performance of this algorithm is sensitive to block size ($N \times M$). For small block size the displacement estimate is affected by noise, and as a result noisy estimates are obtained. If the block size is too large, the assumption of uniform single displacement for the block may no longer be valid as the block may contain more than one object moving in different directions. Fig. 16 and Fig. 17 illustrate the performance of Algorithm II for different block sizes and under different noise conditions for video sequence "Exercise Girl". Note that, in terms of RMSPE and entropy, a block size of 32x16 gives the best results. However the difference in performance between a block size of 32x16 and 16x8 is small. Larger block sizes (64×32) were also used for this sequence, however, no further improvements were noticed.

The effect of varying the change segmentation threshold K_m is shown in Fig. 18. We note that Algorithm II is not as sensitive to this parameter as in the case of Algorithm I.

3.1.3 ALGORITHM III

This algorithm combines both Algorithms I and II. The video signal is divided into virtual blocks of N pels by M lines each. Algorithm II is used to generate the displacement estimate $\underline{d}^{j-1}$ for each block. The displacement estimate is further refined within the block using Algorithm I. In this case $\underline{d}^{j-1}$, obtained at field $j-1$, is used as an initial displacement estimate value for the block at field j. At each pel within the block this displacement estimate is refined to produce a new displacement estimate d^i using Algorithm I. This process is illustrated in Fig. 19. Note that, in this algorithm (III), two sets of displacement estimates are needed as shown in Fig. 20. As a result, the gradient and displaced frame difference calculations could not be shared. Therefore, implementation of this algorithm requires a complexity level equivalent to the implementation of both Algorithm I and Algorithm II combined as illustrated in Fig. 20. This algorithm, though it gives good performance, is considered impractical and is provided for comparison purposes only. The performance of this algorithm under different noise conditions is

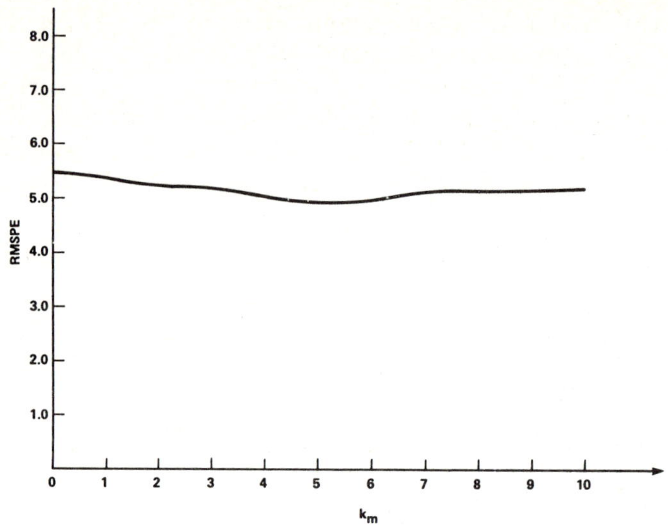

Fig. 18. Relation Between Threshold Value k_m and RMSPE for Algorithm II, Video Sequence "Exercise Girl", Original Luminance Only, 36 Fields

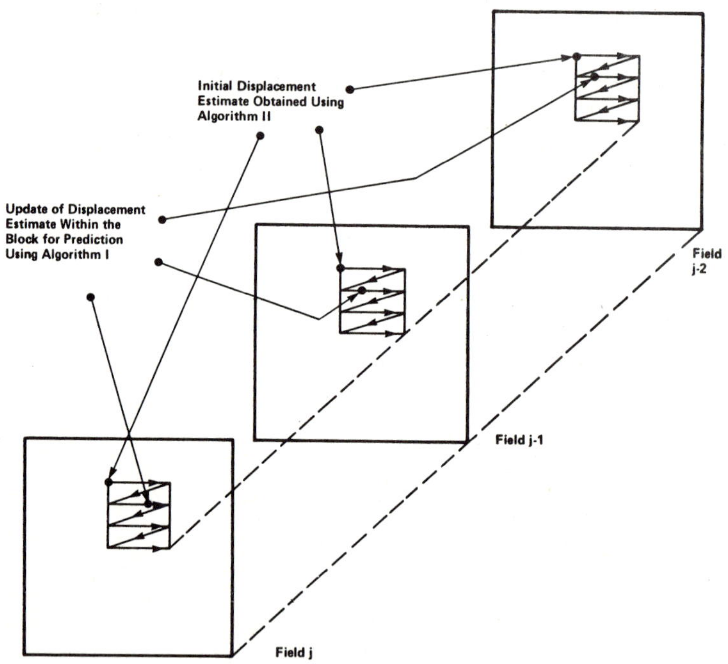

Fig. 19. Illustration of Displacement Estimation Algorithm III

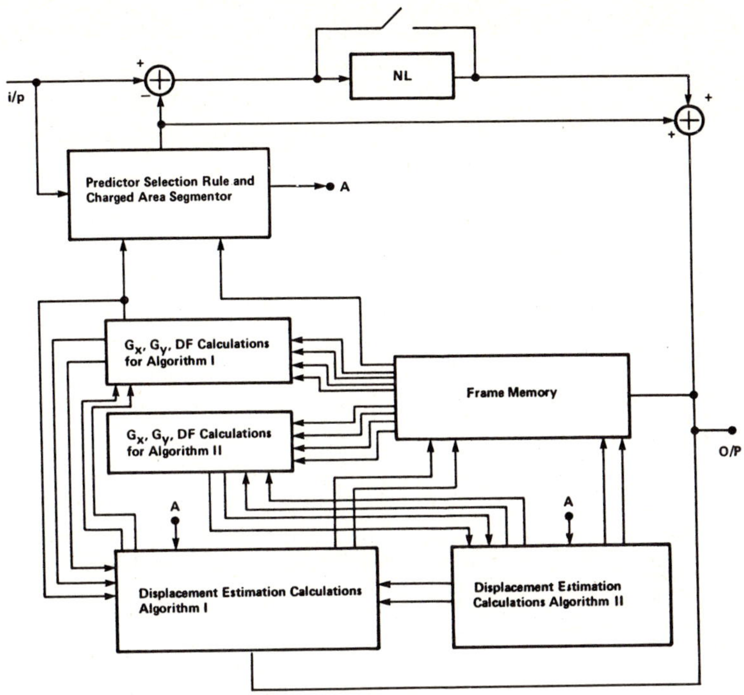

Fig. 20. Implementation of Displacement Estimation Algorithm III for the Component Approach

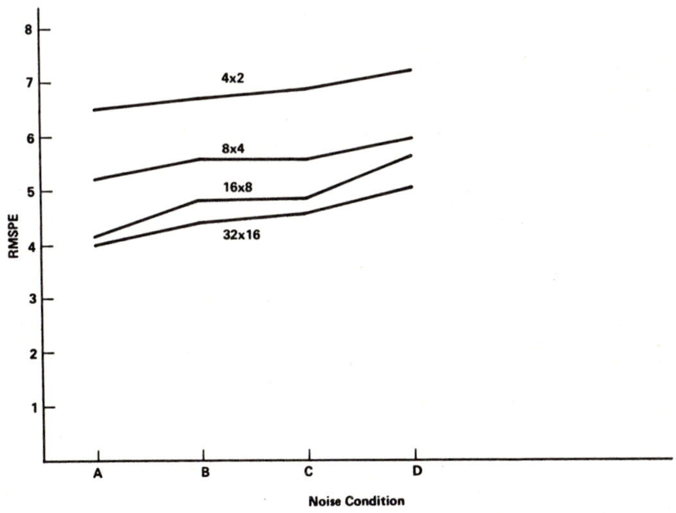

Fig. 21. Performance of Algorithm III Under Different Noise Condition for Different Block Sizes. Video Sequence "Exercise Girl", 36 Fields, Luminance Only $k_m = 4$

illustrated in Fig. 21 for different block sizes for video sequence "Exercise Girl". We note here that the best block size is 32x16.

3.1.4 ALGORITHM IV

This algorithm is based on the pel recursive displacement estimation technique as described in [1]. It alleviates some of the drawbacks of Algorithm I. As discussed in section 3.1.1, Algorithm I updates the displacement estimation in the scan direction, within the whole field. It can be seen that a situation of several objects moving at different speeds in different directions might exist within a field. Therefore the performance of Algorithm I will deteriorate due to its slow convergence property.

In Algorithm IV, the signal is subdivided into virtual blocks of N pels by M lines, the displacement estimates are calculated and updated only within the block. The initial displacement estimate for the block is taken as the last estimate obtained at the end of the same block in the previous field. This process is illustrated in Fig. 22.

Algorithm IV offers several advantages; namely:

(i) Convergence process: The convergence process is significantly speeded up. At the first field to be processed, possible displacements in the range of $\pm$ 8 pels or lines can be obtained (block size of 16×8). Therefore, for most of the practical displacements in TV imagery, the algorithm should converge within one field interval. On subsequent fields, the use of an accurate initial value for displacements insures that convergence is maintained. In addition, Algorithm IV alleviates the problems that result when the picture have multiple objects moving in different directions.

(ii) Newly exposed background and edge problems: Another advantage of this algorithm is that it limits divergence or inaccurate displacement estmates to smaller areas of the picture. To illustrate this point, consider the case of newly exposed background in TV imagery. These areas are classified as changing (or moving) and as a result most of the displacement estimation algorithms will update these displacements accordingly. This will result in inaccurate displacements and divergence problems. In Algorithm IV these problems are limited only to the small block areas where such conditions exist. Also these problems do not propagate to adjacent blocks as the displacement estimation process is carried out within each block only.

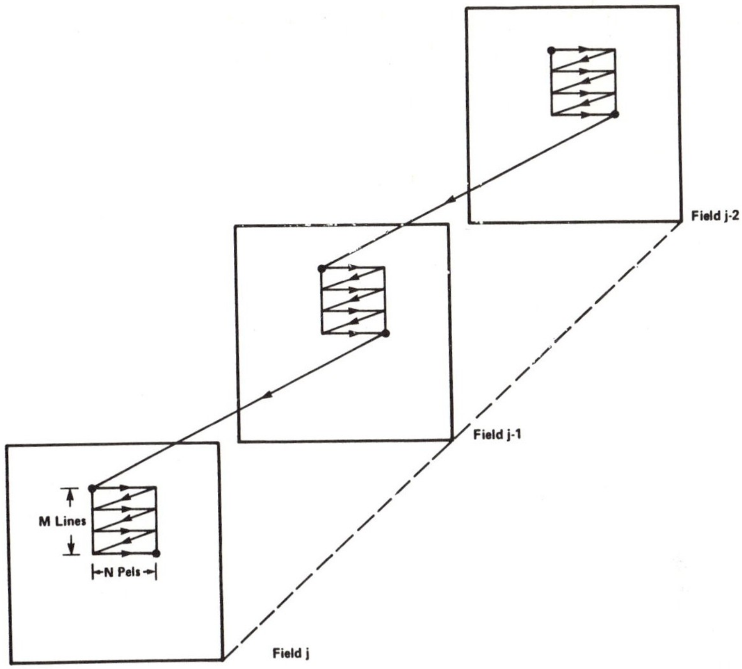

Fig. 22. Illustration of Displacement Estimation Algorithm IV

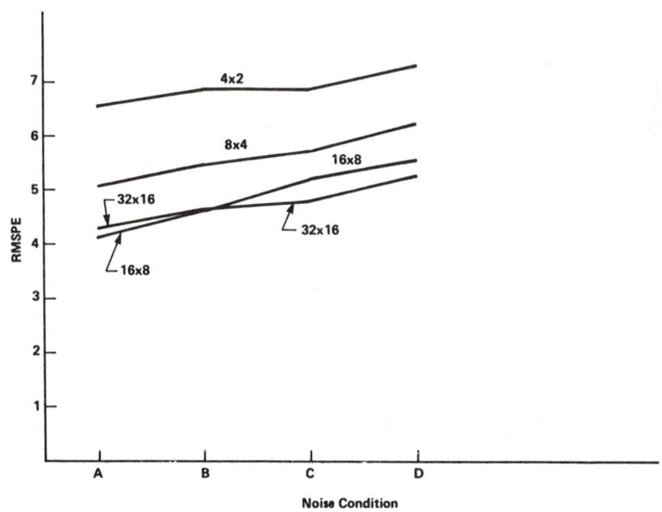

Fig. 23. Performance of Algorithm IV Under Different Noise Conditions, for Different Block Sizes. Video Sequence "Exercise Girl", 36 Fields, Luminance Only, $k_m = 4$

A possible implementation of Algorithm IV is shown in Fig. 24. This realization is similar to that of Algorithm I, except that storage of previous field displacement estimates is required.

The performance of this algorithm is illustrated in Fig. 23 for video sequence "Exercise Girl" under different noise conditions. The optimum block size for this case is 16 pels by 8 lines.

3.1.5 ALGORITHM V

So far we have discussed algorithms where the displacement estimation process is carried out inside the feedback loop and use previously processed data. Therefore, for the coding application, the displacement estimates need not be transmitted. In some applications it might be desirable to place the displacement estimator outside the feedback loop and transmit these displacements together with the coded visual information. For recursive displacement estimation algorithms, where the update of the displacements is carried out on a pel by pel basis, this will require significant overhead and might not be practical. If it is desirable to maintain the pel recursive method for calculation of the displacement estimates and placing the displacement estimator outside the feedback loop, then the algorithm discussed below will meet such objective.

In this algorithm, Algorithm V, the displacement estimation process is carried out in the same way as described in the previous section. However, for the prediction process,i.e., forming the displaced previous frame (or field) value, only one displacement value is used for the whole block. This value can be either the displacement obtained at the end or center of the block. Therefore, a single displacement per block needs to be transmitted. Obviously this algorithm can also be placed inside the feedback loop, if desired.

The performance of this algorithm is illustrated in Fig. 25 for different block sizes and under different noise conditions. We note that a block size of 16 x 8 gives the best performance. In order to insure the same conditions for evaluation of different algorithms, the results shown in Fig. 25 are for the case where the displacement estimator is placed inside the feedback loop.

4. PERFORMANCE OF DISPLACEMENT ESTIMATION ALGORITHMS

In the previous sections, we discussed the performance of each algorithm. In the following sections, issues common to all algorithms will be discussed and a comparative evaluation will be presented. Description of the different

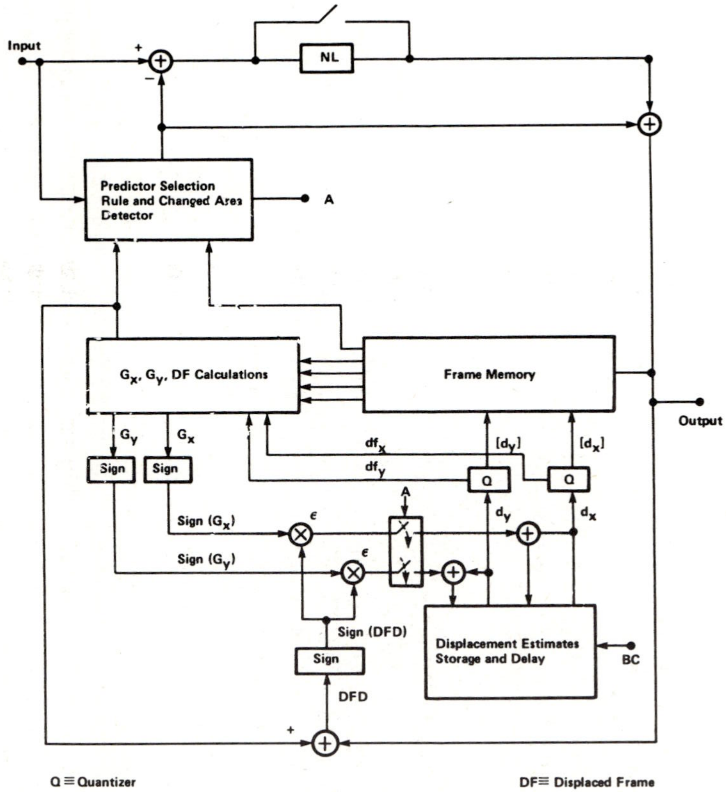

Fig. 24. Implementation of Displacement Estimation Algorithm IV for the Component Approach

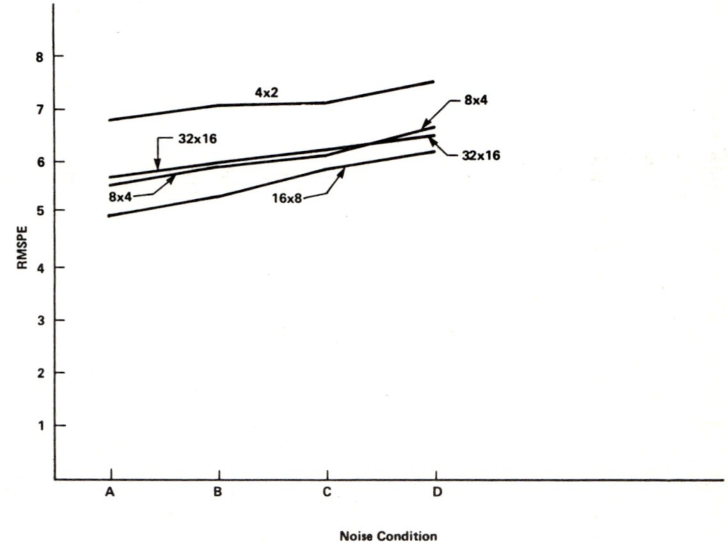

Fig. 25. Performance of Algorithm V (RMSPE) Under Different Noise Conditions for Different Block Sizes. Video Sequence "Exercise Girl", 36 Fields, Luminance Only, $k_m = 4$

video sequences and the facilities used to carry out the simulation are given in the Appendix.

4.1 Previous Frame and Movement-Compensated Prediction

It is clear from results presented so far that significant gains are achieved by using movement-compensated prediction. These gains are quantified by the three parameters: entropy, RMSPE, and percentage of predictable picture area. In terms of entropy, utilization of movement compensation resulted in reduction of the entropy of the prediction error. The RMSPE is reduced by a factor of more than two in most cases. Also the percentage of predictable picture area has increased.

4.2 Composite and Component Prediction

As presented earlier in this paper, the displacement estimation and movement-compensated prediction could be carried out on the video signal in either composite or component forms. For the component approach, we shall restrict our discussion to the luminance signal as it contains most of the energy in the video signal and has much larger bandwidth than the chrominance signals.

A comparison of performance between the composite and component approaches for all the cases under investigation is tabulated in Tables 1 to 6. If we closely examine these results, it can be seen that for video sequences "Moving Letters" and "Bathing Beauty" the performance is quite comparable. In terms of entropy, the composite prediction error signal is 0.01 to 0.20 bits/pel higher than the luminance prediction error. The RMSPE and percentage of predictable picture area are comparable for both cases. For video sequence "Mad Dash" the difference between the two cases is slightly larger. This is mainly due to the fact that this sequence has a high colour saturation and chrominance activity.

It is concluded that the composite approach gives a comparable performance to the component approach (luminance only). Since the chrominance signals (I and Q) will contribute to the bit rate, it is clear that the composite approach has a slightly better performance than the component approach. Another advantage of the composite approach is that the memory requirements are less than that required for the component approach. If we assume that the same resolution is to be maintained, i.e., full NTSC signal resolution, and similar demodulation processes (explicit or implicit) are used, the component approach might require up to 50 percent more memory than the composite approach. It is also concluded that the movement-compensated prediction approach is as efficient for the composite approach as it is for the component approach.

4.3 Effects of Noise

Noise inherent in the video signal influences the performance of the displacement estimator and the prediction process. Evaluation of the performance of each algorithm under different noise conditions has been discussed earlier in this paper. It can be stated that the performance deteriorates due to noise in two ways; the entropy of the prediction error increases substantially and the percentage of predictable picture area decreases. This is true for both movement-compensated prediction and previous frame prediction. For example, if we examine Figs. 11, 17 and 23, we can see that noise could contribute to the entropy as much as the signal itself. For video sequence "Exercise Girl" and using Algorithm I, the entropy of the prediction error signal increases from 1.75 bits/pel (when noise is suppressed) to 4.2 bits/pel when noise is added to the signal (36 db SNR). For other sequences similar results are obtained. For example if we examine Table 3 and Table 4 for video sequence "Bathing Beauty" it can be seen that when no noise suppression is used the entropy is more than 3.4 bits/pel for all algorithms. When noise supression is utilized the entropy goes down by more than 2 bits/pel to the 1-1.4 bits/pel range (algorighms I to V).

It can by argued that such improvements in entropy are due to signal modification by the nonlinear temporal filtering. While this is true, it is stressed that the nonlinearities used in these experiments have been designed so as not to introduce visible degradations in the picture under normal viewing conditions.

Noise also influences the percentage of predictable picture area. This is illustrated in Tables 1 and 2 for video sequence "Moving Letters". This parameter increases from less than 1 percent for the original sequence to more than 40 percent when noise reduction is used. This fact is also illustrated in Tables 3,4,5 and 6, for video sequences "Bathing Beauty" and "Mad Dash".

The RSMPE increases slightly with the increase of noise levels. As shown in Figs. 10,16,18 and 21 the mean square prediction error is influenced by noise and the slight increase in RMSPE is due to the noise itself. These results are also confirmed in Tables 1 to 6.

Several conclusions could be drawn from the previous discussions:

1) For a properly chosen and adjusted displacement estimator, the effect of noise on RMSPE can be made minimal.

2) Noise will have a significant contribution to the bit rate for the coding application, this is illustrated for entropy measurements and percentage of predictable picture area. Therefore, noise reduction is essential to achieve efficient data rate reduction [19].

Process Type	Signal Type	Entropy bits/pel	RMSPE	Percentage Predictable
Previous Frame	Composite	6.81	42.48	0.039
	Luminance	6.76	41.82	0.039
Algorithm I	Composite	5.79	22.84	0.077
	Luminance	5.71	22.22	0.082
Algorithm II	Composite	6.13	29.89	0.071
	Luminance	6.08	29.51	0.074
Algorithm III	Composite	5.99	27.45	0.073
	Luminance	5.91	26.70	0.079
Algorithm IV	Composite	5.83	24.00	0.075
	Luminance	5.71	22.47	0.082
Algorithm V	Composite	5.92	25.59	0.073
	Luminance	5.84	24.39	0.077

Table 1. Comparison between the composite and the component approaches

Video Sequence	:	Moving Letters	K_m	:	4
Nonlinearity	:	none	Sampling frequency	:	4 f_{sc}
Number of fields	:	36	Block size	:	16 x 8

Process Type	Signal Type	Entropy bits/pel	RMSPE	Percentage Predictable
Previous Frame	Composite	5.86	42.36	30.2
	Luminance	5.77	41.71	31.5
Algorithm I	Composite	4.33	22.36	46.6
	Luminance	4.16	21.26	49.1
Algorithm II	Composite	4.84	29.83	43.1
	Luminance	4.17	28.87	43.0
Algorithm III	Composite	4.62	27.2	44.6
	Luminance	4.49	26.53	46.8
Algorithm IV	Composite	4.42	24.14	46.2
	Luminance	4.18	21.95	49.3
Algorithm V	Composite	4.57	25.99	44.6
	Luminance	4.39	24.06	46.9

Table 2. Comparison between the composite and the component approaches

Video Sequence	:	Moving Letters	K_m	:	4
Nonlinearity	:	none	Sampling frequency	:	4 f_{sc}
Number of fields	:	36	Block Size	:	16 x 8

Process Type	Signal Type	Entropy bits/pel	RMSPE	Percentage Predictable
Previous Frame	Composite	2.88	6.77	59.7
	Luminance	2.89	7.0	59.7
Algorithm I	Composite	1.23	2.25	83.8
	Luminance	1.08	2.03	86.0
Algorithm II	Composite	1.35	3.00	82.8
	Luminance	1.29	2.96	83.6
Algorithm III	Composite	1.13	2.38	85.6
	Luminance	0.97	2.26	87.9
Algorithm IV	Composite	1.04	1.97	86.6
	Luminance	0.87	1.83	89.0
Algorithm V	Composite	1.16	2.44	85.2
	Luminance	1.04	2.01	87.4

Table 3. Comparison between the composite and the component approaches

Video Sequence	:	Bathing Beauty	K_m	:	4
Nonlinearity	:	3,10,0.1	Sampling frequency	:	$4\ f_{sc}$
Number of fields	:	36	Block size	:	16 x 8

Process Type	Signal Type	Entropy bits/pel	RMSPE	Percentage Predictable
Previous Frame	Composite	4.63	7.23	0.089
	Luminance	4.69	7.50	0.082
Algorithm I	Composite	3.64	3.27	18.0
	Luminance	3.49	3.00	19.4
Algorithm II	Composite	3.69	3.65	17.8
	Luminance	3.60	3.57	18.7
Algorithm III	Composite	3.62	3.50	18.1
	Luminance	3.52	3.37	19.2
Algorithm IV	Composite	3.57	3.28	18.4
	Luminance	3.45	3.10	19.5
Algorithm V	Composite	3.65	3.57	18.0
	Luminance	3.57	3.48	18.8

Table 4. Comparison between the composite and the component approaches

Video Sequence	: Bathing Beauty	K_m	: 4
Nonlinearity	: none	Sampling frequency	: 4 f_{sc}
Number of fields	: 36	Block size	: 16 x 8

Process Type	Signal Type	Entropy bits/pel	RMSPE	Percentage Predictable
Previous Frame	Composite	2.19	6.50	73.2
	Luminance	1.92	4.93	75.6
Algorithm I	Composite	1.54	4.77	82.0
	Luminance	1.20	2.94	85.3
Algorithm II	Composite	1.60	4.84	81.2
	Luminance	1.25	3.03	84.8
Algorithm III	Composite	1.53	4.73	82.2
	Luminance	1.14	2.80	86.2
Algorithm IV	Composite	1.53	4.76	82.2
	Luminance	1.17	2.86	85.8
Algorithm V	Composite	1.59	4.89	81.3
	Luminance	1.27	3.15	84.5

Table 5. Comparison between the composite and the component approaches

Video Sequence	:	Mad Dash	K_m	:	4
Nonlinearity	:	3,10,0.1	Sampling frequency	:	4 f_{sc}
Number of fields	:	36	Block size	:	16 x 8

Process Type	Signal Type	Entropy bits/pel	RMSPE	Percentage Predictable
Previous Frame	Composite	4.31	6.89	13.5
	Luminance	4.25	5.55	11.1
Algorithm I	Composite	3.79	5.15	22.7
	Luminance	3.59	3.72	21.7
Algorithm II	Composite	3.84	5.3	22.0
	Luminance	3.64	3.72	21.1
Algorithm III	Composite	3.80	5.19	22.2
	Luminance	3.58	3.57	21.5
Algorithm IV	Composite	3.79	5.24	22.6
	Luminance	3.57	3.63	21.9
Algorithm V	Composite	3.82	5.29	22.4
	Luminance	3.64	3.89	21.5

Table 6. Comparison between the composite and the component approaches

Video Sequence : Mad Dash K_m : 4
Nonlinearity : none Sampling frequency: 4 f_{sc}
Number of fields : 36 Block size : 16 x 8

3) Among the algorithms examined,Algorithm II is the least sensitive to noise. This point is illustrated by comparing Fig. 13 and Fig. 18 which show the behaviour of Algorithms I and II as the value of K_m is varied. Therefore, in situations where noise in the signal varies widely, Algorithm II may be preferable. However, if noise level could be determined and the parameter K_m adjusted accordingly this advantage of Algorithm II becomes insignificant.

4.4 Comparison of Performance:

The performance of all displacement estimation algorighms for different block sizes and different video sequences is illustrated in Fig. 26 to Fig. 32 for the composite approach. After close examination of these figures the following picture emerges:

1. For Algorithms II and III, the best block size is 32 x 16. For the video sequence "Moving letters" smaller block sizes gave better results (Fig. 28 and Fig. 29) mainly due the fact that these smaller block sizes matched the letter sizes that exist in this sequence. Since this situation is not common, the larger block size (32 x 16) is preferable.
2. For Algorithms IV and V, block size of 16 x 8 gave the best results.
3. Taking algorithm complexity into consideration, Algorithm III performance does not justify serious consideration for implementation.
4. Algorithm IV gave the best overall performance.
5. Algorithm I, II and V demonstrated a comparable performance.

5. SOME COMPLEXITY ISSUES

Algorithm III which is a combination of Algorithms I and II is discarded as its performance does not justify the added complexity needed for its implementation. Algorighm II, IV and V require memories to store previous field (or frame) displacement estimates. Algorithm I does not require such memory. Algorithm II requires two divisions per block and four memories to store the intermediate summation products. The size of these memories is a function of the block size and the accuracy of representation of the displacement estimates.

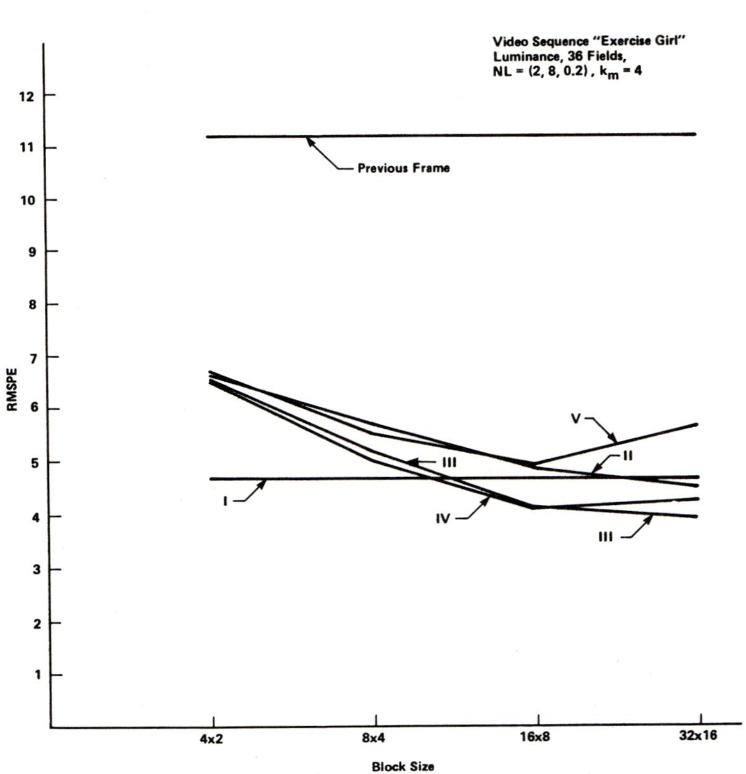

Fig. 26. Comparison Between Different Algorithms for Video Sequence "Exercise Girl"

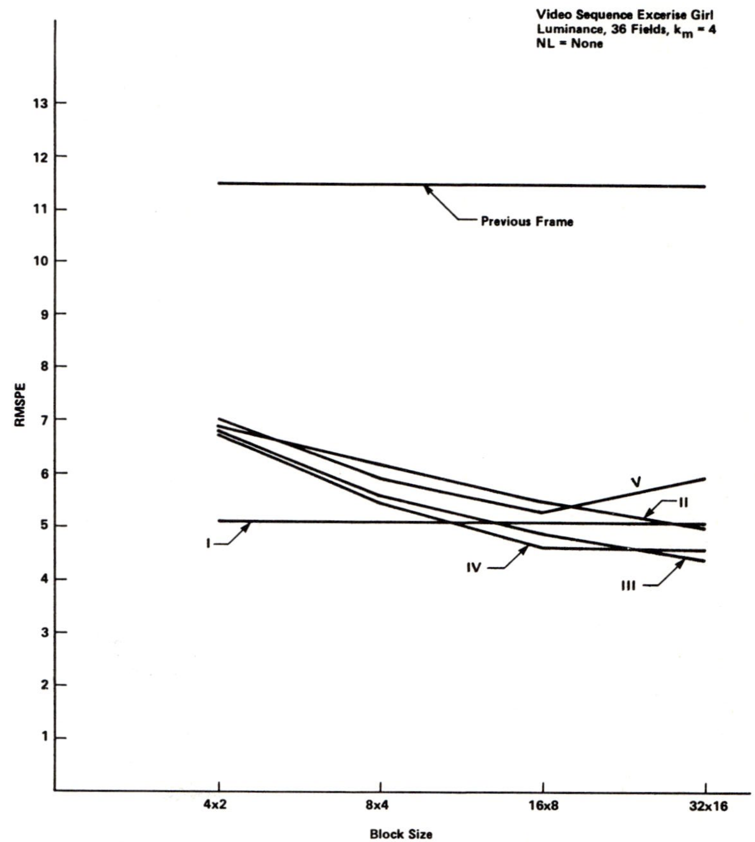

Fig. 27. Comparison Between Different Algorithms for Video Sequences "Exercise Girl"

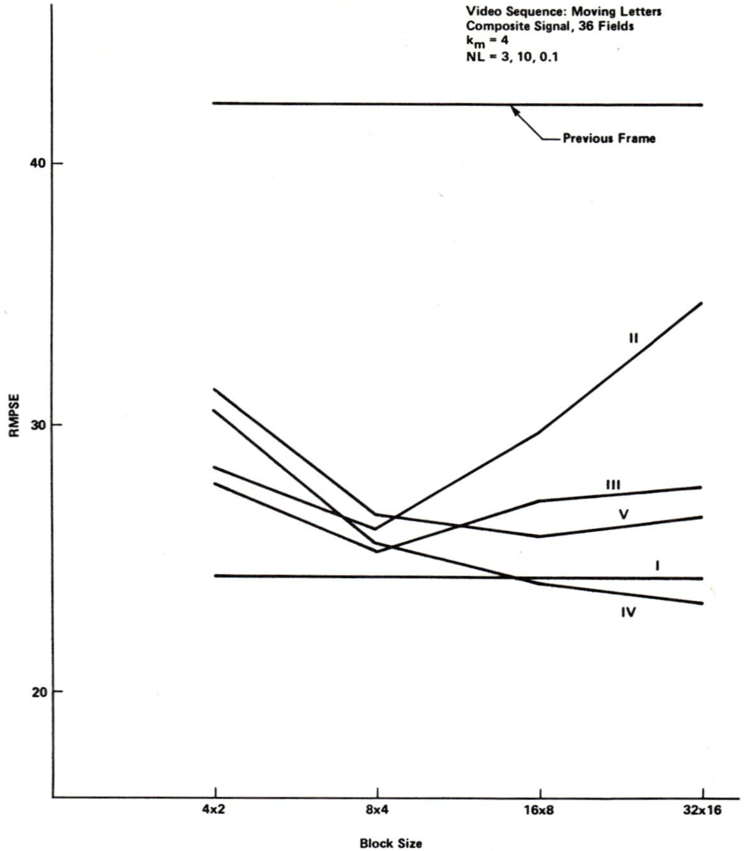

Fig. 28. Comparison Between Different Algorithms for Video Sequence "Moving Letters"

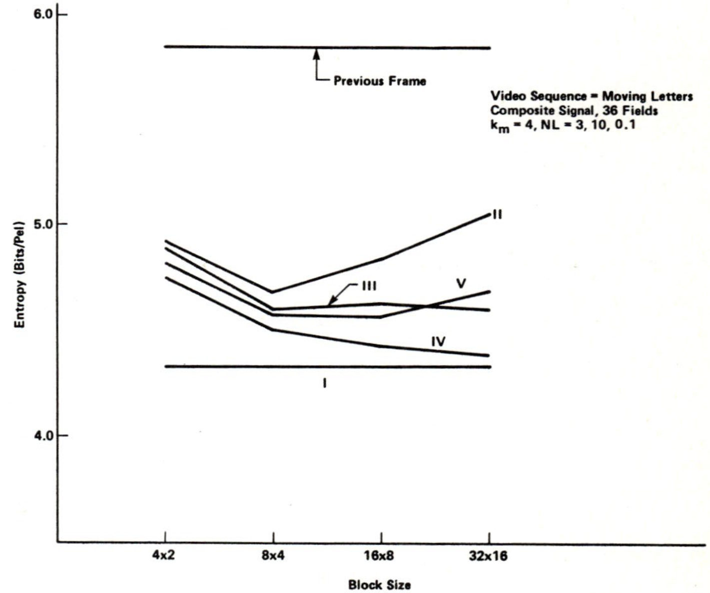

Fig. 29. Comparison Between Different Algorithms for Video Sequence "Moving Letters"

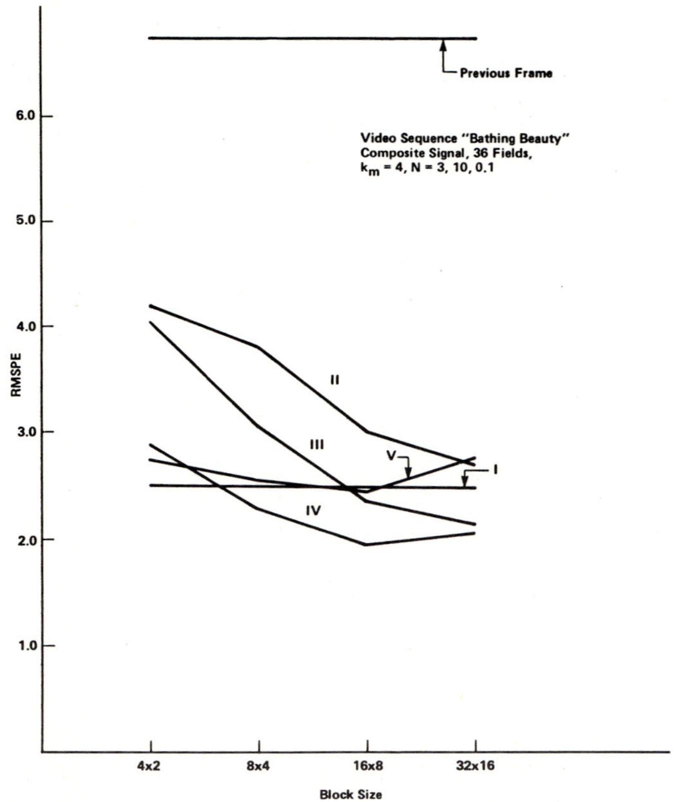

Fig. 30. Comparison Between Different Algorithms for Video Sequence "Bathing Beauty"

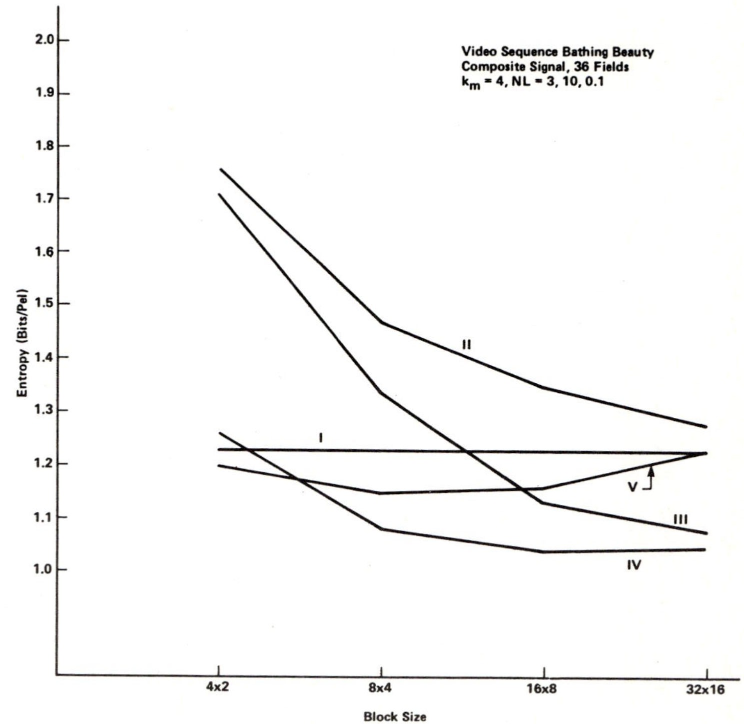

Fig. 31. Comparison Between Different Algorithms for Video Sequence "Bathing Beauty"

For the component and composite approaches, it is evident that the composite approach requires less memory than the component approach. In addition, the composite predictor can be less complex (lower order) than the demodulation filters needed to separate the composite signal into its components. However, for the composite approach, the composite predictor is placed in the feedback loop of the DPCM (or the recursive temporal filter). This fact makes the timing and speed considerations more stringent for this case.

6. CONCLUSIONS

In this paper an approach for composite prediction of colour NTSC video signals has been presented. It has been shown that the approach is applicable to both standard interframe and movement-compensated predictions. Several displacement estimation algorithms have been presented and discussed. A comparative evaluation of performance for both the composite and component approaches has been presented. Some of the main conclusions can be stated as follows:

1. For interframe prediction, the composite prediction approach gives a comparable performance to the component approach provided that a proper sampling frequency is selected ($4^* f_{sc}$ or $2^* f_{sc}$)
2. The gains resulting from utilizing the movement compensation techniques are comparable for both the component and composite approaches.
3. Among the displacement estimation techniques discussed in this paper Algorithm IV, which is based on a modification of the pel recursive method, gives the best overall performance. This result is true for both the composite and component approaches.

ACKNOWLEDGMENTS

The author would like to thank his colleagues Dr. B. Prasada and Dr. E. Dubois for the discussions we had during the course of this work and also for their comments and suggestions during the preparation of this paper.

APPENDIX

Several broadcast TV sequences have been used in the simulation work reported in this paper. These sequences were digitized and stored on BNR/INRS Digital Video Sequence Store (DVS) image processing facility [14]. The sequences are sampled in composite format at 14.3 MHz ($4^* f_{sc}$) and quantized uniformly using 8 bits per picture element.

Description of the content different sequences is given below.

(*i*) Exercise Girl: contains a picture of a girl exercising with medium amount of motion.

(*ii*) Bathing Beauty: sequence contains a complex zoom from an angle

(*iii*) Moving letters: sequence contains electronically generated letters moving in the vertical direction, while the background contains two persons moving while the camera is panning.

(*iv*) Mad Dash: sequence with very high colour saturation and chrominance activities.

REFERENCES

[1] A.N. Netravali and J.D. Robbins, "Motion-Compensated Television Coding: Part l," Bell System Technical Journal, 58, March 1979, pp. 631-670.

[2] A.N. Netravali and J.D. Robbins, "Motion-Compensated Coding: Some New Results," Bell System Technical Journal, 59, November 1980, pp. 1735-1745.

[3] J.D. Robbins and A.N. Netravali, "Interframe Television Coding using Movement Compensation," ICC 79 Conference Record, Vol. 1, pp. 23.4.1-23.4.5, 1979.

[4] E. Dubois, "A Multimode Interframe Coder Using Movement Compensation," presented at the 1981 Picture Coding Symposium, Montreal, Canada, June 3-5, 1981.

[5] S. Brofferio et. al., "Redundancy Reduction of Video signals Using Movement Compensation," Alta Frequenza, Vol. 43, pp. 836-843, October 1974.

[6] S. Brofferio and F. Rocca, "Interframe Redundancy Reduction of Video Signals Generated by Translating Objects," IEEE Trans. on Communication, Vol. COM-25, pp. 448-455, April 1977.

[7] C. Cafforio and F. Rocca, "Methods for Measuring Small Displacements of Television Images", IEEE Transactions on Information Theory, IT-22, September 1976, pp. 573-579.

[8] C. Cafforio and F. Rocca, "Tracking of Moving Objects in Television Images," Signal Processing, 1, 1979, pp. 133-140.

[9] E. Dubois, B. Prasada and S. Sabri, "Image Sequence Coding, in Image Sequence Analysis," (T.S. Huang Ed.), pp. 229-287, Springer-Verlag, Berlin, 1981.

[10] C. Cafforio and F. Rocca, "Detection and Tracking of Moving Objects in Television Images," in Report of the International Workshop on Image Processing: Real-time Edge and Motion Detection/ Estimation, Document technique CCETT: CTN/T/1/80, pp. 19.1-19.10, September 1979.

[11] T.S. Huang and Y.P. Hsu, "Image Sequence Enhancement, in Image Sequence Analysis", (T.S. Huang Ed.), pp. 289-309, Springer-Verlag, Berlin, 1981.

[12] J.O. Limb and J.A. Murphy, "Estimating the Velocity of Moving Images in Television Signals", Computer Graphics and Image Processing, 4, 1975, pp. 311-321.

[13] Y. Ninomiya, "Motion Correction for Interframe Coding Systems", Image Engineering Group Meeting, Record IE-78-6 (in Japanese).

[14] R. Johnston, J. Mastronardi and G. Mony, "A Digital Television Sequence Store," IEEE Transactions on Communications, COM-26, May 1978, pp. 594-600.

[15] J.O. Limb, C.B. Rubinstein, and J.E. Thompson, "Digital Coding of Color Video Signals", IEEE Transactions on Communications, vol. COM-25, No. 11, November 1977, pp. 1349-1385.

[16] Y. Hatori and H. Yamamoto, "Direct Frame to Frame Coding of NTSC Color Television Signals (Simulation Results)", Institute of Television Engineers, (Japan) Monograph, September 1975.

[17] T. Ishiguro, et. al. "Composite Interframe Coding of NTSC Color Television Signals", National Telecommunications Conference, Dallas, Texas, December 1976.

[18] S. Sabri and E. Dubois, "Noise Reduction in TV Imagery Using Motion Compensation Techniques", to be published

[19] S. Sabri and B. Prasada, "Multi-mode Interframe Video Coders for Broadcast TV Signals at 15-20 *Mb/s*", to be published.

[20] S. Sabri, "Movement-Compensated Interframe Coding and Noise Reduction for Broadcast Television Signals", Final Report prepared

by Bell-Northern Research for INTELSAT under contract INTEL-114, May 1981.

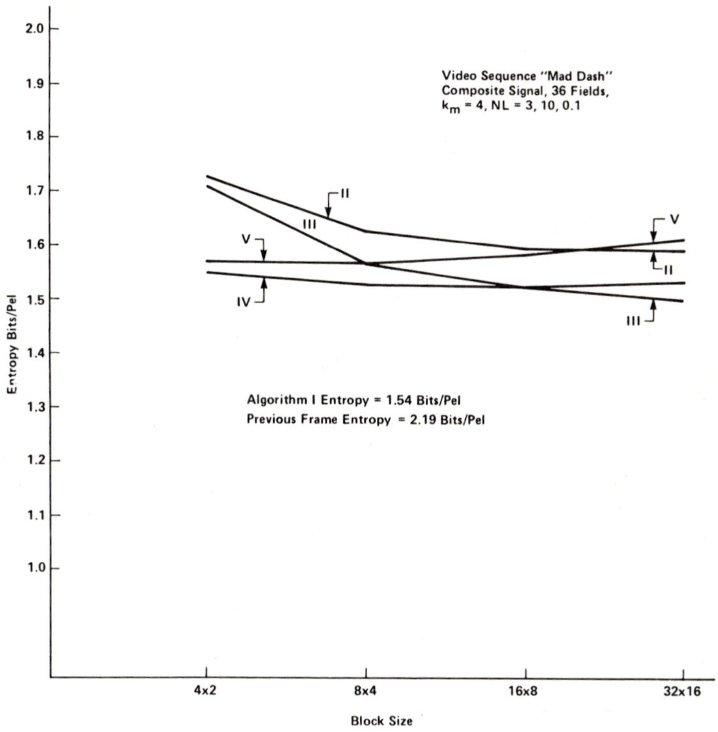

Fig. 32. Comparison Between Different Algorithms for Video Sequence "Mad Dash"

CODING OF COLOUR TV SIGNALS WITH 34 MBIT/S TRANSMISSION RATE

H. G. Musmann

Lehrstuhl für Theoretische Nachrichtentechnik
und Informationsverarbeitung
Universität Hannover
Callinstr. 32
3ooo Hannover 1
West Germany

ABSTRACT

A digital coding scheme for colour television signals is presented that uses a transmission rate of 34 Mbit/s and does not show any picture quality degradations when compared with pictures of the 625-line PAL standard. The luminance and chrominance signals are coded separately by extended differential pulse-code modulation (DPCM). For the example of the luminance decoder an implementation in ECL gate array technology is described.

1. INTRODUCTION

For digital processing of colour television signals in the studio separate coding of the luminance signal Y and the chrominance signals R-Y, B-Y is recommended as a standard /1/. Pulse-code modulation (PCM) of these component signals results in a bit rate of 216 Mbit/s if the proposed sampling frequencies of 13.5 MHz, 6.75 MHz, 6.75 MHz and a uniform quantization with 256 quantizing levels corresponding to 8 bit per sample are applied. By the choice of these relatively high sampling frequencies a head room has been provided for possible distortions introduced by digital signal processing in the studio.

Outside the studio a digital broadcast colour television signal must at least meet the picture quality requirements of the today's analogue PAL standard /2,3/. There a source coding technique /4/ may be applied in order to save storage or transmission channel capacity. Considering these quality requirements the bit rate of the digital colour television signal can be reduced to 34 Mbit/s which is the data rate of

NATO ASI Series, Vol. F2
Image Sequence Processing and Dynamic Scene Analysis
Edited by T.S. Huang

the third stage of the European PCM hierarchy.

To reach this goal the PCM standard of the studio is first converted into a new PCM representation with reduced sampling frequencies which are closely adapted to the signal bandwidths of the analogue PAL standard. According to the specifications presented in /2,3/ the required bandwidth is 4.5 MHz for the luminance and 1.o MHz for the chrominance components. With respect to a simple conversion, the ratio of the new and the studio sampling frequency should be a fraction m/n with small integers m,n /5/. Table 1 shows two possible PCM representations A,B that meet these requirements and that will be used here for coding a colour television signal with 34 Mbit/s data rate.

	STUDIO	TRANSMISSION	
		A	B
SAMPLING FREQUENCY IN MHZ			
Y	13.5	10.125(3/4)	9.00(2/3)
R-Y, B-Y	6.75	3.375(1/2)	2.25(1/3)
SAMPLES PER ACTIVE LINE			
Y:R-Y:B-Y	720:360:360	540:180:180	480:120:120
CODE BITS PER SAMPLE	8	8	8
PCM BITRATE IN MBIT/S	216	135	108

Table 1. PCM representations for the studio and for transmission of colour television signals.
The fractions in the brackets indicate the ratios m/n of the proposed and the studio sampling frequency

For mainly two reasons predictive coding is preferred to transform coding techniques here, in order to reduce the PCM data rates. The implementation of predictive coding techniques as e.g. DPCM is less complex. Furthermore, useful methods have been developed to adapt the parameters of a DPCM system to the properties of the human visual perception /6/ for improving the data reduction. Especially, a DPCM system can be optimized to reach a minimum bit rate without exceeding the visibility threshold of the quantizing error. Starting with the PCM representations shown in Table 1, the DPCM coding is optimized using the design technique by Pirsch /7,8/ and then extended by an adaptive quantizer. After a discussion of the obtained results and of the possible solutions for a 34 Mbit/s codec in chapter 3 the realization of a DPCM decoder in gate array technology is outlined in chapter 4.

2. OPTIMIZATION OF THE DPCM CODING FOR THE LUMINANCE AND CHROMINANCE SIGNALS

Figure 1 shows the block diagram of a DPCM system with a PCM input signal. In a DPCM system the difference e between a prediction value $\hat{s}$ and the amplitude of the actual picture element s is quantized, coded and transmitted. In the past the predictor and the quantizer have been optimized to yield a minimum mean square prediction error e and a minimum mean square quantizing error $q = e' - e$ for a given number of quantizing levels /9/.

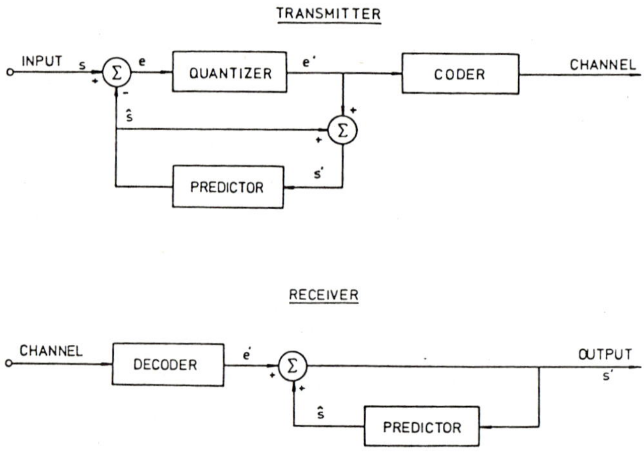

Fig. 1. Block diagram of a DPCM system

In case of the minimum mean square error (MMSE) criterion the optimum coefficients a_j of a linear predictor with

$$\hat{s}_o = \sum_{j=1}^{N} a_j \, s_j \tag{1}$$

can be calculated from the statistics of a television signal. Table 2 presents the results obtained by using the sampling structures shown in Figure 2.

PREDICTION COEFFICIENTS a_1	a_2	a_3	a_4	SAMPLING FREQUENCY IN MHz	DESIGN CRITERION
$\frac{7}{8}$	$-\frac{5}{8}$	$\frac{6}{8}$	–	10	MINIMUM VARIANCE OF PREDICTION ERROR
$\frac{6}{8}$	$-\frac{3}{8}$	$\frac{5}{8}$	–	8.867	
$\frac{4}{8}$	$\frac{1}{8}$	$\frac{2}{8}$	$\frac{1}{8}$	10	MINIMUM NUMBER OF QUANTIZER LEVELS
$\frac{2}{4}$	$\frac{1}{4}$	$\frac{1}{4}$	–	8.867	

Table 2. Optimum coefficients for a predictor with minimum variance of the prediction error and for a predictor with minimum number of quantizing levels

With respect to the CCIR-Recommendation /1/ the sampling structure should be orthogonal also for a 9 MHz sampling rate in a future coding system.

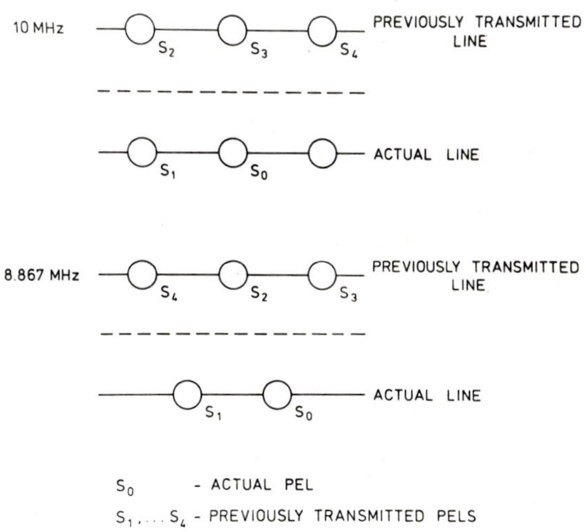

Fig. 2. Sampling structures and position of picture elements $s_o, \ldots s_4$

To design a quantizer with minimum number of quantizing levels and without visible quantizing errors the masking function M_o is required. The masking M_o depends on the luminance activity A_o and can be interpreted as the maximum tolerable quantizing error. Figure 3 shows a masking function where the absolute value of the prediction error is used as a measure for the luminance activity

$$A_o = \left| \sum_{j=1}^{4} a_j \, (s_o - s_j) \right| = |e_o| \tag{2}$$

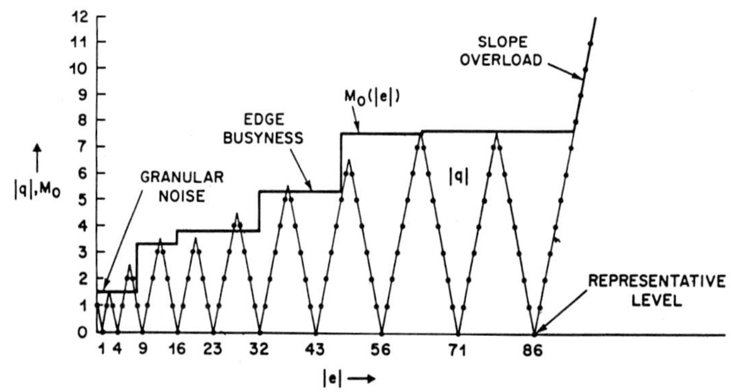

Fig. 3. Masking function M_o versus luminance activity $|e|$ in units 1 out of 256 (from Pirsch /8/)

This special masking function allows a simple geometrical construction of the optimum quantizer. The construction algorithm has been developed by Sharma /1o/ and later on modified by Pirsch /8/. In Figure 3 the zigzag line represents the magnitude of the quantization error as a function of the quantizer input signal e. The inclination of the quantizing error line is $\pm$ 45° when abscissa and ordinate are represented in the same scale 1 out of 256. The construction of the quantizer begins in the overload area of the masking function. The highest representative quantizing level results from the intersection of a line of + 45° inclination with the abscissa, whereby this line is positioned in such a way that the quantizing error does not exceed the masking. Starting from this representative level a line of inclination -45° is drawn. Its intersection with the masking function gives the decision level to the neighbouring quantizing interval and so on. In this way a quantizer is constructed with a minimum number of quantizing levels that does not exceed the visibility threshold for the quantizing error as given by the masking function.

From the construction algorithm it is obvious that the required number of quantizing levels decreases if the level or the slope of the masking function increases. Thus Figure 4 shows that a predictor with only positive coefficients according to Table 2 is superior to a predictor with minimum variance. Further it can be seen that for a given activity a computer generated resolution chart renders less masking than a natural test picture and therefore requires more quantizing levels.

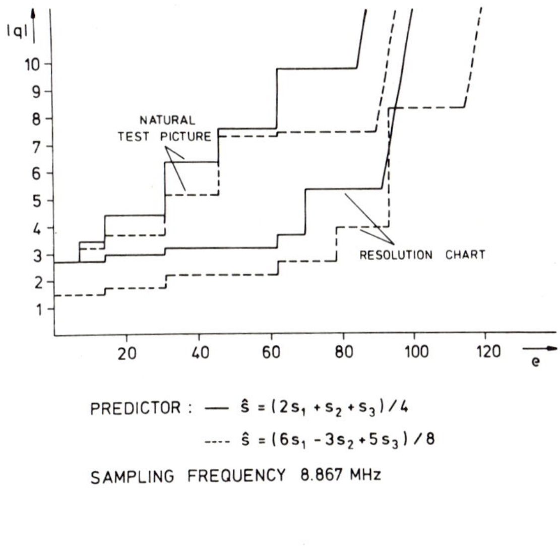

Fig. 4. Masking functions measured for a predictor with minimum variance and for a predictor with subjectively optimized only positive coefficients

Table 3 presents several quantizers which have been found using the described construction algorithm. The predictor with subjectively optimized coefficients allows to reduce the number of quantizing levels from 2o to 17 for 9 MHz and from 15 to 13 for 1o MHz sampling frequency when compared to a predictor with minimum variance.

PREDICTOR NO.	SAMPLING FREQUENCY	NO. OF LEVELS	POSITIVE REPRESENTATIVE LEVELS	PICTURE TYP
1	10 MHz	15	0,5,10,17,28,39,52,67	NATURAL TEST PICTURE
3	10 MHz	13	0,7,16,27,38,51,66	
2	8.867 MHz	20	1,4,9,16,23,32,43,56,71,86	
4	8.867 MHz	17	0,5,10,17,26,35,46,61,78	
1	10 MHz	29	0,5,10,15,22,29,36,43,50,57,64,73,82,93,104	RESOLUTION CHART
3	10 MHz	21	0,5,12,19,26,33,40,47,54,61,72	
2	8.867 MHz	46	1,4,7,10,13,16,19,22,25,28,33,38,43,48,53,58, 63,68,73,80,89,98,111	
4	8.867 MHz	27	0,5,10,15,20,27,34,41,48,55,62,69,78,89	

Table 3. Quantizers for the luminance signal

1,2 Predictors with minimum variance of the prediction error

3,4 Subjectively optimized predictors with only positive coefficients according to Table 2

Considering the masking contributed by the luminance activity in the surrounding of the actual picture element s_o yields an additional reduction from 17 to 16 quantizing levels in the case of 9 MHz sampling rate. However, this requires the application of an adaptive quantizer, whose quantizing characteristic is adapted to the changing luminance activity in the surrounding of the picture elements to be coded /6,8/. The maximum luminance difference between the neighbouring picture elements $s_1, \ldots s_4$ can be used as a measure for this luminance activity

$$A_n = \max |s_i - s_j| \qquad i,j = 1,2,3,4 \tag{3}$$

Figure 5 shows the masking M_n contributed by A_n. The dominating component M_o or M_n determines the masking M for s_o

$$M = \max \{M_o, M_o\} \tag{4}$$

From the composite masking function the quantization characteristics for an adaptive quantizer can be constructed /6/. In Figure 5 three quantizing characteristics for luminance activities of $A_n<32$, $32<A_n<64$, $64<A_n$ are presented. Depending on the actual value A_n always one quantizer out of these three quantizers is switched in.

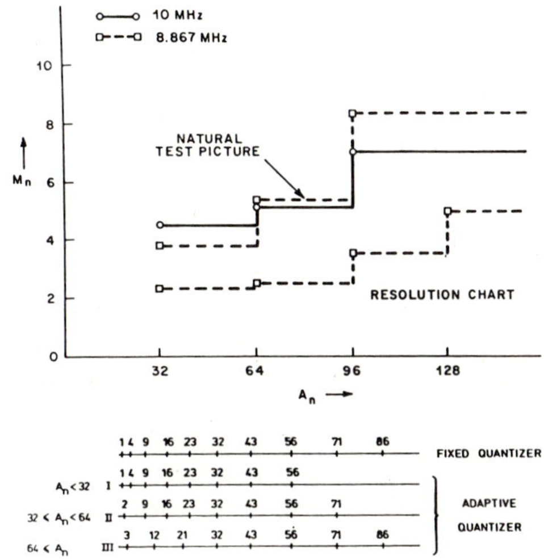

Fig. 5. Masking function M_n versus luminance activity A_n and quantizing characteristics for an adaptive quantizer

In case of the chrominance signals after vertical filtering only every second line is transmitted as described in /8/. The parameters of the optimized chrominance predictor and quantizer are presented in Table 4.

PREDICTION COEFFICIENTS a_1	a_2	a_3	SAMPLING FREQUENCY IN MHz	DESIGN CRITERION
$\frac{3}{4}$	$-\frac{2}{4}$	$\frac{3}{4}$	2.217	MINIMUM VARIANCE OF PREDICTION ERROR

NO. OF LEVELS	POSITIVE REPRESENTATIVE LEVELS	PICTURE TYPE
8	1,4,11,28	NATURAL TEST PICTURE

Table 4. Optimized predictor and quantizer for the chrominance signals R-Y, B-Y

To assess the picture quality of a coding system the viewing conditions according to CCIR-Recommendation 5oo have to be considered. Photos of a displayed resolution test chart, as shown in Figure 6, allow only a restricted comparison of picture qualities in a paper.

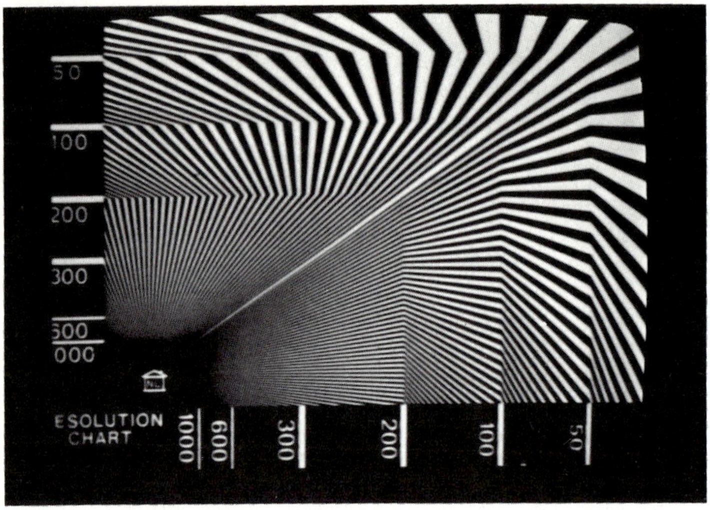

Fig. 6. Television picture of the PCM coded resolution test chart. Sampling frequency 1o MHz

Figure 7 demonstrates the advantages of a DPCM coding system with a subjectively optimized predictor for an enlarged critical section of the resolution test chart. From this photo it can not be recognized that a 13-level quantizer still produces a visible background noise in these areas with diagonal high-frequency structures. Table 3 indicates that 21 quantizing levels would be required to avoid the visible noise in the resolution chart. However, if this picture quality of a component coded colour television signal is compared with that of the composite PAL standard, which is shown in Figure 8, the described remaining background noise is much less annoying than the non-stationary cross colour distortions of a PAL signal. Both distortions appear in the same areas.

(a)

(b)

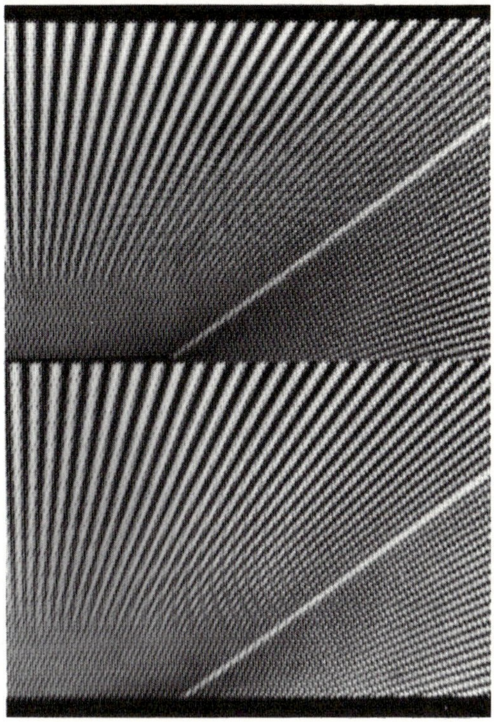

Fig. 7. Enlarged critical section of the DPCM coded resolution test chart. Sampling frequency 1o MHz

(a) Predictor 1 with minimum variance, 15 quantizing levels

(b) Subjectively optimized predictor 3, 13 quantizing levels

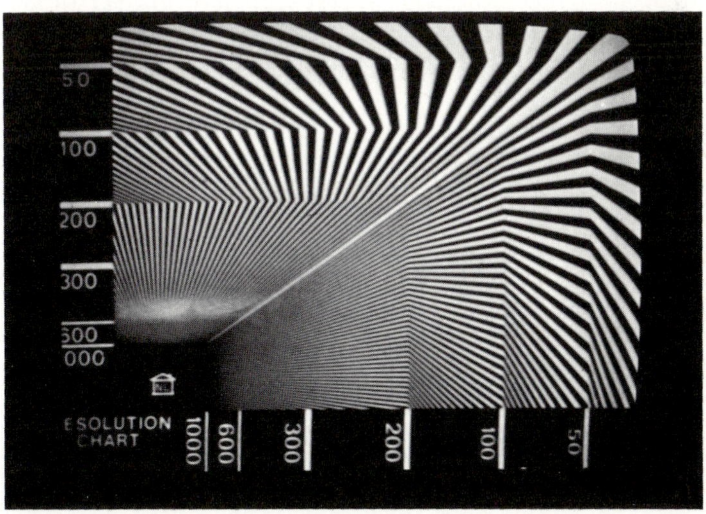

Fig. 8. Television picture of the resolution test chart demonstrating the cross colour distortion of the PAL standard

3. CODING WITH 34 MBIT/S TRANSMISSION RATE

The described optimized DPCM coding techniques yield the transmission bit rates given in Table 5 when applied to the PCM signals as defined in Table 1. For calculating these bit rates a line alternating transmission of the chrominance signals has been assumed.

VIDEO SIGNAL	SAMPLING FREQUENCY	REQUIRED NUMBER OF QUANTIZING	BITS PER SAMPLE	TRANSMISSION RATE IN MBIT/S
U_Y	10.125	13 (3.7 BIT)	4	38.88
U_{R-Y}, U_{B-Y}	3.375	8 (3.0 BIT)	3	
U_Y	9.00	17 (4.1 BIT)	4*	32.83
U_{R-Y}, U_{B-Y}	2.25	8 (3 BIT)	3	

* REALIZED BY AN ADAPTIVE QUANTIZER

Table 5. Transmission bit rates of a digital colour video signal encoded by the described DPCM techniques (Line alternating transmission of the chrominance signals)

In addition to the video signal a digital sound signal of 1o24 Mbit/s rate has to be transmitted and a rate of about 512 kbit/s must be reserved for synchronization etc. Thus in a standard PCM channel of 34.368 Mbit/s only a rate of 32.832 Mbit/s is available for transmitting the video signal.

Table 5 demonstrates that the desired transmission rate of 34 Mbit/s for a complete television signal is achieved only in the case of 9 MHz, 2.25 MHz, 2.25 MHz sampling rate and by the use of an adaptive quantizer. The resulting digital television picture shows no visible coding distortions when compared to pictures of the PAL standard.

In case of 1o.125 MHz, 3.375 MHz, 3.375 MHz sampling an improved chrominance resolution can be provided. However, this sampling structure requires additional techniques for coding the luminance signal to achieve a data rate of 34 Mbit/s. The quantization of the luminance samples with 13 levels according to Table 3 corresponds to 3.7 bit coding per sample. There are the following two possibilities to reduce this rate to 3.2 bit per sample to allow a transmission with 34 Mbit/s. A first solution could be block coding whereby the DPCM code words of one line are represented by variable length code words and are transmitted within one block of constant length /11/. A second possible way is interframe prediction. Both approaches are investigated at the present time and it has to be proven, if the given picture quality requirements can be fulfilled in this way.

4. DESIGN OF A DPCM LUMINANCE DECODER IN GATE ARRAY TECHNOLOGY

Because of the relatively high sampling frequency of the luminance signal only gate arrays in ECL technology can be used at the present to realize the recursive system of a DPCM decoder. The main parameters of an ECL gate array are presented in Table 6.

Number of equivalent gates	96o
Number of cells	36
Delay time of a gate	o.5 ns
Maximum dissipation power	2.3 W
Number of package pins	64

Table 6. Parameters of a gate array in ECL technology

The pre-manufactured gates are arranged in form of cells. Only the wiring has to be designed and realized. Figure 9 shows the geometrical structure of a gate array with 36 cells as an example. The input-output drivers are positioned at the borders of the chip.

A design has been made for the case of a one-dimensional and a two-dimensional predictor with coefficients according to Table 7. This table also presents some dates of the layout which indicate that a DPCM decoder with one-dimensional predictor can be realized using only one gate array. In case of a two-dimensional predictor two gate arrays and an additional line memory are required. When a CMOS technology with 1 ns delay time is available in the future, this predictor could also be implemented in one gate array /12/.

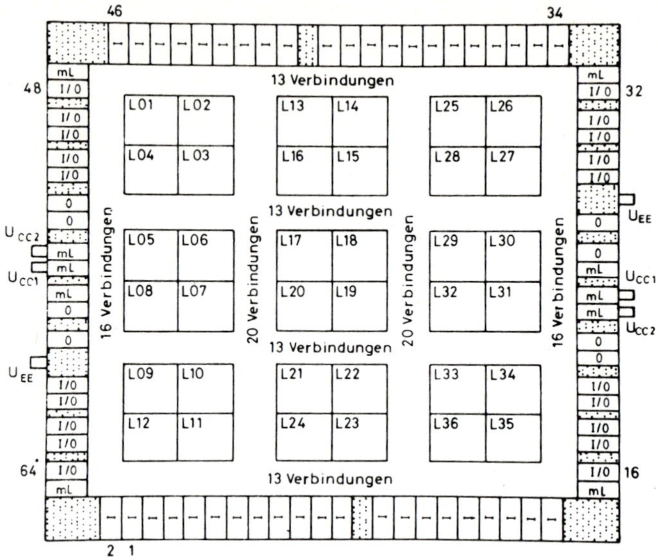

Fig. 9. Structure of a gate array with 36 cells

	ONE-DIMENSIONAL PREDICTOR	TWO-DIMENSIONAL PREDICTOR
COEFFICIENTS	31/32	1/2 1/8 1/4 1/8
NO. OF GATES IN THE LOOP	41	40
DELAY TIME OF THE LOOP	20 NS	20 NS
TOTAL NUMBER OF GATES	332	693 *
REQUIRED NUMBER OF CELLS	27	28 + 24
POWER DISSIPATION	1.59 W	1.66 W + 1.44 W

* WITHOUT LINE BUFFER

Table 7. Parameters of a DPCM decoder in gate array technology

5. CONCLUSION

A digital intraframe DPCM coding technique for colour television signals is described that reduces the transmission rate down to 34 Mbit/s and does not show any picture quality degradation when compared to pictures of the composite PAL standard. By the use of constant word length coding a uniform output data rate of the coding system is maintained. For an evaluation of the implementation expense a design of the luminance decoder demonstrates that depending on the technology such a decoder can be realized by the use of only 3 or 1 gate array chip.

To achieve the relatively low bit rate of 34 Mbit/s without generating visible distortions, the effects of masking have been exploited to an extend that does not allow to introduce further quantizing noise. Also, subsequent processing of the digital video signal might lead to visible distortions if rounding or truncation algorithms are applied. Therefore the described coding technique appears useful for efficient digital recording or digital transmission of television signals where no main subsequent processing is expected as e.g. in video conferencing and video telephone or in digital broadcasting systems. The coding technique could be used to improve the picture quality of a video cassette recorder by digital recording /13/.

6. REFERENCES

/1/ CCIR-Recommendation AA/11 (MODF): Encoding parameters of digital television for studios, October 1981

/2/ G. Wengenroth: "Pass-bands for television signal components considered for digitized coding", EBU Doc COM.T. (C1) 59, Oct.76

/3/ CCIR-Study Groups, Document 11/18-E: "Necessary bandwidth for digitally encoding the luminance component of a 625-line television signal", May 198o

/4/ A.N. Netravali and J.O. Limb, "Picture Coding: A review", IEEE Trans. Commun., Vol.68, pp.366-4o6, March 198o

/5/ CCIR-Study Groups, Document 11/243-E: "Changing the sampling rate of the video signals by non-binary numbers", May 1981

/6/ H.G. Musmann, "Predictive Coding" in Image Transmission Techniques, Ed. by W.K. Pratt, Academic Press, New York 1979

/7/ P. Pirsch, "A new predictor design for DPCM coding of TV signals", ICC Conference Record, pp.31.2.1-31.2.5, Seattle 198o

/8/ P. Pirsch, "Design of DPCM quantizers for video signals using subjective tests". IEEE Trans. Commun., Vol.29, pp.99o-1ooo, July 1981

/9/ J.B. O'Neal, "Predictive quantizing systems", Bell Syst. Techn. Journ., Vol.45, pp.689-721, 1966

/1o/ D.K. Sharma, "Design of absolutely optimal quantizers for a wide class of distortion measures", IEEE Trans. Inform. Theory, Vol.24, pp.693-7o2, Nov. 1978

/11/ P. Pirsch, "Block coding of colour video signals", NTC Conference Record, pp.1o.5-1 to 1o.5-5, Los Angeles 1977

/12/ M. Axmann, "Untersuchung über die Realisierung von DPCM-Systemen für Fernsehsignale durch Gate Arrays", Diplomarbeit an der Universität Hannover, Juni 1982

/13/ U. Fiedler, D. Westerkamp, "Digital video and audio recording on a modified video cassette recorder," Proc. of the fourth Intern. Conf. on Video and Data Recording, pp.231-24o, Southampton, April 1982

ANALYSIS OF DIFFERENT DISPLACEMENT ESTIMATION ALGORITHMS FOR DIGITAL TELEVISION SIGNALS

Hans C. Bergmann

Lehrstuhl für Theoretische Nachrichtentechnik und
Informationsverarbeitung
Universität Hannover
Callinstr. 32
3 Hannover 1
West Germany

ABSTRACT

In this contribution five algorithms for two-dimensional displacement estimation in digital television scenes based on the differential method are mathematically analyzed. The estimation criterion and the mathematical formulation for each algorithm are derived. The estimation procedures and the simplifications are explained. Based on the analytic results a new estimation algorithm is derived that estimates the cross correlation peak of translatory displaced luminance signals within small image segments. This analysis shows the advantage of the new algorithm with respect to the convergency.

1. INTRODUCTION

Several algorithms which estimate the displacement of moving objects between two successive television frames have already been published. With respect to the mathematical formulation these algorithms are divided into three classes /1/:

- Fourier methods
- correlation methods
- methods of differentials.

The third class of algorithms is particularly important for real-time digital television signal processing, because, in many cases, signal transformations or matching procedures are too complex. Concerning calculation complexity the algorithms based on differential methods

NATO ASI Series, Vol. F2
Image Sequence Processing and Dynamic Scene Analysis
Edited by T.S. Huang

can be subdivided into three groups of algorithms, where

1. One displacement estimate is calculated for the entire moving area /2/.
2. The displacement estimates are calculated within the moving area for small rectangular image segments /3,4/.
3. The displacement estimates are calculated recursively pel by pel /5,7/.

When weighting the advantages and drawbacks of each, it is of interest to point out the differences between the algorithms with respect to the formulation and the simplifying assumptions. In this contribution four displacement estimation algorithms are compared on the base of a mathematical analysis. In addition a new algorithm is presented and compared to the others by some experimental results.

2. DISPLACEMENT ESTIMATION BY THE METHOD OF DIFFERENTIALS

2.1 NONRECURSIVE DISPLACEMENT ESTIMATION ALGORITHM OF LIMB AND MURPHY

Limb and Murphy /2/ have proposed an algorithm that calculates one displacement estimate for the entire moving area. The algorithm is based on the two assumptions that the television scene contains one moving object and that the movement can be described by pure horizontal translation. The moving area is determined by a simple frame difference thresholding algorithm. By a heuristic approach the sum over the frame differences within the moving area is found to be a function of the displacement. However, this function depends upon the size of the moving area and the amount of detail within. In order to eliminate these influences the sum over the frame differences is divided by the sum over the spatial luminance differences of adjacent pels, i.e. the element differences. Using this normalization, the amount $|\mu|$ of the horizontal displacement estimate is determined by

$$|\mu| = \frac{\sum\limits_{M} |\Delta_t I(m,n)|}{\sum\limits_{M} |\Delta_m I(m,n)|} \tag{1}$$

$\Delta_t I$: frame difference
$\Delta_m I$: element difference
M : moving area

In order to determine the direction of the displacement, each sum in (1) is divided into two parts in which the summation is carried out with respect to the sign of the element and frame differences. The difference of both parts specifies the estimate of the horizontal displacement

$$\mu = \frac{\sum\limits_{M} |\Delta_t I(m,n)| \cdot sb(\Delta_t I) \not\equiv sb(\Delta_m I) - \sum |\Delta_t I(m,n)| \cdot sb(\Delta_t I) \equiv sb(\Delta_m I)}{\sum\limits_{M} |\Delta_m I(m,n)| \cdot sb(\Delta_t I) \not\equiv sb(\Delta_m I) - \sum |\Delta_m I(m,n)| \cdot sb(\Delta_t I) \equiv sb(\Delta_m I)} \tag{2}$$

$$sb(\cdot) = \begin{matrix} 0 \text{ , if } (\cdot) \text{ is negative} \\ 1 \text{ , if } (\cdot) \text{ is positive} \end{matrix}$$

$\equiv$ LOGICAL EQUIVALENCE
$\not\equiv$ LOGICAL ANTIVALENCE

which has the advantage of a relatively simple hardware realization. An experimental evaluation by the authors results in a good performance for displacements of up to 2.5 pel per frame.

In Figure 1 an example for displacement estimation using this algorithm is demonstrated.

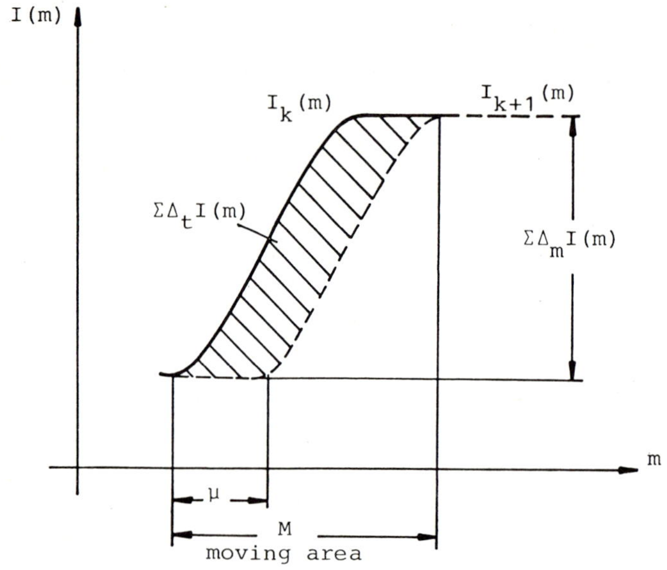

Figure 1. One-dimensional example for displacement estimation using equation (2)

The line denoted by I(m) represents the one-dimensional luminance function of a moving object, the dashed line shows the function in the succeeding frame. Using the displacement estimation algorithm (2) the frame differences are summed over the hatched part in Figure 1 and represent approximately a measure of the area of a parallelogram. Dividing this value by the height of the parallelogram, i.e. the sum over the element differences along m, we get the displacement $\hat{\mu}$.

The problem of displacement estimation is led back to the calculation of the base of a parallelogram knowing its area and height. This implies the requirement that the integral over the frame differences is a linear function in the displacement μ. In this case equation (2) gives an exact estimate for μ. However, in most cases Figure 1 does not hold for real television signals. To take into account the statistical properties of television signals the expected values should be used for an analytic evaluation. Before going into further detail the algorithm of Cafforio and Rocca is derived because it is similar to that of Limb and Murphy.

2.2 NONRECURSIVE DISPLACEMENT ESTIMATION ALGORITHM OF CAFFORIO AND ROCCA

A displacement estimation algorithm for small image segments has been proposed by Cafforio and Rocca /3/. Considering a moving object in a television scene, the two-dimensional luminance signal of two succeeding frames is denoted by

$$I_k(m,n) = I_{k-1}(m-\mu,n-\nu) \qquad (3)$$

$$\mu,\nu : \text{horizontal and vertical displacement}$$

if the movement consists of pure translation. Calculating the frame differences we get

$$\Delta_t I(m,n) = I_k(m,n) - I_{k-1}(m,n) \qquad (4)$$

and approximating (4) by the truncated Taylor series expansions we get

$$\begin{aligned} \Delta_t I(m,n) &= \mu\cdot\frac{\partial}{\partial_m} I_{k-1}(m,n) + \nu\cdot\frac{\partial}{\partial_n} I_{k-1}(m,n) + \varepsilon(m,n) \\ &\approx \mu\cdot\Delta_m I_{k-1}(m,n) + \nu\cdot\Delta_n I_{k-1}(m,n) + \varepsilon(m,n) \qquad (5) \end{aligned}$$

where the partial derivatives are approximated by the spatial differences. The expression $\varepsilon(m,n)$ accounts for the error due to the truncation of the Taylor series expansion of $I_{k-1}(m,n)$ around the location (m,n). Assuming that the error term is not correlated with the signal and has an even probability density function, the optimal estimates for μ and ν are obtained by linear regression. For example the horizontal estimate $\hat{\mu}$ is

$$\hat{\mu} = \frac{\sigma_{\Delta_t I}}{\sigma_{\Delta_m I}} \cdot \frac{\rho_{\Delta_t I \Delta_m I} - \rho_{\Delta_m I \Delta_n I}\cdot\rho_{\Delta_t I \Delta_n I}}{1 - \rho^2_{\Delta_m I \Delta_n I}} \qquad (6)$$

with the variance σ^2 and the correlation coefficient ρ

$$\sigma^2_{\Delta_m I} = \frac{1}{K} \sum_{i,j} \Delta_m I^2(m_i,n_j) \qquad (7)$$

$$\rho_{\Delta_t I \Delta_m I} \cdot \sigma_{\Delta_t I} \cdot \sigma_{\Delta_m I} = \frac{1}{K} \sum_{i,j} \Delta_t I(m_i,n_j) \cdot \Delta_m I(m_i,n_j) \tag{8}$$

The summation is carried out over all the picture element locations (m_i,n_j) within the image segment. If the correlation between $\Delta_m I$ and $\Delta_n I$ is assumed to be negligible and the approximation

$$\rho_{\Delta_t I \Delta_m I} \cdot \sigma_{\Delta_t I} = \frac{1}{K} \sum_{i,j} \Delta_t I(m_i,n_j) \cdot \mathrm{sign}\ \Delta_m I(m_i,n_j) \tag{9}$$

is valid, the estimate is simplified to

$$\hat{\mu} = \frac{\sum_{i,j} \Delta_t I(m_i,n_j) \cdot \mathrm{sign}\ \Delta_m I(m_i,n_j)}{\sum_{i,j} |\Delta_m I(m_i,n_j)|} \tag{1o}$$

which is similar to equation (2). Writing equation (1o) without using the approximation (9) we get

$$\hat{\mu} = \rho_{\Delta_t I \Delta_m I} \cdot \frac{\sigma_{\Delta_t I}}{\sigma_{\Delta_m I}} \tag{11}$$

$$= \frac{1}{N} E\left[\{I_{k-1}(m,n) - I_{k-1}(m+1,n)\} \cdot \{I_k(m,n) - I_{k-1}(m,n)\}\right] \tag{12}$$

$$\begin{aligned} &= \frac{1}{N} \Big[E\left[I_{k-1}(m+1,n) \cdot I_{k-1}(m,n)\right] - E\left[I^2_{k-1}(m,n)\right] \\ &\quad + E\left[I_{k-1}(m,n) \cdot I_k(m,n)\right] - E\left[I_{k-1}(m+1,n) \cdot I_k(m,n)\right]\Big] \end{aligned} \tag{13}$$

with

$$N = E\left[\{I_{k-1}(m,n) - I_{k-1}(m+1,n)\}^2\right] \tag{14}$$

The first two expressions in equation (13) are independent of the displacement and represent an approximation of the first derivative of the autocorrelation function around the peak. A model of the correlation function based on the measurements of real-world television signals is used for further evaluation of equation (13). Figure 2 shows the two-dimensional correlation function of a monochrome television signal. If pure translation is assumed Figure 3 shows the relation between the actual displacement and the estimate according to equation (13) . For simplicity in this case only the horizontal displacment μ and its estimate $\hat{\mu}$ have been evaluated.

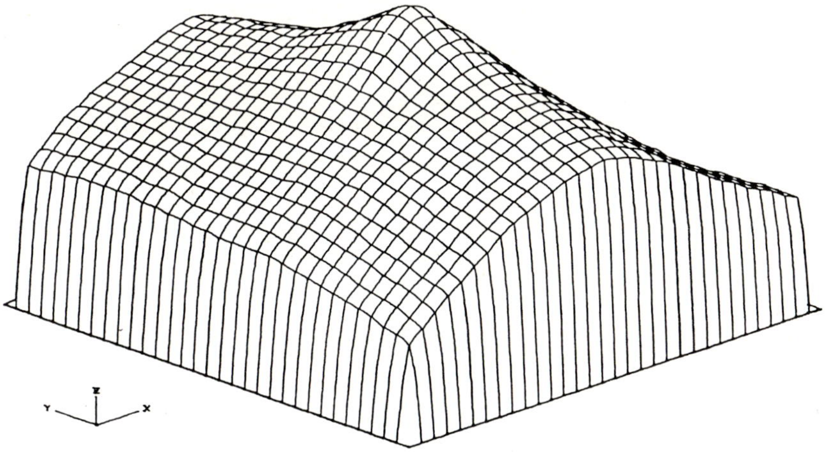

<u>Figure 2.</u> Cross correlation function of two-dimensional luminance signals

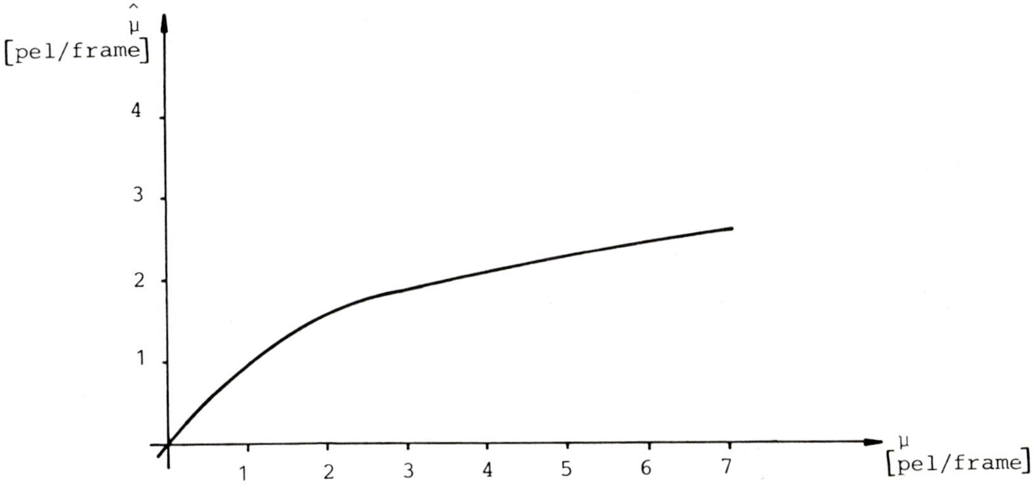

<u>Figure 3.</u> Relation between the actual displacement μ and the estimate $\hat{\mu}$ according to equation (13)

Figure 3 shows a linear relation between μ and $\hat{\mu}$ up to displacements of about 1.5 pel per frame. Exceeding this value equation (13) obviously fails. This may be the reason why the authors of /3/ have evaluated a good performance up to 2.5 pel per frame. For estimating larger displacements they propose spatial low pass filtering of the luminance signals. This would result in a deformation of the correlation function slope in Figure 2, so that the linear range of the function in Figure 3 is enlarged.

Another displacement estimation algorithm which is essentially identical to Rocca's has been proposed by Schalkoff and McVey /4/ and evaluated by Dinse, Enkelmann and Nagel /8/.

One component of the estimate is determined by

$$\hat{\mu} = \frac{1}{\sum_{i,j} \Delta_m I^2 \cdot \sum_{i,j} \Delta_n I^2 - 2 \cdot \sum_{i,j} \Delta_m I \Delta_n I} \cdot$$

$$\left(\sum_{i,j} \Delta_n I^2 \cdot \sum_{i,j} \Delta_m I \Delta_t I - \sum_{i,j} \Delta_m I \Delta_n I \cdot \sum_{i,j} \Delta_n I \Delta_t I \right) \tag{15}$$

which is identical to equation (6).

This algorithm has been implemented and evaluated in /8/. The measurements have shown that the sums over the mixed derivatives in (15) are between one and two orders of magnitude less than the principal diagonal elements and can be neglected in the calculation of estimates. The reason is that these expressions represent the correlation between the partial derivatives in the m- and n-directions which can be assumed to tend to zero. In this case equation (15) becomes

$$\hat{\mu} = \frac{\sum_{i,j} \Delta_m I \Delta_t I}{\sum_{i,j} \Delta_m I^2} \tag{16}$$

$$= \rho_{\Delta_t I \Delta_m I} \cdot \frac{\sigma_{\Delta_t I}}{\sigma_{\Delta_m I}} \tag{17}$$

which is identical to equation (11).

2.3 RECURSIVE DISPLACEMENT ESTIMATION ALGORITHM OF NETRAVALI AND ROBBINS

An elegant method for recursive displacement estimation has been published by Netravali and Robbins /5/. In order to use a pel recursive gradient technique the quadratic error function is given by the squared displaced frame difference DFD^2 which is to be minimized,

$$DFD^2_{m,n} \rightarrow \min \tag{18}$$

with

$$DFD_{m,n} = I_k(m,n) - I_{k-1}(m-\mu_{i-1}, n-\nu_{i-1}) \tag{19}$$

and the displacement estimates from the previous pel using vectorial representation

$$\vec{D}_{i-1} = (\mu_{i-1}, \nu_{i-1}) \quad . \tag{2o}$$

This formula leads to the recursion formula

$$\hat{\vec{D}}_i = \vec{D}_{i-1} - \varepsilon \cdot DFD \cdot \nabla I_{k-1}(m-\mu_{i-1}, n-\nu_{i-1}) \tag{21}$$

∇ : NABLA Operator

where ∇I_{k-1} approximates the spatial gradient of the displaced luminance in frame k-1 by differences.

In the following equation the statistic expectancies of DFD and I_{k-1} are used instead of just one value, because luminance signals generally represent non-deterministic source signals. If we calculate the update term of (21) with an initial estimate $\mu_{i-1} = \nu_{i-1} = 0$, we get for instance the horizontal displacement estimate

$$\hat{\mu} = -\varepsilon \cdot E\left[DFD_{m,n} \cdot \Delta_m I_{k-1}(m,n)\right]$$

$$= -\varepsilon \cdot E\left[\{I_k(m,n) - I_{k-1}(m,n)\} \cdot \{I_{k-1}(m,n) - I_{k-1}(m+1,n)\}\right] \tag{22}$$

which is identical to equation (12), if we choose $\varepsilon = \frac{1}{N}$. The negative sign is due to the definition of DFD in (19). In contrast to /3/ the recursive procedure tries to keep the estimate within the linear range of the curve in Figure 3.

2.4 NEW DISPLACEMENT ESTIMATION ALGORITHMS

Let us consider again the estimation criterion which demands for minimizing the displaced frame differences

$$E[DFD^2_{m,n}] \rightarrow \min$$

$$\begin{aligned} E[DFD^2_{m,n}] &= E[\{I_k(m,n) - I_{k-1}(m-\mu_{i-1}, n-\nu_{i-1})\}^2] \\ &= E[I^2_k(m,n)] + E[I^2_{k-1}(m-\mu_{i-1}, n-\nu_{i-1})] \\ &\quad - 2\,E[I_{k-1}(m-\mu_{i-1}, n-\nu_{i-1}) \cdot I_k(m,n)] \\ &= R_{I^2_{k-1}}(0) + R_{I^2_k}(0) \\ &\quad - 2\,R_{I_{k-1}I_k}(\mu-\mu_{i-1}, \nu-\nu_{i-1}) \end{aligned} \tag{23}$$

R : correlation function

Using the gradient technique for recursive minimization we get the recursion formula for the displacement vector $\vec{D}$

$$\begin{aligned} \hat{\vec{D}}_i(\mu,\nu) &= \vec{D}_{i-1}(\mu,\nu) - \frac{\varepsilon}{2} \cdot \nabla\, E[DFD^2_{m,n}] \\ &= \vec{D}_{i-1}(\mu,\nu) + \varepsilon \cdot \nabla\, R_{I_{k-1}I_k}(\mu-\mu_{i-1}, \nu-\nu_{i-1}) \end{aligned} \tag{24}$$

and, for instance, for the horizontal component

$$\hat{\mu}_i = \mu_{i-1} + \varepsilon \left. \frac{\partial R_{I_{k-1}I_k}}{\partial \mu} \right|_{\mu_{i-1}} \tag{25}$$

If we exchange the summation and the derivation we get for the update term in equation (25)

$$\frac{\partial R_{I_{k-1}I_k}}{\partial \mu} = \frac{\partial}{\partial \mu} \frac{1}{K} \sum I_{k-1}(m-\mu,n-\nu) \cdot I_k(m,n)$$

$$= - \frac{1}{K} \sum \frac{\partial}{\partial (m-\mu)} I_{k-1}(m-\mu,n-\nu) \cdot I_k(m,n)$$

$$= - E[I'_{k-1}(m-\mu,n-\nu) \cdot I_k(m,n)]$$

$$= - R_{I'_{k-1}I_k} \qquad (26)$$

An approximation of the first derivative of the luminance signal by spatial differences leads to the formula

$$R_{I'_{k-1}I_k} = E[\{I_{k-1}(m,n) - I_{k-1}(m+1,n)\} \cdot \{I_k(m,n)\}] \quad . \qquad (27)$$

For an initial value of the displacement $\mu_{i-1} = \nu_{i-1} = 0$ we get

$$\hat{\mu}_i = - \varepsilon [E[I_{k-1}(m,n) \cdot I_k(m,n)] - E[I_{k-1}(m+1,n) \cdot I_k(m,n)]] \qquad (28)$$

While equation (28) contains only a displacement dependent expression which can be interpreted as an approximation of the first derivative of the cross correlation function $R(\mu,\nu)$ in direction of μ the formulation according to equation (22) additionally contains two expressions which are independent from the displacement

$$- E[I^2_{k-1}(m,n)] + E[I_{k-1}(m,n) \cdot I_{k-1}(m+1,n)]$$

$$= - R_{I^2_{k-1}}(o) + R_{I^2_{k-1}}(1) \qquad (29)$$

The algorithm according to equation (24) requires the calculation of the gradient of the cross correlation function within a small image segment. If a minimum window size of only one pel is used, the displacement calculation would require one subtraction and multiplication, while equation (21) requires one subtraction additionally. Thus the gradient technique (24) has the advantage of a more simple hardware realization.

For a more detailed analysis we discuss the convergence behaviour of the gradient algorithm. If we have a symmetric cross correlation function the algorithm converges around the peak up to the two next locally adjacent minima. Assuming an actual displacement μ^* convergence is provided if the update term in equation (25) is smaller than the amount of $2|\mu_i - \mu^*|$,

$$2 \cdot |\mu_i - \mu^*| > \varepsilon \cdot \left| \frac{\partial}{\partial \mu} R_{I_{k-1} I_k} \right| . \qquad (30)$$

Equation (30) gives an upper limit for the constant ε in the neighbourhood of the actual displacement,

$$\varepsilon < 2 \cdot \lim_{\mu_i \to \mu^*} \left| \frac{\mu_i - \mu^*}{\frac{\partial}{\partial \mu} R_{I_{k-1} I_k}} \right| \qquad (31)$$

In order to determine the convergence rate we calculate the estimation error for successive recursions, i.e. the difference between estimate and actual displacement,

$$e(i) = \mu_i - \mu^* \qquad (32)$$

$$\begin{aligned} e(i+1) &= \mu_{i+1} - \mu^* \\ &= \mu_i + \Delta\mu_i - \mu^* . \end{aligned} \qquad (33)$$

In /9/ it has been shown that gradient methods are linear convergent. This means that the limit value of the ratio of successive estimation errors can be determined by

$$c = \lim_{i \to \infty} \frac{e(i+1)}{e(i)} \qquad (34)$$

By inserting (32) and (33) we get

$$c = \lim_{i \to \infty} \left(1 + \varepsilon \frac{\frac{\partial}{\partial \mu} R_{I_{k-1} I_k}}{\mu_i - \mu^*} \right) , \qquad (35)$$

and, because $\lim_{i \to \infty} \mu_i = \mu^*$, we obtain

$$c = 1 + \varepsilon \frac{\partial^2}{\partial \mu^2} R_{I_{k-1} I_k} \Bigg|_{\mu = \mu^*} \tag{36}$$

Equation (36) determines the convergence rate /9/ of the linear convergent recursion sequence (25) in dependence of ε. The smaller the amount of c the faster the algorithm converges to the optimum estimate μ^*. If we choose, for instance, an ε which is equal to the right side of relation (31) we will get a convergences rate

$$c = 1 + 2 \lim_{\mu \to \mu^*} \left| \frac{\mu_i - \mu^*}{\frac{\partial}{\partial \mu} R_{I_{k-1} I_k}} \right| \cdot \frac{\partial^2}{\partial \mu^2} R_{I_{k-1} I_k} \Bigg|_{\mu = \mu^*} \tag{36a}$$

Using the rule of l'Hospital we get

$$c = 1 + 2 \left| \left(\frac{\partial^2}{\partial \mu^2} R_{I_{k-1} I_k} \right)^{-1} \right|_{\mu = \mu^*} \cdot \frac{\partial^2}{\partial \mu^2} R_{I_{k-1} I_k} \Bigg|_{\mu = \mu^*} \tag{36b}$$

Because the second order derivative of the cross correlation function at the location $\mu = \mu^*$ is negative, equation (36b) becomes

$$c = -1 . \tag{36c}$$

In this case the estimates oscillate between the initial estimates $+\mu_0$, $-\mu_0$, in the case $|c|>1$ divergence is provided. For every other chosen ε, whereby relation (31) is valid, equation (36) yields a convergence factor within the limits

$$-1 < c < 1 . \tag{36d}$$

For a graphical demonstration of the convergence rate in dependence of ε we approximate the estimation error as a function of the number of recursion steps using equation (33):

$$|c| = \frac{e(i+1)}{e(i)} \qquad i = 0,1,2,\ldots \quad .$$

This leads to an exponentially decreasing estimation error, for instance

$$e(i) = \alpha \cdot \exp\{\beta i\} . \tag{37}$$

Then we can determine β by

$$|c| = \exp\{\beta\} \tag{38}$$

and with a normalized estimation error $\alpha = 1$ we get

$$e(i) = \exp\{\, \ln|c| \cdot i \,\} \tag{39}$$

where c is determined by equation (36) for a given ε.

In Figure 4a the estimation error with parameter $|c|$ is graphically shown. For a large number of recursion steps both curves $|c| = 0.5$ and $|c| = 0.25$ show sufficient small errors. Especially for only one or two steps the parameter curve $|c| = 0.25$ shows a better convergence concerning the remaining estimation error after truncation.

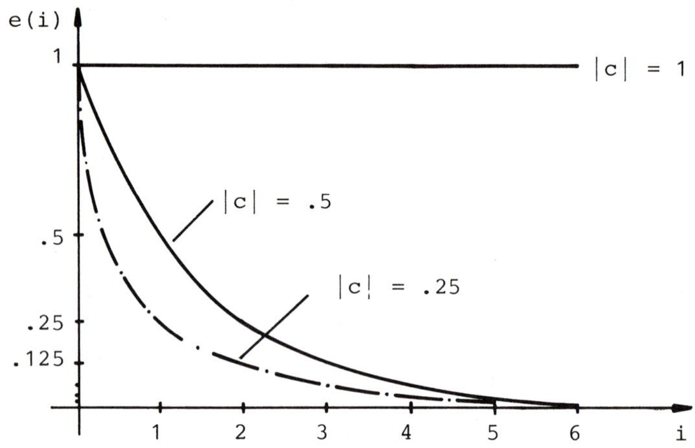

Figure 4a. Estimation error in dependence of the convergence factor $|c|$

The essential disadvantage of the gradient estimation technique is the constant stepwidth ε. By defining the stepwidth the convergency behaviour is determined only for the considered error function, in this case the squared displaced frame differences. Minimizing this function is traced back to maximize the cross correlation function where the direction of the recursive search is derived from its gradient. If the error function is slightly varying its shape the stability and the convergence of the algorithm are influenced. In order to avoid oscillations the stepwidth normally has to be chosen small enough. This would result in a large number of iterations for reaching the zero crossing. A faster convergence is obtained by adapting the stepwidth to the slope of the error function. If the second derivative of the cross correlation function is used for the determination of ε,

$$\varepsilon = \frac{1}{R''_{I_{k-1}I_k}} \tag{4o}$$

the algorithm is similar to the known Newton search method for one-dimensional signals. In Figure 4b an example is shown for the gradient technique and the Newton search technique.

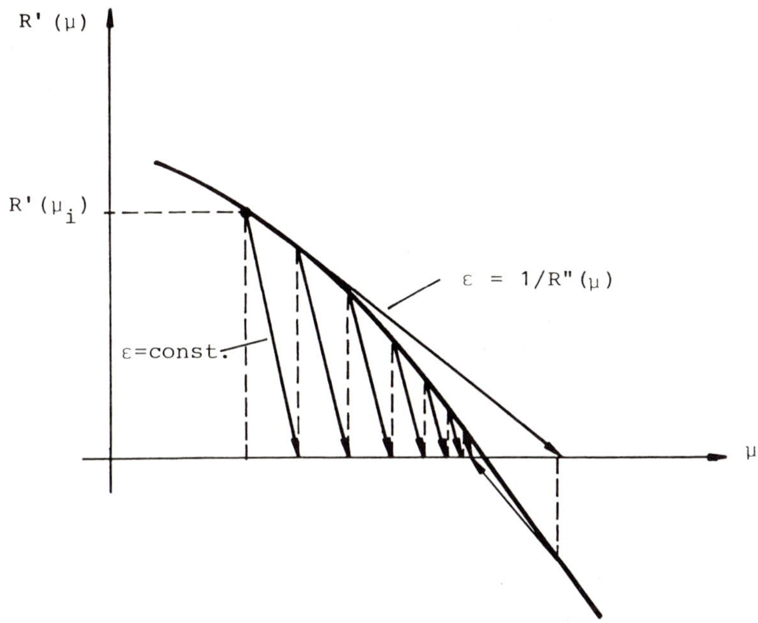

Figure 4b. Example of recursive zero crossing approximation using a constant and an adaptive step width

In /7/ a displacement estimation technique utilizing the derivatives of the cross correlation function for correlation peak estimation is proposed for two-dimensional monochrome television signals. The estimate for the horizontal displacement is given by

$$\begin{aligned}\hat{\mu}_i &= \mu_{i-1} - \frac{\frac{\partial}{\partial\mu} R_{I_{k-1}I_k}}{\frac{\partial^2}{\partial\mu^2} R_{I_{k-1}I_k}} \\ &= \mu_{i-1} + \frac{\sum\limits_{i,j} \Delta_m I_{k-1} I_k}{\sum\limits_{i,j} \Delta_m{}^2 I_{k-1} I_k}\end{aligned} \tag{41}$$

where the derivatives are approximated by differences and calculated within small image segments. The recursion is carried out pel by pel, using the estimate from the previous pel as displacement prediction for the current pel. In general the Newton search technique has a quadratic convergence. If the cross correlation function is symmetrical, i.e. the derivative is S-shaped, it can be shown that the Newton algorithm has a third order convergence /9/. The convergence rate according to /6/ is then given by

$$c = \lim_{i \to \infty} \frac{e(i+1)}{e(i)^3} \tag{42a}$$

and in our case

$$c = \frac{1}{3} \left. \frac{\left(\frac{\partial}{\partial \mu} R_{I_{k-1} I_k}\right)'''}{\left(\frac{\partial}{\partial \mu} R_{I_{k-1} I_k}\right)'} \right|_{\mu = \mu^*} \tag{42b}$$

For the measured cross correlation function according to Figure 2 we get a convergence rate $c = -0.35$. This shows that the new algorithm converges relatively fast because the stepwidth is adapted to the slope of the cross correlation derivative.

In order to compare the performance of both algorithms /5/ and /7/ an experimental evaluation has been done by means of computer simulations. The input signal consists of a digitized monochrome television signal containing a swinging pendulum of 20 x 10 pel size. The known horizontal displacement can be descibed by a sinusoidal function and serves as reference function. Figure 5 shows the actual displacement as reference function and the estimate values. In Figure 6 the RMSE of all estimates within one field is shown. For displacements up to 2.5 pel per frame the RMSE of the gradient algorithm lies between 0.7 and 1.3 pel per frame. For displacements above 2.5 pel per frame the algorithm does not converge fast enough due to the small constant stepwidth ε. The RMSE of the Newton correlation peak estimation remains small and varies only slightly with the amount of the displacement.

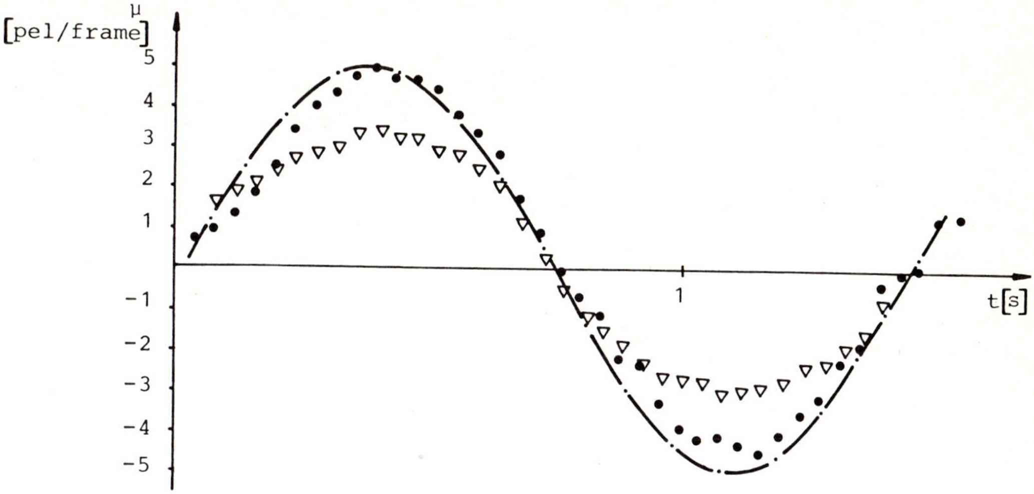

▽ : gradient algorithm

• : correlation peak estimation algorithm

μ : horizontal displacement component

f : field number

Figure 5. Estimates and reference displacement for the horizontal displacement of the pendulum swing

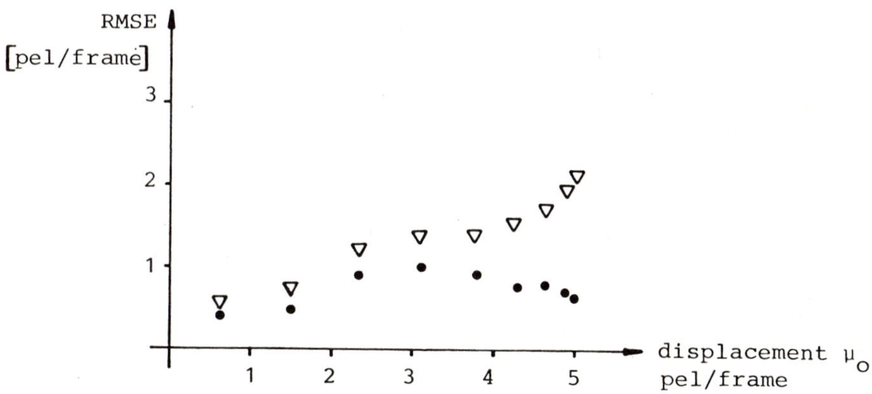

▽: gradient algorithm

•: correlation peak estimation algorithm

Figure 6. RMSE as a function of the reference displacement μ_o

3. CONCLUSION

The analyzation has shown that four of the discussed displacement algorithms basically utilize equation (11). For small displacements each algorithm has a good performance. For larger displacements the recursive gradient technique is expected to have a better performance. But as the stepwidth has to be chosen small in order to avoid oscillations the algorithm does not converge fast enough, i.e. it requires too many iterations for large displacements. This may be the reason for the relatively high root mean square error which has been measured on real television sequences. This analysis has shown that the calculation complexity of the gradient algorithm can be reduced. An improved convergence can be achieved by using the Newton search method for displacement estimation. In this case the stepwidth is derived from the quotient of the first and second order derivatives of the cross correlation function of the displaced luminance signals. A new displacement estimation algorithm which works on the base of the Newton search method has been proposed. The improvement with respect to the convergence could be proved by some experimental evaluations. However, in special cases the range of convergence of the new algorithm may be decreased. Therefore further investigations are provided for optimizing the stepwidth adaptation with respect to a fast convergence within an enlarged range of stability.

REFERENCES

/1/ Huang, T.S. — Image Sequence Analysis, ed. by T.S. Huang Springer Verlag, Berlin 1981, Chapt. 1

/2/ Limb, J.O. Murphy, J.A. — "Measuring the Speed of Moving Objects from Television Signals," IEEE Trans. Comm., Vol. 24, No. 4, April 1975

/3/ Cafforio, C. Rocca, F. — "Tracking Moving Objects in Television Images", Signal Processing, Vol. 1, No. 2, April 1979

/4/ Schalkoff, R.J. McVey, E.S. — "A Model and Tracking Algorithm for a Class of Video Targets", IEEE Trans. Patt. Anal. Mach. Int., Vol. 4, No. 1, January 1982

/5/ Netravali, A.N. Robbins, J.O. — "Motion-Compensated Television Coding: Part I", Bell Syst. Techn. Journ., Vol. 58, No. 3, March 1979

/6/ Burkardt, H. Moll, H. — "Ein modifiziertes Newton-Raphson-Schema zur modelladaptiven Identifikation von Laufzeiten," Mitteilung aus dem Institut für Meß- und Regelungstechnik, Universität Karlsruhe, 1978

/7/ Bergmann, H.C. — "Displacement Estimation Based on the Correlation of Image Segments", IEE Proceedings Int. Conf. Electronic Image Proc., York 1982, UK

/8/ Dinse, T.
Enkelmann, W.
Nagel, H.-H.
"Untersuchung von Verschiebungsvektorfeldern in Bildfolgen", Informatik-Fachberichte Bd. 49, Springer-Verlag, Berlin 1981

/9/ Strubecker, K.
"Einführung in die Höhere Mathematik", R. Oldenbourg Verlag, München, Wien, 1967

An Adaptive Gradient Approach To Displacement Estimation

Richard A. Jones Carl D. Bowling

University of Arkansas
Fayetteville, Arkansas

Abstract

An algorithm for estimating the displacement of objects in successive frames of images is derived. The adaptive gradient algorithm that is utilized in the displacement estimation is discussed from the point of view of Kalman Filtering. It is shown that the residual, or error, term produced by this proceedure is equivalent to the gradient utilized in the coefficient recursive method.

The research reported here is supported by AFOSR grant 77-3456.

NATO ASI Series, Vol. F2
Image Sequence Processing and Dynamic Scene Analysis
Edited by T.S. Huang

I. Introduction

Motion detection and displacement estimation is rapidly becoming a major area of scientific research. This upsurge in activity is based on the need for more effective interframe data compression systems as well as on the theoretical interest in the subject. The detection of significant changes in a sequence of scenes has been a topic of some interest for many years.

Tracking of cloud movements in a satellite picture sequence is an on-going problem in weather forecasting. The need for data compression in picture phone systems does not need amplification and the application of the motion detection and displacement estimation is directly apparent. The military applications, particularly in guidance and tracking, are also obvious.

In the recent past, several methods of detecting motion and estimating displacement have been proposed. The motion detection procedures are usually quite simple. An example is frame to frame threshold examination. On the other hand, estimation of true displacement is usually quite complicated, especially if displacement needs to be determined to an accuracy comparable to the resolution of the original imaging data.

A few techniques, generally various solutions to correpondence problems, are computationally too complex for present day implementation, Examples of these are point by point correlation [3] and pattern matching [9] . Simpler techniques which utilize the relation between spatial and temporal differences have been proposed by Limb and Murphy [8] and Cafforio and Rocca 2.

Recently, the research effort in motion detection and displacement estimation has been directed towards the development of recursive estimation procedures. These procedures attempt to minimize the displaced frame errors. Most of these techniques operate in the time domain. The most recent of the time domain algorithms is a pel (picture element) recursive displacement estimation algorithm by Netravali and Robbins [10].

An algorithm that operates in the transform domain has been discovered by Stuller in Netravali [14]. The algorithm is called the coefficient recursive displacement estimation algorithm. Both these algorithms utilize the steepest descent gradient technique, and the pel-recursive displacement estimation algorithm is a special case of the more general coefficient recursive estimation algorithm.

In this paper, a new approach to the problem of detecting motion and estimating displacement in a sequence of time varying images is described. It is presumed that the objects in the sequence of scenes are subjected to purely translational motion. The algorithm is generalized in the sense that it operates in both the pel and in the transform domain, and it exhibits better stability and convergence properties than the algorithms mentioned above.

The necessary preliminary theory and definition is presented in Section II. It includes a brief description of both the pel-domain and the transform domain algorithms. A description of Adaptive Hybrid Picture Coding (AHPC) is contained in Section III. This procedure serves the dual purpose of data compression and gradient derivation. The main result is discussed in Section IV. The adaptive gradient is utilized in the derivation of the residual recursive displacement estimation algorithm. The results are presented and discussed qualitatively.

II. Preliminary Theory and Defintions

The class of time varying image sequences that are considered in this paper is limited to successive scenes of objects that are undergoing pure translation. The performance of algorithms that detect and estimate the displacement of objects in a sequence of scenes with this description have been investigated in both the picture element (pel) and transform domain.

These algorithms are known as the pel recursive displacement estimation method and the coefficient recursive displacement method respectively.

The main result of this paper is the development of a displacement estimation algorithm that operates in both the pel domain and the transform domain jointly. The algorithm is named the residual recursive displacement estimation algorithm, where the residual term is derived from an image data compression procedure known as Adaptive Hybrid Picture Coding [5], [6] . This procedure will be described in due course. Presently, it is useful to briefly outline both the pel-recursive and the coefficient recursive displacement estimation algorithms. These motions, along with a description of Adaptive Hybrid Picture Coding, will provide the necessary groundwater for development of our main result, the residual recursive displacement estimation algorithm.

In the pel domain, the individual pels (picture elements) form the input data for estimation. If $I(x_k, t)$ is the intensity of the sample point $[x_{1k}, x_{2k}]$ at time t, and $I(\underline{x}_k, t-\tau)$ is the intensity of the same spatial location in the previous frame, then if the frame has an object undergoing pure translation, $\underline{D}$.

$$I(\underline{x}_k, t) = I(\underline{x}_k-\underline{D}, t-\tau \tag{1}$$

This expression neglects any background errors.

The intensity of every point in the current frame differs from the intensity of the previous frame by some amount termed the displaced frame difference (DFD). This quantity is defined as

$$DFD(\underline{x}_k, \underline{D}) = I(\underline{x}_k, t) - I(x_k-\underline{D}, t-\tau) \tag{2}$$

An estimate of the displacement $\underline{D}$, say $\hat{\underline{D}}$, may be obtained by recursively minimizing the quantity $DFD(\underline{x}_k, \hat{\underline{D}})^2$ using a steepest descent iteration of the form.

$$\underline{\hat{D}}_{k+1} = \underline{\hat{D}}_k - \frac{\varepsilon}{2}\nabla_{\underline{\hat{D}}_k} \quad DFD(\underline{x}_k, \underline{\hat{D}}_k)^2 \tag{3}$$

where $\nabla_{\underline{\hat{D}}_k}$ is the two-dimensional gradient operator with respect to $\underline{\hat{D}}_k$. This equation upon expansion becomes

$$\underline{\hat{D}}_{k+1} = \underline{\hat{D}}_k - \varepsilon \, DFD\,(x_k, \underline{\hat{D}}_k) \, \nabla_{\underline{x}} I(\underline{x}_k - \underline{\hat{D}}_k, t-\tau) \tag{4}$$

The coefficient recursive displacement estimation approach is different from the pel-recursive approach in the sense that the estimate is obtained from the two dimension unitary transform coefficients.

Each image frame is partitioned into blocks of size $N = N_r \times N_c$, to obtain the N transform coefficients

$$C_n(q) = \underline{I}(X_q, t)\,\phi_n \tag{5}$$

where $I(X_q, t)$ is the column scaned vector of intensities, $C_n(q)$ is nth transform coefficient and ϕ_n is the nth linear transform basis vector. Similarly

$$C_n(q, \underline{\hat{D}}) = \underline{I}(X_q - \underline{\hat{D}}, t-\tau)\,\phi_n \tag{6}$$

in the nth transform coefficient in the previous (displaced) frame. When $C_n(q)$ is predicted by $C_n(q\ \underline{\hat{D}})$, the prediction error is

$$e(q, \hat{D}) = C_n(q) - C_n(q, \underline{\hat{D}}) \tag{7}$$

The squared prediction errors are minimized by a steepest descent interation as before. The recursive relation is

$$\underline{\hat{D}}_{n+1}(q) = \underline{\hat{D}}_n(q) - \varepsilon/_2 \, \nabla_{\underline{\hat{D}}(q)} e^2(q, \underline{\hat{D}}_n(q)) \tag{8}$$

which becomes

$$\underline{\hat{D}}_{n+1}(q) = \underline{D}_n(q) - \varepsilon \, e_n(q, \underline{\hat{D}}_n(q))\underline{G}_n(q) \tag{9}$$

where

$$\underline{G}_n(q) = \nabla \underline{I}^T(\underline{x}_q - \underline{\hat{D}}_n(q), t-\phi)\,\underline{\phi}_n \tag{10}$$

The vector $G_n(q)$ is called the coefficient gradient vector.

These methods have been investigated and found to be very beneficial in reducing the required data rate for frame to frame image transmission. The subject of this paper is a procedure that possesses the ability to improve upon the above procedure from the point of view of pel-to-pel computational requirements.

III. Adaptive Gradient Derivation

The adaptive gradient is based on an autoregressive model for the transformed image process. Specifically, the original pixels are modeled as a two dimensional non-stationary autoregressive process of unknown statistics, so that

$$p(j,k) = \sum_{m=1}^{p} \sum_{n=1}^{p} b(m,n)\, p(j-m, k-n) + g(j,k) \tag{11}$$

Then the transformed pixel in column "k" is

$$f(j,k) = \sum_{j=1}^{N} p(j,k)\, T(j,k) \tag{12}$$

where T(j,k) indicates a unitary transform.

The precise definition of the distribution of the transform process is difficult to obtain since the distribution of the original process is unknown. However, since the transform operation forms a well behaved weighted sum of all the pixels in the original image, qualitative arguments based upon the Central Limit Theorem can be invoked to make reasonable assumptions regarding the distribution of the transformed pixels. The Central Limit Theorem cannot be rigorously applied in this instance, since the original samples are not independent; however, the nature of the source indicates that the statistics of the transformed picture elements will approach the Gaussian distribution. The distribution should at least be unimodal. These arguments are strengthened as the derivation continues.

The transformed source can now be viewed as a one dimensional process, with the transformed pixels in each column comprising the source samples, viz:

$$s_n(k) = \sum_{i=1}^{p} a_i(k) s_n(k-i) + e_n(k) \tag{13}$$

when n denotes the nth learning period. In vector notation

$$s_n(k) = S_p^T(k-1)A(k) + e(k) \tag{14}$$

With this model, it is evident that the s(k) in the columns are the process observations. Also, since predictor law is constant over a learning period, the message model for the identification is

$$A(k+1) = A(k) \tag{15}$$

With this vector format, the predictor coefficients can be viewed as states. Then sequential estimation theory can be used to obtain an estimate of the state vector by viewing the actual transformed pixels as the observation space. By requiring the estimator to be linear in observations, minimum error variance, and unbiased, the problems associated with the lack of statistical knowledge are partially circumvented.

The observations are also generated according to the equation

$$s(k) = R_p^T(k-1)\, \hat{A}(k-1) + e(k) \tag{16}$$

In this derived observation equation, all noise terms are enclosed in a transmitters feedback loop.

Each element of the vector R_p^T (k-1) contains the corresponding transformed pixel as well as the total noise in the reconstructed sample. Concisely

$$R_p(k-1) = S_p(k-1) + Q_p(k-1) \tag{17}$$

The Vector $\hat{A}_n(k-1)$ is the identified coefficient vector that has undergone element quantization prior to its presentation to the projecting filter. Notationally, the hat (^) refers to identification and the bar (-) indicates quantization.

The identification error is defined as

$$\tilde{A}_n(k) = A_n(k) - \bar{\hat{A}}_n(k) \tag{18}$$

An estimate of the predictor coefficient state is desired that is linear, minimum covariance of error, and sequential. In addition, the estimate will be both conditionally and unconditionally unbiased. The filter form is

$$\hat{A}(k + 1 \mid S\ (k + 1)) = p(k + 1,\ k)\ \ \bar{\hat{A}}(k \mid S\ (k) \\ + K\ (k + 1)s(k + 1) \tag{19}$$

where the subscript "n" has been dropped for convenience.

Each new estimate is a combination of the prediction from the previous quantized coefficient estimate and a correction of this prediction with the information contained in the new observations. Using the conditions stated above and equation (16), it follows directly that

$$P(k + 1,\ k) = \ I - K(k + 1)\ \ R_p^T\ (k) \tag{20}$$

which yields the desired identification algorithm,

$$A(k + 1) = A(k) + K(k + 1)\ \ s(k + 1) - R_p^T(k)\ A\ (k) \tag{21}$$

Combining equation (21) and equation (16) it is possible to express the identification equation in the simpler form

$$A(k + 1) = A(k) + K(k + 1)\ e(k + 1) \tag{22}$$

The residual represents that portion of the observation s(k + 1) that is orthogonal to the observation space S(k). The quantized version of the coefficient vector identified from S(k) is used to form a linear prediction of s(k + 1). The derivation of the gain term, K(k + 1), can be accomplished by making use of the Orthogonal Projection Lemma. The coefficientestimate is then given by,

$$\hat{A}(k + 1) = E\{A(k + 1) \mid S(k)\} + E\{A(k + 1) \mid e(k + 1)\} \tag{23}$$

Or

$$\hat{A}(k + 1) = E\{A(k + 1) \mid S(k)\}$$
$$+ E\{A\ (k + 1)\ e(k + 1)\}\ \ E\{e^2(k + 1)\}^{-1}\ e(k - 1) \tag{25}$$

Comparing (25) and (21), it is apparent that

$$K(k + 1) = \{E\ A(k + 1)\ e(k + 1)\}\ E\ \{e^2(k + 1)\}^{-1} \tag{26}$$

By assuming $\hat{A}(k)$ is orthogonal to $\tilde{A}(k)$, it follows that

$$E\ \{A(k + 1)\ e(k + 1)\} = V_{\tilde{A}}(k)\ R_p(k) \tag{27}$$

Similarly,

$$E\ \{e^2(k + 1)\} = V_z(k + 1) + R_p^T(k)\ V_{\tilde{A}}(k) R_p(k) \tag{28}$$

The matrix $V_{\tilde{A}}(k)$ is the identification error covariance. The scalar term $V_z(k + 1)$ is the variance of the term

$$z(k + 1) = G\ v(k + 1) - Q_p^T(k) A(k) \tag{29}$$

and cannot be determined recursively. It is taken to be a constant, the value of which is determined experimentally. With this stipulation,

$$K(k + 1) = \frac{V_{\tilde{A}}(k) R_p(k)}{V_z + R_p^T(k) V_{\tilde{A}}(k) R_p(k)} \tag{30}$$

The remaining requirement is the error covariance equation. This is obtained by (46) in recognizing that $A(k)$ and $z(k + 1)$ are both independent of $D(k + 1)$, and utilizing (29) to obtain

$$V_{\tilde{A}}(k + 1) = I - K(k + 1)\ R_p^T(k)\ \ V_{\tilde{A}}(k) \tag{31}$$

The equations (14), (15), (22), (30), and (31) describe the identification procedure.

The residual term is utilized in the displacement estimation procedure as a gradient. The justification for this method is that the minimization of $e_n^2(k)$ is accomplished by correcting the vector $a_1(k)$ in the direction of the negative of the negative of the above gradient. If the weighting is done in a general form by the matrix $G(k+1)$, then the estimation equation becomes

$$\hat{A}(k+1) = \hat{A}(k) + G(k+1)\ S_p(k)\ \ e_n(k+1) \tag{32}$$

If the matrix is inversely proportional to the observation noise, the correction will decrease as the variance of the noise increases. Also, if $G(k+1)$ is directly proportional to the error covariance matrix, the elements of $A(k)$ with larger errors will receive larger corrections. Using these notions, it is possible to show that

$$K(k+1) = G(k+1)\ S_p(k) \tag{33}$$

Therefore, this adaptive estimation algorithm is a method of weighting the gradient. This fact will be utilized in Section IV.

IV. Residual Recursive Displacement Estimation

A few essential definitions are worth repeating at this point. First, since the premise is that any uncovered background is ignored, the condition

$$\underline{I}(x_q, t) = \underline{I}(x_q - \underline{D}, t-\tau) \tag{34}$$

remains unchanged.

After a one-sided transform operation on this vector of pels, the result is

$$C_n(q) = \underline{I}^T(\underline{x}_q, t)\ \phi_n \tag{35}$$

where ϕ_n is the nth basis vector of an arbitrary real unitary transform. The definition of $\underline{I}(\circ,\circ)$ is the same as ( ). Obviously, the same operation on the displaced frame is

$$C_n(q, \underline{D}) = \underline{I}^T\ (\underline{x}_q - D, t-\tau)\ \phi_n \tag{36}$$

The procedure described here for estimating the displacement vector depends on a quantity called the displaced residual difference (DRD) due to the fact that it results from a utilization of the residual term produced by the data compression algorithm previously described. The procedure exhibits advantages over either the displaced frame difference (DFD) or the coefficient recursive displacement estimation. These advantages will be enumerated in the sequel. For the present, the derivation of the algorithm will proceed.

At the outset it is recognized that the AHPC procedure yields

$$C_n(q) = s_n(q) \tag{37}$$

and

$$C_n(q, \hat{\underline{D}}) = s_n(q, \hat{D}) \tag{38}$$

Therefore, the AHPC residual is

$$r_n(q) = C_n(q) - \sum_{k=1}^{k} a_n\ C_{n-k}(q) \tag{39}$$

and for the displaced frame

$$r_n(q, \hat{D}) = C_n(q, \hat{\underline{D}}) - \sum_{k=1}^{k} a_k(\hat{D})\ C_{n-k}(q, \hat{\underline{D}}) \tag{40}$$

Where the $a_k(\hat{D})$ are the displaced predictor coefficients

The displaced residual difference is defined as

$$\mathrm{DFD}(q, \underline{\hat{D}}(q)) = r_n(q) - r_n(q, \hat{D}(q)) \tag{41}$$

It is this quantity squared that must be minimized with a steepest descent algorithm of the form

$$\underline{\hat{D}}_{n+1}(q) = \underline{\hat{D}}_n(q) - \varepsilon/_2 \, \nabla_{\underline{\hat{D}}_n(q)} \, [\mathrm{DRD}(q, \underline{\hat{D}}_n(q))]^2 \tag{42}$$

or

$$\underline{\hat{D}}_{n+1}(q) = \underline{\hat{D}}_n(q) - \varepsilon \mathrm{DRD}_n(q, \underline{\hat{D}}_n(q)) \, \nabla_{\hat{D}_n(q)} \mathrm{DRD}(q, \underline{\hat{D}}_n(q)) \tag{43}$$

An examination of the term $\mathrm{DRD}(q, \hat{D}_n(q))$ reveals that it is comprised of the term

$$e_n(q, \underline{\hat{D}}_n(q)) = C_n(q) - C_n \, (q, \underline{\hat{D}}_n(q)) \tag{44}$$

and the past coefficient difference terms

$$\sum_{k=1}^{k} a_k(\hat{D}_{n-k}(q)) \, e_{n-k}(q, \underline{\hat{D}}_{n-k}(q)) \tag{45}$$

Therefore from (41)

$$\nabla_{\underline{\hat{D}}_n(q)} \mathrm{DRD}_n(q \, \underline{\hat{D}}_n(q)) = \underline{r}_n(q \, \underline{\hat{D}}_n(q)) - \nabla_{\underline{\hat{D}}_n(q)} \sum_{k=1}^{k} a_k(\underline{\hat{D}}_{n-k}(q)) \, C_{n-k}(q, \underline{\hat{D}}_{n-k}(q))$$

That is, the gradient term in the adaptive gradient approach consists of two terms: 1) The AHPC residual and 2) the gradient of the prediction term,

$$\underline{P}_n(q, \underline{\hat{D}}_n(q)) = \nabla_{\underline{D}_n(q)} \sum_{k=1}^{k} a_k(\hat{D}_{n-k}(q)) \, C_{n-k}(q, \underline{\hat{D}}_{n-k}(q)) \tag{47}$$

The algorithm that results when the vector $\underline{P}_n(q, \underline{\hat{D}}_n(q))$ is ignored has the same form as the coefficient recursive displacement algorithm. It is this form that is the subject of this paper. Therefore the iterative steepest descent algorithm has the structure

$$\underline{\hat{D}}_{n+1}(q) = \underline{\hat{D}}_n(q) - \varepsilon \mathrm{DRD}_n(q, \underline{\hat{D}}_n(q)) \, \underline{r}_n(q, \underline{\hat{D}}_n(q)) \tag{48}$$

with

$$\underline{\hat{D}}_0(q) = \underline{\hat{D}}_{M-1} \, (q-1) - \varepsilon \mathrm{DRD}_{M-1}(q-1, \hat{D}_{M-1}(q-1) \, \underline{r}_{M-1}(q-1, \underline{\hat{D}}(q-1)) \tag{49}$$

The counting process proceeds with $n=0, 1, \ldots, M-1$ and $q = 0, 1, 2, \ldots$

For conceptual purposes, the image field is considered to be partitioned into rectangular blocks of picture elements, each with N_r rows and N_c columns. This generalized block is viewed as an $(N_r \times N_c)$ matrix which is called the qth matrix. The coordinates of the upper left element are $\underline{X}_q = [x_{1_q}, x_{2_q}]^T$. This procedure for pel-location allows the above counting process. For instance, the $N_n \times N_c$ matrices are sequenced from left to right as $q = 0, 1, 2, \ldots$.

The displaced residual difference algorithm is shown in Figure 2. The gradient is obtained from the AHPC procedure shown in Figure 3.

The algorithm was tested on frames of a synthetically generated pattern. The pattern is a radia decaying cosine function having a radius of 60 pixels and peak to peak amplitude of 228 at its center and decreasing to 130 at the circumference. The period of oscillation is 20 pixels at the center to 10 pixels at the circumference. The pattern is moving in the hoizontal direction at the rate of 2 pixels per frame. It is described mathematically by the intensity function

$$F(R) = 100 \exp(-.01R) \cos(2\pi R/P) + 128 \qquad 0 \leq R \leq 60$$

where

$$P = (1-R/60)\,10 + 10$$

In these frames the background has uniform illumination of 128. (All these intensities are out of an intensity range of 0 to 255).

Typical results indicating the convergence of the displacement error as a function of iteration numbers are shown in Figures 4 and 5. The upper and lower values of the gain term that produced acceptable results are given. Outside these values either oscillation or slow convergence occurred.

Apparently the convergence is better for this particular method because the gradient utilized produces independent elements in the residual gradient sequence. The reason for the importance of this becomes clear when the convergence properties of the displacement error are investigated. In determining the convergence of $\underline{D}_n(q)$, it is presumed that the Euclidian distance $||\underline{D} - \hat{\underline{D}}_n(q)||$ is sufficiently small to insure against an extraneous displacement estimate, then with

$$\underline{P}(\underline{X}_q, \phi) = \underline{P}(\underline{X}_q - \underline{D}, \phi - \tau)$$

The displaced residual difference becomes

$$DRD_n(q, \hat{\underline{D}}_n(q)) = \{C_n(q, \underline{D}) - \sum_{k=1}^{k} a_k C_{n-k}(q, \underline{D})\}$$

$$-\{C_n(q, \hat{D}_n(q)) - \sum_{k=1}^{k} a_k(d) C_{n-k}(q, \hat{\underline{D}}_n(q))\}$$

Using the Taylor's expansion to represent $DRD_n(q, D_n(q))$ about the point $\underline{D} - \hat{\underline{D}}_n(q)$, and assuming that frame rate is fast enough to insure that $a_k - a_k(D)$, $k = 1, 2, \ldots,$, is negligible, then the linear approximation to $DRD_n(q, \underline{D}(q))$ is

$$DRD_n(q, \hat{\underline{D}}(q)) = R_n^T(q, \hat{\underline{D}}(q))\, \underline{\Delta}_n(q)$$

where

$$\hat{\underline{D}}_n(q) - \underline{D} = \underline{\Delta}_n(q) \text{ and } R_n^T(q, \hat{\underline{D}}(q))$$

is as defined previously. Then, using the approximation above

$$\underline{\Delta}_{n+1}(q) = \underline{\Delta}_n(q) - \varepsilon DRD_n(q, \hat{D}(q)) \underline{R}_n(q, \hat{\underline{D}}(q))$$
$$= \underline{\Delta}_n(q) - \varepsilon \underline{R}_n(q, \hat{\underline{D}}(q)) \underline{R}_n^T(q, \hat{\underline{D}}(q)) \Delta_n(q)$$
$$= [I - \varepsilon \underline{R}_n(q, \hat{\underline{D}}(q)) \underline{R}_n^T(q, \hat{\underline{D}}(q))] \Delta_n(q)$$

where I is a 2 x 2 identity matrix. The residual gradient sequence $\{\underline{R}_n(q, \hat{\underline{D}}(q))\}$ consists of independent elements. This follows from the past that the AHPC residuals are statistically independent. Therefore

$$E[\underline{\Delta}_{n+1}(q)] = \{I - \varepsilon E[R_n(q, \hat{\underline{D}}(q)) R_n^T(q, \hat{\underline{D}}(q))]\} E[\underline{\Delta}_n(q)]$$

For completeness, we point out that correlation may exists from block to block even though the within block residual gradients are independent. It is easily shown that the mean displacement error converges to zero by formulating

$$C = \prod_{n=0}^{M-1} [I - \varepsilon E \underline{R}_n(q \hat{\underline{D}}_n(q)) \underline{R}_n^T (q, \hat{\underline{D}}_n(q))]$$

and demonstrating that the eigenvalues of C are inside the unit circle. It follows directly that this is the case when the AHPC predictor coefficient identification scheme produces a system model with poles outside the unit circle.

Summary

An algorithm for estimating the displacement of objects in a sequence of time varying scenes has been derived. The algorithm utilizes a gradient term that is derived from a data compression procedure that was designed to operate on a nonstationary class of images. Therefore, the computational load for the displacement estimate is reduced. The algorithm is expected to be utilized in motion compensated interframe hybrid image coders. The convergence of the mean displacement estimate to the true displacement was established without resorting to assumptions regarding the gradient sequence. The independence is established by virtue of the manner in which the gradients were obtained.

References

[1] Aggarwal, J. K. and Duda, R. O., "Computer Analysis of Moving Polygonal Images," IEEE Transactions on Computers, Vol. C-24, No. 10, November 1975, pp. 966-976.

[2] Cafforio, C. and Rocca, R., "Methods for Measuring Small Displacements of Television Images," IEEE Transactions Information Theory, IE-22, No. 7, September 1976, pp. 573-579.

[3] Haskell, B. G., Mounts, F. W. and Candy, J. C., "Interframe Coding of Videotelephone Picture," Proceedings IEEE, 60, No. 7 (July 1972), pp. 792-800.

[4] Haskell, B., "Entropy Measurements for a Non Adaptive and Adaptive Frame-to-Frame, Linear-Predictive Coding of Videotelephone Signals," BSTJ, Vol. 54, pp. 115-1174, July-August 1975.

[5] Jones, R. A., "Adaptive Hybrid Picture Coding," SPIE, Vol. 87 Advances in Image Transmission Techniques (1976) pp. 247-255.

[6] Jones, R. A. and Mix, D. F., "A Rate Distortion Analysis of Adaptive Hybrid Picture Coding," SPIE 22nd International Symposium, Aug. 1978, San Diego, Ca.

[7] Kailath, T., "An Innovations Approach to Least-Squares Estimation. Part I: Linear Filtering in Additive White Noise," IEEE Trans. on Automatic Control, Vol. AC-13, No. 6, Dec. 1968 pp. 646-655.

[8] Limb, J. O. and Murphy, J. A., "Estimating the Velocity of Moving Images from Television Signals," Computer Graphics and Image Processing, 4, 1975, pp. 311-327.

[9] Martin, W. N. and Aggarwal, J. K., "Dynamic Scene Analysis: The Study of Moving Images," Technical Report No. 184, Information Systems Research Laboratory, University of Texas, Austin, Texas, January 1977.

[10] Netravali, A. N. and Robbins, J. D., "Motion-Compensated Television Coding: Part I," Bell System Technical Journal, Vol. 58, No. 3, March 1979, pp. 631-670.

[11] Netravali, A. N. and Stuller, J. A., "Motion-Compensated Transform Coding," Bell System Technical Journal, Vol. 58, No. 7, September 1979, pp. 1703-1718.

[12] Paul, R. "Robots, Models and Automation," IEEE Computer, Vol. CM-12, No. 7, Jul. 1979, pp. 19-27.

[13] Robbins, J. D. and Netravali, A. N., "Interframe Television Coding Using Movement Compensation," Proceedings of the International Conference on Communications, ICC /79, IEEE Catalog #79CH1435-7, Vol. 2, pp. 23.4.1-23.4.5.

[14] Stuller, J. A. and Netravali, A. N., "Transform Domain Motion Estimation," Bell System Technical Journal, Vol. 58, No. 7, September 1979, pp. 1673-1702.

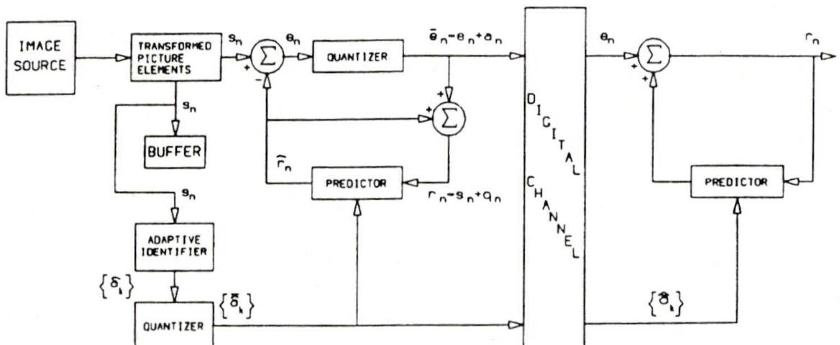

Source Coding by Adaptive Prediction

Figure 2

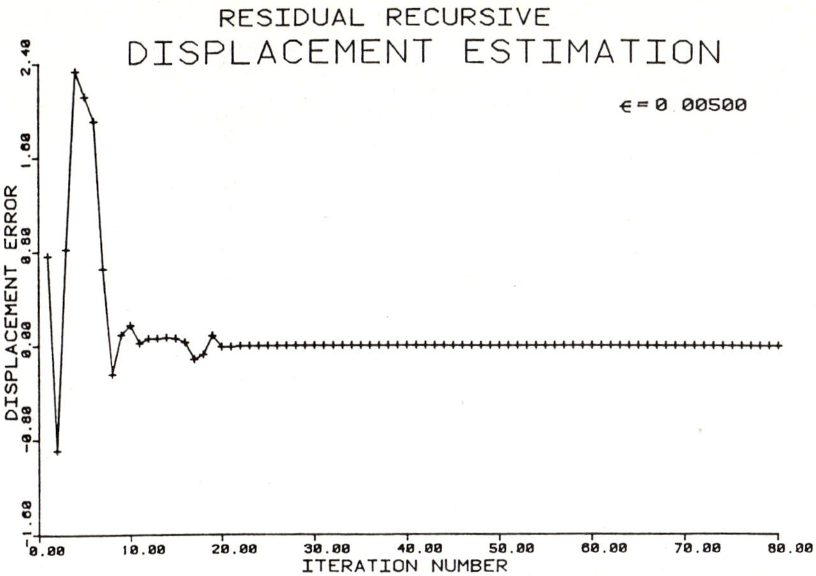

Figure 1

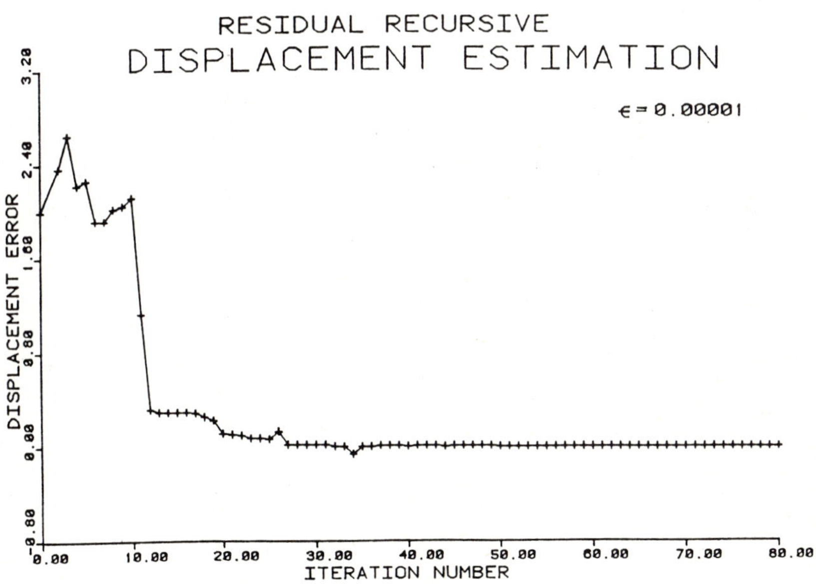

Figure 3

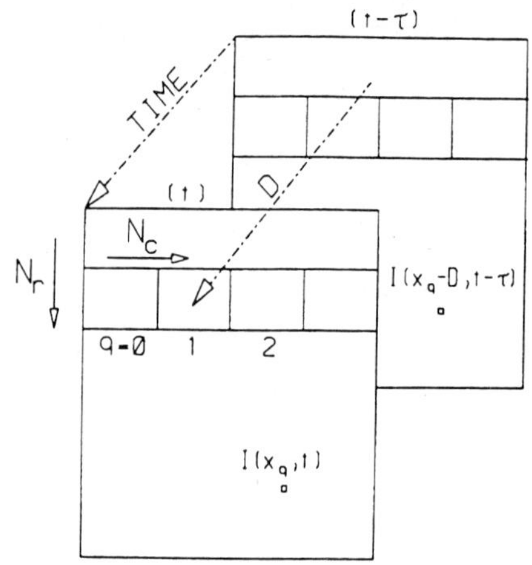

Time Varying Displacement Model
(No Background)

Figure 4

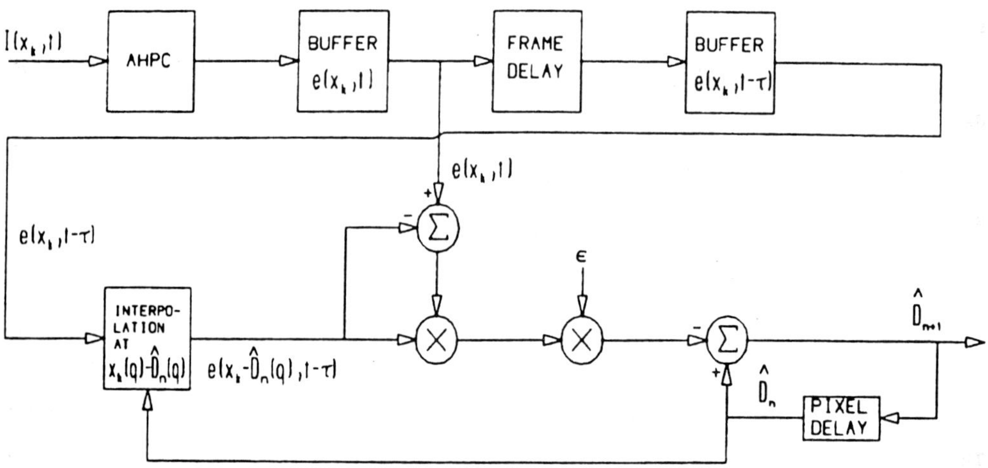

Residual Recursive Displacement Estimation Algorithm

Figure 5

MOTION PARAMETER ESTIMATION IN TV-PICTURES

Z.Houkes

Control and Automation Group

Department of Electrical Engineering

Twente University of Technology

P.O. Box 217, 7500 AE Enschede, The Netherlands

Abstract

Sometimes it may be advantageous or the only way to determine quantities like displacement, rotation angle and scale factor of a moving object from consecutive TV-pictures. A method is presented that estimates the motion parameters using an iterative minimum mean square linear estimator.

The estimation process has been worked out for translations, rotations and scaling. The estimator has been tested for translations and rotations.

First the problem is stated as a non-linear estimation problem. Next is being described how the iterative linear estimator has been obtained.

Results obtained with the estimator, using a picture processing system, and suggestions for further investigations are given.

Introduction

An estimation problem is characterized by two relations that have to be known. The relations in question are known as system and observation model. In this particular case, the system model corresponds with the mathematical description of object motion and is denoted here with "path-model". The observation model determines the relation between observations and the quantities to be estimated.

The principles of the described procedure are taken from a procedure described

NATO ASI Series, Vol. F2
Image Sequence Processing and Dynamic Scene Analysis
Edited by T.S. Huang

in Liebelt [1] for "orbit determination" problems. In common with orbit determination, the motion estimation problem is a non-linear estimation problem. To mold it into a form so that linear theory can be applied, it is linearized by expending about a previous guess of the state-vector. Iterative computations are used to improve the last guess. Hereafter the problem is expressed in terms of the orbit determination problem. Vectors will be designated as lower case with an underscore, matrices as upper case. There has been made one exception for the luminance I, which is denoted with upper case.

Object motion in (2-dimensional) TV-pictures is usually a combination of translation, rotation and scaling. The differential equation

$$\frac{d}{dt}(\underline{p}_i(t)) = \underline{f}_i(\underline{p}_i(t_0),t) \tag{1}$$

representing the movement of the i-th point of a rigid and finite-measured object defines some class of motion estimation problems. The vector $\underline{p}_i(t)$ represents the position vector of the i-th point.

The problem now is to estimate the unknown parameters of this path model. To solve this problem observations and an observation model are needed. The set of observations $\{\theta_i\}$ consists of the luminances I of object pixels in the TV-frame at time $t=t_n$,

$$\theta_i = I\,(\underline{p}_i\,(t_n),\,t_n) + w_i \tag{2}$$

where w_i stands for Gaussian-distributed noise. Because an analytical observation model is missing, the set of luminances of object pixels at $t_{n+1}=t_n+\Delta t$ is taken. With Δt is denoted the frame time, being 40 ms for European standards. From this set the observation model is computed, using numerical methods. This will be explained later on. Using the last guess of the path-parameters in equation (1), the position $\underline{p}_i(t_{n+1})$ is computed. This position is used in the set of luminances of object pixels at $t=t_{n+1}$, to determine an estimate $\hat{\theta}_i$ of the observation θ_i. The equation for $\hat{\theta}_i$ is:

$$\hat{\theta}_i = I\,(\underline{p}_i\,(t_{n+1}),\ t_{n+1}) \tag{3}$$

Because of the non-linear character of the models there will be some difference between θ_i and $\hat{\theta}_i$, denoted with $\delta\theta_i$:

$$\delta\theta_i = \theta_i - \hat{\theta}_i \tag{4}$$

The difference $\delta\theta_i$ is input to a linear estimator whose estimate is used as a correction of the path model parameters. The corrected parameters are used to compute new estimates $\hat{\theta}_i$. The procedure is carried out iteratively, until some error function or its change will be small enough. To determine the set of positions $\{\underline{p}_i(t_n)\}$ the moving area concept [2] is used.

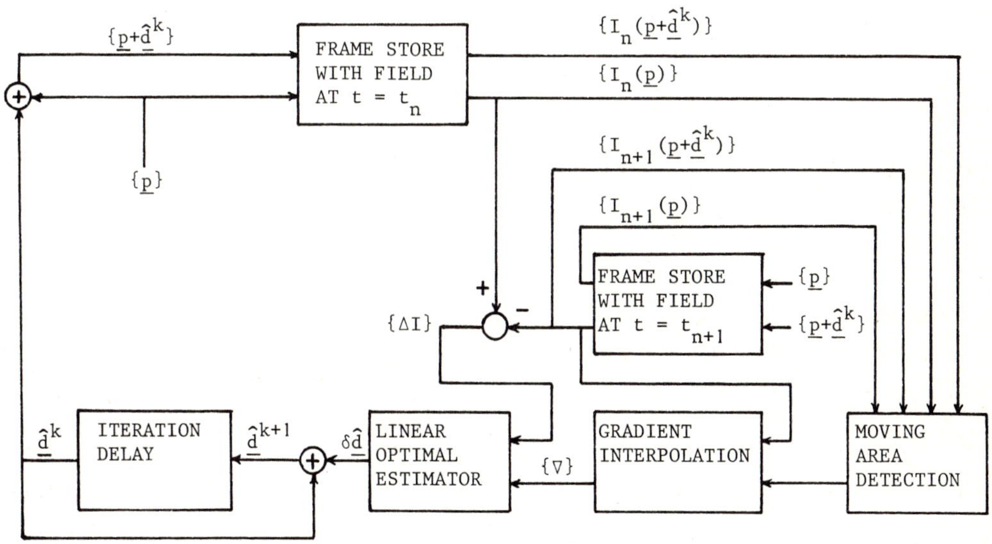

Fig.1 . Block-diagram of the estimation process

Figure 1 illustrates this procedure for the translation of an object. The path-parameter to be estimated is the displacement $\underline{d}$. The index k is the number of the iteration step. The linear estimator output $\delta\hat{\underline{d}}$ is the correction to the current guess $\hat{\underline{d}}^k$. The gradients ∇, computed from the intensities of object-pixels in the frame at t_{n+1}, are used in the linear estimator for the observation model. The components $\delta\theta_i$ of $\delta\underline{\theta}$ are as given by equation (4). $\underline{p}$ indicates the position of the actually scanned pixel.

The next description of the main features of the procedure is restricted to the translation of just one object in a TV-picture. These constraints don´t affect the general principle of the procedure, and they allow comparison with results obtained by Netravali and Robbins [2].

Linear Estimator

As described before, the path model is generally non-linear and the observation model consists of a set of measured luminances. It will be explained how a linear estimator has been obtained.

For the sake of simplicity the problem is restricted to translations, which means that the displacement is equal for all points of the object. The equation (1) becomes for every point of the object:

$$\underline{p}\,(t_{n+1}) = \underline{p}\,(t_n) + \underline{d} \tag{5}$$

with:

$$\underline{d} = \int_{t_n}^{t_{n+1}} \underline{v}\,(t)\,dt \tag{6}$$

$\underline{v}(t)$ = speed at time t

The path parameter to be estimated is the displacement vector $\underline{d}$. An estimate for $\delta\underline{d}$ can be obtained from expressions for the "a priori form" of the minimum mean square error linear estimator given by Liebelt [1]. This assumes that $\delta\underline{d}$ and noise $\underline{w}$ are uncorrelated, and no information is available concerning the true $\delta\underline{d}$. If an "a priori" estimate is missing the following equations will give an estimate $\delta\hat{\underline{d}}$ of $\delta\underline{d}$:

$$\delta\hat{\underline{d}} = C_e B^T (C_w)^{-1} \delta\underline{\theta} \tag{7}$$

with:

$$C_e = (B^T (C_w)^{-1} B)^{-1} \tag{8}$$

C_e: Error matrix

C_w: Noise covariance matrix

B : Observation matrix

In these formulas the transposed of the observation vector $\delta\underline{\theta}$, is given by the following expression,

$$(\delta\underline{\theta})^T = (\delta\theta_1, \delta\theta_2, \ldots\ldots, \delta\theta_m) \tag{9}$$

and with $\delta\theta_i$ given by equation (4). The index m indicates the number of pixels belonging to MA. The covariance matrix C_w of the noise is assumed to be a diagonal matrix with equal and constant terms q. The actual value of q, being the expectation of w_i^2, is only of importance for the error matrix C_e. The observation matrix B, defined by $\delta\underline{\theta} = B\,\delta\underline{d}$, is obtained from the equations (3) and (5) by linearization. To simplify notation, the field to be used for computation of luminances I will be indicated by an index n or n+1.

$$\delta\theta_i = (\nabla I_{n+1}(\underline{p}_i(t_n) + \hat{\underline{d}}))^T(\delta\underline{d}) + w_i \tag{10}$$

with:

$$(\delta\underline{d})^T = (\delta d_x, \delta d_y) \tag{11}$$

$$\nabla I_{n+1}(\underline{p}_i(t_n)+\hat{\underline{d}}) = \begin{pmatrix} \frac{\partial}{\partial x_i^{n+1}} I_{n+1}(\underline{p}_i(t_n)+\hat{\underline{d}}) \\ \frac{\partial}{\partial y_i^{n+1}} I_{n+1}(\underline{p}_i(t_n)+\hat{\underline{d}}) \end{pmatrix} \tag{12}$$

The observation matrix B is now given by the expression:

$$B = \begin{pmatrix} \frac{\partial}{\partial x_1^{n+1}} I_{n+1}(\underline{p}_1(t_n)+\hat{\underline{d}}) & \frac{\partial}{\partial y_1^{n+1}} I_{n+1}(\underline{p}_1(t_n)+\hat{\underline{d}}) \\ \vdots & \vdots \\ \frac{\partial}{\partial x_m^{n+1}} I_{n+1}(\underline{p}_m(t_n)+\hat{\underline{d}}) & \frac{\partial}{\partial y_m^{n+1}} I_{n+1}(\underline{p}_m(t_n)+\hat{\underline{d}}) \end{pmatrix} \tag{13}$$

The gradients in the B-matrix are computed from the luminances of the object pixels in the frame at t_{n+1}. For that purpose, use is made of a simple algorithm with linear interpolation. This will be described next.

Computation of Gradients

Figure 2 shows a pixel P belonging to the moving area MA in the TV-frame at $t=t_{n+1}$. Because the estimate of $\underline{d}$ generally doesn't have an integer value, the point with position vector $\underline{p}_i(t_n)+\hat{\underline{d}}$ will not coincide with a grid point. For that reason the luminance of displaced pixels is estimated by linear interpolation from the luminances of the 3 nearest grid points. So the luminances of four new points, indicated by U,D,L,R are computed. The gradients at the point $\underline{p}_i(t_n)+\hat{\underline{d}}$ are obtained with:

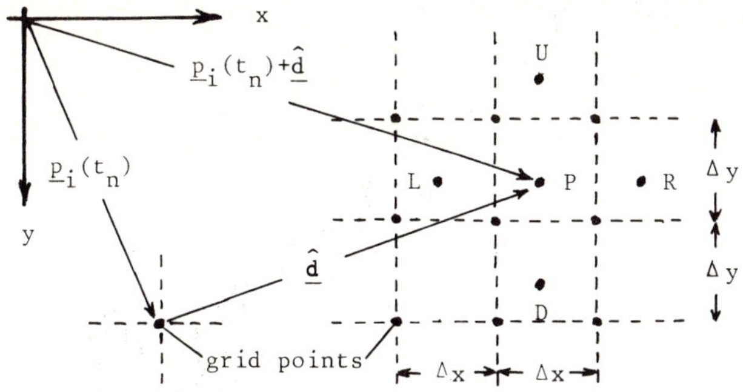

Fig. 2. Configuration for computation of gradients

$$\frac{\partial}{\partial x_i^{n+1}} I_{n+1}(\underline{p}_i(t_n)+\hat{\underline{d}}) \approx (I_R - I_L)\ /2\Delta x \tag{14}$$

$$\frac{\partial}{\partial y_i^{n+1}} I_{n+1}(\underline{p}_i(t_n)+\hat{\underline{d}}) \approx (I_U - I_D)\ /2\Delta y \tag{15}$$

with Δx , Δy being the grid distances for the x- and y-direction. The four points U,D,L,R are taken as:

$$x_U = x_P \qquad y_U = y_P - \Delta y$$

$$x_D = x_P \qquad y_D = y_P + \Delta y$$

$$x_L = x_P - \Delta x \qquad y_L = y_P$$

$$x_R = x_P + \Delta x \qquad y_R = y_P$$

Moving Area

Without explanation it was assumed earlier in this paper that the positions of the object pixels $\underline{p}_1(t_n),\underline{p}_2(t_n),\ldots,\underline{p}_m(t\ )$ are known. This set of pixels is only an approximation. To know which pixels are to be taken to compute the observation matrix B, the idea of the "moving area" MA [2] is used. The pixel with position vector $\underline{p}(t_n)$ belongs to MA if the following equations are satisfied.

$$\left| I_n(\underline{p}(t_n)) - I_{n+1}(\underline{p}(t_n)) \right| \geq \Delta_{th} \quad (16)$$

$$\left| I_n(\underline{p}(t_n) + \hat{\underline{d}}) - I_{n+1}(\underline{p}(t_n) + \hat{\underline{d}}) \right| \geq \Delta_{th} \quad (17)$$

with Δ_{th} is a threshold to eliminate noise.

When parts of the moving object have equal greylevel (difference within the threshold Δ_{th}), it will happen that pixels within these parts don't satisfy the equations (16) and (17). When this phenomenon is ignored, the following can be said. When the estimate of $\underline{d}$ is zero, the moving area MA0 is formed by the pixels of the object at $t=t_n$ and $t=t_{n+1}$. When the estimate of $\underline{d}$ is a good estimate, the positions $\underline{p}$ that satisfy these equations will denote pixels that belong to the object at $t=t_n$. Figure 3 illustrates this for a rectangular shaped object.

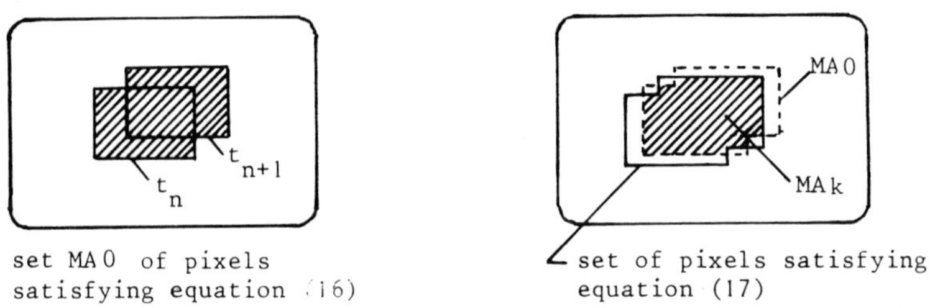

Fig. 3. Construction of moving area MA_k

Estimation of Rotation Angle and Scale Factor

In case of rotations and scaling, the motion of the object is characterized by the rotation angle and scale factor, which both are equal for every point of the object. So it can be proved that the following equation holds:

$$\underline{p}_i(t_{n+1}) = \underline{p}_c + sF(t_{n+1}, t_n)(\underline{p}_i(t_n) - \underline{p}_c) + \underline{d} \quad (18)$$

with:

$\underline{p}_c$: position vector of rotation centre

F : transition matrix

s : scale factor

$\underline{d}$: displacement vector

The transition matrix is given by the equation:

$$F\ (t_{n+1},\ t_n) = \begin{pmatrix} \cos\phi & -\sin\phi \\ \sin\phi & \cos\phi \end{pmatrix} \tag{19}$$

with: $\phi = \int_{t_n}^{t_{n+1}} \omega(t)dt$

$\omega(t)$ = rotation speed at time t

The path parameters to be estimated are the elements of the displacement vector $\underline{d}$, the scale factor s, and the rotation angle ϕ. To simplify notation, $\underline{d}$, s and ϕ are combined to a vector $\underline{\alpha}$ according to:

$$\underline{\alpha} = \begin{pmatrix} d_x \\ d_y \\ s \\ \phi \end{pmatrix}$$

On the same conditions and in the same way it is possible to derive a linear estimator that will estimate corrections $\delta\underline{\alpha}$ to a previous guess $\hat{\underline{\alpha}}$. As explained for translations we need an observation matrix B. This matrix can be obtained from equation (3) and (18) by linearization around a previous guess. This yields:

$$\delta\theta_i = \left(\frac{\partial}{\partial\underline{\alpha}}\ I_{n+1}\ (\underline{p}_i\ (t_{n+1}))\right)^T \delta\underline{\alpha} + w_i \tag{20}$$

with: $(\delta\underline{\alpha})^T = (\delta d_x, \delta d_y, \delta s, \delta\phi)$

$$\left(\frac{\partial}{\partial \underline{\alpha}} I_{n+1}(\underline{p}_i(t_{n+1}))\right)^T = \left(\frac{\partial}{\partial \underline{p}_i^{n+1}} I_{n+1}(\underline{p}_i(t_{n+1}))\right)^T \frac{\partial}{\partial \underline{\alpha}} \underline{p}_i(t_{n+1}) \qquad (21)$$

If we denote the i-th row of the observation matrix B with B than:

$$B_i = \left(\frac{\partial}{\partial x_i^{n+1}} I_{n+1}(\underline{p}_i(t_{n+1})), \frac{\partial}{\partial y_i^{n+1}} I_{n+1}(\underline{p}_i(t_{n+1}))\right) \begin{pmatrix} \frac{\partial x_i^{n+1}}{\partial d_x} & \frac{\partial x_i^{n+1}}{\partial d_y} & \frac{\partial x_i^{n+1}}{\partial s} & \frac{\partial x_i^{n+1}}{\partial \phi} \\ \frac{\partial y_i^{n+1}}{\partial d_x} & \frac{\partial y_i^{n+1}}{\partial d_y} & \frac{\partial y_i^{n+1}}{\partial s} & \frac{\partial y_i^{n+1}}{\partial \phi} \end{pmatrix}$$

The derivatives to d_x, d_y, s and ϕ are computed from equation (18) and are found to be:

$$\frac{\partial x_i^{n+1}}{\partial d_x} = 1 \qquad \frac{\partial y_i^{n+1}}{\partial d_x} = 0$$

$$\frac{\partial x_i^{n+1}}{\partial d_y} = 0 \qquad \frac{\partial y_i^{n+1}}{\partial d_y} = 1$$

$$\frac{\partial x_i^{n+1}}{\partial s} = (x_i^n - x_c) \cos\phi - (y_i^n - y_c) \sin\phi$$

$$\frac{\partial y_i^{n+1}}{\partial s} = (x_i^n - x_c) \sin\phi + (y_i^n - y_c) \cos\phi \qquad (23)$$

$$\frac{\partial x_i^{n+1}}{\partial \phi} = -s(x_i^n - x_c) \sin\phi - s(y_i^n - y_c) \cos\phi$$

$$\frac{y_i^{n+1}}{\partial \phi} = s(x_i^n - x_c) \cos\phi - s(y_i^n - y_c) \sin\phi$$

The derivatives of $I_{n+1}(\underline{p}_i(t_{n+1}))$ to x_i^{n+1} and y_i^{n+1} are computed in the same way as has been done for translations. With the same assumption for C_w, the equations (7) and (8) yield an estimator for $\delta\underline{\alpha}$. The complete formulas are omitted, because of the lot of space needed for it. The reader who is interested in the complete set of formulas, is invited to contact me.

This estimator is used to obtain the results in the following paragraph. The movements of the object (a motor vehicle) are respectively a translation and a combination of rotation and translation.

Results

The described algorithms are implemented on an INTELLECT picture processing system, consisting of a framestore of 512*512 bytes, with input and output processing facilities and interfaced with an LSI-11 micro computer. The pixels are digitized into one byte words, so that 256 different greylevels can be distinguished.

To determine the performance of the iterative linear estimator, it is sufficient to store two consecutive and corresponding half frames (2 * 256 lines * 512 pixels). As is shown in figure 4 , the 2-dimensional picture of a moving car serves as the moving object. The movement is nearly a translation.

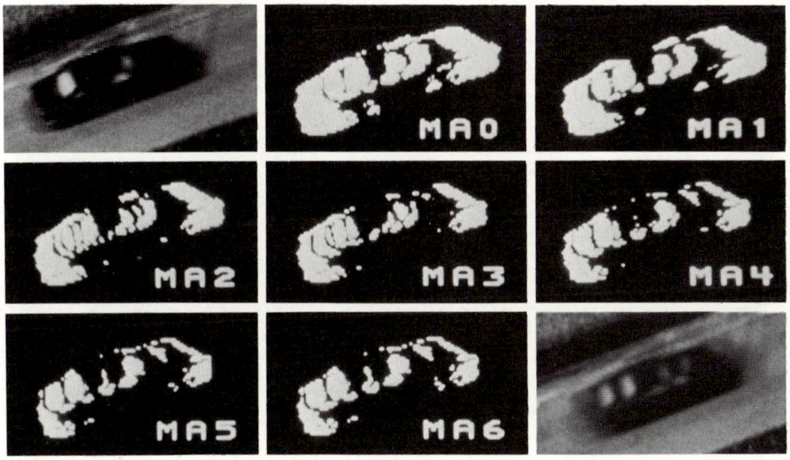

Fig. 4. Moving area after successive iteration steps

The photographs show only a detail of the stored frames (143*126 pixels). This method allows a thorough study of the operation of the procedure. Also is shown the moving area after different numbers of iteration steps.

In figure 5 the estimated displacement is drawn as a function of the number of iteration steps. The rotation angle and scale factor are omitted , because they are near 0 and 1 respectively. After about 5 steps the estimate is stationary. The procedure was started with an initial estimate $\underline{\hat{d}} = \underline{0}$, $\hat{s}=1$ and $\hat{\phi}=0$.

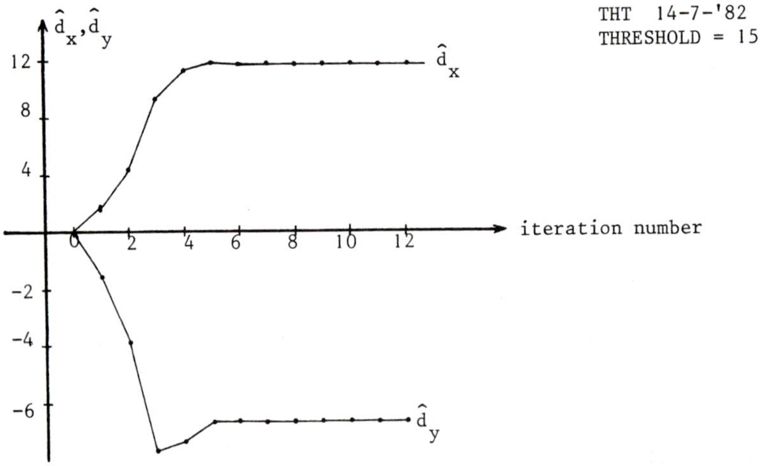

Fig.5 . Estimated displacement for a nearly translating object

The estimated displacement is not very sensitive for the threshold level, as can be derived from Table 1. This table gives for different thresholds Δ_{th}, the number of pixels IMA0 and IMA, respectively belonging to the moving area with $\underline{\hat{d}}=0$ and with $\underline{\hat{d}}=\underline{\hat{d}}^{12}$ the value listed in the table. The quantity NMA is the difference between IMA0 and IMA. The last column shows the estimated displacement for an other choice of the rotation centre (0,0). There is a small difference in the estimated displacement caused by $\hat{s}$ and $\hat{\phi}$, being not equal to 1 and 0 respectively. This introduces small extra displacements, without changing the total displacement per pixel.

The lowest value of Δ_{th} (=8) just doesn't add pixels to the moving area caused by noise. The highest value (=30) causes an essential loss of object pixels, which results in a decrease of NMA. But nevertheless, the estimate of $\underline{d}$ is not affected very much.

x_c	90	90	90	0	
y_c	90	90	90	0	
Δ_{th}	8	15	30	15	
IMA	1051	664	226	664	
NMA	785	756	577	756	
IMAO	1836	1420	803	1420	
$\hat{d}_x^{12}$	11.748	11.744	11.747	11.468	
$\hat{d}_y^{12}$	-6.629	-6.592	-6.610	-6.497	
$\hat{s}$	1.001	1.001	1.001	1.001	
$\hat{\phi}$	-0.001	-0.002	-0.004	-0.002	Table 1

The estimator has been tested for a translating and rotating car. The results are shown in figure 6.

Final Remarks

First it must be understand that the procedure only works for moving objects. When there is no moving object, the moving area MA will be an empty set. The iterative linear estimator, whose derivation has been based on (linear) estimation theory, has proved to operate correctly. For pure translations the results correspond (not shown in this paper) to those obtained by Netravali and Robbins [2]. But from now on, it will be possible to include other types of motion such as rotation and scaling. The estimator needs about 6 steps to give a stationary estimate starting from a zero estimate. If there is a reasonable initial estimate then less steps will be satisfactory. Such an initial estimate could be obtained from the displacements of the moving area.The thresholding of the successive pictures is nearly a differentiating operation. So the low frequency components are more attenuated then the high frequency components. This will make the estimator unstable for high thresholds.

It must be said that the used picture has good contrast and texture. Also the displacement is smaller than the measures of the object. Objectives of future investigations will be to determine the influence of these factors.

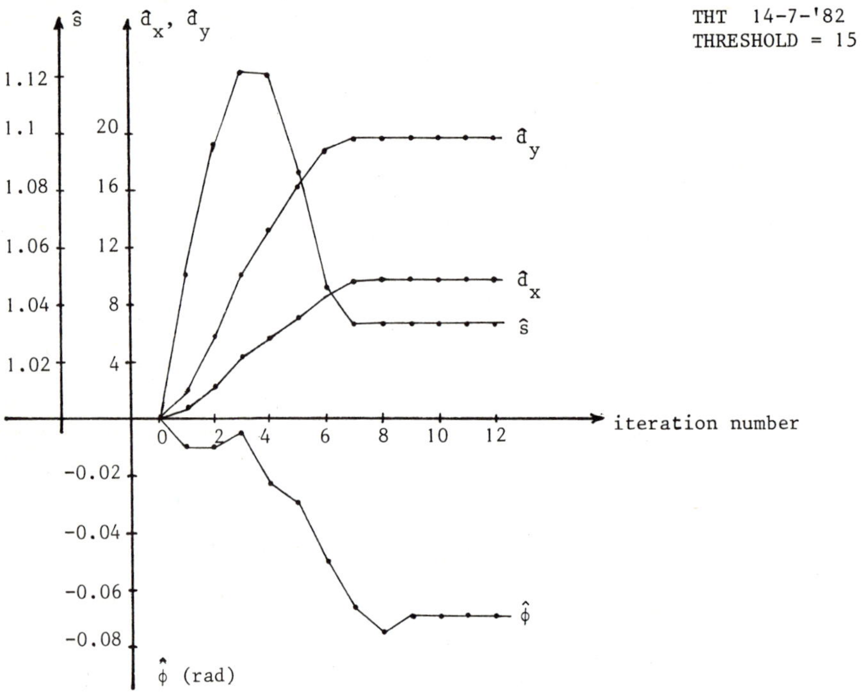

Fig.6 . Estimated displacement, scale factor and rotation angle for a translating and rotating object.

The described estimation proces can be applied to problems who accept a slow estimation of motion. In cases where a fast implementation is needed, a realisation in dedicated hardware has to be thought about. Besides it is not superfluous to find methods that will accelerate the execution of the procedure. An acceleration can be obtained from a division of the moving area in (rectangular) overlapping segments. The estimate of a previous segment is used as the initial estimate in the next section. This as well as realization of dedicated hardware are objectives of near future investigations.

References

[1] Paul B.Liebelt, An Introduction to Optimal Estmation, Addison Wesley Publishing Company, pp. 169-171 and pp. 161-163.

[2] A.N.Netravali, J.D.Robbins, Motion Compensated Television Coding, The Bell System Technical Journal, March 1979

IMAGE SEQUENCE CODING USING SCENE ANALYSIS AND SPATIO-TEMPORAL INTERPOLATION

R. Lenz, A. Gerhard
Lehrstuhl für Nachrichtentechnik
(Institute for Telecommunications)
Technische Universität München
Arcisstrasse 21, D-8000 München 2
Federal Republic of Germany

Abstract

In video-conferencing or monitoring applications it might be desirable to use data channels with very low bandwidth. In order to achieve bit rates of 64kBit/s and less in seems inevitable to subsample the input scene in time. Therefore the reciever will have to perform a spatio-temporal interpolation of the transmitted key-frames in order to reconstruct the missing in-between images. To do this properly, further information about the moving parts of the scene, derived by scene analysis, has to be transmitted. The segmentation procedure as well as the interpolation procedure are presented.

Keywords
Coding, Scene Analysis, Interpolation, Occlusion, 3-D Model, Motion Estimation

Introduction

A moderate picture quality can be achieved at about 0.2 bit/pel when adaptive Hybrid-DCT coding techniques are employed. Using a frame size of 256 by 256 pels and limiting the data rate to 64kBit/s implies that only about every fifth to tenth frame can be transmitted. Since subsampling of the sequence violates the sampling theorem, linear reconstruction procedures for the missing frames, i.e. interpolation only in time, will not result in the original sequence and cause image blurring or motion jerkiness. It is therefore necessary to perform spatio-temporal interpolation (movement compensation), which has to take into account the non-linear effects resulting from the fact that the background may be occluded by moving objects and these in turn may be occluded by other moving objects. To enable the receiver to do this kind of interpolation, the transmitter has to supply a description of the scene in terms of moving objects, their boundary lines, trajectories and occlusion priority levels (OPL).

NATO ASI Series, Vol. F2
Image Sequence Processing and Dynamic Scene Analysis
Edited by T.S. Huang

Image Model

The image model used takes into account the occlusion occurring when a 3-D scene is projected into two dimensions and is limited in size by the frame boundaries. Each part of the scene is labelled with its occlusion priority level (OPL): a lower number indicates that it may be occluded by an object with higher OPL. Figure 1 shows the scetch of an image, the background is labelled with OPL 0, the moving car with '1', holes in the car i.e. windows with '0' and a bird in front of the car with '2'. In order to have a consistent description of the 3-D scene, the frame boundaries which may occlude parts of objects moving into the visible image area are automatically labelled as a stationary object with maximal OPL, in this example '3'.

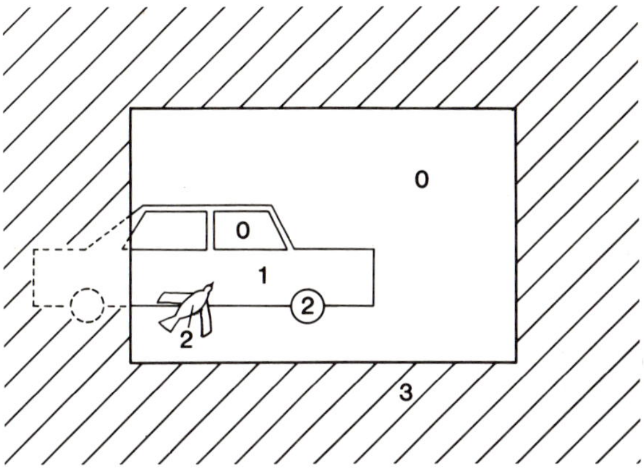

Fig. 1. Pseudo 3-D image model with indicated occlusion priority levels

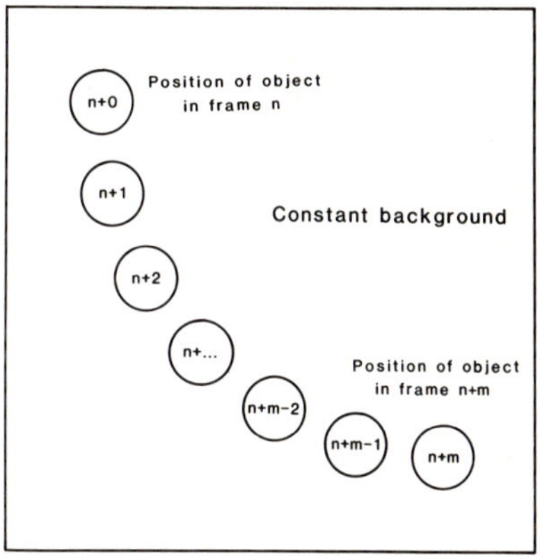

Fig. 2. Superimposed sequence with one moving object and stationary background

Interpolation

Figure 2 shows a superimposed scene consisting of one rigid object with translational motion in front of a stationary background. Every m-th frame, called key-frame, is transmitted to the receiver. If the receiver would interpolate only in time without taking into account the movement in space, the object would gradually disappear at position n+0 and gradually appear at position n+m. To reproduce the missing frames correctly, the receiver has to interpolate the object along its trajectory and fill up the uncovered background with the pictorial data from that key-frame which has the data available (extrapolation), or alternatively from a background memory if it were supplied (figure 3).

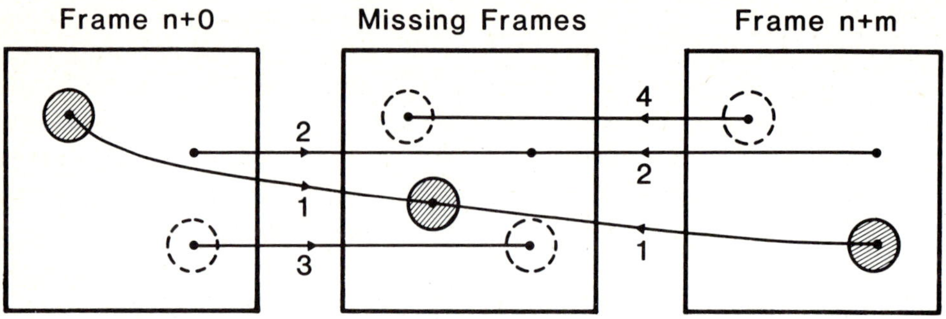

1: Interpolation of moving object along trajectory

2: Interpolation of stationary background (along zero-trajectory)

3: Extrapolation of covered background from first key-frame

4: Extrapolation of uncovered background from second key-frame

Fig. 3. Reproduction procedure for the missing frames by use of the key-frames

Segmentation

It is the transmitter's task to provide a description of the scene, i.e. to perform a segmentation into stationary and moving parts and to track the latter through the scene. One possibility would be to determine the displacement vector field on a pel-by-pel basis and to cluster regions of similar displacement. The segmentation algorithm presented in this paper is object-based and attemps to track moving objects as a whole. A block-diagram of the segmenter is shown in figure 4.
The segmentation and tracking process is separable into three stages:

1) Derivation of a primary mask
2) Region growing of that mask
3) Feature extraction and object tracking

Stage 1:
The first stage uses frame-to-frame differences in order to derive a primary mask of the moving objects. A symmetrical arrangement of difference pictures on each side of the investigated frame ensures a one-to-one relationship from input image to output mask which is not provided in the case of single-sided differencing. The difference pictures between previous and current frame as well as between current and next frame are thresholded and logically AND-ed. As one can immediately see from figure 5, the

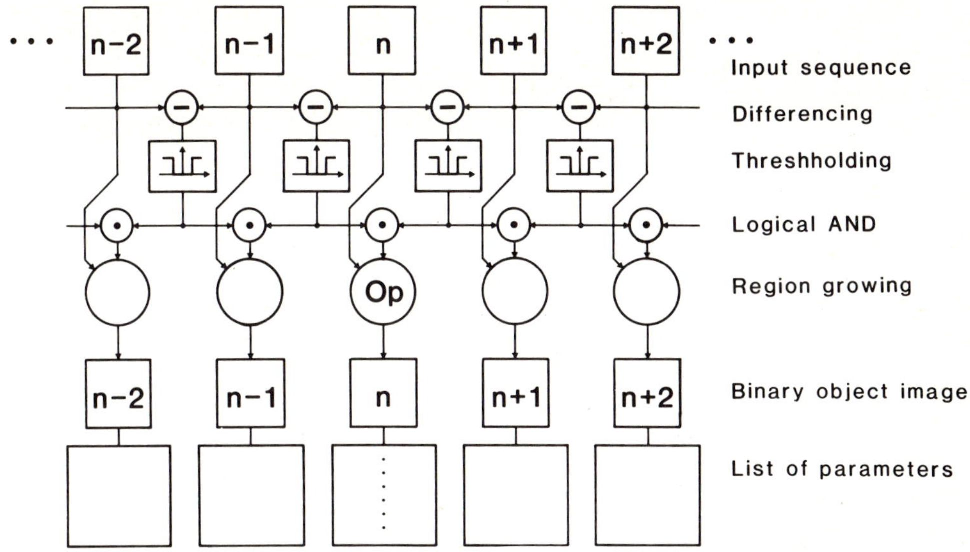

Fig. 4. Block-diagram of the segmenter for image sequences

primary mask will not completely cover the object in the input frame if it has a fairly homogenious greylevel and does not completely displace itself from one frame to the next. It is not possible to lower the binarization threshold too much, because then camera noise would be detected also. Therefore an adaptive threshold is used (figure 6), directing it's attention to regions of interest, i.e. of change. This adaptive threshold may be considered as a low-pass filtered version of the absolute value of the difference picture subtracted from a relatively high fixed threshold. This procedure will not blurr the object boundaries but still suppress noise effectively. Another measure to obtain a better primary mask is to take more than one difference picture on both sides of the investigated frame. Figure 7 shows an example for 2+2 DPs. The thresholded DPs on each side are OR-ed independently and then both groups are AND-ed.

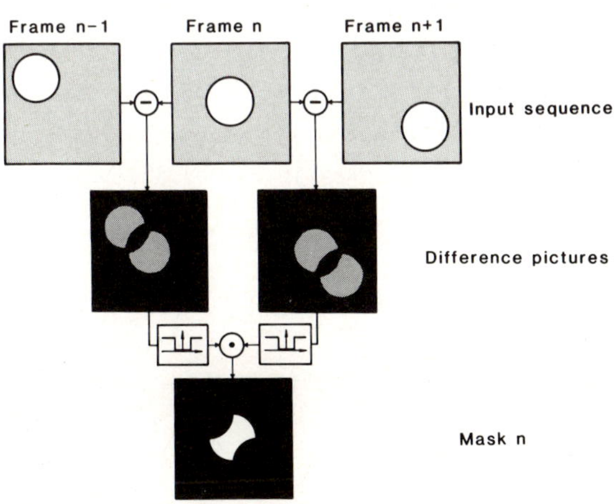

Fig. 5. Symmetrical differencing

This differencing procedure is easy to implement and quite effective, but it will still not deliver a complete mask of the moving objects. Therefore another processing step is added.

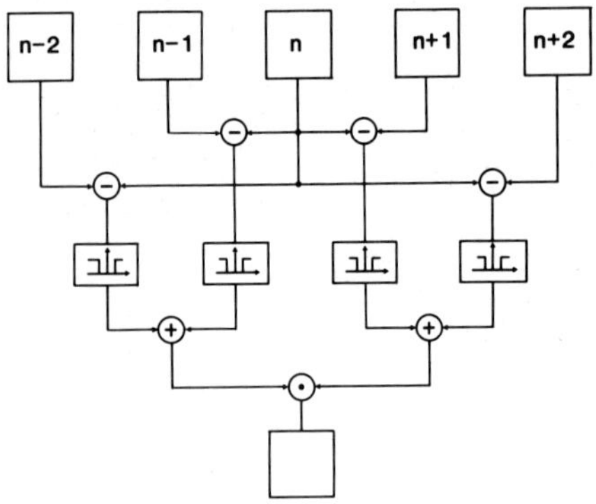

Fig. 6. Extension of figure 5 for multiple difference pictures

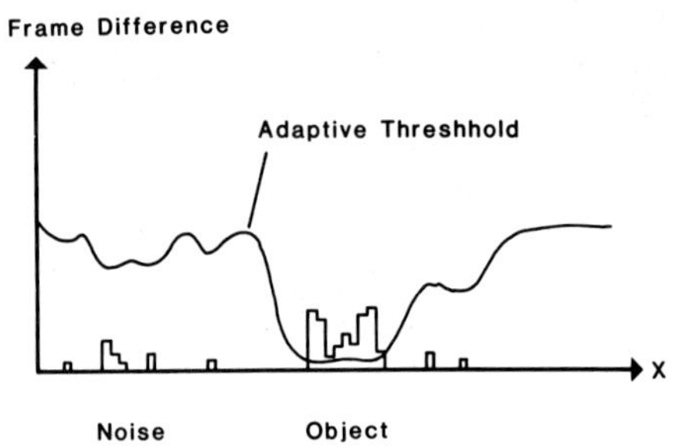

Fig. 7. Adaptive threshold for noise suppression

Stage 2:

In the second stage a region growing algorithm is applied on the primary mask. It can take advantage of the fact, that the greyvalue is relatively constant in those parts of the object where the difference picture was nearly zero. The 8-connected neighborhood of the investigated pel is searched for pels already marked by the primary mask. If marked neighbors exist, then their average greyvalue is calculated and compared with the greyvalue of the pel in question. If their difference lies within a certain confidence interval the pel is marked as part of the mask. This is done twice in a recursive manner, once scanning the image from the upper left to the lower right and once in the opposite direction. More elaborate region growing techniques might lead to better results but have not been tested by the authors yet.

Results of the first two stages applied on a scene with a battery-driven toy car, climbing up a small hill on the left are shown in figures 9 through 12. The scene was taken directly into RAM in real time, the frame size is 256 by 256 pels with 64 greylevels.

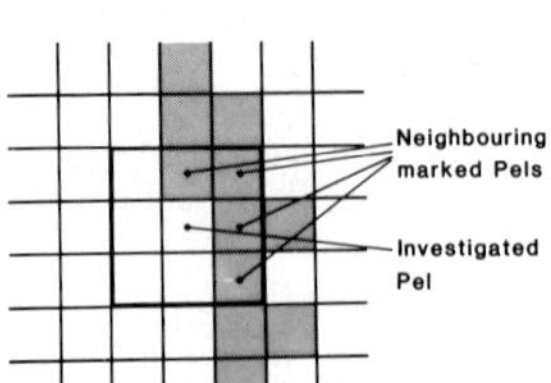

Fig. 8. Region growing window

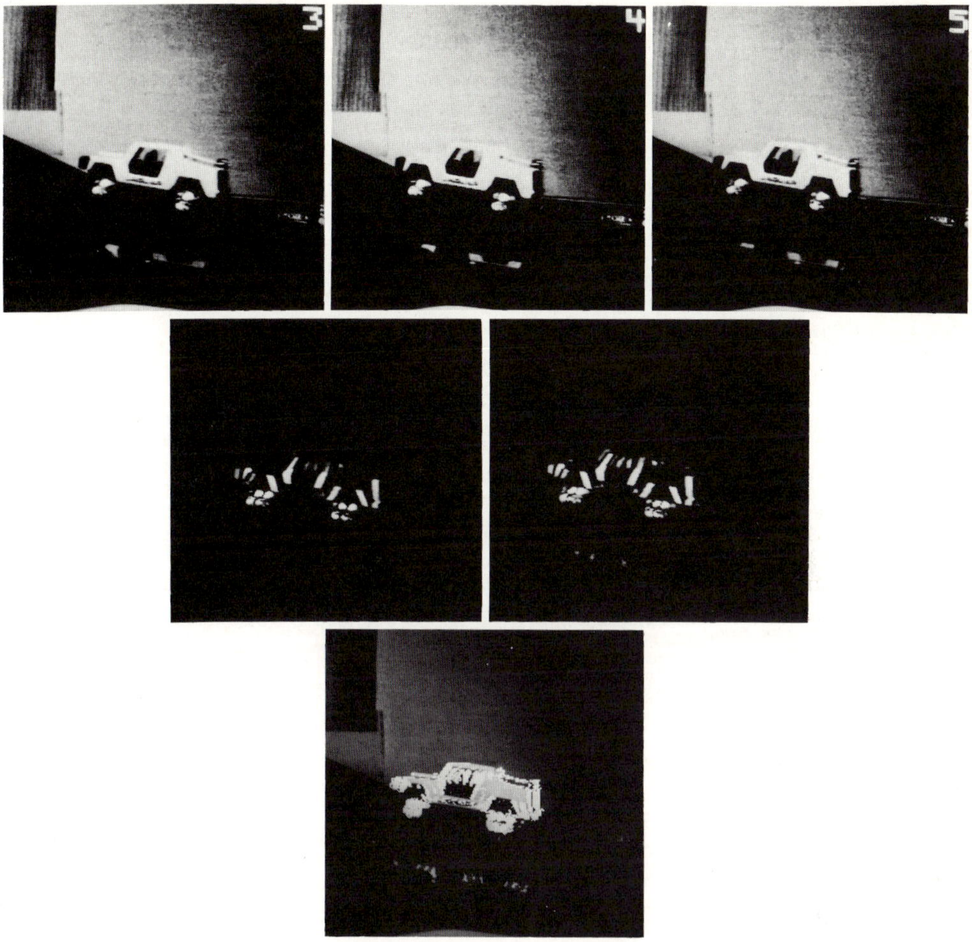

Fig. 9 Derivation of primary mask for image no. 4 with one DP on either side

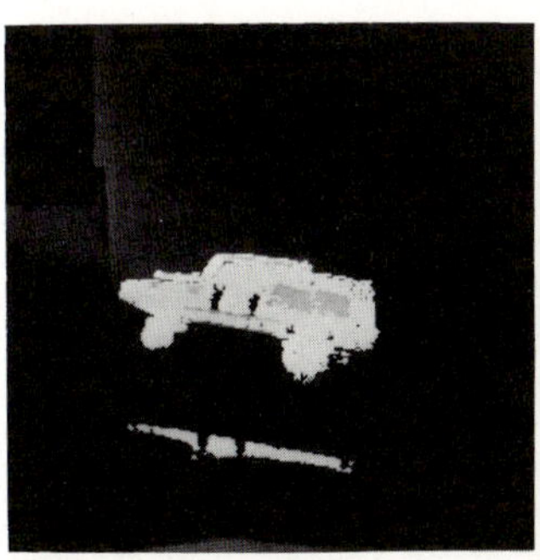

Fig. 10 Improved mask with 3 DPs/side

Fig. 11 Mask after region growing

Fig. 12 Boundary line and center of gravity

Stage 3:
In the third stage, object features are extracted and used to track the individual objects through the scene. The features are object area, average greyvalue, center of gravity, boundary line and occlusion priority. When these parameters are calculated, one has to consider that the objects might be partly occluded, like the car in fig. 1 which is occluded by the frame border and the bird. The displacement of the center of gravity of the visible portion of the car, for example, is not equivalent to its true displacement.

Recursive Differencing:
One can easily see that by simple differencing only it is not possible to detect moving objects in front of moving objects, i.e. objects of OPL 2 like the bird in figure 1. One pass through the three stages of the segmenter will only deliver objects of OPL 1. If objects of higher OPL shall be detected, it is necessary to repeat the differencing and tracking procedure, this time with the objects of OPL 1 being registered, not the background with OPL 0 (figure 13). Here the technique of adaptive thresholding proves to be particularly useful, since it is very difficult to register moving objects exactly, especially when they are distorted by occlusion.
The number of recursions required is determined by the number of priority levels desired. For coding applications two passes seem to be sufficient.

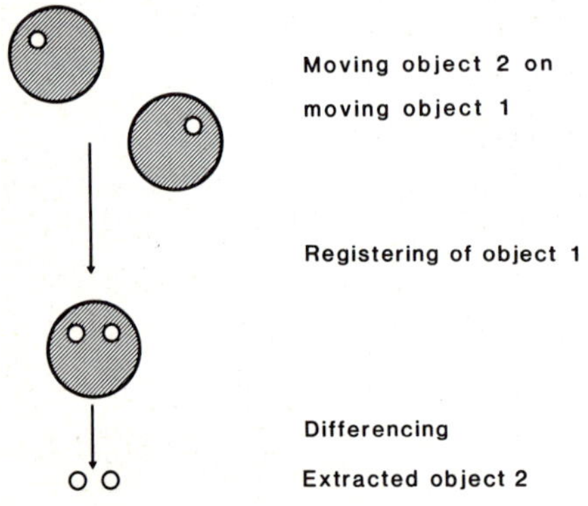

Fig. 13. Recursive differencing

Interpolation Examples

In figure 14 a reconstructed scene of an object moving over a stationary background is shown. Only the first (No. 1) and the last (No. 16) frame (errorless keyframe) were transmitted, together with the trajectory and boundary line of the object (figure 15). The little white dots appearing in the last few frames (except for the keyframe itself) result from a non-perfect segmentation of the original scene. The dots lie along the boundary line of the objects position in the last frame and were con-

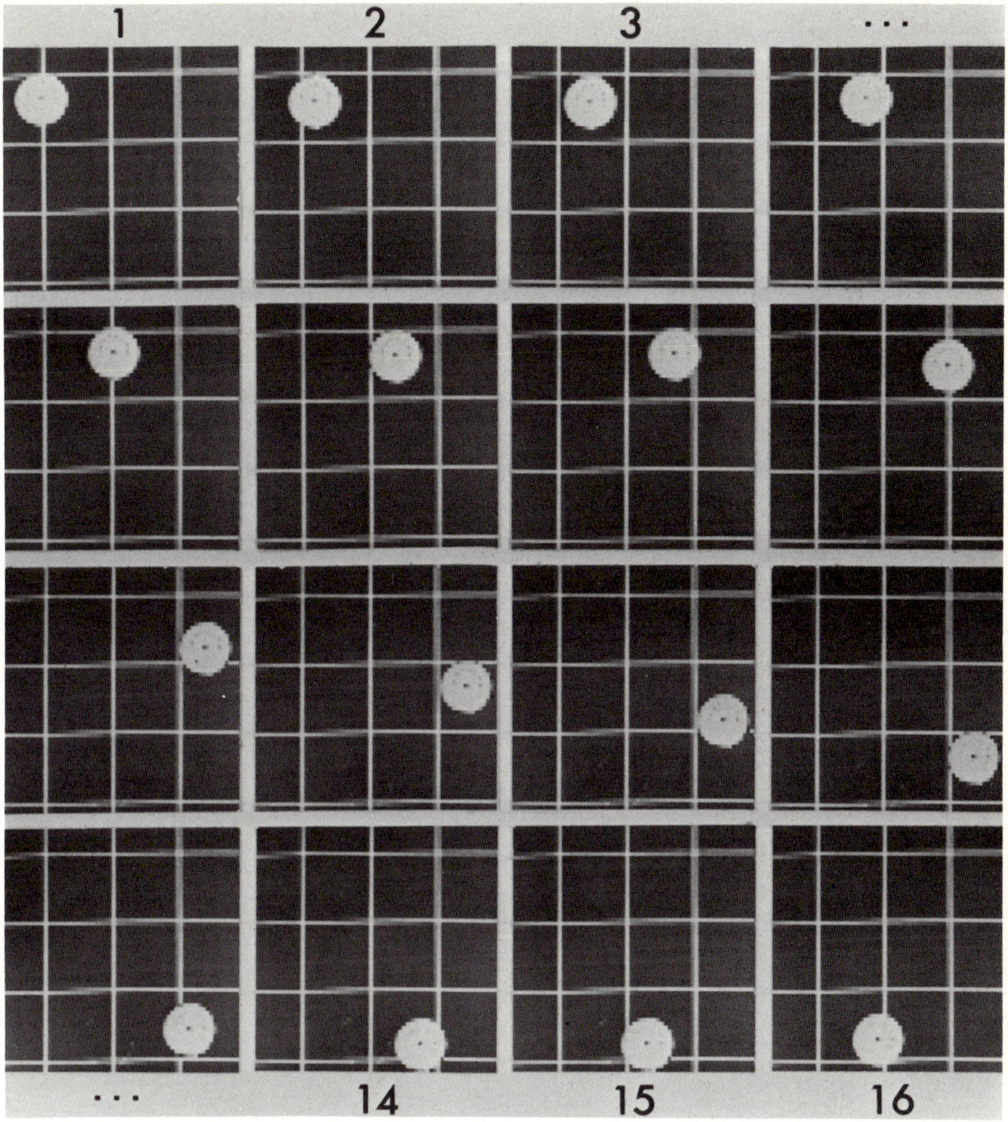

Fig. 14. Reconstructed sequence, only frame 1 and frame 16 were transmitted

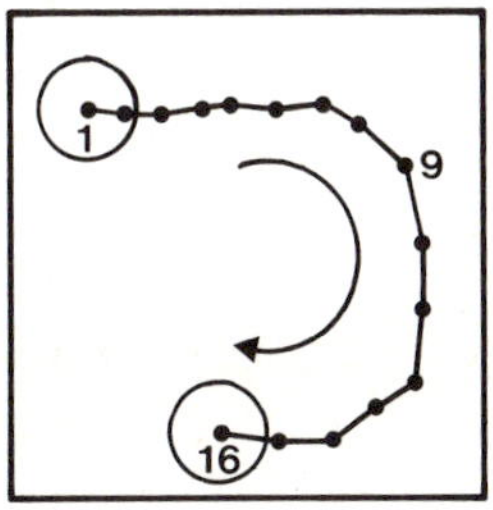

Fig. 15. Trajectory and boundary lines of object in fig. 14

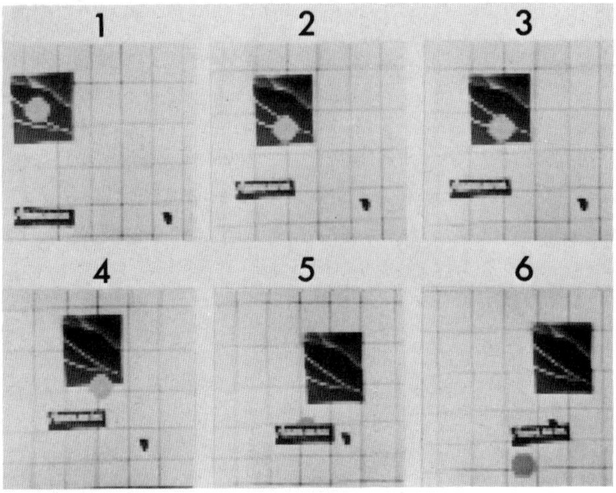

Fig. 16. Original sequence with moving background

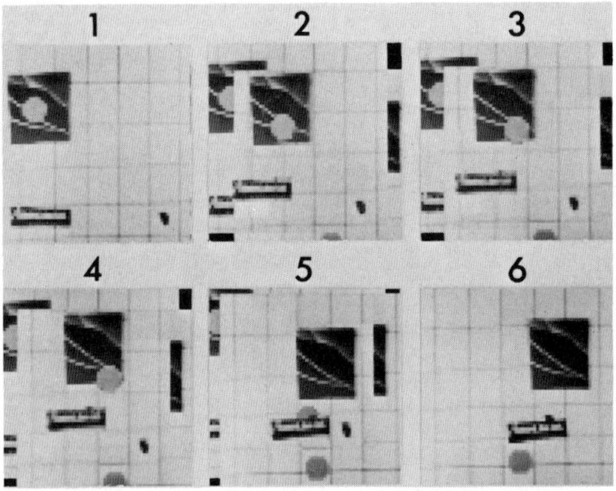

Fig. 17. Interpolated sequence with regions of missing information showing up

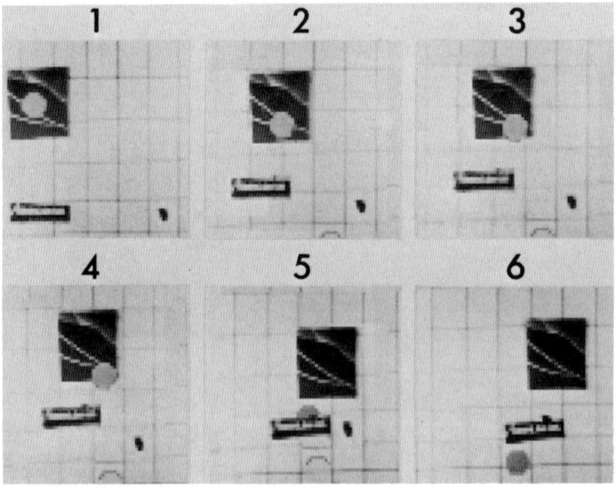

Fig. 18. Interpolated sequence with regions of missing information filled up with their neighborhood

sidered by the segmenter to be part of the background. Therefore they were interpolated only in time and not in space (see also figure 3). One can immediately see the importance of the correct function of the segmenter.

Another interpolation example can be seen in figures 16 through 18, this time with moving background and several objects occluding each other. The segmentation was done by hand in order to cope with the moving background. Figure 16 shows the original 6-frame-sequence, the black blob in the lower right corner is fixed to the background and indicates its diagonal movement. Again only the first and the last frame were transmitted together with the background and object movement, trajectories and OPLs. Figure 17 shows the effect when the receiver does not take into account that diagonally moving background exhibits image regions where no pictorial data exists. These regions, the lower left and upper right corner, therefore appear black in the reconstructed frames. A similar problem of lacking information occurs when an objects occludes parts of the background in the second key-frame which are not visible in the first key-frame and are exhibited by background movement. This is the case for the little grey circular object near to the lower margin of the second key-frame and it therefore appears twice in the interpolated in-between frames.

Figure 18 shows the interpolation results when the receiver takes measures to let the areas of missing or invalid information not show up as much. Since by means of the transmitted scene description the receiver knows where data is invalid, it can fill up these regions by taking the average greyvalue of the surrounding picture elements.

Conclusion

A coding scheme is proposed which uses low-level scene analysis to isolate moving objects and to describe their motion and occlusion properties. Recursive application of the segmentation procedure allows the measurement of relative motion. This information is used to derive a pseudo 3-D model of the scene and is made available to the receiver to interpolate the missing frames of the subsampled input sequence.

The image model makes it possible to describe the effects resulting from occlusion and suggest a straightforward approach to build up a background memory with a size greater than the actual visible image. Up to now the segmentation algorithm is limited to stationary background and rigid objects with purely translational motion. Future work will persue to overcome these constraints. The output of the segmenter might then also prove useful as a low-bitrate input for computer vision in robotics applications.

References

/1/ J. K. Aggarwal, R. Duda: Computer Analysis of Moving Polygonial Images IEEE Trans. Comp., vol. C-24, Oct. 1975, pp. 966-976

/2/ H. C. Bergmann: Motion Adaptive Interpolation of Eliminated TV-Fields Proc. Picture Coding Symposium, Montreal 1981

/3/ T. S. Huang (ed.): Image Sequence Analysis Springer Series in Information Sciences, Vol. 5, 1981, ISBN 3-540-10919-6

/4/ R. Jain, H. H. Nagel: On the Analysis of Accumulative Difference Pictures from Image Sequences of Real World Scenes, PAMI, vol. 1, no. 2, Apr. 79, pp. 206-214

/5/ F. May: Hybrid Coding of Picture Sequences for Transmission over Narrowband Mobile Radio Channels, Signal Processing: Theories and Applications, North-Holland Publishing Company, Eurasip 1980, pp. 283-288

/6/ S. Yalamanchili, W. N. Martin, J. K. Aggarwal: Extraction of Moving Object Descriptions via Differencing, Comp. Graph. and Image Proc., vol. 18, 1982 pp. 188-201

TWO MOTION ADAPTIVE INTERFRAME CODING TECHNIQUES FOR AIR TO GROUND VIDEO SIGNALS

Siegfried Beyer

Lehrstuhl für Theoretische Nachrichtentechnik und
Informationsverarbeitung
Universität Hannover
Callinstr. 32
3000 Hannover 1
West Germany

ABSTRACT

Two motion adaptive interframe coding techniques for video signals of air to ground surveillance systems are described. Using motion adaptive techniques both coders take advantage that successive fields of air to ground video signals are mainly displaced depending on the flight velocity of the camera. The first coding technique is based on motion adaptive interpolation and results in a data rate of about 1 Mbit/s. The second coder uses motion adaptive extrapolation and achieves a data rate of about 0.1 Mbit/s. The efficiency of both coders with respect to picture quality and hardware requirement is discussed.

1. INTRODUCTION

The transmission of air to ground video signals requires a data link which is robust against jamming. This can be achieved by reducing the transmission data rate and in addition by channel coding techniques. This paper is concerned with the reduction of the transmission data rate using motion adaptive interframe coding techniques.

Intraframe coding techniques as predictive coding, transform coding or hybrid coding allow a data reduction from 8 bit per pel to 4 bit per pel without visible degradation /1,2,3,4,5/. These techniques remove intraframe redundancy but not take advantage of the correlation between successive frames. Using interframe coding techniques without any motion adaptation a further reduction of the data rate

NATO ASI Series, Vol. F2
Image Sequence Processing and Dynamic Scene Analysis
Edited by T.S. Huang

can be achieved mainly in stationary parts and in parts with moderate movement /4,6,7,8/. New results indicate that 3 bit per pel are required for coding natural TV-signals without visible degradation /9/. Motion adaptive transform coding techniques as well as predictive coding techniques provide further reduction of interframe redundancy /10,11,12/. For special applications motion adaptive interpolation techniques can be utilized to reduce the frame rate at the transmitter and to interpolate the omitted frames at the receiver /13,14/.

The investigations described in this paper are specially concerned with applications of air to ground surveillance. For these applications the coder has to be simple whereas the receiver can be more complex. For a typical surveillance mission the flight track is straight forward and the TV-camera is directed to the ground whereby the direction of the lines is vertically to the flight direction. In this situation successive fields are mainly displaced against each other depending on the flight velocity and only a few lines of each field contain new information. Figure 1 shows the scanning of the ground in two successive fields.

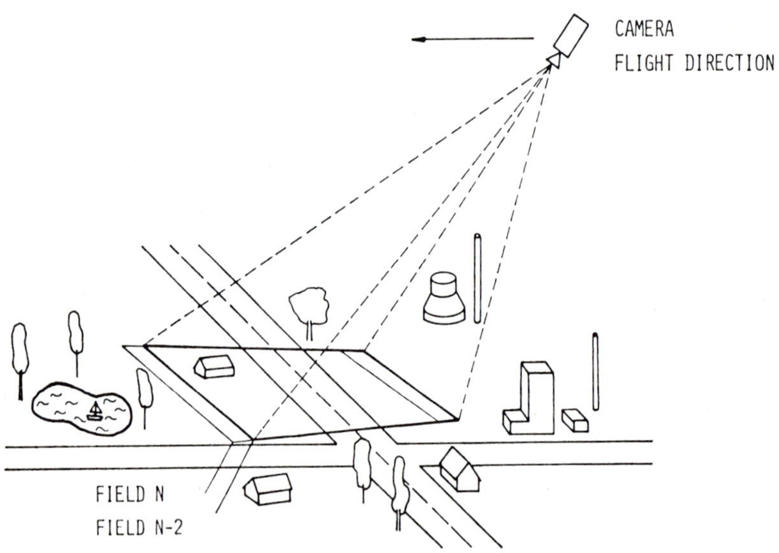

Figure 1. Scanning of the ground in successive fields

This a priori knowledge about the special scanning situation is taken into account designing a motion adaptive interpolation and extrapolation coding technique.

2. INTERFRAME CODING WITH MOTION ADAPTIVE INTERPOLATION

The motion adaptive interpolation technique is based on the idea to reduce the field rate at the transmitter and to interpolate the omitted fields at the receiver. Due to the camera movement the scanned ground is displaced from field to field. Therefore the interpolation at the receiver has to be carried out with regard to the displacement between transmitted fields. Figure 2 shows how this interpolation is performed in the case that only every L-th field is transmitted. Transmitted picture elements are marked by solid points.

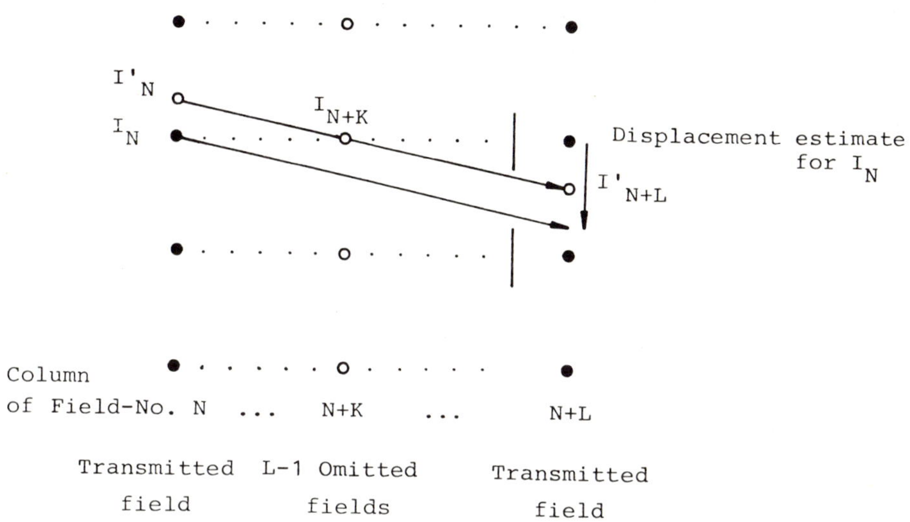

Figure 2. Motion adaptive interpolation of picture element I_{N+K} shown by columns of the transmitted fields N, N+L and one column of field N+K to be reconstructed

For the special case of air to ground TV it can be assumed that each picture element of field N is displaced to field N+L along a linear trajectory. In this case it is obvious to reconstruct the picture elements of each omitted field by linear interpolation in direction of the trajectories between the transmitted field N and N+L. The direction of each trajectory depends on the displacement between field N and N+L. From the displacement estimate for I_N the trajectory originating at I_N is given. For reconstructing picture element I_{N+K} the trajectory crossing this picture element must be known. This trajectory has to be calculated from the displacement estimate respectively from the neighbouring trajectory. For simplification it is assumed that the trajectory crossing I_{N+K} is parallel to the trajectory with foot at picture element I_N. In the case the trajectory crossing I_{N+K} connects the picture elements I'_N and I'_{N+L}. Because these picture elements in general are not placed on rasterpoints, each luminance is calculated by spatial bilinear interpolation from four adjacent picture elements /15/. The reconstruction of picture element I_{N+K} is then determined by

$$I_{N+K} = \frac{L-K}{L} I'_N + \frac{K}{L} I'_{N+L} \qquad , K = 1,2,3,\ldots,L-1 \qquad (1)$$

For hardware simplification this formula is rearranged to

$$I_{N+K} = \frac{K}{L} (I'_{N+L} - I'_N) + I'_N \qquad (2)$$

Due to the displacement between the transmitted fields a certain part in the top and the bottom of the reconstructed fields cannot be interpolated from two transmitted fields, because luminance information from only one field is available. Reconstructing field N+K three areas have to be distinguished. The top of the field N+K is extrapolated from field N+L, the middle is interpolated from field N and N+L, and the bottom is extrapolated from field N, see Figure 3.

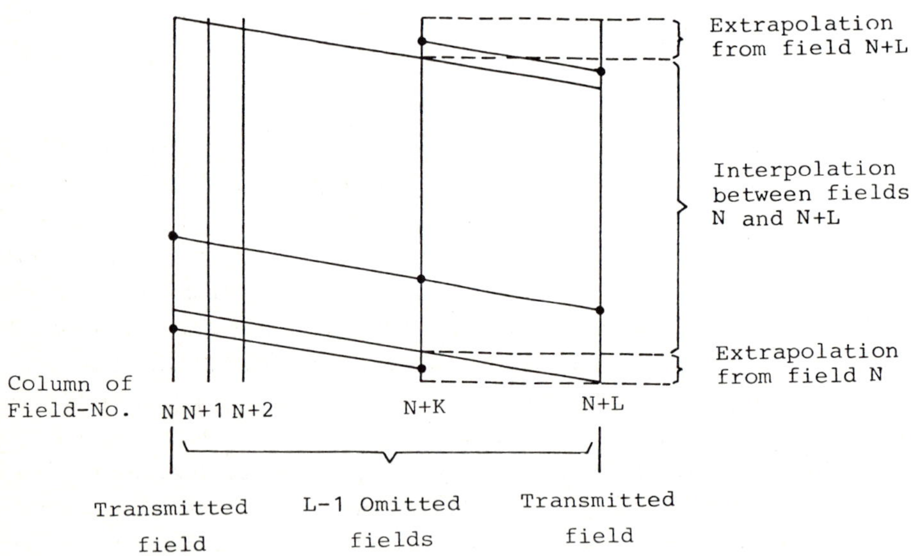

Figure 3. Areas of extrapolation and interpolation for reconstructing field N+K

An additional reduction of the transmission data rate is achieved by an intraframe DPCM-coder. Figure 4 shows a blockdiagramm of the complete source coder and decoder.

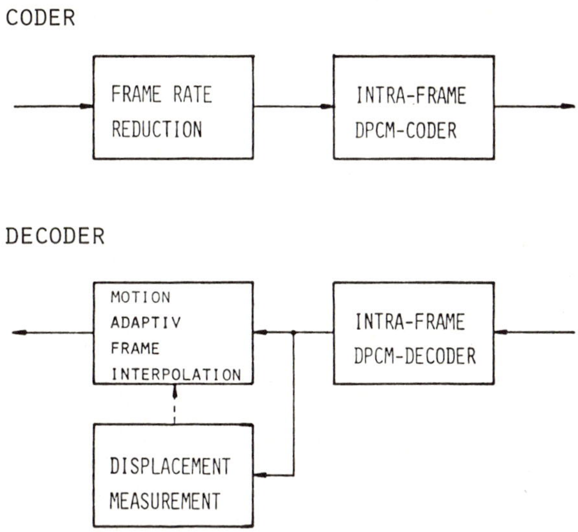

Figure 4. Blockdiagram of the codec including DPCM-coder and decoder with motion adaptive frame interpolation

The two components of the digital coder at the transmitter are simple to implement in hardware. The frame rate reduction can be performed by a switch and the intraframe DPCM-coder requires a line memory for the DPCM loop. At the receiver the transmitted codewords are decoded to obtain the luminance information of the transmitted fields. The displacement estimation for the motion adaptive frame interpolation is performed for blocks of 15 by 15 pel based on a special correlation algorithm using a window size of 15 by 15 pel /16/. This algorithm calculates the normalized correlation function of the spatial differentials and requires only a 3 level quantization before multiplication is carried out retaining good accuracy. The value of the normalized correlation peak defined in the range of -1 and +1 is taken as a measure of accuracy. Only if this value is greater than 0.4 the position of the correlation peak is taken as a displacement estimate and assigned to the corresponding block. For correlation peaks less or equal 0.4 the displacement for this block is interpolated from the four adjacent estimates. For displacement measurement and motion adaptive interpolation two field memories are required.

This coding concept has the advantage of a simple transmitter because the displacement measurement is performed at the receiver. Transmitting F fields per second and using a n bit DPCM-coder the resulting data rate R for a field size of M pels per line and N lines per field is

$$R = M \cdot N \cdot n \cdot F \quad (3)$$

in bit per second.

The spatial resolution depends on the field size $M \cdot N$, the choice of the DPCM-codeword length determines the picture quality of the transmitted fields. The temporal resolution of moving objects or changing background depends on the field frequency F. For a given fields size $M \cdot N$ and a codewordlength n the data rate depends only on the field frequency F. To achieve a low data rate it is necessary to reduce the field rate as much as possible. Of course there is a lower bound for the minimum field rate which is reached when the area of interpolation vanishes (s. Figure 3). In this case the ground would be scanned only once by successive transmitted fields and a displacement measure at the receiver is possible no longer. For applications where the temporal resolution can be reduced to a single scan of the ground a coding concept based on motion adaptive extrapolation is proposed.

3. INTERFRAME CODING WITH MOTION ADAPTIVE EXTRAPOLATION

The motion adaptive extrapolation technique is based on the model that successive fields are vertically displaced against each other. Then the main part of a field is identically with its preceding field, only a few lines of each field contain new information. Therefore it is sufficient to transmit only these lines from each field and to extrapolate the rest of each field at the receiver from already transmitted information. In the case of additional horizontal displacement also some columns at the border contain new information with regard to the preceding field. For simplicity of the coder and decoder this information is not transmitted. Figure 5 shows this segmentation for two fields N and N-2.

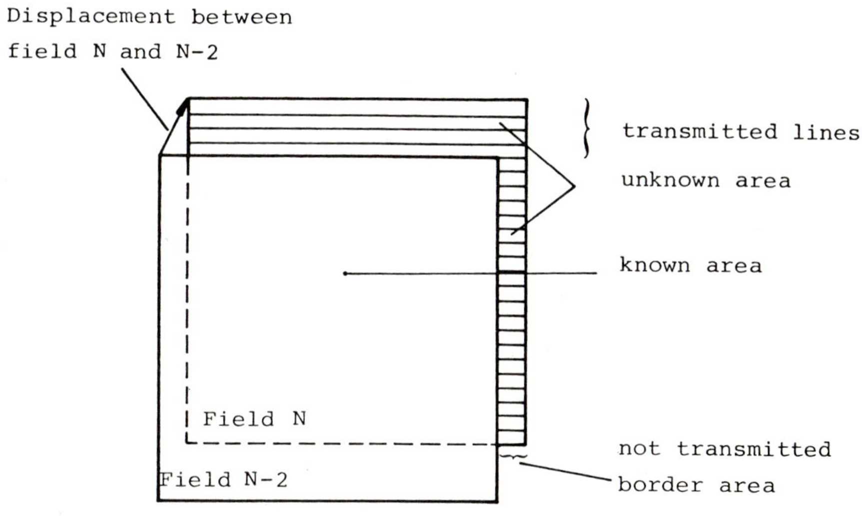

Figure 5. Segmentation of field N in known and unknown areas and marking the lines to be transmitted

For selecting the lines with new information the vertical component of the displacement between field N and N-2 must be measured. In order to reconstruct a field at the receiver both components of the displacement are required. Because the transmitted segments of each field do not overlap, displacement estimation is not possible at the receiver and therefore must be performed at the transmitter. Figure 6 shows the composition of field N at the receiver.

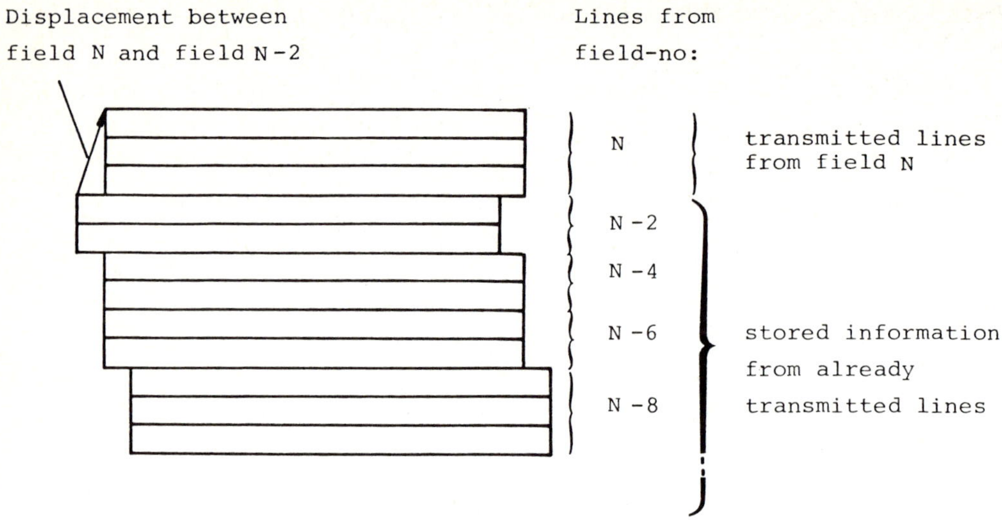

Figure 6. Reconstruction of field N from transmitted lines of field N and from already transmitted and stored lines

In case of a perpendicular looking camera the resolution of the ground is constant within one field. Then each picture element has the same displacement if the flight path is straight forward. Therefore the displacement estimate can be performed in any position of a field. However, in the case of a forward looking camera the spatial resolution of the ground and displacement increases from top to bottom within one field due to the perspective viewing conditions. In this situation the top of a field contains information of the sky or ground in far distance. Therefore the lines to be transmitted are taken from the bottom part of a field, starting at an eligible reference line r. At this reference line r also the displacement measurement is performed.

Figure 7 shows the scanning situation using a forward looking camera. Assuming a vertical displacement of 3 lines between field N and field N-2 at reference line r, the three lines r-2, r-1 and r of field N have to be transmitted. The motion adaptive extrapolation at the receiver is the same as for a perpendicular looking camera.

For displacement measurement seperate displacement estimates are carried out along the reference line r in distance of 15 pels using the same correlation technique explained before. Only if the peak of the normalized correlation function is greater than 0.4 a displace-

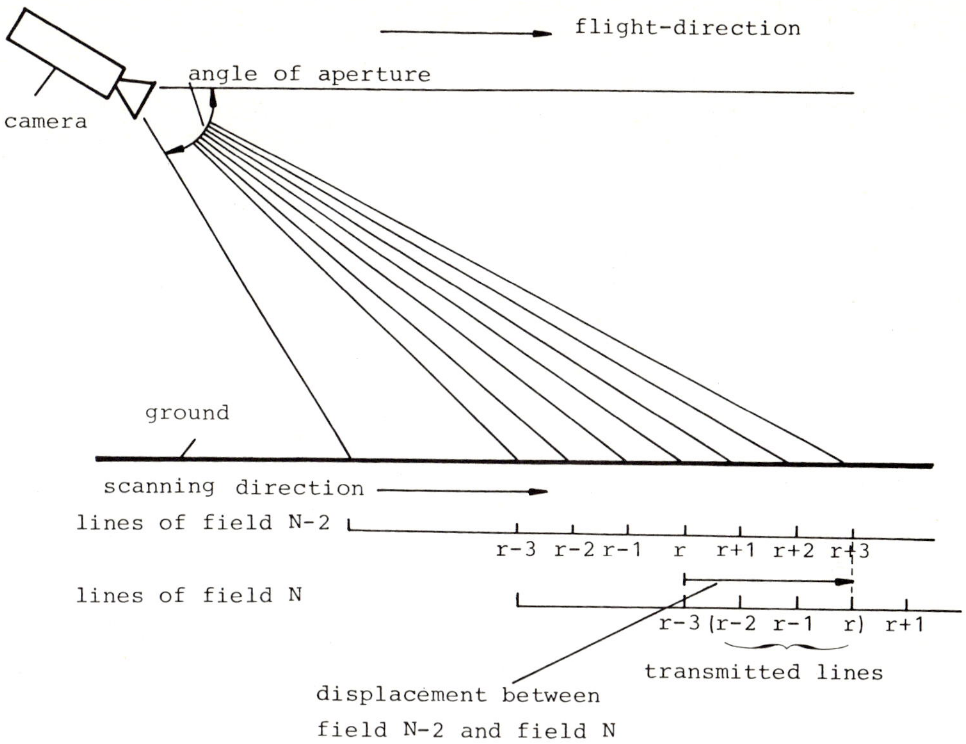

Figure 7. Line selection in case of a forward looking camera

ment estimate is taken into account. The displacement estimate between two successive fields is calculated as the mean value of all selected displacement estimates along the reference line r. For hardware simplification the horizontal and vertical components are rounded to integer values. The error due to the rounding is taken into account calculating the next integer displacement by error feed back. In case of a constant noninteger vertical displacement of for instance 1.5 pel between two fields the error feed back prevents transmitting continously 2.0 lines per field which would magnify the picture in vertical direction. Due to the error feed back alterning 1 line and 2 lines per field are transmitted which allow a better geometric reconstruction. Figure 8 shows the block diagram of the coder and decoder with motion adaptive extrapolation.

Closing switch "A" 15 lines and additional lines depending on the maximum displacement in vertical direction are stored. The displace-

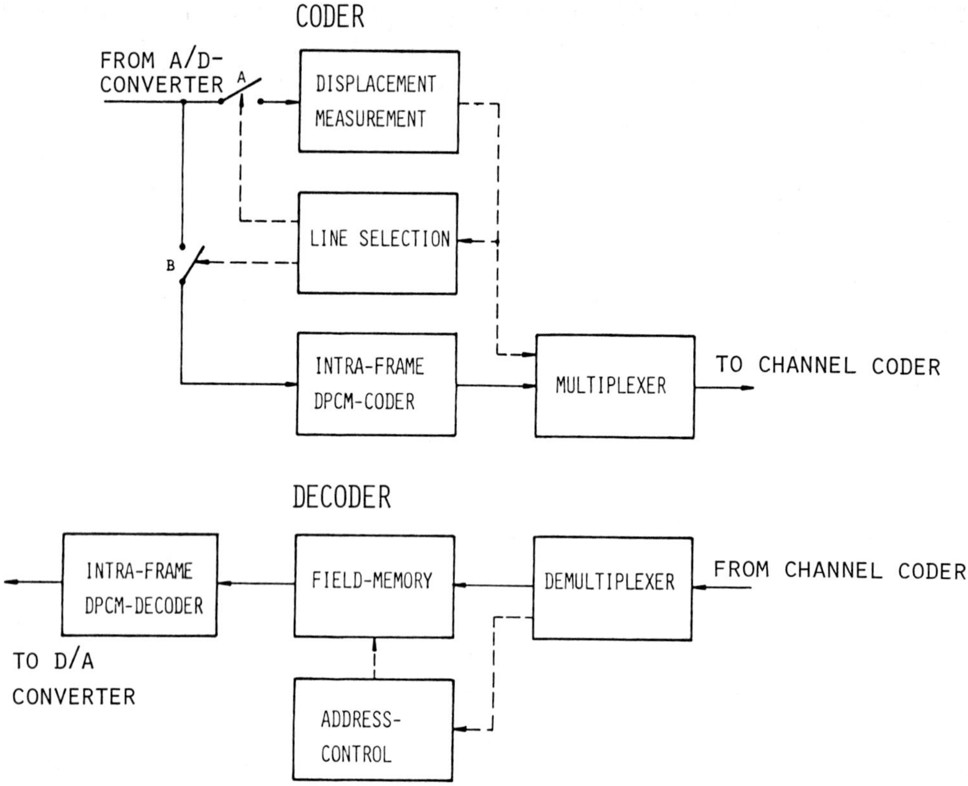

Figure 8. Block diagram of the DPCM-coder with motion adaptive extrapolation

ment estimate is calculated from this data and the stored lines from the preceding field. The integer component of the displacemnt estimate in vertical direction is used to control switch "B" for selecting the lines to be transmitted. These lines are coded using a 3 bit intraframe DPCM-Coder. For continous coding of the new lines to be transmitted it is necessary that the processing of the lines within each field is from bottom to top. The DPCM-codewords of each field are transmitted together with the integer components of the vertical and horizontal displacement estimates.

At the receiver the integer displacement components are separated from the DPCM-codewords. The stored information in the field memory is shifted according to the integer displacement components and the new transmitted lines are inserted at the top of the field memory. This operation requires mainly the calculation of new addresses and is provided by a special address control. For reconstruction of the TV-signal the content of the field memory is DPCM-decoded in direction from bottom to top which is the same processing direction as at the transmitter.

The integer value of the displacement in vertical direction determines the number of lines to be transmitted. The integer horizontal displacement specifies how to compose lines from successive fields in horizontal direction. The data rate of this coder depends on four parameters:

$$R = a \cdot v \cdot M \cdot n \qquad (4)$$

in bit per second, where 'a' specifies by how many lines per meter the ground is scanned and 'v' is the flight velocity in meter per second. 'M' is the number of pels per line and 'n' the codewordlength of the DPCM-coder in bit per pel.

The described coding technique is based on the assumptions that the flight height is great compared with object altitudes at the ground, the camera is directed vertically to the ground and the displacement between successive fields is translatoric. In this case the achieved picture quality depends mainly on the DPCM codewordlength and the accuracy of the displacement measurement. In the case of rotary movements or forward looking camera this simple technique produces geometric distortion which can be prevented using more complex reconstruction algorithms which take into account the nontranslatoric movement. This coder is more complex than the coder with motion adaptive interpolation but has the advantage to achieve a lower data rate.

4. SIMULATION RESULTS

Both coding techniques were simulated on a digital picture processing system. The TV-scene was taken from a forward looking standard TV-camera mounted on an airplane and recorded on a videorecorder. The flight velocity was 200 km/h and the altitude about 200 m. Due to the maximum data rate of the picture processing system only TV-sequences of 1/6 of the whole frame size could be processed and displayed. The videosignal was samples with 10 MHz and quantized with 8 bit per pel.

Figure 9 shows one frame out of the TV-sequence which was taken from a flight over a bridge.

Figure 9. Single frame from the TV-sequence
Size of frame segment: 192 pels by 256 lines
Coding: 8 bit PCM

(Publication authorization no. Brg. 5232/41, released by the Braunschweig district authorities, Federal Republic of Germany)

The motion adaptive interpolation coder has been simulated for the case that only every 16th field is transmitted and a 3 bit-DPCM-coder is used. Figure 10 shows a segment of the interpolated 7th and 8th field out of 15 interpolated fields. According to formula 3 the resulting data rate is $R = 1,4$ Mbit/s.

Using the motion adaptive extrapolation technique with 3 bit DPCM the resulting data rate for the same TV-sequence is $R = 96$ kbit/s. A single reconstructed frame is shown in Figure 11.

The resulting data rates according to the simulation results shown in Figure 9, 10 and 11 are composed in Table 1. These data rates are related to the processed picture segment of 192 pels per line and 256 lines per frame and in addition recalculated for the full frame size of 576 pels per line and 512 lines per frame.

Figure 10. Interpolated 7th and 8th field of a motion adaptive interpolation coding technique
Coding: 8 bit PCM

Figure 11. Reconstructed frame of a motion adaptive extrapolation coding technique
Coding: 3 bit DPCM

Coding technique	Data rate in Mbit per second	
	picture segment 192 pels x 256 lines per frame	full frame size 576 pels x 512 lines per frame
8 bit PCM	9.8	59
Reduction of the field frequency: F = 16 Motion adaptive interpolation 3 bit DPCM	0.23	1.4
Motion adaptive extrapolation 3 bit DPCM	0.032	0.096

Table 1. Data rates of the presented coding techniques

Analyzing the picture quality degradations generated by the motion adaptive interpolation coder, two types of errors can be recognized. The first type of error is produced by the quantizing noise of the 3 bit intraframe DPCM coder and is noticeable as slope overload and granular noise. The known edge business of DPCM coders is masked by the motion adaptive field interpolation and is not visible. The second type of error occurs at small moving objects and at the border of moving objects. This error is due to the window size of 15 by 15 pel performing the displacement estimation.

Using the motion adaptive extrapolation technique with a 3 bit intraframe DPCM coder again slope overload and granular noise are recognized. Because the picture content of successive fields is only shifted all errors are static and edge business errors do not occur. Since only lines with completely new information are selected by the coder moving objects are displayed as fixed. Therefore it is not possible to detect moving objects at the receiver. Since the TV-scene was taken from a forward looking camera the resolution is changing from the top to the bottom within a field. The described simple reconstruction technique does not take into account these changes of resolution. Therefore the perspective view of the reconstructed fields is not the same as in the original fields. To avoid this geometric reconstruction error a more complex reconstruction technique is necessary which considers the changing resolution within a field.

5. CONCLUSION

In this paper two TV coding techniques are presented for the special application of air to ground surveillance using motion adaptive coding techniques. The two coding techniques lead to different data rates and involve different kinds of restriction with regard to an evaluation of the TV-signals at the receiver. Applying motion adaptive interpolation techniques a data rate of 1.4 Mbit/s is achieved. A certain degradation of the picture quality occurs if moving objects are interpolated. The motion adaptive extrapolation technique achieves an extremly low data rate of about 100 kbit/s but involves more restrictions concerning the evaluation. For a straight forward flying airplane the perspective view is lost and moving objects are displayed unmoved. With respect to the special application a coding technique with a simple coder or transmitter is preferred. The complexity of the two coding techniques depends mainly on the displacement measurement. Contrary to the motion adaptive interpolation technique which requires a displacement estimator at the receiver,the motion adaptive extrapolation technique needs a displacement estimator at the transmitter.

ACKNOWLEDGEMENT

Special thanks are due to R. Lippmann from DFVLR, Braunschweig, West Germany, for providing the air to ground TV-scene.

REFERENCES

/1/ Musmann, H.G., "Digital Coding of Television Signals" in Advances in Digital Image Processing, edited by P. Stucki, 1979 Plenum Press, New York

/2/ Pirsch, P., "Design of DPCM Quantizers for Video Signals Using Subjective Tests", IEEE Trans. on Communications, Vol. Com-29, No. 7, July 1981

/3/ Wintz, P.A., "Transform Picture Coding", Proc. IEEE 60, 1972

/4/ Roese, J.A., "Interframe Coding of Digital Image Using Transform and Hybrid Transform/Predictive Techniques", in Image Transmission via Spread Spectrum Techniques, ARPA Annual Technical Report ARPA QR7 Jan. 1977, Naval Undersea Center, San Diego

/5/ Pratt, W.K.,"Digital Image Processing", 1978, John Wiley and Sons, Inc., New York

/6/ Haskell, B.G., " Entropy Measurements for Nonadaptive and Adaptive, Frame-to-Frame, Linear-Predictive Coding of Videotelephone Signals", BSTJ, Vol. 54, 1975

/7/ Musmann, H.G., and Klie, J., " TV-Transmission Using a 64 kbit/s Transmission Rate", Int. Conf. on Communications 1979, Boston, USA

/8/ Pirsch, P., " Adaptive Intra-Interframe DPCM-Coder", BSTJ, Vol. 61, 1982

/9/ Westerkamp, D., "An Adaptive Three-Dimensional Predictor for DPCM Coding of Studio Television Signals", Proc. of the Intern. Conf. on Electronic Image Processing, 1982, York, UK

/10/ Netravali, A.N. and Robbins, J.D., "Motion Compensated Television Coding Part I", BSTJ, Vol. 58, No. 3, March 1979

/11/ Dubois, E. and Sabri, S., "Video Teleconferencing: Sequenzing the Signals Pays Off", Telesis, Vol. 9, No. 1, 1982

/12/ Netravali, A.N. and Stuller, J.A., "Motion Compensation Transform Coding", BSTJ, Vol. 58, No. 7, Sept. 1979

/13/ Bergmann, H.C., "Motion-Adaptive Interpolation of Eliminated TV Fields", Proc. of the Picture Coding Symposium, Montreal (Canada) 1981

/14/ Lippmann, R., "Video Transmission of Areal Scenes at Reduced Frame Rates Using Motion Compensation", Int. Conf. on Communications 1980, Seattle, USA

/15/ Castlemann, K.R, "Digital Image Processing", Prentice-Hall 1979, Inc. Englewood Cliffs, USA

/16/ Beyer, S., "Displacement Measurement with Fast Video Correlators", Proc. of the Picture Coding Symposium, Montreal (Canada) 1981

MOTION ESTIMATION IN A SEQUENCE OF TELEVISION PICTURES

C. LABIT *A. BENVENISTE*

IRISA

Campus universitaire de Beaulieu - Av. Gén. Leclerc
35042 RENNES CEDEX - FRANCE

KEYWORDS : Interframe coding, motion compensation, adaptive estimation, ruptures, edge, automata

ABSTRACT : We present in this paper a pel-by-pel motion estimation which must be fitted any type of scene or motion in a sequence of television pictures.

The basic assumption of this algorithm is that abrupt changes in motion are located along boundaries of objects. These boundaries are spatial edges modelized by state automata including motion parameters.

The pel-by-pel propagation of the estimator uses adaptive algorithms and instantaneous detection of motion changes. When a rupture in motion is detected, the estimator is reinitialized using local motion informations computed in a causal neighbourhood of the current pixel and specific motion estimation computed along the rupture edge.

I. INTRODUCTION

Motion estimation in a sequence of pictures may have two different purposes (see [5] for a survey on this topic) :

- *a pattern recognition purpose*, where detection or displacement estimation of moving objects is a tool for scene understanding and analysis ; this task generally involves a-priori information on the moving objects to be detected and tracked (i.e target tracking or traffic control) ; however a global processing of the current frame may be used before segmentation and estimation be done.

- *a coding purpose*, by taking into account the temporal redundancy present in a picture sequence. In this context, motion estimation is a tool for motion compensated prediction of the intensity in a picture sequence. The following constraints are to be taken into account :

NATO ASI Series, Vol. F2
Image Sequence Processing and Dynamic Scene Analysis
Edited by T.S. Huang

. *the causality constraint :* the estimation must be based only upon informations available at the receiver ; hence no global processing of a frame can be used.

. *the real-time constraint :* this means that all the algorithms are subject to the rate of the input data flow.

Several methods of determining the motion vector have been performed but most of them are devoted to "*videoconferencing*" sequences where only "*head-an-shoulders*" images are involved with a smaller bandwith than in T.V broadcast applications.

In T.V. sequences, motions often are much morecomplex (several superposed moving objects with different amplitudes, several kinds of motions : translation, rotation, zoom, travelling ...) and no a-priori information on them.

Moreover *abrupt changes in the motion* frequently occur. This leads to the need of the detection of motion *ruptures* along a line. Because of the delay for detection [4] arising in classical sequential detection techniques, we tried to obtain an instantaneous detector (see [1]) of this abrupt change by using an a priori information : as basic assumption, the abrupt changes in motion only occur on the boundaries of objects. Hence we are interested in estimating the motion along edges.

All the experiments have been performed using two type of sequences :

*) a "head and shoulders" sequence of videoconferencing pictures : "Judith" : 50 pictures 8 bits (256 lines of 256 points each) (from Bell Labs.)

*) a television sequence : 50 pictures, intensity 8 bit (576 lines of 624 points each). (from FTZ, Germany) : "Castle"

II - MOTION ESTIMATION ALONG THE EDGES. (detailled in [1] and [2])

I.1. Basic principles.

A sequence of pictures can be considered as a three-dimensional data cube (the third dimension consisting of the temporal axis).

In this space, three types of section are available (cf. Fig. 2).

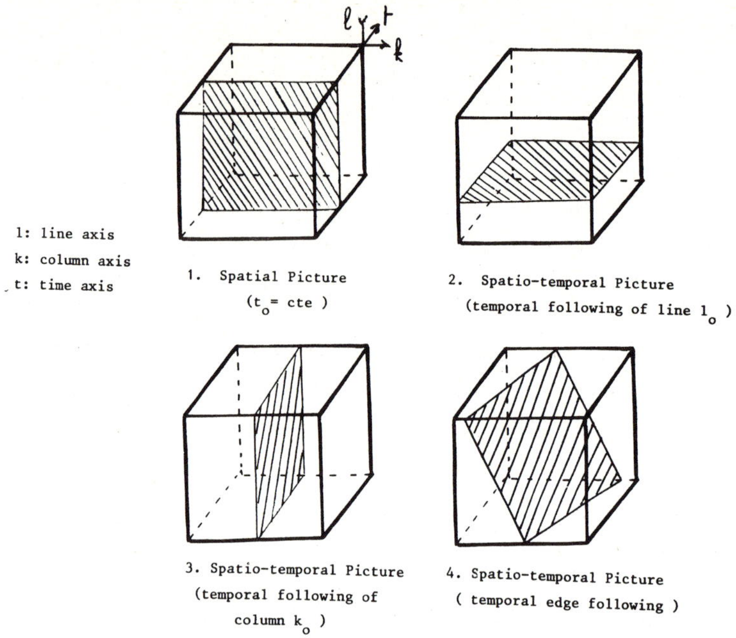

Fig.2

Especially, if we follow the same line at each picture, so we create a spatio-temporal image where classical procedures for edge detection and parametrization can be used. The procedure of following spatio-temporal edges provides us with motion information along the edges.

I.2. Spatial edge modelling and following.

. Edge model : An edge is a regular (hence well predictable) line with an orientation which is nearly constant.

$$\theta = ct + \text{noise}$$

. Edge following : At each spatial edge is associated a state automaton which summarizes the whole past of the edge (set of filtered parameters). Therefore the edge detector/estimator processes according to the following scheme :

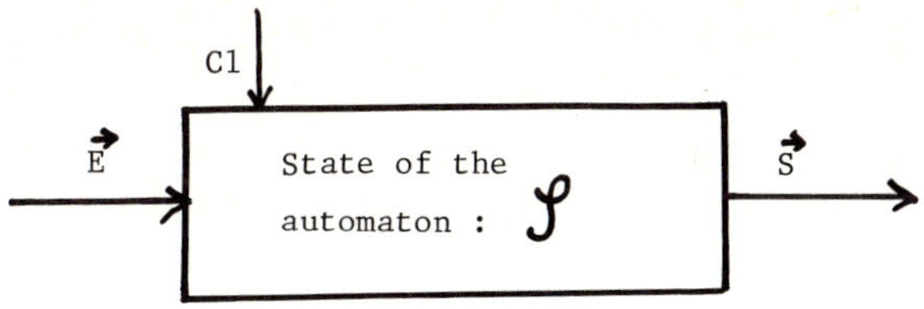

Fig. 3

$\vec{E}$: input vector of local observations at the current line.

$\mathcal{S}$: internal state vector of the automaton. State vector consists of filtered variables of edge lenght, thickness and sign.

$\vec{S}$: output vector of estimated parameters (for example, estimation of the orientation, the sign or motion of edges).

Cl : running clock of the automaton. It runs at the T.V. line- frequency according to the scanning process.

These methods for edge following have been used in [3] to determine an ADPCM intraframe coding scheme and tested on a hardware processor [9].

I.3. Edge motion estimation

Only the final version of the algorithm is proposed in this paper. (see[1,2)]

As for the other state parameters, motion estimation consists of :

- a local estimation which is a temporal processing and which is included in the input vector of the automaton,
- a recursive filtering of these local observations

- <u>Local motion estimation (Fig. 4)</u>

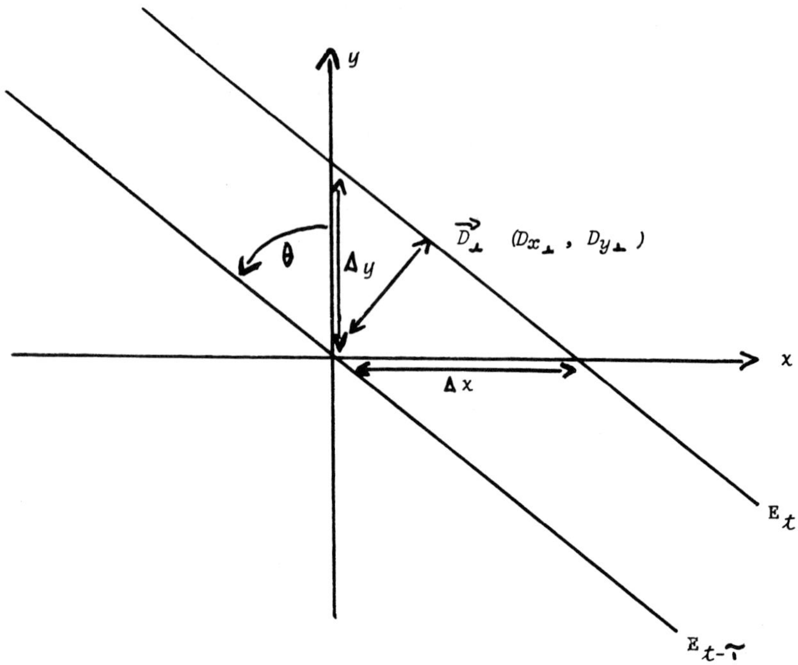

Fig. 4

If E_t and $E_{t-\tau}$ are two edges detected respectively at picture (t) and (t−τ), by computing temporal gradients between these two edges, we determine at each line an information Δ_X which is a local motion estimation between $E_{t-\tau}$ and E_t. It is outstanding to notice that the information Δ_X is biased and does not take into account the sliding of the edge on itself. So it is better to determine the local motion as $\vec{D}_\perp$ (minimal norm vector).

$$\vec{D}_\perp = \begin{cases} \cos^2\theta \ \Delta_X \\ \sin\theta.\cos\theta \ \Delta_X \end{cases}$$

- <u>validation rules for determining Δ_X</u> :

In presence of texture gradients, the estimation of Δ_X (hence $D_\perp$) is too noisy and inefficient. Therefore it is necessary to estimate motion parameter only along very regular edge.

Hence temporal gradients between $E_{t-\tau}$ and E_t are detected only if the edge $E_{t-\tau}$ verifies some rules to be valid. These rules are constructed using different components of the state vector associated to $E_{t-\tau}$ and define constraints about length, thickness of the corresponding edge.

The edge validation is only used at the next picture and so can be determined with a delay of some T.V. lines. We introduce an a-posteriori validation process which also validate the beginning of edges.

- filtering of local motion :

$\vec{D}_{\perp}$ is derived from two components of the state vector of the edge automaton. So $\vec{D}_{\perp}$ is filtered line-by-line (clock Cl) with the same processing detailed above ; we simply extend the state of the automaton associated to the given edge.

III. PEL-BY-PEL MOTION ESTIMATION

III.1. Introduction

The aim of this second part is to estimate a displacement vector at each pixel. For that we use a local approach well suited for :

- a coding scheme with a pel-by-pel motion compensated coding scheme
- the estimation of any kind of motion, without any global and physical interpretation.

The algorithms used, are the following :

$$\hat{D}_i = \hat{D}_{i-1} - \varepsilon\, f(\hat{D}_{i-1}, x_{i-1}, t)$$

where $\hat{D}_i$: estimation of the displacement at pixel x_i

ε : constant gain

f : error function which depends on time, previous pixel and previous estimated value and minimizes a given error criterion.

The propagation ((i-1) → (i)) is according to the T.V. scanning process.

III.2. Test of different functions f.

These algorithms have been introduced in [6], [7] and tested on videoconferencing pictures sequence.

. Complete version : propagation P_c.

To minimize the error criterion $E(DFD^*(x,D^i)^2)$, the adaptive algorithm

---

Notations ([6]) : "DFD" : displaced frame difference

$$DFD(x,D) = I(x,t) - I(x-D, t-\tau)$$

results in the following recursion :

$$\hat{D}_i = \hat{D}_{i-1} - \varepsilon\, DFD(x_i, \hat{D}_{i-1})\ \nabla I(x_i - \hat{D}_{i-1}, t-\tau)$$

$(\varepsilon = 1/1024)$

(∇I : spatial gradient)

. Simplified version : propagation P_s.

The previous error function provides great jumps in estimation. To limit these different jumps, a very rough quantization of the adaptive function is performed as following :

$$\hat{D}_i = \hat{D}_{i-1} - \varepsilon \text{ sign } DFD \times \text{sign } \nabla I \qquad (\varepsilon = 1/8)$$

where "sign" is the symetric function of the sign value.

III.3. Propagation with rupture along edges.

The basic principle is to use the motion information of edges computed above for reinitializing the estimator along boundaries of objects where abrupt change in motion is detected. This reinitialization can be made systematically or conditionnally to rupture tests.

- systematic reinitialization :

We use for that the displacement vector $\vec{D}_\perp$ computed with the two filtered parameters Δ_X and θ extracted from the output vector of the edge automaton

$$\left.\begin{array}{l}\Delta_X : \text{horizontal motion estimation along edges}\\ \theta : \text{orientation}\end{array}\right\} \rightarrow \vec{D}_{rupt}$$

- rupture/initialization tests :

It is necessary :

1) to detect real abrupt changes in motion
2) to solve the difficult problems of motion assignment or objects superposition.

Using the available motion informations (part II), as described in Fig 5, we can decide whether the current edge belongs to the left and/or right area, or to none of them.

- <u>propagation</u> :

We use the recursive algorithm P_s to propagate the estimation between the rupture pixels.

The structure of the final algorithm (P_r : rupture + propagation) is described in Fig. 5.

Fig. 5

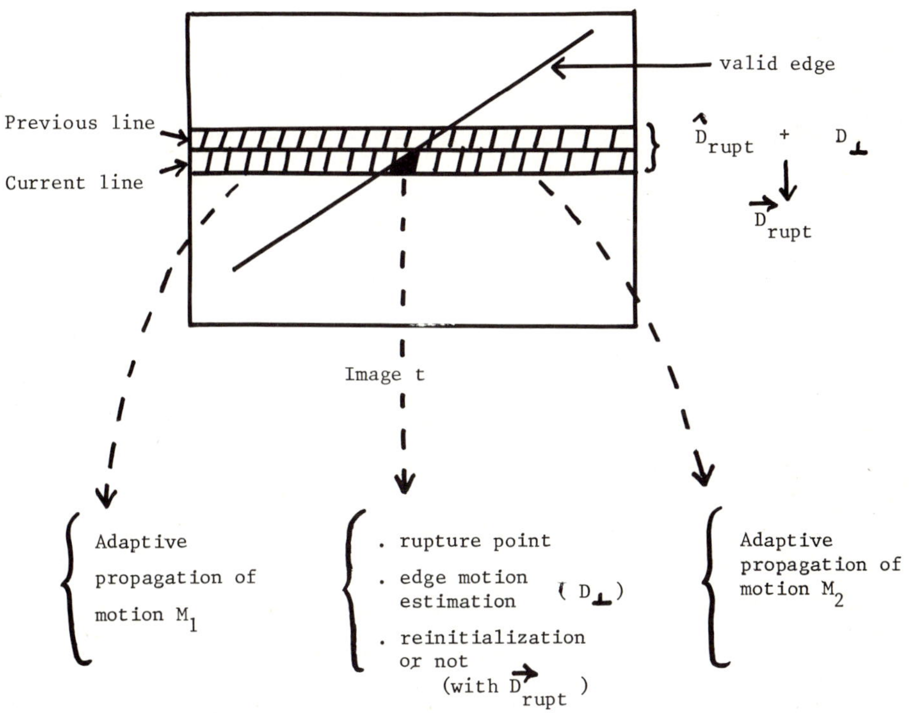

CONCLUSION

This paper presents the main basic ideas of several proposals for determining a pel-by-pel motion estimator in pictures. Complete comparison and discussion are developped in [1].

For summarizing, we can briefly observe that :

- when a sequence of pictures contains much amount of motion (in variety or in amplitude), the main problem in a coding approach we have to solve, are located on spatial edges.

- the estimation using propagation P_c allows too great jumps (and sometimes very far from convergence point) in displacement estimation.

- the estimation using P_s reduces these jumps but in presence of very different motion, the estimator does not converge fast.

- it appears necessary to introduce a specific processing along rupture configurations where the estimator has to jump quickly close to a new convergence point. Hence we need to estimate the value of rupture edge motion which has to be computed precisely and used it to reinitialize the pel-recursive estimator.

REFERENCES.

[1] C. Labit : Estimation de mouvement dans une séquence d'images de télévision Thèse de docteur-Ingénieur. IRISA-RENNES (Fev. 82)

[2] F. Kretz : Edges in visual scenes and sequences; Application to filtering, sampling and adaptive DPCM coding.
Lecture on Advance Study NATO Institute, June 82 Braunlage (this book)

[3] C. Richard : Thèse de 3ème cycle RENNES I. IRISA Rapport n° 124 (nov. 79). Estimation récursive de l'état local des contours d'image et application à la prédiction adaptative en codage différentiel des signaux de télévision.

C. Richard, A. Benveniste, F. Kretz : (to be published)
Recursive estimation of local caracteristics of edges in T.V pictures.

[4] M. Basseville : Edge detection using methods for change in level IEEE/ASSP Fev. 81

[5] H.H Nagel : Analysis techniques for image sequences.

[6] A.N Netravali, J.O Robbins: BSTJ Vol. n°58-3 (March 79)
Motion compensated Television coding (part I)

[7] A.N Netravali, J.O Robbins: BSTJ Vol. n°59-9 (Nov. 80)
Motion compensated coding: some new results.

[8] R. Jain, W.N Martin, J.K Aggarwal : Computer graphics and image processing n°11 (1979)
Segmentation through the detection of changes due to motion.

[9] C. Wagner, G. Michel, F. Jorgensen, C. Labit : Proceedings IEEE/ICASSP 82
PREDITE: a real time processor for bandwith compression in T.V transmission.

[10] P. Robert, C. Labit : A comparative study between intra and interframe prediction schemes. (NATO Institute JUNE 82 Braunlage)

APPENDIX :

- . Fig 1 : four consecutive pictures of videoconferencing sequence "Judith"
- . Fig 2 : a) valid edges for temporal following

  b) detection/following of all the edges
- . Fig 3 : one T.V picture of sequence "Castle"
- . Fig 4 : edge detection/following
- . Fig 5 : a) visualisation of edges present in a little area of "Castle"
  - -) continuous line : picture t-1
  - -) dashed line : picture t

  b) visualisation of the same edges after edge motion estimation between picture (t-1) and picture (t)
- . Fig 6 : prediction errors histograms

  a) Frame difference b) "DFD" with propagation P_c

  c) "DFD" with propagation P_s d) "DFD" with final algorithm P_r (rupture + propagation P_s)
- . Fig 7 : id as Fig 6 with prediction errors pictures

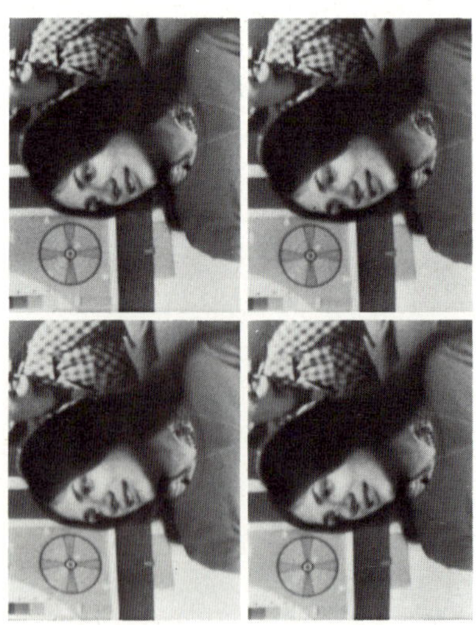

Fig. 1

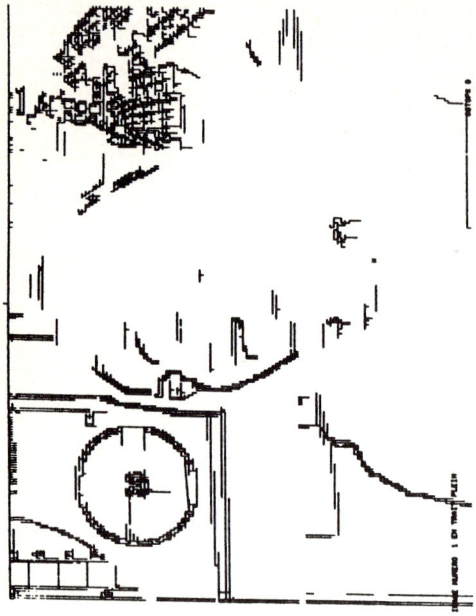

Fig. 2 a)

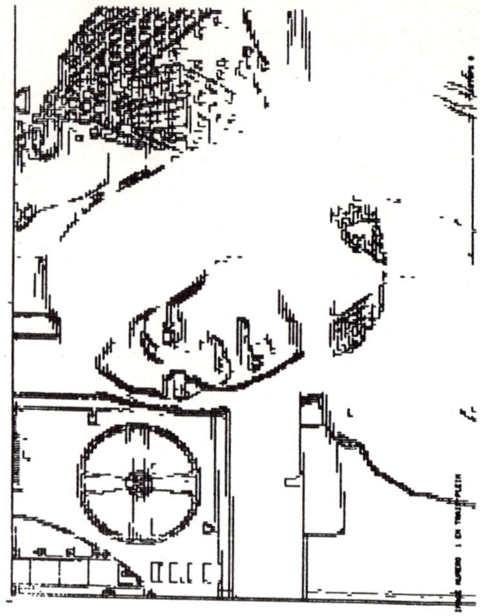

Fig. 2 b)

Fig .3

Fig. 4

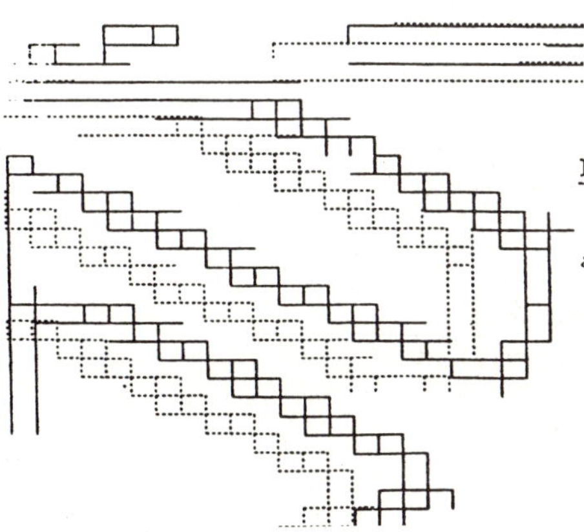

Fig. 5

a)

b)

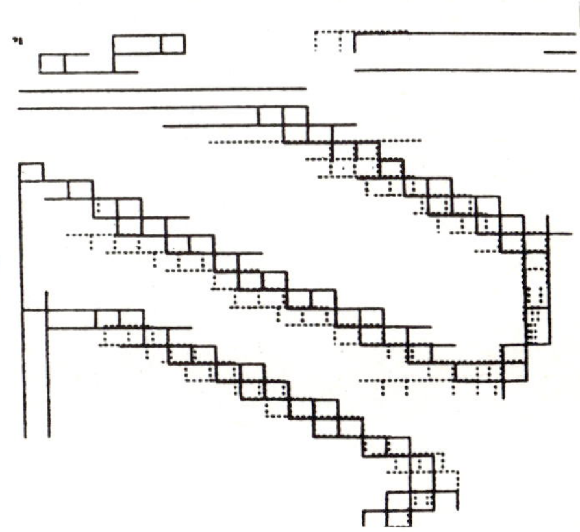

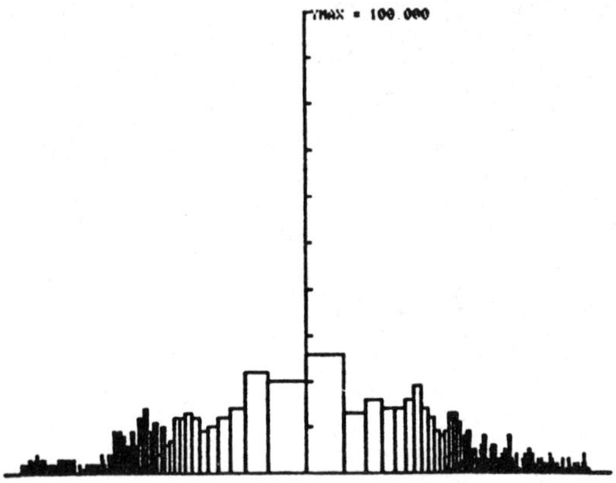

Fig. 6 a)

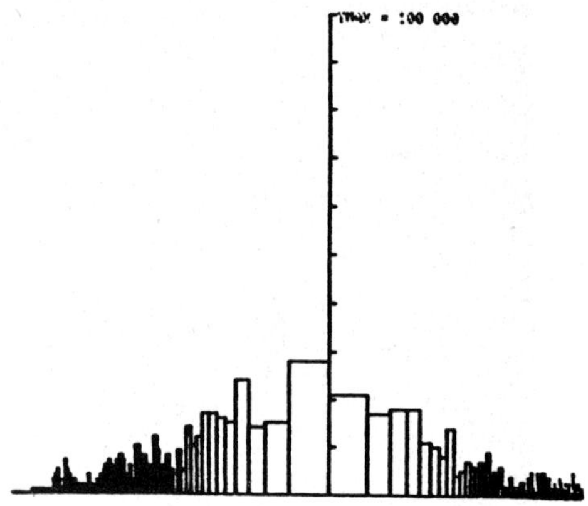

Fig. 6 b)

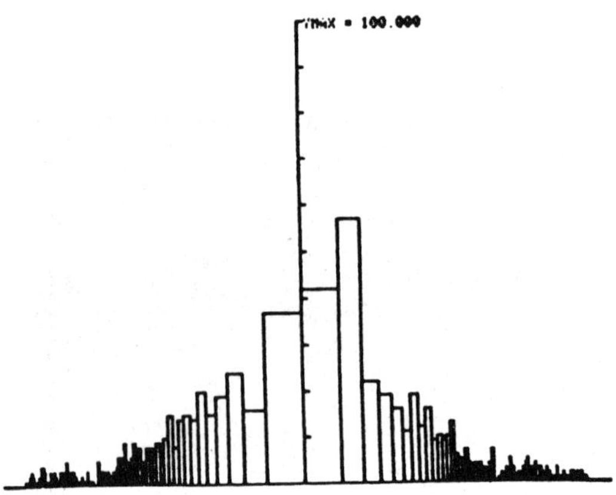

Fig. 6 c)

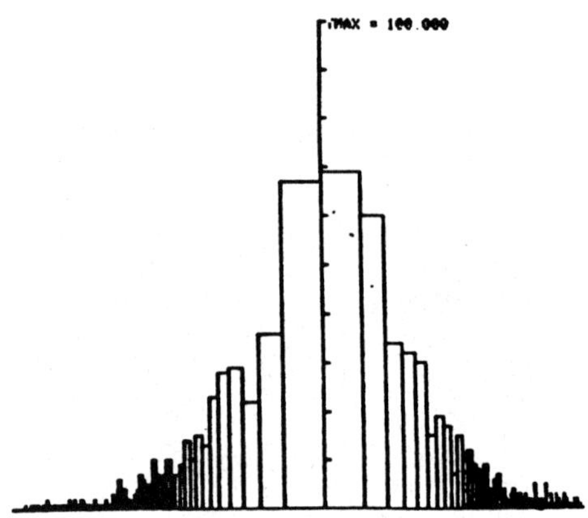

Fig. 6 d)

Fig. 7 a)

Fig. 7 b)

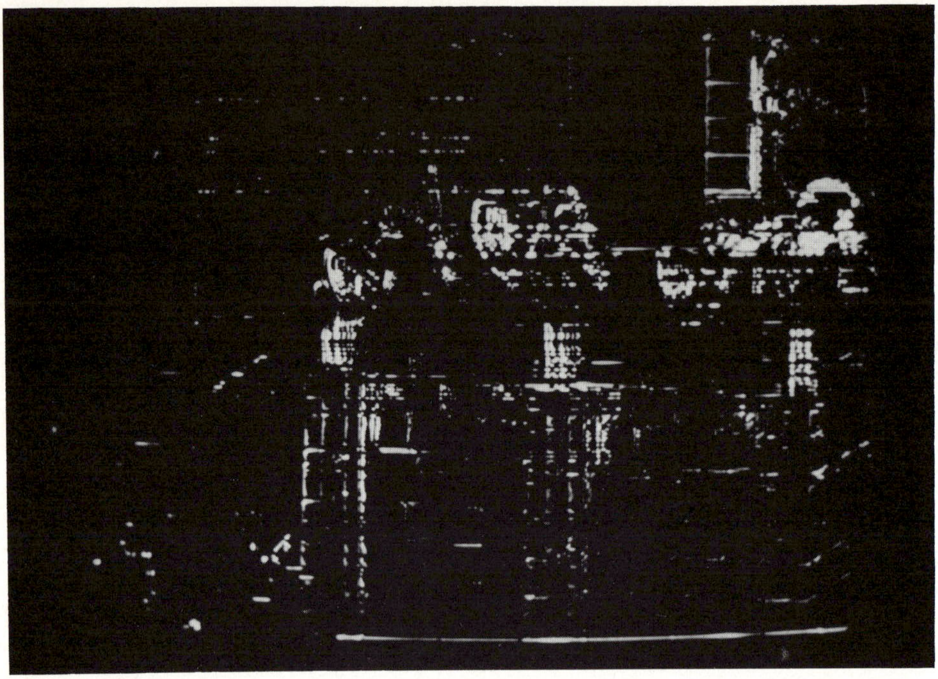

Fig. 7 c)

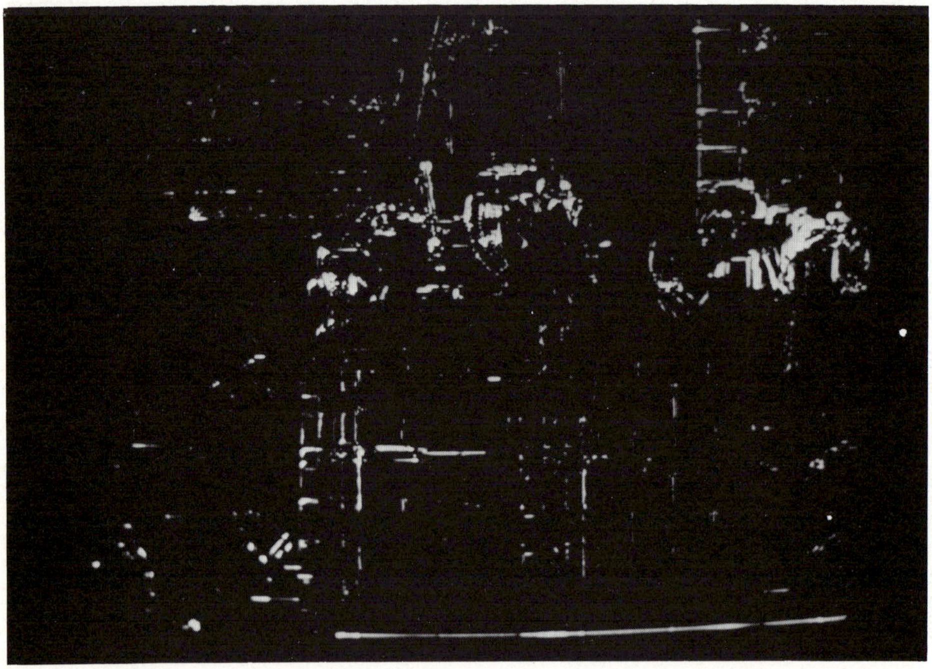

Fig. 7 d)

COMPARATIVE STUDY BETWEEN INTRA AND INTERFRAME PREDICTION SCHEMES

P. ROBERT *C. LABIT*

IRISA

Campus universitaire de Beaulieu - Av. Gén. Leclerc

35042 -RENNES CEDEX- FRANCE

KEYWORDS: intraframe,interframe coding, motion estimation, error probability, entropy.

ABSTRACT: This paper describes some statistical results on intraframe and interframe prediction schemes. We compare spatial and temporal correlations within a whole image (or within two consecutive images for temporal domain) and within some smallest areas: homogeneous,edge, or moving areas. These comparative results will be used to define a combining intra/interframe coder well fitted to T.V broadcast pictures sequences.

I. INTRODUCTION:

Some years ago, after studies on intraframe coding (DPCM or ADPCM coding), some new researches have been developped in interframe domain mainly in order to use temporal correlations by estimating some motion parameters. However, although it appears quite natural, very few coding schemes combine these two approaches. Obviously, a simple comparison between intra and interframe coding is not convenient because spatial and temporal domains are characterized with quite different sampling, blurring, masking, and correlation properties.

We present in this paper some statistical results which will be used to define a complete and combining intra/interframe coder fitted to T.V broad cast applications. We refer here to some recent studies on:

- pel-recursive algorithms for motion estimation (see [1], [2], [3])
- some intraframe coding, to get comparison. (see [4], [5], [6])

All these results have been computed using the T.V sequence "Castle" which involves rotation, zoom, and panning motions (sometimes about 3 to 5 pels per frame.) It was digitized (12 Mhz) at FTZ. Darmstadt, Germany ([2], [4])

NATO ASI Series, Vol. F2
Image Sequence Processing and Dynamic Scene Analysis
Edited by T.S. Huang

II. COMPARATIVE STUDY ON INTRA AND INTERFRAME CORRELATIONS.

This section is devoted to a comparison between spatial and temporal correlations which could be computed respectively in one picture (in both horizontal and vertical dimensions) and in two consecutive pictures.

If $\hat{I}(x,t)$ indicates the luminance estimate at current pixel (x,t), three configurations have been considered to test spatial correlations:

$\hat{I}= A,\ C,\ \frac{A+C}{2}$ (see notations below)

In temporal domain, we get an idea on interframe correlations by using:

- frame differences: "FD": in this case, $\hat{I}(x,t)$ equals to $I(x,t-1)$.
- displaced frame differences: "DFD(x,D)" (see notations below).We need motion estimations for these temporal differences and we refer here to three pel-recursive motion estimation detailed in [1]-[4]:
  - . Algorithm P_c (complete propagation):

    $$D^i = D^{i-1} - \varepsilon.\nabla I(x-D^{i-1}).DFD(x,D^{i-1})$$

  - . Algorithm P_s (simple propagation): (see [1])

    $$D^i = D^{i\ 1} - \varepsilon.\mathrm{sign}(\nabla I).\mathrm{sign}(DFD)$$

  - . Algorithm P_r : we can summarize it as a specific motion estimation on edges which is used for reinitializing algorithm P_s along motion break areas. (see [2];[3],[4])

We compare all these "prediction errors" in Table 1, 2, 3 and 4.

Table 1 shows statistical results such as entropy and error probabilities ($P_\lambda = \Pr(|I-\hat{I}| \leq \lambda)$) on a whole field of T.V picture "Castle" (624x288 pixels)

Table 1: comparison between spatial and temporal correlations on image n°41 from sequence "Castle"

$P_\lambda = \Pr(|I(x,t)-\hat{I}(x,t)| \leq \lambda)$

Entropy E

	A	C	$\frac{A+C}{2}$	FD	P_c	P_s	P_r
Entropy E	3.71	3.99	3.75	4.46	4.31	3.63	3.51
P_{60}	98.9%	96.5%	99.2%	92.8%	90.2%	96.9%	97.7%
P_{40}	96.8%	92.4%	97.2%	86.0%	83.7%	94.2%	95.5%
P_{20}	89.4%	83.2%	88.8%	72.7%	73.5%	88.0%	89.8%
P_{10}	78.6%	72.0%	75.8%	58.3%	63.8%	79.9%	82.0%

Table 2, 3, 4 give us error probabilities P_λ ($\lambda \in [0,100]$) respectively in the whole image, edge pixels (spatial gradient threshold = 25), and moving area (temporal gradient threshold =25).

All the results lead to prove that motion compensation with algorithms P_r and in a less degree P_s allows to reduce quite well the temporal errors amplitude. If we compare with spatial domain (where, for this sequence "Castle", pixels correlation indeed seem to be greater statistically than in temporal domain), we observe that it is clearly possible to obtain similar and even better results when we use motion compensation P_r or P_s.

Obviously all these improvements would be degraded if we compared not only estimation errors but real prediction errors, so, for example, if we replaced real motion estimation values in "DFD" by an approximated prediction of them.

Notations: - $DFD(x,D)=I(x,t)-I(x-D,t-1)$

-

	C	← previous line
A	X	← current pixel

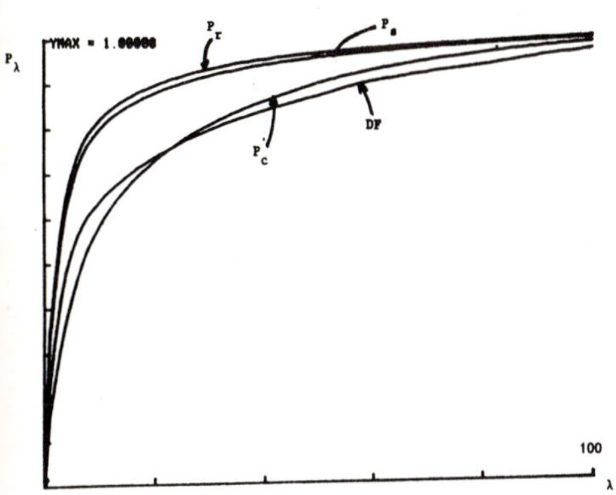

$\lambda \in [0,100]$

Comparison between temporal differences.

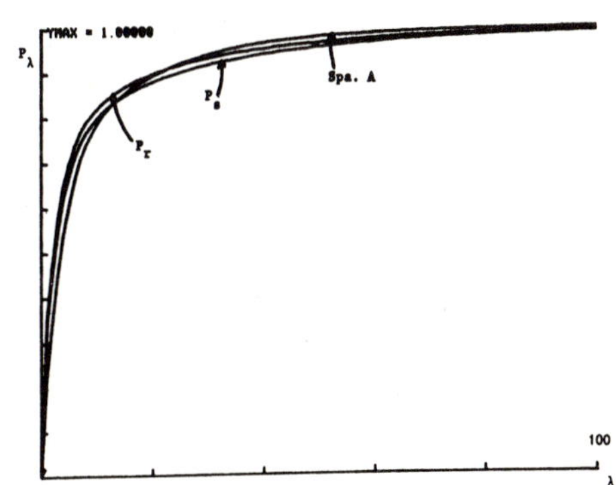

$\lambda \in [0,100]$

Comparison between temporal and spatial differences.

Table 2

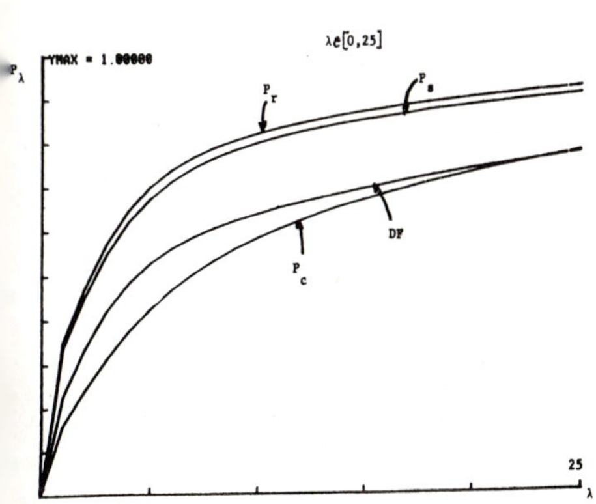

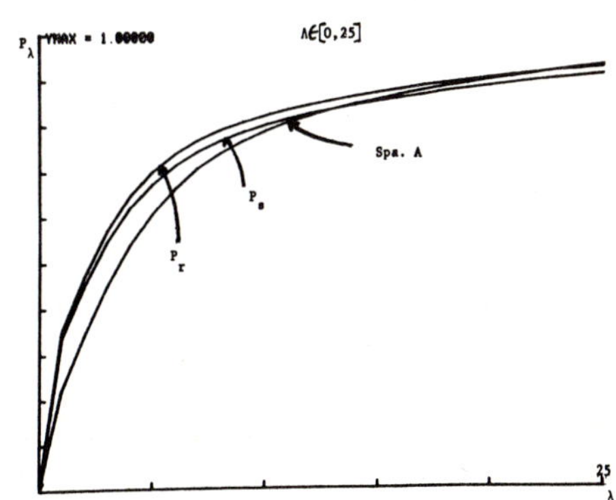

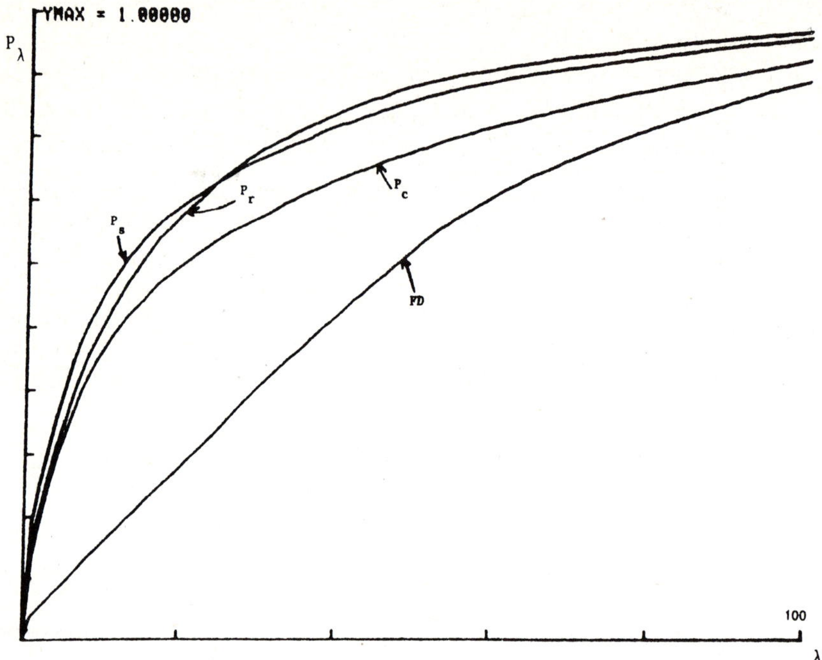

Comparison between temporal differences.

Table 3 : edge pixels (8.25% of all the pixels)

Comparison between temporal ans spatial differences.

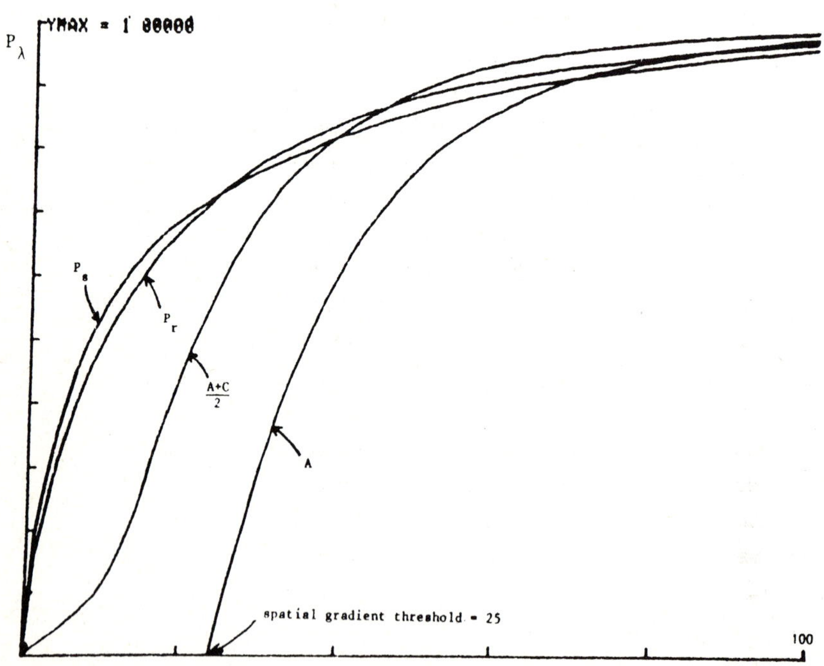

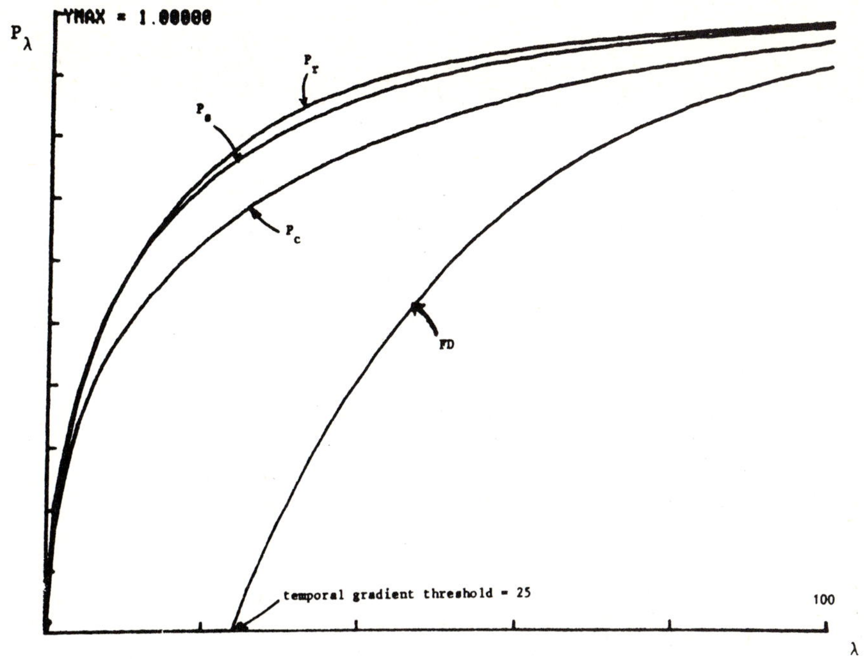

Comparison between temporal differences.

Table 4: moving area (23.7% of all the pixels

Comparison between temporal and spatial differences.

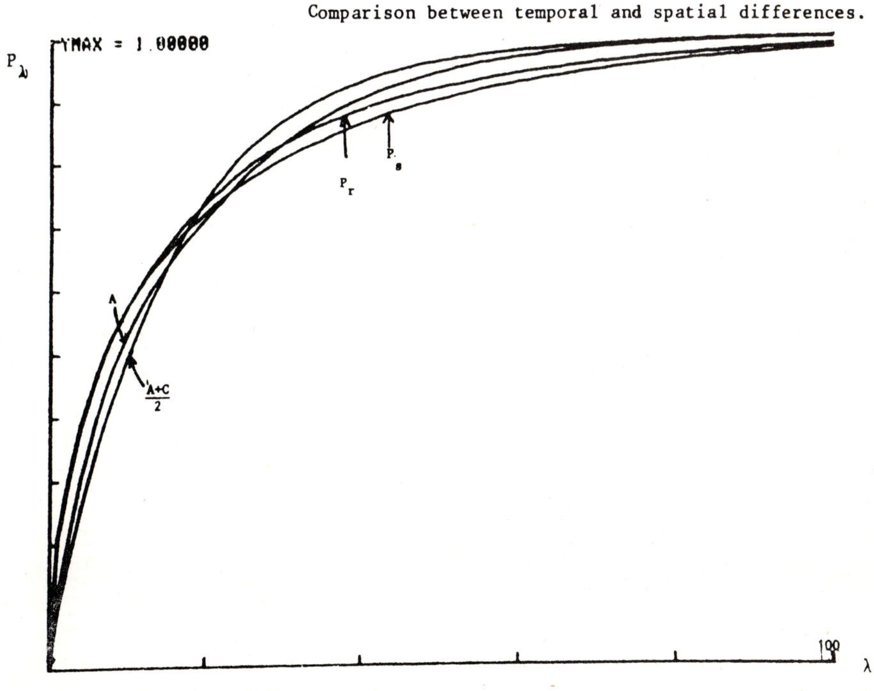

III. BASIC PRINCIPLES FOR A COMBINING INTRA/INTERFRAME CODER.

As we observed in previous section, an efficient way would be to define a coder using both spatial and temporal redundancies. This study is just beginning. Our approach for this coder is based on three prediction modules:

- Module 1: adaptive spatial prediction module

The spatial predictor we use is an adaptive one which varies with local picture configurations and properties (see [5], [6]) . This adaptive prediction sheme has to be compared with usual fixed predictors.

- Module 2: interframe prediction module based on conditional refresment. We need here no motion estimation but only a moving area detector.

- Module 3: motion compensation coding module. In this case, we use motion estimation algorithm detailed in [2],[3] ,[4].

Table 5 and 6 show some comparative prediction errors statistics respectively on edge and homogeneous areas. Other experiments (on fixed, moving or noisy pictur areas) have been performed but presently without any quite efficient results. We can observe that in homogeneous area, spatial and temporal predictions have similar results. However in edge area (and statistically on the whole image) spatial predictors work better than simplest temporal ones. So if we want to take into account temporal redundancies , we have to use a complete motion compensation scheme. All these remarks are obv iously quite depending on specific characteristics of the sequence: amount of noise, contrast, motion....

CONCLUSION:

This paper has been devoted to a comparative study about improvements we can expect by using together intra- and interframe predictors in a general coding scheme. It leads to prove that on one hand,adaptive spatial prediction and on the other hand, motion compensation prediction separately but all the more simultaneously must reduce significantly the transmission bit rate. An efficient way would be to define a prediction strategy based on local and combined spatio-temporal pixel configurations and therefore on global properties of the sequence.

Table 5: edge area

Entropy E and error probabilities $P_\lambda = Pr(|X-\hat{X}| > \lambda)$

Prediction	E	P_{60}	P_{40}	P_{20}
Fixed spatial A	4.55	5.53%	16.6%	34.7%
Fixed spatial $\frac{A+C}{2}$	4.22	3.28%	7.89%	21.9%
Adaptive spatial	4.06	1.08%	3.43%	17.4%
Frame difference (FD)	4.82	13.0%	24.6%	40.7%
Motion compensation (DFD)	4.03	3.43%	7.22%	16.7%

Table 6: homogeneous area

Entropy E and error probabilitie $P_\lambda = Pr(|X-\hat{X}| > \lambda)$

Prediction	E	P_{10}	P_7	P_4
Fixed spatial A	2.69	.56%	4.2%	21%
Adaptive spatial	2.47	.15%	1.2%	10%
Frame difference (FD)	2.63	.21%	2.8%	17.8%
Motion compensation (DFD)	2.57	.16%	2.5%	14.7%

REFERENCES:

|1| : A.N Netravali, J.O Robbins: BSTJ Vol n°58-3 (March 79) Motion Compensated Television Coding

|2| : C. Labit: Thèse de docteur-ingénieur IRISA-Rennes (Fev. 82) Estimation de mouvement dans une séquence d'images de télévision.

|3| : C. Labit, A. Benveniste: Motion estimation in a sequence of TV pictures. (This book)

|4| : F. Kretz: Edgrs in visual scenes and sequences. Application to filtering, sampling and adaptive DPCM coding. (This book)

|5| : F. Kretz: Codage ADPCM en télévision. Peport CCETT/EAS/T/3/81 (May 81)

|6| : C. Richard: Thése 3ème cycle RennesI. IRISA Report n°124 (Nov. 79) Estimation récursive de l'état local des contours d'images et application à la prediction adaptative en codage DPCM des signaux TV

A NARROW-BAND VIDEO COMMUNICATION SYSTEM FOR THE TRANSMISSION OF SIGN LANGUAGE OVER ORDINARY TELEPHONE LINES

by J.F. ABRAMATIC, PH. LETELLIER and M. NADLER

I.N.R.I.A.
Domaine de Voluceau, Rocquencourt
B.P. 105, 78153 Le Chesnay Cêdex
FRANCE

ABSTRACT

After a brief review of the origin and present state of sign-language as the principal means of communication among the deeply deaf, the main existing proposals for enabling telecommunications for this sub-community of handicapees are reviewed. Arising from this analysis, a new system is proposed, based on the transmission of a severely simplified image sequence, using extraction of contours to obtain the effect of a real-time animated cartoon drawn from the actual scene. Some preliminary experimental results are presented, together with a discussion of the future stages of the investigation. An appendix describes a contour extraction algorithm applied in this work.

The present project is situated within the general framework of aids to the handicapped, in particular to the hard of hearing and deaf (1-3). The deaf communicate with each other or with the hearing by two widely used visual means :lip reading and sign language.

This project is designed to evaluate the feasability of a system that would allow the deeply deaf to communicate over ordinary telephone lines by one or the other of these means.

This paper is a progress report on the initial phases of our work, giving a statement of the problem, pointing out the lines of investigation, and presenting the initial experimental results.

In section I we present sign language and its characteristics. Section II gives a critical discussion of previous work in this field and the available equipement. Section III presents our proposals, consisting in the transmission of sequences of contour images. Our two main problems are discussed : contour detection in an image sequence and the coding of these contour images. The first results obtained on image sequences are presented.

I) SIGN LANGUAGE

Deaf children express themselves spontaneously by signs, the only means of communication available to them. Arising out of that, two formalised means of communication have been developed : signed French (English, etc.), which is a direct translation into signs of the oral language, and a true, independent language, French (American, etc.) sign language.

After an initial flourishing of signed communication, developed by the Abbe de l'Epee, the Abbe Sicard, Bebian, and others, there succeeded in France almost a century of negation in order to favorise oralisation, starting with the Milan Congress of 1880. It is only recently that we observe a renewal of French sign language (LSF). On the other hand, after its introduction into the United-States early in the 19th century by French Immigrants, American Sign Language (ASL) was able to continue to enrich itself and is today a model for the other sign languages because of the numerous studies that have been made of it.

Sign language expresses itself by gestures situated in a volume bounded by the signer's waist, the top of his head, and extended laterally to about 30cm from each shoulder. The signs are differentiated by (cf.(5,6)) :

- the hand configuration(s) ;
- motion of the hands or change of configuration ;
- the positions of the hands with respect to the rest of the body.

Furthermore, many nuances are transmitted by facial expression (affirmation, negation, etc.).

The deaf sub-community is a minority, but possesses its own culture as a result of the strong interaction between the mode of thought and language. It is very important for the development of each individual to be able to communicate with others and, in particular, with other members of the deaf sub-community.

Thus, the need for a telecommunications system, analogous to the telephone for the hearing, is obvious. In order not to impair communication, the transmission means should be adapted to the rate of the language used. Now sign language is as fast as spoken language (for more detailed information on sign-language, see (4-7)).

II) DISCUSSION OF PRIOR WORK AND AVAILABLE EQUIPMENT

1) Prior work

i) Tartter and Knowlton

Tartter and Knowlton have carried out experiments at Bell Laboratories (8) on communication between two deaf people fluent in sign language. They have shown that with 27 luminous spots placed strategically on the hands (special gloves !) and nose the participants could understand each other fairly well, using only the sight of these 27 points in motion. The experiments showed that the simple transmission of the positions of these 27 points with a resolution of about 1% (7 bits along each axis) could be accomplished at an adequate rate over a telephone line.

Poizner, Bellugi, and Lutes-Driscoll (9) have come to similar conclusions by psychometric experiments.

ii) Sperling and Pearson

Sperling and Pearson have studied the minimum information rate necessary for intelligibility of ASL. to this end, they have varied the following parameters :

- number of images/s

- number of lines/images
- number of points/line

Sperling (10) comes to a minimum bandwith of 21 khz (30 images/s of 38 lines of 50 points) in order that the intelligibility not be too reduced.

As for Pearson (11), he describes experiments with videophone over a period of 6 years, showing the value of such a communication means. In addition, he shows the sensitivity of sign language to the number of images transmitted per second and gives results comparable to those of Sperling concerning the minimum rate. For him 100 kbits/s seems to be the threshold between comfortable and difficult communication ; below 5kbits/s communication becomes impossible.

iii) Wallis and Pratt

Wallis and Pratt (12) have built a video-conference system (not intended for the deaf) at 9600 baud. The resolution of the transmitted binary image varies from 120*120 points for 8 images/s to 240*240 for 2 images/s as a function of the amount of motion between images.

2) Available material

At the present time hard of hearing can use amplifiers, directly connected to the telephone receiver, which do not interfere with use by hearing people. There are also terminals with screen and keyboard, in particular, the Portatel system, under evaluation by the French PTT. This system is being experimented by deaf people in France with a view to its eventual adoption there. The cost is 5700 Francs per terminal in France.

3) Conclusions

The technique used by Tartter and Knowlton has the advantage of allowing fast communication between deaf people but unfortunately is not very comfortable and requires a certain training time. Furthermore, it cannot transmit the information furnished by the facial expression.

The work of Sperling and Pearson shows that it is not possible to reach 4800 baud by simple degradation of the image. However they do show that the allowable image degradation can be very extensive.

The material of Wallis and Pratt gives encouraging results, but it would need an additional factor of two while the number of images/s may be inadequate to signed communication, requiring the signers to slow down.

As for the available material, amplifiers are a good solution for the hard of hearing, but are absolutely out of the question for the deeply deaf ("In France there are 400,000 deeply deaf out of 3 millions with impaired hearing" - Mr. Veit, President of the French Committee for Audiophonology).

Portatel has the great drawback of slowing down communication enormously because of the use of the written word (the equivalent of writing letters for the hearing).

This brief review of the previous work and the available equipment, as well as our discussions with deaf people, has convinced us of the need for a communication system for deaf and the requirement that this system not impede communication by slowing down the rate of "talking".

III) OUR PROPOSAL : THE USE OF CONTOURS

Since a video sequence cannot be transmitted over a telephone channel, we propose to extract its significant characteristics, namely, a contour image. This type of image is perfectly adapted to the experience of deaf people, who often learn sign language from drawings (Fig.1,taken from (6)). Rather than the direct transmission of a sequence of contour images, the changes in contour between images will have to be encoded in order to reach the limit of 4800 baud. We therefore have two main problems :

- contour detection ;
- encoding a sequence of contour images.

1) Contour detection

We propose to use classical contour detection techniques, such as the Sobel, the pseudo-Laplacean (cf.Appendix II), etc. (13-17), which we adapt to sequences of images. Figure 4 gives a comparison of the quality to be expected from the Sobel algorithm and that obtained by the pseudo-Laplacean of (16). For a sequence obtained by the Sobel, see Fig.3a.

In order to obtain a suitable quality of contour image, our first experiments indicate that the following procedure is advisable :

1) extraction of contours in a high-resolution image (e.g. 256*256) ;

2) reduction of resolution by sub-sampling.

This procedure gives a better quality of image than the direct extraction of contours at the final resolution for the following reason : contours are detected over a finite width. Thus, the quality of contour in a sub-sampled image is practically that of the original high definition image with the width of line proportional to the resolution ; on the other hand, contour extraction at the final definition results in loss of detail due to the thickness of the contours (Fig.2).

In order to speed up contour extraction and to improve the stability of contours over a sequence, Yachida, Ikeda, and Truji (19), using Kelly's work on plans (18), propose to search for contours exclusively in a zone about the contours of the preceding image. We have implemented this algorithm. In the presence of rapid movement certain contours located outside of the selected zone disappear. Furthermore, if an entire contour should disappear in an image, the surrounding zone would no longer be validated, and no search would be made for the contour. If it should reappear in further frames, it would take several frames to reconstitute it. Figure 3b gives a sequence obtained with this algorithm.

To eliminate this defect, we have associated to the extraction of contours within a zone, contour following starting from "contour points" found at the edge of this zone. This effectively eliminates the defect (Fig.3c).

2) Coding

As yet we have not considered this problem in detail, but we expect to use a coding of the type of "run-length" or block coding, modified to take into account the preceding images in a sequence. In a first approximation, we shall consider only the contour points that have changed ; to reduce the number of points to be encoded and to reduce flickering of contours, we can eliminate isolated contour points (cf.(28-30)).

3) Direction of research

i) In order to evaluate the effectiveness of the various algorithms, we intend to verify the results of our work by deaf people. We must therefore define a plan of psychovisual experimentation to validate or reject our experimental results.

ii) We are also considering the use of motion detection in order to better predict the zone of search, in order to limit the number of points in which contours are followed (cf.Appendix I), thereby accelerating the algorithm.

iii) The coding problem has not yet been approached. Motion prediction should be of great help in limiting the rate to 4800 baud, which is our objective.

APPENDIX I

In this Appendix we present some quantitative data on the number of points involved in contour detection. A deaf person explaining our experiments in sign language was recorded on digital video tape for a total duration of 1 minute of signing. Our experiments have been carried out on a sequence of 50 images of 256*256*8 bits, representing 2 seconds at 25 images/s.

4) The number of contour points in an image (256*256)

The different algorithms used give about 4000 contour points for a "good" image quality (cf.Fig.4) ; this number of points is practically constant over the entire sequence.

i) The number of points tested in the different parts of an algorithm using planning

If simple planning is used, with a 9*9 neighborhood taken about each contour point found, the number of points tested for contour is of the order of 16k, i.e., a quarter of the total number of points in the image (cf.Fig.3b). However, the improvement obtained by contour following at the edges of the planning zone allow us to reduce the latter to the 8 neighbors about a contour point. In this case the number of points is reduced to 8000 out of the 64k. On the other hand, the number of contour points found at the edges of the planning zone and the number of points tested in following the contour outside of this zone varies extensively. The former is found in the interval (620-1250), the latter in (4500-9500), a variation of two to one.

The magnitude of the number of points tested in contour following (close to the number in the original plan) justifies the idea that the detection of motion can improve the algorithm by predicting in a more exact way the region in which the contour is sought.

5) The number of contour points varying from one image to the next

The number of points of difference between two successive contour images (Fig.5) is found in the interval (2200-2700), i.e., about a quarter of the contour points have moved. This number can, however, be reduced by elimination of isolated points, which only give a flickering of the contours.

APPENDIX II

The Pseudo-Laplacean

The Pseudo-Laplacean is a non-linear finite-difference operator that is the combination of a gradient operator and a Laplacean in the limit as the non-linearity goes to zero. Aside from the application described here, operators of this type have been applied to the binarisation of poor print quality images in OCR and to restoration of medical images (echocardiographic sections) input to a computer via an image dissector (*). In this note we shall present the general philosophy of the operator and then describe the actual operator used in the present work.

A.1 General Philosophy

Edge detection and contrast enhancement algorithms have generally been limited to the use of gradient operators in one or another form ; unfortunately, they yield thick contours since all points on a contrast gradient are detected as part of the above-threshold gradient.

The Laplacean has the theoretical advantage of extracting one or the other extremity of a contrast slope depending on the sign of the response, or the locus of maximum rate of contrast variation when the zero crossings are used. Thus the resulting contours are thinner, but, unfortunately, the Laplacean, as a second-difference operator, is quite sensitive to noise.

We shall illustrate these considerations by a series of diagrams in one dimension, although the gradient, as a vector operator, is only meaningful in two or more dimensions.

Consider, then, Fig.A.1. It shows a piecewise-linear function (a), its finite-difference "gradient" (b), and its finite difference "Laplacean" (c) in two variants : the usual direct form (dashed line) and a form obtained as a difference operator applied to the "gradient" (full line). Examination of this diagram shows the well-known properties of the two operators, namely : the gradient displays extrema at points of most rapid change of function (maximum "contrast gradient"), whereas the Laplacean marks the "edges" of the zones of contrast variation. By suitable choice of polarity, we can place the detected contours just at the inside edges of the darker areas or

(*) It is an unfortunate fact that the physical arrangement of the image dissector is such as to obscure fine detail as a result of glare produced on the photoelectric target by internal reflection from the electrode strucutre (25). This property of the image dissector does not seem to be as widely known as it merits.

the lighter. It is this property of the Laplacean that has attracted repeatal attempts to exploit it for edge detection.

However, just as soon as we comme up against real images and scenes the situation is quite different. Figure A2 illustrates the problem, even though it is far from a "real" image function. (It was obtained by a freehand drawing of the curve shown). The + signs indicate the nearest discrete values on the assumed a/d conversion grid, applying the same rules as for Fig.1. Further, applying the same rules to the function (a) as in Fig.1, we obtain the corresponding functions (b) and (c). Now, whereas the gradient behaves more or less as before , except for the parasitic peak marked, the Laplacean is totally unusable, giving responses all over the place. The slight variations in slope, simulating high-frequency spatial noise in the function (a), produce false peaks in the Laplacean response, independently of the mode of calculation (L or L').

It is this property of the Laplacean that has induced its experimenters to abandon it...or to introduce modifications that are, effectively, nonlinearities to suppress the noise responses. The pseudo-Laplacean is one such modification.

For the moment we continue with our one-dimensional qualitative explanation. Figure A3 reproduces the "function" of Fig.A2. In (b) the gradient has been thresholded and clipped at one unit, giving the solid line curve. In the two dimensional case this operation is equivalent to thresholding the gradient modulus and retaining only the unit vector, giving the direction of above-threshold modulic (an operation analogous to retaining only the phase information in the Fourier transform : cf.(21)).

Now, if we take second differences on this thresholded gradient, we obtain the pseudo-Laplacean response shown at (c).

The positions of the responses are in agreement with the intuitive notion of Laplacean, as illustrated by the piecewise-linear function of Fig.A1.

The original algorithm for computing the two-dimensional pseudo-Laplacean (16) used as its first step the edge vector, which is the gradient vector rotated by $\pi/2$ (20-21). Certain operations were effected on a 5x5 window of edge vectors to obtain the pseudo-Laplacean (16). It was shown subsequently (22) that the combination of modulus-thresholded edge vector and the operations thereon were equivalent in the limit, as the non-linearity and threshold go to zero (i.e., with retention of the modulus information in the algorithm), to a Laplacean operator, with the weights shown in Fig.A4.

A.2 The Algorithm

Examination of Fig.A.4. suggests a more direct way to calculate the pseudo-Laplacean (23).

It is sufficient to construct a system of radially symmetrical differences (Fig.A.5), to estimate the gradient in each direction ; each arrow represents the difference of gray values between the head and the foot, where the value at the foot is subtracted from the value at the head. These differences are then processed by a non-linear filter : the absolute value of each such difference is compared to a threshold t and, if above threshold, the sign of the difference is retained. This is equivalent to a quantisation to three levels (Fig.A.6). The signs are counted separately : a counter CP for the + signs, a counter CN for the - signs. The final edge/no-edge decision (16-23) is based on two further threshold operations :

$$CP \geq t_1 \tag{1}$$

and

$$CP - CN \geq t_2. \tag{2}$$

Threshold t_1, as we have seen, eliminated the noise variations of high spatial frequency ; threshold t_1 is equivalent to requiring that the gradient exceed the first threshold in at least t_1 directions (i.e.,the t_1-th maximum),instead of simply requiring that the max.(gradient) be above threshold. It is the difference CP-CN that corresponds to the Laplacean, and t_2 selects the contour points. Another alternative for condition (2), could be to look for zero crossings of the pseudo-Laplacean :

$$CP-CN = 0$$

but it would be difficult to apply here ; in contradistinction to the Marr-Hildreth (26) or Keskes (27) algorithms the operator does not filter the image by a large-diameter convolution, and thick contours would result.

The "Zero crossings of the second directional derivative" due to Haralick (24) present a certain analogy to the pseudo-Laplacean, as it is also the combination of a gradient operator with a second-difference operator. Haralick filters the image by representing it locally with a linear combination of polynomials ; the gradient is computed and thresholded, using the simple maximum ; however, in place of the Laplacean, he takes the zero crossings of the second directional derivative in a sector about the direction of the gradient maximum.

Conclusions

The pseudo-Laplacean is essentially the combination of a gradient operator, a non-linear transformation, and a Laplacean operator. It yields thin contours with good noise-suppression properties, while retaining fine detail, unlike operators using large-diameter convolutions.

REFERENCES

AIDS TO THE HANDICAPPED

(1) COLBY, CHRISTINAZ, GRAHAM. "A computer-driven, personal, portable and intelligent speech prosthesis". (Computer and Biomed. Research, 11,4,337-343 , 1978).

(2) M.C. HATON and J.P. HATON. "Sirene, a system for speech training of deaf-people". (Proc. ICASSP, pp.482-5, Washington D.C., USA , April 1979).

(3) CHOUARD, MAC LEOD, MEYER, PIALOUX. "Appareillage électronique implanté chirurgicalement pour la réhabilitation des surdités totales et des surdimutilés. (Ann. O.L. 94,7-8,353-363 , 1977).

SIGN LANGUAGE

(4) LANGAGES
(La langue des signes revue LANGAGES N 56 Décembre 1979, Larousse).

(5) STOKOE, CASTERLINE, CRONEBURG. "A dictionary of American Sign Language on linguistic principles". (Gallaudet College Press, Washington DC , 1965).

(6) KLIMA, BELLUGI. "The signs of language". (Harvard Univer. Press , Cambridge Mass. , 1979).

(7) CUXAC. "L'éducation des sourds en France depuis l'Abbé de l'Epee". (Thèse de Doc. 3ème Cycle Université de Paris V

COMMUNICATION SYSTEM FOR THE DEAF

(8) TARTTER, KNOWLTON. "Perceiving sign language from an array of 27 moving spots". (Bell Laboratorie Murray Hill). (1980).

(9) POIZNER, BELLUGI, LUTES-DRISCOLL. "Perception of American Sign Language in dynamic point-light displays". (Journal of Experimental Psychology, Vol.7, n° 2, pp.430-440 , 1981).

(10) SPERLING. "Video transmission of American Sign Language and Finger Spelling : Present and Projected Bandwidth Requirements". (I.E.E.E. Transactions on Communications, Vol. Com. 29, n° 12 , Dec. 1981).

(11) DON PEARSON. "Visual Communication Systems for the Deaf". (I.E.E.E. Transactions on Communications, Vol.COM29, n° 12, Dec. 1981).

(12) WALLIS, PRATT. "Video-teleconferencing at 9600 baud (I.E.E.E. preprint). (ibid).

CONTOUR DETECTION

(13) DAVIS. "A survey of edge detection techniques". (Computer Graphics and Image Processing, Vol.4 , 1976).

(14) RICHARD. "Estimation récursive de l'état locale des contours d'images et application à la description adaptative en codage différentiel de signaux de télévision. (Thèse d'Ingénieur-Docteur, Rennes , Novembre 1978).

(15) LABIT. "Estimation de mouvement dans une séquence d'images de télévision". (Thèse d'Ingénieur-Docteur, Rennes, Février 1982).

(16) M. NADLER, A. OISEL, and G. ADAMOFF. "Procédé et dispositif d'analyse élémentaire d'une surface imprimée". French Patent No 2 163 815. (2 July 1973). See also : M. NADLER. "Effective and Cost-Effective Real-Time Picture Operators for Medical Imagery". (in : Decision Making and Medical Care, F.T. DOMBAL and F. GREMY, Eds, North-Holland. 1976).

(17) RAMER. "The transformation of photographic images into stroke arrays". (I.E.E.E. Trans on Circuits and System, Vol.22, n° 4, April 1975).

(18) KELLY. "Edge detection in pictures by computer using Planning". (in : Machine Intelligence, n° 6, Edinburgh, 1971).

(19) YACHIDA, IKEDA, TSUJI. "A plan-guided analysis of cineangiograms for measurement of dynamic behavior of heart wall". (I.E.E.E. Pami-2, n° 6, Novembre 1980).

(20) M. NADLER. "Un système analogique-digital pour la reconnaissance de caractères". (in : Information Processing 1962, North-Holland Publ. Co., Amsterdam, 1963, pp.456-461). Also : "An Analog-Digital Character Recognition System".(IRE Trans EC-12, n° 5, pp.814-821, December 1963).

(21) J. KENT BOWKER. "Edge Vector Image Analysis".(in : Second International Joint Conference on Pattern Recognition, Copenhagen, pp.520-525, 1974).

(22) ASHOK AGRAWALA, Private Communication.

(23) M. NADLER. "European Patent Application 81402085.5. (28 December 1981).

(24) R.M. HARALICK. "The Digital Step Edge". (In Press).

(25) A.S.J. FARROW. "Image Dissectors Dissected".(in : Proceedings of the Oxford Conference on Computer Scanning, 2nd-5th April 1974, Nuclear Physics Laboratory, Oxford, 1974).

(26) MARR and HILDRETH. "Theory of Edge detection". MIT A.I. Memo, n° 518, April 1979).

(27) N. KESKES and O.D. FAUGERAS. "Technique de Traitement Numérique d'Images et de Reconnaissance des Formes Appliquées aux Signaux Sismiques".(Proc. 2e Congrès AFCET-IRIA de Reconnaissance des Formes et d'Intelligence Artificielle, Toulouse, Septembre 1979).

PICTURE CODING

(28) JAIN. "Image data compression : a review". (Proceedings of the I.E.E.E., Vol. 69, n° 3, March 1981).

(29) NETRAVALI, LIMB. "Picture coding : a review". (Proceedings of the I.E.E.E., Vol.68, n° 3, March 1980).

(30) KUNT, JOHNSEN, DE COULON. "Réduction de redondance d'images par codage de blocs". (7ème Colloque sur le Traitement du Signal et ses Applications, Nice, 28 Mai - 2 Juin 1979).

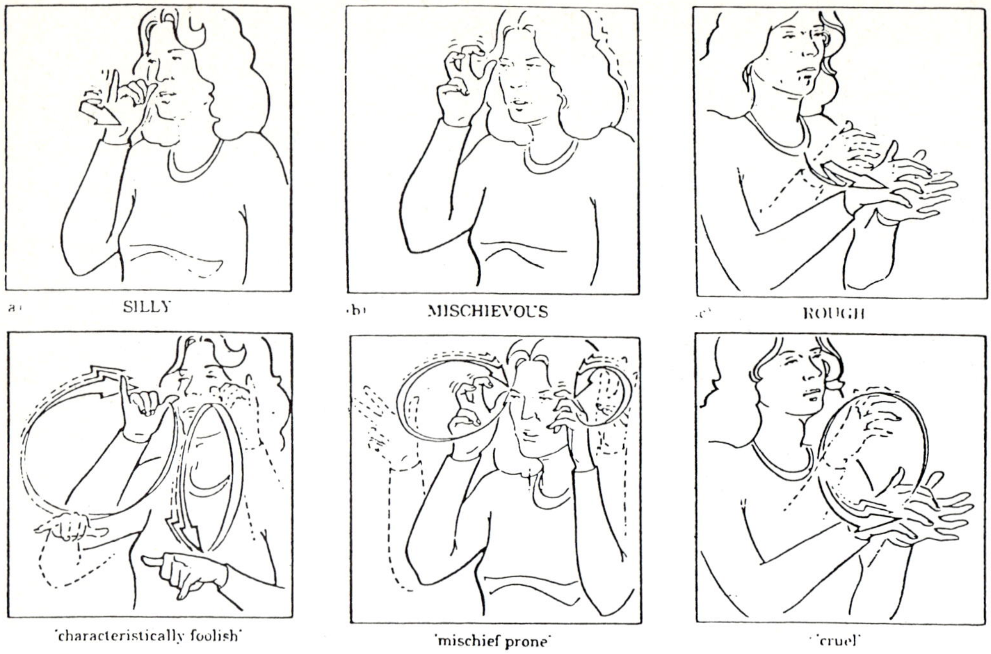

Figure 1.

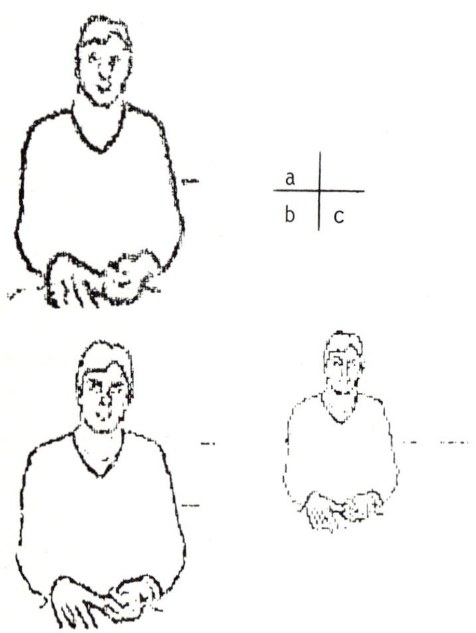

a) Contour image 128x128, computed on the half-tone image 128x128.
b) Contour image 128x128, computed on the half-tone image 256x256, then sampled at 1 point out of 2 and 1 line out of 2.
c) Contour image 85x85, computed on the half-tone image 256x256, then sampled at 1 point and 1 line out of 3.

Figure 2.

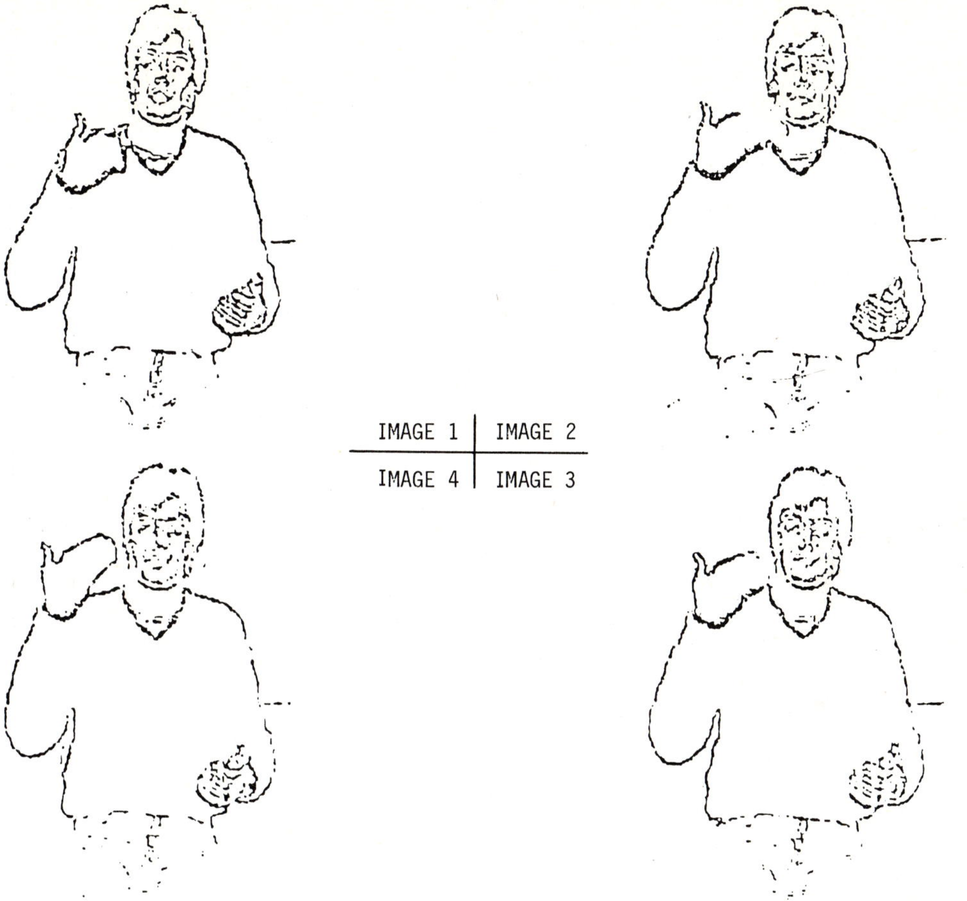

Figure 3a. The Sobel algorithm computed over the entire image 256x256 ; note that the hand disappears in image 2.

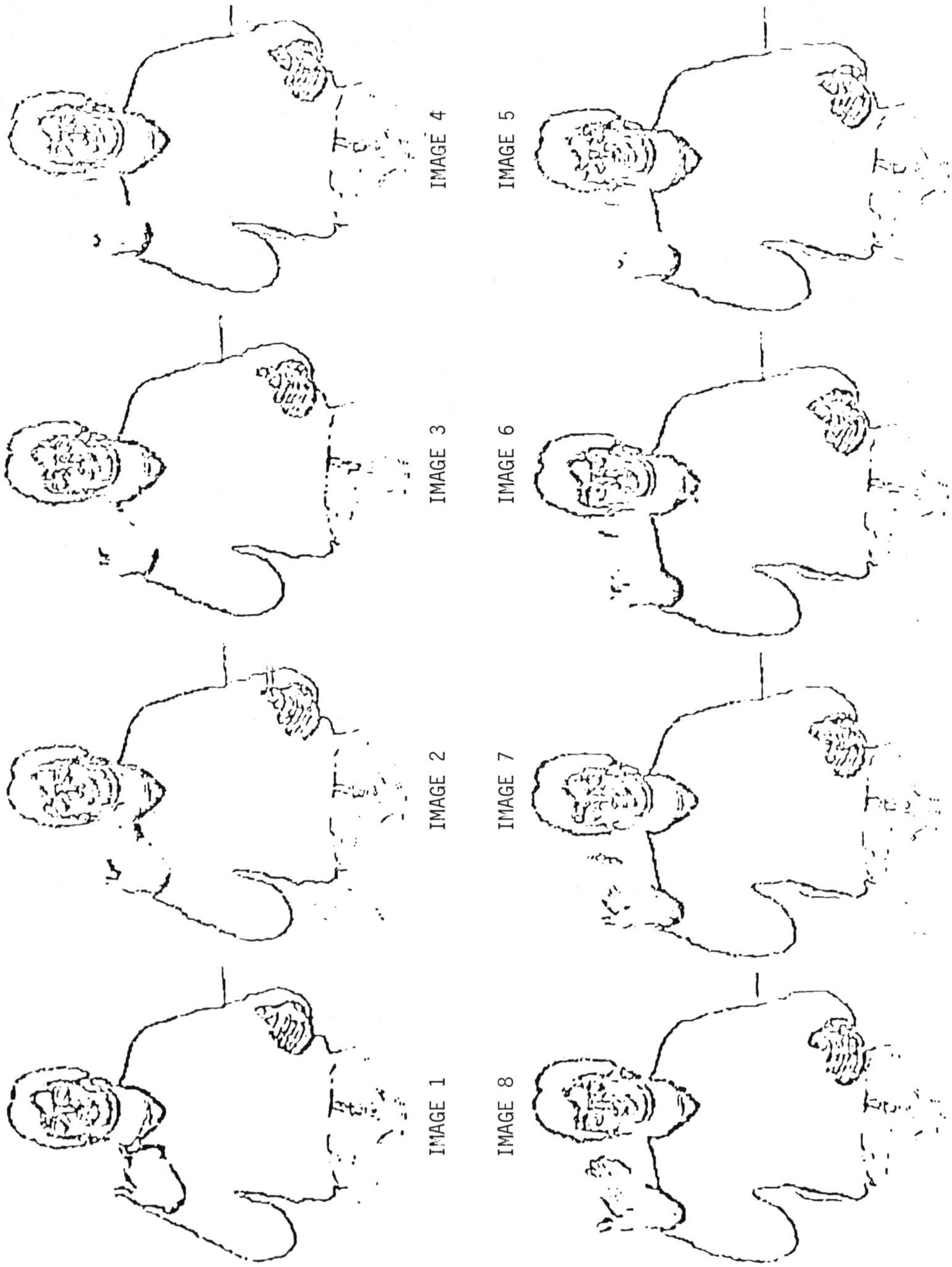

Figure 3b. The Sobel algorithm, computed in the planning zone (16000 pts) ; note that the hand disappears in 7 of the images.

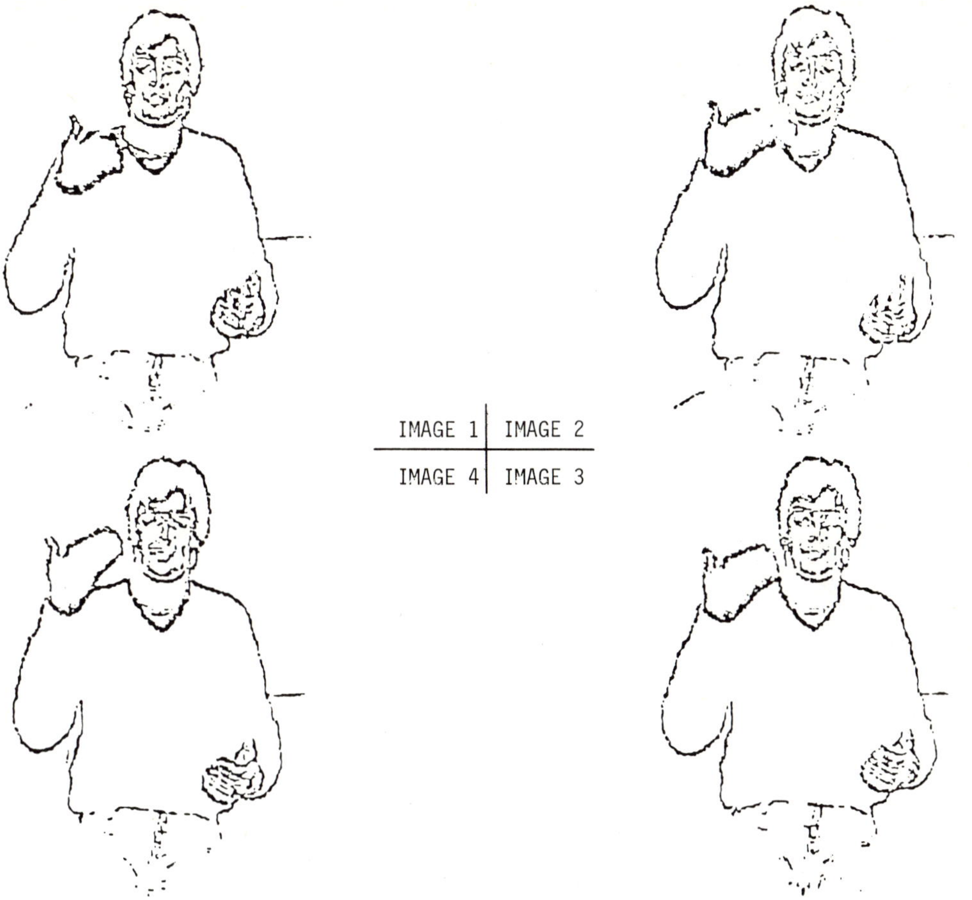

Figure 3c. The Sobel algorithm computed i- the planning and contour following zones (16000 pts) ; note that the hand disappears in 1 image.

SOBEL PSEUDO-LAPLACIEN (Ref.16)

Figure 4. Contour detection over the image 256x256 points with the Sobel (4155 pts) and the pseudo Laplacean (4284 pts) algorithms.

Figure 5. Difference image between two successive contour images.

Figure 6. The planning zone for a 9x9 neighborhood (15823 pts tested).

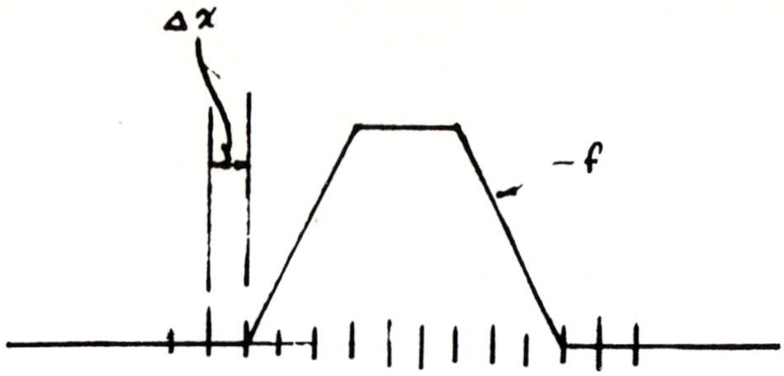

(a) Original function.

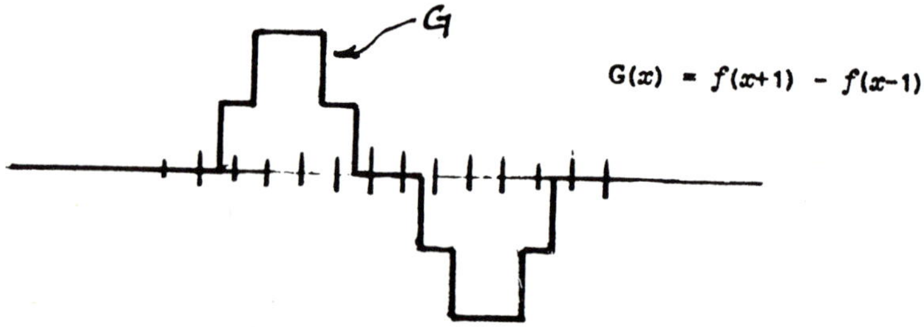

(b) Discrete gradient calculated on the function (a).

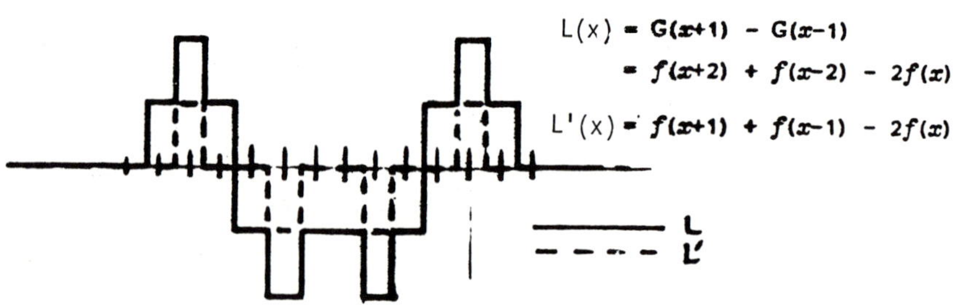

(c) Laplacean calculated on the function (a).
L' - by the usual scheme
L - by second differences on the gradient (b).

Figure A1. Gradient and Laplacean of a piecewise-linear function.

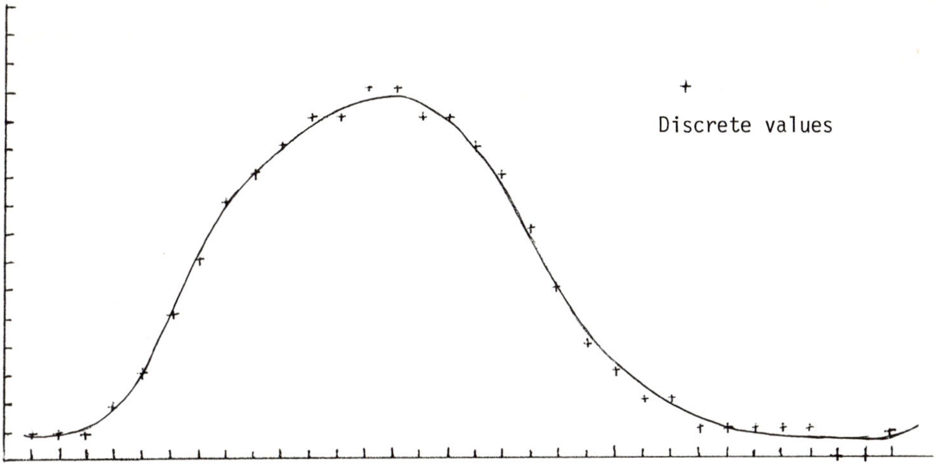

(a) Original function.

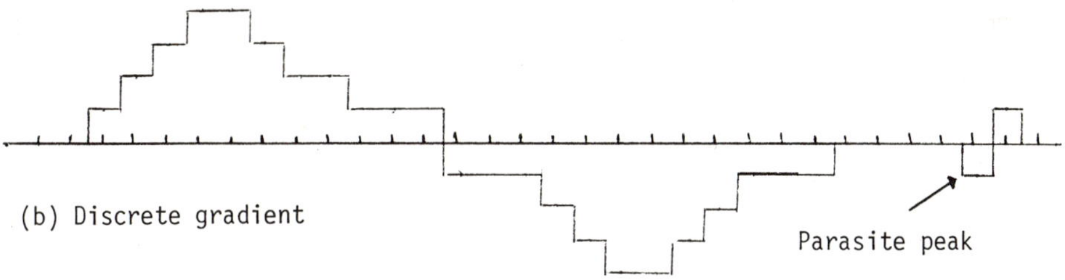

(b) Discrete gradient

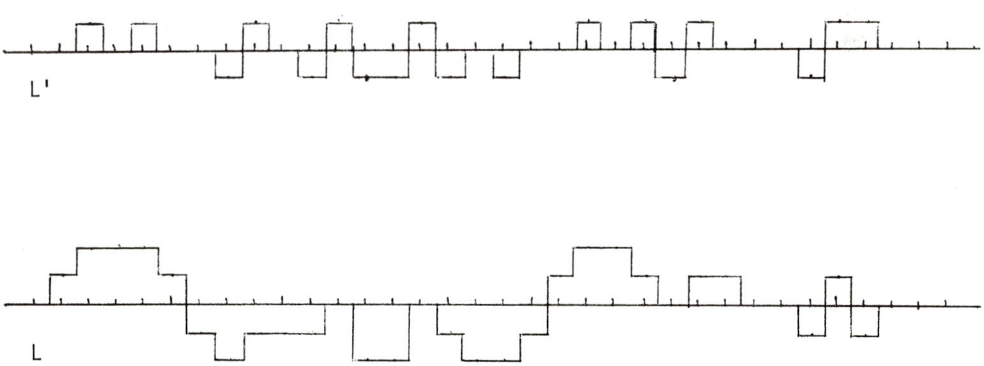

(c) Laplacean : L' - by the usual scheme.
L - by second differences on the gradient (b).

Figure A2. Gradient and Laplacean of a continuous function with high-frequency disturbances.

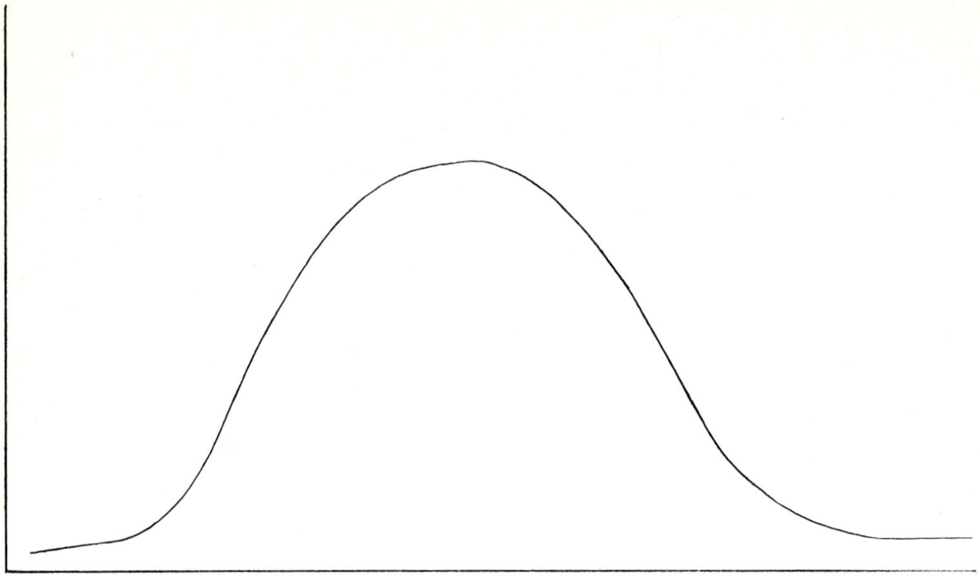

(a) The function (a) of Fig.A2.

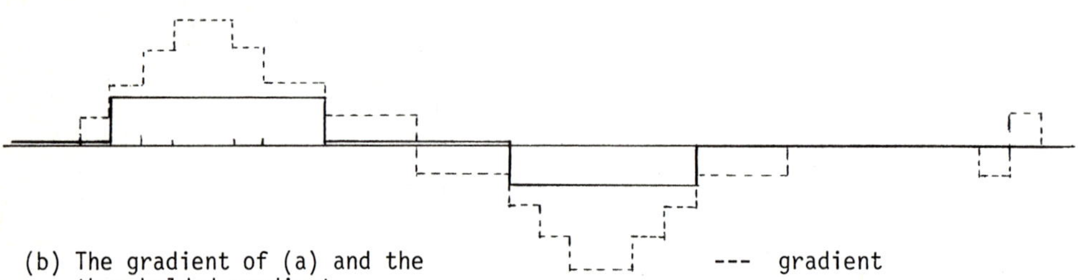

(b) The gradient of (a) and the thresholded gradient.

--- gradient

— thresholded gradient

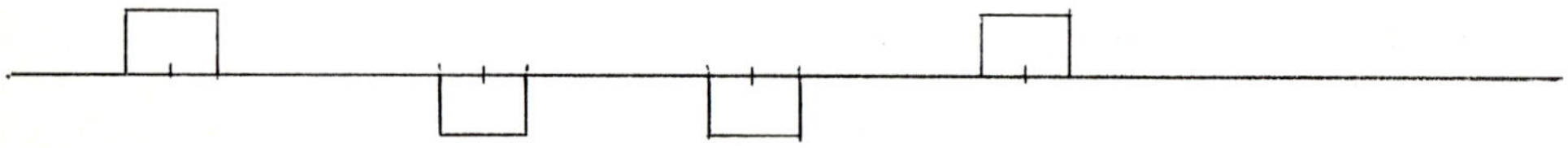

(c) The pseudo-Laplacean.

Figure A3. The pseudo-Laplacean calculated by second differences on the function of Fig.A2.

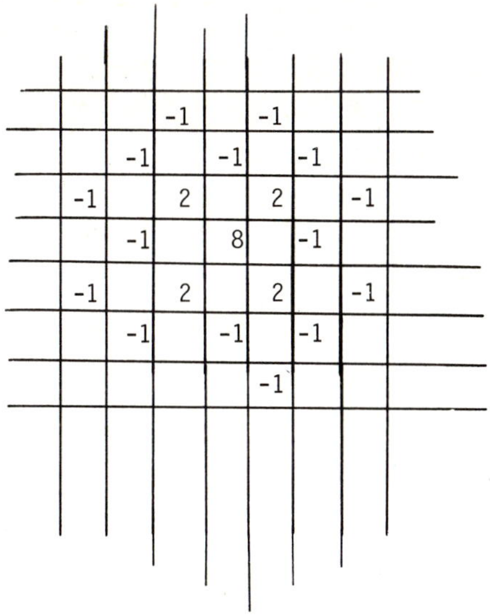

Figure A4. Weights of "equivalent" Laplacean.

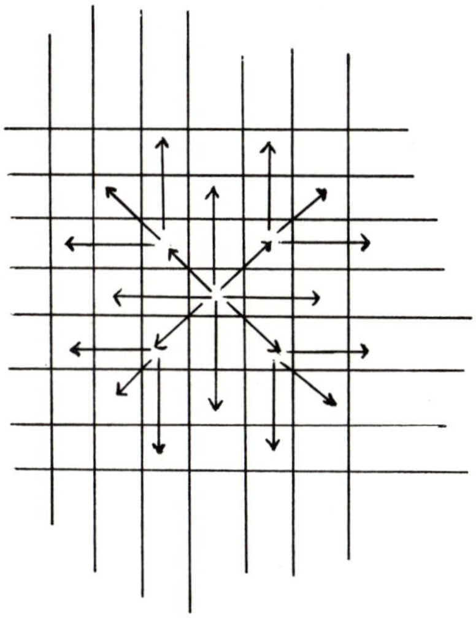

Figure A5. Alternative Pseudo-Laplacean.

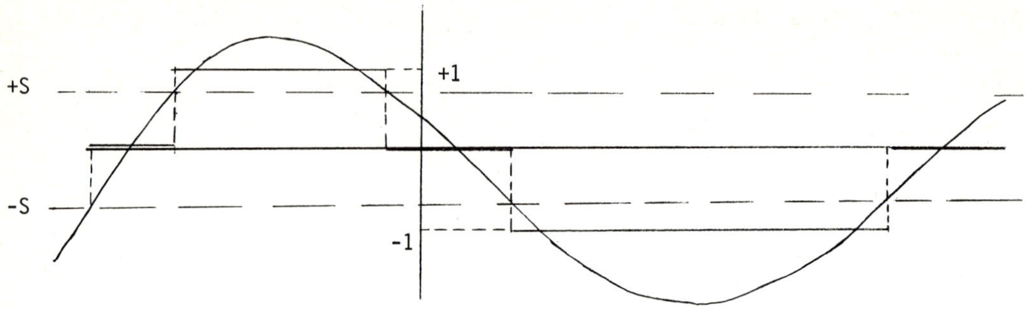

Figure A6.

CLASSIFICATION AND BLOCK CODING OF THE FRAME DIFFERENCE SIGNAL

Giovanni Zarone
Istituto Elettrotecnico, Facoltà d'Ingegneria, Università di Napoli
21 via Claudio, 80125 Napoli, Italy

An interframe-intraframe predictive coder which uses an adaptive subdivision of the frame difference signal into homogeneous regions (blobs), the conditional replenishment technique, an adaptive quantization, and the block coding of the quantizer output, is investigated. In order to recognize the homogeneous regions, three different classification algorithms are applied to a videotelephonic image sequence and compared. The performance of this coding scheme, in terms of Normalized Mean Squared Error and bit rate, are then evaluated for several values of the coder parameters. The cost of blob boundary coding is evaluated too.

1. INTRODUCTION

A lot of coders have been developed to reduce the video signal redundancy by predictive coding. Moreover, since the video signal is highly non stationary, it is advantageous to adapt the coding to the signal local properties. The adaptivity may be implemented either by making the predictor adaptive or by making the quantizer adaptive. We decided to investigate an interframe predictive coding scheme |1| provided with an adaptive quantizer.

The simplest way (but not the only one) to attain this aim is to subdivide every frame in regions whose shapes and sizes are fixed once and for all and to quantize the intensities of the pels included in the same region using some statistical features of the region itself. Of course, the values of parameters which cause the choice of quantization characteristic to be used need to be transmitted as a side information.

This approach, however, presents some disadvantages. If the region sizes are very large, the adaptivity to the local features is, in fact, very poor: little heterogeneous subregions (i.e. areas having different statistical features) may be included in the same region. On the other hand, if the region sizes are very small, the bit rate pertinent to the side information is high (it may happen that many neighbouring regions have the same statistics).

Therefore we decided to test out a scheme where the frame subdivision in regions is adaptive, i.e. a coding scheme which everytime subdivides the picture into "homogeneous" areas (called "blobs") which have neither a pre

NATO ASI Series, Vol. F2
Image Sequence Processing and Dynamic Scene Analysis
Edited by T.S. Huang

fixed shape nor prefixed sizes.

At the beginning, the frame difference signal (fds) is subdivided into equally shaped and equally sized elementary blocks, each one characterized by the local mean and standard deviation values. A special clustering algorithm then processes these elementary regions by grouping those one which are statistically similar together. So the output signal is subdivided into homogeneous and uncorrelated regions, having thus removed partially the spatial redundancy existing in the video signal.

The same classification algorithm provides, moreover, as a fall out, a simple and interesting way to segment the image into background and moving area: in fact, if both the absolute value of fds mean and fds standard deviation of a blob are below two appropriate thresholds, then that blob can be considered as the background. Therefore this algorithm allows us to exploit the "conditional replenishment" technique, consisting in skiping from the transmission the background (at the receiving end it is reconstructed by reading the pel intensities stored in a frame memory) and in transmitting only the moving area.

Moreover we sudied the coding of the quantizer output both on a pel-by-pel basis and through a block coding, with a further reduction of intraframe redundancy.

In summary, the investigated coding scheme presents the following properties:

Cons: i) The subdivision into homogeneous regions needs to be computed at each frame; ii) The map of blob boundaries needs to be transmitted (however even a conditional replenishment scheme needs the transmission of moving or fixed areas addresses).

Pros: i) The homogeneity of blocks included in the same region results in an efficient adaptivity; ii) The blob sizes are as large as possible (compatibly with the condition that the included blocks are homogeneous): the transmission cost of adaptivity parameters is low; iii) The free segmentation between fixed and moving areas results in a free conditional replenishment.

Our test signal was obtained by taking the most critical part of the typical videotelephonic sequence named Judy. It is 64 pels x 64 pels wide and consists of 16 images. Some statistics about this signal can be found in |2|.

The investigated coder is described in section 2; the study concerning the appropriate algorithm of block classification is presented in section 3; finally

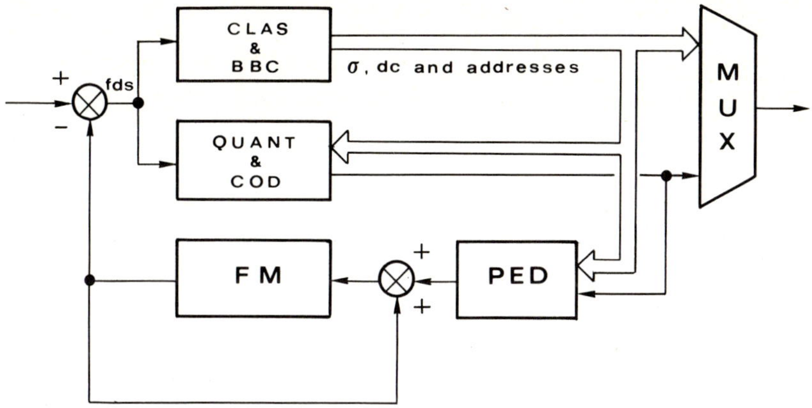

Fig. 1. Block scheme of the investigated coder.
Legenda: CLAS Classifier
QUANT Quantizer of the frame difference signal
COD Pel-by-pel or block coder of QUANT output
BBC Blob Boundary Coder
FM Frame memory
PED Prediction error decoder
MUX Multiplexer

the performance of this coding scheme (tested by Judy sequence) is evaluated by means of NMSE (normalized mean squared error) and average bit rate and the results are given by section 4.

2. CODING SCHEME

The coding scheme concept can be inferred from fig. 1.

The previous frame, stored in the frame memory FM, is subtracted from the present one. The resulting fds is subdivided into elementary blocks on which mean and standard deviation are computed. A classification algorithm compares then the adjacent blocks and merges together those ones which are statistically similar; the similarity measure is a special feature of the used algorithm. The fds is therefore subdivided into a certain amount of no longer elementary regions ("blobs"), whose local mean (dc) and standard deviation (σ) values are known.

If the clustering algorithm has processed the fds accurately enough, it is correct to estimate that the homogeneous region having the absolute dc value and the σ value below two particular threshold levels, should coincide with the background of the image.

All the dc and σ values of the moving area are uniformly quantized and transmitted to the receiver.

BBC means Blob Boundary Coder. In order to code the blob map we tried run length coding, block coding and a simple contour tracing algorithm. This algorithm consists in Huffman coding the angle that a segment of the traced contour makes with the previous segment of the same contour. This quantity can assume, in a sampled image, only three values (-90°, 0°, 90°). The position of the first segment of every contour was coded by run length coding.

The QUANT block quantizes the differences between the fds value pertinent to a generic pel of the moving area and the dc value of the blob containing that pel; the quantization step is proportional to the σ value of the same blob. It is interesting to note that, in this way, the quantization becomes dynamically matched to the local features of the region to be transmitted; this is in agreement with the psychophysics of the visual system: in fact the signal results quantized coarsely in those regions having higher σ values, i.e. where the errors are less noticeable since the movement content is greater. Moreover a step proportional to σ reduces the bit rate non uniformities. This either makes the buffering problem less dramatic or allows us to make the channel capacity closer to the long term average data rate.

The quantizer output is coded by COD block. We tried both a pel-by-pel Huffman coding and a Huffman coding of pel blocks. In fact, if the quantizer representative levels are M and the block sizes are SH and SV, every bidimensional block can assume M^S different patterns, where S=SH•SV. Since these different patterns present different occurrence probabilities because of high spatial correlation, their Huffman coding is expected to reduce the bit rate.

The reconstruction of the prediction error is performed by the PED block by utilizing the blob map information, the quantized values of dc and σ, and the COD output.

3. INVESTIGATED CLUSTERING ALGORITHMS

The first algorithm applied to Judy sequence fds is the Blob algorithm introduced in 1975 by Gupta and Wintz to obtain data compression on aircraft multispectral images. The similarity criterion is based on the concept of hypothesis testing depending on the first and second order statistical properties of the elementary regions ("blocks").

In order to overcome some restrictions presented by Blob algorithm, a new algorithm to be added to the previous one was developed.

The third processing, finally, is a typical maximum likelihood clustering algorithm, the Isodata, introduced in 1965 by Ball and Hall and then widely employed in pattern recognition.

The three algorithm logic schemes were partially modified in order to match them to the fds properties.

To evaluate the clustering performance, we used the quantities ε and e_5, defined as follows:

$$\varepsilon^2 = \overline{(d - d_b)^2} \qquad\qquad \text{Prob}\{|d-d_b|>e_5\} = 0.05$$

where d is the fds value pertinent to the generic pel and d_b is the dc value of the blob including that pel.

a) Blob algorithm

Blob algorithm, employed in this work, processes the fds in the same way as the single channel procedure utilized in the multispectral analysis introduced by Gupta and Wintz |3|. Adjacent subsets having similar first and second order statistics are merged into blobs (a blob is a connected set of pels all of which have some common characteristics). Hypothesis testing (F-test and student t-test) are used to measure the statistical similarity of pel groups. Performing this operation over the entire image, the fds is subdivided into homogeneous areas having shapes and sizes depending only on the local properties of the signal.

The algorithm output is an array of blob identifiers and of dc and σ values. The final results depend appreciably on the significance levels α_F and α_t used, respectively, in the F-test and t-test; in the hypothesis testings, in fact, the significance level represents the error probability of false rejection of the hypothesis. The H_0 null hypothesis is therefore accepted more easily as the corresponding α-level decreases; in this situation, however, the error probability of false alarm of the test increases. Consequently, in our application, even not very homogeneous pel groups are merged into blobs. On the other hand, in order to avoid the fds subdivision into a very large number of blobs, it is not convenient to use too restrictive α values.

In order to come to the best arrangement between classification accuracy and blob number, Blob algorithm was tested on a large amount of α_F and α_t values. ε, e_5 and final number of homogeneous areas were computed per each pair (α_F, α_t). The most significant results of this analysis are reported in table I, implementing Blob algorithm on the first fds of Judy sequence.

Table I. - Blob algorithm performance

			α_t 0.005	0.010	0.025	0.050
α_F	0.00010	Blobs	33	32	42	56
		ε	3.49	3.56	3.54	3.16
		e_5	8.02	8.11	7.92	7.38
	0.00025	Blobs	32	36	42	56
		ε	3.66	3.60	3.49	3.15
		e_5	8.34	8.15	7.89	7.35
	0.00050	Blobs	40	44	59	75
		ε	3.53	3.42	3.39	3.09
		e_5	8.14	7.96	7.66	7.20
	0.00075	Blobs	46	55	63	79
		ε	3.58	3.42	3.36	3.05
		e_5	8.23	8.02	7.65	7.20
	0.00100	Blobs	54	47	65	86
		ε	3.73	3.61	3.30	3.04
		e_5	8.55	8.33	7.48	7.08
	0.00500	Blobs	87	93	105	133
		ε	3.45	3.20	3.15	2.93
		e_5	7.98	7.66	7.30	6.75

Table III. - Isodata algorithm performance

No of classes	Monocon. areas	ε	e_5
2	23	3.58	7.58
3	71	3.07	6.95
4	81	3.09	6.89
5	174	2.86	6.53
6	178	2.84	6.40
7	177	2.84	6.29
8	180	2.80	6.22
9	202	2.74	6.18
10	212	2.74	6.12
11	214	2.74	6.10
12	215	2.73	6.03
13	221	2.72	6.14

Table V. - Cost of Blob Boundary Coding (bits/pel)

N	Block Coding 2x8	8x2	4x4	Run-length	Contour Tracing a	b	a+b
9	0.365	0.342	0.356	0.439	0.208	0.196	0.404
6	0.309	0.277	0.266	0.343	0.126	0.128	0.254

b) Blob II algorithm

The Gupta and Wintz algorithm respects some priorities in choosing the scan paths to be followed in order to carry out the test of statistical likelihood between blocks and blobs. These paths fit particularly well aircraft multispectral images. Moreover these pictures show a blob shape range that is much more narrow than that presented by a video sequence.

The strict application of the suggested procedure (even if it were optimized on the fds) implies sometimes the aggregation of homogeneous blocks to distinct blobs. In fact, implementing F-test and t-test directly on two distinct and adjacent blobs resulting from Blob processing, the two null hypotheses sometimes have been accepted. During the processing, growing size samples (i.e. the blobs) are indeed compared with fixed size sample (i.e. the blocks). The blob statistical features are therefore averaged on a greater number of pels; in a region where there are occasional but sharp transitions between adjacent pels, the growth of a blob can be stopped even if the homogeneous region is actually wider.

Therefore we decided to process again the output signal of the Gupta and Wintz algorithm by comparing all the blobs, one with the other, even the non adjacent ones, and by merging together the statistically similar ones. The comparison criterion is still hypothesis testing; also note that this time the sample population consists of a large number of pels. Hence, from the central limit theorem, it follows that the distribution approaches the Gaussian one and so the tests are more and more accurate as such a number increases.

Table II shows the most significant results of some experiments carried out by using different significance levels α_F and α_t for the Blob algorithm and α_{II} for the Blob II algorithm (we assumed equal the two levels of Blob II algorithm: in fact an excessive precision on these values is unnecessary, since the Blob II clustering has not to be very selective).

In order to make a right choice of α_F, α_t and α_{II}, one must bear in mind that Blob II just reduces the number of areas to that of the actually homogeneous ones; it is useful, therefore, to make the Blob tests more selective than the Blob II tests. In fact it is necessary to implement the Blob algorithm with a low "false alarm" probability, since the numerous, distinct but correlated blobs, which are formed, are then associated by the Blob II algorithm.

Note, finally, that the comparison of non adjacent blobs gives rise to

Table II. - Blob II algorithm performance

α_F	α_t		α_{II} 0.1	0.025	0.00075	0.0001
0.005	0.075	Multicon. areas	46	34	22	19
		Monocon. areas	144	142	113	109
		ε	2.91	2.98	3.16	3.23
		e_5	6.74	6.88	7.43	7.57
0.005	0.100	Multicon. areas	44	34	21	19
		Monocon. areas	172	163	122	111
		ε	2.91	2.95	3.15	3.23
		e_5	6.88	7.03	7.47	7.60
.0075	0.075	Multicon. areas	51	36	21	18
		Monocon. areas	154	150	123	115
		ε	2.90	2.97	3.10	3.26
		e_5	6.81	6.86	7.33	7.57
.0075	0.100	Multicon. areas	47	32	22	17
		Monocon. areas	180	165	126	117
		ε	2.89	2.95	3.15	3.25
		e_5	6.85	6.87	7.49	7.64
0.010	0.075	Multicon. areas	52	33	20	20
		Monocon. areas	159	153	131	127
		ε	2.89	2.97	3.11	3.19
		e_5	7.06	6.65	7.05	7.42
0.010	0.100	Multicon. areas	48	37	21	19
		Monocon. areas	184	181	139	129
		ε	2.89	2.96	3.15	3.18
		e_5	6.88	6.70	7.20	7.51

Table IV - Pel-by-pel coding

(x, y)	$10^4 \cdot$ NMSE	bits/pel dc	σ	Δ/σ
(2, 2.0)	11.56	0.02	0.01	0.90
(2, 1.0)	2.59	0.01	0.01	1.34
(2, 0.5)	1.93	0.01	0.01	1.70
(2, 0.2)	1.41	0.02	0.01	2.02
(1, 2.0)	11.90	0.02	0.01	0.85
(1, 1.0)	3.53	0.02	0.01	1.34
(1, 0.5)	1.93	0.02	0.01	1.62
(1, 0.2)	1.34	0.02	0.01	2.10

x: σ and dc quantization step
y: Δ/σ quantization step

the forming of homogeneous multiconnected regions. This peculiarity brings down the cost which is necessary to transmit the quantized σ and dc values, since the number of multiconnected areas at the output of Blob II algorithm is appreciably smaller than the number of monoconnected regions at the output of Blob algorithm, ε and e_5 being equal.

c) Isodata algorithm

The Blob II algorithm can be seen as a classification algorithm which is applied to non elementary image parts, in a way that is absolutely non dependent on the geometric position inside the frame.

Therefore we deemed it useful to compare the previously obtained results with those which can be achieved through a classification algorithm directly applied to elementary 2x2 blocks, without any "a priori" knowledge of the image.

The Isodata algorithm |4|, a typical example of unsupervised clustering, matches perfectly this requirement; the adopted criterion to perform the comparison is that of maximum likelihood. The features, used in the "measurement space", were, in our study, the average value and the standard deviation of each sample (an elementary block). The position inside this space, of the barycentre of the samples belonging to the same class is called "centroid".

During the processing, the generic sample is assigned to the class including the nearest centroid. Then the centroid coordinates are up to dated, by computing again the different classes. This procedure is repeated until just a small percentage of samples changes class from an iteration to the next one.

By summarizing, we can say that the algorithm searches for that particular elementary block classification which makes minimum, in the measurement space, the Euclidean distances between the points (representing elementary blocks) and the corresponding centroids. If one uses the Isodata algorithm, one cannot assign at random the starting centroid positions; this would imply, in fact, too long convergence times. On the other hand, some studies about the dependence of the Isodata convergence on the choice of the starting centroids show that this algorithm can stop at a relative minimum position. We decided to overcome this problem by adopting a solution borrowed from iterative optimization systems through hierarchical clustering.

The starting centroids, for the c-class problem, are the final ones for the (c-1)-class problem and, as additional centroid, the sample which is the

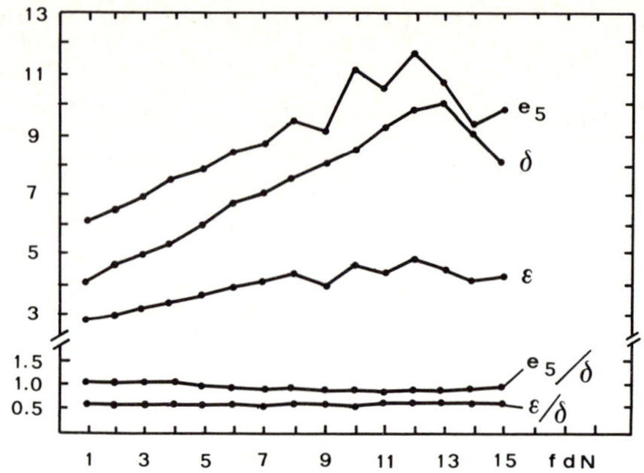

Fig. 2. Behaviour of Isodata accuracy vs. frame difference number (fdN).

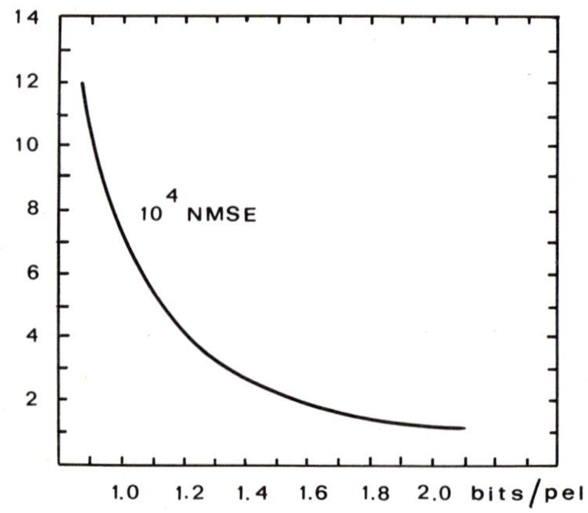

Fig. 3. Pel-by-pel coding. Normalized mean squared error (NMSE) vs. bit rate.

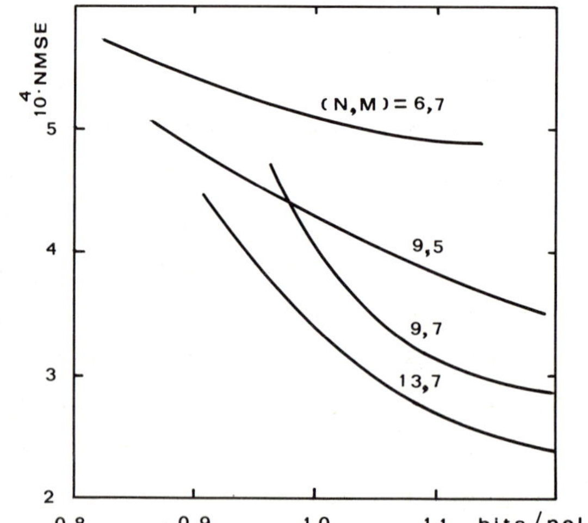

Fig. 4. 2x2 Block coding. Normalized mean squared error (NMSE) vs. bit rate for some values of N (number of Isodata classes) and M (number of quantizer representative levels).

farthest from the nearest class centroid. Such a complicated beginning procedure is fortunately necessary just after sudden changes of the scene; usually the previous frame final centroids can be adopted directly.

Table III shows the results obtained by applying the Isodata algorithm to the first frame difference of Judy sequence for different number of classes and by adopting the described procedure. Note that the error ε is small enough even if one uses a small number of classes.

4. RESULTS

Among the different investigated algorithms, the Isodata one proved the most efficient. In fact, as we can infer from tables I, II and III, it is able to classify the fds through the minimum number of distinct pairs (dc, σ), ε being equal. The Isodata was therefore selected to try the above described coder.

Fig. 2 shows the behaviour of ε and e_5 all over the Judy sequence; the same figure shows the r.m.s. value (δ) of the fds, as well as the ratios ε/δ and e_5/δ. It has to be noted that the trends of ε, e_5 and δ are similar (the ratios ε/δ and e_5/δ, in fact, do not change significantly all over the sequence); the algorithm is therefore very robust as regards the signal variations. By adopting the Isodata algorithm in our coder, the largest classification errors occur where the activity is greater, i.e. where coding errors are less noticeable.

Fig. 3 shows the NMSE (normalized mean squared error) between the processed images and the original ones vs. bit rate for the coder utilizing the 13 class Isodata and a pel-by-pel coding. Table IV reports, for the same coder, the transmission cost of the three quantities dc, σ and Δ/σ = (fds-dc)/σ.

The results obtained applying the 2x2 block coding to the quantizer output for some values of M (number of quantizer representative levels) and N (number of Isodata classes) are given by figure 4.

The investigated coding scheme showed the possibility to achieve a NMSE = 0.0003 at 1 bit/pel. One image of the original sequence (8 bits/pel) and some corresponding processed ones (1.05 bits/pel) are shown by figure 5. The degradation, introduced by the investigated coder, is not noticeable.

In table V we give the bit rate (bits/pel) necessary to transmit the map of blob boundaries for some value of N and using different techniques: run

a)

b)

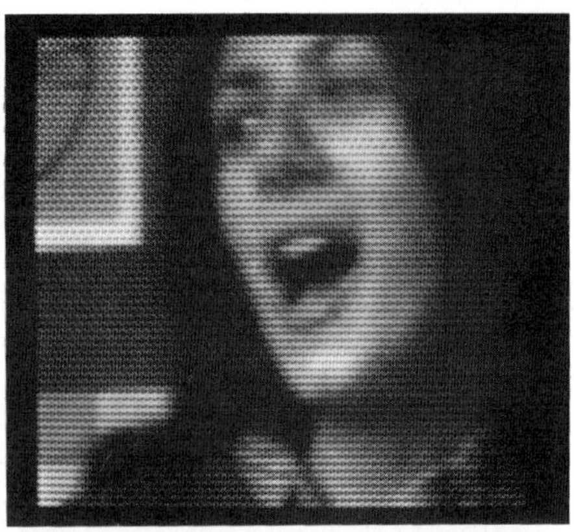

c)

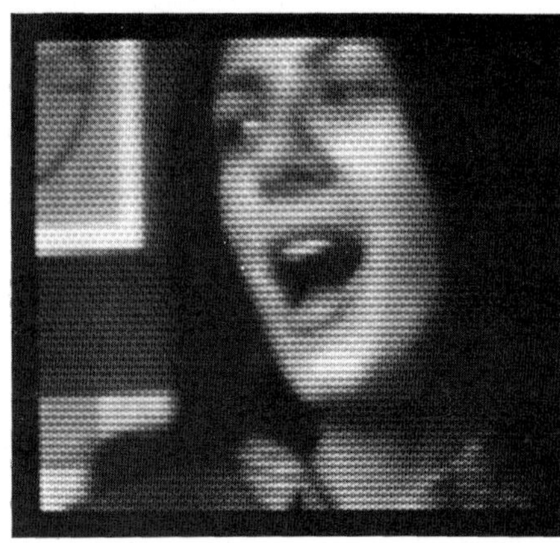

d)

Fig. 5

a) Original image (8bits/pel)
b) Processed image (1.05 bits/pel). Pel-by-pel coding; M=3.
c) Processed image (1.05 bits/pel). Pel-by-pel coding; M=7.
d) Processed image (1.05 bits/pel). Block coding; M=5.

length , block coding and contour tracing. In the last case,the coding of the segments following the first one of each contour needs the bit rate "a", while the coding of first segment position needs the bit rate "b". As N decreases, the number of blobs decreases, the cost necessary to transmit their map de creases.

5. CONCLUDING REMARKS.

Although the coder parameters (N, M, SH, SV) need to be still optimized, we deem that the fds classification can significantly improve the overall per formance in frame-to-frame coding. The subjective results (see the pictures pro cessed at 1.05 bits/pel) and the objective results are, in fact, good enough.

Actually an objective parameter, as NMSE, is ill-suited to evaluate the quality of a sequence of pictures coded using a quantization partially matched to the property of vision; it would be more appropriate to evaluate this qua lity by some parameters more correlated to the subjective opinion of the obser vers, particularly by a parameter related to the temporal masking phenomenon. Unfortunately such a parameter does not exist.

A further study could be devoted to code the elementary blocks by Tran sform Coding; this, indeed, would make the coder a hybrid coder.

The author thanks prof. C.Cafforio and prof. F.Rocca who allowed him to display the processed pictures.

REFERENCES

|1| P.Gustuti, O.Telese, G.Zarone, "Frame difference signal classification and co ding", Internat. Conf. on Digital Signal Processing, Firenze (Italy), Sept. 2-5, 1981, pp. 1096-1102.

|2| L.Arena, G.Zarone, "3-D filtering of television signal", Alta Frequenza, No 2, 1977, pp. 108-116.

|3| J.Gupta, P.A.Wintz, "A boundary finding algorithm and its applications", IEEE Trans. on Circuits and Systems, No 4, April 1975, pp. 351-362.

|4| G.H.Ball, D.J.Hall, "Isodata, a novel method of data analysis and pattern clas sification", Standford Research Institute, Technical Report, chapt. 11.

HISTOGRAMS OF IMAGE SEQUENCE SPECTRA

A.W. Lohmann, Ch. Thum
Physikalisches Institut
Universität Erlangen-Nürnberg
Erwin-Rommel-Straße 1
8520 Erlangen, West Germany

Abstract

We investigated histograms of the Fourier transform amplitudes of image sequence spectra. Our results show that the Fourier amplitudes of images within the same sequence do not differ very much if the scenery in general remains the same throughout the sequence. As a useful consequence of this diagnosis we replaced the Fourier amplitude of every image in the sequence by an estimate. The estimate was obtained by averaging over the Fourier amplitudes of the first few images. We reconstructed images from their correct Fourier phase and from the estimated Fourier amplitude. This concept can be used in transform coding, where we can achieve a data compression of approximately 1/2.

I. Introduction

It has been known for some time that the Fourier amplitude of an image is often less important than the Fourier phase. A recent paper on this subject was published by Oppenheim and Lim /1/. There it was shown how two different images can be converted into each other simply by exchanging their Fourier phase, whereas the Fourier amplitude remains unchanged. In some earlier experiments, the fact that the detailed structure of the Fourier amplitude is relatively unimportant was used also, e.g. for code translation /2/. It is therefore reasonable to expect that if the Fourier amplitudes of a group of images are similar, then the images can be reconstructed fairly accurately from some estimated Fourier amplitude and the correct Fourier phase. If the images in an image sequence have the same general features (e.g. a moving object on a fixed background), it can be expected that their Fourier amplitudes do not differ very much, too. We want to show how far these assumptions are true, and especially, how well an image sequence can be reconstructed using the same estimated Fourier amplitude for all images.

II. Evaluation of histograms of image sequence spectra

First, we wanted to know how much the Fourier amplitude changes within an image sequence. The single images of an image sequence were recorded on magnetic tape and Fourier transformed digitally by a PDP 11/34 with a picture processing memory for storage. In the Fourier spectrum of the n-th image in the sequence, at every spatial frequency we calculated the relative amplitude difference $\Delta A_n = |A_n - A_1|/A_1$ of the Fourier amplitude A_n of the n-th image and the Fourier amplitude A_1 of the first image. From these values the histogram $h(\Delta A_n, n)$ was calculated, i.e. the frequency of occurrence of the values ΔA_n in the spectrum of the n-th image.

The histogram $h(\Delta A_n, n)$ of an image sequence (fig. 1) is shown in fig. 2. For every n, the histogram has a peak near $\Delta A_n = 0$ and then decreases for larger ΔA_n. This shape does not change very much for different images n. About 80-90% of the values of ΔA_n are in the range $0 < \Delta A_n < 1$. We can gain more insight if we divide the Fourier spectrum of every image according to the amplitude values A_1 of the first image (fig. 3). For high values of A_1 (fig. 3a), which generally occur at low spatial frequencies, the histogram is rather narrow, i.e. the amplitude A_n is generally not very different from A_1, whereas for

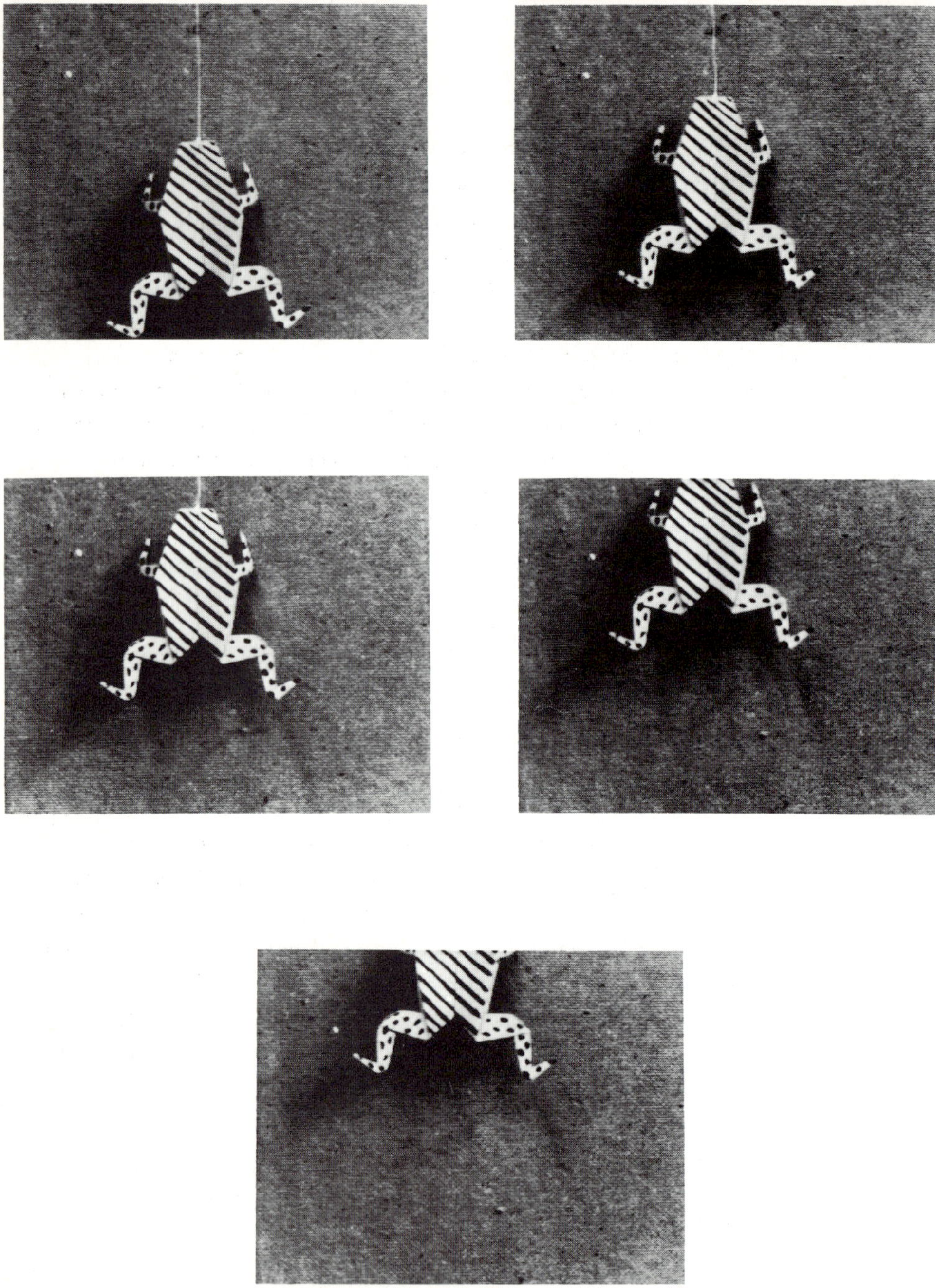

Fig. 1. Image sequence: moving paper frog
images no. 1,30,60,90,120 are shown

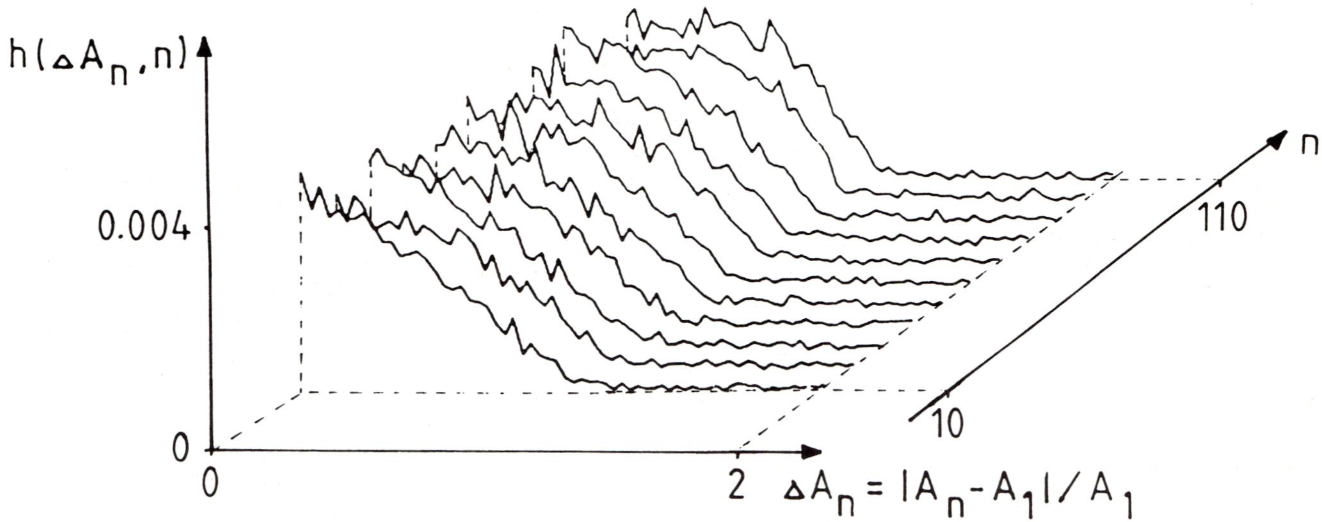

Fig. 2. Histogram $h(\Delta A_n,n)$ of the relative differences $\Delta A_n = |A_n - A_1|/A_1$ of the Fourier amplitude A_n of the n-th image and the Fourier amplitude A_1 of the first image in the sequence of fig.1.

a)

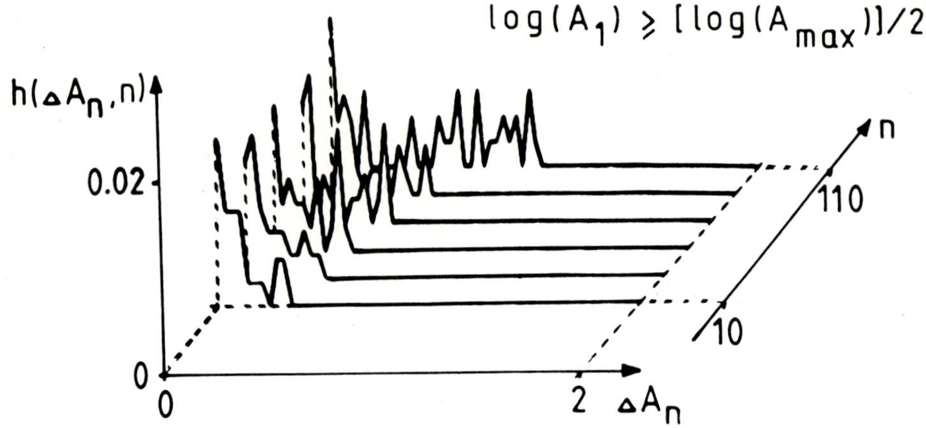

b)

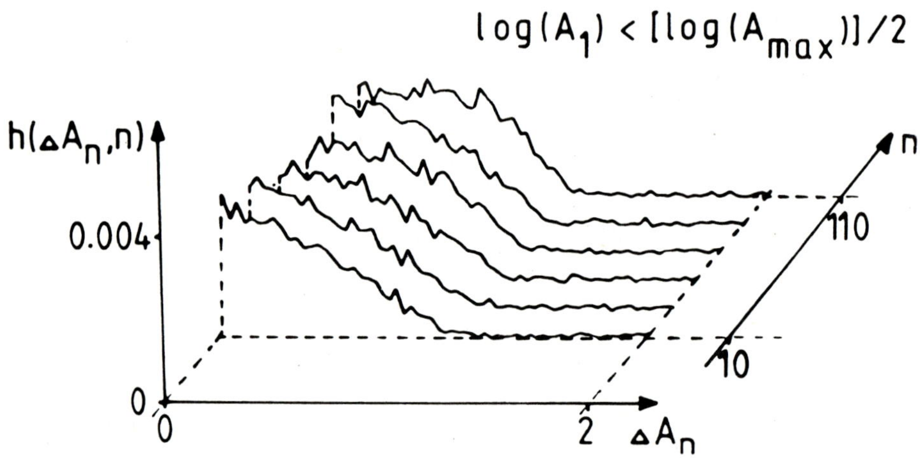

Fig. 3. Histograms $h(\Delta A_n,n)$ of the image sequence of fig. 1, where the Fourier plane is divided according to the magnitude of the Fourier amplitude A_1 in relation to the maximum Fourier amplitude A_{max} of the first image:
a) $h(\Delta A_n,n)$ for $\log (A_1) > (\log(A_{max}))/2$
b) $h(\Delta A_n,n)$ for $\log (A_1) < (\log(A_{max}))/2$

low values of A_1(fig. 3b), which generally occur at higher spatial frequencies, the changes in the relative amplitude are larger. So if the Fourier amplitude of the n-th image is replaced by the Fourier amplitude of the first image, we can expect that the coarse structures of the image are reconstructed fairly well, whereas possibly the finer structures with low contrast are not so well reconstructed.

III. Reconstruction of image sequences from the correct Fourier phase and an estimated Fourier amplitude

For different image sequences, we tried to reconstruct every single image from its correct Fourier phase and an estimated Fourier amplitude. This estimated amplitude was calculated by averaging over the Fourier amplitudes of the first few images in the sequence, or even by simply taking the Fourier amplitude of the first image in the sequence.

Our image sequences consisted of an object $u_o(x)$ moving on a fixed background $b(x)$. If the moving object keeps its shape, every image in such a sequence can be described in the following way (calculations are shown here for one-dimensional images only):

$$v_n(x) = b(x) \cdot [1-m(x-x_n)] + u_o(x-x_n)$$

where $v_n(x)$ is the n-th image in the sequence, $m(x)$ is a mask which is unity within the boundary of the moving object and zero outside, and x_n is the distance which the moving object has travelled from the first to the n-th image. The Fourier transform is:

$$\tilde{v}_n(\nu) = \tilde{b}(\nu) * [\delta(\nu) - \tilde{m}(\nu) \cdot e^{-2\pi i \nu x_n}] + \tilde{u}_o(\nu) \cdot e^{-2\pi i \nu x_n}$$

where '~' denotes Fourier transform, '*' denotes convolution, ν is the spatial frequency coordinate, and $\delta(\nu)$ is the Dirac delta function.

The quality of the reconstructed image depends heavily on the spatial structure of the image sequence, especially on the structure of the background. This can be illustrated by a simple example: if the background is uniform, then the reconstructed images are exactly equal to the original images, if the Fourier amplitude of every single image is replaced by the Fourier amplitude of the first image in the sequence. The first image $v_1(x)$ consists of a constant background and a moving object $u(x)$:

$v_1(x) = c + u(x)$

The amplitude $v_n(x)$ of the n-th image is:

$v_n(x) = c + u(x-x_n)$

This expression, however, is related to the first image simply by:

$v_n(x) = v_1(x-x_n)$.

Hence, the Fourier amplitudes are identical: $|\tilde{v}_n(\nu)| = |\tilde{v}_1(\nu)|$. Under these circumstances, if $|\tilde{v}_n(\nu)|$ is replaced by $|\tilde{v}_1(\nu)|$, nothing is changed at all, and the reconstructed image is exactly equal to the original image $v_n(x)$.

However, things are not so easy if the background has a varied structure, and/or if the moving object changes its shape (e.g. a walking person moving his legs). This can be seen in the following image sequences:

The sequence of fig.4 shows a paper frog containing high spatial frequencies (stripes and dots) moving on a background of coarse cardboard that varies only slowly in space. First, the Fourier amplitudes of the single images of the sequence were replaced by the Fourier amplitude of the first image (fig.5). It can be seen that the moving object is reconstructed rather well, whereas some of the high spatial frequencies of the object appear now in the background, too. Especially when the shape of the object changes (only part of the paper frog is visible in the last two images of the sequence), some kind of weak "ghost image" appears (lower half of the last two images). All these effects are much less noticeable if the estimate of the Fourier amplitude is not taken from the Fourier amplitude of the first image alone, but is averaged over the Fourier amplitudes of the first three images (fig.6). It is to be noted that in spite of the errors in the reconstructed images, the structure of the background is retained. E.g. the little white spot in the upper left corner of the original images appears in all reconstructed images, too.

Our next image was a moving person on a background containing many low frequencies and a considerable amount of high frequencies, too (edges, fig.7). The walking person moves his legs and arms and thereby changes his shape. The Fourier amplitudes of the first three images were averaged and used as an estimate for the Fourier amplitudes of

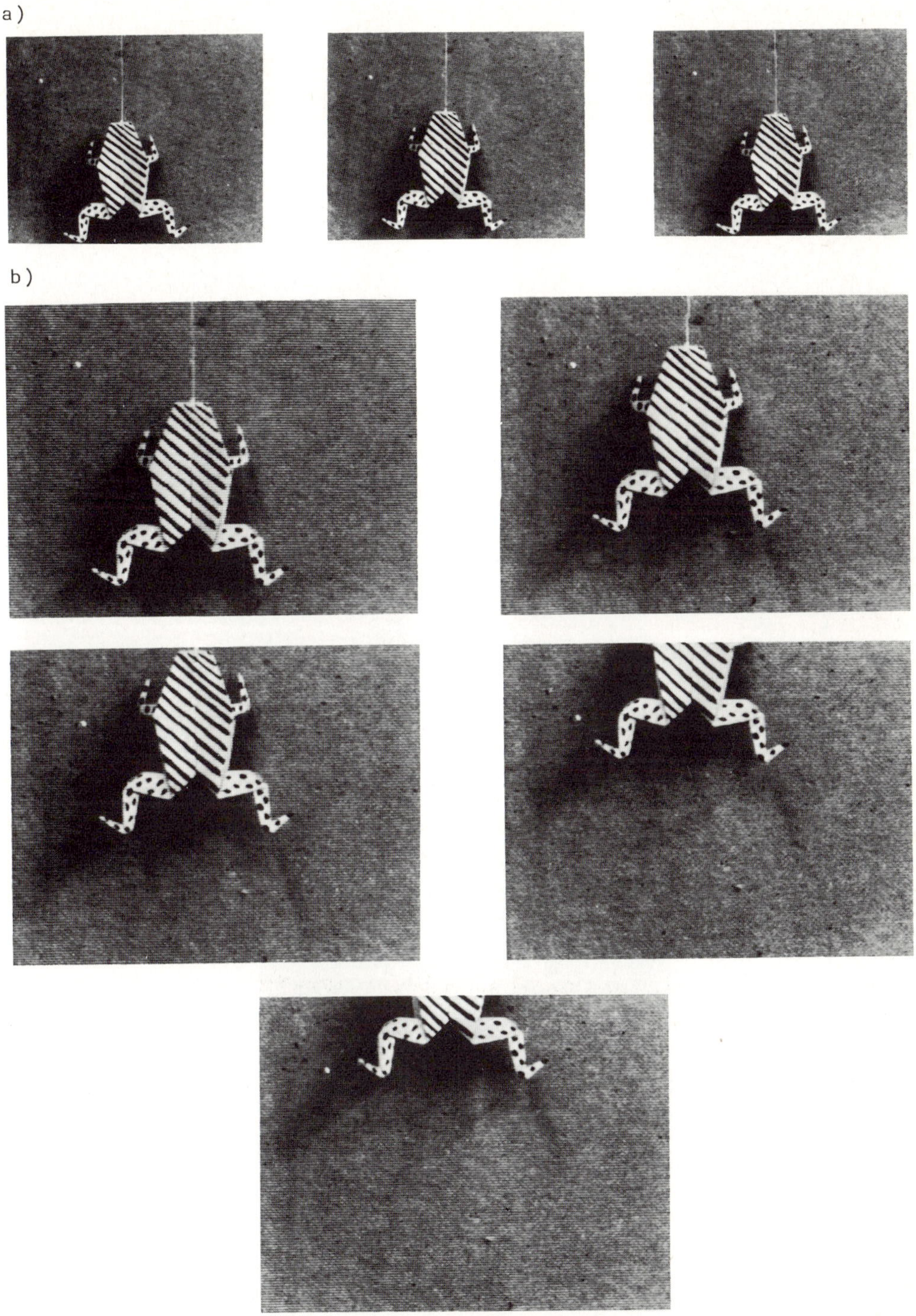

Fig. 4. Image sequence: moving paper frog
a) first three images in the sequence
b) images to be reconstructed: no. 20,50,80,120,140

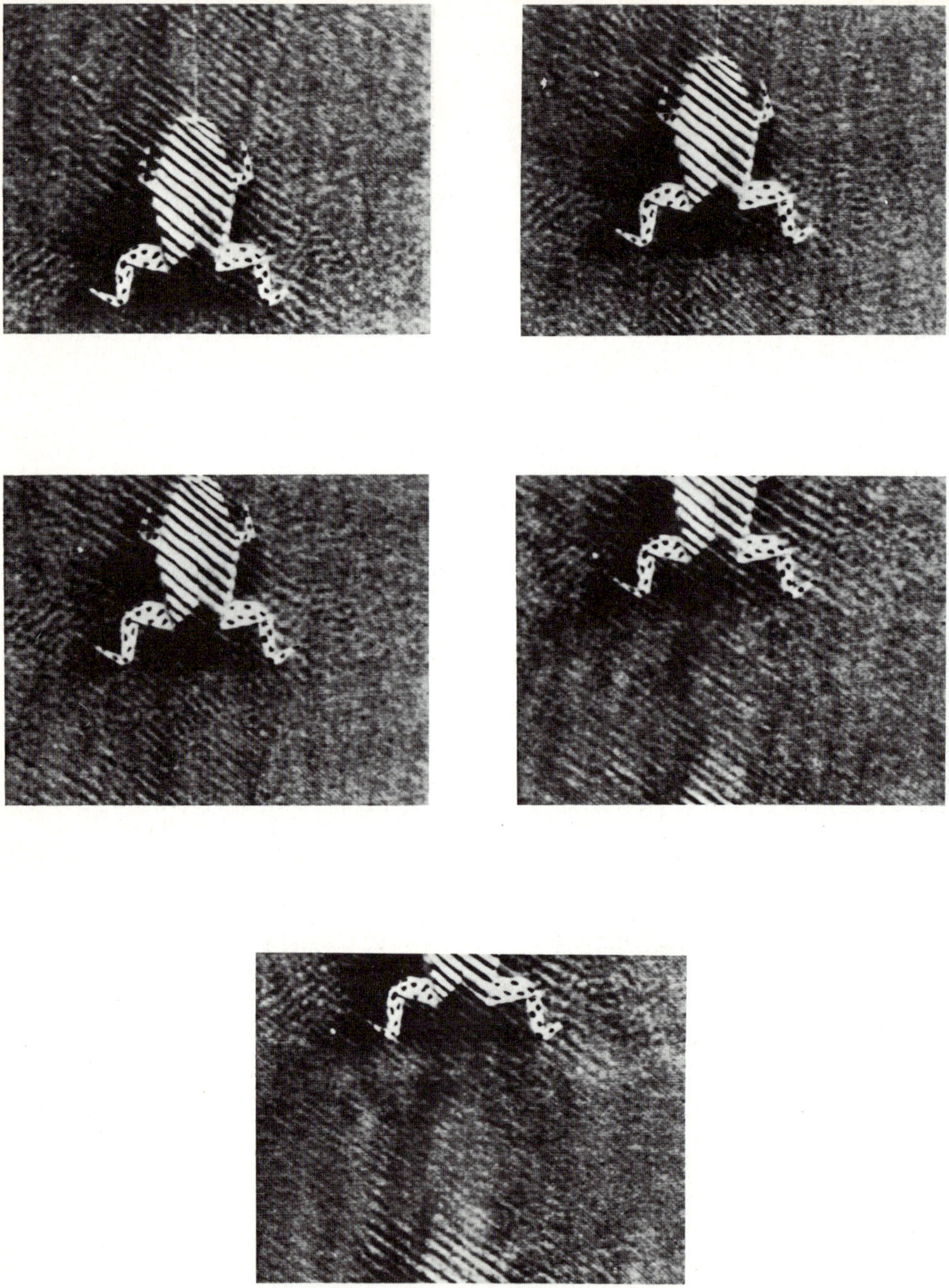

Fig. 5. Reconstructed images, estimated Fourier amplitude taken from the first image in the sequence

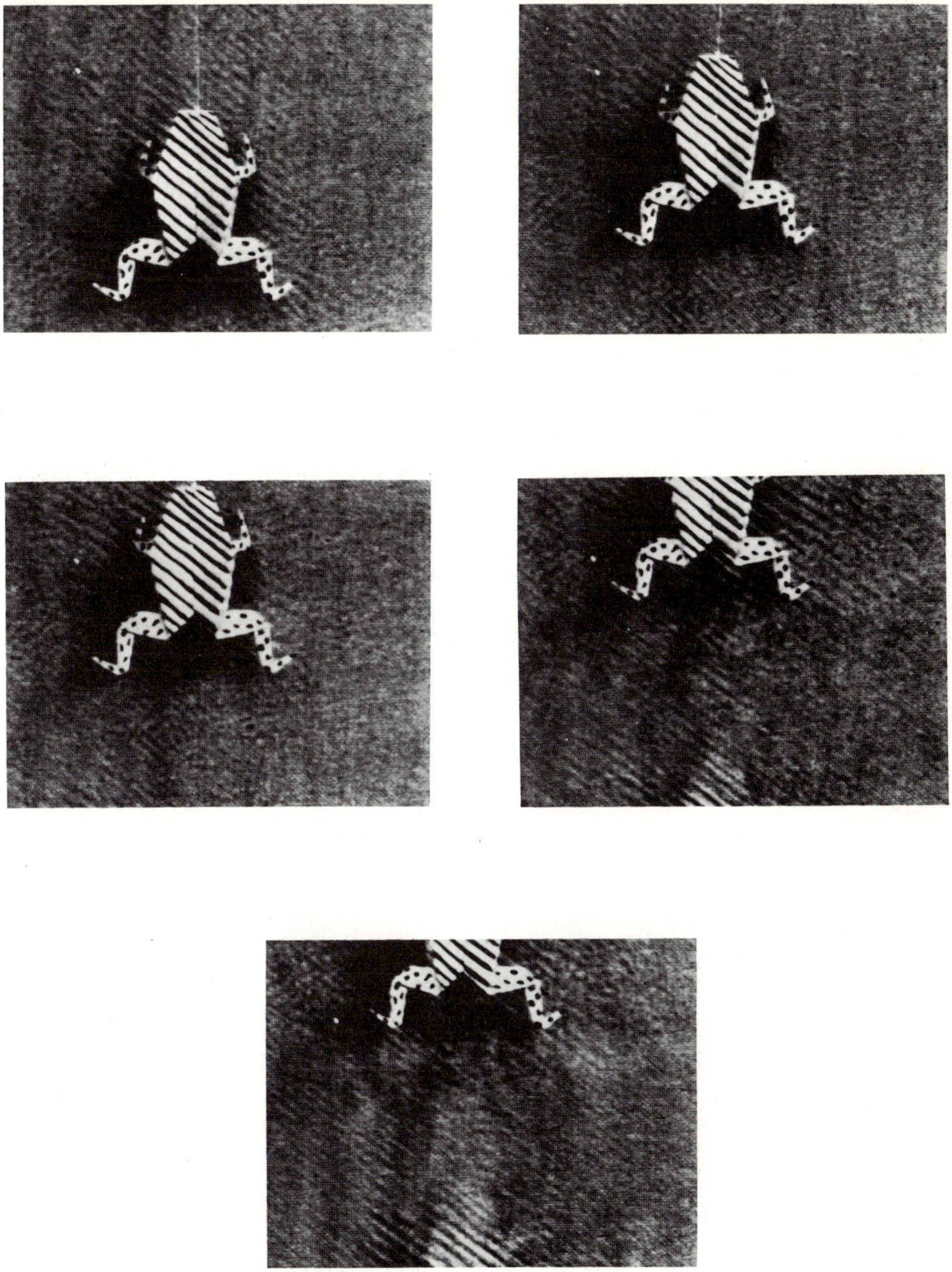

Fig. 6. Reconstructed images, estimated Fourier amplitude taken from the first three images in the sequence

a)

b)

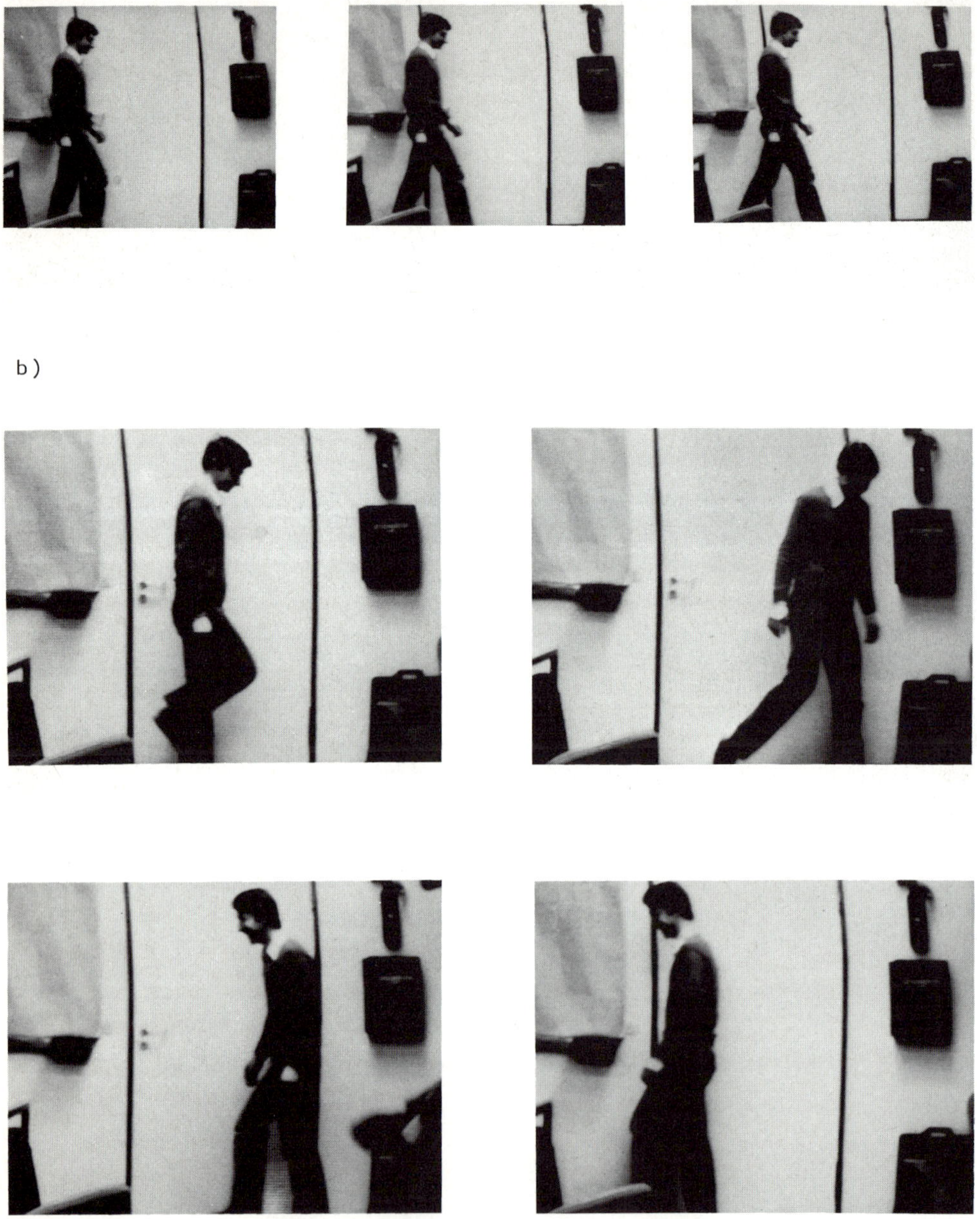

Fig. 7. Image sequence: walking person
a) first three images in the sequence
b) images to be reconstructed: no. 12,25,58,78

Fig. 8. Reconstructed images, estimated Fourier amplitude taken from the first three images in the sequence

the other images. The reconstructed images (fig.8) show noisy structures, especially in the background. However, the important features of the images are retained: For the person, the shape of the legs and arms is reconstructed although it changes in every image. In the background, the shape of the objects is retained, although there are some dark noisy structures, especially in the upper left corner of the first reconstructed image. It is to be noted however that even the left vertical border of the door and the doorknobs beside it can be seen in the reconstructions, although they were concealed in the first three images from which the Fourier amplitude was taken.

IV. Conclusion

We have reconstructed images of image sequences from an estimated Fourier amplitude and the correct Fourier phase. The results seem encouraging, as all essential features of the reconstructed images are retained, although in some cases there is considerable noise in the image. If the noise gets too strong, it may be necessary to modify the method, e.g. by coding the Fourier amplitudes at some of the higher spatial frequencies more exactly than it was done here. However, if the method is applied e.g. to a video film, the eye of the observer will follow mainly the moving object, so that noisy structures in the background will not be noticed as much as in a still picture.

The transform coding technique we investigated here yields a data compression of approximately 1/2 as only the Fourier phase of every single image and one Fourier amplitude for all images are to be transmitted. It remains to be seen how far the factor 1/2 can be reduced further by working on the Fourier phase, e.g. by using prediction and interpolation methods between frames. One can hope that the errors introduced by coding in the Fourier domain will spread out all over the spatial domain. The result should be a rather general degradation of the images which is more tolerable for the human eye than the introduction of noise in specific image details.

References:

/1/ A.V.Oppenheim, J.S.Lim: The importance of phase in signals, IEEE Proc. 69 (1981) 529

/2/ A.W.Lohmann, H.W.Werlich: Holographic production of spatial filters for code translation and for image restoration, Phys. Lett. 25A (1967) 570

PART III

SCENE ANALYSIS AND INDUSTRIAL APPLICATIONS

DETERMINING 3-D MOTION AND STRUCTURE OF A RIGID BODY USING STRAIGHT LINE CORRESPONDENCES

B. L. YEN T. S. HUANG

Coordinated Science Laboratory
University of Illinois
1101 W. Springfield
Urbana, Illinois 61801

ABSTRACT

This paper investigates the determination of 3-D motion and structure of a rigid body containing straight line segments, given a sequence of (2-D) perspective views. Central projections of 3-D lines over the image sequence are used, and line correspondences are assumed to be given. The approach is to use central projection on the unit sphere in the analysis. Three basic types of rigid motion are considered - pure rotation about an axis through the origin, pure translation, and a general rigid motion.

For the two frame case, the determination of a general rigid motion requires that the 3-D lines lie on set(s) of planar surfaces. The methods are analogous to methods based on point correspondences. In some cases, a combination of point and line correspondences can be used. For the three frame case, no restriction on the 3-D line configuration is required for the determination of a general rigid motion. Also described briefly are some 3-D methods.

NATO ASI Series, Vol. F2
Image Sequence Processing and Dynamic Scene Analysis
Edited by T.S. Huang

I. Introduction

The determination of 3-D motion and structure of rigid bodies from image sequences has important applications in many areas, such as target tracking and robotic vision. Any 3-D motion of a rigid body can be decomposed into a rotation about an axis through the origin, followed by a translation. The 3-D motion parameters are the 3 rotational and the 3 translational components. The structure of an object can be described with a relative depth map of points on its surface.

Methods for determining 3-D motion of rigid bodies from image sequences have been based exclusively on the geometrical and transformation properties of point projections on the image plane (or the unit sphere). For small frame to frame displacements of the object image, the instantaneous velocity field of points (optical flow) is computed and used [1-2]. Alternatively, image space point shifts may be used [3]. For general displacements, point projections in the initial frame and their correspondences in succeeding frames are used [4-10].

In this paper it is assumed that the 3-D object has features in the form of straight line segments. In the man-made environment such objects are widespread - polyhedra for example. Central projections on the image plane of corresponding 3-D lines (image space line correspondence (LC)) are parameterized by corresponding pseudo points on the unit sphere. For general displacements of the object image, a set of image space LCs are located. In this paper, it is assumed LCs are given - the problem is then the determination of 3-D motion and a relative depth map of object surface points/lines.

The analysis is based fundamentally on the use of the unit sphere as the surface of central projection. This is a continuation of the spherical projection technique, which exposed the basic geometry of point correspondence (PC) based motion estimation [9-10]. This technique is crucially fundamental to a higher degree for LC based motion estimation [11], which is now described.

II. Parameterization of a 3-D Line and its Central Projection

A 3-D line or plane will be classified as containing the origin O (CO) or not containing O (NCO). The projection of a line (CO) on the image plane degenerates to a point, indistinguishable from the projection of a point. Thus, only lines (NCO) are considered. In reality, only a 3-D line segment l_s is visible. It is represented by a 3-D line l with infinite extension (fig. 1).

A 3-D line l (NCO) is parameterized by $\hat{m}$ and $p_{\perp}$ (fig. 1). $\hat{m}$ is a unit vector aligned with l, and represents one of 2 senses. $p_{\perp}$ is a vector which terminates on

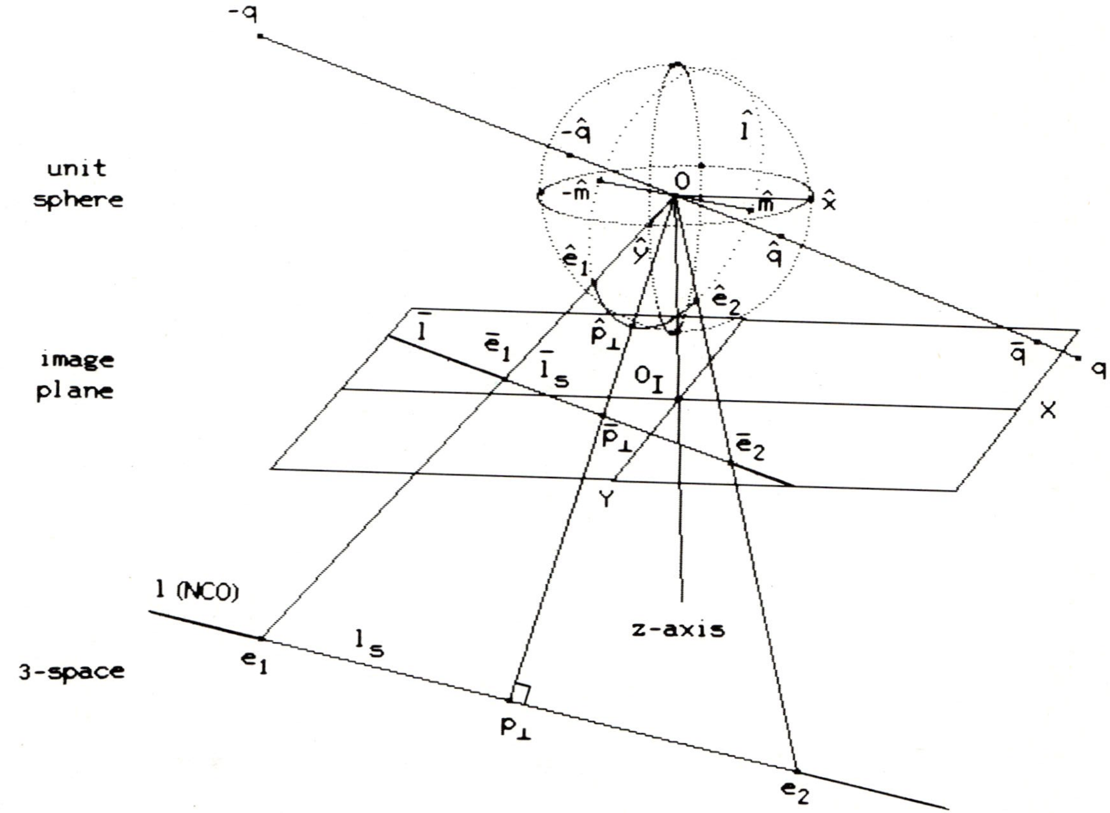

<u>fig. 1.</u> parameterization of 3-D line ℓ (NCO) by ($\hat{m}$, $p_{\perp}$), line $\bar{\ell}$ (on the image plane) by $\bar{q}$, and great circle $\hat{\ell}$ (on the unit sphere by $\underline{\pm}\hat{q}$

1 and is perpendicular to it. Note that $||p_\perp||$ is the perpendicular distance of 1 from 0 (depth of 1). Given 2 distinct points p_1 and p_2 on 1, $\hat{m}$ and $p_\perp$ can be determined from

$$\hat{m} = \pm \frac{p_2 - p_1}{||p_2 - p_1||} \qquad (1)$$
$$p = \frac{1}{||p_2 - p_1||^2}[(||p_2||^2 - p_1 \cdot p_2)p_1 + (||p_1||^2 - p_1 \cdot p_2)p_2]$$

A line 1 (NCO) defines a plane π (CO) containing it, with normal q (fig. 1) defined by

$$q = \hat{m} \times p_\perp \quad \text{where} \quad ||q|| = ||p_\perp|| \qquad (2)$$

The sense of 1 is represented by q, in terms of which side of π it lies. q is fundamental to the parameterization of the central projection of 1 on the image plane and unit sphere.

The central projection of 1 on the image plane is a planar line $\bar{1}$ (fig. 1). It is uniquely parameterized by its planar point dual $\bar{q}$, the central projection of q on the image plane. In general, it is a finite image space point. However, it degenerates to a directed point at infinity when $\bar{1}$ contains O_I, the image plane origin.

The central projection of 1 on the unit sphere is half of a great circle $\hat{1}$ (fig. 1). It is parameterized by one of 2 poles $\pm\hat{q}$ of the great circle, where $\hat{q}$ is the central projection of q on the unit sphere. Though there is a 2 way ambiguity in parameterization, there is no degeneracy. For this reason, $\pm\hat{q}$ is used to parameterize $\hat{1}$ (and $\bar{1}$). Representation of $\pm\hat{q}$ is in terms of xyz coordinates, which is equivalently a homogeneous coordinate representation of $\bar{q}$. It is important to preserve the character of the central projections of points (which occupy a 2-D surface) as a 3-D vector, both theoretically and computationally.

III. Determination of Correspondences ($\pm\hat{q}$, $\pm\hat{q}'$) of LC ($\bar{1}$, $\bar{1}'$)

In terms of the 3-D parameters of (1, 1') (NCO), the corresponding parameterizations ($\hat{q}$, $\hat{q}'$) of LC ($\bar{1}$, $\bar{1}'$) are given by

$$\hat{q} = \hat{m} \times \hat{p}_\perp \qquad (3)$$
$$\hat{q}' = \hat{m}' \times \hat{p}'_\perp$$

where ($\hat{m}$, $\hat{m}'$) represent corresponding senses of (1, 1'). The corresponding senses

$(\hat{s}, \hat{s}')$ of $(\bar{l}, \bar{l}')$ (respectively) are given by

$$\hat{s} = \widehat{q \times \hat{z}} = \widehat{(\hat{m} \times p_{\perp}) \times \hat{z}}$$
$$\hat{s}' = \widehat{q' \times \hat{z}} = \widehat{(\hat{m}' \times p_{\perp}') \times \hat{z}} \qquad (4)$$

Thus, $(\hat{m}, \hat{m}')$ also represent corresponding senses of $(\bar{l}, \bar{l}')$ (fig. 2). Given the LC $(\bar{l}, \bar{l}')$, there are 4 possible correspondences $(\pm\hat{q}, \pm\hat{q}')$. $(\hat{q}, \hat{q}')$ and $(-\hat{q}, -\hat{q}')$ are termed "sensed" correspondences. They parameterize $(\bar{l}, \bar{l}')$ and also their corresponding senses. $(\hat{q}, -\hat{q}')$ and $(-\hat{q}, \hat{q}')$ are termed "non-sensed" correspondences, which represent only $(\bar{l}, \bar{l}')$.

The 2 sensed correspondences can be selected from $(\pm\hat{q}, \pm\hat{q}')$ by determining the corresponding senses of $(\bar{l}, \bar{l}')$. The following are examples of "sense" indicators for image space lines. One indicator is corresponding ends of a line. The necessary computation is given by

$$\hat{q} = \widehat{\bar{p}_1 \times \bar{p}_2} \qquad \hat{q}' = \widehat{\bar{p}_3' \times \bar{p}_4'} \quad \text{where} \qquad (5)$$

$$\bar{p}_1 - \bar{p}_2 = \lambda(\bar{e}_b - \bar{e}_a) \qquad \bar{e}_a, \bar{e}_b \text{ - endpoints of } \bar{l}_s$$
$$\bar{p}_3' - \bar{p}_4' = \lambda'(\bar{e}_b' - \bar{e}_a') \qquad \bar{e}_a', \bar{e}_b' \text{ - endpoints of } \bar{l}_s'$$
$$\operatorname{sgn}(\lambda) = \operatorname{sgn}(\lambda')$$

A second sense indicator is corresponding intensity gradients in the direction perpendicular to $(\bar{l}, \bar{l}')$ (fig. 2). Pairs of corresponding directed points $\bar{p}_1 \rightarrow \bar{p}_2$ and $\bar{p}_3' \rightarrow \bar{p}_4'$ can be found and $(\hat{q}, \hat{q}')$ determined using (5).

There are 2 LC types. $(\bar{l}, \bar{l}')$ is termed a fixed LC if $\bar{l} = \bar{l}'$, i.e. (l, l') lie in the same plane (CO). $(\bar{l}, \bar{l}')$ is termed a non-fixed LC if $\bar{l} \neq \bar{l}'$, i.e. (l, l') do not lie in the same plane (CO). This distinction is important for the 3 basic types of motion.

IV. Pure Rotation About an Axis (CO)

The 3-D motion of the rigid body relative to the camera over a sequence of images is assumed to be a rotation Θ about an axis $\hat{n}$ (CO). Though such motion rarely occurs, it does in special cases (e.g., a camera panning about its focal point). The parameters $(\hat{m}, p_{\perp})$ of l transform by a rotation R (fig. 3).

$$\hat{m}' = R\hat{m} \qquad p_{\perp}' = Rp_{\perp} \quad \text{where} \qquad (6)$$

R = 3 x 3 orthogonal matrix and det R = 1.

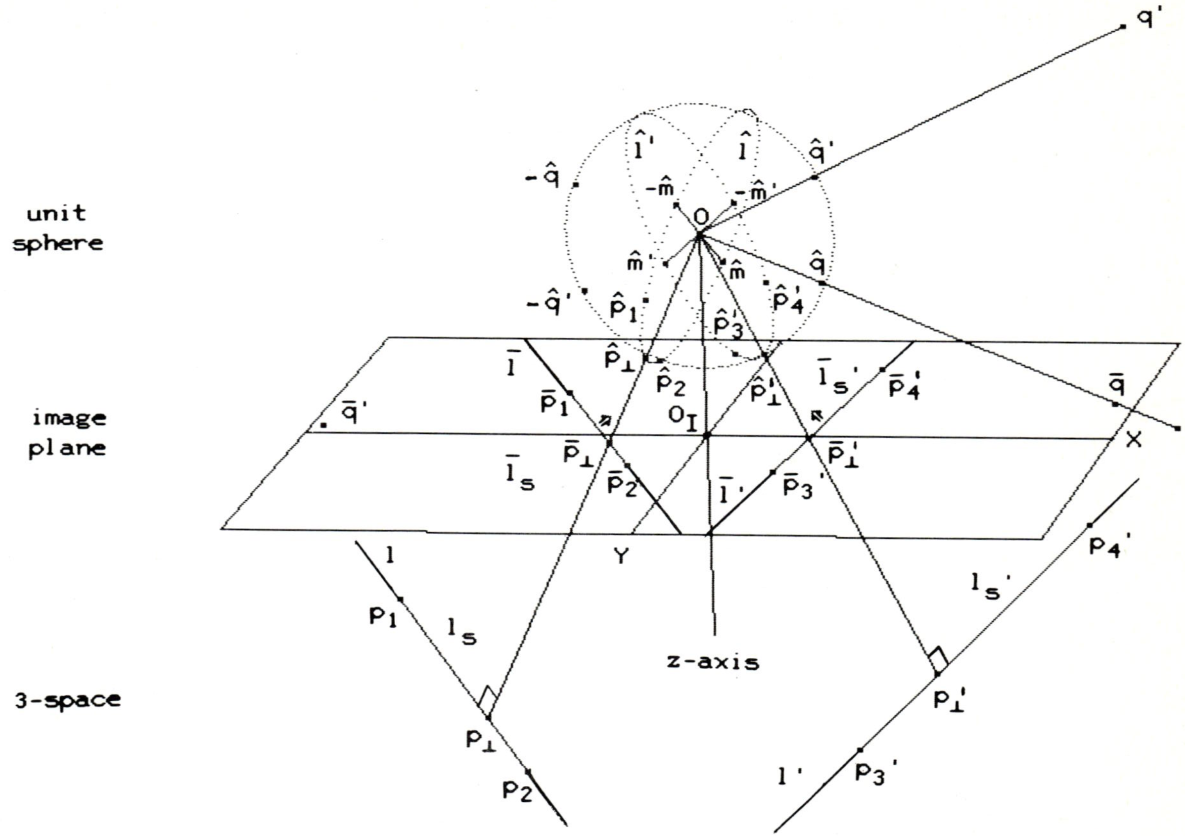

fig. 2. determination of sensed correspondences $(\hat{q},\hat{q}')$ $(-\hat{q},-\hat{q}')$ from LC $(\bar{\ell},\bar{\ell}')$

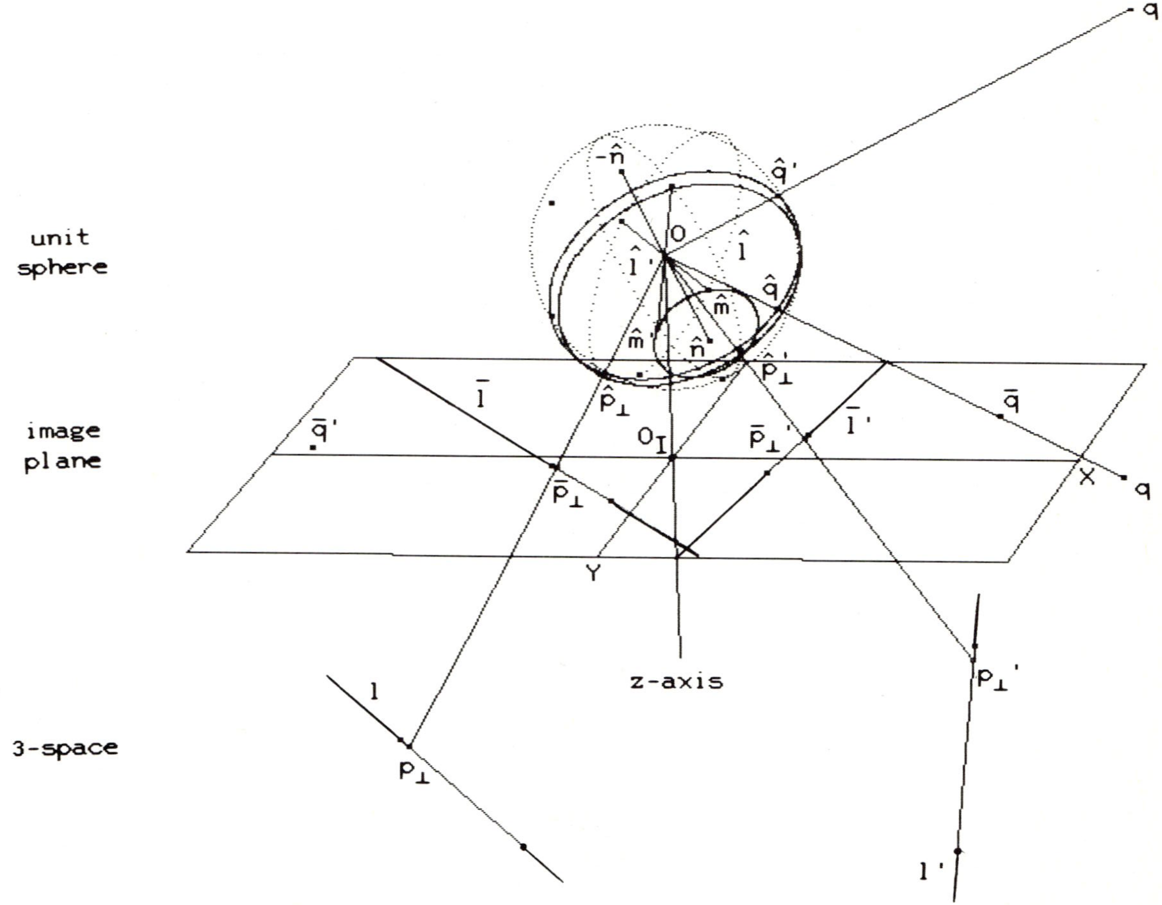

<u>fig. 3.</u> pure rotation R about axis (CO) - rotation of ($\hat{m}$, $p_\perp$, $\hat{p}_\perp$, q, $\hat{q}$) under rotation of ℓ

In fact, from (2), q also transforms by a rotation R.

$$\hat{q}' = R\hat{q} \tag{7}$$

Thus, the PC based methods for determining the rotation R in [9-10] are applicable.

For image space methods, the point duals $(\bar{q},\bar{q}')$ of LC $(\bar{l},\bar{l}')$ are not required to be associated with sensed correspondences. The point $\bar{q}$ transforms by an orthogonal collineation R. The rotation R can be determined from a total of 4 LCs $(\bar{l}_i,\bar{l}_i')$ i = 1,..,4, where no 3 of the lines in the 1st frame are concurrent [11]. In general, the rotation R can be determined from a total of 4 PCs and LCs, where no 3 of the points and point duals of lines in the 1st frame are collinear.

For methods based on the unit sphere, sensed correspondences are required. The point q transforms by a rotation R (fig. 3). Consider the 2 frame case. The rotation R can be determined from any 2 LCs $(\bar{l}_i,\bar{l}_i')$ i = 1,2 [11]. In general, the rotation R can be determined from any total of 2 PCs and LCs. Note that a unique solution is guaranteed. Consider the 3 frame case. The rotation R can be determined from the correspondences of a single line over 2 different frame pairs. In general, the rotation R can be determined from a combined total of 2 correspondences of lines and points, each over some frame pair.

There is no constraint on the configuration of 3-D lines on the object. It is impossible to determine a relative depth map of object surface lines, i.e. line structure of the object cannot be determined.

V. Pure Translation

The 3-D motion of the rigid body relative to the camera over a sequence of images is assumed to be a translation in a constant direction. This occurs not only in situations where there are translating objects, but also where a camera is purposely translated in a static environment. Under a translation t, the parameters $(\hat{m}, p_\perp)$ of l transform by (fig. 4)

$$\hat{m}' = \hat{m}$$
$$p_\perp' = p_\perp + t - (\hat{m}\cdot t)\hat{m} \qquad \text{where } t = \begin{bmatrix} \Delta y \\ \Delta x \\ \Delta z \end{bmatrix} \tag{8}$$

Note that (l, l') are parallel. From (2), q transforms by

$$q' = q + (\hat{m} \times t) \tag{9}$$

The mapping of q is a function of $\hat{m}$ in addition to t, and is not a translation t.

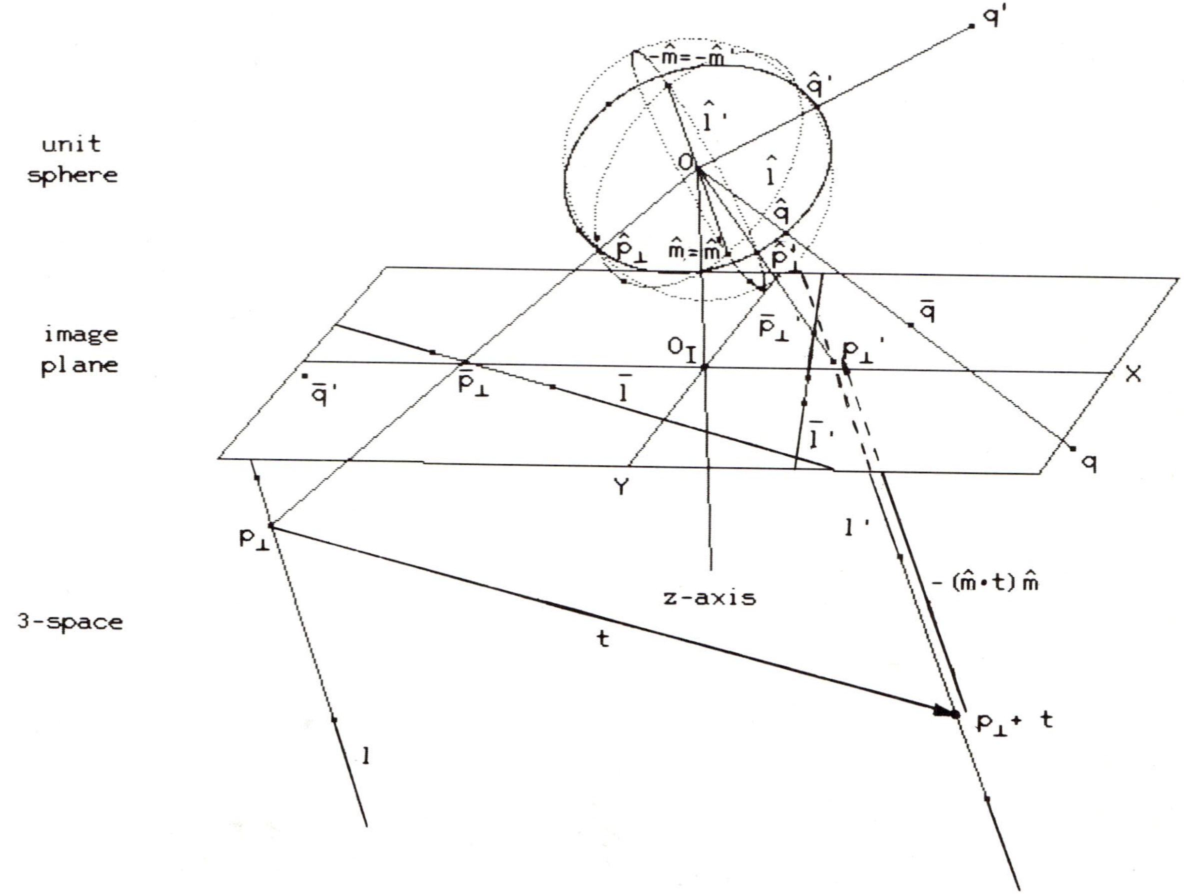

<u>fig. 4.</u> pure translation t – transformation of $(\hat{m},\ p_\perp,\ \hat{p}_\perp,\ q,\ \hat{q})$ under translation of ℓ

On the unit sphere, $\hat{q}$ rotates on a great circle with poles $\pm\hat{m}$. In general, $\hat{q}\cdot\hat{t} \neq 0$ (t does not lie in the plane (CO) containing l) and thus $\hat{q}' \neq \pm\hat{q}$ (($\bar{l},\bar{l}'$) is not fixed). In rare cases, $\hat{q}\cdot\hat{t} = 0$ (t lies in the plane (CO) containing l) and thus $\hat{q}' = \pm\hat{q}$ (($\bar{l},\bar{l}'$) is fixed).

Given a sensed correspondence ($\hat{q}$, $\hat{q}'$) of ($\bar{l},\bar{l}'$), various quantities can be determined. If ($\bar{l},\bar{l}'$) is not fixed ($\hat{q}\cdot\hat{t} \neq 0$) then from (8) and (9)

$$\hat{q} \times \hat{q}' = \frac{\|t\|}{\|p'_\perp\|}(\hat{q}\cdot\hat{t})\hat{m} \tag{10a}$$

$$\frac{\|t\|}{\|p'_\perp\|} = -\frac{\hat{q}'\cdot\hat{p}_\perp}{\hat{q}\cdot\hat{t}} \tag{10b}$$

$$\frac{\|p_\perp\|}{\|p'_\perp\|} = \frac{\hat{q}'\cdot\hat{t}}{\hat{q}\cdot\hat{t}} \tag{10c}$$

Given ($\hat{q},\hat{q}'$) of ($\bar{l},\bar{l}'$), ($\hat{m},\hat{p}_\perp$, $\hat{p}'_\perp$) can be uniquely determined [11]. Given $\hat{t}$, the ratios $\|p_\perp\|/\|p'_\perp\|$ (relative depth of l to l') and $\|t\|/\|p'_\perp\|$ (amount of translation relative to the depth of l') can be determined from (10b,c). This is particularly useful for determining object structure in the situation where the camera is purposely translated within a static environment, in a known direction. Also,

$$\begin{array}{llll} \text{case 1} & \hat{q}\cdot\hat{t} > 0 & \hat{q}'\cdot\hat{t} > 0 & \text{where } \hat{q}'\cdot\hat{p}_\perp < 0 \\ \text{case 2} & \hat{q}\cdot\hat{t} < 0 & \hat{q}'\cdot\hat{t} < 0 & \text{where } \hat{q}'\cdot\hat{p}_\perp > 0 \end{array} \tag{11}$$

The solution region for $\hat{t}$ can be determined as the intersection of 2 hemispheres - a hemispherical portion (fig. 5). If ($\bar{l},\bar{l}'$) is fixed, then ($\hat{m}$, $\hat{p}_\perp$, $\hat{p}'_\perp$) cannot be uniquely determined. But,

$$\hat{q}\cdot\hat{t} = 0 \tag{12}$$

The solution region for $\hat{t}$ is a great circle with poles $\pm\hat{q}$ (fig. 6).

A. Case when 3-D Lines are of an Arbitrary Configuration

Given the LCs ($\bar{l}_i$, $\bar{l}'_i$) i = 1,..,n_δ the solution region for $\hat{t}$ is found as the intersection of the hemispherical portions (11) and/or great circles (12). This is continued for 3 or more frames, to further isolate the solution region. In general, $\hat{t}$ cannot be uniquely determined. However, from 2 fixed LCs, $\hat{t}$ can be determined to within a sign. With an additional non-fixed LC, $\hat{t}$ can be uniquely determined. Note that each possible $\hat{t}$ has a corresponding map of relative line depths (10b,c).

B. Case when 3-D Lines Lie on Planar Surfaces

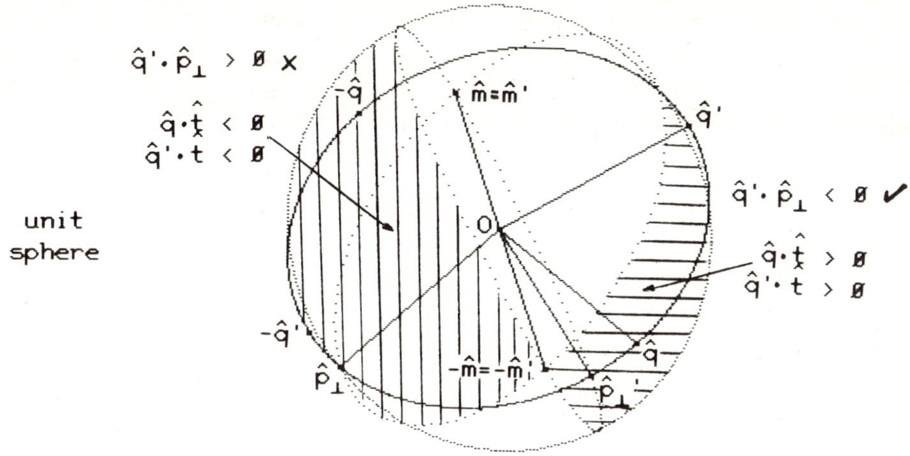

fig. 5. (pure translation) for non-fixed LC $(\bar{\ell},\bar{\ell}')$ - solution region for $\hat{t}$ is a hemisphere portion

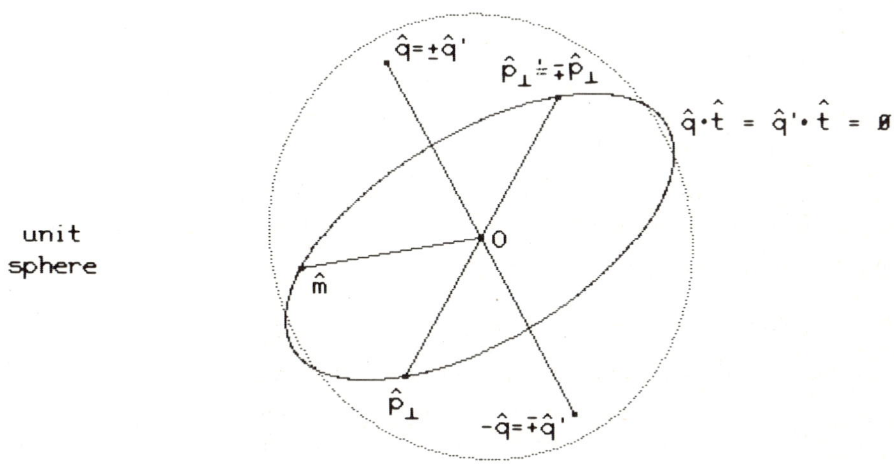

fig. 6. (pure translation) for fixed LC $(\bar{\ell}, \bar{\ell}')$ - solution region for $\hat{t}$ is a great circle

It is assumed that the 3-D lines l_i $i = 1,..,n_\delta$ lie on planes π_i $i = 1,..,n_\pi$. There are intersection points o_i $i = 1,..,n_o$ ($n_o \neq 0$) among the n lines. Such a configuration will be denoted by (n_π n_δ n_o). The degenerate case (* n_δ n_o), which describes an arbitrary line configuration, is excluded. An example is a polyhedral object, where there are n_π faces, n_δ edges, and n_o vertices.

Let o be the intersection of (l_1, l_2) and o' be the intersection of (l_1', l_2') (fig. 7). Given the correspondences ($\pm\hat{q}_i$, $\pm\hat{q}_i'$) i = 1,2 (not required to be sensed) of non-fixed LCs ($\bar{l}_i,\bar{l}_i'$) i = 1,2, where each of the line pairs (l_1,l_2) (l_1',l_2') do not lie in a plane (CO), the o-point correspondence (OPC) ($\hat{o},\hat{o}'$) can be determined [11]. This holds for any general rigid motion.

There are 2 classes of methods for the solution of $\hat{t}$. In Class 1, two homogeneous linear equations in $\hat{t}$ can be determined by two methods.

method 1 $$\begin{bmatrix} a_{12}^T \\ a_{34}^T \end{bmatrix} \hat{t} = 0 \quad \text{where } a_{ij} = \frac{\|p_{\perp j}\|}{\|p_{\perp i}\|}(\hat{q}_j' \cdot \hat{p}_{\perp j})\hat{q}_i - (\hat{q}_i' \cdot \hat{p}_{\perp i})\hat{q}_j \tag{13a}$$

method 2 $$\begin{bmatrix} (\hat{o}_1 \times \hat{o}_1')^T \\ (\hat{o}_2 \times \hat{o}_2')^T \end{bmatrix} \hat{t} = 0 \quad \text{where} \quad \begin{matrix} o_1 = l_1 \cap l_2 & o_1' = l_1' \cap l_2' \\ o_2 = l_3 \cap l_4 & o_2' = l_3' \cap l_4' \end{matrix} \tag{13b}$$

Both methods are equivalent from a theoretical standpoint, where method 2 is the typical PC approach. The difference is purely computational. Method 1 makes use of the fact that $\|p_{\perp j}\|/\|p_{\perp i}\|$ (relative depth of l_j wrt l_i) can be determined, where l_i and l_j are coplanar. $\hat{t}$ can be uniquely determined from configurations (n_π n_δ n_o) n_o = 2 (there are 2 o-points) where the o-points in the 1st frame and t are not contained in a plane (CO). In Class 2, a non-homogeneous set of linear equations in a scaled version of t is determined.

method 1 $$\begin{bmatrix} a_i^T \\ b_1^T \\ b_2^T \end{bmatrix} t_o = \begin{bmatrix} 0 \\ \gamma_1 \\ \gamma_2 \end{bmatrix} \quad \text{where} \quad \begin{matrix} l_o, l_1 \text{ are coplanar} \\ l_o, l_2 \text{ are coplanar} \end{matrix} \tag{14a}$$

method 2 $$\begin{bmatrix} b_1^T \\ b_2^T \\ b_3^T \end{bmatrix} t_o = \begin{bmatrix} \gamma_1 \\ \gamma_2 \\ \gamma_3 \end{bmatrix} \quad \text{where} \quad \begin{matrix} l_o, l_1 \text{ are coplanar} \\ l_o, l_2 \text{ are coplanar} \\ l_o, l_3 \text{ are coplanar} \end{matrix} \tag{14b}$$

$$b_i = \frac{\|p_{\perp o}'\|}{\|p_{\perp i}'\|}(\hat{q}_o' \cdot \hat{q}_o)\hat{p}_{\perp i} - (\hat{q}_i' \cdot \hat{q}_i)\hat{p}_{\perp o}$$

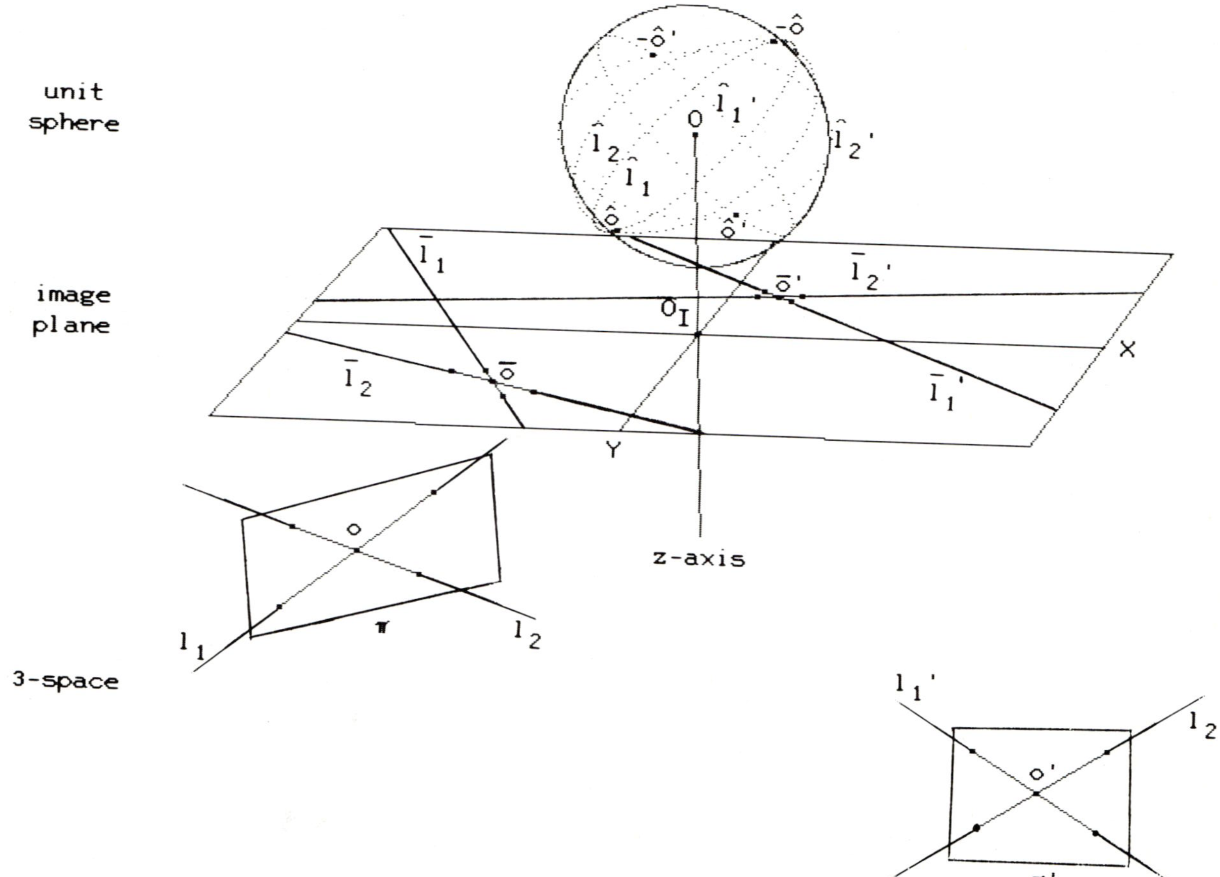

<u>fig. 7.</u> determination of OPC ($\hat{o}$, $\hat{o}'$) from 2 LCs ($\bar{\ell}_1, \bar{\ell}_1'$) ($\bar{\ell}_2, \bar{\ell}_2'$) where $o = \ell_1 \cap \ell_2$
$o' = \ell_1' \cap \ell_2'$

$$\gamma_i = (\hat{q}_i'\cdot\hat{q}_i) - \frac{\|p'_{\perp o}\|}{\|p_{\perp o}\|}\frac{\|p_{\perp i}\|}{\|p'_{\perp i}\|}(\hat{q}_o'\cdot\hat{q}_o)$$

$$t_o = \frac{\|\ t\ \|}{\|p_{\perp o}\|}\hat{t}$$

Configurations where there are at last 2 lines coplanar with a reference line are required.

In addition, the relative depths of lines $\{l_i\}$ on a plane π_i and of lines $\{l_i'\}$ on a plane π_j' can be determined. Given $\hat{t}$, the ratios in (10b,c) can be determined. In general, a relative depth map of lines (and o-points) can be determined.

In general, $\hat{t}$ can be uniquely determined from a combination of PCs and LCs for an $(n_\pi\ n_\delta\ n_o)$ configuration. The Class 1 LC based methods can be extended to 3-frame analysis, as the standard PC based methods were [10]. In general, $\hat{t}$ is determined from a combined total of 2 correspondences of points and o-points, each over a different frame pair. In general, a relative depth map of points, o-points, and lines can be determined.

D. 3-D Methods

It is assumed that the xyz coordinates of 3-D points on a line can be determined, by a stereo camera configuration or a rangefinder. Given the 3-D points (p_1, p_2) on l and (p_3', p_4') on l', the 3-D parameters $(\hat{m},p_\perp)$ of l and $(\hat{m}',p_\perp')$ of l' can be determined from (1) (the 3-D configuration of lines can be determined). t can be uniquely determined from two 3-D LCs (l_i,l_i') i = 1,2 by

$$t = \frac{a_i\cdot b_i}{\|b_i\|^2}\hat{t} \qquad \text{where}$$

$$\hat{t} = \pm\ \widehat{(\hat{m}_1 \times a_1) \times (\hat{m}_2 \times a_2)} \tag{15}$$

$$a_i = p'_{\perp i} - p_{\perp i}$$

$$b_i = \hat{t} - (\hat{m}_i\cdot\hat{t})\hat{m}_i \quad i \in 1,2 \qquad \text{such that } a_i\cdot b_i > 0$$

where $\hat{m}_1$, $\hat{m}_2$, and $\hat{t}$ are not contained in a plane (CO).

The following is an alternative (stronger) method, using only two 3-D correspondences (q_i,q_i') i = 1,2. Then $\pm\hat{t}$ and $\hat{m}_i' = \hat{m}_i$ are found for some i, from which t is determined.

$$\alpha\hat{t} = (q_1' - q_1) \times (q_2' - q_2) = \hat{t}[\hat{t}\cdot(\hat{m}_1 \times \hat{m}_2)]$$

$$\beta\hat{m}_i' = \beta\hat{m}_i = q_i \times q_i' = \hat{m}_i(\hat{q}_i\cdot\hat{t}) \quad \text{such that} \tag{16}$$

$$\| t \| = \frac{(q_i' - q_i)\cdot(\hat{m}_i \times \hat{t})}{(\hat{m}_i \times \hat{t})\cdot(\hat{m}_i \times \hat{t})} > 0 \quad i \in 1,2$$

where $\hat{t}\cdot(\hat{m}_1 \times \hat{m}_2) \neq 0$.

VI. Rotation plus Translation - A General Rigid Motion

It is assumed that the body does not undergo a pure rotation about an axis (CO) or a pure translation, but a general rigid motion, over a sequence of images. The parameters $(\hat{m}, p_\perp)$ of 1 transform by

$$\begin{aligned} \hat{m}' &= R\hat{m} \\ p_\perp' &= Rp_\perp + t - (R\hat{m}\cdot t)R\hat{m} \end{aligned} \tag{17}$$

Note that the mapping of $\hat{m}$ is a pure rotation R. Consequently, q transforms by

$$q' = Rq + (R\hat{m} \times t) \tag{18}$$

The mapping of q is a function of $\hat{m}$ in addition to R and t, and is not a rotation R followed by a translation t (as for the case with points).

On the unit sphere, the induced motion of q can be decomposed as the following. $\hat{q}$ rotates by R to $\hat{q}''$ on a circle, contained in a plane perpendicular to $\hat{n}$. $\hat{q}''$ then rotates to $\hat{q}'$, on a great circle with poles $\pm\hat{m}'$ (fig. 8). If $\hat{q}''\cdot\hat{t} = 0$ (t is contained in the plane (CO) containing 1″, i.e. $(\bar{1}'',\bar{1}')$ is fixed), $\hat{q}' = \pm\hat{q}''$. If $\hat{q}''\cdot\hat{t} \neq 0$ (t is not contained in the plane (CO) containing 1″, i.e. $(\bar{1}'',\bar{1}')$ is not fixed), $\hat{q}' \neq \pm\hat{q}''$. In this case only, $\hat{m}'$ is given by

$$\hat{m}' = \pm \frac{R\hat{q} \times \hat{q}'}{\| R\hat{q} \times \hat{q}' \|} \tag{19}$$

That is, from the pure translation case, a solution for R immediately gives the orientation of the 3-D lines (1,1′).

A. Case when the Relative Orientations of 3-D Lines are Known

It is assumed that the relative orientations between $(1_i,1_j)$ $i \neq j$ $i,j = 1,\ldots,n_\delta$ (also the relative orientations between $(1_i',1_j')$ $i \neq j$ $i,j = 1,\ldots,n_\delta$) are known (fig. 9). Define θ_{ij} as the inclusive angle between $(\hat{m}_i,\hat{m}_j)$ (also between $(\hat{m}_i',\hat{m}_j')$).

$$\cos\theta_{ij} = \hat{m}_i\cdot\hat{m}_j = \hat{m}_i'\cdot\hat{m}_j' \tag{20}$$

Since $(\hat{m}_i,\hat{m}_i')$ have a $\pm$ ambiguity, $\cos\theta_{ij}$ has a $\pm$ ambiguity, i.e. the relative angle

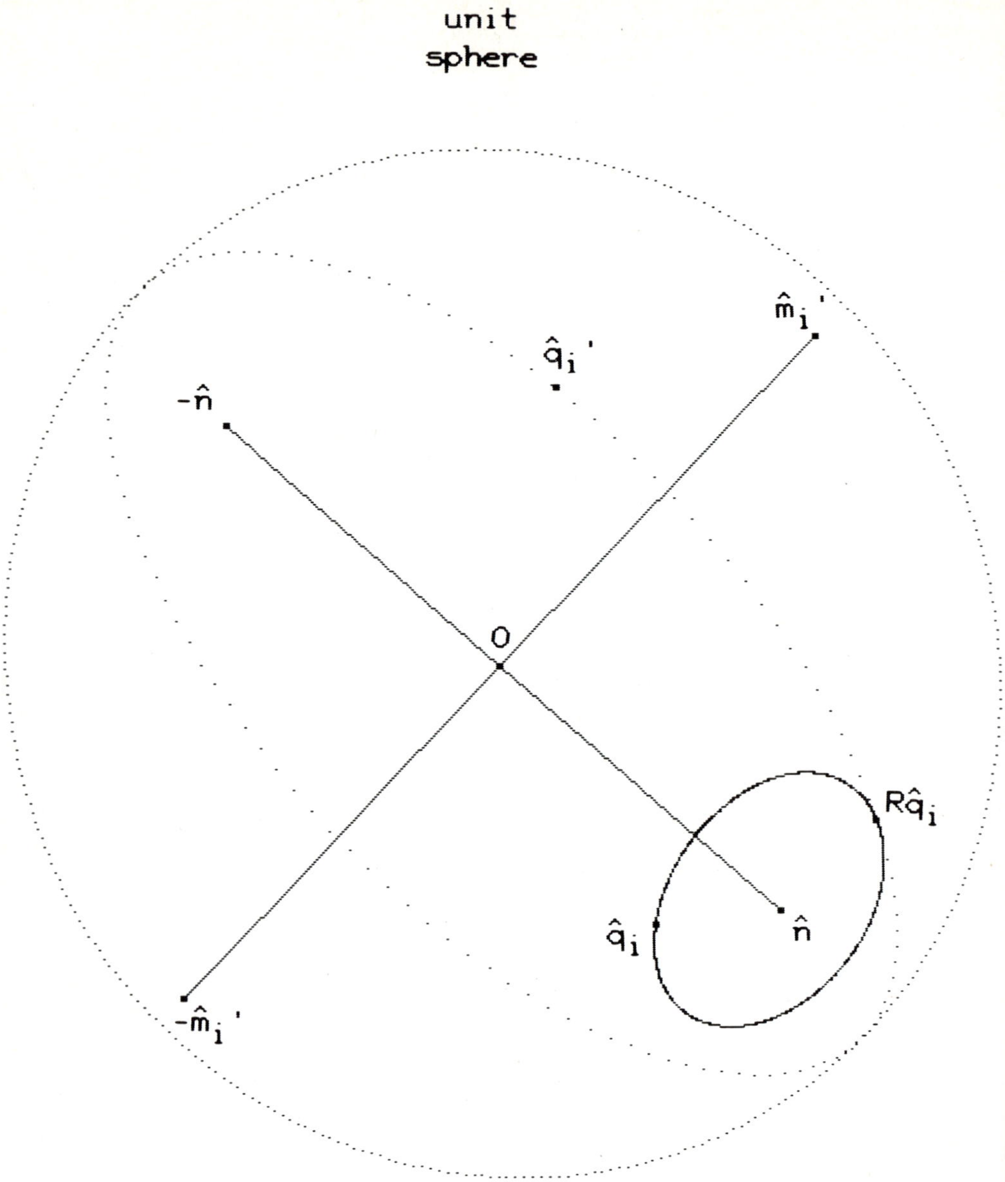

fig. 8. geometry on unit sphere of $(\hat{q}_i, \hat{q}_i')$ with respect to $(\hat{n}, \hat{m}_i')$

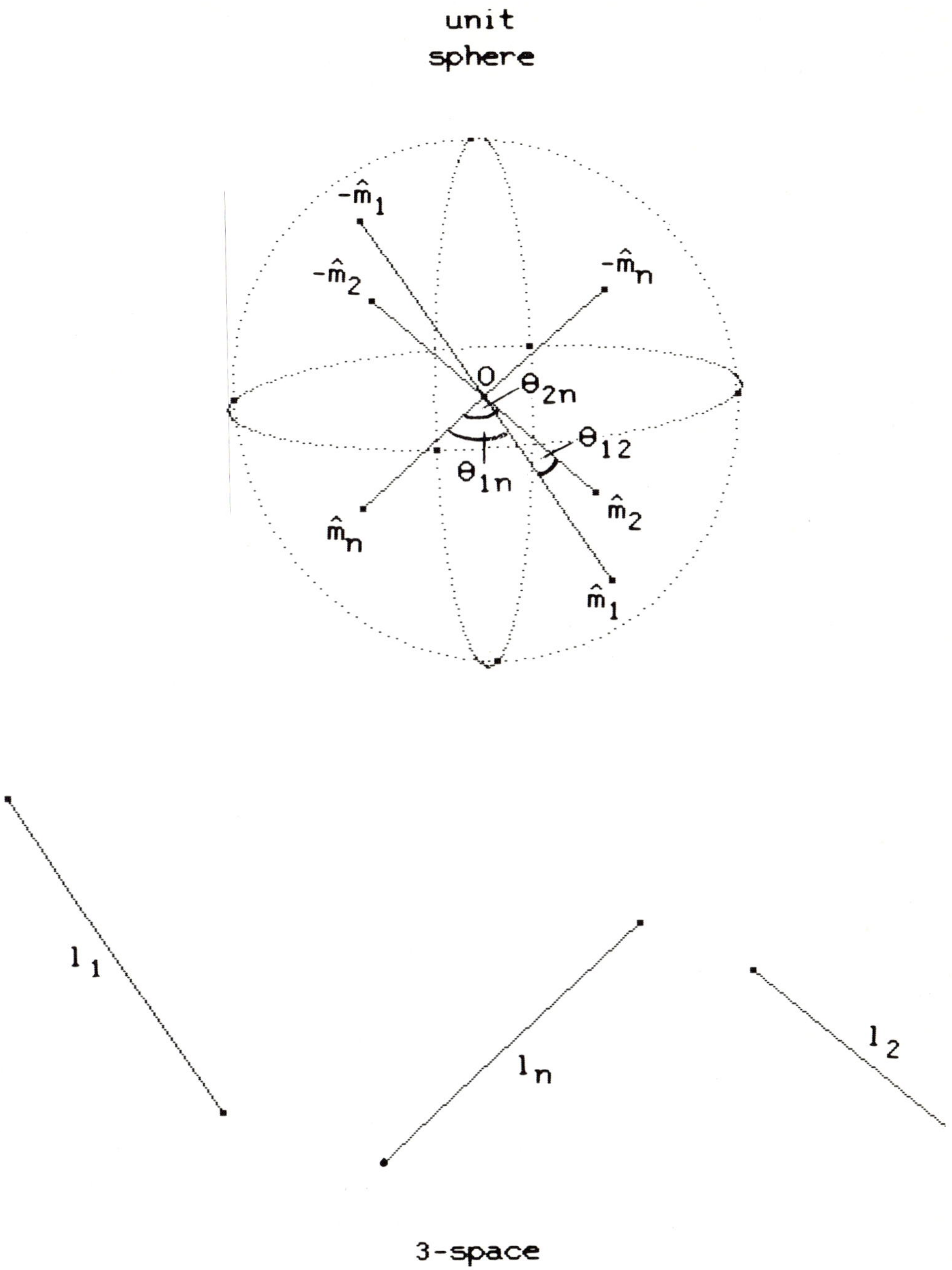

fig. 9. relative configuration among 3 lines ℓ_1, ℓ_2, ℓ_n is unique up to a rotation and/or reflection given the relative angles θ_{12}, θ_{1n}, θ_{2n}

between $\hat{m}_i$, $\hat{m}_j$ can be described by θ_{ij} or $180 - \theta_{ij}$.

Given a set of correspondences $(\pm\hat{q}_i, \pm\hat{q}'_i)$ $i = 1,..,n_\delta$ (not required to be sensed) of $(\bar{l}_i, \bar{l}'_i)$ $i = 1,..,n_\delta$ 2 sets of equations in the rotational parameters can be determined [11].

set 1
$$\frac{\sigma_{12}^2}{\tau_1^2\tau_2^2} = \cos^2\theta_{12}$$
$$\frac{\sigma_{i1}^2}{\tau_i^2\tau_1^2} = \cos^2\theta_{i1} \qquad \frac{\sigma_{i2}^2}{\tau_i^2\tau_2^2} = \cos^2\theta_{i2} \qquad i = 3,..,n_\delta \tag{21a}$$

set 2
$$\frac{\sigma_{12}\sigma_{2i}\sigma_{i1}}{\tau_1^2\tau_2^2\tau_i^2} = \cos\theta_{12}\,\cos\theta_{2i}\,\cos\theta_{i1} \qquad i = 3,..,n_\delta \tag{21b}$$

where
$$\sigma_{ij} = (\hat{q}_i\cdot\hat{q}_j)(\hat{q}'_i\cdot\hat{q}'_j) - (R\hat{q}_i\cdot\hat{q}'_j)(R\hat{q}_j\cdot\hat{q}'_i)$$
$$\tau_i = \| Rq_i \times q'_i \|$$

where it is assumed that t is not contained in any of the planes (CO) containing l''_i $i = 1,..,n_\delta$. The equations in each set are independent, based on the fact that the relative configuration of 3 unit vectors is fixed (up to a rotation and/or reflection) by their 3 relative angles (fig. 9). Set 1 contains $(2n_\delta - 3)$ equations and set 2 contains $(n_\delta - 2)$ equations, where in both cases 2 lines are used as the reference. At least 3 equations in the 3 rotational parameters are required (4 LCs for set 1 and 5 LCs for set 2). Solution of these equations and the uniqueness question are currently under investigation.

Given R, the $\bar{l}_i$ $i = 1,..,n_\delta$ are mapped to $\bar{l}''_i$ $i = 1,..,n_\delta$. From $(\bar{l}''_i, \bar{l}'_i)$ $i = 1,..,n_\delta$, the solution region for $\hat{t}$ is determined by the methods for pure translation. The (absolute) orientations $(\hat{m}_i, \hat{m}'_i)$ $i = 1,..,n_\delta$ of the lines (l_i, l'_i) $i = 1,..,n_\delta$ can be determined from (19).

B. Case when 3-D Lines Lie on Planar Surfaces

An $(n_\pi\ n_\delta\ n_o)$ configuration is assumed. First, general $(n_\pi\ n_\delta\ n_o)$ configurations are considered. Second, $(n_\pi\ n_\delta\ n_o)$ $n_\pi = 1$ configurations are considered. Third, $(n_\pi\ n_\delta\ n_o)$ configurations where the n_π planes are parallel are considered.

1. <u>General</u> (n_π n_δ n_o) <u>Configurations</u>

A general (n_π n_δ n_o) configuration has n_δ lines and n_o ($\neq 0$) o-points scattered among n_π planes. Given ($\pm\hat{q}_i$, $\pm\hat{q}_i'$) i = 1,..,n_δ (not required to be sensed) associated with($\bar{l}_i$, $\bar{l}_i'$) i = 1,..,n_δ, the OPCs ($\hat{o}_i$, $\hat{o}_i'$) i = 1,..,n_o can be determined, in general (fig. 7). The 2 principal PC based methods [5-6], [7-8] for determining the motion can be applied. Then a relative depth map of o-points can be determined [9-11].

2. (n_π n_δ n_o) n_π = 1 <u>Configurations (Single Planar Surface)</u>

A configuration (n_π n_δ n_o) n_π = 1 is the situation where n_δ lines lie on a single planar surface. There are 2 basic methods. One is to determine 4 OPCs over 2 to 3 frames and apply the planar patch methods in [7]. In general, a total of 4 PCs and OPCs can be used to determine the motion. The other method is to use ($\bar{q}_i$, $\bar{q}_i'$) i = 1,..,n_δ (not required to be sensed) of ($\bar{l}_i$,$\bar{l}_i'$) i = 1,..,n_δ directly [11]. The collineation of a point dual $\bar{q}$ of an image space line $\bar{l}$ is $(A^{-1})^T$, where A is the (non-singular) collineation of the image space point $\bar{p}$ [12-13]. The motion can be determined over 2 to 3 frames from 4 LCs ($\bar{l}_i$,$\bar{l}_i'$) i = 1,..,4, where no 3 of $\bar{l}_i$ i = 1,..,4 are concurrent.

3. (n_π n_δ n_o) <u>Configurations where Planar Surfaces are Parallel</u>

It is assumed that the n_π planar surfaces are parallel, i.e. they have common normals $\pm\hat{c}$. The orientations $\hat{m}_i'$ i = 1,..,n_δ of l_i' i = 1,..,n_δ lie on a great circle with poles $\pm R\hat{c}$ (fig. 10). Geometrically, $R\hat{q}_i$ and $\hat{q}_i'$ lie on a great circle C_i with poles $\pm\hat{m}_i'$. Thus, the great circles C_i have a common diametrical axis about $\pm R\hat{c}$. That is, ($R\hat{q}_i$, $\hat{q}_i'$, $\pm R\hat{c}$) lie on the great circle C_i.

Now, the geometry of ($\hat{p}_i$, $\hat{p}_i'$) on the unit sphere is reviewed. $R\hat{p}_i$ and $\hat{p}_i'$ lie on a great circle T_i which contains $\pm\hat{t}$. The great circles T_i i = 1,..,n_p have a common diametrical axis about $\pm\hat{t}$. That is, ($R\hat{p}_i$, $\hat{p}_i'$, $\pm\hat{t}$) lie on the great circle T_i (fig. 11).

The geometry of ($R\hat{q}_i$, $\hat{q}_i'$, $\pm R\hat{c}$) i = 1,..,n_δ (in the case where $\hat{m}_i'$ i = 1,..,n_δ lie on a great circle) is similar to the geometry of ($R\hat{p}_i$, $\hat{p}_i'$, $\pm\hat{t}$) i = 1,..,n_p. The equations used in the 2 principal PC based methods for curved surfaces [5-6], [8] are both based fundamentally on the fact that ($R\hat{p}_i$, $\hat{p}_i'$, $\pm\hat{t}$) lie on a great circle [9-10]. There are 2 sets of homogeneous equations which can be obtained from PCs ($\pm\hat{p}_i$, $\pm\hat{p}_i'$) i = 1,..,n_p.

set 1 $$\hat{t}\cdot(R\hat{p}_i \times \hat{p}_i') = 0 \qquad i = 1,..,n_p \tag{22a}$$

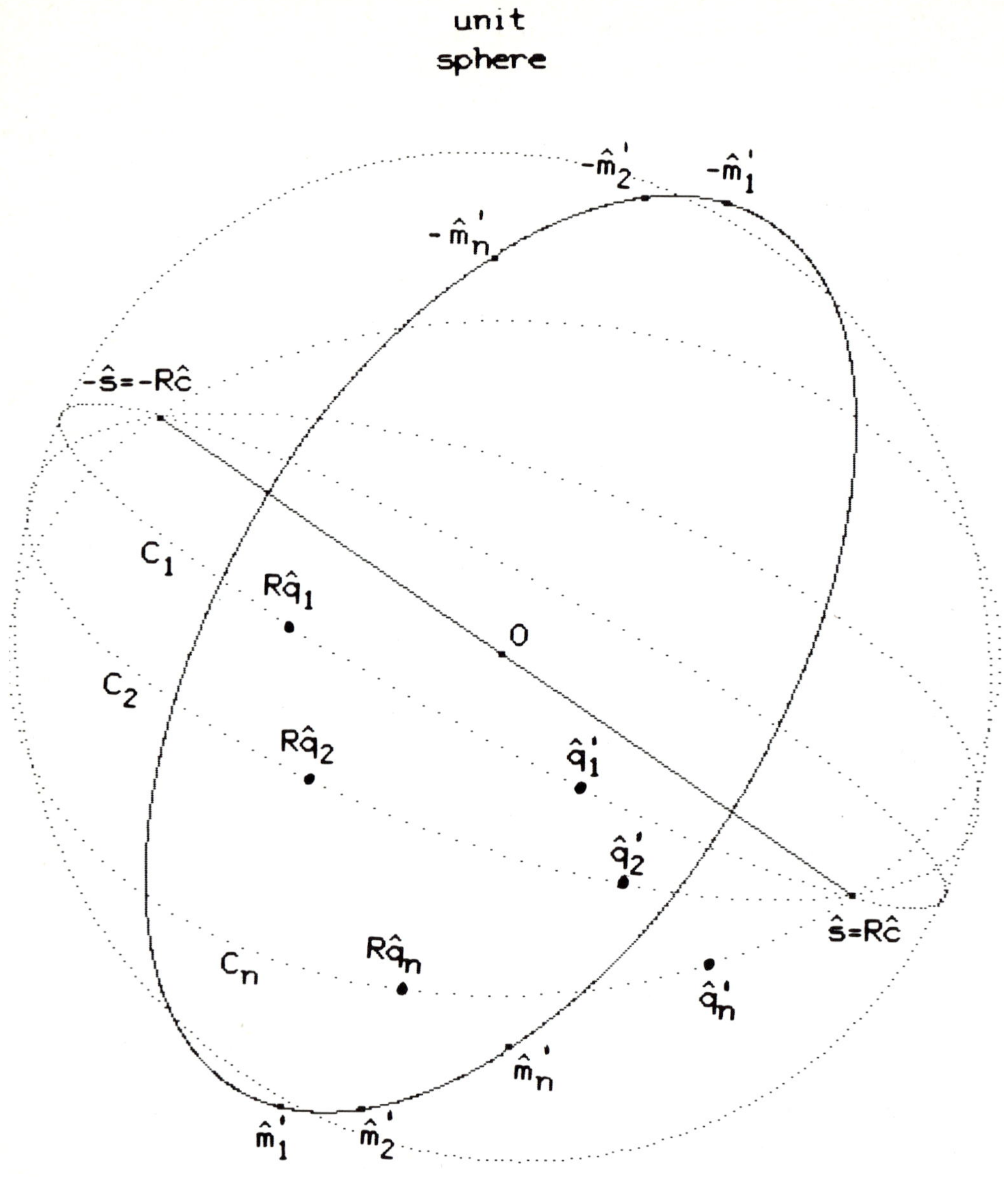

fig. 10. geometry on unit sphere of $\{(\hat{q}_i, \hat{q}_i')\}$ of LCs $\{(\overline{\ell}_i, \overline{\ell}_i')\}$ wrt $\hat{s}$ (where 3-D lines $\{\ell_i\}$ are contained in parallel planes): great circles C_i have a common diameter $\pm\hat{s}$

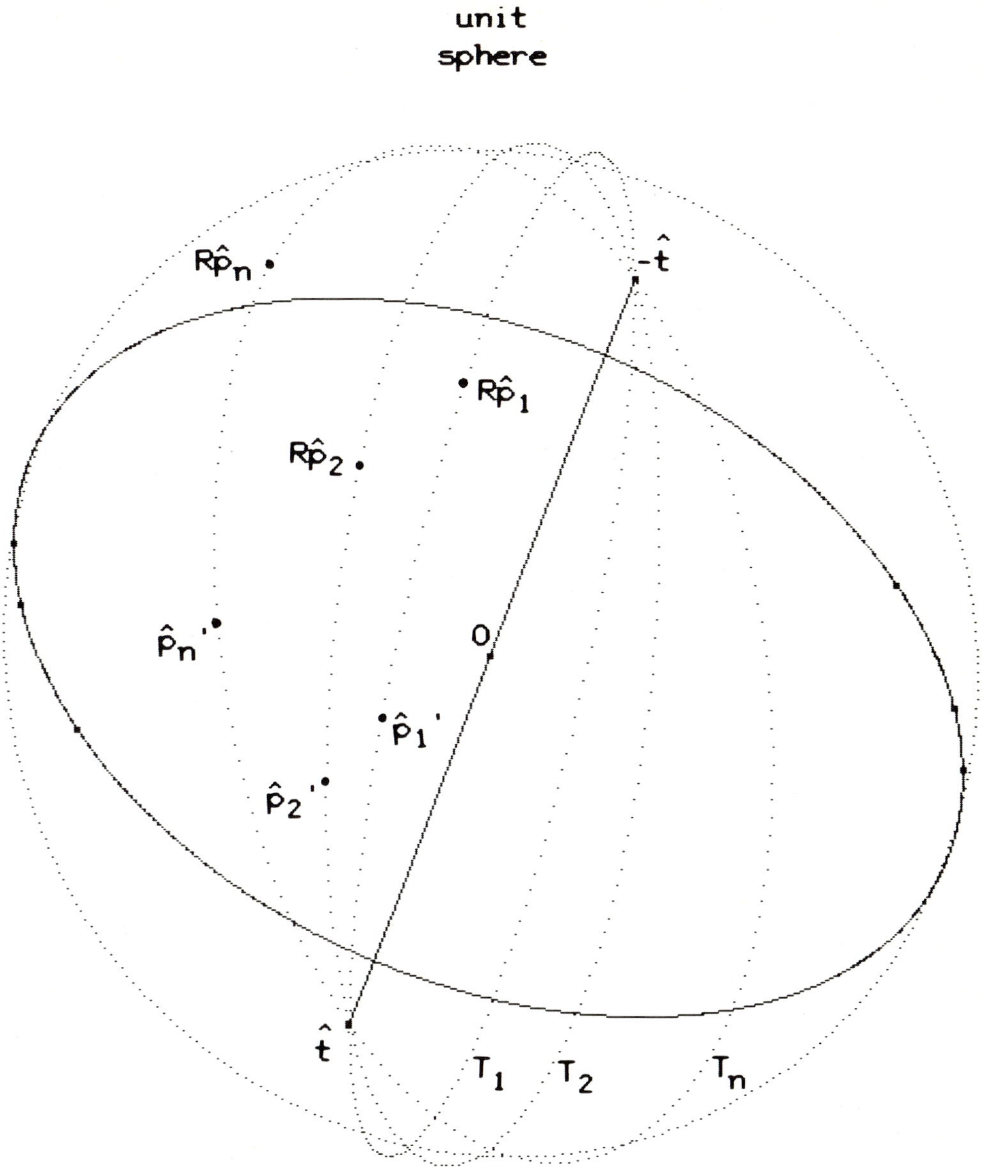

fig. 11. geometry on unit sphere of PCs $\{(\hat{p}_i, \hat{p}_i')\}$ wrt $\pm\hat{t}$: great circles T_i have a common diameter $\pm\hat{t}$

set 2 $$(Rp_i \times p_i')\cdot[(Rp_1 \times p_1') \times (Rp_2 \times p_2')] = 0 \qquad i = 3,..,n_p \tag{22b}$$

Equivalently,

$$p_i'^T Ep_i = 0 \qquad i = 1,..,n_p \qquad \text{where}$$

set 1

$$E = [t \times c_1 \quad t \times c_2 \quad t \times c_3] \qquad R = [c_1 \quad c_2 \quad c_3] \tag{23a}$$

$$[R^{-1}p_i'\cdot (p_1 \times p_i)][Rp_2\cdot (p_1' \times p_2')] - [Rp_i\cdot (p_1' \times p_i')][R^{-1}p_2'\cdot (p_1 \times p_2)] = 0 \tag{23b}$$

set 2 $$i = 3,..,n_p$$

(23a) are the equations used in [8] and (23b) are the equations used in [5-6]. An analogous set of equations to (23a,b) can be derived given $(\pm q_i, \pm q_i')$ $i = 1,..,n_\delta$ associated with LCs $(\bar{l}_i,\bar{l}_i')$ $i = 1,..,n_\delta$

set 1 $$s\cdot(Rq_i \times q_i') = 0 \qquad i = 1,..,n_\delta \tag{24a}$$

set 2 $$(Rq_i \times q_i')\cdot[(Rq_1 \times q_1') \times (Rq_2 \times q_2')] = 0 \qquad i = 3,..,n_\delta \tag{24b}$$

where $s = Rc$

Equivalently,

set 1 $$q_i'^T G\, q_i = 0 \qquad i = 1,..,n_\delta \qquad \text{where}$$

$$G = [s \times c_1 \quad s \times c_2 \quad s \times c_3] \qquad R = [c_1 \quad c_2 \quad c_3] \quad s = Rc \tag{25a}$$

$$R^{-1}q_i'\cdot(q_1 \times q_i)][Rq_2\cdot(q_1' \times q_2')] - [Rq_i\cdot(q_1' \times q_i')][R^{-1}q_2'\cdot(q_1 \times q_2)] = 0 \tag{25b}$$

set 2 $$i = 3,..,n_\delta$$

The advantage in (25a,b) is that ambiguous, non-sensed correspondences $(\pm q_i, \pm q_i')$ can be used.

From equations (25a), using 8 LCs, $\pm s$ can be determined uniquely, but there are two solutions for R [11]. Consider the solution of equations (25b). At least 5 LCs are required to obtain 3 equations in the 3 rotational parameters. Analytical solutions and the uniqueness question are currently being investigated. Given a solution for R, $\bar{l}_i$ $i = 1,..,n_\delta$ are mapped to $\bar{l}_i''$ $i = 1,..,n_\delta$. From $(\bar{l}_i'', \bar{l}_i')$ $i = 1,..,n_\delta$ a solution region for t and a relative depth map of o-points/lines is determined by the methods for pure translation.

In general, a combination of PCs and LCs can be used to determine the rotation. A combined total of PCs and LCs are required to obtain at least 3 equations of the type in (23b), (25b). In general, a solution region for $\hat{t}$ and a relative depth map of points, o-points, and lines can be determined.

The (n_π n_δ n_o) configuration can be generalized to lines on b subsets of parallel planes. Let b_i ($\geq$ 3) be the number of lines on the ith subset. Given n_δ LCs $\sum_{i=1}^{b} b_i - 2b$ equations of the type in (25b) can be obtained. Additional n_p PCs can be used to obtain additional n_p-2 equations of the type in (23b). At least 3 equations in the rotational parameters are required. Given the rotation R, and solution region for a $\hat{t}$ and a relative depth map of o-points and lines can be determined.

C. 3-D Methods

It is assumed that the xyz coordinates of 3-D points on the line can be determined by a stereo camera configuration or a rangefinder. Given the 3-D points p_1, p_2 on l and p_3', p_4' on l', the 3-D parameters ($\hat{m}, p_\perp$) of l and ($\hat{m}', p_\perp'$) of l' can be determined from (1) (the 3-D configuration of lines can always be obtained). R and $\hat{t}$ are uniquely determined from the parameters ($\hat{m}_i$, $p_{\perp i}$) of 2 3-D LCs (l_i, l_i') i = 1,2 where $\hat{m}_1$, $\hat{m}_2$, $\hat{t}$ are not contained in a plane (CO). R can be determined from the correspondences ($\hat{m}_i, \hat{m}_i'$) i = 1,2 (always). Then ($\hat{m}_i, p_{\perp i}$) i = 1,2 are mapped by R to ($\hat{m}_i''$, $p_{\perp i}''$) i = 1,2. From ($\hat{m}_i''$, $p_{\perp i}''$) ($\hat{m}_i', p_{\perp i}'$) i = 1,2, t is determined by the 3-D methods for pure translation.

VII. 3-Frame Analysis

The 2-frame methods to this point are characterized by various restrictions on the object 3-D line configuration. There are no such constraints for 3-frame methods, in general. The n-frame case can of course be handled by application of the well known 2-frame methods over n - 1 frame pairs [3-8]. To date, only pure rotation about an axis (CO) and pure translation have been seen to lend themselves intrinsically to 3-frame analysis. It is evident that this is also the case for arbitrary motion using LCs over 3 frames, which is now described.

First, a word on notation. The correspondences (and related quantities) in the ith frame are denoted by the conventional symbol, with a superscript i. For example, the parameters of corresponding lines ($\bar{l}_i^1, \bar{l}_i^2, \bar{l}_i^3$) over 3 frames f^1, f^2, f^3 will be denoted by ($\hat{q}_i^1, \hat{q}_i^2, \hat{q}_i^3$). The motion parameters (and related quantities) from f^i to f^j are denoted by their conventional symbols with double subscript ij. For example, the rotation matrix and translation from f^i to f^j are denoted by R_{ij},

t_{ij}. Note that $R_{ji} = R_{ij}^{-1}$ and $t_{ji} = -R_{ij}^{-1}t_{ij}$. Note also that given the rotation/translation of any 2 distinct frame pairs over f^1, f^2, f^3, the rotation/translation, is uniquely determined for the third remaining frame pair. Without loss of generality, the rotation/translation pairs (R_{12}, t_{12}) and (R_{23}, t_{23}) are chosen as solution variables.

From the 2-frame analysis from f^1 to f^2, the geometry on the unit sphere is such that $R_{12}\hat{q}_i^1$ and $\hat{q}_i^2$ lie on a great circle with poles $\pm\hat{m}_i^2$ (fig. 8). There are 2 cases. In Case 1, $R_{12}\hat{q}_i^1 \cdot \hat{t}_{12} = 0$ (t_{12} lies in the plane (CO) containing l_i rotated by R_{12} - a rare occurrence), and $R_{12}\hat{q}_i^1$ and $\hat{q}_i^2$ are aligned. In Case 2, $R_{12}\hat{q}_i^1 \cdot \hat{t}_{12} \neq 0$ (t_{12} does not lie in the plane (CO) containing l_i rotated by R_{12} - the case in general), and $R_{12}\hat{q}_i^1$ and $\hat{q}_i^2$ are not aligned. Making use of this fact for the successive frame pairs 1 to 2 and 3 to 2 (frame 2 is chosen as the reference), (q_i^2 $R_{12}\hat{q}_i^1$, $R_{23}^{-1}\hat{q}_i^3$) lie on a great circle with poles $\pm\hat{m}_i^2$ (fig. 12). Thus, the following set of equations can be obtained from LCs over 3 frames.

$$\hat{q}_i^2 \cdot (R_{12}\hat{q}_i^1 \times R_{23}^{-1}\, \hat{q}_i^3) = 0 \qquad i = 1,..,n_\delta \tag{26}$$

It is assumed that case 1 does not occur for either of the frame pairs $1 \rightarrow 2$ and $3 \rightarrow 2$ ($i = 1,..,n_\delta$), otherwise the equations in (26) are always trivially satisfied. Note that the unknowns are strictly the interframe rotations R_{12}, R_{23} - the interframe translations t_{12}, t_{23} have been nulled out. Each equation in (26) is a 4th order polynomial in the 6 independent rotation parameters. In general, 6 LCs would be required to produce 6 equations. The solution of (26) by numerical methods and the uniqueness question are currently under investigation. Given the determination of R_{12}, R_{23} $\{l_i^1\}$ are mapped by R_{12} and $\{l_i^2\}$ are mapped by R_{23}. Then the 2-frame pure translation methods can be applied to determine the solution region for $\hat{t}_{12}$, $\hat{t}_{23}$ (which cannot be uniquely determined in general).

The uniqueness question and direct solution of non-linear equations such as (23b), (25b), (26) for the rotational parameters has been met with little success. The only statement on uniqueness and a method for analytical solution has been with the solution of (23a), as demonstrated by Tsai/Huang [7-8]. The basic approach is the linear solution of an intermediary set of 8 essential parameters using 8 PCs for curved surfaced objects. The motion parameters are subsequently determined. An analogous approach will be taken for the solution of (26).

The equation set (26) can be rewritten as

$$[\hat{q}_i^{2\,T} F_1 \hat{q}_i^3 \quad \hat{q}_i^{2\,T} F_2 \hat{q}_i^3 \quad \hat{q}_i^{2\,T} F_3 \hat{q}_i^3]\; \hat{q}_i^1 = 0 \qquad i = 1,..,n_\delta$$

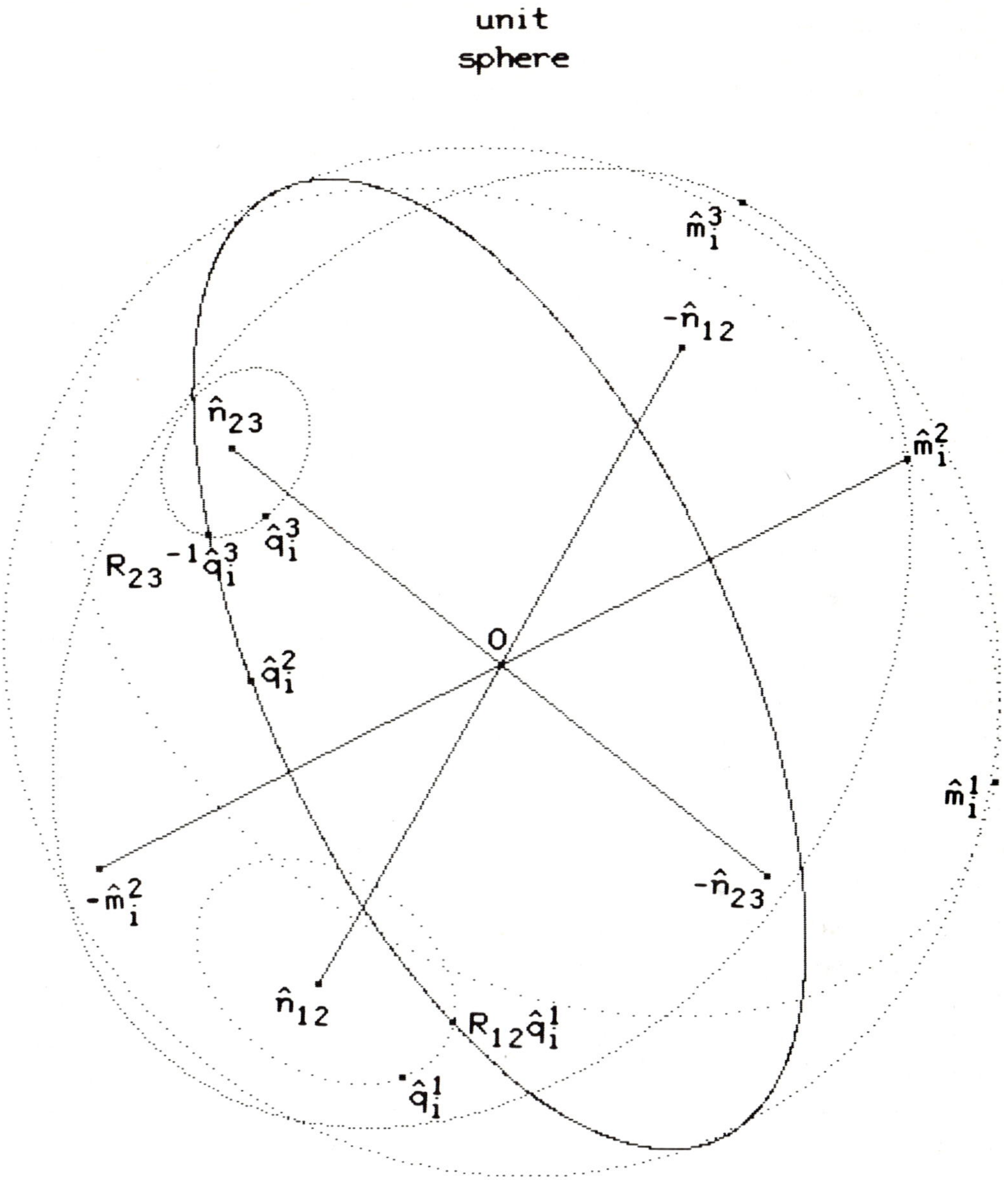

fig. 12. geometry on unit sphere of $R_{12}\hat{q}_i^1$ $R_{23}^{-1}\hat{q}_i^3$ $\hat{q}_i^2$ with respect to $\hat{n}_{12}$, $\hat{n}_{23}$, $\hat{m}_i^1$, $\hat{m}_i^2$, $\hat{m}_i^3$

where

$$F_1 = [(\hat{b}_1 \times \hat{c}_1) \quad (\hat{b}_1 \times \hat{c}_2) \quad (\hat{b}_1 \times \hat{c}_3)] \qquad R_{12} = [\hat{b}_1 \quad \hat{b}_2 \quad \hat{b}_3]$$

$$F_2 = [(\hat{b}_2 \times \hat{c}_1) \quad (\hat{b}_2 \times \hat{c}_2) \quad (\hat{b}_2 \times \hat{c}_3)]$$

$$F_3 = [(\hat{b}_3 \times \hat{c}_1) \quad (\hat{b}_3 \times \hat{c}_2) \quad (\hat{b}_3 \times \hat{c}_3)] \qquad R_{23} = \begin{bmatrix} \hat{c}_1^T \\ \hat{c}_2^T \\ \hat{c}_3^T \end{bmatrix} \tag{27}$$

Clearly, each equation in (27) is linear in the elements of the 3 intermediate parameter matrices (F_1, F_2, F_3). On comparing with (23a), (27) is seen to be a higher level more complex. (F_1, F_2, F_3) can be determined to a scale factor by arbitrarily setting one of the unknowns to 1. 26 LCs are required to generate 26 equations in 26 variables. The uniqueness question requires the analysis of a 3rd order equation. This aspect is currently under investigation.

It is now shown that the interframe rotation matrices R_{12}, R_{23} are uniquely determined given that the intermediary matrices (F_1, F_2, F_3) have been determined to a scale factor as (F_1', F_2', F_3').

$$F_1' = kF_1 = k[f_{11} \quad f_{12} \quad f_{13}] = [f_{11}' \quad f_{12}' \quad f_{13}']$$

$$F_2' = kF_2 = k[f_{21} \quad f_{22} \quad f_{23}] = [f_{21}' \quad f_{22}' \quad f_{23}'] \tag{28}$$

$$F_3' = kF_3 = k[f_{31} \quad f_{32} \quad f_{33}] = [f_{31}' \quad f_{32}' \quad f_{33}']$$

where

$$F_1 = B_1 R_{23}^{-1} \qquad F_2 = B_2 R_{23}^{-1} \qquad F_3 = B_3 R_{23}^{-1} \qquad B_i = \begin{bmatrix} 0 & -(\hat{b}_i)_z & (\hat{b}_i)_y \\ (\hat{b}_i)_z & 0 & -(\hat{b}_i)_x \\ -(\hat{b}_i)_y & (\hat{b}_i)_x & 0 \end{bmatrix}$$

Note that each of (F_1, F_2, F_3) have the same structure as the E matrix used in the 2 frame PC method for curved surfaces. In terms of the interframe rotations (R_{12}, R_{23}), the polar decomposition of F_i is the product of a skew symmetric matrix B_i (function only of b_i, the ith column of R_{12}) and an orthogonal matrix R_{23}^{-1} of the 1st kind.

Thus, the methods for computing the motion from the E matrix by SVD [8] or vector operations [10], [14] can be applied. From F_i, R_{23} is uniquely determined but the columns of R_{12} are determined only to a sign. However, R_{12} can be determined from

$$G_1' = kG_1 = k\,[f_{11} \quad f_{21} \quad f_{31}] = [f_{11}' \quad f_{21}' \quad f_{31}']$$
$$G_2' = kG_2 = k\,[f_{12} \quad f_{22} \quad f_{32}] = [f_{12}' \quad f_{22}' \quad f_{32}'] \tag{29}$$
$$G_3' = kG_3 = k\,[f_{13} \quad f_{23} \quad f_{33}] = [f_{13}' \quad f_{23}' \quad f_{33}']$$

where

$$G_1 = -C_1 R_{12} \qquad G_2 = -C_2 R_{12} \qquad G_3 = -C_3 R_{12} \qquad C_i = \begin{bmatrix} 0 & -(\hat{c}_i)_z & (\hat{c}_i)_y \\ (\hat{c}_i)_z & 0 & -(\hat{c}_i)_x \\ -(\hat{c}_i)_y & (\hat{c}_i)_x & 0 \end{bmatrix}$$

(G_1, G_2, G_3) are simply 3 matrices derived from the 1st, 2nd, 3rd columns of (F_1, F_2, F_3) (respectively). The same above techniques uniquely determine R_{12}. In summary, 2 applications of the $E \rightarrow (R, \pm\hat{t})$ computation are required.

Another proof of uniqueness of R_{12}, R_{23} is now shown, which gives a computationally more efficient method than the above. Consider the solution of R_{12}, with columns $(\hat{b}_1, \hat{b}_2, \hat{b}_3)$. Note that $\hat{b}_i$ can be determined to a sign by some 1 of the 3 cross products among the columns of F_i'.

$$f_{i1}' \times f_{i2}' = k^2\hat{b}_i[\hat{b}_i\cdot(\hat{c}_1 \times \hat{c}_2)] = k^2\hat{b}_i(\hat{b}_i\cdot\hat{c}_3) \tag{30a}$$
$$f_{i2}' \times f_{i3}' = k^2\hat{b}_i[\hat{b}_i\cdot(\hat{c}_2 \times \hat{c}_3)] = k^2\hat{b}_i(\hat{b}_i\cdot\hat{c}_1) \tag{30b}$$
$$f_{i3}' \times f_{i1}' = k^2\hat{b}_i[\hat{b}_i\cdot(\hat{c}_3 \times \hat{c}_1)] = k^2\hat{b}_i(\hat{b}_i\cdot\hat{c}_2) \tag{30c}$$

Using (30a), (30b), (30c) requires that $\hat{b}_i\cdot\hat{c}_3 \neq 0$, $\hat{b}_i\cdot\hat{c}_1 \neq 0$, $\hat{b}_i\cdot\hat{c}_2 \neq 0$ (respectively). At least 1 of the 3 (maybe 2, even 3) are non-zero (since $(\hat{c}_1, \hat{c}_2, \hat{c}_3)$ are mutually orthogonal), which gives a non-singular solution. Now consider the solution of R_{23}, with rows $(\hat{c}_1, \hat{c}_2, \hat{c}_3)$. Using a similar analysis above, $\hat{c}_i$ can be determined to a sign by some 1 of the 3 cross products among the columns of G_i'.

$$f_{1i}' \times f_{2i}' = k^2\hat{c}_i(\hat{c}_i\cdot\hat{b}_3) \tag{31a}$$
$$f_{2i}' \times f_{3i}' = k^2\hat{c}_i(\hat{c}_i\cdot\hat{b}_1) \tag{31b}$$
$$f_{3i}' \times f_{1i}' = k^2\hat{c}_i(\hat{c}_i\cdot\hat{b}_2) \tag{31c}$$

Using (31a), (31b), (31c) requires that $\hat{c}_i\cdot\hat{b}_3 \neq 0$, $\hat{c}_i\cdot\hat{b}_1 \neq 0$, $\hat{c}_i\cdot\hat{b}_2 \neq 0$ (respectively). At least 1 of the 3 (maybe 2, even 3) are non-zero (since $(\hat{b}_1,\hat{b}_2,\hat{b}_3)$ are mutually orthogonal), which gives a non-singular solution. Hence, (30a-c) and (31a-c) determine the columns of R_{12} to a sign $(\pm\hat{b}_1, \pm\hat{b}_2\ \pm\hat{b}_3)$ and the rows of R_{23} to a sign $(\pm\hat{c}_1, \pm\hat{c}_2\ \pm\hat{c}_3)$ (respectively).

It is now shown that the signs of $(\pm\hat{b}_1, \pm\hat{b}_2, \pm\hat{b}_3)$ and $(\pm\hat{c}_1, \pm\hat{c}_2, \pm\hat{c}_3)$ can be resolved. In each of (30a), (30b), (30c), $\hat{b}_i$ occurs twice - the alignment of each vector cross product is determined only by $\hat{c}_3, \hat{c}_1, \hat{c}_2$ (respectively). Note that at least one of (30a), (30b), (30c) is a non-zero vector, since $\hat{c}_1, \hat{c}_2, \hat{c}_3$ are mutually orthogonal. Thus, the correct signs of $\hat{b}_i$ $i = 1,2,3$ can be determined. Similarly, the correct signs of $\hat{c}_i$ $i = 1,2,3$ can be determined from (31a-c). Hence, R_{12} and R_{23} are uniquely determined.

It should be noted that any LC based methods automatically produce PC based methods. This is because a LC can be obtained from 2 PCs. Admittedly, 26 LCs is a prohibitive number. But, from 8 PCs, $\frac{8(7)}{2} = 28$ LCs can be obtained. Thus, 8 PCs can be used to determine (F_1', F_2', F_3'), from which the motion over 3 frames can be found. The question of uniqueness of (F_1', F_2', F_3') using PCs (to generate LCs) is currently being studied. This appears to be consistent with applying the 2-frame method for curved surfaces over successive frame pairs, where 8 PCs over 3 frames are required.

VIII. Summary

From a sequence of images from a single camera, the goal is to determine the rotation R and unit translation $\hat{t}$ of the rigid motion of the object (which contains lines in addition to points) and the relative depth of points/lines on it. The basic method is a 2 step procedure. In Step 1, a set of interframe PCs and LCs are obtained. The LC problem is a new area for investigation, but was not considered. In Step 2, the rotation and translation components of the motion are determined. Also, a relative depth map of points/lines is determined. Step 2 was analyzed over 2 frames for the simple rigid motions - pure rotation about an axis (CO) and pure translation - and for a general rigid motion. Preliminary results of 3-frame analysis were also described.

The key operation is the spherical projection, which directed the analysis onto the surface of the unit sphere. The central projection of a 3-D line is represented by a pseudo PC. Also, an OPC can be determined from LCs, where the 3-D lines intersect. As a result, many of the PC based methods were applicable in some situations. In addition, the LC based methods immediately yield PC based methods.

For pure rotation about an axis (CO), there is no constraint on the line configuration on the object. The rotation can be determined from a total of 4 PCs and LCs (not required to be sensed). Alternatively, the rotation can be determined from a total of any 2 PCs and (sensed) LCs. Over 3 frames, the rotation can be determined from a combined total of 2 correspondences of points and lines, each over some frame pair. A relative depth map of points/lines cannot be obtained.

For pure rotation about an axis (CO), there is no constraint on the line configuration on the object. The rotation can be determined from a total of 4 PCs and LCs (not required to be sensed). Alternatively, the rotation can be determined from a total of any 2 PCs and (sensed) LCs. Over 3 frames, the rotation can be determined from a combined total of 2 correspondences of points and lines, each over some frame pair. A relative depth map of points/lines cannot be obtained.

For pure translation, a unique solution for t is impossible in general, for objects with arbitrary line configurations. The lines are required to lie on planar surfaces. t can be determined from a total of 2 PCs and OPCs. Over 3 frames, t can be determined from a combined total of 2 PCs and OPCs, each over some frame pair. In general, a relative depth map of points, o-points, and lines can be obtained.

For a general rigid motion, a unique solution for the motion over 2 frames is impossible in general, where the object line configuration is arbitrary. Several forms of constraints are required. Given the relative orientations of the lines, 2 sets of equations involving the rotational parameters can be solved for. If the lines lie on planar surfaces, the motion can be determined by PC based methods. For arbitrary such configurations, OPCs can be used. For configurations where the lines lie on a single planar surface, OPCs can be used or LCs used directly. For configurations where the lines lie on parallel planar surfaces, LCs can be used directly. Analogous sets of equations of the 2 principal PC based methods can be obtained. In general, for all the previous methods, a combination of PCs and LCs can be used. A relative depth map of points, o-points, and lines can be obtained, in general. Over 3 frames, the motion can be uniquely determined from 26 LCs for objects with arbitrary line configurations.

3-D methods can be used for the pure translation case and a general rigid motion. A stereo camera configuration or rangefinder is required to obtain the 3-D xyz coordinates of points. For the pure translation case, t can be determined from 2 LCs by 2 different methods. For a general rigid motion, R and t can be determined from 2 LCs.

ACKNOWLEDGEMENT

This work was supported by NSF Grant ECS 81-12082.

REFERENCES

1. K. Prazdny, "Egomotion and Relative Depth Map from Optical Flow," Biological Cybernetics, v. 36, 1980, pp. 87-102.
2. W. Clocksin, "Perception of Surface Slant and Edge Labels from Optical Flow: A Computational Approach," Perception 9, 1980, pp. 253-269.
3. T. Huang and R. Tsai, "Three-Dimensional Motion Estimation from Image Space Shifts," Proc. ICASSP, March 1981.
4. N. Badler, "Three-Dimensional Motion from Two Dimensional Picture Sequences," 2nd IJCPR, Aug. 1974.
5. H. Nagel, "Representation of Moving Rigid Objects Based on Visual Observations," Computer, Aug. 1981, pp. 29-39.
6. H. Nagel, "On the Derivation of 3D Rigid Point Configurations from Image Sequences," PRIP, Aug. 1981.
7. R. Tsai and T. Huang, "Estimating Three-Dimensional Motion Parameters of a Rigid Planar Patch," CSL Technical Report R-922, Nov. 12, 1981.
8. R. Tsai and T. Huang, "Uniqueness and Estimation of Three-Dimensional Motion Parameters of Rigid Objects with Curved Surfaces," CSL Technical Report R-921, Oct. 30, 1981.
9. B. Yen and T. Huang, "Determining 3-D Motion and Structure of a Rigid Body Using the Spherical Projection," PRIP, June 1982.
10. B. Yen and T. Huang, "Determining 3-D motion and Structure of a Rigid Body Using the Spherical Projection," 1982, CSL Technical Report.
11. B. Yen and T. Huang, "Determining 3-D Motion and Structure of a Rigid Body Containing Lines Using the Spherical Projection," 1982, CSL Technical Report.
12. O. Veblen and J. Young, Projective Geometry vol. I, Ginn & Co., 1938.
13. F. Ayres, Projective Geometry, Schaum Publishing Co., 1967.
14. H. C. Longuet-Higgins, "A Computer Algorithm for Reconstructing a Scene from Two Projections," Nature, no. 293, pp. 133-135, Sept. 1981.
15. M. Spiegel, Vector Analysis, Schaum Publishing Co., 1959.

COMPARISON OF FEATURE OPERATORS
FOR USE IN MATCHING IMAGE PAIRS

Thomas D. Williams *
Frank Glazer **

ABSTRACT

Various feature operators are applied to image pairs prior to inter-frame matching. Results of the match operation are presented where several areas and interesting points have been selected for comparison.

I. INTRODUCTION

Inter-frame matching of image pairs is an activity essential to all stereo and motion understanding systems. The methods by which the matching is performed, the information used by the matcher, and the context within which the matches are used vary considerably. This paper is focused on the information used by the matcher, but we cannot ignore the other factors.

---

* Corporate Research Group, DIGITAL EQUIPMENT CORPORATION, 77 Reed Road, Hudson, MA 01754

** Computer & Information Sciences, University of Massachusetts, Amherst, MA 01003

NATO ASI Series, Vol. F2
Image Sequence Processing and Dynamic Scene Analysis
Edited by T.S. Huang

The majority of motion analysis systems (as reported in Martin and Aggarwal 77) perform a purely "bottom-up" analysis and interpretation. This means that data in the images, either directly or after some processing, are compared to infer correspondence, and thus motion. A purely "top-down" analysis (and we believe none exist) would hypothesize motion from some a-priori knowledge, and test the hypothesis by inter-frame matching. Systems employing motion estimation (Jain and Nagel 78), region based models (Williams 81), and edge based models (Martin and Aggarwal 78) are a mixture of the two.

If a high level model is available, then interpretation would consist of describing some parameter(s) of motion or spatial disposition. The matching activity would then be categorized as a comparison of many points (an entire surface or object) where such a match in totto would constitute one search point in a parametric search (Williams 81, Barrow et.al. 77). The essential characteristic of the match is that many matches could be averaged together over the area covered by an object or by its border (edges).

Contrast this with more conventional "bottom-up" approaches, where no a-priori model is assumed. The knowledge available is that of image structure. Usually, some operator is employed to select small areas which are somehow unambiguous when matched, that is, areas which tend to match themselves very well only when the area is perfectly registered with itself. These operators are called "interest operators" since the points they find are "interesting", at least more so than other points. The essential characteristic of the match is that there are relatively few points, which individually are expected to derive unambiguous and accurate matches.

II. EXPERIMENTS

Three image pairs were selected from three sequences (See figures 1-3). They have different qualities and display different forms of motion. They by no means exhaust the possible situations, but they are all naturally occuring scenes (were not synthesized).

II.1 Images

The wrench data were obtained by physically moving a wrench horizontally and digitizing from a video camera. The camera has shading and vertical stripe anomolies. The motion was best controlled in this data, but there were no occlusions of interest.

The road data were obtained by driving down a road in an automobile shooting super-8 movie film. Several frames were digitized through red, green, and blue filters with a video camera. A 128 x 128 portion of that data (original image digitized 512 x 512) is used. The gradients in the data are quite small due to the resolution of the film and optics. Occlusions are apparent in several areas, and motion is in the Z axis.

The house data were obtained by photographing the original scene from several places with 35mm color slide film. The slides were

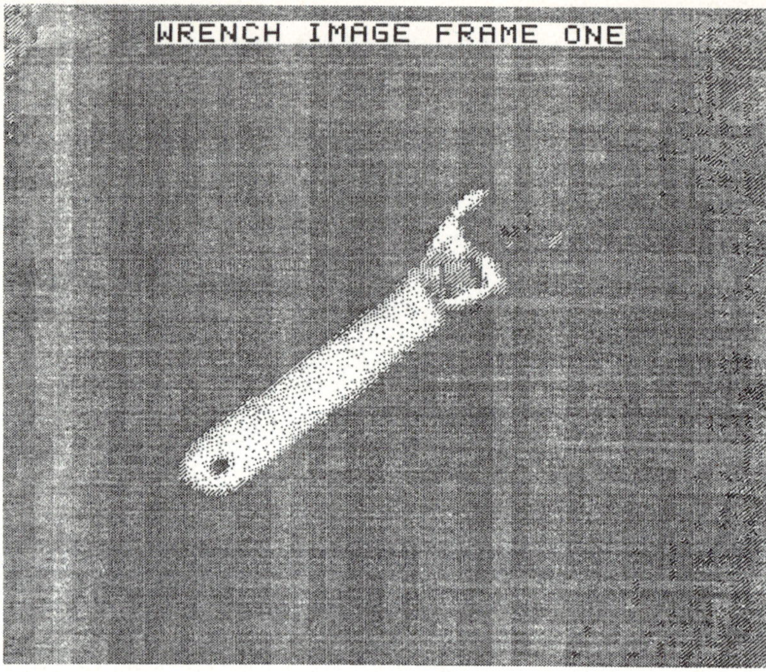

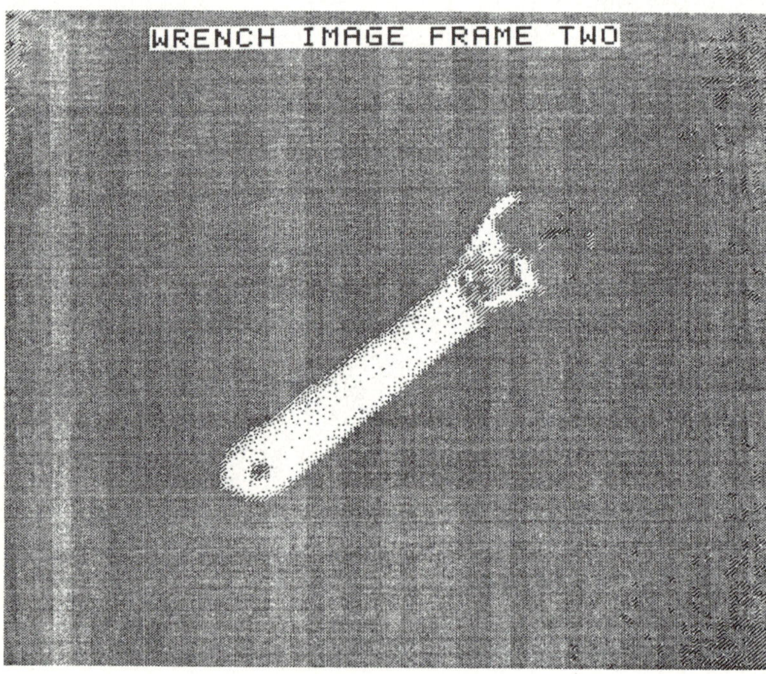

Figure 1

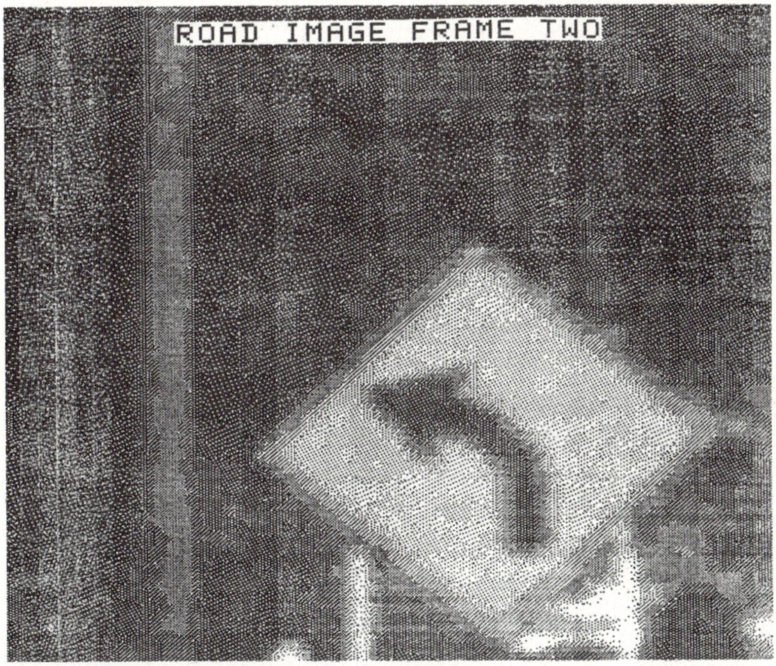

Figure 2

Figure 3

digitized on a high quality drum scanner with three filters. The images exhibit translation, rotation, and multiple occlusions. (Thanks to Terry Weymouth at the University of Massachusetts for this data set).

II.2 Data For Matcher

The image data are sometimes used directly in the match process, and sometimes pre-processed to enhance some attribute or diminish some other. We use the term "feature operator" to describe this pre-processing data selection step. This is not to be confused with the more common term "image feature", where some decision is made based upon the value of a feature operator. We have chosen to compare the match results from single point comparisons of original data values (intensity), and local gradients (inter-pixel difference, and 4-direction Kirsch) and also area comparisons involving 2x2, 4x4, and 8x8 areas using original data values. The result of the matches (details below) are compared to see if there is any obvious reason to choose between point, gradient, and area type.

The "point" feature is considered a special case of the "area" feature. By point we mean the value at the pixel is the feature value. In the case of color data it is a vector (R,G,B). The second type of feature is the two direction local gradient - we call it 1 x 2. It is simply the signed inter-pixel difference vertically and horizontally. Only green data are used for this operator. The last feature is a 4-direction Kirsch, set at 0°, 45°, 90° and 135°. By area we mean that point values about the central point will be used for matching. In the case of color data only the green data are used in area matching.

II.3 Displacements

Displacement fields relating frame pairs were generated for the three image sequences to provide the basis for interframe matching. For a given pixel in frame one, the displacement vector based at that pixel, points to the corresponding pixel in frame two. Displacements were generated by manually comparing areas of the images. Where good visual matches were found, displacement values were deposited into "displacement images". These were then used to generate smooth fields of displacements which agreed with image dynamics. Particular attention was paid to areas where detailed analysis was to be performed. Displacements are integer valued, where each vector begins in the center of a pixel and ends in the center of a pixel.

II.4 Masks

To simulate top-down type of matching, areas of the images that correspond roughly to objects were segmented by hand. Also, several areas corresponding to the boundaries of objects were selected both by hand and automatically. We call all of these areas "masks" and present results of all points' match values averaged together. These results are meant to indicate the ease with which some parametric search might proceed if a model could predict the objects as segmented. In some sense, such a result indicates the "best-case" for search since the interframe displacements and occlusion are accounted for, and because the match is averaged over relatively large areas, reducing the likelihood that problematic points will contribute significantly to the match result.

Masks were generated by several means. For the road scene an automatic segmentation was modified by hand, generating three primary masks. Two other masks were entirely hand selected to investigate the borders of objects (see figure 4).

The wrench was segmented by hand to produce the wrench-only mask, and the background mask. The border was generated by hand, it is one pixel wide. Also, the image was thresholded producing a mask of the brighter pixels, and a piece of the background was selected.

The house scene was segmented by hand into bushes, railing, a piece of the door, and a piece of railing and background (see figure 5).

III. RESULTS

Results of matching must be considered within the framework of application. In order to compare results, they are presented in two major forms, the match array and sharpness values for that array.

ROAD IMAGE, THREE PRIMARY MASKS

ROAD: SIGN EDGE & SIGNPOST EDGE

Figure 4

BUSHES AND DOOR MASKS

RAILING AND PIECE OF RAILING & HOUSE

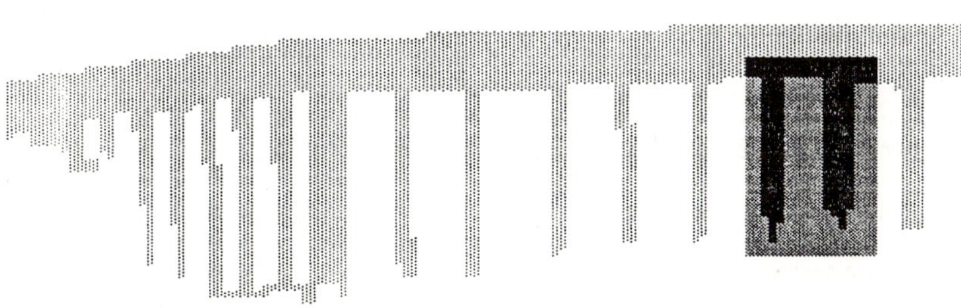

Figure 5

III.1 Match Metric

The match is simply the Euclidean distance between the processed data point values (see figure 6).

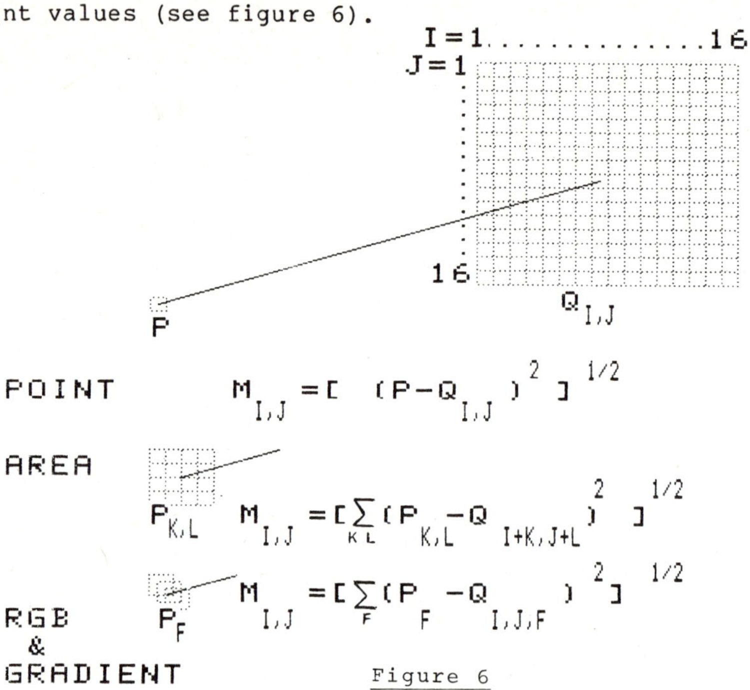

Figure 6

The case of point intensity has only one dimension, so the match value is the absolute value of the difference between the intensity value at the base of a displacement vector and the intensity value at the tip in the second frame. The two direction local gradient is a two-dimensional value, and thus the match value is the magnitude of the vector difference of two two-dimensional vectors, and four-dimensional in the case of the Kirsch values. For the area match the number of dimensions is simply larger. The 2x2 area describes a four-dimensional vector, the 4x4 a 16-dimensional, and the 8x8 a 64-dimensional one. Note that this measure is not equivalent to multiplicative correlation. Such correlation describes the angle between the two vectors, whereas ours is the Euclidean distance between their tips. If intensity were normalized, the two measures would be similar, but not identical. Ours is sensitive to differences in intensity or the "DC" component of the image and sub-images.

Interframe matching is, of course, a means to an end, that being correspondence interpretation. Any system which determines correspondences utilizes some search strategy and tolerates error in the match process. Often some constraints are applied such as smoothness of the resultant field of displacements, or continuation
across several frames.

All of the matching we perform follows one simple algorithm. The point or area value(s) in the first image is compared with the 16x16 array of points surrounding the tip of the displacement vector in the second image. Thus, a two-dimensional array of difference values (16x16) is developed for each match. This match result is easy to display and qualitatively examine, but difficult to compare and analyze when there are large numbers of such arrays. Therefore, we take a measure which has some meaning to most correspondence interpretation systems. We expect that such systems' performance will degrade if minima exist in the match result array at places other than the correct (central) point. Furthermore, as those minima occur further away from the center, they would reduce the likelihood of successfully choosing the center as the correct point. Thus, as the form of the result becomes "sharper", with steeper slopes from the center, the performance of an interpretative system should improve. This intuition led to the sharpness measures as follows:

$$1)\quad S_{min} = \frac{\sum_r r\,(M_r - M_o)}{(\sum_r r)(MAX - M_o)}$$

$$2)\quad S_{ave} = \frac{\sum_r r\,(A_r - M_o)}{(\sum_r r)(MAX - M_o)}$$

Where r is the distance (8 direction) from the central minimum M_o in pixels, and M_r is the minimum in set of match result values in the r-distance square annulus. A_r is the corresponding average. MAX is the maximum value in the entire array.

Because the displacements used in the experiments are integers (are only accurate to ± 1 pixel), we select the center to be the minimum of the 3x3 area around the actual center of the array.

Several examples of S_{min} are given in section III.2. Note that all displays are inverted, that is, minima appear as peaks, maxima as valleys, to ease viewing and comprehension. Note also that the sharpness value ranges from 0 to 1 when the central value is a peak, and goes negative when it is a valley. Also a "ridge" will give a low value, since we are using minima within each annulus. We call this measure the "min rule" sharpness value. We interpret it as a strong measure because its value decreases quickly when there are any other good match values in the 16x16 array.

The second sharpness function S_{ave} measures the average value for each annulus prior to weighted summation. The effect of this is primarily to improve the sharpness value when the result array has only a few points or a ridge of points which are similar to the central minimum. Thus, smooth and radially symmetric arrays will have similar sharpness values, and those where a few bad points exist will degrade the first measure much more quickly than the second. Several figures demonstrate the difference. We will often present both measures.

III.2 Masked Areas

In the first set of results (figures 7, 8, 9 and 10) we present the masked averages. This gives us examples of the approximate search space for parmeterized top-down processing. The surface plots have been inverted so that minima (best match) become peaks and large differences (poor match) become valleys. Because of display scaling, one must compare sharpness numbers as part of evaluation. Also, where the two sharpness values are given, M denotes the value S_{min} and A denotes the value S_{ave}. Two views are given of the match array, where the right-hand one is rotated to the right about a vertical axis by 90°.

The results shown confirm intuition about the interframe matching:

1) Where enough contrast and/or texture exists there are unique

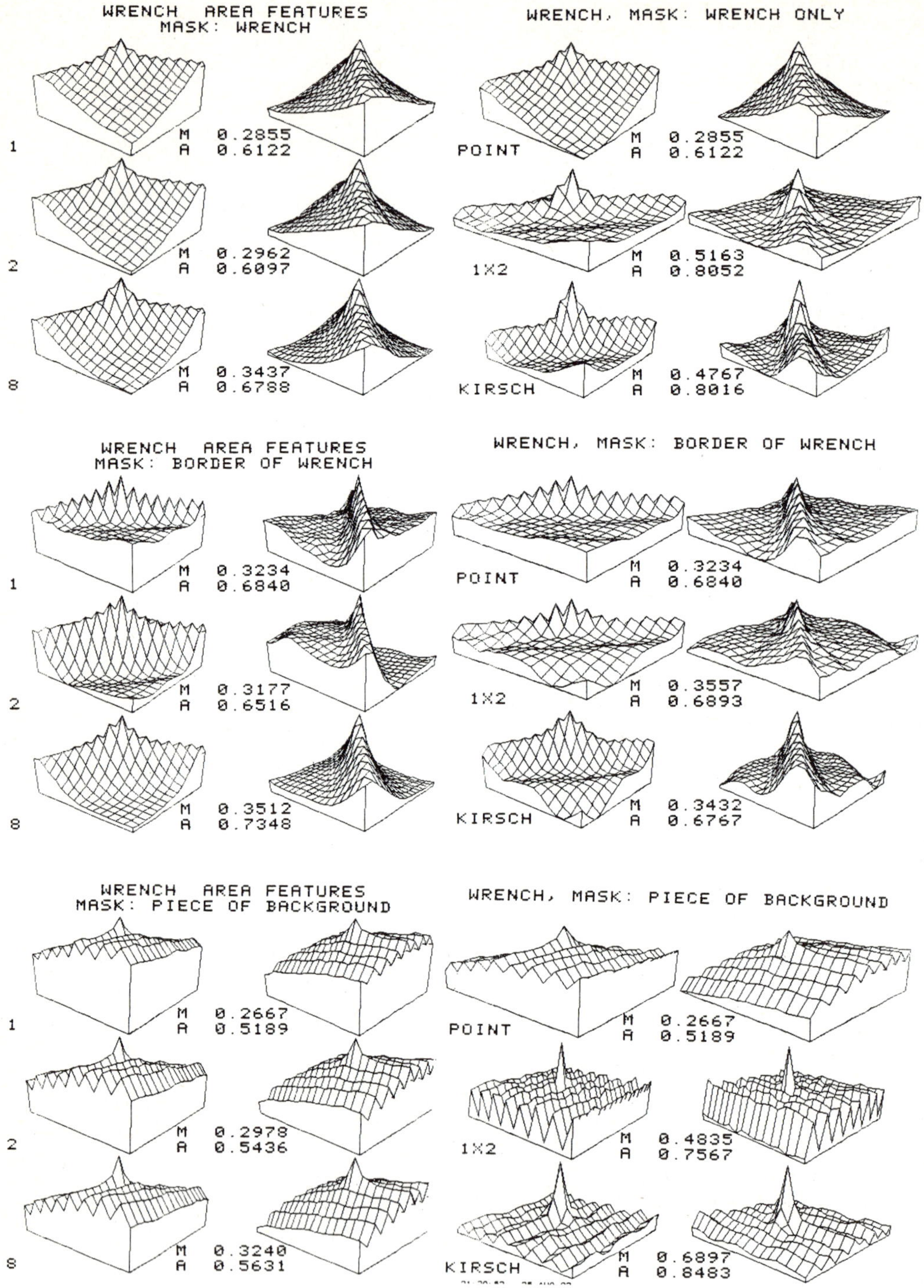
WRENCH AREA FEATURES
MASK: WRENCH
1
M 0.2855
A 0.6122
2
M 0.2962
A 0.6097
8
M 0.3437
A 0.6788
WRENCH, MASK: WRENCH ONLY
POINT
M 0.2855
A 0.6122
1X2
M 0.5163
A 0.8052
KIRSCH
M 0.4767
A 0.8016
WRENCH AREA FEATURES
MASK: BORDER OF WRENCH
1
M 0.3234
A 0.6840
2
M 0.3177
A 0.6516
8
M 0.3512
A 0.7348
WRENCH, MASK: BORDER OF WRENCH
POINT
M 0.3234
A 0.6840
1X2
M 0.3557
A 0.6893
KIRSCH
M 0.3432
A 0.6767
WRENCH AREA FEATURES
MASK: PIECE OF BACKGROUND
1
M 0.2667
A 0.5189
2
M 0.2978
A 0.5436
8
M 0.3240
A 0.5631
WRENCH, MASK: PIECE OF BACKGROUND
POINT
M 0.2667
A 0.5189
1X2
M 0.4835
A 0.7567
KIRSCH
M 0.6897
A 0.8483

Figure 7

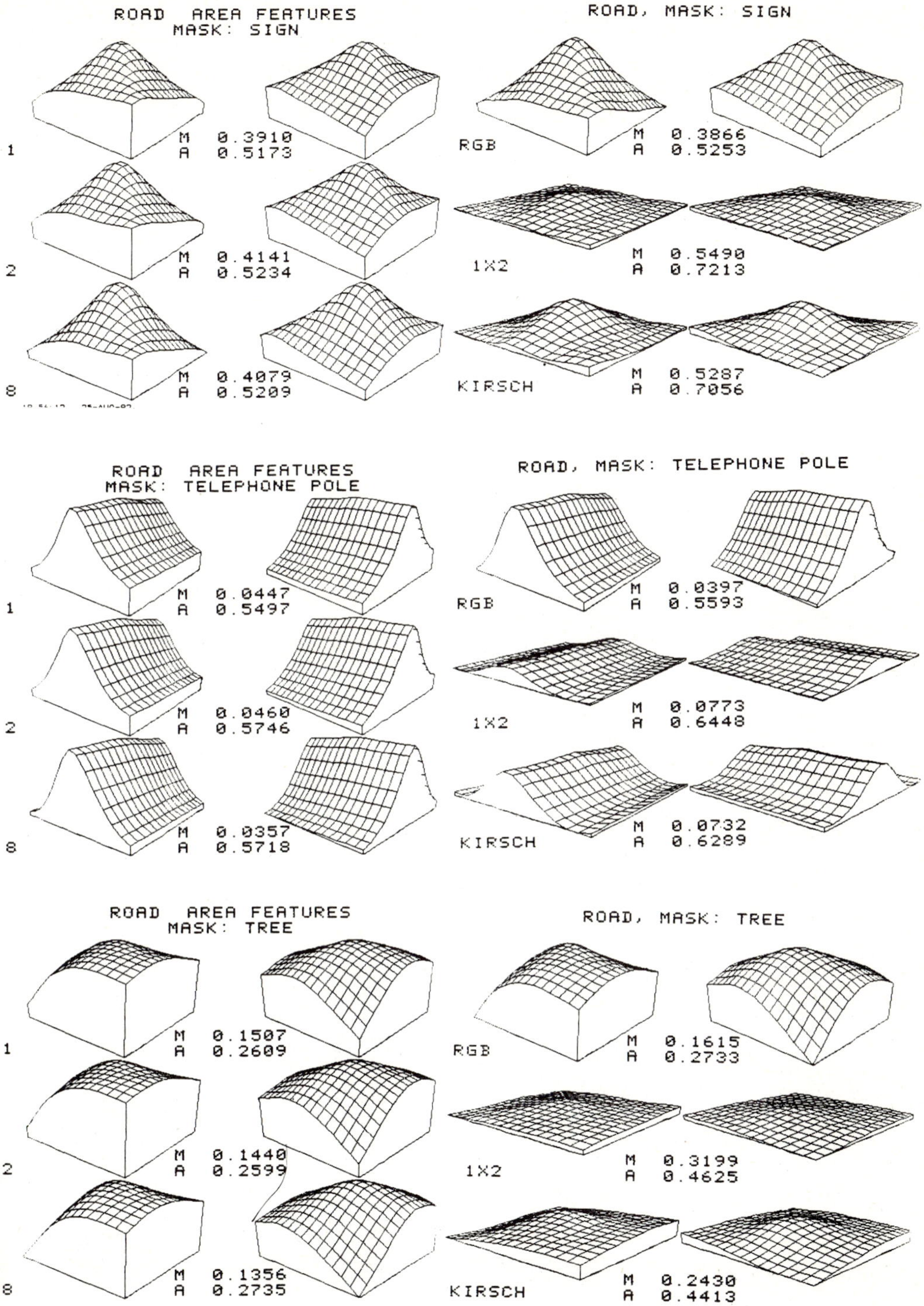
ROAD AREA FEATURES
MASK: SIGN
1
M 0.3910
A 0.5173
2
M 0.4141
A 0.5234
8
M 0.4079
A 0.5209
ROAD, MASK: SIGN
RGB
M 0.3866
A 0.5253
1X2
M 0.5490
A 0.7213
KIRSCH
M 0.5287
A 0.7056
ROAD AREA FEATURES
MASK: TELEPHONE POLE
1
M 0.0447
A 0.5497
2
M 0.0460
A 0.5746
8
M 0.0357
A 0.5718
ROAD, MASK: TELEPHONE POLE
RGB
M 0.0397
A 0.5593
1X2
M 0.0773
A 0.6448
KIRSCH
M 0.0732
A 0.6289
ROAD AREA FEATURES
MASK: TREE
1
M 0.1507
A 0.2609
2
M 0.1440
A 0.2599
8
M 0.1356
A 0.2735
ROAD, MASK: TREE
RGB
M 0.1615
A 0.2733
1X2
M 0.3199
A 0.4625
KIRSCH
M 0.2430
A 0.4413

Figure 8

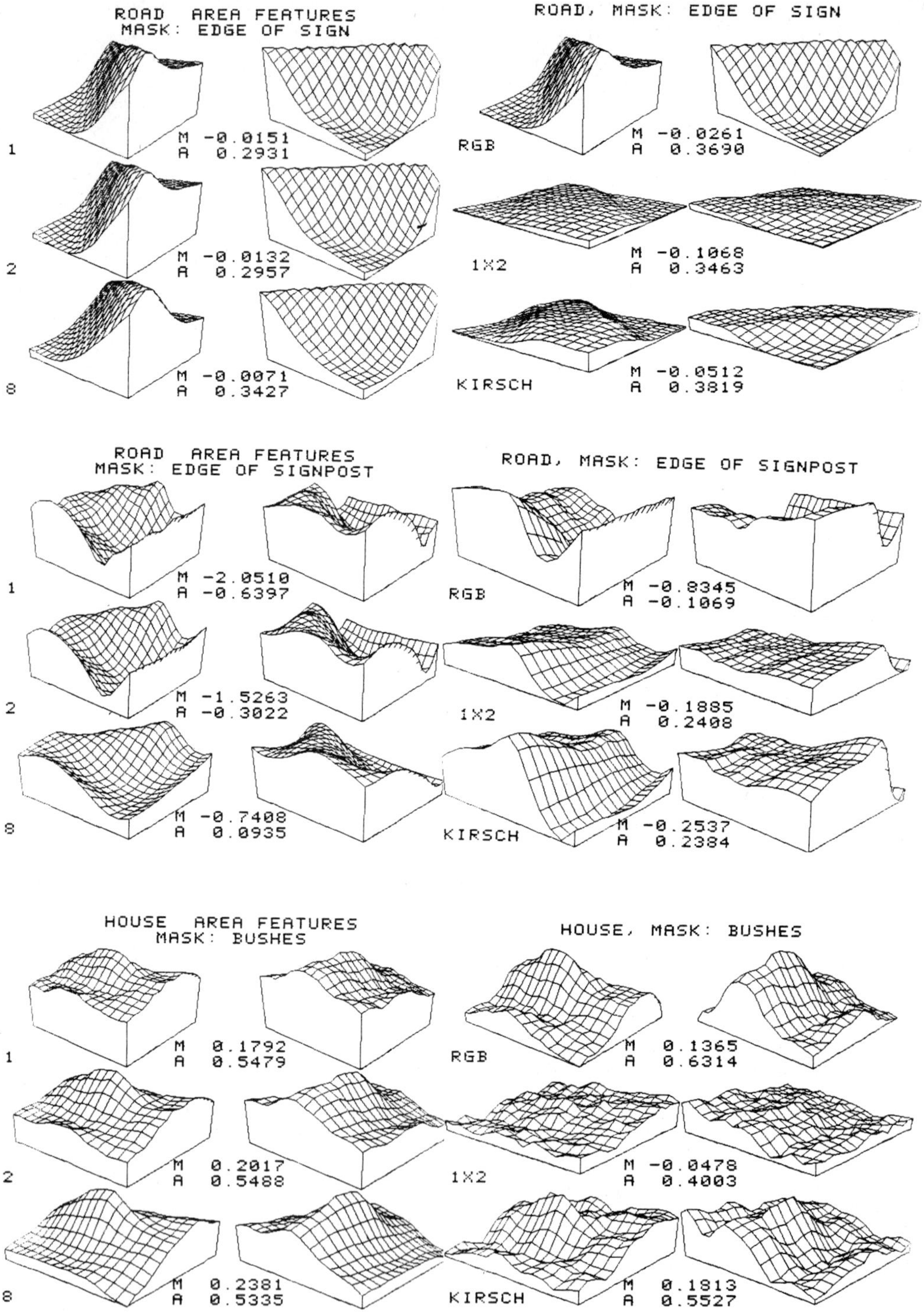
ROAD AREA FEATURES
MASK: EDGE OF SIGN
1
M -0.0151
A 0.2931
2
M -0.0132
A 0.2957
8
M -0.0071
A 0.3427
ROAD, MASK: EDGE OF SIGN
RGB
M -0.0261
A 0.3690
1X2
M -0.1068
A 0.3463
KIRSCH
M -0.0512
A 0.3819
ROAD AREA FEATURES
MASK: EDGE OF SIGNPOST
1
M -2.0510
A -0.6397
2
M -1.5263
A -0.3022
8
M -0.7408
A 0.0935
ROAD, MASK: EDGE OF SIGNPOST
RGB
M -0.8345
A -0.1069
1X2
M -0.1885
A 0.2408
KIRSCH
M -0.2537
A 0.2384
HOUSE AREA FEATURES
MASK: BUSHES
1
M 0.1792
A 0.5479
2
M 0.2017
A 0.5488
8
M 0.2381
A 0.5335
HOUSE, MASK: BUSHES
RGB
M 0.1365
A 0.6314
1X2
M -0.0478
A 0.4003
KIRSCH
M 0.1813
A 0.5527

Figure 9

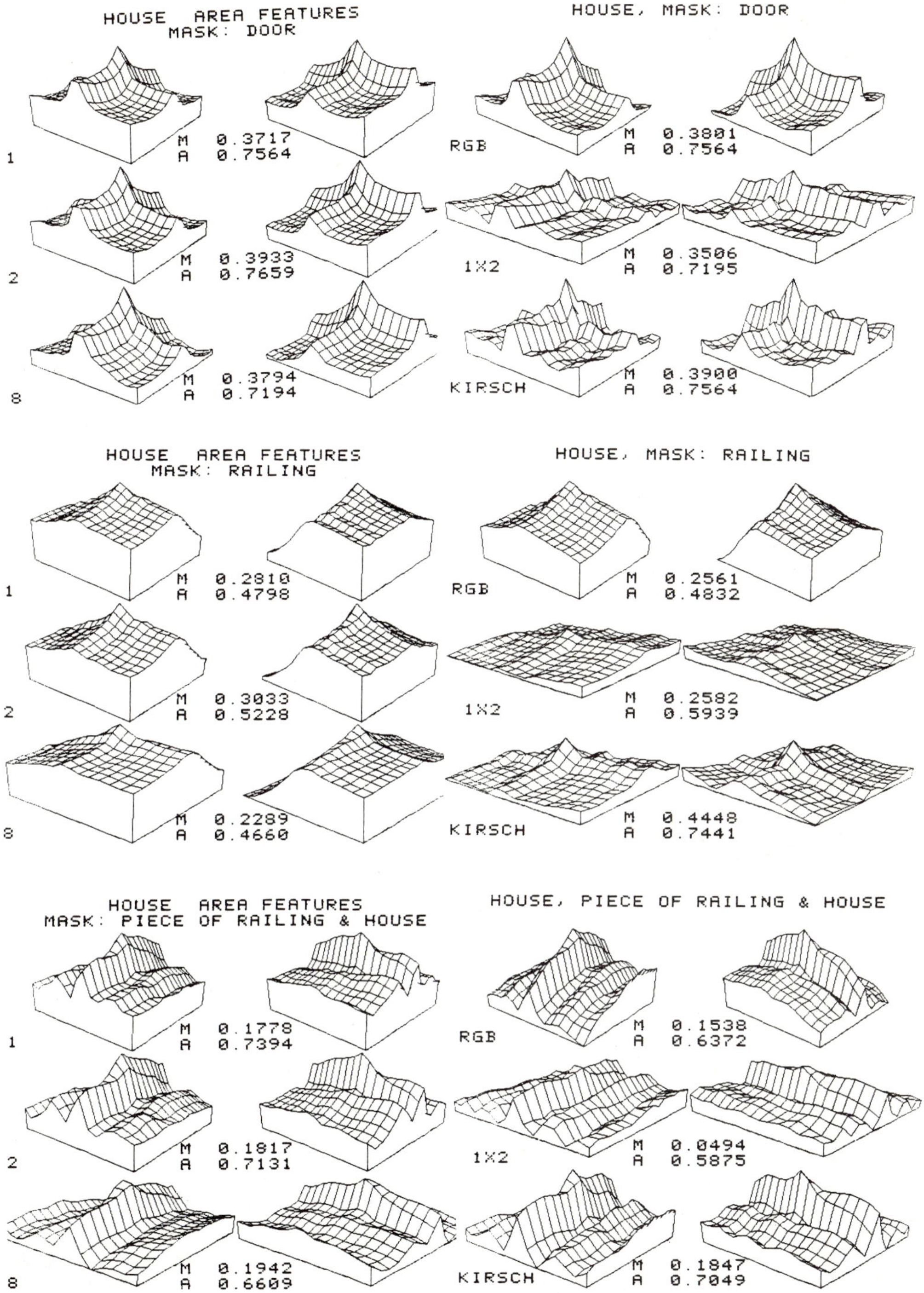
HOUSE AREA FEATURES
MASK: DOOR
1
M 0.3717
A 0.7564
2
M 0.3933
A 0.7659
8
M 0.3794
A 0.7194
HOUSE, MASK: DOOR
RGB
M 0.3801
A 0.7564
1X2
M 0.3506
A 0.7195
KIRSCH
M 0.3900
A 0.7564
HOUSE AREA FEATURES
MASK: RAILING
1
M 0.2810
A 0.4798
2
M 0.3033
A 0.5228
8
M 0.2289
A 0.4660
HOUSE, MASK: RAILING
RGB
M 0.2561
A 0.4832
1X2
M 0.2582
A 0.5939
KIRSCH
M 0.4448
A 0.7441
HOUSE AREA FEATURES
MASK: PIECE OF RAILING & HOUSE
1
M 0.1778
A 0.7394
2
M 0.1817
A 0.7131
8
M 0.1942
A 0.6609
HOUSE, PIECE OF RAILING & HOUSE
RGB
M 0.1538
A 0.6372
1X2
M 0.0494
A 0.5875
KIRSCH
M 0.1847
A 0.7049

Figure 10

and clear peaks.

2) Where there is one primary direction of contrast (an edge) there is a resulting ridge of ambiguity.

3) Where there is considerable self occlusion (the house bushes) there are generally poor matches, which improve with larger area matching.

4) The edge of the sign post shows a combination of occlusion and low-pass filtering (due to resolution of the film) where a gradient dissapears. In this situation the data changes so drastically that no feature accurately recovers the correct displacement.

III.3 Interest Points

Interest operators usually operate in two steps. Firstly, they compute some value we call the "interest value", and secondly, a decision is made (usually on absolute value, and local maximum) to determine which points are "interesting". The result is called "interest points".

Two sets of interest points were selected. They are 1) (Moravec 81) and 2) (Hannah 80), so named after the researchers who developed them. The following are the equations for the interest values:

Moravec: $\min_{j=1,4} \sum_i (X_i - X_j)^2$, j is an index of nearest neighbor relative to i.

Hannah: where DVj is Moravec evaluated at index j:

$$(\text{Local variance}) * \min \left(\frac{DV_1}{DV_3}, \frac{DV_3}{DV_1}, \frac{DV_2}{DV_4}, \frac{DV_4}{DV_2} \right)$$

To understand the utility of these operators at selecting points which are good for matching we utilize 2d histograms.

This analysis was conducted on the House image. The first set of histograms (figures 11 and 12) show the relation between sharpness and interest values. The X axes of the histograms displays are the square root of the absolute value of interest. Since the Moravec (and therefore the Hannah) values are based on the square of pixel units, taking the square root of the interest value displays the data in a linear format. In these figures the first house frame is matched with itself. Thus, the effects of motion are removed, and we can clearly see the ability of interest operators to imply high sharpness. The primary objective of this paper is to observe differences arising from the use of different matching (point, gradient, area), and histograms are presented in each category. Note that the strongest effect occurs where the matches are based on larger (4x4) areas. Although this paper is not intended to compare interest operators, the Hannah operator seems to give superior results.

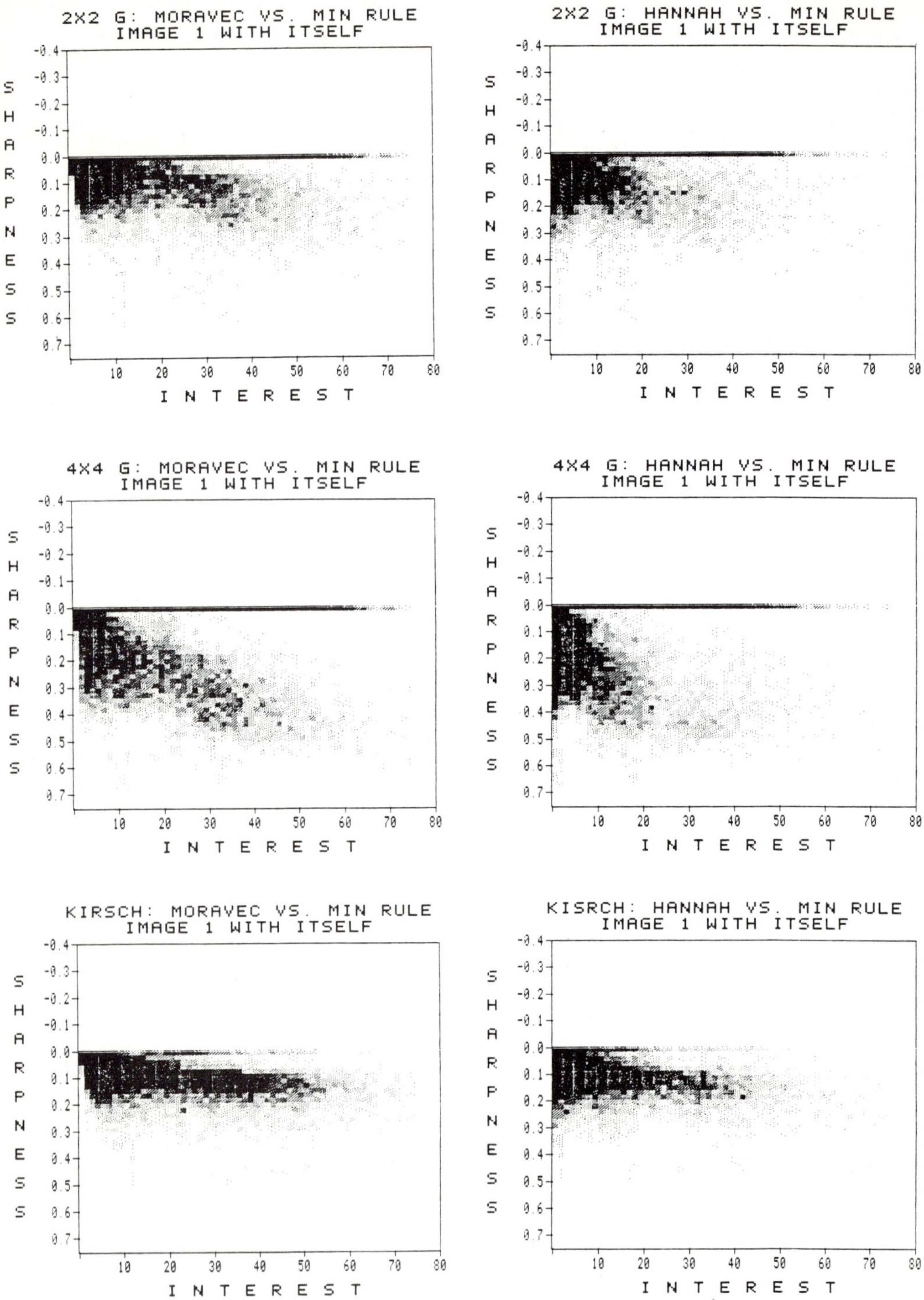
2X2 G: MORAVEC VS. MIN RULE
IMAGE 1 WITH ITSELF
2X2 G: HANNAH VS. MIN RULE
IMAGE 1 WITH ITSELF
4X4 G: MORAVEC VS. MIN RULE
IMAGE 1 WITH ITSELF
4X4 G: HANNAH VS. MIN RULE
IMAGE 1 WITH ITSELF
KIRSCH: MORAVEC VS. MIN RULE
IMAGE 1 WITH ITSELF
KISRCH: HANNAH VS. MIN RULE
IMAGE 1 WITH ITSELF
SHARPNESS
INTEREST
-0.4 -0.3 -0.2 -0.1 0.0 0.1 0.2 0.3 0.4 0.5 0.6 0.7
10 20 30 40 50 60 70 80

Figure 11

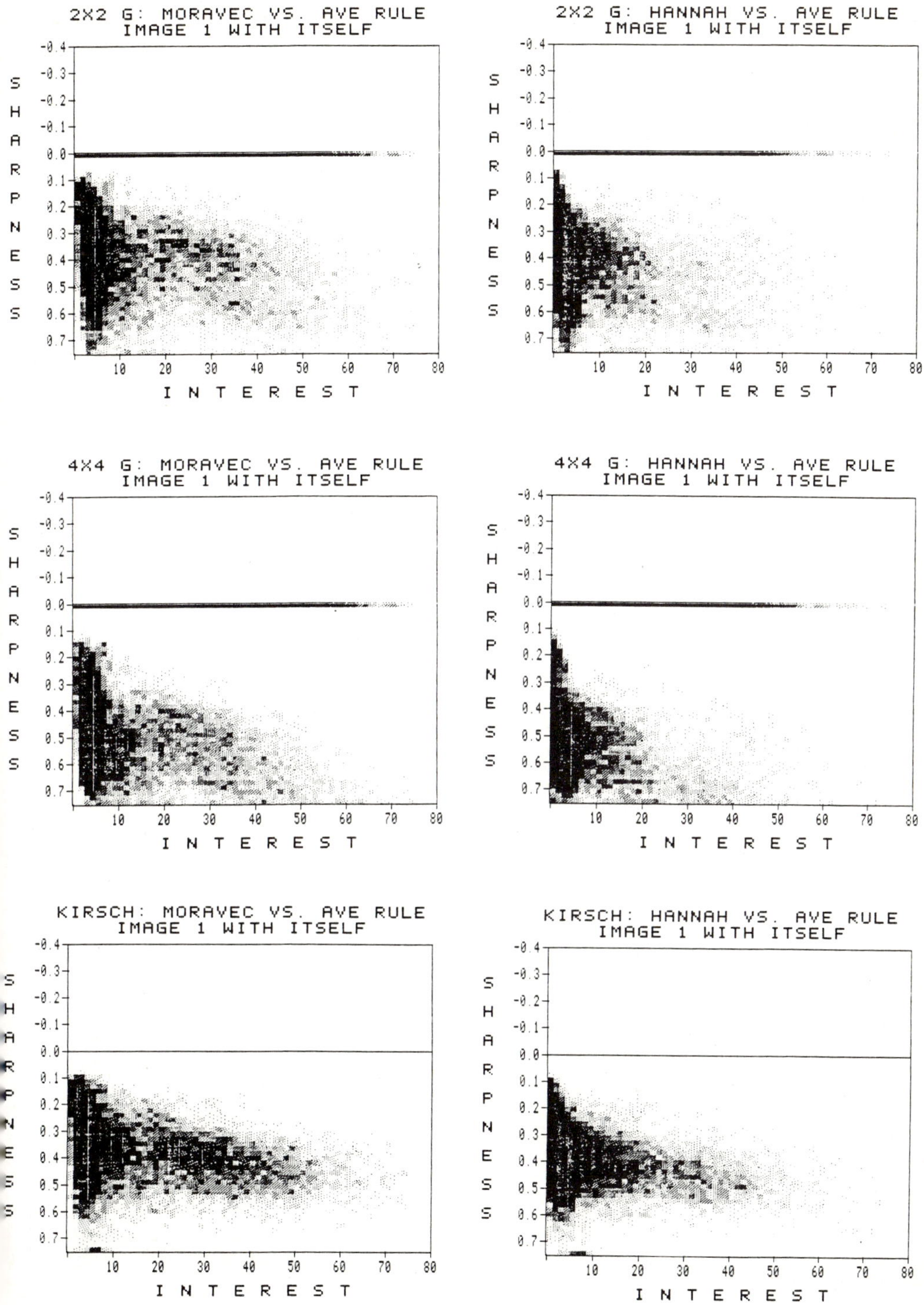
2X2 G: MORAVEC VS. AVE RULE
IMAGE 1 WITH ITSELF
2X2 G: HANNAH VS. AVE RULE
IMAGE 1 WITH ITSELF
4X4 G: MORAVEC VS. AVE RULE
IMAGE 1 WITH ITSELF
4X4 G: HANNAH VS. AVE RULE
IMAGE 1 WITH ITSELF
KIRSCH: MORAVEC VS. AVE RULE
IMAGE 1 WITH ITSELF
KIRSCH: HANNAH VS. AVE RULE
IMAGE 1 WITH ITSELF
SHARPNESS
INTEREST
-0.4
-0.3
-0.2
-0.1
0.0
0.1
0.2
0.3
0.4
0.5
0.6
0.7
10
20
30
40
50
60
70
80

Figure 12

To illustrate the utility of interest operators on the image pair (motion present), the next set of figures 13 and 14 are presented. The most obvious change from the intra-image matches is the presence of negative values and a dispersion from the diagonal. By observation we conclude that although the interest operators do not guarantee sharpness, some correlation still exists. Of particular import however, are the great numbers of image points that result in high sharpness, but have low interest values (are good at matching but not identified as such). This motivates us to consider (in future work) further research in interest operators, and matching in general.

III.4 Categories

Interest points were classified by manual inspection into nine categories according to image and scene characteristics. Firstly, a decision was made, based upon whether the point is on or near the occluding border of objects. In each of these categories, subcategories were made to classify the point as being on or near a straight edge, a curved edge, or a sharp corner. If the point is within an object, and not one of these sub-classes, it was classified as homogeneous, somewhat textured, or strongly textured. All of these categories are subjectively described and determined. The purpose is to assist in understanding the utility of features and matching under different situations. Figures 15 and 16 depict the number of points in each category according to sharpness values. In these figures categories 1, 2 and 3 represent homogeneous, somewhat textured and strongly textured respectively. Categories 4, 5 and 6 represent corner, curved and straight edges within an object, while 7, 8 and 9 are corner, curved and straight at a border.

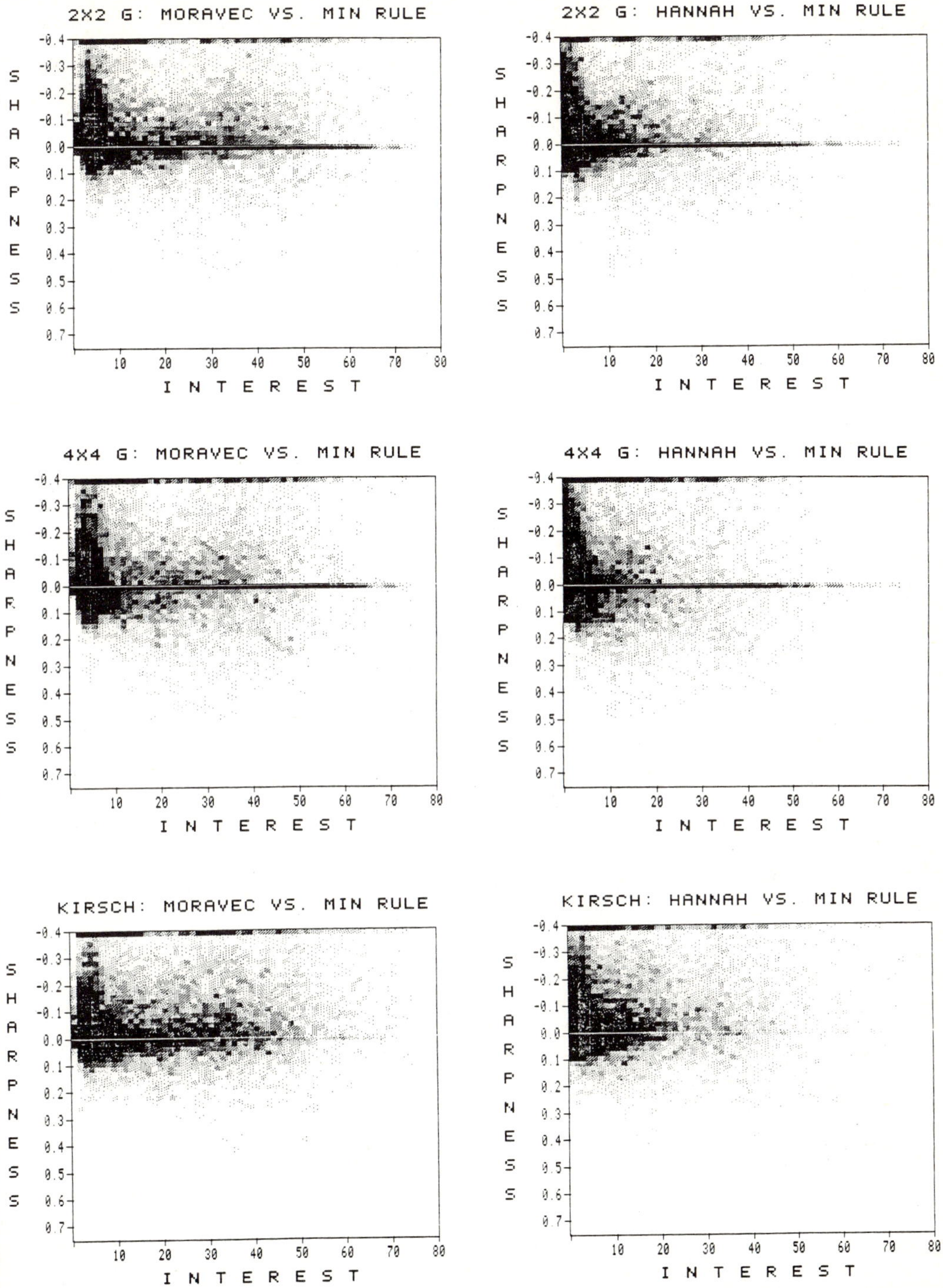
2X2 G: MORAVEC VS. MIN RULE
2X2 G: HANNAH VS. MIN RULE
4X4 G: MORAVEC VS. MIN RULE
4X4 G: HANNAH VS. MIN RULE
KIRSCH: MORAVEC VS. MIN RULE
KIRSCH: HANNAH VS. MIN RULE
SHARPNESS
INTEREST
-0.4
-0.3
-0.2
-0.1
0.0
0.1
0.2
0.3
0.4
0.5
0.6
0.7
10
20
30
40
50
60
70
80

Figure 13

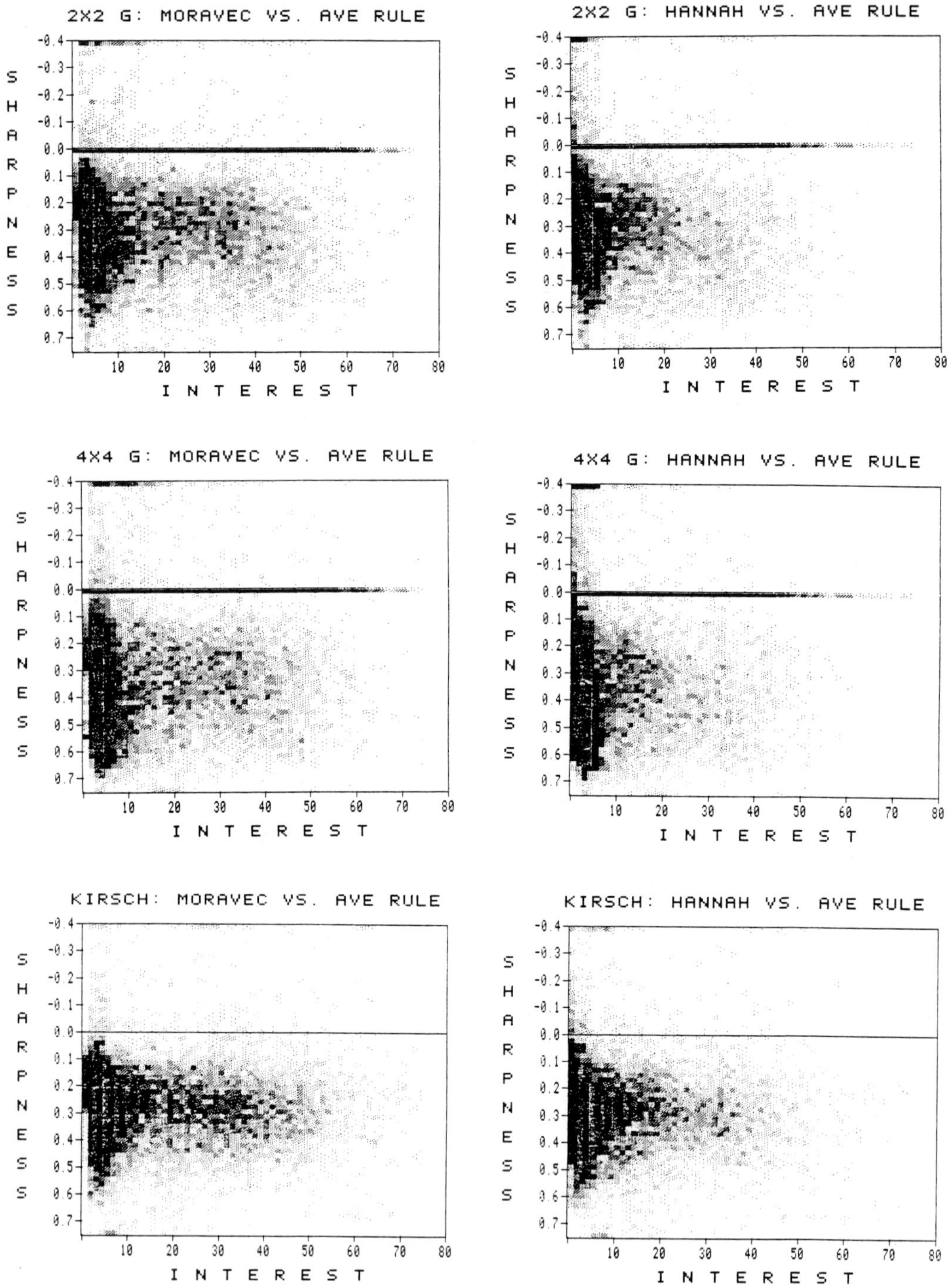
2X2 G: MORAVEC VS. AVE RULE
2X2 G: HANNAH VS. AVE RULE
4X4 G: MORAVEC VS. AVE RULE
4X4 G: HANNAH VS. AVE RULE
KIRSCH: MORAVEC VS. AVE RULE
KIRSCH: HANNAH VS. AVE RULE
SHARPNESS
INTEREST
-0.4
-0.3
-0.2
-0.1
0.0
0.1
0.2
0.3
0.4
0.5
0.6
0.7
10
20
30
40
50
60
70
80

Figure 14

SHARPNESS : MIN RULE
4X4 AREA, BY CATEGORY

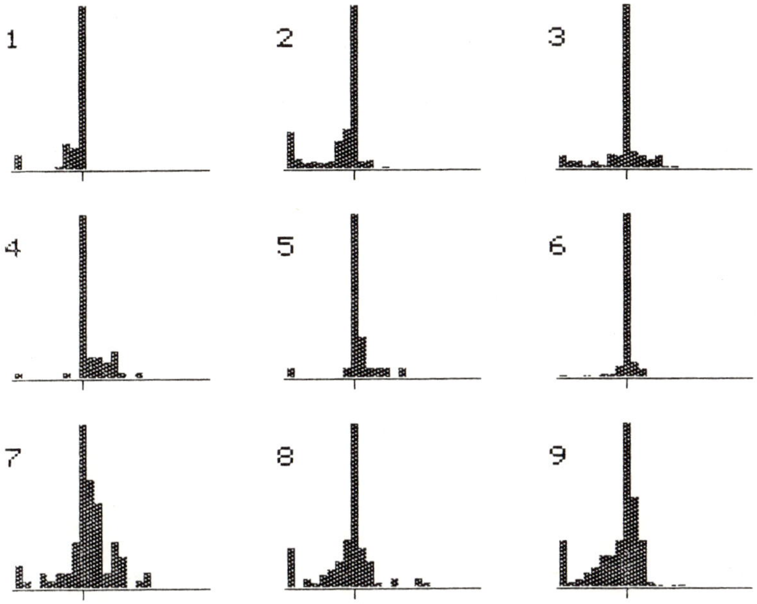

SHARPNESS : AVE RULE
4X4 AREA, BY CATEGORY

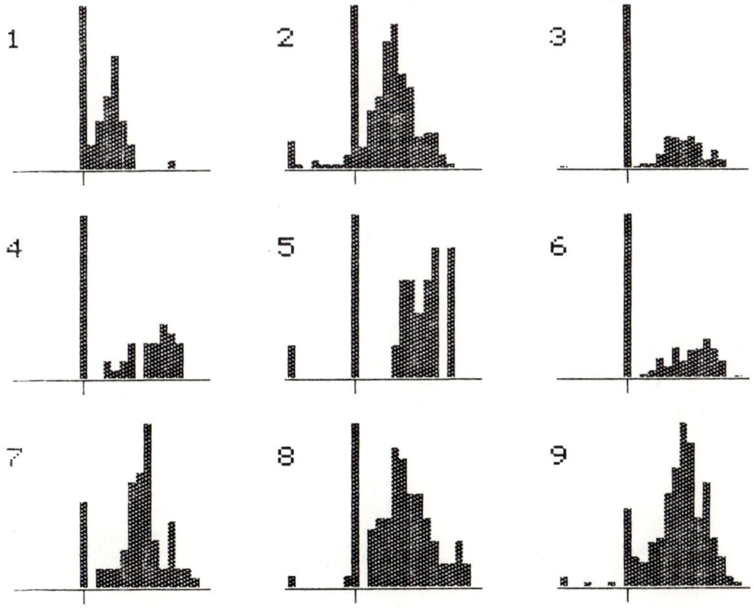

Figure 15

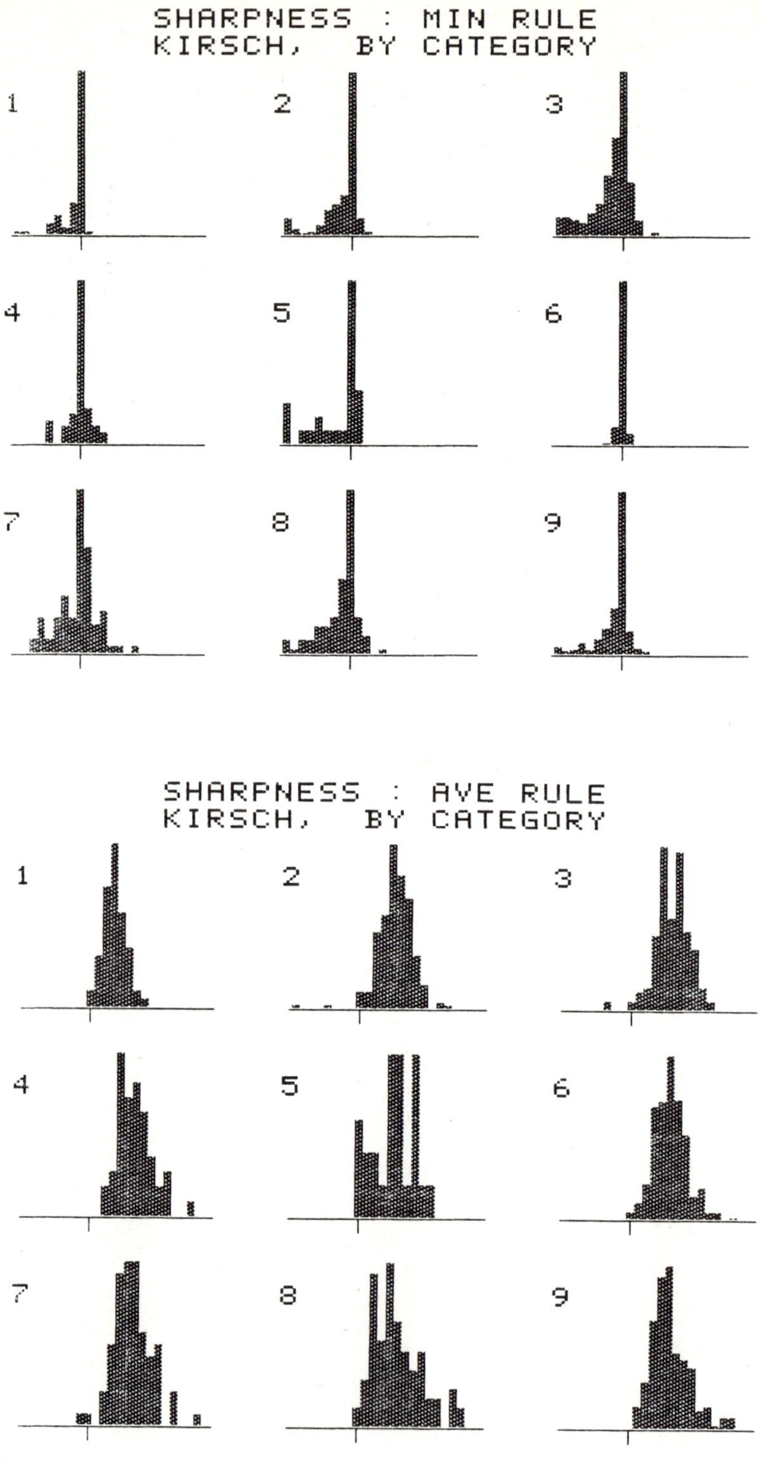
SHARPNESS : MIN RULE
KIRSCH, BY CATEGORY
1
2
3
4
5
6
7
8
9
SHARPNESS : AVE RULE
KIRSCH, BY CATEGORY
1
2
3
4
5
6
7
8
9

Figure 16

REFERENCES

Barrow, H.G., J.M. Tennenbaum, R.C. Bolles, and H.C. Wolfe, "Parametric Correspondence and Chamfer Matching: Two New Techniques for Image Matching," in Proc. IJCAI-5, Cambridge, MA, August 1977.

Hannah, M.J., "Bootstrap Stereo," in Proc. Image Understanding Workshop, DARPA Conference, pp. 201-208, April 1980.

Hanson, A., E. Riseman, "Segmentation of Natural Scenes," from Hanson and Riseman (ed.), Computer Vision Systems, Academic Press, 1978.

Jain, R. and H-H Nagel, "On a Motion Analysis Process for Image Sequences from Real World Scenes," Tech Report 48-78, Dept. of Information Science, University of Hamburg, Hamburg, Germany, 1978.

Martin, W. and J. Aggarwal, "Dynamic Scene Analysis: The Study of Moving Images," Tech Report 184, Information Systems Research Laboratory, Univ of Texas at Austin, January 77.

Moravec, H.P., "Rover Visual Obstacle Avoidance," in Proc. IJCAI-7, Vancouver B.C., August 1981.

Williams, T., "Computer Interpretation of a Dynamic Image from a Moving Vehicle", (Ph.D. Thesis) COINS Tech Rpt 81-22, University of Massachusetts, Amherst, MA., 1981.

DISPLACEMENT ESTIMATION FOR OBJECTS ON MOVING BACKGROUND

Peter Spoer

Lehrstuhl für Theoretische Nachrichtentechnik
und Informationsverarbeitung

Universität Hannover
Callinstr. 32
3 Hannover 1
West Germany

ABSTRACT

In a sequence of TV-pictures moving objects can be detected by their frame-to-frame displacements. In the case of a moving TV sensor also the non moving background will show a frame-to-frame displacement. Here a method is presented which allows the estimation of the displacement relative to the background and an automatic detection of relatively moving objects. The main problem is the estimation of the background displacement. A method is described to estimate the background displacement with low computational complexity for a background displacement varying spatially in the picture. For simplification the assumption of a flat surface of the ground and a small size of relatively moving objects has been made. Two laboratory generated sequences have been used to determine the accuracy of the background displacement estimate and the amount of relative displacement needed to separate moving objects from background. The error of the estimated background displacement shows a standard deviation of 0.50 pel per frame if only background motion occurs. In case of a relatively moving object this error increases. For the simulated TV-sequence a separation between relatively moving area and background has been achieved if the relative displacement exceeds a threshold of 2.4 pel per frame.

NATO ASI Series, Vol. F2
Image Sequence Processing and Dynamic Scene Analysis
Edited by T.S. Huang

1. INTRODUCTION

In air-to-ground surveillance a forward looking TV-sensor mounted on an airborne vehicle scans a certain region of the ground. For this application it is desired to detect moving objects and to estimate their velocity. In the case of a moving sensor the detection of moving objects by simply evaluating the frame difference signal is not possible because the whole image plane is displaced between successive frame of the sequence. Tracking methods may be used to renew information about position and velocity of moving objects, but for starting the tracking algorithm must be furnished with data of position and shape of objects of interest /1/.

This paper presents first results of a work concerned with a method to detect moving objects in aerial TV-sequences and to measure their velocity. The method used here is based on the evaluation of displacement vector fields estimated between successive frames of a sequence. No knowlegde about the position of moving objects is assumed for the operation of the algorithm. Also no information about the actual flight data of the TV-sensor such as flight velocity, flight height etc. is needed applying the presented method.

2. GENERAL APPROACH

The aerial TV-sequence contains a certain region of the ground and additionally objects moving on the ground. Both the motion of the TV-sensor and the motion of the objects on the ground cause displacements of picture elements between successive frames of the sequence. Generally the resulting displacement vector $\vec{D}$ is a function of the image coordinates X and Y. The horizontal and vertical components of the vector are denoted by D_X and D_Y.

The motion of the TV-sensor causes a displacement component called background displacement and will be denoted by $\vec{D}_b$. The angle of inclination between the optical axis of the TV-sensor and the vertical axis is different from zero and the background displacement becomes a function of the image coordinates. If an object is moving on the ground a second displacement component is superimposed to the background displacement. This component is called the relative displacement $\vec{D}_r$ of moving objects.

$$\vec{D}(X,Y) = \vec{D}_b(X,Y) + \vec{D}_r(X,Y) \qquad 2.1$$

The blockdiagram of the method to detect and to estimate the velocity of moving objects is shown in Fiqure 1. The digitized luminance signal I(X,Y) of two fields n and n+2 is used to estimate a displacement vector $\vec{D}$ at every position X,Y of the image plane. An approximation of the spatial function of the background displacement $\vec{D}_b$ is calculated from the displacement vector field. For that purpose assumptions concerning the flatness of the ground and the size of moving objects are made. The difference between the displacement $\vec{D}$ and the background displacement $\vec{D}_b$ represents the relative displacement $\vec{D}_r$. The vector of relative displacement estimated in the image plane is a measure for the velocity of corresponding points on the ground. A detection and clustering process extracts information about position and shape of moving objects from the vector field of relative displacement. The velocity for each moving object is then determined by averaging the vectors of relative displacement in the region covered by the object.

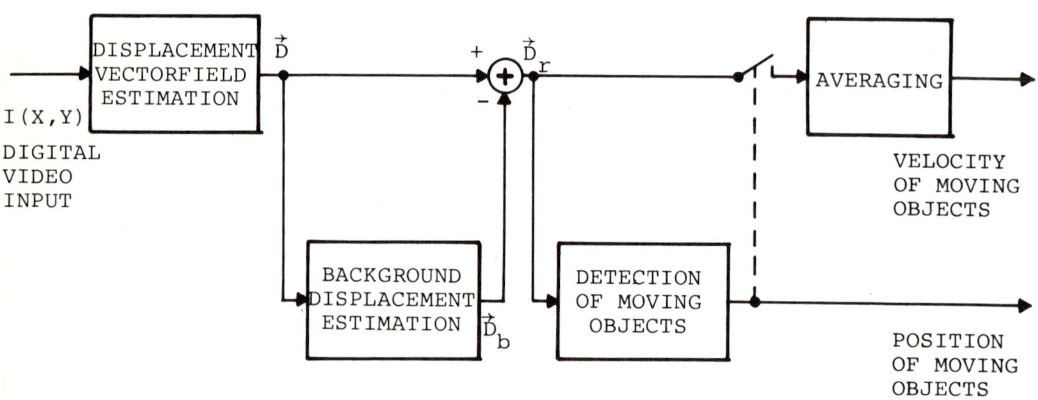

Figure 1. Blockdiagram of the detection and velocity estimation algorithm

3. DISPLACEMENT ESTIMATION

Three basic concepts for displacement estimation are known from the literature /2/: the method of differentials, the Fourier method and the correlation method.

In this work a correlation algorithm developed by Beyer /3/ is used for displacement estimation.

To estimate the displacement vector for one position X,Y the similarity between the luminance signal of a reference window centered at the current position in field n and the luminance signal of displaced search windows in field n+2 is determined. As a measure of similarity the cross correlation coefficient is calculated for all search window positions within a certain search area. The position where the correlation coefficient achieves his maximum value is the result of the displacement estimation $\vec{D}(X,Y)$ for the current position.

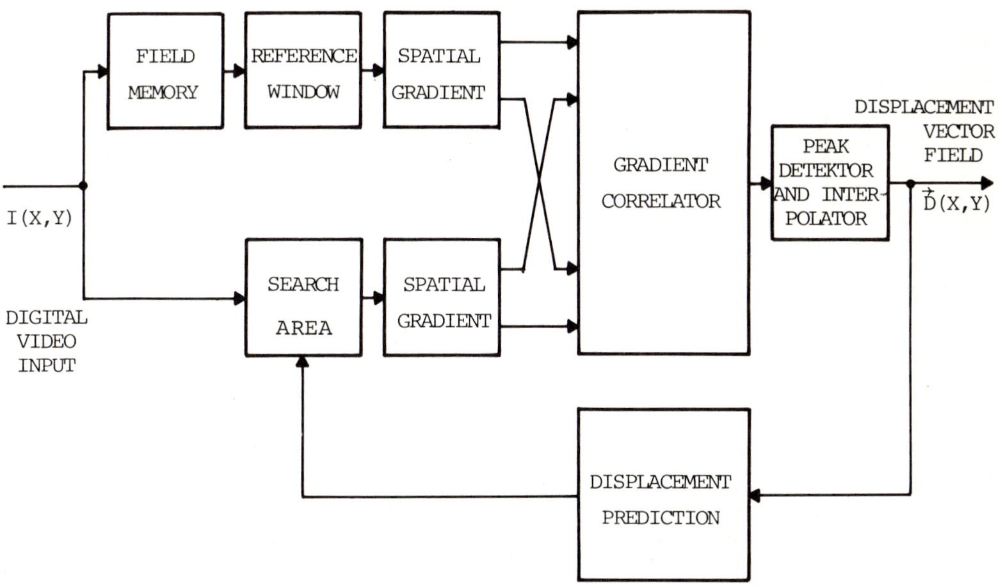

Figure 2. Blockdiagram of the displacement vectorfield estimator

The blockdiagram of the displacement estimator used for this application is shown in Figure 2. By the aid of a field memory the luminance of two successive fields is available for the displacement estimation. The reference window with a size of 5 by 5 picture elements is constructed around the actual position in the field n. The search area which covers 9 by 9 picture elements is constructed in the field n+2.

The computational expense of the correlation method is determined by the number of search window positions which are compared with the reference window i.e. by the size of the search area. For the application of the presented method it may be assumed that the variation of displacement estimates of corresponding picture elements in successive fields is small. The assumption allows a reduction of the size of the search area. For this purpose the displacement predictor derives a displacement prediction $\hat{\vec{D}}_n = (\hat{D}_{x,n};\hat{D}_{y,n})$ from the estimates of the previous field. If the search area is displaced corresponding to the predicted displacement the size of the search area can be reduced because the center of the displaced search area is already close to the position of best match.

The displacement prediction is concerned with special problems at regions moving relatively to the background. The predictor

$$\hat{\vec{D}}_n(X,Y) = \vec{D}_{n-2}(X,Y) \qquad 3.1$$

gives wrong results in the case of an object moving with a displacement which is large compared to the object size because this type of predictor does not take into account that a moving point will appear at a displaced position in the next field. An adaptive displacement prediction will avoid this disadvantage:

$$\hat{\vec{D}}_n(X+D_{x,n-2}(X,Y),\ Y+D_{y,n-2}(X,Y)) = \vec{D}_{n-2}(X,Y) \qquad 3.2$$

where $\vec{D}_{n-2}(X,Y) = (D_{x,n-2}(X,Y);D_{y,n-2}(X,Y))$.

However this method can cause ambiguity if background is covered by a moving object and can give no prediction for background which is decovered from an object as illustrated in Figure 3.

To avoid the difficulties concerned with the two methods mentioned above the information about the background displacement of the whole image plane is used to improve the prediction. From those points of the image where the displacement differs from the background displacement an adaptive prediction according to equation 3.2 is derived. For those points to whom no displacement prediction has been assigned in this way the background displacemnt is used according to equation 3.1 as prediction.

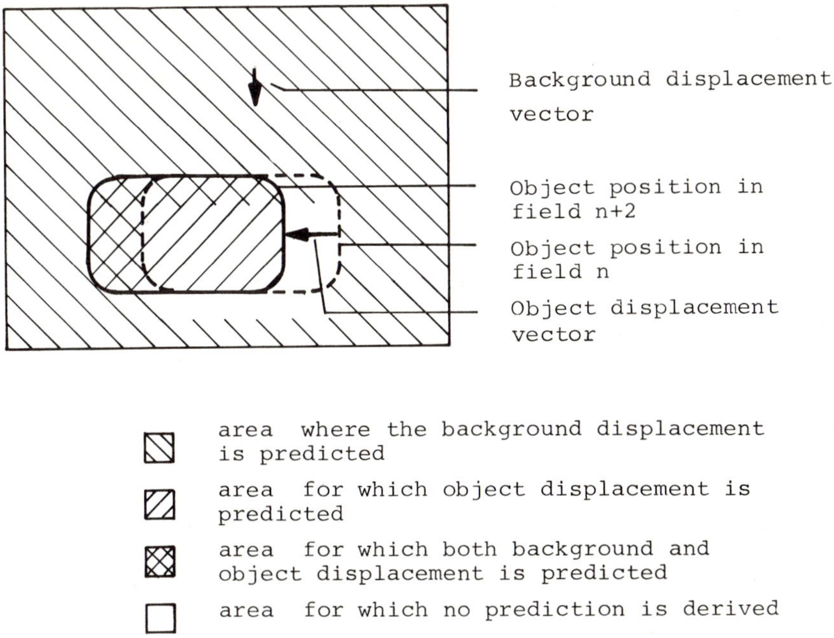

Figure 3. Influence of an object moving relatively to the background on the adaptive displacement prediction according to equation 3.2 in the case of uniform background and object displacements

It has been shown /3/ that preprocessing of the luminance signal by differentiation increases the peak-to-sidelobe ratio of the correlation i.e. the ratio of the maximum of the correlation coefficient and the standard deviation of the correlation coefficients at the sidelobes. For this purpose a gradient vector $\vec{G}$ of the luminance is used for correlation instead of the luminance itself.

$$\vec{G}(X,Y) = \begin{pmatrix} \partial I/\partial X \\ \partial I/\partial Y \end{pmatrix} \approx \begin{pmatrix} I(X+1,Y)-I(X-1,Y) \\ I(X,Y+1)-I(X,Y-1) \end{pmatrix} \qquad 3.3$$

In the search area the correlation coefficient R of the preprocessed luminance signal $\vec{G}$ is determined for every search window displaced by d_x, d_y.

$$R(d_x,d_y) = \frac{\sum_M \sum_N \vec{G}_n(X,Y) * \vec{G}_{n+2}(X+d_x,Y+d_y)}{\{\sum_M \sum_N \vec{G}_n^2(X,Y) \cdot \sum_M \sum_N \vec{G}_{n+2}^2(X+d_x,Y+d_y)\}^{1/2}} \tag{3.4}$$

M,N : shape of the windows

X,Y : current position in the image

d_x,d_y : spatial offset between the center of the search and the reference window

The peak detector searches for the position with the maximum correlation coefficient. The spatial offset D'_x,D'_y of this position is the estimated displacement which is resolved to within 1 pel.

$$R(d_x,d_y) = \text{Max.} \Big|_{d_x=D'_x,\ d_y=D'_y} \tag{3.5}$$

To improve the resolution of the displacement estimation the correlation function R is approximated at the peak position separately in horizontal and vertical direction by a polynome of second order:

$$R(X,D'_y) = u_x + v_x\ (X-D'_x) + w_x\ (X-D'_x)^2 \tag{3.6}$$

$$R(D'_x,Y) = u_y + v_y\ (Y-D'_y) + w_y\ (Y-D'_y)^2 \tag{3.7}$$

Each set of the parameters u,v and w of the polynomes can be derived from 3 values of the correlation function. For that purpose the peak value and the two values of the correlation function next to the peak in each direction are used. The positions of the maxima of the polynomes 3.6 and 3.7 serve as the corrected values D_x,D_y for the initial estimate D'_x,D'_y.

$$D_x = D'_x - \frac{v_x}{2 \cdot w_x} \tag{3.8}$$

$$D_y = D'_y - \frac{v_y}{2 \cdot w_y} \tag{3.9}$$

4. BACKGROUND DISPLACEMENT ESTIMATION

The outcome of the displacement estimation is a displacement vector field for every field of the TV-sequence. This vector field describes the displacement $\vec{D}$ for each picture element of the field. The background displacement is now estimated in order to separate each image into picture elements belonging to the background and picture elements belonging to moving objects. For this purpose an approximation of the vector field of the background displacement is derived from the displacement vector field. Then the difference between the displacement and the background displacement is determined for every picture element. The difference is denoted as the relative displacement. It is a measure of the motion relative to the background and allows the separation between background and moving objects.

To find out the background displacement for every picture element of the field it is assumed that the TV-sensor moves in constant flight height above a ground with flat surface. The model of the geometry for the image generation is shown in Figure 4. Starting with the general formulas presented in /2/, which describe the mapping between a 3-dimensional objectspace and the 2-dimensional imageplane the two components $D_{b,x}$ and $D_{b,y}$ of the background displacement $\vec{D}_b$ have been derived as a function of the image coordinates X,Y for the geometry and the flight parameters shown in Figure 4:

$$D_{b,x} = -\frac{\Delta}{h} \cdot \frac{F\cos\Phi\sin\Theta - X\sin\Phi\cos\Phi\cos\Theta - Y\sin\Phi\sin\Theta + \frac{X \cdot Y}{F}\sin\Phi\cos\Theta}{1 - \frac{\Delta}{h}\sin\Phi\cos\Phi\cos\Theta\left(1 + \frac{Y}{F}\tan\Phi\right)} \qquad 4.1$$

$$D_{b,y} = \frac{\Delta}{h} \cdot \frac{F\cos^2\Phi\cos\Theta - 2Y\sin\Phi\cos\Phi\cos\Theta\left(1 - \frac{Y}{2 \cdot F}\tan\Phi\right)}{1 - \frac{\Delta}{h}\sin\Phi\cos\Phi\cos\Theta\left(1 + \frac{Y}{F}\tan\Phi\right)} \qquad 4.2$$

The equations 4.1 and 4.2 describing the two components of the background displacement can be approximated by bilinear functions

$$D_{b,x} = a_x + b_x \cdot X + c_x \cdot Y + d_x \cdot X \cdot Y \qquad 4.3$$

$$D_{b,y} = a_y + b_y \cdot X + c_y \cdot Y + d_y \cdot X \cdot Y \qquad 4.4$$

if $\frac{Y}{F} \cdot \tan\Phi \ll 1$.

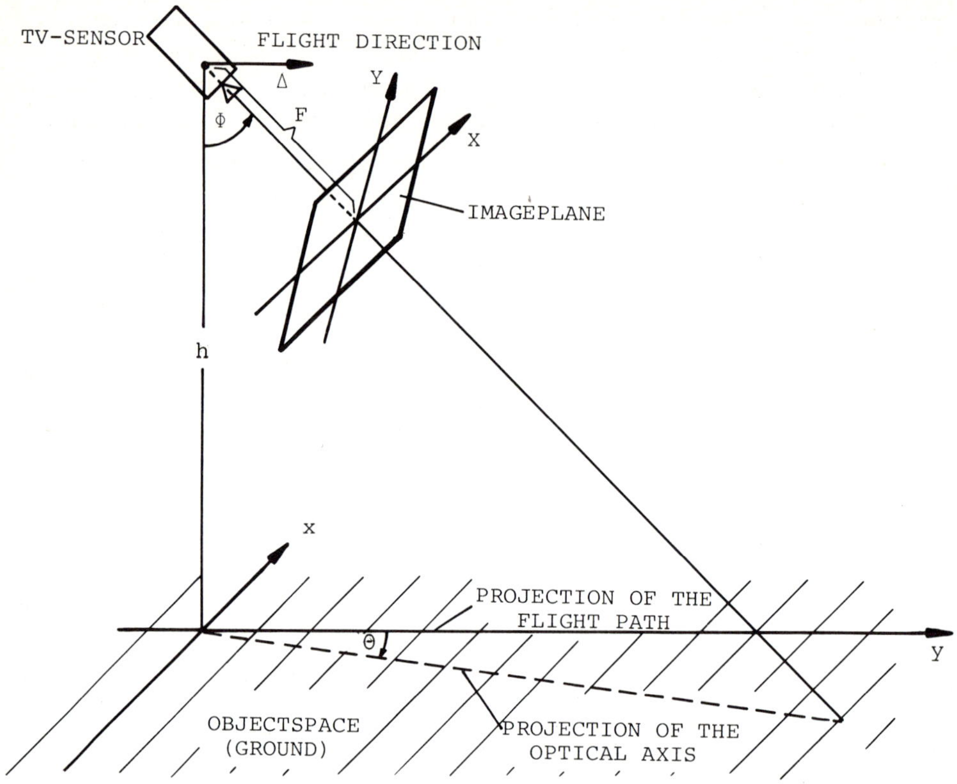

Δ : distance covered between two fields used for estimation
x,y : coordinates of the objectspace on the ground
X,Y : coordinates of the imageplane
F : focal length of the TV-sensor
h : flight height
Φ : angle of inclination between the vertical axis and the optical axis of the TV-sensor
Θ : angle of inclination between flight direction and the optical axis of the TV-sensor

Figure 4. Geometrical model for the image generation

The maximum of the quotient Y/F is a measure of the optical aperture of the TV-sensor. Thus the bilinear approximation of the background displacement is valid at least if both the angle of aperture and the angle of inclination between the optical axis of the sensor and the vertical axis are small. The evaluation of the equations 4.3 and 4.4 to calculate the background displacement requires low computational complexity because the bilinear function reduces to a linear function if one coordinate is fixed /4/.

The bilinear functions contain 8 parameters which have to be calculated from 4 background displacement vectors. Since the displacement vector field contains both vectors belonging to background and vectors belonging to moving objects 4 single estimates of the displacement vector $\vec{D}$ can not be used. Nevertheless 4 local background displacement estimates can be obtained by utilizing the assumption that the size of moving objects is small. Therefore the image is divided into 4 rectangular regions of the same size M·N. For each region the mean value $\overline{D_{x,i}}$, $\overline{D_{y,i}}$ of the two components of the vector field $\vec{D}$ is computed.

$$\overline{D_{x,i}} = \frac{1}{M \cdot N} \sum_{M_i} \sum_{N_i} D_x(X,Y) \quad , \; i=1,\ldots,4 \qquad 4.5$$

$$\overline{D_{y,i}} = \frac{1}{M \cdot N} \sum_{M_i} \sum_{N_i} D_y(X,Y) \quad , \; i=1,\ldots,4 \qquad 4.6$$

where M_i,N_i denote the i-th region. As the size of movong object is small the amount of object displacement contributing to the average is negligible. It is a property of the bilinear function that the mean value computed in a region of retangular shape represents the value of the function at the center of the region. For this reason the averages computed in equation 4.5 and 4.6 are assigned to the center of each region as local background displacement estimate. Then the 8 parameters $a_x,\ldots,d_x,a_y,\ldots,d_y$ of the bilinear representation can be determined by solving a set of 8 linear equations. Introducing these parameters into equations 4.3 and 4.4 the background displacement can be computed for every position of the imageplane.

The vector field obtained by calculating the difference between the estimated displacement $\vec{D}$ and the background displacement $\vec{D}_b$ at each picture element of the current field is the vector field of relative displacement. It delivers information if the point on the ground corresponding to a picture element is moving or not. Every field of the TV-sequence is segmented into background and moving objects using the relative displacement as a feature for segmentation. The simple segmentation procedure uses a threshold operation with a fixed threshold of 2.4 pel/frame applied to the absolute value of the relative displacement to detect relatively moving pictures elements. Neighboring picture elements which both are moving relatively to the background are then combined to a cluster if the absolute value of the difference of their relative displacements does not exceed the threshold of 2.4 pel per frame. For each moving object obtained by segmentation the mean value of the relative displacements of all picture elements within an object is determined as the relative displacement of the object.

5. EXPERIMENTAL RESULTS

Two laboratory generated TV-sequences have been used to determine the accuracy of the displacement and background displacement estimation and to show that a detection of objects moving relatively to the background is possible by evaluation of the relative displacement. The TV-sequence 1 shows a pendulum swinging from the right to the left in front of a chequerboard which rotates clockwise around the lower left corner of the frame. The rotation of the chequerboard represents the background motion. The motion of the swinging pendulum contains a component which moves relatively to the background. The TV-sequence 2 shows the rotating chequerboard without pendulum and thus represents the background motion only. For both TV-sequences a number of 20 frames has been taken up using a sampling frequency of 10MHz and has been digitized with a resolution of 8 bit per pel. Frame segments of 96 by 128 pels have been processed. Photo 1.a shows one of the processed frame segments. The spatial function of the background displacement has been determined from the motion parameters of the rotation. For the processed frame segments the absolute value of the background displacement spans a range of 1.0 pel per frame to 8.3 pel per frame.

For both sequences the background displacement $\vec{D}_b$ has been estimated and compared to the known background displacement. The standard deviation of the estimation error has been calculated for all processed frame segments. For sequence 2 showing only background motion the standard deviation amounts to 0.5 pel per frame. For sequence 1 showing background and moving object the standard deviation of the background displacement estimation error amounts to 1.85 pel per frame. The deviation increases for sequence 1 because the assumption of negligible moving object size is not fulfilled. The pendulum occupies about 10% of the processed frame size.

To separate moving objects from background a threshold corresponding to 2.4 pel per frame has been applied to the absolute value of the relative displacement estimate for sequence 1. Photo 1.b shows the result of the separation.

Photo 1.a

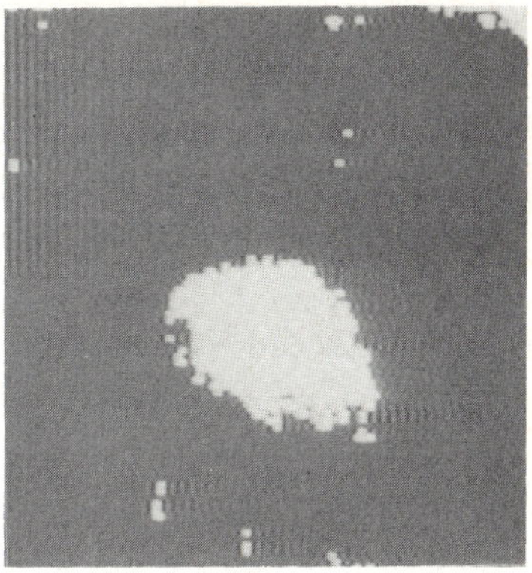

Photo 1.b

Photo 1.a Processed frame segment from the original TV-sequence 1 showing a swinging pendulum in front of a rotating chequerboard

Photo 1.b Separation of a processed frame segment into relatively moving area and background obtained by thresholding the absolute value of the relative displacement at 2.4 pel per frame, light regions of the photo representing the moving area

The area which moves relatively to the background is represented by light regions on the photo. The swinging pendulum can be recognized as a cluster of large size while the areas detected falsely have a much smaller size. It can be seen that the boundary of the pendulum is not detected correctly. This error is caused by the displacement estimator which assumes a uniform displacement in a window. The assumption is not valid at the boundary of regions moving with different displacements. The relatively high size of 2.4 pel per frame for the selected threshold is mainly required because of the errors in the background displacement estimate which are introduced by the large size of the moving object.

6. CONCLUSION

A method has been presented to estimate the relative displacement of objects in front of moving background for TV-sequences taken from an airborne vehicle. The method is based on the determination of the background displacement from the frame-to-frame displacements of the sequence. The correlation method used to estimate the frame-to-frame displacements has been described. The assumptions of a flat surface of the ground and a small size of relatively moving objects are the only limitations for the determination of the background displacement. A bilinear model of the background displacement has been used to achieve a low computational complexity.

The estimates of the relative displacement have been used to detect automatically objects moving relatively to the background. For a laboratory generated TV-sequence imitating the intended application regions moving relatively to the background have been separated from the background. To enable a detection of relatively moving objects at low amounts of relative displacement further investigations are needed to reduce the influence of the object size on the background displacement.

REFERENCES

/1/ Bohner, M. "Objektermittlung in Bildfolgen" in Erfassung und maschinelle Verarbeitung von Bilddaten ed. by H. Kazmierczak Springer Verlag, Wien 1980

/2/ Huang, T. S., Tsai, R. Y, "Motion Estimation" in Image Sequence Analysis ed. by T. S. Huang Springer Verlag, Berlin 1981

/3/ Beyer, S. "Displacement Measurement with Fast Video Correlators" in Proceedings of the Picture Coding Symposium Montreal (Canada) 1981, pp. 56-57

/4/ Castleman, K. R. Digital Image Processing Prentice Hall, Inc., Englewood Cliffs 1979

LINEAR FILTERING IN IMAGE SEQUENCES

Ulrich Boes

Deutsche Forschungs- und Versuchsanstalt fur Luft- und Raumfahrt e.V.
Insitut fur Dynamik der Flugsysteme
D-8031 Wessling/Obb.

Abstract

It is shown that linear and shift-invariant filters, acting on time and space simultaneously, can discriminate between image regions moving uniformly with different velocities, irrespective of contrast. A general scene ist formulated mathematically and its three-dimensional Fourier transform is calculated. In frequency space, the energy of a moving image region is concentrated on a plane. A filter, the frequency response of which is fitted to this plane, can detect the object and separate it from the differently moving background. Design techniques for recursive and nonrecursive filters are described, and simulation results demonstrate the ability of this method.

1. Introduction

In scenes, which are recorded by a moving imaging sensor, often an object is to be recognized and separated from the differently moving environment or background. This task has to be performed in a variety of applications, as tracking a target or controlling a robot by a TV-camera. Frequently a bad contrast deteriorates the visibilities, so that only velocity information can be used. This problem is often approached by estimating the displacement and thus the velocity of pixels or image regions from one frame to the next. Here, on the contrary, it is attempted to incorporate the dynamics of the whole process by using a model for the two-dimensional motion. A scene with one object and a background, both moving differently, is to represent a general case. A mathematical model is

NATO ASI Series, Vol. F2
Image Sequence Processing and Dynamic Scene Analysis
Edited by T.S. Huang

developed and the effects of motion and relative motion on the three-dimensional Fourier spectrum is investigated.

Differently moving image regions can be discriminated by linear and shift-invariant, spatio-temporal filters irrespective of contrast, if they move uniformly. All other temporal changes are neglected. No assumptions are necessary concerning the textures and shapes of image regions, especially object and background may have identical textures. The filter is matched to a specific velocity and will respond maximally at the site of an image region moving with this velocity, independent of spatial frequencies. A filter is to be represented by its discrete unit sample response or, correspondingly, by its z-transform. The presupposed knowledge of the velocity may be dropped, if the filter is adapted. It will be outlined, how to manage this.

Two-dimensional filters are treated as a special case, which is to illustrate the proposed method more obviously. In practice, two two-dimensional filters may be paralleled in order to process the projections of image rows and columns separately. This might be advantageous, for only translatory motion can occur, and less time is needed for the filtering.

The ability of discriminating between differently moving image regions is demonstrated by simulation results. Artificially generated input scenes were used, where object and background have identical intensity distributions. As it was expected, the filter responds maximally to the object, which moves with the velocity the filter is matched to, showing a resonance phenomenon. These results and the possibility of an exact mathematical analysis of the system's behaviour seem to convince of this linear filtering approach.

2. A Mathematical Formulation of Dynamic Scenes

By a moving imaging sensor, the surroundings are mapped onto a planar, time-dependent intensity distribution s(x,y,t), which is defined to be the dynamic scene. It may have a Fourier transform S(u,v,f), where u and v are spatial frequencies and f is the temporal frequency. For the sake of a manageable mathematical analysis, a general scene is to be represented by a scene containing one object and the background, both moving uniformly with different velocities. The textures of both object and background are arbitrary. All variables in the following derivation will refer to a coordinate system fixed to the sensor.

The textural function of the object, centered at the origin, be $m(x,y)$. It is a spatially limited function and assumed to be square-integrable, which guarantees the existence of its two-dimensional, spatial Fourier transform $M_0(u,v)$. The textural function of the background $b(x,y)$ is extended over the entire space and also assumed to be square-integrable; it has the two-dimensional Fourier transform $B_0(u,v)$. Following [1], a mask function $\lambda(x,y)$ is introduced, which is equal to one where the object exists in the scene and zero elsewhere. It moves in the same manner as the object and has the Fourier transform $\Lambda_0(u,v)$. If (c_x,c_y) and (d_x,d_y) are the velocity vectors of object and background, respectively, the total scene is modeled as follows:

$$s(x,y,t) = m(x-x_0-c_xt,y-y_0-c_yt) + \left[1-\lambda(x-x_0-c_xt,y-y_0-c_yt)\right] \cdot b(x-d_xt,y-d_yt), \tag{1}$$

where x_0 and y_0 denote the position of the object's centre at $t = 0$.

Now, the three-dimensional Fourier spectrum of (1) is to be calculated. The transforms of each term are given first. Then they will be combined to yield the total spectrum. The three-dimensional transform of the moving object is:

$$M(u,v,f) = M_0(u,v)\cdot e^{-2\Pi i(ux_0+vy_0)}\cdot\delta(c_xu+c_yv+f). \tag{2}$$

The phase factor is caused by the initial position. Due to the delta-function, spectral values exist only on the velocity-plane

$$c_xu+c_yv+f = 0 \tag{3}$$

in the frequency space spanned by u,v and f. Thus, the total energy of the object is concentrated on this plane. Similarly, the transform of the mask function is:

$$\Lambda(u,v,f) = \Lambda_0(u,v)\cdot e^{-2\Pi i(ux_0+vy_0)}\cdot\delta(c_x u+c_y v+f). \tag{4}$$

The Fourier spectrum of the background, finally, exists only on a second velocity-plane:

$$B(u,v,f) = B_0(u,v)\cdot\delta(d_x u+d_y v+f). \tag{5}$$

Since a product in the original domain is equivalent to a convolution in the frequency domain, the total transform of (1) is:

$$S(u,v,f) = M(u,v,f)+B(u,v,f)-\Lambda(u,v,f)*B(u,v,f). \tag{6}$$

The sign * denotes the convolution operation irrespective of the dimension.

As a special case, spatially one-dimensional signals are regarded. The spectrum results from (6) for - say - $v = 0$ and depends therefore on the spatial frequency u and the temporal frequency f:

$$\begin{aligned} S(u,f) = {} & M_0(u)\cdot e^{-2\Pi iux_0}\cdot\delta(cu+f)+B_0(u)\cdot\delta(du+f)- \\ & -e^{2\pi i\frac{x_0}{c-d}(du+f)}\cdot B_0\left(\frac{cu+f}{c-d}\right)\cdot\frac{\sin\left[2\Pi\frac{\ell}{c-d}(du+f)\right]}{\Pi(du+f)}, \end{aligned} \tag{7}$$

where c and d are the velocities of object and background, respectively (the index x is omitted), and 2ℓ is the size of the object.

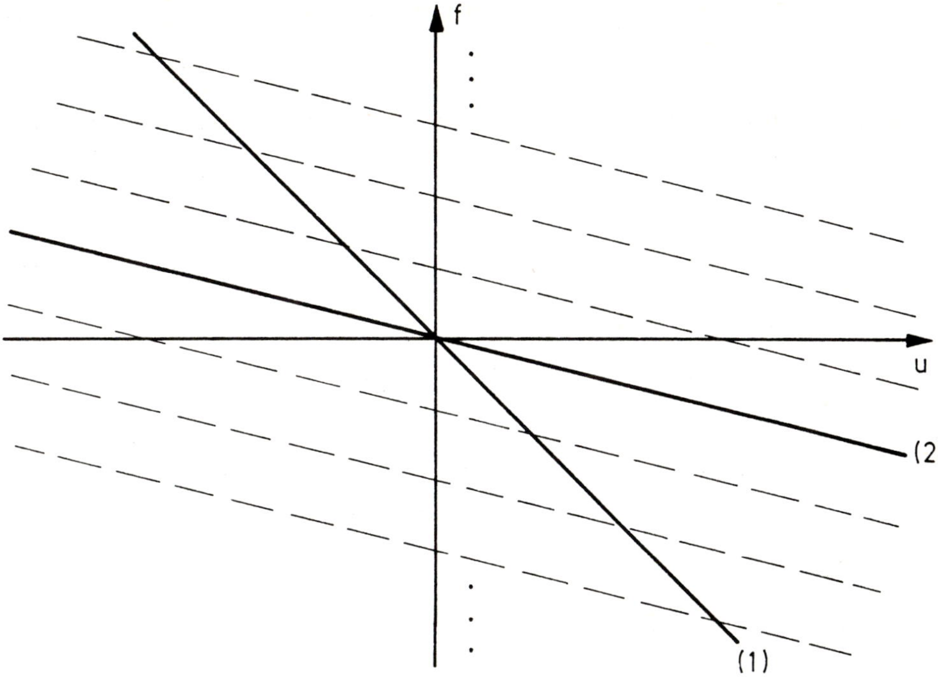

Fig.1
Contour lines of the Fourier spectrum of the representative scene. f is the temporal, u the spatial frequency. (1) velocity-line of the object, (2) that of the background. The dashed lines, parallel to (2), are caused by relative motion.

This spectrum is characterized by a variety of straight lines, as depicted in fig. 1. It is equal to zero on the velocity -line of the background and on lines parallel to it, which are caused by the relative velocity. Their intercepts are equal to the ratio of the relative velocity and the size of the object. The major part of the spectral energy due to relative motion is concentrated in the region between the two lines nearest to the background's velocity line. The energy of the object is concentrated on another line, which is clearly discernible in the total spectrum. Similar relations hold in the three-dimensional case, but here planes instead of straight lines arise.

3. On the Design of Velocity-Matched Filters

Considering the results of the previous section, it may be inferred that image regions with different velocities can be discriminated by linear spatio-temporal filters, the frequency response of which is fitted to one velocity-plane. That is, the filter must suppress all frequencies except those lying on the velocity-plane. In this section, two techniques are outlined, to design both recursive and nonrecursive, velocity-matched, digital filters in the frequency domain. The non-recursive filters will have zero phase with respect to the spatial coordinates and linear phase temporally.

The pass-region of a filter is defined to be a region in the multidimensional frequency space, inside of which a frequency response is greater than $1/\sqrt{2}$. A velocity-matched filter should have a pass-region on a plane through the origin and perhaps in a small neighbourhood of it, in order to tolerate small deviations from the design velocity. Yet, the origin must not belong to the pass-region.

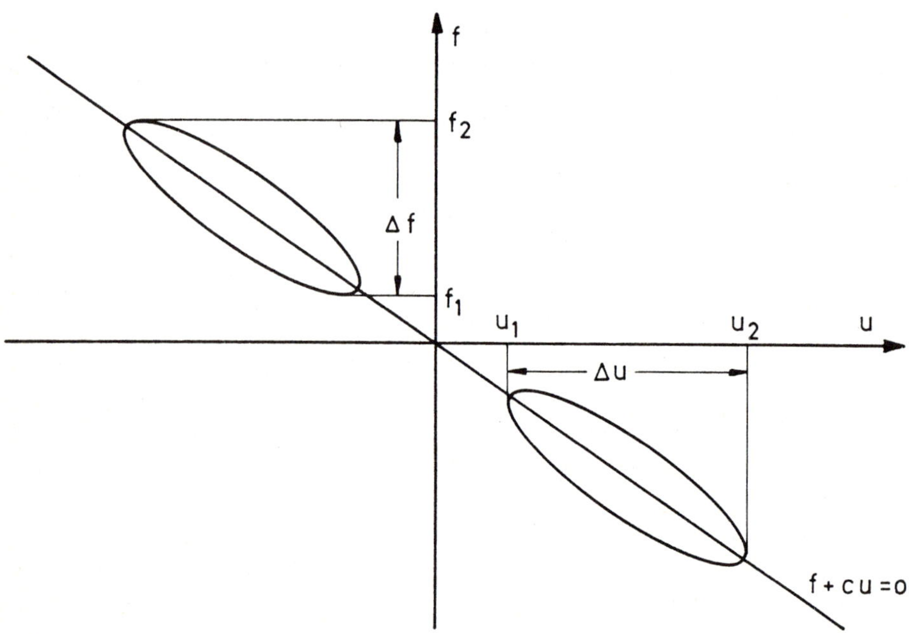

Fig. 2
Specification of the frequency response of a velocity-matched filter.

In two dimensions, a possible pass-region is depicted in fig. 2 as two ellipses whose major axes lie on the velocity-line. An elliptical pass-region favours the design. Two pass-regions symmetrical to a line perpendicular to the velocity-line are necessary in order to obtain a system function with real coefficients [2]. In three dimensions, the pass-region may consist of two oblate spheroids, the major axes of which lie on the velocity-plane.

According to methods known in two-dimensional signal processing [2], a recursive filter is designed as follows. A one-dimensional, continuous lowpass filter $H_0(u)$ is rotated onto the velocity-plane and hereafter multiplied by a bandpass filter $H_1(v,f)$. A pass-region like a distorted ellipsoid is yielded. Finally, the digital filter is obtained via a multiple bilinear transformation. Some drawbacks have to be noted. First, the multiple bilinear transformation causes a distortion of the frequency response at higher frequencies. Second, the stability of the filter is mainly determined by the rotation and therefore by the sign of the velocity. The above outlined technique will yield a stable filter only for positive velocities. Also, this recursive filter has a nonlinear phase, the effect of which on the input spectrum is hardly to estimate.

These drawbacks shall be avoided by designing a nonrecursive filter. In two dimensions, an ideal frequency response is defined according to fig. 2: it is equal to one in the interior of the ellipses and zero in the exterior. The coefficients of the filter are then obtained by minimizing a mean-square error criterion, which was mainly chosen in order to facilitate the three-dimensional approximation. The unit sample response is given by:

$$h(\nu,\kappa) = 1/c \iint_{\mathbb{D}} \cos(2\Pi\nu u) \cos(2\Pi\kappa f) \; du \, df, \qquad \nu = 0,\dots,N; \quad \kappa = 0,\dots,K. \tag{8}$$

$\mathbb{D}$ is the interior of both ellipses of fig. 2, and c is a constant. The integral can be evaluated analytically. Each coefficient is independent of each other and of N and K, the number of coefficients. The three-dimensional approximation yields a similar, but more complex relation for the coefficients of the unit sample response.

4. Simulation Results

In this section, first simulation results are presented, which illustrate the theory and demonstrate the ability of the proposed filters to discriminate between differently moving image regions. The scenes were artificially generated, using a gaussian distribution; object and background possess identical statistics. Fig. 3 shows a line of an input scene for a fixed instant. It is clearly seen that the object cannot be recognized in one frame only. Object and background move uniformly, but with different velocities.

First, a two-dimensional, recursive filter was designed. A first order lowpass filter was rotated and multiplied by a temporal bandpass filter of second order. The resulting digital filter has a region of support of four pixels in the temporal direction and two pixels in the spatial direction. It was designed for a velocity of one pixel per sampling period.

Fig. 4 and 5 show the response of the filter to a scene consisting of one object and the background, both moving with different velocities; the object moves with the design velocity. Fig. 4 shows the temporal response at one fixed spatial position. After the system was turned on, the output increases and decays hereafter. At the site of the moving object the output oscillates strongly and decays quickly to approximately zero again, after the object has passed. Fig. 5 shows the output at a fixed instant, after the transient response has decayed completely.

Various simulations were performed with different sizes of the object and also different statistical distributions. The results demonstrated that the output of the filter is not influenced by these factors.

As in the two-dimensional case, a three-dimensional filter was designed. It has a region of support of four pixels in the temporal and in the horizontal spatial direction and two pixels in the vertical spatial direction; the number of coefficients amounts to 64. Fig. 6 shows the response to an input scene, which was generated in the same manner as the image line of Fig. 3. Here, the object - not distinguishable from the background - was a circular disc of radius five pixels. At the site of the object, the filter responds clearly maximally.

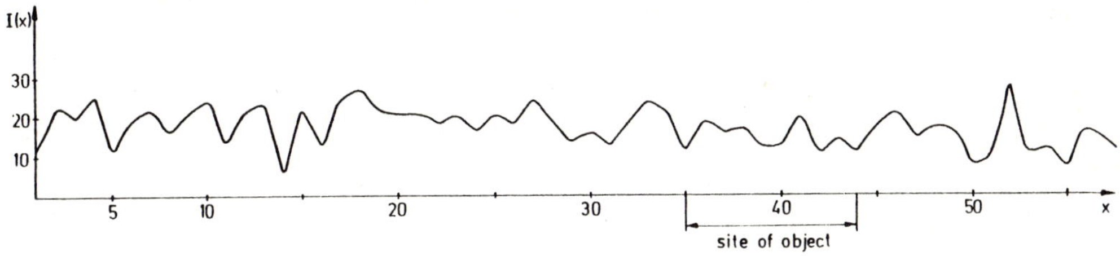

Fig. 3
Spatially one-dimensional input scene; the intensity is depicted versus the spatial coordinate x for a fixed instant. Object and background are gaussian distributed with equal mean and variance. Both move uniformly, with velocities v_0 and v_B, respectively.

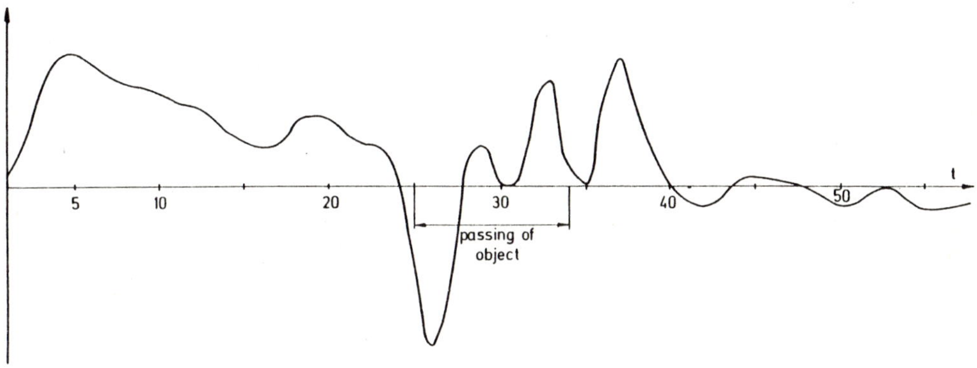

Fig. 4
Transient response of the filter at a fixed spatial position to the scene of fig. 3. The passing of the object in the input scene is marked.

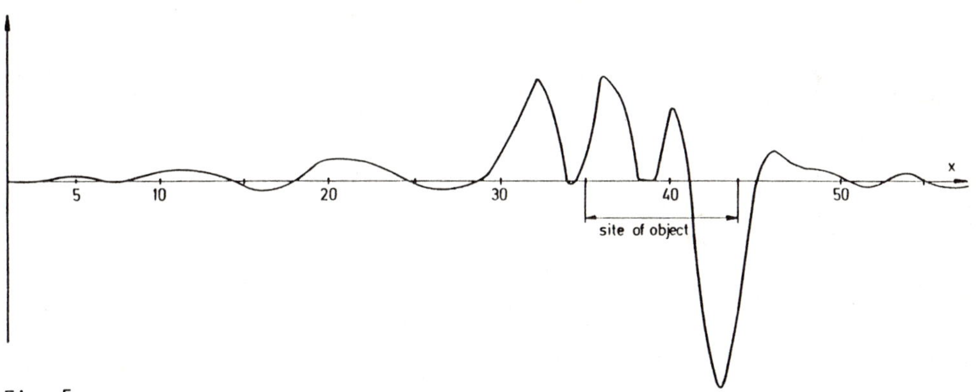

Fig. 5
Response of the filter to the same scene as in fig. 4, plotted versus the spatial coordinate x. The site of the object in the corresponding input scene is marked.

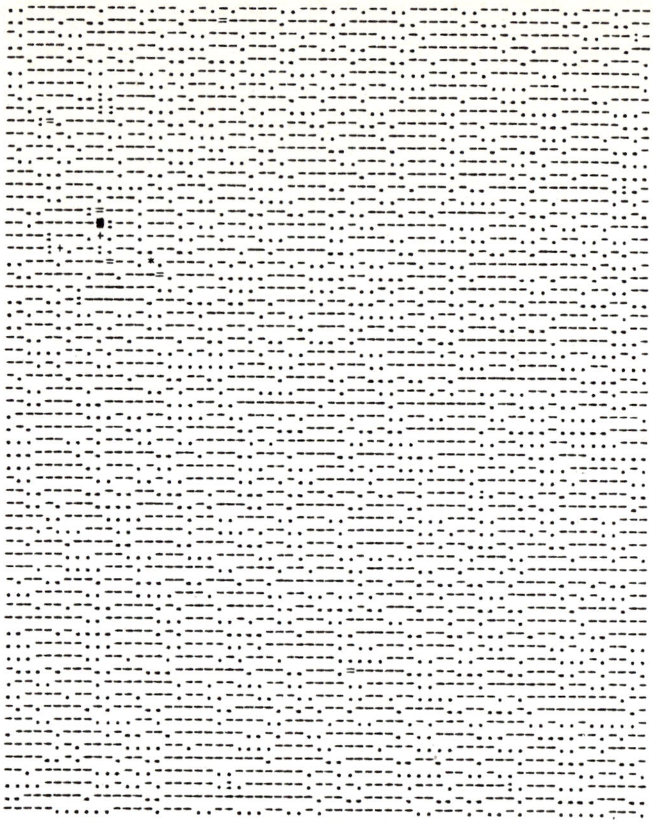

Fig.6
Response of the filter to a spatially two-dimensional input scene, one line of which is depicted in fig. 3.

The maximum of the output is about three or four pixels behind the object. This can be seen clearly in fig. 5. Yet, this delay is considered small. The slowly decaying transient response seems to be more important; the filter output takes about forty sampling periods until it has decayed completely, if a stationary image has been presented. It is expected, however, that this behaviour could be considerably improved using a FIR filter with linear phase.

5. Conclusions

The ability of linear, shift-invariant, multidimensional filters to recognize and discriminate image regions due to their velocity was demonstrated. The basic assumption was uniform motion. The intention was to recognize the moving object. The simulation results showed that a velocity-matched filter responds maximally at the sites of image regions moving with the velocity the filter is designed for. Segmentation of the image, that is, separation of the moving object from the also moving background may be completed by simply thresholding the output of the filter.

It may be regarded as a drawback that the filters proposed here are fixed to a velocity, which must be specified prior to the design. In reality, a system often has to deal with changing or unknown velocities. Small changes can be tolerated due to the pass-region defined in section three (cf. fig. 2). But changing velocities or accelerated movements during a longer period of time require an adaptation of the filter's coefficients. If the velocity of an object is slightly different from the design velocity, the transmission bandwidth reduces (Δf and Δu in fig. 2), and the output decreases, too. To make the filter follow the velocity can therefore be accomplished by maximizing a local output energy. The auxiliary condition is that the frequency response of the filter must always lie on the velocity-plane, that is, it has to be rotated in the frequency space. The adaptation procedure is facilitated by the known dependence of the filter's coefficients on the velocity. Also, determination of unknown velocities might be possible with an adaptive, velocity-matched filter.

References

[1] R.J. Schalkoff, and E.S. McVey:
'A Model and Tracking Algorithm for a Class of Video Targets',
IEEE Transactions on Pattern Analysis and Machine Intelligence, Vol. PAMI-4, No. 1, Jan. 1982.

[2] R.R. Read, J.L. Shanks, and S. Treitel:
'Two-Dimensional Recursive Filtering',
in: T.S. Huang (ed.): 'Picture Processing and Digital Filtering',
2nd ed., Topics in Applied Physics, Vol. 6 (Springer, Berlin, Heidelberg, New York 1979).

PHOTOMETRIC STEREO FOR MOVING OBJECTS

H. WESTPHAL

Fachbereich Informatik
Universität Hamburg
Schlüterstraße 70
D-2000 Hamburg 13

ABSTRACT

A method for inferring local properties of surfaces (orientation, albedo) of moving rigid objects from image sequences by exploiting image formation knowledge is presented. Two representations of the reflectance characteristics of surfaces - the reflectance map and a parameterized analytic function - are compared. An improved photometric stereo method allowing an unknown, locally varying diffuse fraction of the illumination is described.

1. INTRODUCTION

In a sequence of images moving objects can often be seen from different vantage points. This gives rise to special 3D-modelling capabilities. One of the few approaches so far available selects prominent points and establishes correspondences between their projections in the images of the sequence [Dreschler 81, Dreschler&Nagel 81a]. When these 2D-projections are known, the trajectory of the moving object and the 3D-coordinates of the points in an object-centered coordinate system can be calculated [Nagel 80]. The convex hull of these points is a polyhedral 3D-model of the moving object [Dreschler&Nagel 81b, Dreschler&Nagel 82].

NATO ASI Series, Vol. F2
Image Sequence Processing and Dynamic Scene Analysis
Edited by T.S. Huang

In the present paper, we propose a method to improve this polyhedral description by replacing the plane surfaces of the convex hull by curved ones. In this way, the intensity distribution off the prominent points can be described more accurately. For this purpose we additionally need to integrate into the model the locally changing reflectance of the surface.

An approach will be described that aims to infer the local orientation and reflectance of the moving object's surfaces from the image sequence by exploiting image formation knowledge.

2. *ASSUMPTIONS*

Several assumptions are necessary:

1. The moving object is rigid and opaque.

2. The trajectory of the moving object is known.
   [Neumann 80] points out that the grayvalues of a rotating object do not supply enough constraints to infer the orientation of a surface patch if no additional knowledge (e.g. the rotation angles between the images) is given.

3. The illumination of the scene is known and constant over the entire path of the object; i.e. neither the intensity of the illumination changes while the object is in motion (e.g. as a result of shadows) nor does its direction. If we assume a scene illuminated by the sun without any other objects close to the moving object's path, this assumption is reasonable.

4. The photometric properties of the surfaces are known.

5. The projections of fixed points on the surface of the object are approximately predicted by the convex hull model and the trajectory.

The last assumption provides the intensities of a patch on the surface under different, known illumination directions. This is the basic idea of photometric stereo.

3. PHOTOMETRIC STEREO

Photometric stereo is a method for inferring local properties of a surface such as its orientation or albedo by exploiting image formation knowledge for the evaluation of measured grayvalues. Mainly a known directional illumination distribution and reflectance characteristics of the surface are taken into account.

Measuring only one grayvalue for a pixel is insufficient to determine the various parameters involved in image formation. It is therefore necessary to add constraints derived either from the local neighborhood of the point under consideration [Horn 77, Woodham 78] or from images taken under different illuminations [Horn et al. 78, Ikeuchi&Horn 79, Coleman&Jain 82]. Changing illuminations can also occur in scenes with moving objects [Westphal&Nagel 81].

4. THE REFLECTANCE MAP

A crucial problem of all these methods is the knowledge assumed about the reflectance function and the illumination distribution. Horn introduced the reflectance map for this purpose [Horn&Sjoberg 78]. This is a table which maps surface orientation into intensity given the knowledge about the illumination and the reflectance characteristics. It is either measured in a calibration step for each illumination distribution or calculated from a model.

The reflectance map has the advantage of allowing almost any kind of illumination distribution and reflectance characteristics. Even locally changing albedo values (fraction of total reflected light and illumination) can be determined using multiple illuminations and the corresponding reflectance maps [Horn et al. 78].

5. THE ANALYTIC REFLECTANCE FUNCTION

One disadvantage of the reflectance map is its inflexibility. It is set up for one special illumination distribution. If it were a parameterized function it would be no problem to adapt it to new light conditions. Even if some of the illumination parameters are unknown, they can be determined if enough data are available from redundant images. An unparameterized reflectance map in the form of a table is not a proper tool for this task.

Furthermore, because of errors introduced by quantization and noise, one can be forced to use more than the minimal necessary number of light directions and fit a model to these overdetermining data. If we use some parameterized analytic function instead of a table to describe the reflectance characteristics, we can fit the parameters of this function to the measured data with known methods such as least-squared error approximation.

A 'Lambertian' surface is a totally diffuse reflector, i.e. the incident radiation is scattered uniformly over the hemisphere above the surface. This results in a cosine dependency between the emitted intensity and the incidence angle. The cosine can be expressed by a dot product of two unit vectors. Additionally we assume a fraction (1-a) to be absorbed by the surface, where a is called the albedo. Then the intensity sensed by the camera can be described analytically by

$$I = a \cdot \vec{L} * \vec{N},$$

where I is the measured intensity (range 0 to 1), a is the albedo, $\vec{L}$ is the unit vector in the direction of the light source (which is assumed to have unit radiance), $\vec{N}$ is the unit normal of the surface, and '*' denotes the vector dot product. This reflectance function is used, for example, by [Woodham 80] or [Coleman&Jain 82].

6. *A REFLECTANCE FUNCTION FOR DIFFUSE ILLUMINATION*

The disadvantage of this approach is the very constrained illumination model used. There must be a point light source very far away without any diffuse illumination component. But we always have some fraction of diffuse light, especially if the scene is not totally under control, as in outdoor scenes. Additionally, there is a kind of diffuse light due to mutual illumination if the object is not convex.

For all these reasons, a reflectance model is proposed using an analytic reflectance function and considering an unknown, locally varying diffuse fraction of illumination d:

$$I = a \cdot (\vec{L} * \vec{N} + d).$$

In contrast to [Westphal&Nagel 81] the diffuse fraction is not given but is an additional unknown.

As the object rotates while in motion, the normal $\vec{N}$ measured in an object-centered coordinate system is rotated by a time-dependent known rotation matrix R_t. We thus get for every time step t an equation of the form

$$I_t = a \cdot (\vec{L} * (\vec{N} * R_t) + d).$$

Small rearrangements show that the problem is formally equivalent to the photometric stereo with fixed object-camera relation and changing illumination direction:

$$I_t = a \cdot ((\vec{L} * R_t^{-1}) * \vec{N} + d)$$

or

$$I_t = a \cdot (\vec{L}_t * \vec{N} + d).$$

This model is not linear. We introduce two new variables $\vec{N}' = a.\vec{N}$ and $d' = a.d$:

$$I_t = \vec{L}_t * \vec{N}' + d'.$$

This makes the equation linear, but the dependencies between $\vec{N}'$ and d' are neglected.

There are 4 unknowns (d' and the 3 components of $\vec{N}'$), so we need 4 intensity values for every point measured under 4 different light directions $\vec{L}_t$. The result is a system of 4 nonhomogeneous linear equations. The length of $\vec{N}'$ is the albedo a (because of $\vec{N}' = a.\vec{N}$) and d' divided by a is the diffuse fraction d.

7. EXISTENCE OF A UNIQUE SOLUTION

A unique solution to the system of equations exists if the determinant of the coefficient matrix does not vanish, i.e.:

$$\det \begin{vmatrix} L_{1x} & L_{1y} & L_{1z} & 1 \\ L_{2x} & L_{2y} & L_{2z} & 1 \\ L_{3x} & L_{3y} & L_{3z} & 1 \\ L_{4x} & L_{4y} & L_{4z} & 1 \end{vmatrix} \neq 0.$$

The first 3 row vectors can always be made linearly independent by choosing 3 non-coplanar light directions L_1, L_2, and L_3. In the case of a rotating object, this is true if the rotation axis is not perpendicular or parallel to the light direction.

But the fourth light direction L_4 must be dependent on the others in 3-space. So let

$$L_4 = \sum_{i=1}^{3} \mu_i \; L_i .$$

The row vectors have four components. The fourth row vector is a linear combination of the first three iff the sum of the three μ_i is 1 (look at the fourth component). Thus we can choose any unit vector L_4 as long as the sum of the μ_i is not 1, e.g.

$$\begin{aligned} L_1 &= (1\ 0\ 0), \\ L_2 &= (0\ 1\ 0), \\ L_3 &= (0\ 0\ 1), \\ L_4 &= 1/3\ (1\ 1\ 1). \end{aligned}$$

So if the direction of the sun and the rotational component of the movement fulfill the condition of the independence of the equations, we can calculate the local normal and the local albedo of many points between the prominent points and fit a curved surface to these data. This can be done iteratively, introducing more points if the fitting error is above some threshold.

8. IMPLEMENTATION

The photometric stereo algorithm has been implemented and tested on artificial data. It produces exactly correct results as long as the grayvalues used as input are real numbers. But the roundoff errors introduced by quantization into 256 graylevels cause errors of a few degrees in the orientation of the surface normal to be determined. Neglecting the dependencies between $\vec{N}'$ and d' for the linearization results in a good approximation, but it should be improved.

To cope with additional errors due to noise we have to use more data and perform a least-squared error approximation to this overdetermined system. The analysis of moving objects usually supplies many more than only four images. The additional problem of correspondence, in particular a method for tackling the problems induced by the errors due to the approximation by the convex hull, is treated in [Westphal&Nagel 81].

For the minimization task we cannot neglect the dependencies between $\vec{N}'$ and d' and have to apply an iterative scheme to the nonlinear problem. The actual implementation uses the Levenberg-Marquardt Algorithm [Marquardt 63]. The problem of finding a good starting value is solved by using the result of the linear method.

9. DISCUSSION

For shape calculations in scenes which are not totally controlled the parameterized, analytic reflectance function has some advantages compared to the reflectance map. The main advantages are its flexibility to adapt to illuminations not known in advance and the possibility of minimizing errors in an overdetermined case. The method described overcomes the very restricted illuminations that were used so far for reflectance functions by allowing a part of the illumination to be diffuse.

The use of all images of a sequence to determine local properties of a surface has the advantage of supplying a global interpretation (in a temporal sense) of the whole image sequence rather than only of a pair of consecutive images. This makes it possible for the approach to be more robust and accurate.

Many methods for the analysis of time-varying imagery (e.g. [Schalkoff&McVey 79]) assume that the intensity of a surface patch fixed to the moving object does not change. But the approach discussed here exploits the changes that do occur.

The photometric approach works best in regions with little local grayvalue changes. In these regions, errors in the convex hull approximation, which cause errors in the assumed projections of a hypothesized fixed point on the surface (cf. assumption 5 of paragraph 2), do not result in major errors in the intensities fed into the linear equations.

This approach is complementary to the feature based method used for the prominent points.

The main limitation of this approach consists in the restrictive assumptions (cf. paragraph 2) that must be made.

10. ACKNOWLEDGEMENTS

I wish to thank H.-H. Nagel and B. Neumann for many helpful discussions on this topic.

11. REFERENCES

Coleman & Jain 82
E.N. Coleman and R. Jain: Obtaining 3-dimensional shape of textured and specular surfaces using four-source photometry, Computer Graphics and Image Processing 18, pp. 309-328 (1982).

Dreschler 81
L. Dreschler: Ermittlung markanter Punkte auf den Bildern bewegter Objekte und Berechnung einer 3D-Beschreibung auf dieser Grundlage, Dissertation, Fachbereich Informatik, Universität Hamburg, 1981.

Dreschler & Nagel 81a
L. Dreschler and H.-H. Nagel: On the frame-to-frame correspondence between greyvalue characteristics in the images of moving objects, 5. GI-Fachtagung GWAI-81 (German Workshop on Artificial Intelligence) Bad Honnef, Jan. 26-30, 1981, in J. Siekmann (ed.), Springer Verlag Berlin-Heidelberg-New York, series Informatik Fachberichte, pp. 18-29.

Dreschler & Nagel 81b
L. Dreschler and H.-H. Nagel: Volumetric model and 3D-trajectory of a moving car derived from monocular TV-frame sequences of a street scene, IJCAI 1981, Vancouver, Canada, pp. 692-697.

Dreschler & Nagel 82
L. Dreschler and H.-H. Nagel: Volumetric model and 3D-trajectory of a moving car derived from monocular TV-frame sequences of a street scene, Mitteilung 90, Fachbereich Informatik, Universität Hamburg (September 1981), to appear in Computer Graphics and Image Processing 1982.

Horn 77
B.K.P. Horn: Understanding image intensities, Artificial Intelligence 8 (1977), pp. 201-231.

Horn et al. 78
B.K.P. Horn, R. Woodham, W.M. Silver: Determining shape and reflectance using multiple images, M.I.T. A.I.Memo 490, August 1978.

Horn & Sjoberg 78
B.K.P. Horn, R.W. Sjoberg: Calculating the reflectance map, M.I.T. A.I.Memo 498, Oct. 1978.

Ikeuchi & Horn 79
K.Ikeuchi and B.K.P. Horn: An application to the photometric stereo method, IJCAI-79, pp. 413-415.

Marquard 63
D.W. Marquardt: An algorithm for least-squares estimation of nonlinear parameters, SIAM J. Appl. Math. 11 (1963), pp. 431-441.

Nagel 80
H.-H. Nagel: From digital picture processing to image analysis, Int. Conf. on Image Analysis, Pavia, Oct. 22-24, 1980, pp. 27-40.

Neumann 80
B. Neumann: Exploiting image formation knowledge for motion analysis, IEEE-PAMI-2 (1980), pp. 550-554.

Schalkoff & McVey 79
J.R. Schalkoff and E. McVey: Algorithm development for real-time automatic video tracking systems, Proc. 3rd Int. Computer Software and Applications Conf., Chicago/IL, Nov. 1979, pp. 504-511.

Westphal & Nagel 81
H. Westphal and H.-H. Nagel: Zur Ermittlung gekrümmter Oberflächen bewegter Objekte aus Bildfolgen, in B. Radig (ed.): Modelle und Strukturen, 4. DAGM Symposium, Hamburg, Oct. 6-8, 1981, Informatik-Fachberichte 49, Springer-Verlag Berlin-Heidelberg-New York 1981, pp. 123-129.

Woodham 78
R.J. Woodham: Reflectance map techniques for analyzing surface defects in metal castings, AI-TR-457, M.I.T., June 1978.

Woodham 80
R.J. Woodham: Photometric method for determining surface orientation from multiple images, Optical Engineering 19 (1980), pp. 139-144.

On the Selection of Critical Points and Local Curvature Extrema of Region Boundaries for Interframe Matching

L.S. Dreschler and H.-H. Nagel
Fachbereich Informatik
Universität Hamburg
Schlüterstr. 70
D-2000 Hamburg 13

1.0 ABSTRACT

Points of high boundary curvature can be used as critical points for displacement estimation in interframe matching. The mathematical relations between the two-stage interest operator of DRESCHLER+NAGEL 81, the zero-crossing contours of the "Mexican Hat"-operator of MARR+HILDRETH 80 and the zero-lines of the gaussian curvature are discussed.

2.0 INTRODUCTION

Local disparity measurement is an important problem in stereo-vision as well as in motion analysis. There are several approaches suggesting small local image features like corners or isolated spots as basic elements for disparity measurements [BARNARD+THOMPSON 80, DRESCHLER+NAGEL 81, HANNAH 80, KITCHEN+ROSENFELD 80, MORAVEC 77+80+81]. We shall call these small features critical points and algorithms to select critical points interest operators.

NATO ASI Series, Vol. F2
Image Sequence Processing and Dynamic Scene Analysis
Edited by T.S. Huang

The interest operators mentioned above do not use image segmentation but detect critical points by local evaluation of the image intensity function. On one hand, this is an advantage because segmentation errors may occur especially in the interesting area around critical points. On the other hand global knowledge about the image components sometimes might be helpful to interprete more complex image features.

Fig. 1 represents one frame of a sequence recorded and buffered in real-time from a street scene, using an analog video disc. Subsequently, these TV-frames have been digitized and processed without real-time constraints. Fig. 2 shows the locations of critical points selected by the two-stage interest operator of DRESCHLER+NAGEL 81. Although the set of selected points is satisfying compared to the results of other local interest operators, there are visually prominent points which seem to be difficult to recognize by local operators, e.g. points where three or more regions are connected or endpoints of very thin lines.

We therefore investigate alternative means to characterize visually prominent points.

3.0 CRITICAL POINTS

3.1 "Greyvalue Corners"

KITCHEN+ROSENFELD 80 developed an interest operator which detects points with significant changes of gradient direction along a curve of local maximum gradient.

This characterization can be formulated mathematically in the following way. Consider a local coordinate system at the position of a "greyvalue corner". It should be aligned with the directions of principle curvature for the greyvalue distribution. Let I_x and I_y denote the gradient components and I_{xx}, I_{xy}, I_{yy} the second order partial derivatives of the greyvalue distribution. Then we may assume $I_{xy} = 0$ due to the coordinate axes being aligned with the principle curvature directions. A maximum in the gradient requires that the gradient is aligned with a principle curvature direction, for example I_x = maximum, $I_y = 0$, $I_{xx} = 0$. The premise of a maximal rate of change in the gradient direction has the consequence that I_{yy} should be a local maximum. A rigorous derivation of these conditions can be found in NAGEL 82.

It has been shown by NAGEL 82 that the concept of KITCHEN+ROSENFELD 80 is equivalent to one of the methods investigated by DRESCHLER+NAGEL 81 for the localization of a subclass of critical points, namely "greyvalue corners".

3.2 Corner Points On Edges

In images of high contrast and with significant region boundaries, corners may be detected as points of high boundary line curvature. Boundary points of high curvature have often been used as breakpoints for the piecewise approximation (by straight lines, arcs,..) of boundary contours [DUDA+HART 73, PAVLIDIS 77]. YAM+DAVIS 81 use an interest operator which selects areas of high edge-line curvature from a binary edge picture as "control points" for image registration. The binary edge is derived by a simplified version of an algorithm described by MILGRAM 79 in combination with the "Mexican Hat"-operator of MARR+HILDRETH 80.

Fig. 3 shows lines of constant greyvalue for the car image selected from Fig. 1. Fig. 4 shows zero-crossings of the "Mexican Hat"-operator of MARR+HILDRETH 80 reflecting very well the boundary of the car image. Since these zero-crossing contours form closed lines and are not too ragged it is easy to detect points of high boundary curvature by the algorithm of PAVLIDIS 77 [p. 168]. Local extrema of the curvature for the boundary line are marked by short lines perpendicular to the boundary. The zero-crossing contours have been computed using a circular gaussian filter with a bandwidth corresponding to a standard deviation of one pixel in a 5x5 pixel operator.

One problem with the "Mexican Hat"-operator is that zero-crossing contours are not only detected at edges but also inside almost homogeneous regions. This can easily be corrected by suppression of zero-crossings with low gradient.

Fig. 4 shows also the locations of the critical points from Fig. 2, which have been selected by the two-stage interest operator of DRESCHLER+NAGEL 81. Several critical points are very close to a point of strong curvature in the zero-crossing contour, but they are never exactly in the same position. In fact the critical (corner) points cannot lie exactly on the zero crossing contour because the interest operator is choosing corner points at the zero-crossing between a maximum and a minimum of the <u>gaussian</u> <u>curvature</u> of the image intensity function. Near a corner point, zero-crossings of the "Mexican Hat"-operator cannot coincide with zero-crossings of the gaussian curvature. The "Mexican Hat"-operator is a convolution of the Laplacian and a gaussian filter. Both Laplacian and gaussian curvature can be expressed in terms of the two principle curvatures σ^{+} and σ^{-}. The Laplacian is the sum of σ^+ and σ^- and the gaussian curvature the product of σ^+ and σ^-. So zero values of the Laplacian and the gaussian curvature can only coincide in points where both principle curvatures are zero. Near corner points, however, at least one principle curvature must be different from zero or there would be no corner.

Along straight edges this is different. The principle curvature in direction of the edge vanishes identically while the principle curvature in gradient direction has high values of different sign on both sides of the edge and is crossing zero at the point of steepest gradient. So along straight edges the zero-crossings of the Laplacian are zero-crossings of one principle curvature. In these points the gaussian curvature is zero, too.

3.3 Zero-crossing Contours Of A Gaussian-Filtered Laplacian

Without a well-founded mathematical concept for greyvalue characteristics which are visually recognized as acceptable critical points, it appears to be difficult to judge the actual influence of the gaussian filter employed by MARR+HILDRETH 80 based on theoretically and biologically motivated considerations. In order to convey at least a visual impression regarding the effect of a circular symmetric gaussian filter with a standard deviation of one pixel for a 5x5 pixel operator, Fig. 5 presents the zero-crossing contour obtained for the Laplacian with ("Mexican Hat") and without gaussian filter.

In the following, we use the greyvalues without gaussian filter. The influence of the gaussian filter will have to be studied in detail by a separate investigation.

3.4 Laplacian Versus Gaussian Curvature

Since the points where the gaussian curvature becomes zero - without vanishing identically within an entire 2D environment - are points of extreme gradient it seems to be more interesting to track the zero-crossing contours of the gaussian curvature instead of the zero-crossing contours of the Laplacian. Fig. 6 shows the zero-crossing contours of the gaussian curvature (straight line) and the zero-crossing contours of the Laplacian (dashed line). Laplacian and gaussian curvature have been computed using the 5 by 5 operators of BEAUDET 78. It is surprising at first glance that the zero-crossing contours of the gaussian curvature do not coincide with the zero-crossing lines of the Laplacian along the straight edges of the car image. The explanation is that the gaussian curvature becomes zero along these edges but is not always passing through zero. In places where both principle curvatures change their sign simultaneously the gaussian curvature is only touching zero. A simple algorithm which inspects the sign of the gaussian curvature at two neighboring raster locations will in general not be able to detect such zero-touchings. It is, however, possible to detect this case by looking at the zero-crossings of the two principle curvatures separately.

3.5 Zero-crossings Of Principle Curvatures

Fig. 7 shows the zero-crossings of the principle curvature which is parallel to the direction of the gradient. As expected, these lines coincide with the zero-crossing lines of the Laplacian along straight edges and differ in corners (see Fig. 8: dashed line Laplacian, straight line gaussian curvature).

The different results near corner points are important for the selection of critical points as points of high boundary curvature. In areas where at least one principle curvature is different from zero, the zero-crossing line of $(\sigma^+ + \sigma^-)$ is forced to deviate from the zero-crossing contours of the principle curvature related to the steepest greyvalue slope into the area where the gaussian curvature is negative, eg. $-\sigma^+ = \sigma^-$. This may be helpful for the detection of corners because it tends to enhance the plane curvature of the zero-crossing contour and, therefore, emphasizes the corner. On the other hand, the zero lines of the gaussian curvature are lines of extreme gradient and it might be better to select critical points at a location of maximum greyvalue slope. In clear cut situations like "greyvalue corners", the differences do not appear to be great. In more complicated situations of many closely neighbored extrema and zeros of the gaussian curvature, more care appears to be required.

4.0 CONCLUSION

For the disparity estimation algorithm described in NAGEL+ENKELMANN 82 it is very important to select critical points near locations where only one principle curvature passes through zero - implying an extremum of the gradient - and the other principle curvature is different from zero.

The results discussed in section 3 show that it is possible to find critical points by detecting points of high curvature along the zero-crossing contours of the "Mexican Hat"-operator, the Laplace-operator or the principle curvature in gradient direction. Corner points detected by the latter operator are similar to a subclass - i.e. the "greyvalue corners" characterized in section 3.1 - of those localized through the two-stage interest operator by DRESCHLER+NAGEL 81. In addition to "greyvalue corners", greyvalue extrema - i.e. bright or dark spots - and more complex combinations of at least three extrema of the gaussian curvature have been used by DRESCHLER+NAGEL 81. Further investigations are necessary to establish the precise relation between these heuristically introduced critical points and the related mathematical structure of the greyvalue distribution. More complex image features which are no simple "greyvalue corners" require an analysis of the greyvalue distribution which comprises a larger environment, i.e. is less local than the analysis required in the case of mathematically well understood "greyvalue" corners. Such investigations are in progress at our laboratory.

5.0 AKNOWLEDGEMENT

We thank Y.-S. Hsu for many interesting discussions on this subject.

6.0 REFERENCES

BARNARD+THOMPSON 80

Disparity Analysis of Images, S.T. Barnard and W.B. Thompson, IEEE Trans. Pattern Analysis and Machine Intelligence PAMI-2 (1980) 333-340

BEAUDET 78

Rotationally Invariant Image Operators, P.R. Beaudet, IJCPR-78, pp.579-583

DRESCHLER+NAGEL 81

Volumetric Model and 3D-Trajectory of a Moving Car Derived from Monocular TV-Frame Sequences of a Street Scene, L. Dreschler and H.-H. Nagel, IJCAI-81, pp. 692-697

HANNAH 80

Bootstrap Stereo, M.J. Hannah, Proc. Image Understanding Workshop, April 1980, pp. 201-208

KITCHEN+ROSENFELD 80

Gray-Level Corner Detection, L. Kitchen and A. Rosenfeld, TR-887 (April 1980), Computer Science Center, University of Maryland, College Park/MD

MARR+HILDRETH 80

Theory of Edge Detection, D. Marr and E. Hildreth, Proc. R. Soc. Lond. B 207 (1980) 187-217

MILGRAM 79

Region Extraction Using Convergent Evidence, D. Milgram, Comp. Graphics and Image Procesing 11 (1979) 1-12

MORAVEC 77

Towards Automatic Visual Obstacle Avoidance, H.P. Moravec, IJCAI-77, p. 584

MORAVEC 80

Obstacle Avoidance and Navigation in the Real World by a Seeing Robot Rover, H.P. Moravec, Ph.D. Thesis, Department of Computer Science, Stanford University, available as CMU-RI-TR-3 (September 1980) Robotics Institute, Carnegie-Mellon University Pittsburgh/PA

MORAVEC 81

Rover visual Obstacle Avoidance, H.P. Moravec, IJCAI-81, pp. 785-790

NAGEL 82

Displacement Vectors Derived from Second Order Intensity Variations in Image Sequences, H.H. Nagel, to appear im Computer Graphics and Image Processing, Fall

NAGEL+ENKELMANN 82

Investigation of Second Order Greyvalue Variations to Estimate Corner Point Displacements, H.-H. Nagel and W. Enkelmann, ICPR-82

PAVLIDIS 77

Structural Pattern Recognition, T. Pavlidis, Springer-Verlag Berlin Heidelberg New York 1977

YAM+DAVIS 81

Image Registration using Generalized Hough Transform, s. Yam and L.S. Davis, Proc. IEEE Conf. Pattern Recognition and Image Processing PRIP-81, Dallas/TX, August 3-5, 1981, pp. 526-533

Fig. 1. Second frame of a series of 22 frames showing a moving car

Fig. 2. Enlarged section showing the car image from Fig. 1. The little squares indicate the locations of critical points selected by the two-stage interest operator of DRESCHLER+ NAGEL 81

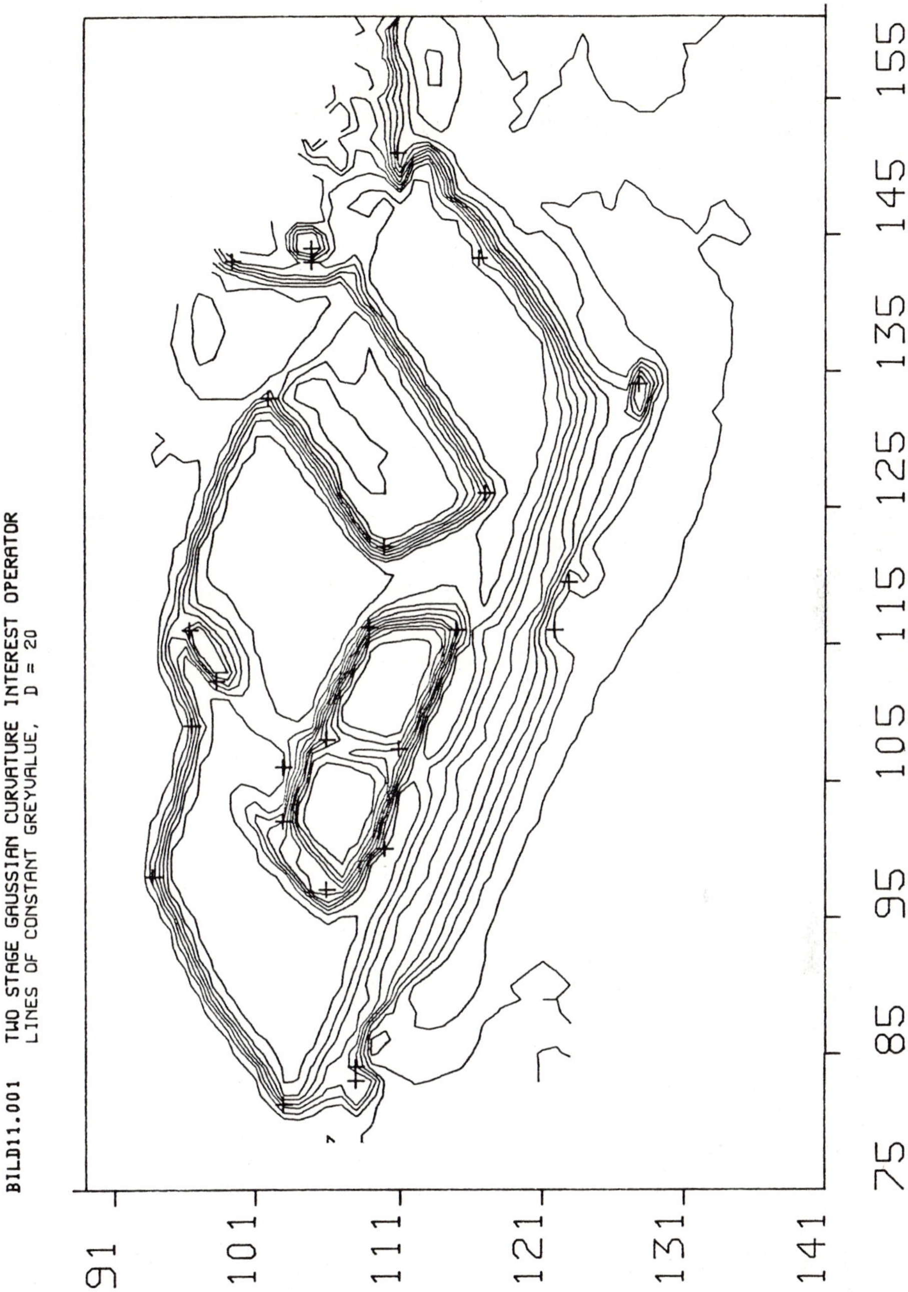

Fig. 3. Lines of constant greyvalue for the image section of Fig. 2.

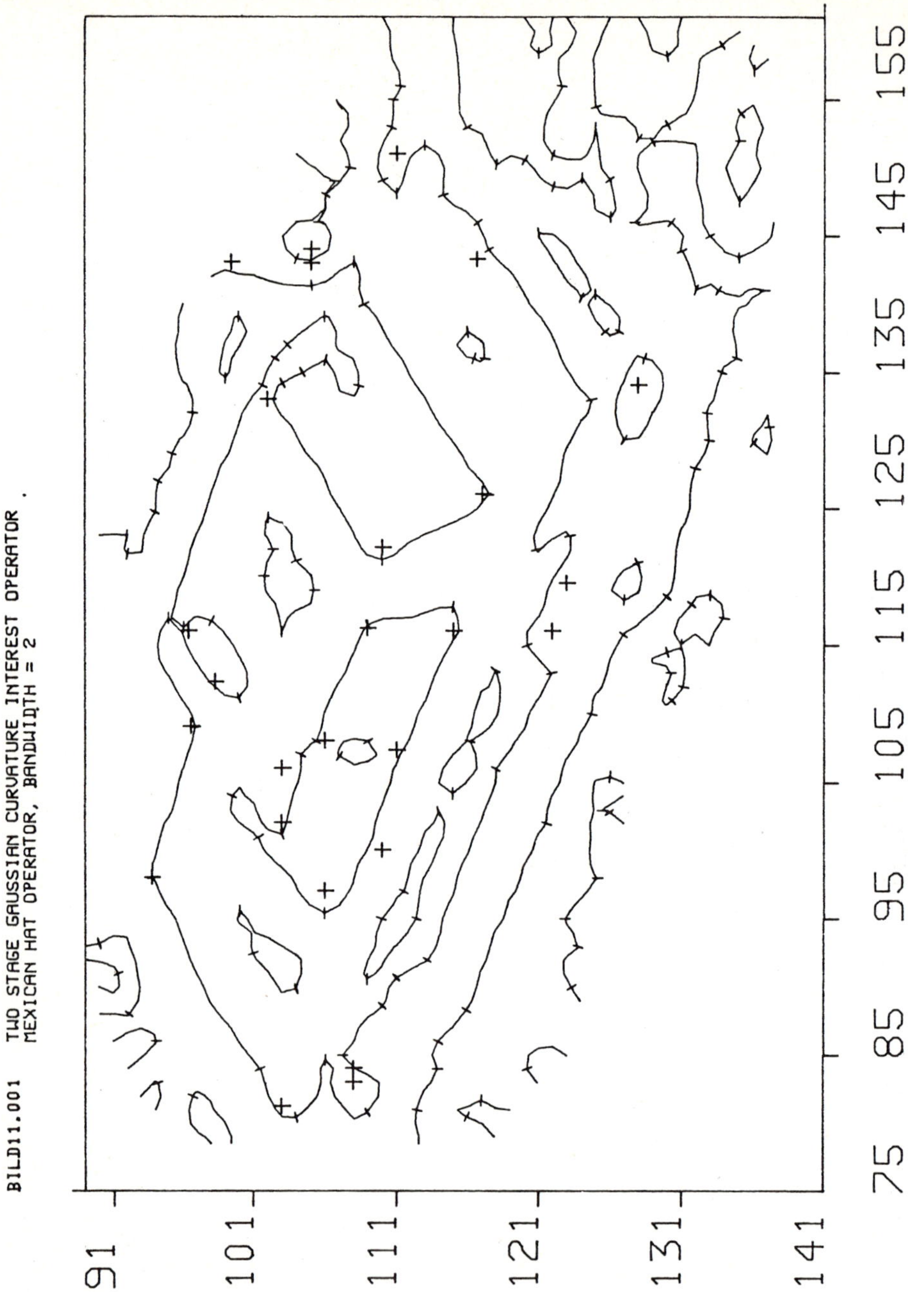

Fig. 4. Zerocrossing contours of the "Mexican Hat"-operator for the image section of Fig. 2. The dashes perpendicular to the zero-crossing contours indicate local extrema of the boundary curvature. Black crosses show the locations of the critical points from Fig. 2.

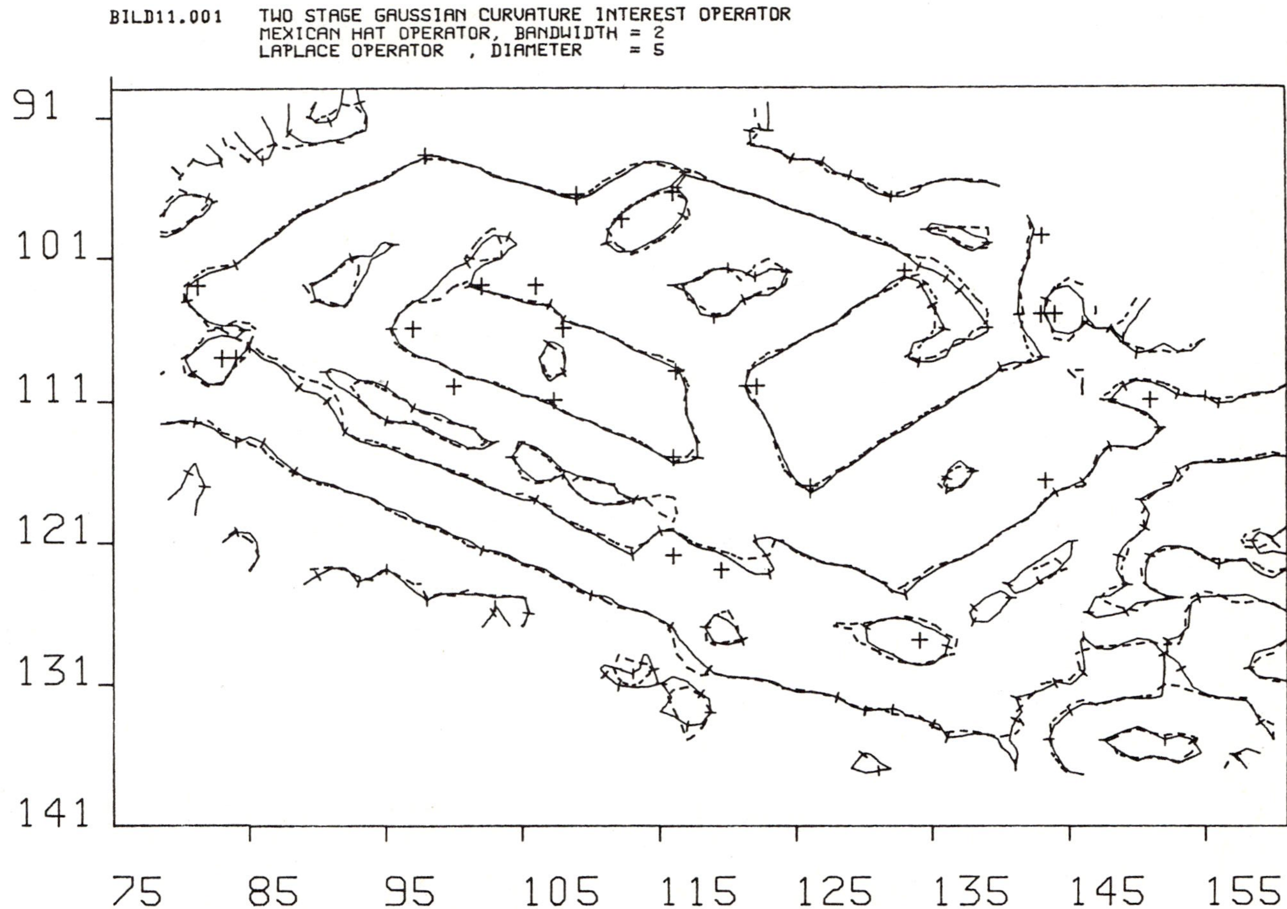

Fig. 5. Zero-crossing contours of the "Mexican Hat" operator (straight line) and of the Laplacian (dashed line)

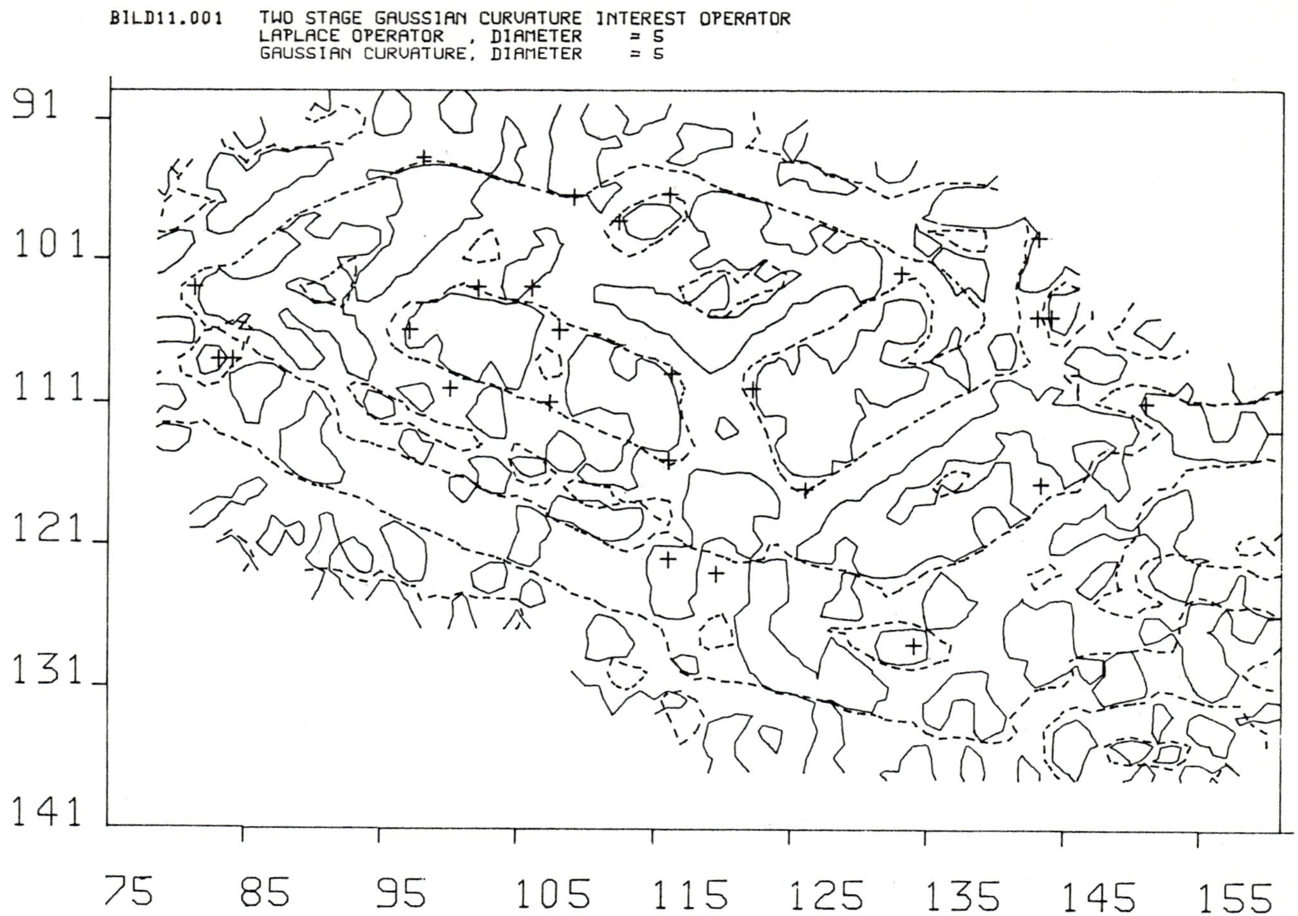

Fig. 6. Zero-crossing contours of the gaussian curvature (straight lines) and of the Laplacian (dashed lines).

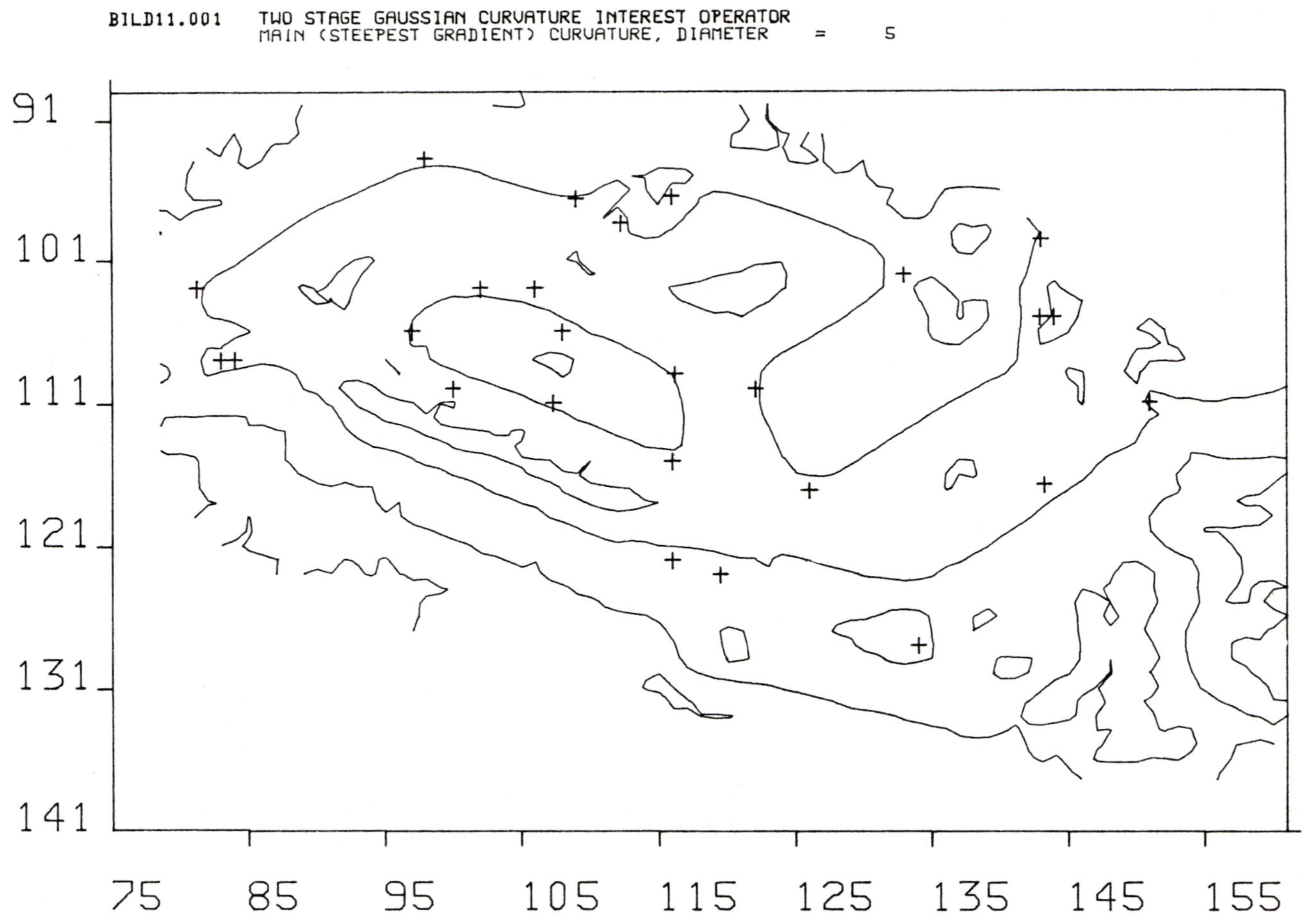

Fig. 7. Zero-crossing contours of the principle curvature in gradient direction

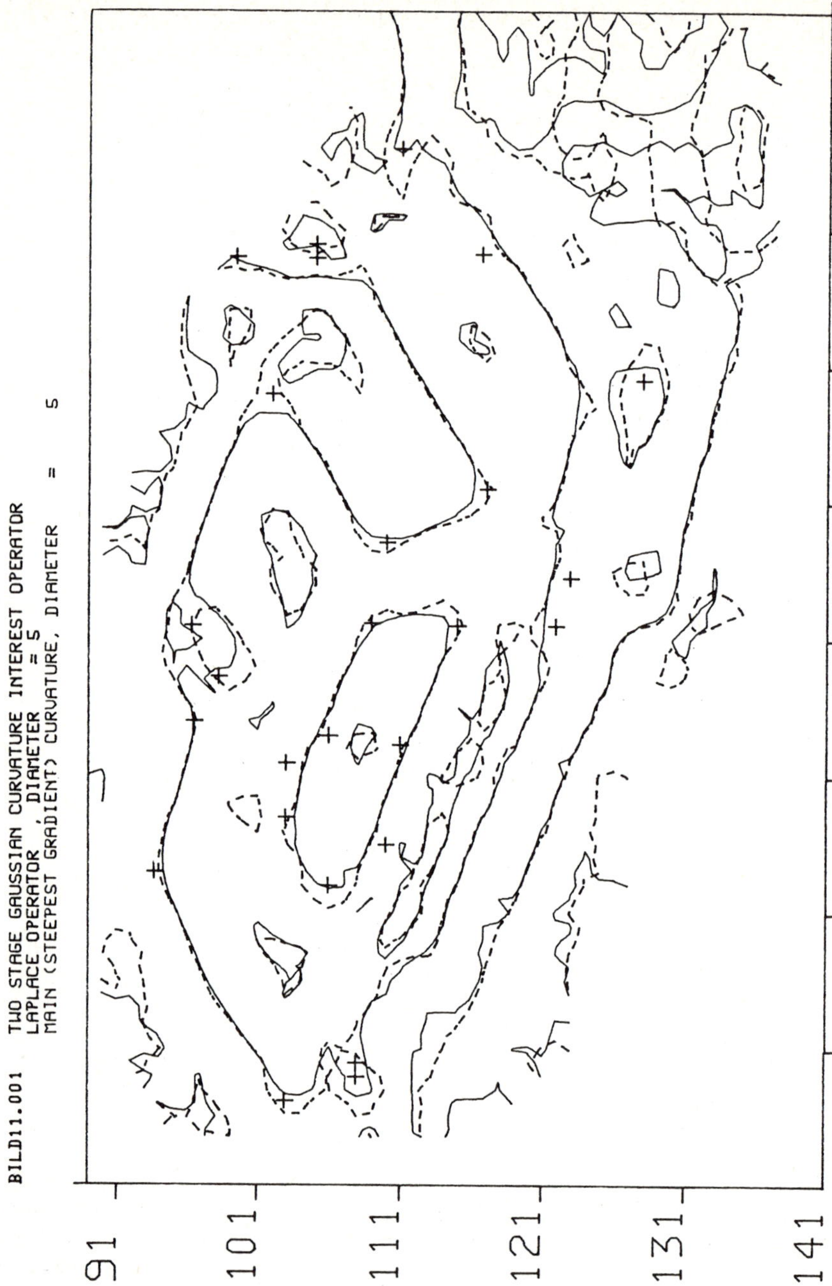

Fig. 8. Zero-crossing contours of the Laplacian (dashed line) and zero-crossing contours of the principle curvature in gradient direction (straight line).

IMAGE SEGMENTATION CONSIDERING PROPERTIES OF THE HUMAN VISUAL SYSTEM

Claus-E. Liedtke

Lehrstuhl für Theoretische Nachrichtentechnik und
Informationsverarbeitung
Universität Hannover
Callinstr. 32
3 Hannover 1
West Germany

ABSTRACT

Basic algorithms for image segmentation like contour detection and texture discrimination have mostly been derived from optimizations of mathematically defined quality criteria. This paper describes new approaches to the development of algorithms for image segmentation based on properties of the human visual system.

1. INTRODUCTION

Image segmentation is the decomposition of a scene into objects contained in the scene and background. For image analysis purposes we understand objects to be those parts of an image, which carry a meaningful label. Approaches to image segmentation are usually dependent on special applications. The algorithms are derived from optimizations of mathematically defined quality criteria and happen to work with much lower success on scenes which are different from those for which they have been developed. A system which works extremely well on a great variety of scenes, but of course not optimal in the mathematical sense, is the human visual system. Therefore we study properties and processing methods of the human visual system in order to arrive at new algorithms and concepts for image segmentation.

Figure 1 shows a blockdiagram of the human visual system which has been published earlier by Foith /1/. Due to nonlinear processes in the retina and lateral inhibition of neighboring cells the visual system performs a bandpass filtering on the incoming image and shows adaptive

NATO ASI Series, Vol. F2
Image Sequence Processing and Dynamic Scene Analysis
Edited by T.S. Huang

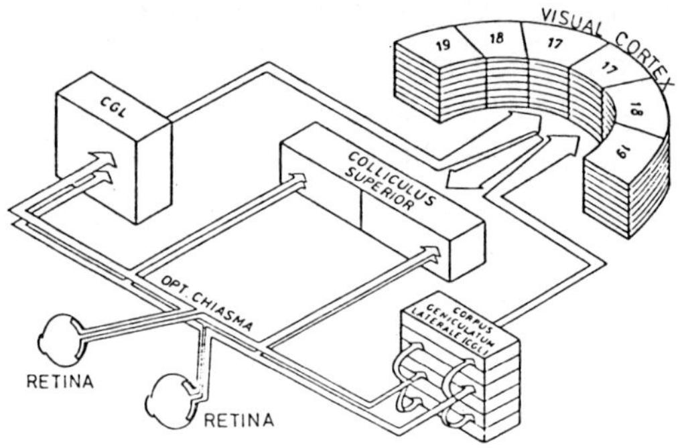

Figure 1. Blockdiagram of the human visual system

properties to slowly changing intensity values. On its way from the retina to the CGL (corpus geniculatum laterale) the original gray level image is step by step reduced to its binary contours. In the visual cortex the binary image is filtered by various two-dimensional filters of different spatial extent and orientation sensitivity in different directions. Further processing is done in higher centers of the brain which are no more accessible to direct physiological measurements where lines and homogeneous areas, which we call blobs, are detected and combined to form meaningful objects partially guided by "experience" from an earlier learning phase.

Various qualitative and quantitative measurements have been made on the visual system on the basis of physiological measurements and subjective tests /2,3,4/.

2. CONTOUR DETECTION

Objects in a scene can be found by detecting the contours surrounding them. Contour detection algorithms are mostly based on mathematical criteria applied to artificial contour models. Our approach is to develop a contour detection algorithm by a different approach, namely by modelling of the most prominent properties of the human visual system. The two properties which we regard to be most important for contour detection are the frequency response and adaptation. The frequency response of the visual system during contour detection has quantitatively been measured by several authors /5,6/. The detection of contour points is achieved by filtering and choosing zero-crossings as possible contour points. The method is explained in more detail by Geuen /7/. Figure 2b shows the zero-crossings which have been detected after applying a bandpass filter to the original parking lot scene of Figure 2a. The frequency response of the bandpass filter has been derived from measurements from Shapley and Tolhurst /5/. Many points in Figure 2b are not perceived as contour points because the threshold of perception depends on the local contrast. This is part of the

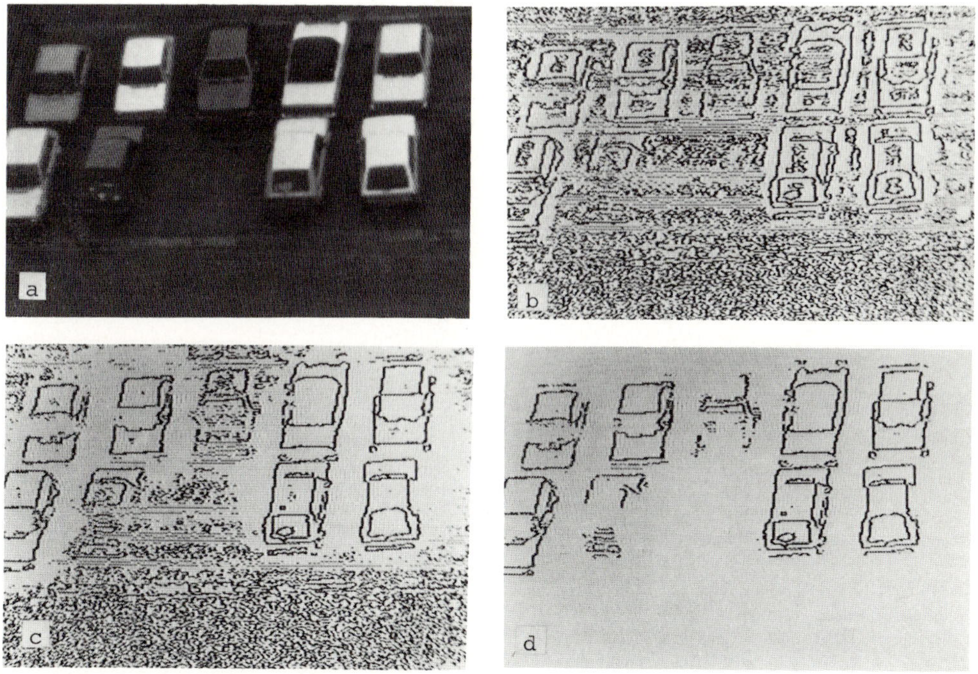

Figure 2. (a) Original scene. Modelling of human contour perception considering (b) frequency response, (c) adaptation, and (d) noise reduction

property called adaptation. A quantitative description of the just visible contrast has been published by Moon and Spencer /8/. On this basis a model for the influence of adaptation on contour detection has been developed. The results are presented in Figure 2c. It becomes clearly visible that in areas of high contrast which appear between the bright car roofs and the dark backgound the threshold for perception of contour points in the background is raised and the number of contour points found by the algorithm has been reduced.

The results are still not always satisfactory. In contrast to human perception noise in the image results in small gaps within contour lines and many extra contour points within textured areas. Errors of this kind have been reduced by introduction of the concept of contour strength. It could be shown from subjective tests that for a given contrast a contour must have a minimum length in order to be perceived as a contour. The elimination of short contours below that threshold results in the contours of Figure 2d.

The model of contour detection based on the visual system has been described in detail by Liedtke and Geuen /9/. A very efficient implementation of parts of the model is described by Geuen /7/.

3. TEXTURE ANALYSIS

In many cases objects in an image can be described efficiently as being those subimages which are homogeneous in some local property. This local property is called texture and we exclude in our experiments colour from being a special subclass of textures in this sense. Objects can then be separated from each other and from the background by the detection of textural differences. Our approach to texture analysis is to develop an algorithm which automatically indicates textural differences in all those cases where the human visual system perceives spontaneously textural differences as well. In order to detect differences in the texture of two local neighborhoods parameters from the two local neighborhoods have to be extracted and be compared with each other. The parameters which we have chosen correspond to parameters which in a similar fashion are available to the visual system as well. The blockdiagram for parameter extraction is shown in Figure 3. Since visual image processing includes gray level and contour perception it can be concluded that the higher cortical

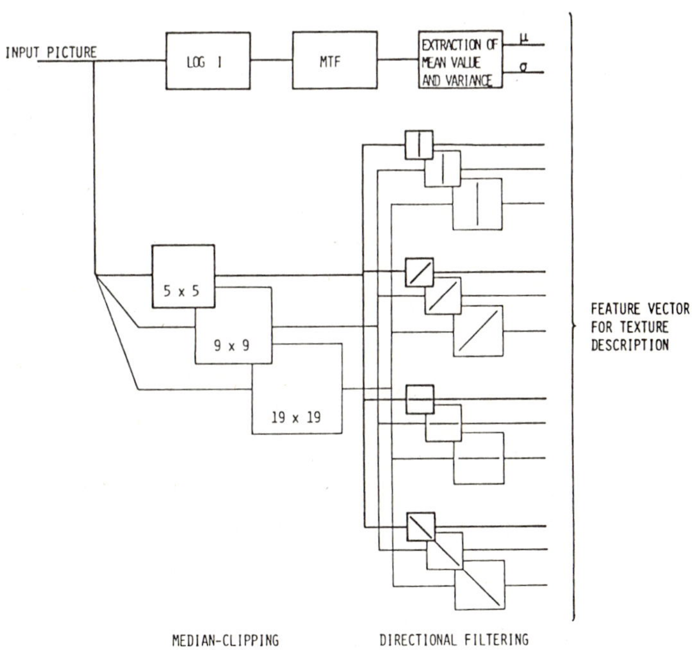

Figure 3. Parameter extraction for texture analysis

centers use simultaneously visual information from different parts of the visual path. This corresponds to the two parallel paths in Figure 3.

The upper path represents the processing of gray level information and the lower path the processing of the image contours. For simplification of the mathematical description and in agreement with physiological measurements, gray level information can be represented by its first order statistics. In the upper path of Figure 3 the description of the first order statistics is approximated by the mean value and the variance corresponding to the subjective properties of brightness and contrast. The processing of gray level information is performed mainly by the retina. Therefore the upper path contains a logarithmic characteristic and a bandpass filter (MTF = Modulation Transfer Function) to model the properties of the retina as mentioned above. The lower path in Figure 3 models the processing of the two-dimensional structural information of texture. For simplification of the modelling the processing steps of the visual system like adaptation, bandpass filtering and contour extraction have been combined in the computationally simple algorithm of "median-clipping". The algorithm of median-clipping works as follows: the gray level of each pixel is compared to the median value of the gray levels in a square local neighborhood surrounding the pixel. If the present pixel shows a higher gray level than the median value of the corresponding neighborhood, the pixel is set to a bright level. Otherwise it is set to dark. The effect of this binarization which is supposed to model the visual path from the retina to the CGL is demonstrated in Figure 4b which has been derived from the original test pattern in Figure 4a. In correspondence with image processing steps in the cortex the binary contour image is further processed by two-dimensional filters with narrow orientation sensitivity in different directions. Subjective properties like coarseness and regularity can probably only be perceived because of the existence of a large number of receptive fields of different spatial extent in the visual system. This can be modeled in Figure 3 by three parallel filters of different window size for the median filtering and its corresponding set of four directional filters. The parameters for the quantitative description of the structural texture properties are calculated as the mean energy at the output of each directional filter. The application of just the structural information, i.e. without using gray level information, for the segmentation of textured areas is shown in Figure 4.

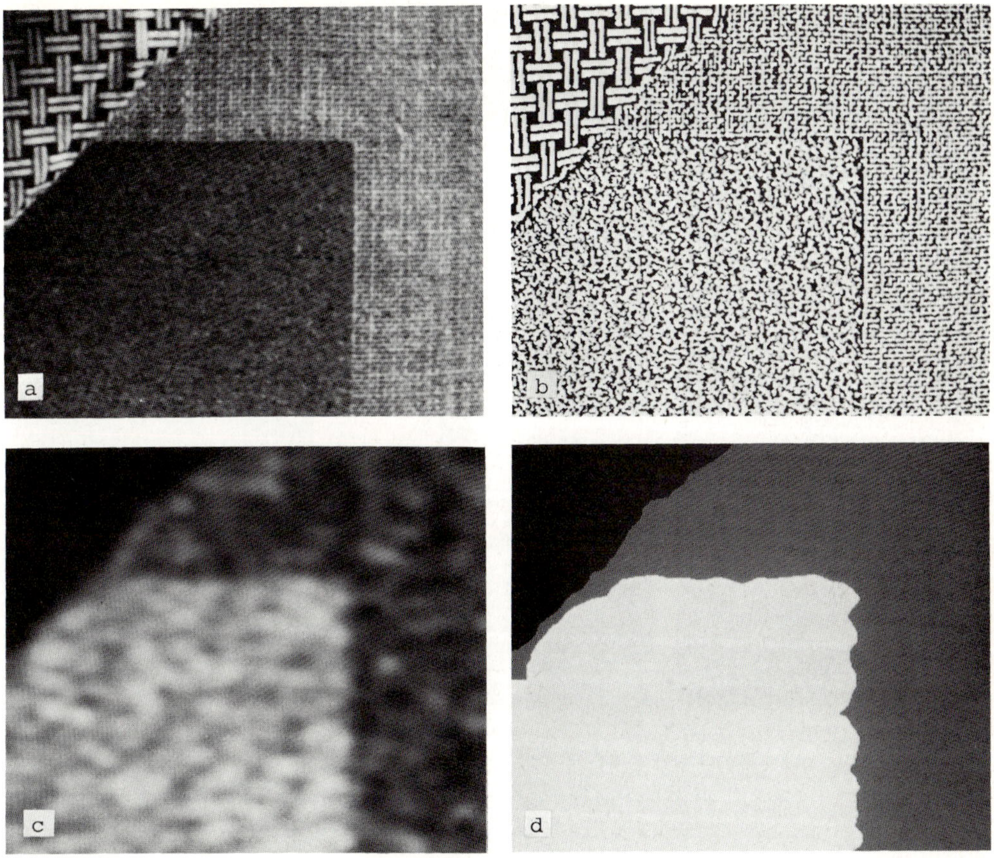

Figure 4. Image segmentation of textured areas. The processing steps show (a) the original test pattern, (b) median clipping, (c) mean energy from the output of one directional filter,and (d) the texture masks obtained from (c) by thresholding

Most segmentation methods for textural subimages in the literature are based on a classification of local neighborhoods into a finite number of pregiven texture classes. They show usually minor performance when applied to image segmentation in the sense of an unsupervised classification. The power of our approach can be demonstrated for the latter application in Figure 5. The original image containing eight different textures in Figure 5a is scanned horizontally in Figure 5b and vertically in Figure 5c. Neighboring local areas are analyzed for significant changes in the textural parameters. The

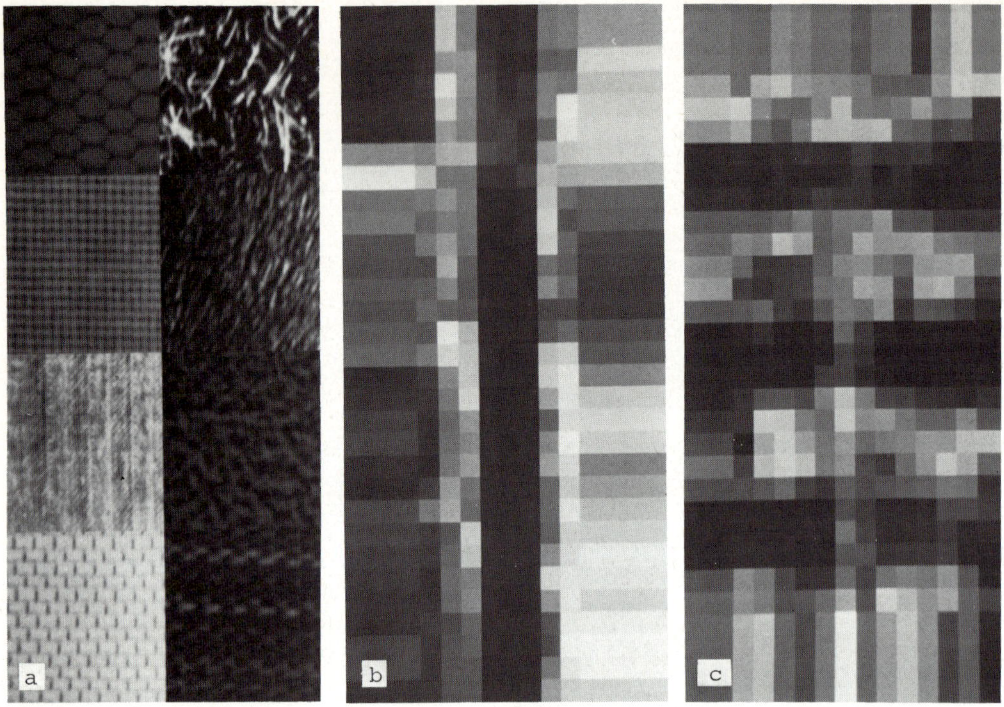

Figure 5. (a) Original image containing eight different textures. Detection of textural differences in scanning the image in horizontal direction (b) and in vertical direction (c). Dark gray levels indicate major and bright gray levels minor textural differences

dissimilarity of neighboring textures is coded in Figure 5b and 5c by gray levels, i.e. dark gray levels indicate major and bright levels minor textural differences. The borderlines between areas of different texture can clearly be detected.

The model and further experiments have been described in more detail by Wermser and Liedtke /10/.

4. USE OF A PRIORI KNOWLEDGE

We understand by image segmentation the subdivision of an image into meaningful objects. Standard image segmentation methods use fixed thresholds on gray levels, colour, or textural features or thresholds on amounts of change of these features. From the observation of natural objects it becomes apparent, that not only the thresholds but even the relevant features may change along the boundary around an object. In this connection changes in the preparation and staining of biological specimen and changes in the illumination of industrial scenes pose major limitations on the applicability of automated image analysis techniques while they have only a minor influence on the recognition capability of a human viewer. One important difference between the human viewer and the machine is that the human has in addition to the pictorial information contained in the image an extensive amount of a priori knowledge about the scene available, something which we usually describe by the term "experience". Automated methods for image segmentation can be expected to become the more reliable and the more insensitive to unrelevant changes in the scene the more a priori knowledge about the scene can be entered into the segmentation process.

In connection with the segmentation of microscopic cell scenes from blood smears we found that in addition to local properties of the scene like gray level, colour, and texture global a priori knowledge about neighborhood relations, shape, and size of the objects has to be used in order to improve segmentation. Our approach to enter the missing a priori knowledge into the segmentation algorithm originates from the observation that in order to perceive an object visually it must either consist of one or a few homogeneous areas or it must at least partically be surrounded by a contour. For our data material, namely leucocytes in a microscopic cell scene, contours seemed to be of minor importance. Therefore in a first step we subdivide the image into a number of areas which appeared to be homogeneous in some local properties like gray level, colour or texture. We call these areas blobs. In a second step the objects are assembled from the blobs by using a priori knowledge about the scene content and well known photometrical, morphological, topological properties and size information which has been collected during an earlier learning phase. The procedure for entering relevant a priori knowledge into the segmentation procedure in order to extract nucleus and cytoplasm from

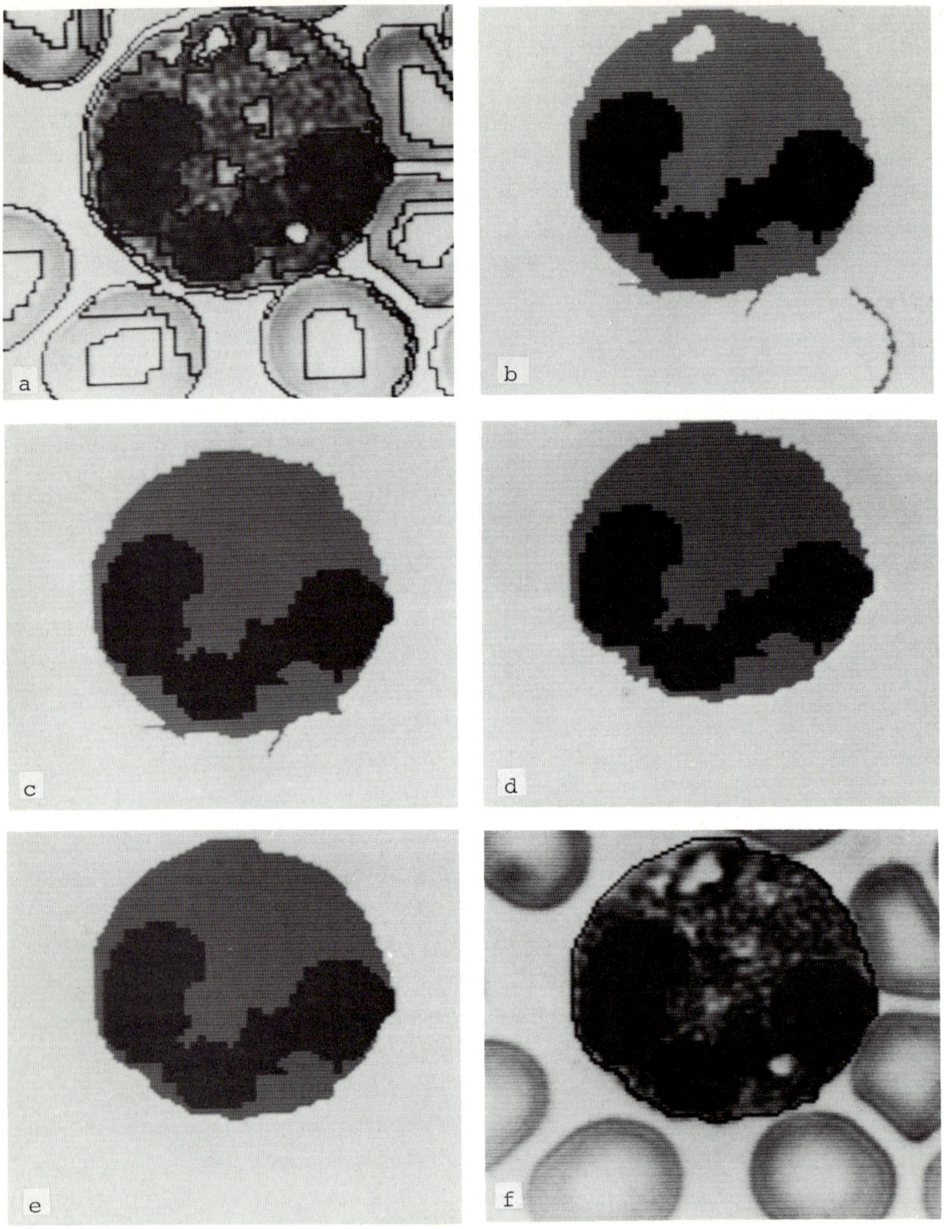

Figure 6. Entering of a priori knowledge (apk) into the segmentation of leucocytes: (a) uniform regions detected by a Split & Merge algorithm, region labelling on the basis of photometri apk (b), topological apk about neighborhood relations (c), and morphological apk (d). (e) shows nuclear and cytoplasmic masks after postprocessing and (f) gives the resulting boundaries of nucleus and cytoplasm superimposed to the original image.

leucocytes in a microscopic cell scene is demonstrated in Figure 6. Figure 6a shows the blobs which have been found during a search for homogeneous areas. In Figure 6b those blobs have been assembled to objects which belong on the basis of photometric information to the nucleus (black) and cytoplasm (gray). In Figure 6c - 6e corrections have been made for well known topological and morphological properties. The resulting boundaries of nucleus and cytoplasm superimposed on the original cell image are shown in Figure 6d. The quality of the result has improved markly over any other segmentation method which we have applied to this type of cell. Details about the segmentation method have been described by Haussmann and Liedtke /11/.

REFERENCES

/1/ J.P. Foith, Prinzipien der Szenenanalyse beim Menschen und im Computer. IITB-Mitteilungen, p. 27 - 34, 1976

/2/ T.N. Cornsweet, Visual Perception, Academic Press, New York, 1970

/3/ D. Jameson and L.M. Hurvich, Visual Psychophysics, Handbook of Sensory Physiology, Springer Verlag, Berlin, Heidelberg, New York, 1972

/4/ O.D. Creutzfeldt and H.C. Nothdurft, Representation of Complex Visual Stimuli in the Brain, Naturwissenschaften 65, 1978

/5/ R.M. Shapley and D.J. Tolhurst, Edge detectors in human vision, J. Physiol. 229 (1973), p. 165 - 183

/6/ G. Hauske, Adaptive filter mechanisms in human vision, Kybern. 16 (1974), p. 227 - 237

/7/ W. Geuen, A Fast Edge Detection Algorithm Matching Visual Contour Perception, Proceedings of the NATO ASI 82/52 on Image Sequence Processing and Dynamic Scene Analysis, 1982

/8/ P. Moon, D.E. Spencer, Visual data applied to lighting design, J. Opt. Soc. Amer. 34 (1944), p. 605 - 617

/9/ C.-E. Liedtke, W. Geuen, Konturfindung unter Berücksichtigung des visuellen Systems des Menschen, Archiv der NTZ 3, 1981

/10/ D. Wermser and C.-E. Liedtke, Texture Analysis Using a Model of the Visual System, Proceedings of the 6th Intern. Conf. on Pattern Recognition, Munich 1982

/11/ G. Haussmann and C.-E. Liedtke, A Region Extraction Approach to Blood Smear Segmentation, submitted for publication to Computer Graphics and Image Processing

A FAST EDGE DETECTION ALGORITHM MATCHING VISUAL CONTOUR PERCEPTION

W. Geuen

Lehrstuhl für Theoretische Nachrichtentechnik und Informationsverarbeitung
Universität Hannover, West Germany

ABSTRACT

For several applications it is very useful to have an edge detection algorithm matching human visual contour perception.

The main characteristic of the frequency response of visual contour perception shows a resemblance to a spatial bandpass filter. We simulated this characteristic by using different low pass filters. The realization of these filters is achieved by calculating the mean value within rectangular windows of different sizes in the spatial domain. The recognition of a contour point requires only the detection of the change in the filter output sign (zero crossing).

One of the advantages of the described algorithm is the implementation. It is quite simple especially for real-time applications.

INTRODUCTION

In all those applications where pictures (e.g. repro-photos, tv-pictures etc.) are processed for ultimate evaluation by a human observer it is very useful to have an edge detection algorithm which matches human visual contour perception. For this purpose an algorithm has been developed based on a simple mathematical model. The aim was to find an algorithm which is rather simple and which approximates the visual contour perception. The model consists of spatial filters matching the frequency response of visual contour perception as have been measured by subjective tests published in literature.

The algorithm which we have developed has been applied to a natural test picture and compared with a number of other edge detection algorithms /1/.

NATO ASI Series, Vol. F2
Image Sequence Processing and Dynamic Scene Analysis
Edited by T.S. Huang

SPATIAL FILTERING

A deterministic characterisation of the visual process is not available, but several visual phenomena are known. One of them is the spatial frequency response, called Modulation Transfer Function (MTF) of the optical system. The MTF does not describe contour perception produced by contours and simple patterns, such as spots, lines or bars. This is due to adaptation and masking effects. The perceptibility of a visual stimulus is therefore reduced, if another stimulus is presented immediately afterwards. In order to describe these effects different sensitivity measurements were carried out e.g. by Shapley and Tolhurst /2/, and Hauske /3/. The characteristic of the sensitivity functions is roughly the same as that of the MTF (see Fig.1) but it differs in respect to the method of measuring.

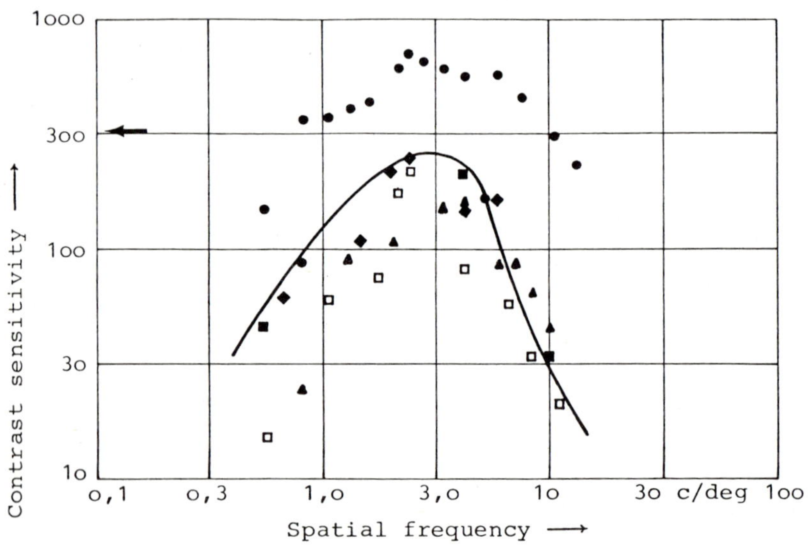

FIGURE 1

Results of Shapley and Tolhurst: Spatial frequency response of the edge detector. Subthreshold sinusoidal gratings were added to several test patterns in the sine phase. The test patterns were simple edges (◆), and low spatial frequency square-wave gratings of 0,4 c/deg (▲) and 0,53 c/deg (□■). The same spatial frequency response fits all the test patterns. The continuous curve is the Fourier transform of the subject's edge detector and it has been adjusted on the sensitivity scale so as to exactly predict the sensitivity to an edge, indicated by the arrow. On the same graph are plotted the results (•) of simple detection of sine-wave gratings (the spatial MTF of the whole visual system).

REALIZATION OF THE SPATIAL FILTERING

The main characteristic of the frequency response of visual contour perception resembles a spatial band pass filter. We approximated this characteristic with different low pass filters, by subtracting the filter outputs from one another (Fig.2).

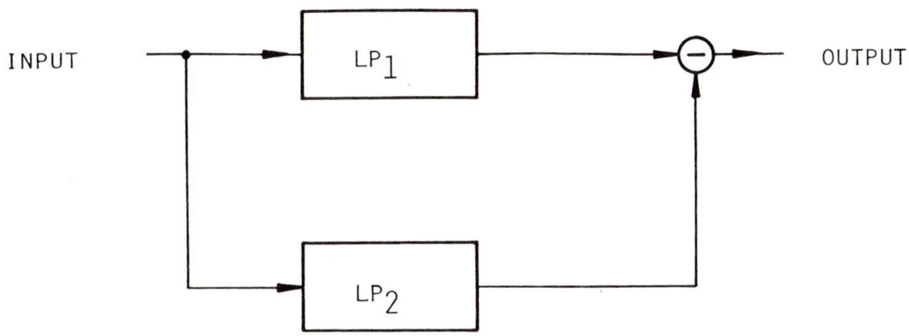

FIGURE 2

Structure of the spatial filter

The low pass filters are realized by calculating the mean value within rectangular windows of different size in the spatial domain. The characteristic of the band pass filter is the result of the structure of the low pass filters, each having a different window size (Fig.3).

The spatial frequency response of the eye is approximately rotationally symmetric, whereas the filter response of the rectangular windows is not. The spatial frequency response of Shapley and Tolhurst is based on special test conditions but human contour perception is naturally much more complex. Therefore our main objective was the performance of the final algorithm judged by human observers /1/ and not to achieve the best fit to the frequency response of Fig.1. For our application we used a TV-system with different spatial resolution in horizontal and vertical direction with a ratio of 5:4. The corresponding frequency response was also different in horizontal and vertical direction. (Fig.4).

In order to calculate the mean value of intensities within a window it is necessary to add all of them for the starting window of a line. Then with successive horizontal shift of the window by one pixel, only the values of the previous first column have to be subtracted and the

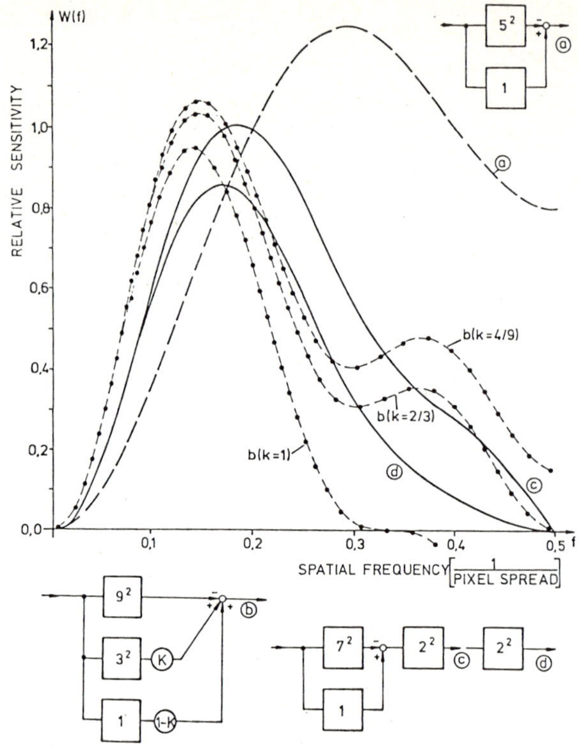

FIGURE 3

One-dimensional frequency response of different spatial band pass filters

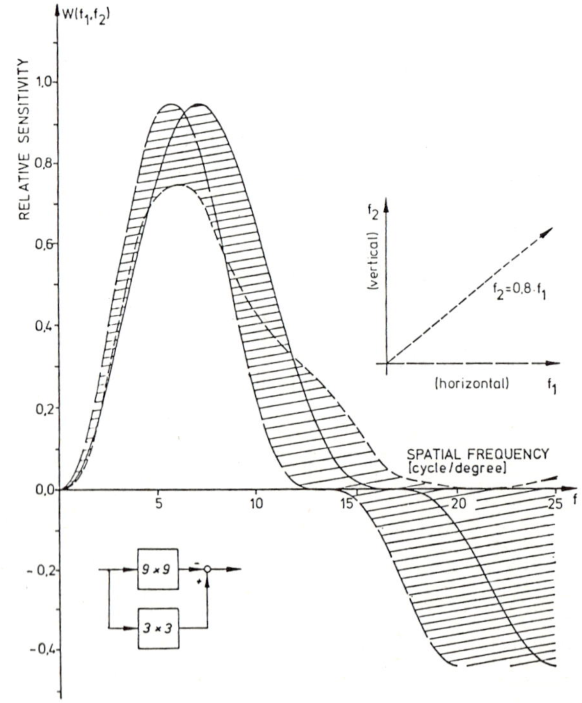

FIGURE 4

Frequency response in vertical and horizontal direction

values of the present last column have to be added to alter the previous output (Fig.5)

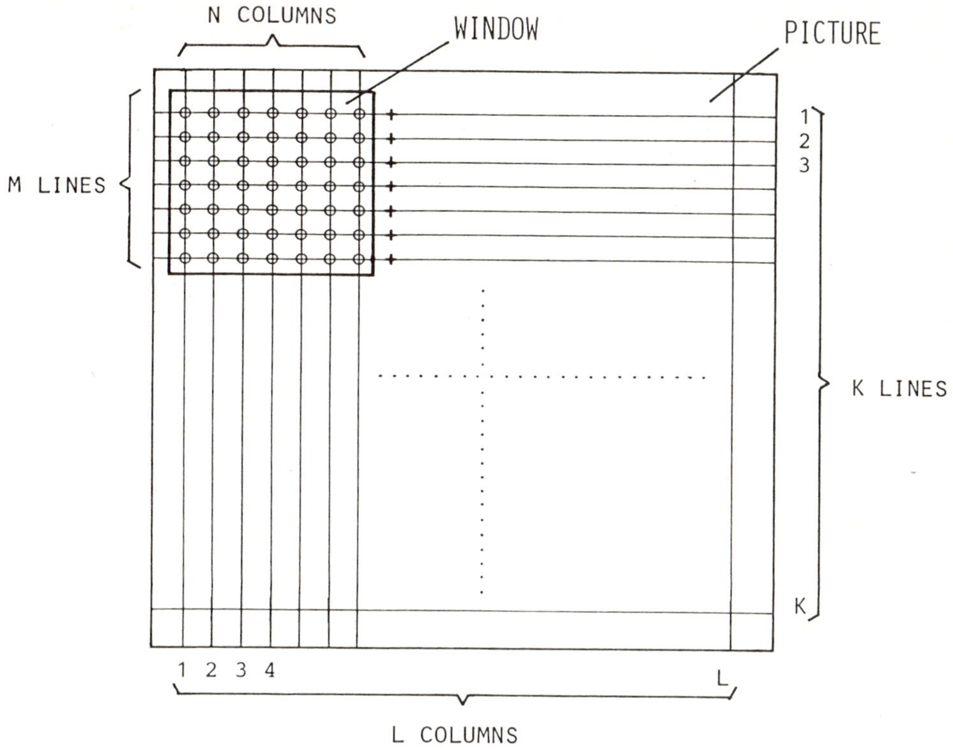

FIGURE 5

Calculation of the mean value of intensities within a window

For a picture with K lines and L columns and a window size of M lines and N columns the number of additions is:

$$A = K\cdot\{M\cdot N-1 \ + (L-1)\cdot 2\cdot M\} = K\cdot M\{N-1/M + 2\cdot(L-1)\}$$

for $L \gg N$ the number of additions is approximately:

$$A \simeq K\cdot L\cdot 2\cdot M$$

The number of additions is therefore proportional to the number of lines in the window.With this technique a considerable reduction in computation time is possible,especially for bigger sized windows.An additional improvement concerning a reduction in computation time is possible by using a technique which is based on the decomposition of a two-dimensional operator into two one-dimensional ones.In our case this decomposition technique is simple because the weighting factors of the two-dimensional operators have always the value one (Fig.6). Therefore the value of the weighting factors of both one-dimensional operators is also one.

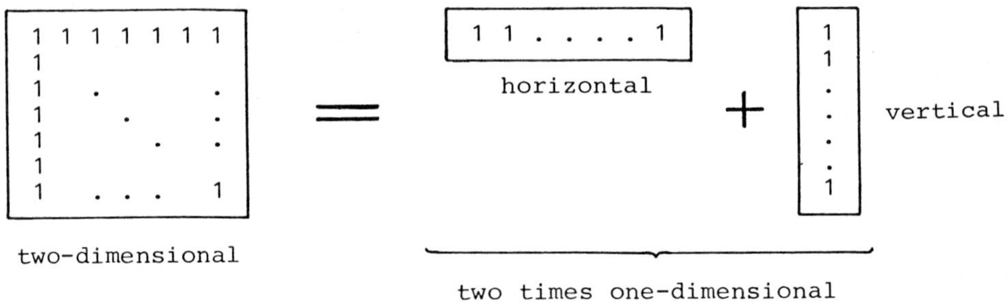

FIGURE 6

Decomposition of a two-dimensional operator into two one-dimensional operators

For this application the number of additions is independent of the window size; an important factor for real-time applications. This technique is especially suited for pipelining in hardware processors.

DETECTION OF CONTOUR POINTS

The above described filter simulates approximately the visual contour perception, although the output of the filter is not a contour picture. In Fig.7 the result of one-dimensional filtering of an input pattern is shown.

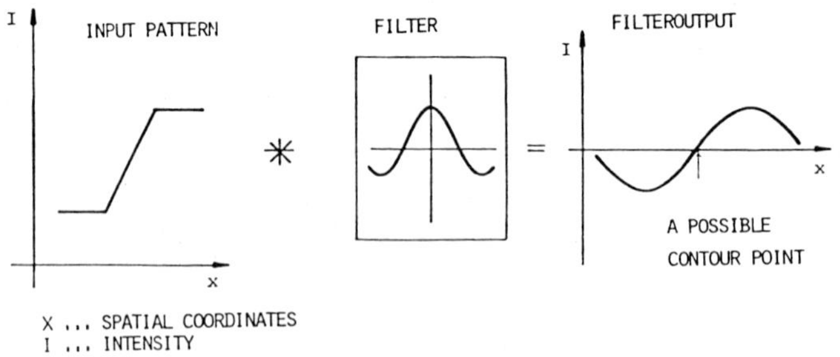

FIGURE 7

One-dimensional filtering of an input pattern

We define the zero crossing of the filter output as a possible contour point. For the detection of zero crossings only a change in the filter output sign, regarding the four-connected-neighbourhood, has to be recognized. For input patterns with low contrast (e.g. noise) the filter also generates zero crossings which are false contour points. In order to reject these we have adapted a threshold for the detection of contour points. The difference between the previous and the current sample has to be greater than the threshold. Figure 8 shows the block diagram of the edge detection algorithm described above.

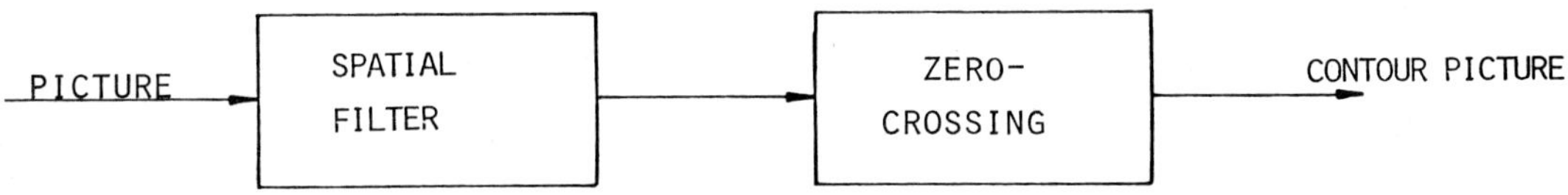

FIGURE 8

Block diagram of the edge detection algorithm

RESULTS

The natural test pictures "parking lot" and "playboy" have been processed by the following three edge detection algorithms:

- Sobel operator /4/
- Robinson operator /5/
- Hueckel operator /6/

The results of these processes are shown in Fig.9. The disadvantages of these algorithms are broad contours, gaps, partially missing contours etc.

The final results of the proposed edge detection algorithm based on the spatial filters of Fig.3 can be seen in the diagram of Fig.1o.

The results of our edge detection algorithm shown in Fig.1o are already very similar to the visual perception of a human observer. Fig.1o d shows a large number of false contour decisions due to the high spatial filter cut-off frequency. Compared with the results of Fig.9 our algorithm has produced thin contours. Also is the average length of the contour segments relatively long, which is an important feature for some applications.

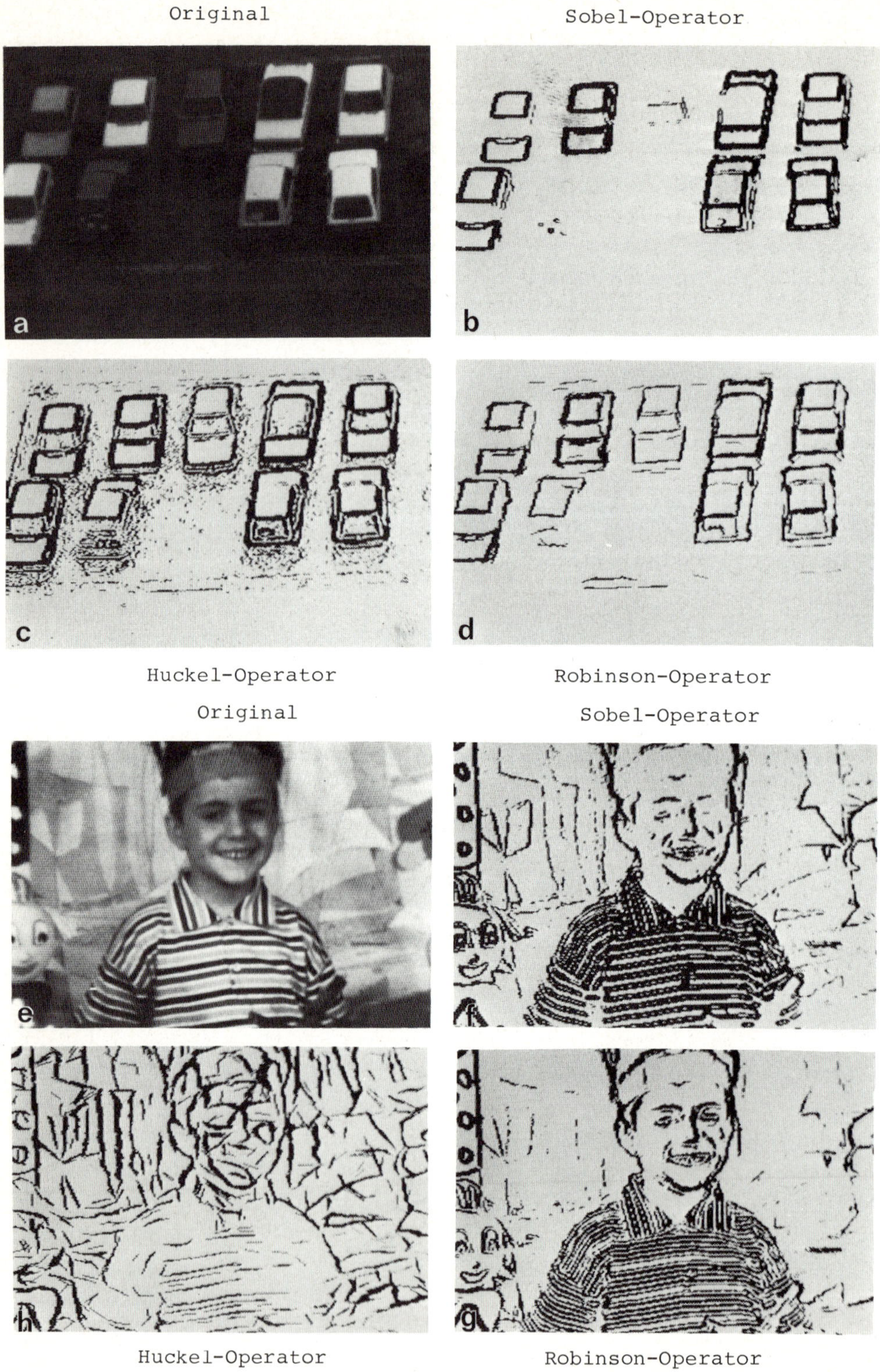

FIGURE 9. Results of the different edge detection algorithms

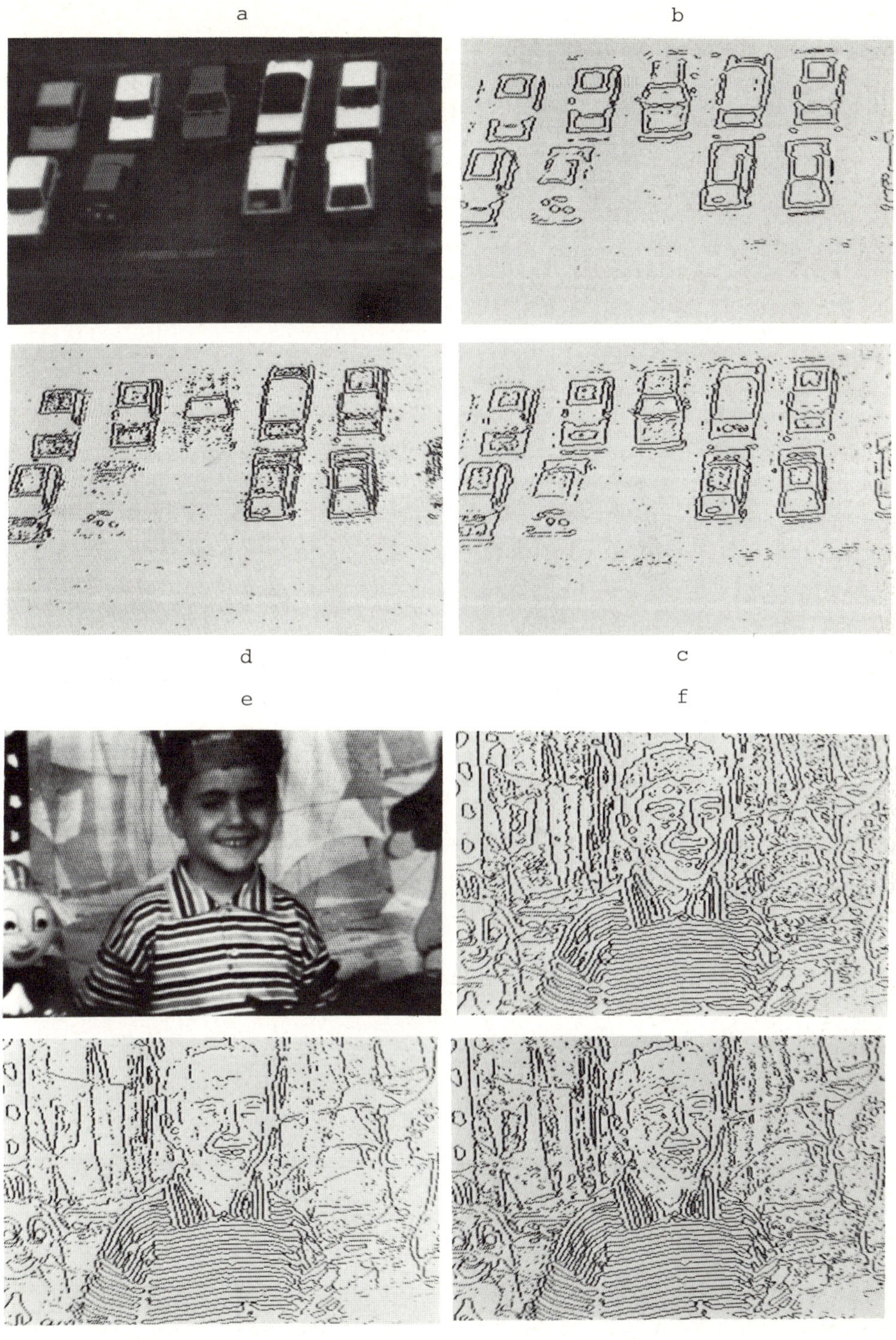

FIGURE 10. (see next page)

FIGURE 10

Results of the edge detection algorithm:
a & e) natural test picture (256 lines with 256 pixel per line and eight bit gray level quantization)

Results applying a spatial filter with the frequency response corresponding to Fig.3:
b) window size 9x9/3x3; k=1
c+g) window size 7x7/1x1 + 2x2
d) window size 5x5/1x1
f) window size 9x9/3x3/1x1; k=2/3
h) wondow size 8x8/2x2

CONCLUSION

The structure of the described edge detection algorithm is quite simple and it also matches human visual contour perception. The results are better than those of other edge detection algorithms already known from literature /1/. The algorithm can easily be implemented on different computer-systems or be used for special hardware applications.

REFERENCES

/1/ Geuen, W.; Preuth, H.G.: New Performance Criteria of Edge Detection Algorithms. ICASSP 82, May 1982, Paris, pp. 1936-1939
/2/ Shapley, R.M.; Tolhurst, D.J.: Edge detectors in human vision. J. Physiol. 229 (1973), pp. 165-183
/3/ Hauske, G.: Adaptive filter mechanisms in human visions. Kybern. 16 (1974) pp. 227-237
/4/ Pratt, W.K.: Digital image processing. New York, Wiley & Sons 1978
/5/ Robinson, E.S.: Edge detection by compass gradient masks. Comp. Graphics and Image Proc. 6 (Oct. 1977) No. 5
/6/ Hueckel, M.: An operator which locates edges in digital pictures. JACM 18 (1971) No. 1, pp. 113-125

IMAGE SEQUENCE ANALYSIS FOR TARGET TRACKING

K.-H. Bers, M. Bohner, P. Fritsche

Forschungsinstitut für Informationsverarbeitung
und Mustererkennung (FIM/FGAN)
(Research Institute for Information Processing
and Pattern Recognition)
Breslauer Straße 48, D-7500 Karlsruhe 1
Fed. Rep. Germany

Summary

This paper describes a procedure for the tracking of moving or stationary objects in image sequences, taken by a moving or stationary sensor. Primarily the objects are located on the ground in their natural environment so that simple detection procedures (e.g. detection by maximum intensity) cannot be applied. For those situations a tracking system has been designed, simulated on a digital computer and tested with an equipment for real time application. At the beginning the object is detected by an human operator. From that moment it will be tracked automatically based on an evaluation of the correlation function between the actual scene and a memory which is continuously updated. The scale (distance between object and sensor) is determined separately and normalized by the location of object parts. The basic shortcomings of a correlation tracker system (e.g. image modification by foreground and background objects) have been eliminated in a system where in addition to the evaluation of the correlation function objects in the foreground or background are detected based on features like contrast, image differences, contour lines, shape and relative speed of the objects.

NATO ASI Series, Vol. F2
Image Sequence Processing and Dynamic Scene Analysis
Edited by T.S. Huang

Introduction

In the past systems have been built where object detection for target tracking was based on a simple evaluation of the intensity values of the object and its environment. After preprocessing of the sensor signals the maximum intensity level or other similar features like centre of gravity are used to detect the position of a selected object (contrast tracking). Those systems only work successfully if a high contrast between object and background is given (e.g. aircraft in the sky in a TV-scene, hot vehicles in a cold environment in thermal imagery). Objects on the ground are often lost especially in the European landscape.

For those situations a tracking system based on correlation has been designed, simulated on a digital computer and tested with an equipment for real time application /1/. The basic shortcomings of a correlation tracker system (e.g. image modification by foreground and background distortions) have been eliminated in combining correlation and target detection methods, where parts of the moving object and mainly foreground and background objects are detected and identified within a period of time.

Tracker System

Fig. 1 shows the principle mode of operation of a system based on the correlation function. An operator detects an interesting target (the car), this target is stored in a memory as a reference and in the next image the correlation function between the intensity values of the actual scene t(n+1) and the reference is computed. The reference image has to be shifted within a search area which results in a field of correlation values K(x,y). These values are used to detect the target in the actual scene by a decision criterion (e.g. maximum of the correlation values) and to update the reference memory according to changes in the representation.

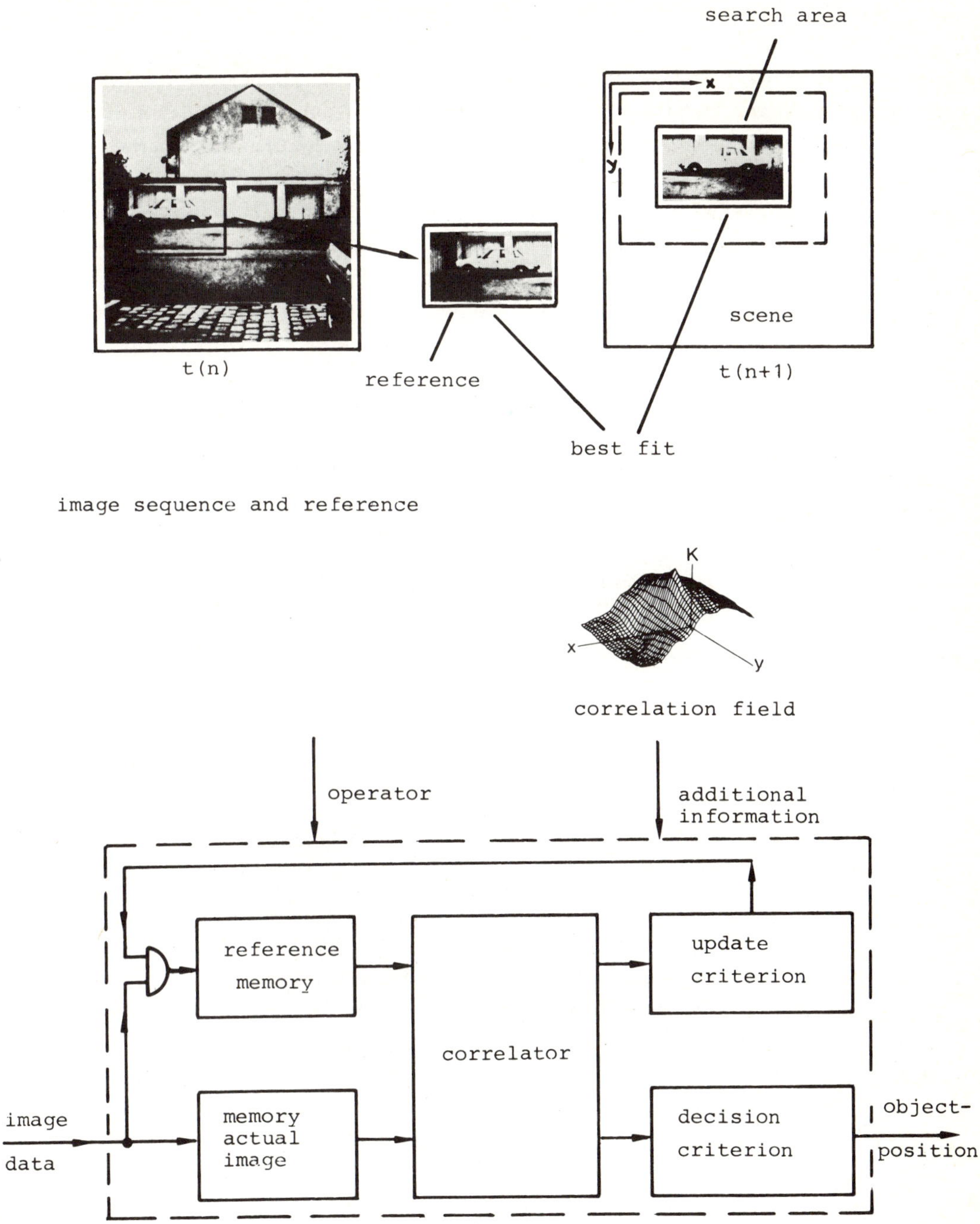

Fig. 1. Correlation-Tracker (Principle)

For such a system a number of different decision and update criterions have been tested and the optimal system parameters have been evaluated. The software simulation at FIM as well as an online hardware system of a German company revealed the basic problem of the method, demonstrated in Fig. 2 in principle. In the moment t(n) the reference memory contains the target without any foreground objects, whereas in the actual scene the car is partially hidden by the trunk of a tree. The maximum of the correlation function decreases but the object position can still be found with a high probability. If the update criterion is active in this moment, the trunk is learned into the memory as a part of the ob-

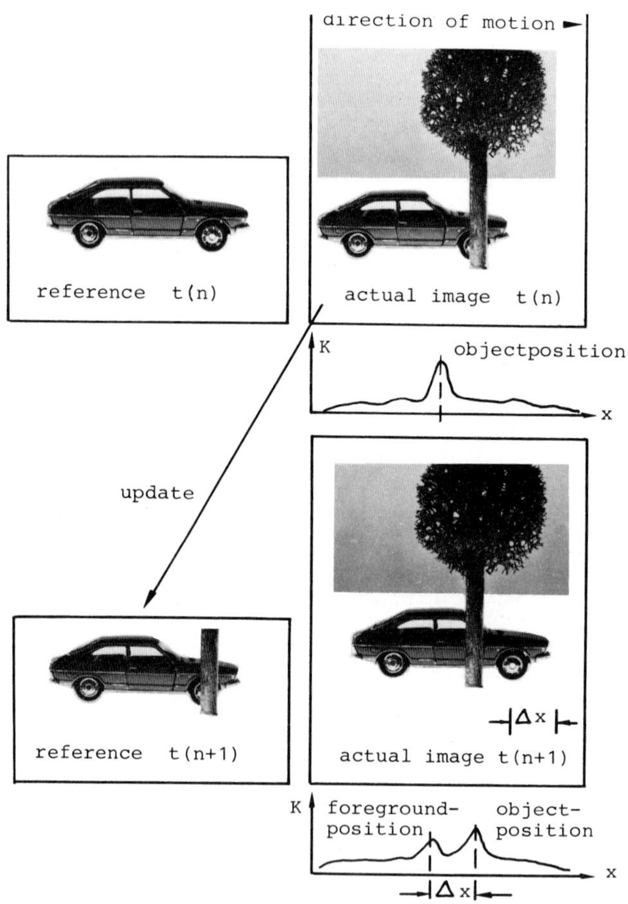

Fig. 2. Influence of foreground objects

ject. Therefore in the moment t(n+1) two peaks in the field of the correlation values are obtained (matches of car and trunk). Dependent on the size and contrast of the trunk a wrong object position may be found. A similar problem arises by big contrast changes in the background of the target.

To solve this problem at FIM a higher sophisticated tracking system has been developed, in which additional features have been added to the original correlation system (pointed out in Fig. 3) /2/:

- A new reference is composed of a part of the old reference and actual image data (indicated by ⊕ in Fig. 3). Because of the relative speed between a moving target and stationary background, only the moving target is well represented, whereas background information is blurred.

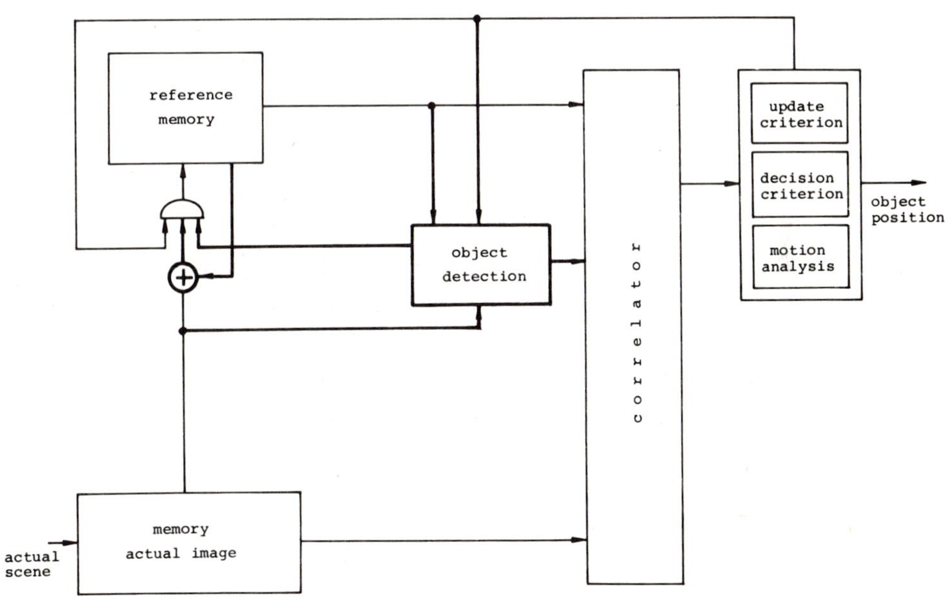

Fig. 3. FIM-Tracker-System (Picture Processing Unit)

- Monitored by an object detection module parts of the moving object and mainly foreground and background distortions are detected and identified.

Thus this system combines correlation and object detection methods. The object to be tracked is mainly located by the correlation of the actual scene and a reference. However the evaluation of the correlation values and especially the decision for the final object position is highly influenced by object detection methods. The detection and identification of the objects or parts of objects are based on features like:

- intensity and contrast levels
- image differences of "consecutive" images within a sequence
- distribution of contour lines
- shape of special areas near the moving object
- relative speed between objects.

Object Approach

When approaching a target (sensor on board) the scale of the object representation is constantly increased.

The homing systems applied until now especially analyse the intensity of the sensor signals and the approach of the missile to the target normally is compensated by a variable amplification of the sensor signal. As for the systems with simple analysation of intensity this kind of tracking system is very susceptible to false targets.

The method developed at FIM is based on the correlation tracker system as described before. A new element has been added which continuously adapts the size of the actual incoming object to a stored reference /3,4/. Because of the sensitivity of the correlation function to changes in scale the object size in the image has to be reduced to an original value. This adaptation can be done by zooming or electronically by a digital size transformation.

The principle of determining the scaling factor is shown in Fig. 4. To evaluate the degree of size reduction it is necessary to detect and classify parts of the object or marked details in the object although the target is changing in size. Therefore a reference including the object is generated at time t(n), which is systematically divided into several subreferences, the so-called drift- references (e.g. 4 drift-references in Fig. 4). Depending on the speed of the missile the size

Fig. 4. Determination of the scaling factor

of the target increases at t(n+1) and the position of the sub-references drifts apart. According to the general system for target tracking the object position in the actual scene is found by evaluating the correlation function between the actual scene and the main reference. The difference between the position of the drift-references and the main reference is a measure for the change in size. The scaling factor is computed by analyzing these geometrical proportions and the results of the evaluation of the correlation values of main reference and drift-references.

Fig. 5 shows the picture processing unit for the realized homing system, pointing out the additional parts which compute and compensate the object size in the actual image.

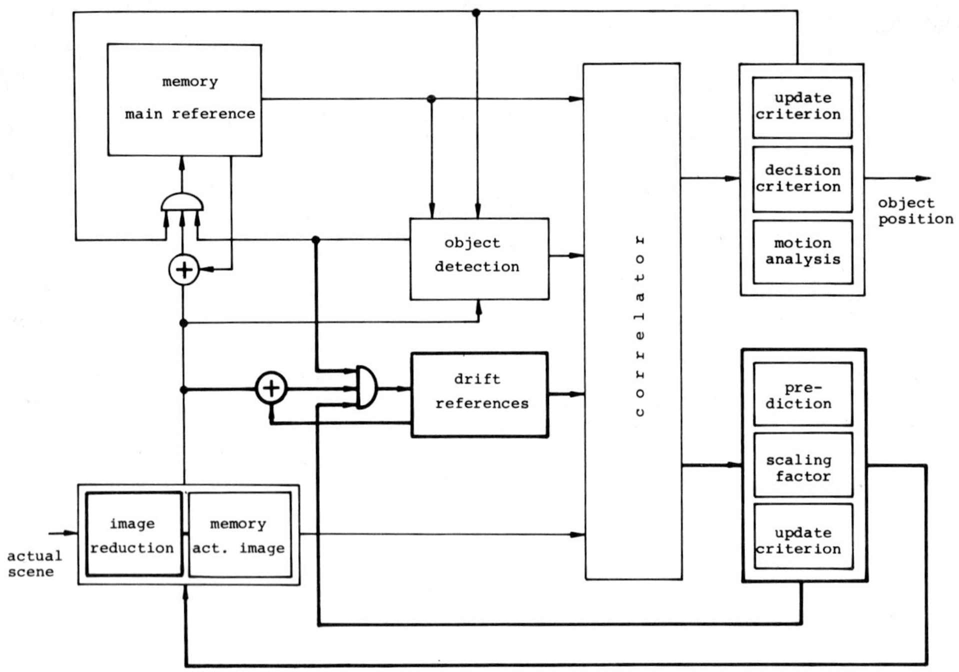

Fig. 5. FIM-Homing-System (Picture Processing Unit)

Results

The simulation with scenes in the visual and IR-range has been very successful and has demonstrated the advantage of the FIM-Tracker-System to other tracking systems when used in a complex scenario.

References

/1/ M. Bohner, H. Gerlach — "Simulation eines Korrelationstrackers" FIM-Bericht Nr. 44, September 1976

/2/ H. Gerlach — "Digitale Bildfolgenauswertung zum Wiederfinden von Objekten in natürlicher Umgebung" FIM-Bericht Nr. 68, Mai 1979

/3/ K.-H. Bers, M. Bohner, H. Gerlach — "Object Detection in Image Sequences" ICPR 1980, Miami, USA

/4/ K.-H. Bers, H. Doll, P. Fritsche, H. Stroh — "Automatische Kompensation der Bildvergrösserung für Zielanflugverfahren" FIM-Bericht Nr. 97, Dezember 1981

TRACK ACQUISITION OF SUB-PIXEL TARGETS

W. E. Snyder
S. A. Rajala

Department of Electrical Engineering
North Carolina State University
Raleigh, North Carolina 27650/USA

1.0 Introduction

In this paper, the problem of automatic identification of sub pixel sized (unresolved) targets moving against a cluttered background is discussed. One possible scenario for such a system is the earth as observed from a geosynchronous satellite. In this scenario, no a priori information is assumed regarding target position or velocity. Hence, such systems are properly called "track acquisition" algorithms rather than "trackers."

Two techniques for track acquisition are discussed in this paper. Both assume targets which are less than one pixel in size, and both assume linear (constant heading) motion of the target.

The algorithms have both been tested on the same simulated data base, and their performance evaluated under disturbances due to noise, spacecraft jitter, and time-varying background.

2.0 General Problem Definition

One can view the problem of tracking in an image sequence as an optimization of a mapping. First, we describe the generic mapping. Let f(x,y,t) represent an image sequence. The set of all such sequences defined on real numbers,

$$F_R = \{f(x,y,t) | x \varepsilon R,\ y \varepsilon R,\ t \varepsilon R\}$$

can be used as the domain of a generalized mapping. This mapping τ, takes images (or image sequences) as input, and produces functions (quasi-images) as output; that is,

$$\tau: F_R \rightarrow F_R^d ,$$

where F^d is the set of output functions.

$$F_R^d = \{f(x_1,x_2,\ldots x_d) | x_1 \ldots x_d \ \varepsilon R,\ f(x_1 \ldots x_d)\ \varepsilon R\}$$

For convenience, we have chosen to deal with only real numbers, however, this is not a theoretical restriction, and many mappings, including the Fourier transform, produce complex outputs. The dimensionality of the output space is denoted by d.

Probably the most familiar such mapping is the Fourier transform

$$\tau: F_R \rightarrow F^3$$

NATO ASI Series, Vol. F2
Image Sequence Processing and Dynamic Scene Analysis
Edited by T.S. Huang

$$\tau(f(x,y,t)) = \iiint f(x,y,t) \exp\{-j(w_x x + w_y y + w_z t)\}\, dx\, dy\, dt$$

$$= F(w_x, w_y, w_t)$$

The Fourier transform preserves the dimensionality of the input, however, this is not a requirement for our generic mappings. In fact, we shall see two instances of mappings which reduce the dimensionality (from 3 to 2 in both cases).

Furthermore, a particular instance of τ may be the composition of two or more mappings, such as a transform followed by a filter.

In all problem environments, the objective is to choose a form for τ so that the resulting output function will have a peak at a point corresponding uniquely to a feature of interest in the original image sequence.

To find necessary conditions for a peak, we could take the gradient in the domain space and set the result to zero. Again, using the simple case of the 1-D Fourier,

$$F(w) = \int f(x)\, e^{-jwx} dx$$

$$\frac{\partial F(w)}{\partial w} = \frac{\partial}{\partial w} \int f(x)\, e^{-jwx} dx \quad ,$$

which can be evaluated and set to zero. Unfortunately, this approach generally fails to produce useful results since the image functions are often ill-behaved and noisy, and the domain space generally has multiple peaks, including peaks at "uninteresting" places, (e.g. a peak at zero velocity, corresponding to the background). Such problems are resolved using a priori knowledge about the nature of f(x), filtering, and hill-climbing techniques.

After the appropriate peak has been found, τ^{-1} may be evaluated at that peak, resulting in the target image, location, velocity, or other parameters of interest.

2.1 Specific Problem Definition: Subpixel Track Acquisition

This problem is motivated by a scenario involving a downward-looking satellite whose mission is to observe the Earth's surface and identify moving objects such as airplanes. If one assumes a geosynchronous orbit, then spacecraft motion (drift) can be reduced to semirandom jitter. The algorithms discussed in this paper are certainly applicable to the moving satellite case, however, a geometric correction preprocessor must be used, and the explanations are clearer if we assume geosynchronous orbits.

2.1.1 The Constant Velocity Assumption

The mapping, τ, mentioned above, is one which searches for consistency in time and space. Different τ's may be appropriate for different problem environments, but in any environment, the concept of "consistency" is critical to the definition of τ. In this problem, we will define consistency to mean constant (two dimensional) velocity. That is, any pixel in S, (the image sequence), having velocity v will map to a point v(s) in the domain space P, and furthermore, all pixels having the same velocity will map to the same point. Note that this means all pixels in the background will map to a single point in P, corresponding to zero velocity for geosynchronous satellites, or to the drift velocity, for other orbits.

In addition, we would like to require that τ be locally continuous. That is, suppose pixels α,β, and λ have velocities V_α, V_β, and V_λ, respectively, and $|V_\alpha - V_\beta| > |V_\alpha - V_\lambda|$, then $d(\tau(\alpha), \tau(\beta)) > d(\tau(\alpha), \tau(\lambda))$, where d is some distance metric defined in P. We will observe that this requirement for continuity is not met in the case of the Hough transform mapping, under perturbations due to jitter, and that this causes failures of the algorithm.

Constant velocity is a restrictive assumption in that it eliminates the possibility of acquiring the tracks of maneuvering or accelerating targets. Our assumption is that the sampling rate is sufficiently high that tracks appear, locally, to be straight lines. Should this assumption be false, the acquisition algorithms can be generalized to detect, say, parabolic tracks at considerable computational expense, or include determination of acceleration along track.

2.1.2 Simulation

Proper testing of the algorithms requires a data base on which to test, and since no satellite exists which gathers the appropriate type of data, it was necessary to develop a quasi-simulation. That simulation was constructed by taking real data (a LANDSAT frame), superimposing a target, and perturbing the resulting image in a way characteristic of satellite performance. In this section, we will describe that simulation in more detail.

A pixel in the LANDSAT image is taken to represent a square region, homogeneous in intensity. The target is likewise a square, 1/4 pixel in area. Depending on where it is located, the target may be located totally within a background pixel, or may overlap pixels. The choice of 1/4 pixel in area is arbitrary, and represents one possible guess at the resolution of the hypothetical satellite.

The detector on board the spacecraft only outputs a single number. This number is derived in simulation by assuming that the target occludes the background, and by passing this characteristic through a Gaussian point spread function, integrated over the detector area. The Gaussian PSF spreads the image of the target over adjacent pixels, as well as mixing the target and background responses. In this way, blur due to optics and atmosphere, as well as the detector's response to subpixel targets, are modeled.

"Clutter" is probably the most ill-defined of all terms in the realm of tracking and track acquisition. After an extensive literature search and discussions with numerous experts, particularly in the realm of radar processing, we concluded that clutter is "whatever gets though the detector that is undesirable." For simulation, however, the concept of clutter needs to be rigorously characterized. We have, thus, defined the term to mean spatially correlated, time-varying perturbations of the background. Our "clutter" therefore, simulates the effect of such events as large flocks of birds on the wing, or cloud shadows.

The satellite undergoes perturbations due to mechanical vibrations, gyro instability, star tracker error, or round off in the geometric correction system. All of these phenomena cause the pointing angle to be different from one frame to another. This effect we define as "jitter" and simulate by adding a small, random, perturbation to the x and y coordinates of the image pixels, varying from one frame to the next. It should be noted that, since Δx and Δy are random, we are not accurately modeling spacecraft vibrations, which, presumably, would be sinusoidal in time.

Finally, white noise is added to the resulting image to model detector dark current. Figure 1 shows the imagery used.

The resulting perturbation model is given in table 1. With this simulation we can examine the performance of algorithms, while adjusting:

1) contrast ratio between target and background;
2) target size;
3) point spread radius;
4) clutter;
5) jitter;
6) noise.

This simulated data base has been used to test both of the τ functions described in this paper.

3.0 Two Approaches to Sub-pixel Tracking

In this paper, we discuss two candidate mappings which function as described in section 2.1. These are τ_C, a mapping based on the Exponential-Area transform, and τ_H, based on the Hough transform. Both of these techniques assume a target moving at constant velocity, and map the image sequence into a space in which the target track is characterized by a peak.

3.1 The Exponential-Area Transform [5]

Like the Hough-tracker, the exponential-area transform assumes a target with constant velocity, thus requiring a track which is a straight line. For pedagogical clarity, we will first define the exponential-area transform of a one-dimensional "image" sequence, and then extend that derivation to two-dimensions.

The basis functions for this particular mapping are one-dimensional, complex-exponential functions. Target motion in the x and y directions is used to generate the target-related one-dimensional complex-exponential sequences which are similarly functions only of x or y. The exponential area transform then maps these target-related sequences into an inverse domain in which target velocity is easily measured. The process which achieves this mapping is a linear transformation that measures the projection of the target-related sequences along each basis function. The largest projection corresponds to the target velocity along the specified coordinate.

Let f(x,t) be a one-dimensional signal with an amplitude boundary $0 \leq x \leq a$ and time duration $0 \leq t \leq c$, which is formed by sampling the signal at regular intervals in time t. Define a general transformation as:

$$F(k_x,t) = \int_0^a f(x,t)\, w(x,k_x)\, dx \qquad (1)$$

where w(x,k) is a weighting function of order k. A properly chosen weighting function will allow us to use this weighted area, g(t), to find the velocity of a moving object while discriminating against a stationary background.

If $w(x,k) = \exp(-j2\pi k_x x/a)$, equation (1) becomes:

$$F(k_x,t) = \frac{1}{a} \int_0^a f(x,t)\, e^{-2\pi k_x x/a}\, dx \quad . \qquad (2)$$

The parameter k_x is chosen such that the ratio (k_x/a) controls the number of complete cycles along the x-axis of the region. If a moving object is present in the processed region, the object extent will move with velocity v_x through the sequence and, more significantly, through the cycles of the spatial exponential function, which remain fixed with respect to object motion. As a result, the sequence $F(k_x,t)$ displays a complex exponential property whose complex frequency is directly proportional to the velocity v_x of the object, (with all stationary objects and the background appearing as a zero frequency component.) Figure 2 illustrates this operation when

$$w(x,k_x) = \mathrm{Re}\,\{\exp(-j2\pi k_x x/a)\} = \cos\,(2\pi kx/a).$$

Figure 1. Imagery Used in the Simulation.

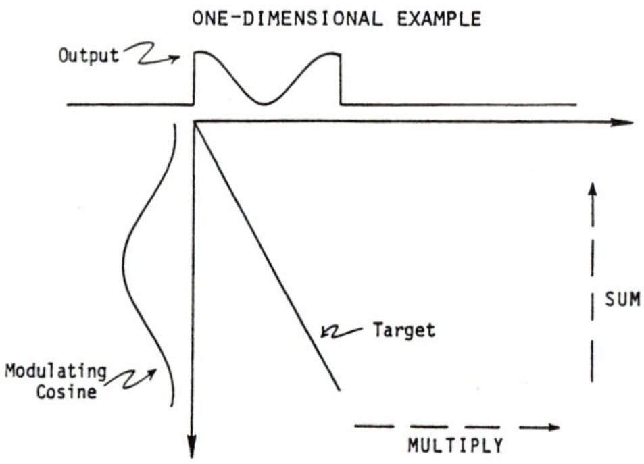

Figure 2.

$$g(x,y,t) = \int (w(x',y',t)T(x',y') + (1-w(x',y',t))\,((B(x'(x',y',t)))*h(x,y) + n(x',y',t)\,dx\,dy$$

where

$B(x,y)$ is the stationary background.

$B'(s,y,t)$ represents "clutter", a time-varying, correlated perturbation of the background due (for example) to cloud shadows.

$x'=x+\Delta x$, and $y'=y+\Delta y$ with Δx and Δy time-varying random variables, representing spacecraft jitter.

h is the point spread function of the optics, assumed Gaussian. No model of the PSF of the atmosphere is included in this model.

$w(x,y,t)$ is a binary valued window function exactly coincident with the target, providing a mechanism for modeling occlusion.

$T(x,y)$ is the intensity profile of the target.

$n(x,y,t)$ is time-varying white noise.

Table 1. Simulation of Sub-pixel Target Motion

Spectral analysis techniques (such as the one-dimensional Fourier transform) may be used to extract the velocity information. Applying the Fourier transform results in:

$$G(k_x,f) = \frac{1}{c} \int_0^c F(k_x,t)\, e^{-j2\pi ft/c}\, dt$$

The velocities of any moving objects in the one-dimensional signal will appear as peaks in the frequency function $G(k_x,f)$.

We can immediately generalize the cosine to two dimensions: Let the time sequence of images be $f(x,y,t)$ for $0 \leq x \leq a$, $0 \leq y \leq b$, and $0 \leq t \leq c$. The velocity desired is then a vector in the image space $V = [v_x,v_y]^T$. The exponential-area transform becomes:

$$F(k_x,t) = \frac{1}{a} \int_0^b \int_0^a f(x,y,t)\, e^{\frac{-j2\pi k_x x}{a}}\, dx\, dy,\quad k_x = 0,\pm 1,+2,\ldots \tag{3}$$

$$F(k_y,t) = \frac{1}{b} \int_0^b \int_0^a f(x,y,t)\, e^{\frac{-j2\pi k_y y}{b}}\, dx\, dy,\quad k_y = 0,\pm 1,+2,\ldots \tag{4}$$

$$\tau_c(k_x,f) = \frac{1}{c} \int_0^c F(k_x,t)\, e^{\frac{-j2\pi ft}{c}}\, dt,\quad f = 0,0,\ \pm 1,\ \pm 2,\ldots \tag{5}$$

$$\tau_c(k_y,f) = \frac{1}{C} \int_0^c F(k_y,t)\, e^{\frac{-j2\pi ft}{c}}\, dt \tag{6}$$

The convolution can thus be performed in n log n time using the fast Fourier transform.

By detecting a peak in the spectrum of $G(k_u,f)$ for a fixed k_u, the speed of an object is determined. At the peak, the frequency is divided by k_u and the object speed in dimension u results. Once the moving object has been found, the sign of the velocity is determined from the phase of $F(k_u,t)$.

Using the complex exponential form also allows us to extract the image of the moving target by notch-filtering out the DC term and applying an inverse Fourier transform. In addition, further processing of the magnitudes of the $\tau(k,f)$ coefficients results in the extraction of acceleration information along x and y.

The results of applying the exponential transform is quite effective in distinguishing target tracks and velocities. Its performance will be discussed in section 4 of this paper, and as one might predict, the algorithm is most sensitive to time-varying noise.

A first step in increasing noise immunity is taken by averaging. (Those frequency spectra indicating a specific velocity over their corresponding, k_i.) Recall that for a target of velocity v, a peak occurs at f = vk where k is the frequency of the modulating complex exponential. Thus, within bounds defined by the sampling theorem, we can choose any modulating frequency and still extract the target velocity. By choosing several such frequencies and averaging the resultant functions, the contributions due to noise are reduced. Several experiments (section 4) have verified this.

3.2 The Hough Transform

In this analysis, we consider motion to be the only attribute of interest. That is, we assume that targets produce well defined tracks. Furthermore, we assume that the targets are sub-pixel in size, and moving at a constant velocity.

The first step in this process is the application of a moving target indicator algorithm, whose output enhances those areas in the image sequence which potentially contain moving targets. The algorithms which perform this change detection function are referred to as moving target indicators (MTI) and will be described in the next section.

A constant velocity target will trace out a straight line in image sequence space. That is, if we apply an MTI algorithm to an image sequence and combine the results into one frame, straight lines in that frame will correspond to target tracks. Thus, with this approach, the target identification task can be broken into several components:

1) moving target indication;
2) track (straight line) detection;
3) track refinement;
4) determination of target position and velocity along track.

3.2.1 Moving Target Indication

If the background is assumed constant over time (e.g. geosynchronous satellite), then a simple frame-to-frame difference will produce an image containing only those pixels which have changed. Clutter, noise, and other image perturbations will cause simple frame-to-frame differences to produce false hits when used as a moving target indicator. The false hit rate can be reduced by using more elaborate MTI schemes [6]. We have chosen to use the third-order difference, because it couples acceptable performance with simplicity. In general, an nth order difference is:

$$D^n (x,y,t) = \sum_{i=0}^{n} (-1)^i \binom{n}{i} f(x,y,k-i) , \tag{7}$$

which for the third order difference is

$$D^3 (x,y,t) = f(x,y,t) -3f(x,y,t-1) +3f(x,y,t-2) -f(x,y,t-3).$$

3.2.2 Track Acquisition

We refer to the output of the MTI algorithm as an "MTI frame." Several MTI frames must be combined to produce a track map. This combination can be accomplished in several different ways:

- Threshold several MTI frames and OR them together
- Add (average) several MTI frames
- Recursively combine MTI frames

The third technique preserves the direction of motion information and eliminates the need for storing more than two frames. In recursive combination,

$$T(x,y,t) = a \mid D(x,y,t) \mid + b \cdot T(x,y,t-1), \qquad (8)$$

where a and b are constants.

In this approach, the MTI frames are weighted in time. The combination of k frames results in

$$T(x,y,t) = a \sum_{i=0}^{k} b^i D(t-i). \qquad (9)$$

$T(x,y,t_0)$, is then a single two dimensional function which reflects the activity in the image during the k frames prior to time t_0. In this function (image), straight lines will correspond to target tracks, with the intensity varying along the track. This intensity variation is shown in figure 3. Those tracks are detected using the Hough transform.

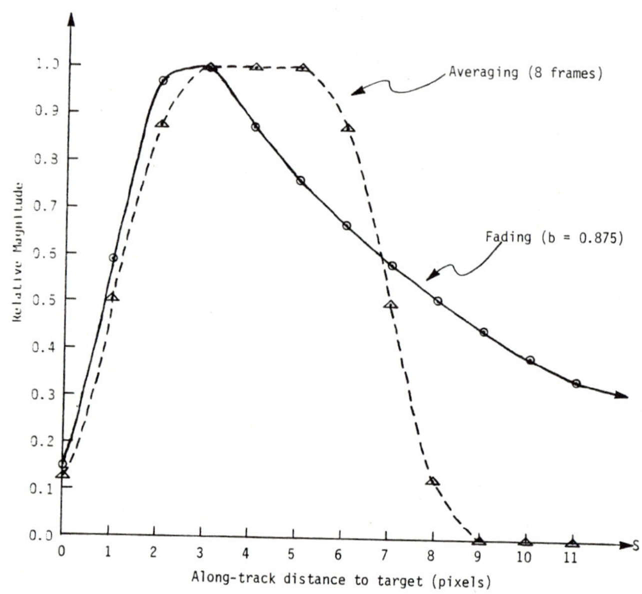

NOTE: Target travels at constant velocity of 1 pixel/frame and induces the same change in the intensity of each pixel in its path. Current target position is at S = 0.

Figure 3. Target Track Intensity Distribution.

The Hough transform is well known [4] and will be described here only briefly.

A line in the track map may be described by the equation

$$\rho = x \cos\theta + y \sin\theta \qquad 0\leqslant\theta\leq\pi. \qquad (10)$$

Then for any single point $\langle x_i, y_i \rangle$ in the track map, an equation may be written:

$$\rho = x_i \cos\theta + y_i \sin\theta , \qquad (11)$$

describing a sinusoid in ρ-θ space. All such sinusoids corresponding to points along the same straight line in x-y space will intersect at the same point in ρ-θ space.

Rather than explicitly computing intersections, the intersections are usually computed by using accumulator arrays. In such an array, ρ-θ space is quantized into an array of accumulator cells. Eq. (11) is applied to each point of interest in the track map, and each accumulator cell lying along the resultant sinusoidal curve is incremented.

Deans [3] has shown that the accumulator array approach makes the Hough transform equivalent to the discrete implementation of the Radon,

$$R(\rho,\theta) = \tau(g(x,y)) = \iint g(x,y)\ \delta(\rho - x\cos\theta - y\sin\theta)\ dx\ dy . \qquad (12)$$

This fact has led to some interesting observation:

Theorem: The Hough transform is a linear operator.

To prove this, we will show that the Hough transform applied to an arbitrary linear combination of functions gives the same result as first transforming the functions and then linearly combining them. Proof will be by induction on the number of terms in the combination. We will first make use two lemmas

LEMMA 1: $\mathcal{H}(af) = a\mathcal{H}(f)$ where f is an arbitrary function of x and y.

Proof of this lemma is trivial from the Radon definition

LEMMA 2: $\mathcal{H}(f(x,y) + g(x,y)) = \mathcal{H}(f(x,y)) + \mathcal{H}(g(x,y))$

Again, using the Radon definition,

$$\mathcal{H}(f(x,y) + g(x,y)) = \int [f(x,y) + g(x,y)]\ \delta(\rho - x\cos\theta - y\sin\theta)\ dx\ dy$$

$$= \int f(x,y)\ \delta(\rho - x\cos\theta - y\sin\theta)\ dx\ dy + \int g(x,y)\ \delta(\rho - x\cos\theta - y\sin\theta)\ dx\ dy$$

$$= \mathcal{H}(f(x,y)) + \mathcal{H}(g(x,y))$$

From lemma 2, the theorem is true for 2 terms. Assume it is true for n terms. That is, let f(x,y) be a linear combination of n functions. Again, from lemma 2, the theorem is true if one more term is added.

Finally, we have the interesting result,

Corollary: The Hough transform of a difference picture (of any order) is the (corresponding) difference of the Hough transforms of the original images.

The impact of this corollary, as it affects computational complexity, architecture, and noise/false hit immunity, has yet to be investigated.

3.2.3 Track Refinement

The Hough transform, applied to the track map, defines the desired mapping function,

$$R(\rho,\theta) = \tau_H\ (f(x,y,t)) = \iint T(x,y)\ \delta(\rho - x \cos\theta - y \sin\theta)\ dx\ dy$$

where T is the track map defined in eq. (9).

A peak in the Hough transform space (accumulator array) indicates a probable target track. The impact of noise can be reduced by maximum entropy quantization [1], rather than uniform quantization of the accumulator array.

Due to noise and quantization error, the actual target track may be located anywhere within a strip on either side of the line corresponding to the peak in the accumulator array (figure 4). We assume, therefore, that the target exists at a point (x_T,y_T) within this strip.

The best estimate of the location of the target can be determined by a matched filter especially tuned to the characteristic intensity profile along track (figure 3). The fact that the target lies within the strip rather than precisely on the proposed track is handled by an algorithm based on a second Hough transform which essentially projects (x_T,y_T) from within the strip onto the proposed track. Details of this algorithm are provided in [2].

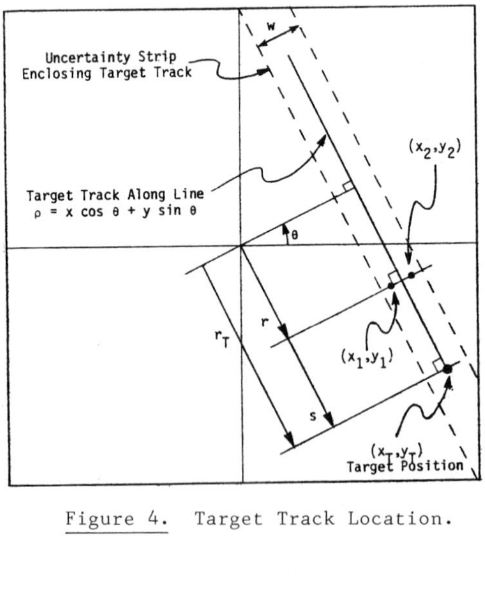

Figure 4. Target Track Location.

4.0 Performance

Both the exponential-area transform technique and the Hough transform technique have been applied to the simulated data base described in section 2. Both techniques work essentially perfectly in the absence of time-varying perturbations. Sensitivity to such perturbations has been the objective of the tests described in this section. Figure 1 shows the image frame on which the targets were superimposed.

4.1 Exponential-area Technique Performance

Figure 5a,b shows the results of applying the generalized area transform to the simulated data base with jitter and clutter suppressed. Figure 5a is the x component of the Fourier transform, and figure 5b is the y component. In this case, the SNR was 23.3dB, and the highest peak accurately corresponded to the target track.

Figures 6-8 show percentage velocity error versus pixels of jitter, SNR, and signal-to-clutter ratio. In all cases, averaging improved the performance. The performance measure should be taken as the point where velocity error deviates from zero. The apparent step function behavior is due to the simple peak search technique used in frequency space. When the peak due to perturbations exceeds the peak due to target motion, the algorithm choses the incorrect peak. No past history and hence no a priori knowledge of probable target velocity is used. Such knowledge would significantly improve performance.

Combining the various perturbations severely complicates the detection process. Figure 8 shows performance with jitter and white noise, as a function of signal to clutter ratio.

4.2 Hough Technique Performance

Figure 9 shows a track map from a sequence containing six different targets. Table 2 lists the target characteristics.

Target	Velocity (pixels/frame)	Relative Magnitude
1	1.0	1.0
2	0.5	1.0
3	1.0	1.0
4	1.0	0.707
5	2.0	1.0
6	1.0	1.0

Table 2. Target Characteristics

The frame sequence was divided into windows in which each window contains exactly one target, and each window was processed independently by the algorithm. Figure 10 shows the Hough transform resulting from a window which does not contain a target while figure 11 shows the result of a window containing a target.

The target location algorithm employing the matched filter was applied to all six target tracks. Only in the case of targets 2 and 4 was there any appreciable error in target position and heading. These errors are plotted in figures 12 and 13. In each of these cases, the heading was 180° in error. Except for frames 11 and 12 of target 2, this heading error resulted in significant position error.

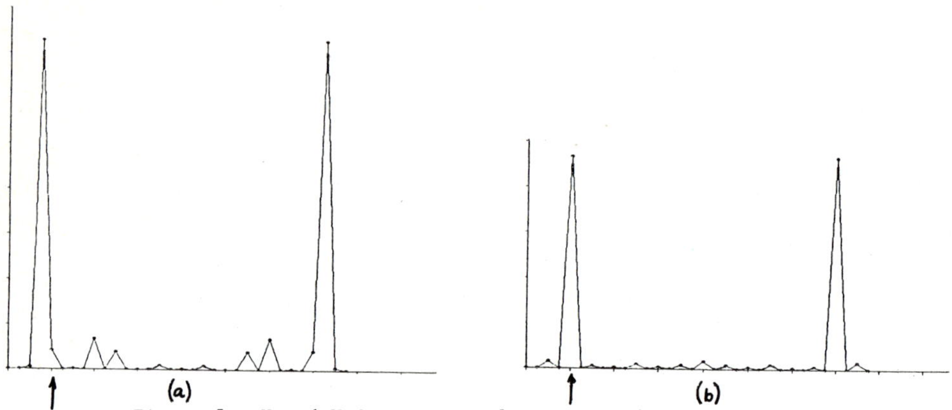

Figure 5. X and Y Components of Exponential Area Transform. Correct peak is denoted by arrows.

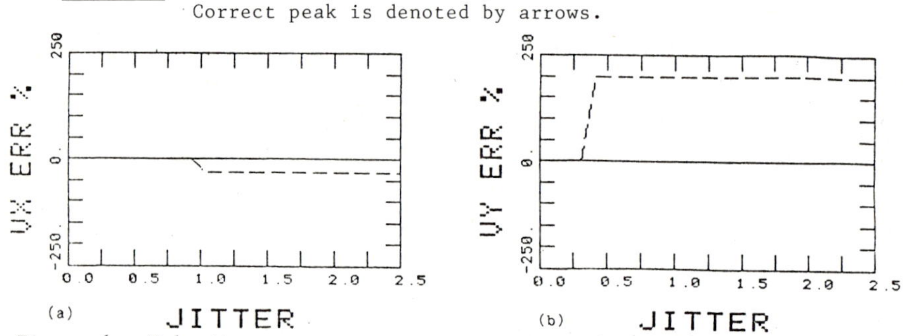

Figure 6. Velocity error percentage versus pixels of jitter in LANDSAT image. Solid line is new averaging algorithm, dashed line is generalized area transform. (a) x direction, k=6 (b) y direction, k=4.

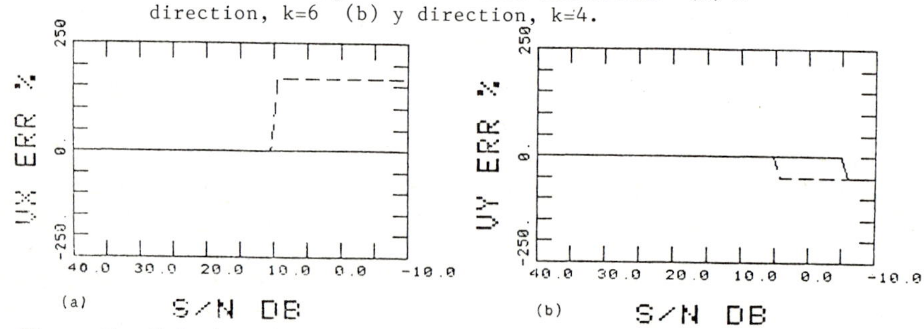

Figure 7. Velocity error percentage versus mean square signal (target) to white noise ratio in LANDSAT image. Solid line is averaging algorithm, dashed line is generalized area transform. (a) x direction, k=6 (b) y direction, k=4.

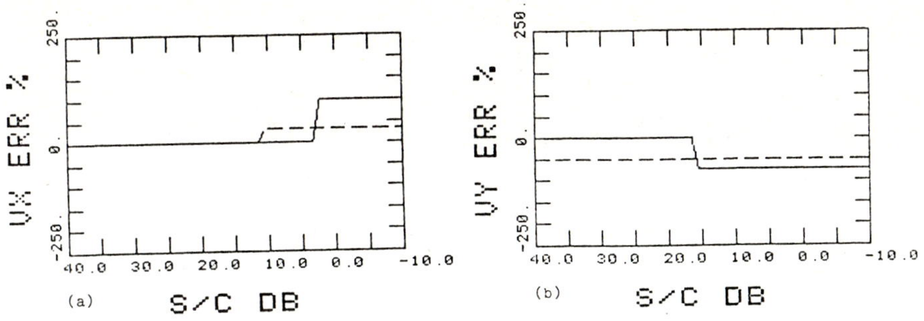

Figure 8. Velocity error percentage versus signal (target) to clutter ratio with .2 pixels of jitter and a 23.3 db signal to white noise ratio in LANDSAT image. Solid line is averaging algorithm, dashed line is generalized area transform. (a) x direction, k=6 (b) y direction, k=4.

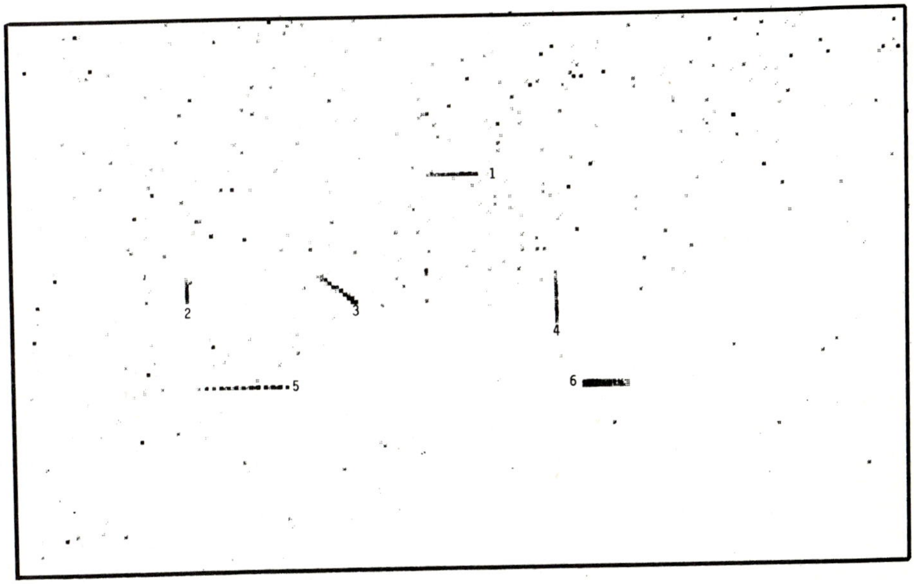

Figure 9. Target track map showing 6 targets. The high velocity of target 5 causes its discontinuous appearance.

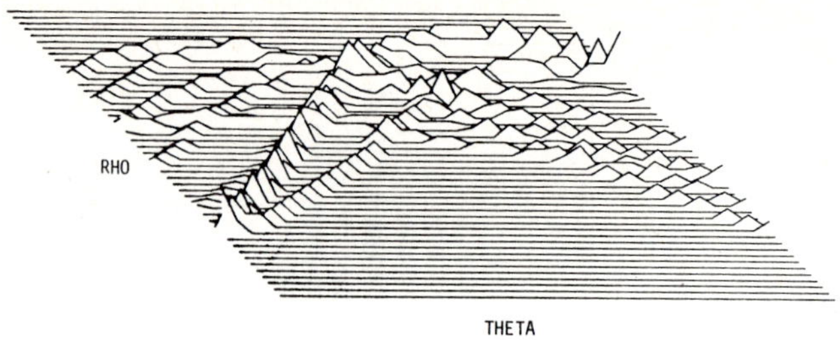

Figure 10. Hough Transform for Window with No Target.

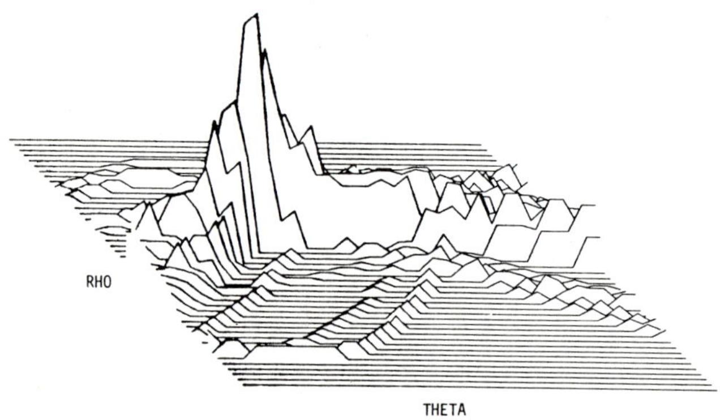

Figure 11. Hough Transform for Window Containing a Target.

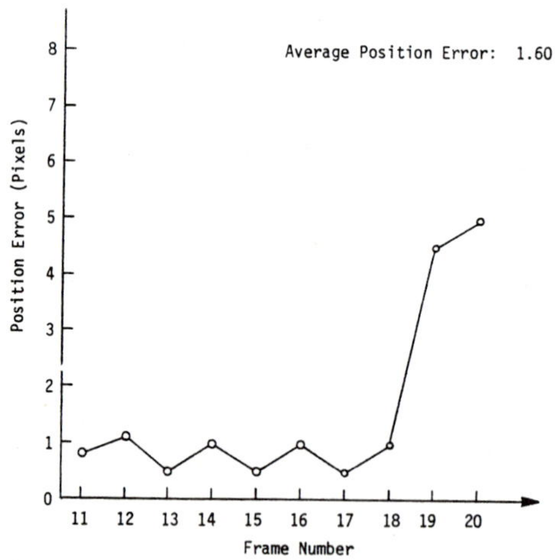

Figure 12. Target 2 Position Error.

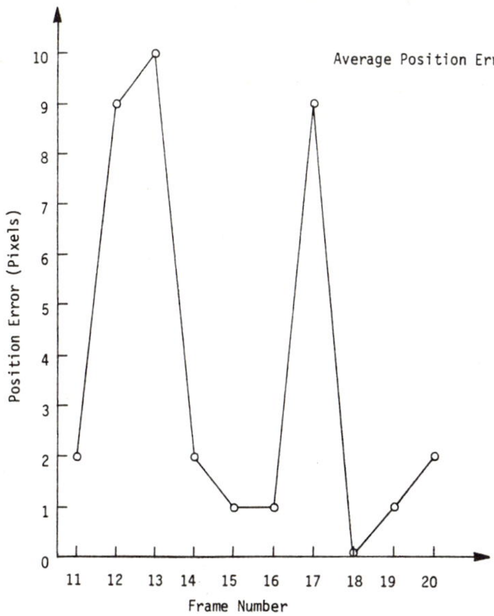

Figure 13. Target 4 Position Error.

5. Conclusion

Two techniques have been presented which detect sub pixel sized targets in sequences of satellite images. Both techniques employ a mapping to another space and a gradient search in that space.

One property which is noteworthy in comparing the two mappling τ_H and τ_C. τ_H is specifically designed for targets of one pixel in extent or smaller, whereas τ_C can accommodate large targets, mapping all pixels with the same velocity together. Thus, has greater potential for motion sequentation.

As of this writing, work is still underway in the analysis and evaluation of these techniques.

Since the intensity profile along a track differs for various target velocities (in the Hough transform case), a bank of filters sensitive to these differences could be used to select the best matching profile and hence determine target velocity. Sampling causes problems with this approach for high velocities, however.

Also under investigation are exact analytic and experimental comparisons of the two techniques to evaluate their relative performance.

Finally, appropriate architectures are being considered for implementation at high speeds.

(1) Cohen, M. and Toussaint, G., "On the Detection of Structures in Noisy Pictures," Pattern Recognition, Vol. 9, 1977.

(2) Cowart, A., Snyder, W., and Ruedger, H., "The Detection of Unresolved Targets Using the Hough Transform," CGIP, October, 1982.

(3) Deans, S. R., "Hough Transform from the Radon Transform," IEEE PAMI-3, 1981.

(4) Duda, R. and Hart, P., "Use of the Hough Transformation to Detect Lines and Curves in Pictures," CACM, Vol. 15, 1972.

(5) Rajala, S., Riddle, A., and Snyder, W., "Application of the One-Dimensional Fourier Transform for Tracking Moving Objects in Noisy Environments," CGIP, October, 1982.

(6) Williams, R., and Fried, D., "Signal Processing for Clutter Rejection in a Quasi-staring Sensor," 23rd SPIE International Technical Symposium, San Diego, CA, August, 1979.

Acknowledgements

The authors are grateful to Alfy Riddle, who did much of the work on the exponential area transform, to Alan Cowart, who likewise did much of the development of the Hough technique, and to the Research Triangle Institute for their support.

A Pre-Processor for the Real-Time Interpretation of Dynamic Scenes

Volker Graefe
German Armed Forces University Munich

Abstract

A pre-processor is described that can be combined with a conventional computer into a powerful instrument for experiments in the field of computer vision. It contains several micro-computers, all of which work in parallel and have access to the digitized video signal at the same time. Each micro-computer executes its own program analyzing the video signal with respect to one particular aspect.

Results of experiments involving the pre-processor are reported.

1. Introduction

In the literature on computer vision there are not many statements on the amount of computing time required to segment an image or to interpret a scene. It appears, however, that times between 100 s and 1000 s are not uncommon (e.g. Reddy 1978, Shirai 1979, Moravec 1980), if general-purpose computers or minicomputers are used. Reddy (1978) estimates that the segmentation of an image alone requires 1000 "operations" per pixel. For an image of 256 by 256 pixels this would amount to $6.5 \cdot 10^7$ operations, or 65 seconds on a computer that executes 10^6 such operations per second.

Obviously, conventional computers are too slow to interpret scenes in real time if "real time" is understood as referring to something comparable to human speed.

But also regarding their input-capabilities, conventional computers are insufficient. Continuous interpretation of a dynamic scene requires that a video signal be continuously monitored. A digitized video signal contains $3 \cdot 10^6$ bytes per second or more (depending on the spatial and temporal resolution of the system), and for a conventional computer it is very difficult to continuously input data at this rate and process them intelligently at the same time.

NATO ASI Series, Vol. F2
Image Sequence Processing and Dynamic Scene Analysis
Edited by T.S. Huang

Uhr (1980) postulates an array of 10^4 to 10^8 special processors for a computer vision system that would probably match human performance in most respects. Although it is possible that in the future such systems may become available and that ways of programming them will be found, much simpler systems are needed if the goal is to do experimental work now.

Our answer to these problems is a hierarchical system containing a pre-processor that is able to handle the input of video-data and permits some degree of parallelism of its internal operations, and a conventional master-computer (figure 1). The system is to be used in experiments where the control of a dynamic system depends on the continuous interpretation of a scene. In such experiments the pre-processor performs various low-level vision tasks while the master-computer uses task-specific knowledge to combine the results of the low-level analysis into an abstract image containing all the task-relevant aspects of the scene. If, for instance, the task were to drive an automobile on a highway, the knowledge involved might include the general layout of roads, vehicle dynamics, and projective geometry.

The amount of time needed for communication between the pre-processor and the master-computer should be minimized in the interest of efficiency. The pre-processor therefore generates a compact intermediate description of the scene, containing only such information as is necessary to accomplish the given purpose, e.g. controlling some mechanism or navigating a vehicle through some environment, perhaps driving a car on a highway.

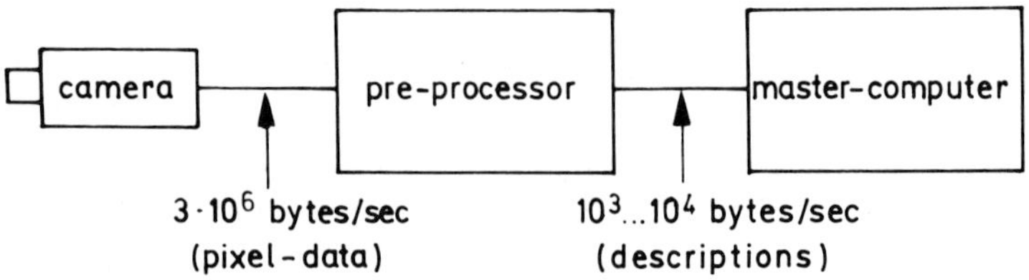

Figure 1. Hierarchical computer vision system

2. The pre-processor

2.1 Basic concepts

One basic concept for the pre-processor stems from the observation that typically only small areas within an image contain information that is relevant for the control of a dynamic system (fig. 2). Therefore only these small areas ("windows") have to be considered in order to generate the intermediate description, and, in the interest of efficiency pixels from other areas should not even be input and stored by the pre-processor*.

Totally ignoring most parts of an image is a large step towards the goal of removing undesired redundancy from the video signal and helps at the same time to sidestep the problem of inputting vast amounts of pixel data; it poses, however, a problem: how does the processor know which parts of an image it should input and analyze before it has looked at the image? The answer is that the system is not looking at a sequence of unrelated images, but at a continuous stream of images of a specific, gradually developing scene. Therefore, if it has successfully analyzed

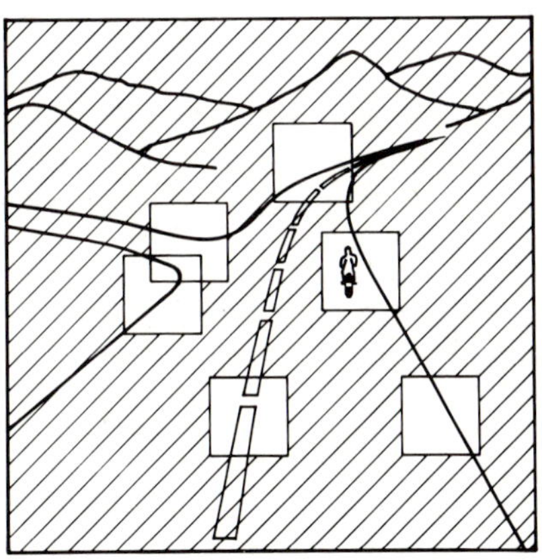

Figure 2. To extract all the information necessary to guide a car safely on a road, a small number of appropriately placed "windows" is sufficient. Small rectangles in the figure indicate where such windows might possibly be placed*.

---

*The idea to use several freely movable windows within an image was proposed by E. D. Dickmanns.

the previous images, then it knows where in the images the essential features have been and how they are moving. General knowledge about the type of scene being analyzed then makes it possible to infer where approximately these features will appear in the next image; as a matter of fact, this prediction is usually quite easy, since the time span covered by the prediction is only 20 or 40 milliseconds.

This process needs, of course, in order to get started, an initialization phase during which the system has ample time to find the relevant features of the scene. This is similar to a person who suddenly finds himself in an unexpected and unfamiliar environment and also needs time to orient himself.

The pre-processor contains several micro-computers (called "picture processors"), one for each window. This makes it possible to analyze all windows simultaneously and in parallel, using either similar or totally different programs for each window. Because the subtasks of analyzing different windows are highly unrelated, there is no synchronization and almost no communication between different picture processors. This makes it easy to program the pre-processor, since each picture processor can be programmed as if it were a stand-alone computer, avoiding almost all the problems generally associated with parallel processing.

Providing each window with its own picture processor makes the pre-processor extremely flexible, because it permits each picture processor to select at any time the size, shape, and position of its window independent of any other window. In particular, windows may overlap; this may happen by coincidence when two windows tracking different objects in the image are moved to similar positions, but it is also possible to lock several windows to each other in order to let several picture processors analyze the same part of the image simultaneously with different programs.

2.2 Physical realization

Figure 3 shows the structure of the pre-processor. The camera delivers a video-signal with 256 visible lines per frame and 50 frames per second. The analog-to-digital converter samples each visible line 256 times and distributes the digital video-signal, together with some auxiliary signals, to all picture-processors via the video-bus. There can be up to 15 picture-processors. This number is more than sufficient to conduct interesting experiments, and it is small enough to let the entire pre-processor fit into a standard card cage.

The picture-processors receive commands from, and report results to, the system controller via the system bus. The system controller controls the system bus and manages the communication between the picture processors and the master-computer.

In a system with many processors, often the connecting bus (in this case the system bus) is a bottleneck. The picture-processors, however, do not have to communicate very much. They typically receive only short messages, e.g. instructing them to reposition their window or to execute a specific program. They also typically transmit only short messages, e.g. a few parameters or the coordinates of some object or feature. If we assume that there are 15 processors and that each of them receives and transmits 20 bytes every 20 ms, then the total bus load is only 30 000 bytes per second. In reality, the bus load is even lower.

The system controller is not a bottleneck either, because usually it just passes on messages, not doing much more than some blocking or unblocking of messages.

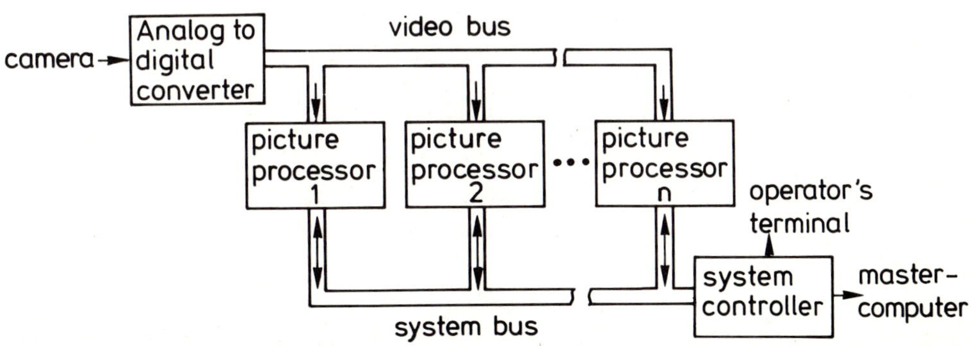

Figure 3. Block-diagram of the pre-processor.

2.3 The picture processor

Figure 4 shows the structure of the picture processor. The CPU is an Intel 8085 A microprocessor, total memory is 16 K bytes. 4 DIP-switches and 4 lamps permit limited communication with the operator. The system bus interface consists essentially of 2 FIFO-buffers of 64 bytes each, one for input and one for output. Most communication between a picture processor and the system controller involves these buffers. If the system controller sends a message to a picture processor, it simply writes it into the picture processor's receiving buffer without waiting for the picture processor to enter a state where it is ready to receive messages. If the picture processor sends a message to the system controller, it just writes it into its own output buffer and proceeds without waiting for the system controller to get ready to read it.

The advantage of this scheme is that no processor has to wait for another processor to start communicating, and no processor is forced to interrupt what it is doing and start communicating instead. This not only improves the efficiency of the system but it also simplifies the development of software.

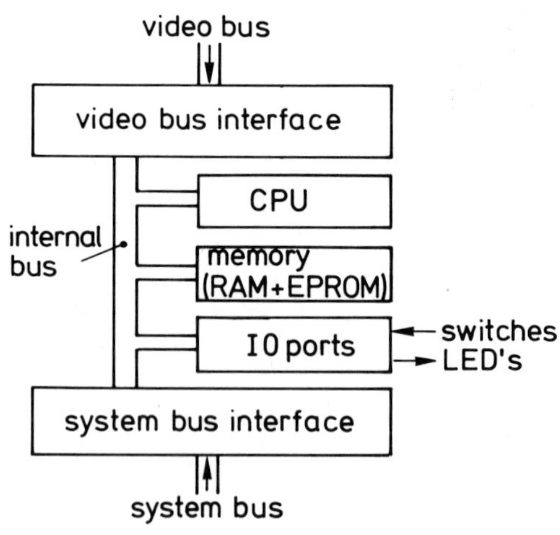

Figure 4. Block-diagram of the picture processor

The task of the video bus interface is to read those data-bytes that belong to the selected window, and to store them in the picture processor's memory. The data rate on the video bus is too high to store pixel data directly in the processor's main memory, therefore data from the video bus are first stored in a fast FIFO-buffer and then transferred from there to the main memory by a DMA-controller (fig. 5).

The window is a rectangle, its sides are parallel to the borders of the image. The size, shape, and location of the window are controlled by the CPU by sending the corresponding parameters to the window controller. There, counters for lines and pixel-columns of the image are initialized appropriately, and they control the FIFO-buffer in such a way that it accepts only those pixel-data from the video bus that belong to the selected window.

The fact that each picture processor has its own window controller explains why the windows of different picture processors are totally independent of each other. This independence greatly simplifies the writing of programs for the picture processors, as these programs normally need not consider the location of any other window when they position their own window. Of course, it is also possible to write programs that let one window track some other window.

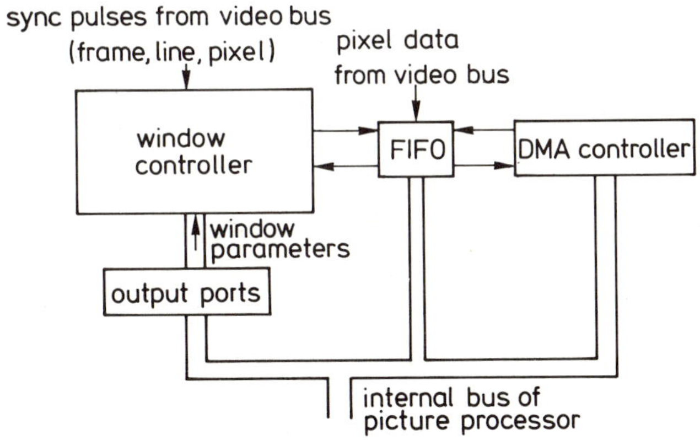

Figure 5. The video bus interface. The window parameters (location, size, and shape of the window) are stored in output ports, and updated by the CPU only when they have to be changed.

2.4 Software

As a result of its simple structure, the pre-processor does not need a large operating system, certainly nothing that resembles a conventional multi-task multi-processor operating system with all its overhead.

Each picture processor has an identical "operating system" of about 4 K bytes, consisting essentially of utility programs for
- setting windows and reading data from the video bus
- communication over the system bus
- loading and starting programs

and some debugging aids.

The system controller has an "operating system" of about 6 K bytes, consisting essentially of utility programs for communication with the master-computer and with the picture processors, and of a multi-processor monitor as a debugging aid. The monitor permits, for instance, to inspect and to modify the contents of memory locations in every processor. It uses a terminal for communication with the operator.

Task-specific application programs are written and compiled on a microcomputer development system. They can be loaded into the picture processors via the system controller, or, if they are of a more permanent nature, they can be stored in EPROM's that are inserted into the picture processors.

3. Experimental Results

In order to evaluate the performance of the pre-processor, experiments were performed in which dynamic systems were controlled and where no sensors other than the computer vision system were used. The systems were inverted pendulums of different lengths on an electric cart (fig. 6). A DEC PDP 11/60 with an RSX 11M operating system was used as the master computer.

The controller that stabilized the pendulum is described by Meissner and Dickmanns (in this volume). It requires as input a time-series of sampled values of the coordinates of two different points on the pendulum.

Within the pre-processor, one picture processor is assigned to each of the selected points. They track the assigned points with their windows and report the horizontal coordinate of their point to the system controller. There, the coordinates of both points are combined into one message that is then sent to the master computer. Both picture-processors have identical application programs; the size of this program is about 4 K bytes.

The inverted pendulum, because of its inherent instability, is very critical regarding the response time of its controller: if the controller is too slow, the pendulum topples over. This characteristic makes the inverted pendulum useful for verifying the real-time-behavior of a computer vision system, as it imposes an absolute limit on the time available for the interpretation of each image. In our case, the picture processors need about 35 ms to input their image data, to find the desired coordinates - this includes some digital filtering to reduce the quantization noise - , and to compute the position of their windows for the next cycle. The camera needs 20 ms to complete a scan of the

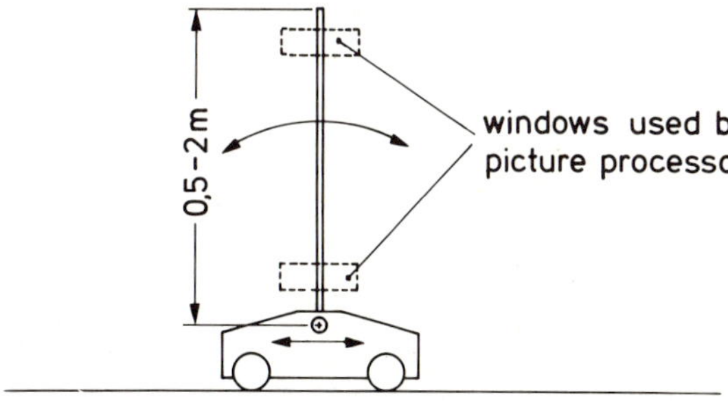

Figure 6. The electro-cart with the inverted pendulum. An inverted pendulum (a straight rod in this case) stands vertically on an electric cart. Its lower end is attached to the cart by means of a hinge. The pendulum is therefore unstable and would normally topple over. It can, however, be balanced by appropriate motions of the cart. This is accomplished by a controller that applies at all times the appropriate voltage to the drive motor of the cart.

image; the picture processors, therefore, use every other frame, sampling the scene every 40 ms. A high sampling rate is desirable, because it facilitates both the tracking of the pendulum by the pre-processor and the control of the system by the master-computer. However, the overhead associated with the communication between the master-computer and the pre-processor made it necessary to communicate data only once every 80 ms. (In the near future we shall use a different master-computer, and we expect the overhead to become negligible then.)

The experiments show that the pre-processor is a sufficiently fast and accurate sensor for the controller to balance the inverted pendulum about as well as a man could do it if he balanced the rod directly on his hand. (Balancing it with the cart is more difficult, by the way; people cannot do it.)

Since the processing-time of the picture processors is only 35 ms, it could be expected that they would be fast enough. The accuracy of the pre-processor was determined separately in a number of experiments. Figure 7 shows as an example the results of one such experiment where a vertical line on the screen of an oscilloscope was translated horizontally at a known speed. The camera was looking at the screen, and the pre-processor measured the distance traveled by the vertical line within

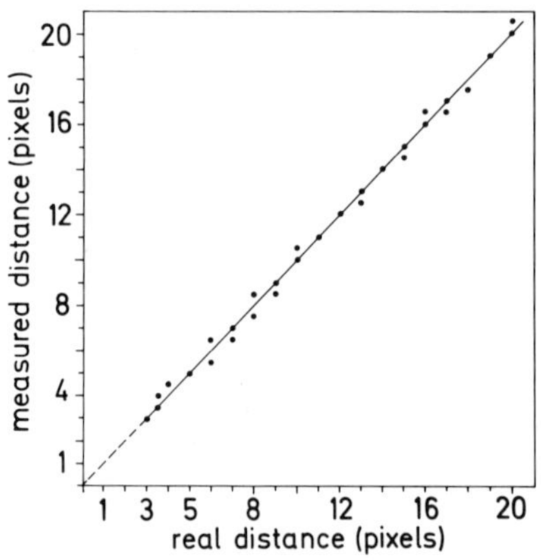

Figure 7. Measurements of the horizontal motion of a vertical line. The line was displayed on the screen of an oscilloscope and moved horizontally at a known, constant velocity. The graph shows the true and the measured distance traveled by the line within 20 ms. The velocity of the motion (referred to the image) was varied in a series of measurements between 150 pixels per second and 1000 pixels per second.

time intervals of 20 ms. Motion blur imposes an upper limit on the permissible velocity of the line; at a speed of 1000 pixels per second the image of the line, which is only 2 pixels wide at rest, is smeared over an interval more than 20 pixels wide; the corresponding loss of contrast then makes it difficult to detect the line. Special, nonlinear filtering algorithms, developed by G. Haas (1982), made it possible for the picture processor to locate its target with good accuracy ($\pm$ 1/2 pixel) even under such adverse conditions.

In other experiments we found that almost always the measured position of a target is in error by not more than $\pm$ 1 pixel; the greatest errors we have found are $\pm$ 2.5 pixels, but these are rare exceptions. The errors are caused by quantization noise and by motion blur. There may be additional errors caused by geometric distortions of the lens or the camera but they are neither specific to the pre-processor nor would they influence the balancing of the pendulum.

The main shortcoming of the present system is that the algorithms that find the pendulum in the image are rather simple-minded. They are, therefore, easily confused by any dark, elongate, approximately vertical object of the same diameter as the rod. The problem is, that more sophisticated algorithms would require much more computing time or a much more powerful picture processor. A partial solution was found by taking advantage of the flexible architecture of the pre-processor, and tracking more than two points on the pendulum using more than two windows. The master-computer knows that all points should lie on the same straight line, and it applies this knowledge to eliminate those points that appear questionable. This redundancy makes the recognition process more robust in the presence of occluding objects or confusing features in the image.

4. Conclusions

The pre-processor has demonstrated its ability to contribute to the interpretation of dynamic scenes in real time. The problem of inputting vast quantities of data has been sidestepped by the implementation of small windows that can be moved freely all over the image. Several subtasks can be executed simultaneously, but the difficulties normally associated with the programming of parallel processes are avoided. This is due to the facts that the task of low-level vision is easily decomposed into rather unrelated sub-tasks, and that the picture processors within the pre-processor are highly independent of each other.

The most serious shortcoming of the pre-processor is the limited computing power of its present picture processors. This restricts it to simple images, excluding, for the time being, many natural scenes from real-time interpretation. The root of this problem is the limited speed of our microprocessor 8085A, in particular the lack of sufficient index registers, which results in large amounts of overhead when accessing elements of two-dimensional arrays. The obvious solution is to redesign the picture processors using one of the much more powerful micro-processors that have now become available; this should provide a gain in processing speed of more than one order of magnitude. An even greater increase in speed can be expected if several superfast processing elements are used within each picture processor.

The basic concepts and the architecture of the pre-processor, however, appear valid and useful for the study of various methods of low-level computer vision in a real-time environment.

Acknowledgements

The pre-processor in its present form is largely the result of work by G. Haas. The experiments have been conducted by G. Haas in close cooperation with E. D. Dickmanns and H. G. Meissner, who also developed and implemented all the necessary software for the master-computer.

References

Haas, G. (1982)
Meßwertgewinnung durch Echtzeitauswertung von Bildfolgen
(Doctoral dissertation, to be published)

Meissner, H.G. and E.D. Dickmanns (1983), in this volume
Control of an unstable Plant by Computer Vision

Moravec, H.P. (1980)
Obstacle Avoidance and Navigation in the Real World by a Seeing Robot Rover
Robotics Institute, Carnegie Mellon University, Pittsburg, PA

Reddy, R. (1978)
Pragmatic Aspects of Machine Vision
in Hanson and Riseman (ed.): Computer Vision Systems, pp. 89-98, Academic Press

Shirai, Y. (1979)
On Application of 3-Dimensional Computer Vision
Bul. Electrotech. Lab., Vol. 43 No. 6, pp. 358-377

Uhr, L. (1978)
"Recognition Cones", and Some Test Results
in Hanson and Riseman (ed.): Computer Vision Systems, pp. 363-377, Academic Press

Control of an Unstable Plant by Computer Vision

Meissner*, H.-G. and Dickmanns¤, E.D.

Abstract

The measurement and control concept for a computer vision system is discussed that works on the basis of conventional TV-signals. A preprocessor system of parallel independently working window processors reduces the data flow. The system is used to stabilize an inverted pendulum on an electro-cart with closed loop corner frequencies up to 1 Hz. The performance achieved by the closed loop system and the sensitivity to parameter changes and disturbances is discussed. The vision system allows some features of higher intelligence as compared to conventional measuring systems.

Content

0. Introduction
1. Plant description: an electro-cart balancing an inverted pendulum
   1.1 linear mathematical model for the balanced rod and the cart
   1.2 A more realistic model with nonlinearities
2. The measurement concept
   2.1 geometry and redundancy
   2.2 initialization
   2.3 time delays and sampling rate
3. The control concept
   3.1 dynamical model for state observation
   3.2 closed loop dynamics
4. Performance analysis of the closed loop system
   4.1 cart track orientation
   4.2 camera orientation
   4.3 camera zoom
   4.4 hilly terrain
   4.5 higher level intelligence
5. Conclusions
Literature

*Dipl.-Ing., scientific coworker
¤Dr.-Ing., professor for control
} Aerospace Department HSBw Munich, Germany

NATO ASI Series, Vol. F2
Image Sequence Processing and Dynamic Scene Analysis
Edited by T.S. Huang

Introduction

A large variety of animals on earth prove by their existence, that vision systems may be very adequate to control complex motions: e.g. a humming bird hovering at the same relative distance to a blossom, while this is moved by the wind. Nature has succeeded in building all necessary equipment for measuring and control into a volume of about 1 cm^3. Much smaller systems are known (e.g. insects). In nature, vision seems to occur only in connection with the ability of active locomotion.

With the electronics evolution taking place right now, human technology is approaching packing densities for electronic systems that become more and more comparable to organic devices. For vision systems, however, computing power is still marginal, but paralleling of more powerful microcomputers may lead to systems that can perform some useful tasks in the domain of motion control in the near future and that may have growth potential to control complex motions of aerospace and other systems in the long run.

As an entry into this interesting field of development, a simple task was designed to become acquainted with some basic problems of this control loop using a visual measuring system. Since realtime control is an essential feature, the use of previous information for the present measuring and control step seems mandatory. To unify this information, a dynamical model of both the measuring system, the system to be controlled and the actuation system has been chosen. This model is part of the control algorithms and allows the determination of state variables of the system that are not directly measured but help to improve the closed loop performance. Information from this model may also be used to adapt parameters of the measuring system.

In order to achieve a vivid demonstration, a familiar unstable plant has been chosen which is easily changed and known to everybody from his own experience.

1. Plant description: An electro-cart balancing an inverted pendulum

Figure 1 shows a physical sketch of the system and the natural state variables. The cart is about 1 ft long and has a mass of about 4 kg. The electric drive (EM) has a maximum power of about 300 W and can accelerate the cart at 0.8 Earth gravity units. The maximum values of φ are $|\varphi_{max}| = 11^{o}$ by hardware constraints to garantee identical initial conditions in the range of linearized trigonometric functions around 0 ($\cos\varphi \approx 1$, $\sin\varphi \approx \varphi$).

The cart is controlled by a voltage from a power amplifier which in turn is controlled from a minicomputer PDP 11/60 with a process interface. The general arrangement of the closed control loop is shown in fig. 2. The only physical connection between the plant and the measuring/controlling devices is the voltage input to the electric drive of the cart (lower left corner). The upper left corner shows a sketch of the viewing area of the camera (tilted by 90° to make it visible in the paper plane). These data are transferred into the TV camera by optical sight (photon flow). Two lines in the image are marked window 1 and 2. They are specified by distance arrows y_u and y_o above the hinge point of the pendulum. Vertically centered around these lines are rectangular windows free to move horizontally.

The position and tilt of the pendulum is measured through these windows (see below); therefore the dynamical model for the pendulum will be described in terms of the window-center coordinates.

1.1 Linear mathematical model for the balanced rod and the cart

The common point of motion is the hinge point, where the lower end of the rod is forced to move horizontally. Introducing vertical and hori-

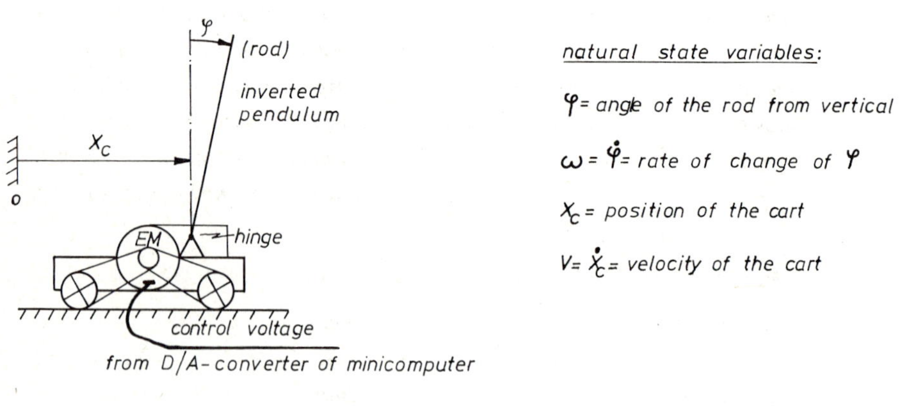

Figure 1. Electro-cart balancing an inverted pendulum

zontal forces at this point and setting up the moment equation for the rod around its center of gravity one obtains the differential equation

$$\ddot{x}_c \cos\varphi - g \sin\varphi + \frac{2}{3} l\ddot{\varphi} = 0 . \tag{1}$$

Here it has been assumed that the rod is homogeneous with constant cross section along its length l. Introducing the "reduced pendulum length" $l_r = 2\ l/3$ and linearizing the trigonometric function around $\varphi = 0$ yields

$$\ddot{\varphi} - g/l_r\ \varphi = -1/l_r\ \ddot{x}_c . \tag{2}$$

Expressing the rod coordinates at the vertical center of the windows in terms of φ and x_c (figure 2 upper left) leads to

$$x_o = x_c + y_o\varphi ; \qquad x_u = x_c + y_u\varphi \tag{3}$$

and

$$v_o = \dot{x}_o = \dot{x}_c + y_o\dot{\varphi} ; \qquad v_u = \dot{x}_u = \dot{x}_c + y_u\dot{\varphi}$$

$$\dot{v}_o = \ddot{x}_c + y_o\ddot{\varphi} ; \qquad \dot{v}_u = \ddot{x}_c + y_u\ddot{\varphi} .$$

Subtracting the last 2 eqs. from each other yields

$$\dot{v}_o - \dot{v}_u = \ddot{\varphi}(y_o - y_u) \quad \text{or} \quad \ddot{\varphi} = (\dot{v}_o - \dot{v}_u)/(y_o - y_u) . \tag{4}$$

Doing the same for (3) and inserting both results into (2) gives

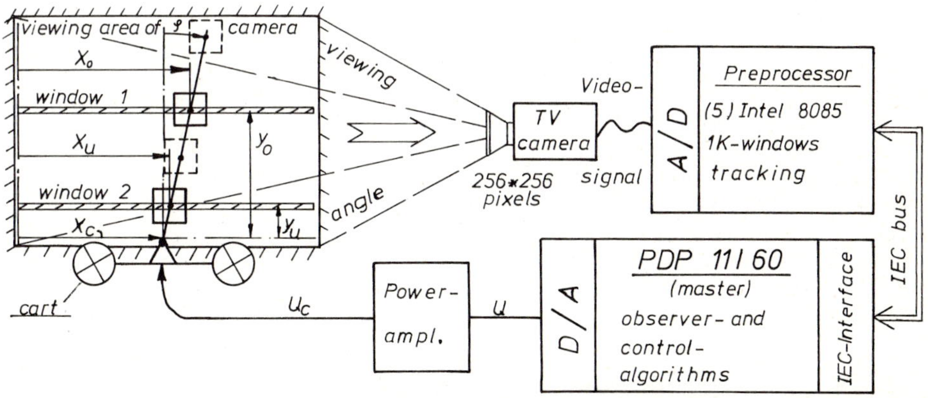

Figure 2. Blockdiagramm of closed loop control system

$$\frac{\dot{V}_o - \dot{V}_u}{y_o - y_u} - \frac{g}{l_r}\frac{x_o - x_u}{y_o - y_u} = -\frac{1}{l_r}\ddot{x}_c \, , \tag{5}$$

which for $y_u = 0$, i.e. $x_u = x_c$, and $y_o = l_r$ simplifies to the diff. eq.

$$\ddot{x}_o - g/l_r \, x_o = - g/l_r \, x_c \, . \tag{6}$$

This indicates, that there are 2 points on the pendulum to be tracked in order to abtain a relatively simple model: the base point at the hinge and the point 2/3 of the rod-length upward. At this point an acceleration of the cart does not show up as an acceleration of the upper observed point of the rod.

From eq. (6) it can be seen, that there are 2 eigenvalues of equal magnitude $|\lambda|$, one stable and one unstable. The following table gives some numbers for different rod lengths l:

$l/[m]$	0,5	1	2
$l_r/[m]$	0,33	0,67	1,33
$\lvert\lambda\rvert/[s^{-1}]$	5,4	3,8	2,7

The cart has essentially the dynamic characteristics of the electric drive with a shifted time constant which may be modelled by the transfer function (u = control voltage)

$$\frac{x_c(s)}{u(s)} = G_c(s) = \frac{k_c a}{s(s+a)} \, , \quad k_c \approx 0{,}14\left[\frac{m/s}{V}\right] \, , \quad a \approx 2[s^{-1}] \tag{7}$$

taking the loads and the transmission gear into account.

This leads to the diff. eq. for the cart

$$\ddot{x}_c + a\dot{x}_c = k_c a \, u(t) \, . \tag{8}$$

With the state vector $\underline{x}^T = (x_o, V_o, x_c, V = \dot{x}_c = \dot{x}_u)$ eqs (7) and (8) may be written in standard state eq. form for linear systems

$$\dot{\underline{x}} = \begin{pmatrix} \dot{x}_o \\ \dot{V}_o \\ \dot{x}_c \\ \dot{V} \end{pmatrix} = \begin{pmatrix} 0 & 1 & 0 & 0 \\ g/l_r & 0 & -g/l_r & 0 \\ 0 & 0 & 0 & 1 \\ 0 & 0 & 0 & -a \end{pmatrix} \begin{pmatrix} x_o \\ V_o \\ x_c \\ V \end{pmatrix} + \begin{pmatrix} 0 \\ 0 \\ 0 \\ k_c a \end{pmatrix} u \tag{9}$$

with the measuring matrix

$$\underline{\underline{C}} = \begin{bmatrix} 1 & 0 & 0 & 0 \\ 0 & 0 & 1 & 0 \end{bmatrix}$$

yielding the output vector

$$\underline{y} = \begin{bmatrix} x_o \\ x_c \end{bmatrix} = \underline{\underline{C}}\,\underline{x}\ . \tag{10}$$

The velocities V_o and V cannot be measured directly.

1.2 A more realistic model with nonlinearities

The model described above has deliberately been derived as a linear one in order to be able to apply the powerful methods of linear control theory. The real system, however, has marked nonlinearities, one of which cannot be linearized and appears around the nominal operating condition of the closed loop system: it is a dry friction (plus an additional stiction for $V = 0$) torque M_F of the cart, which opposes the direction of motion

$$M_F = -\ M_F^{*}\ \mathrm{sign}(V)\ . \tag{11}$$

As a result the cart remains at rest until the input voltage is larger than 0,8 to 1 V; then the acceleration of the cart is achieved by the remaining torque of the motor exceeding M_F^{*}.

Saturation of the motor is reached for commanded voltages beyond 30 V. This, however, is of minor influence under usual operating conditions. In section 4 it will be shown, that the limit cycling occurring around the commanded cart position can be contributed to the dry friction in connection with the unstable inverted pendulum.

2. The measurement concept

Until now the camera has been kept at rest so that motion occurring in the picture is inertial motion in the scene. The effect of angular motion of the camera will be discussed in section 4. Details of the measurement concept are given in [1] and [2].

2.1 Geometry and redundancy

The view of the scene as it is seen from the camera is given in fig. 2 left. Up to 4 windows may be used by the presently available system. Each window may contain up to 1 K pixels arranged in a rectangular pattern. Using a 32 * 32-pixel window with maximum resolution (all neighbouring lines), 8 windows would be required to cover the total picture fully in 1 dimension. Since the camera has a resolution of 256 * 256 pixels, 64 windows would be required to cover the picture entirely.

For the present task with 1 vertical line of interest in the picture, 4 windows are considered sufficient for studying some basic questions of redundancy, 2 are the minimum to obtain reasonable resolution and a large enough operating range in x_c and φ.

With the requirement that the picture covers the reduced pendulum length of the 2 m-rod ($l_{r_2} = 1.33$ m), say 1.4 m, the resolution of the camera is about 5.5 mm. The diameter of the rod is 12 mm. To control the 1 m-rod, resolution may be stepped up to about 3 mm, however, this simultaneously reduces the operating range of the cart to about 75 cm for the fixed camera mount. The camera position nominally is about 3 m off the carttrack at a right angle.

For the present task each window is assigned a line position which is kept constant while tracking is done horizontally. The maximum speed of the cart is about 4 m/s. For a TV-framerate of 25 per second (40 msec) a point may move about 16 cm (or 29 pixels for a viewing area of 1.4 m) from one frame to the next. Each picture processor tries to keep its window centered at the rod. Since motion blur distorts the image of the rod, special algorithms [1] take care of the effects of the direction of motion and the imaging properties of the camera. This is essential for handling these high dynamics. From the last 2 frames an estimate for the velocity of the observed points is computed by numerical differencing and this is used to position the window for the next measurement. The actual position of the observed point is obtained by adding the x-coordinate of the window to the measured coordinate within the window. This number is transferred to the master computer.

If the master computer receives the results of only 2 windows, it has no choice but to accept them for good or to stop. If it receives the results of more windows, the situation is much more favorable, since the known line structure of the rod allows some redundancy checking. This may be used, exploiting the dynamical model described below, to redirect a window of the preprocessor which has been trapped by some other feature in the scene.

2.2 Initialization

Before getting started with the fast real time control loop, the master computer checks several operating conditions.

It directs the preprocessor to search for the rod using one window around the center line of the picture. If a feature is met which may be part of the rod, the preprocessor tracks it upward and downward to verify it as a rod. If the candidate line leaves the picture at one of the sides, this is reported to the master computer which then applies an appropriate voltage to slowly drive the cart into the picture. If a lower end is found, this is assumed to be the hinge point.

After the rod has been recognized, the desired number of windows is positioned and the system is switched into the tracking mode.

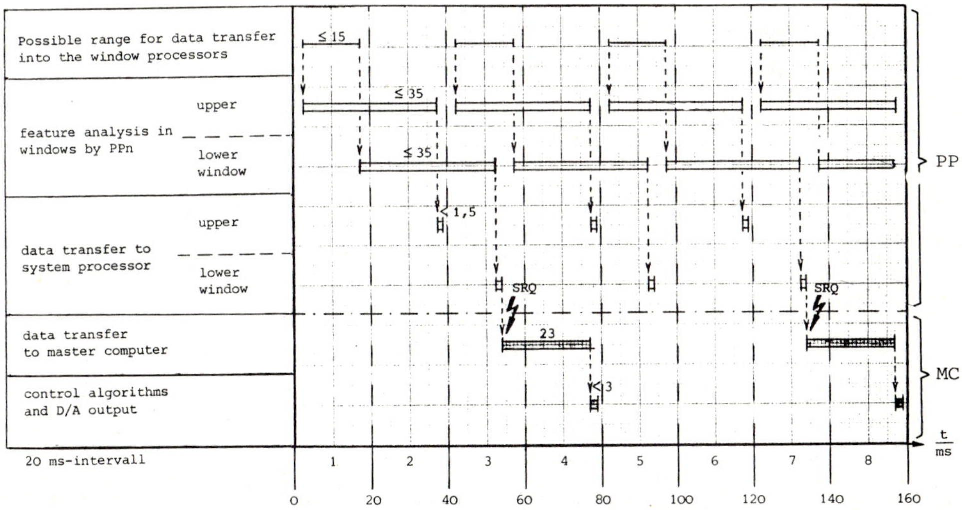

Figure 3. Time delays for a 2-window system from measurement of single features until output of control voltages (see also [5])

2.3 Time delays and sampling rate

The video beam requires a finite amount of time to scan the picture (~64 μs per line). Therefore, features in the image are not sampled simultaneously. If these features are moving, there is a systematic error connected to the line difference and the speed of motion. If the upper and lower windows are e.g. 160 lines apart, the time difference is about 10 ms. Within this time a feature with a velocity of 2 m/s travels 2 cm. Depending on the optical resolution, this may correspond to up to 6 pixels. Therefore, a correction to this is done in the window processors.

Figure 3 shows the time delays occurring from video frame start at 0 until output of the commanded voltage to the power amplifier at 80 ms. Therefore 80 ms have been chosen as sampling period.

This yields the most simple sampling control system and has proven to be sufficient at least for the 1 m and 2 m-rods.

3. The control concept

To optimize system performance, a state feedback controller has been selected which requires all state variables to be known. However, only the positions can be measured directly. A wellknown method to obtain the missing variables is state observation by a dynamical model [3].

3.1 Dynamical model for state observation

This model is not just given by eq. (9), but has to take the digital controller and the time delay into account. With a digital controller,

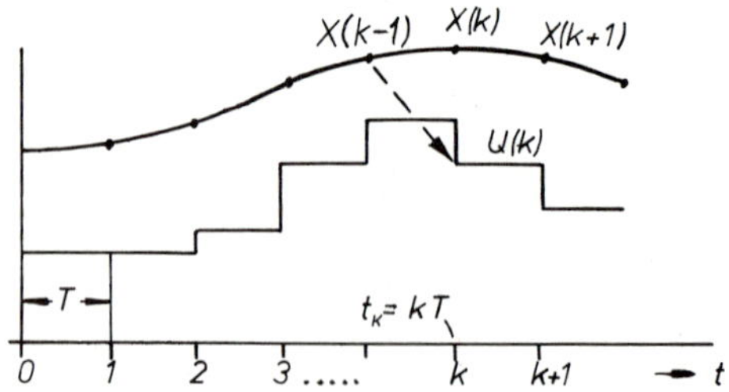

Figure 4. Sampling control

u is discontinuous at the sampling times and constant in between (figure 4). While the state x is a continuous function of time, it is measured only at the sampling points. Its value at point (k+1)T can be obtained from its value at kT and the applied constant control u(kT) by integration of eq. (9) over T. The resulting difference equation is

$$\underline{x}(k+1) = \underline{\underline{\Phi}}'(T)\ \underline{x}(k) + \underline{g}(T)\ u(k)\ , \tag{12}$$

where the transition matrix $\Phi'(T)$ and the control coefficient vector g(T) are constants depending on the sampling period T and the coefficients of eq. (9). In the closed loop system, the control u(k) is computed from the state variables; since the measuring and data transfer takes time T, one can write

$$u(k) = f\left(\underline{x}(k-1)\right)\ . \tag{13}$$

It is an established procedure (e.g. [4]) to deal with this time delay by introducing one additional state variable for each sampling period delay time. In the case at hand, the observation state vector is defined as (see [5])

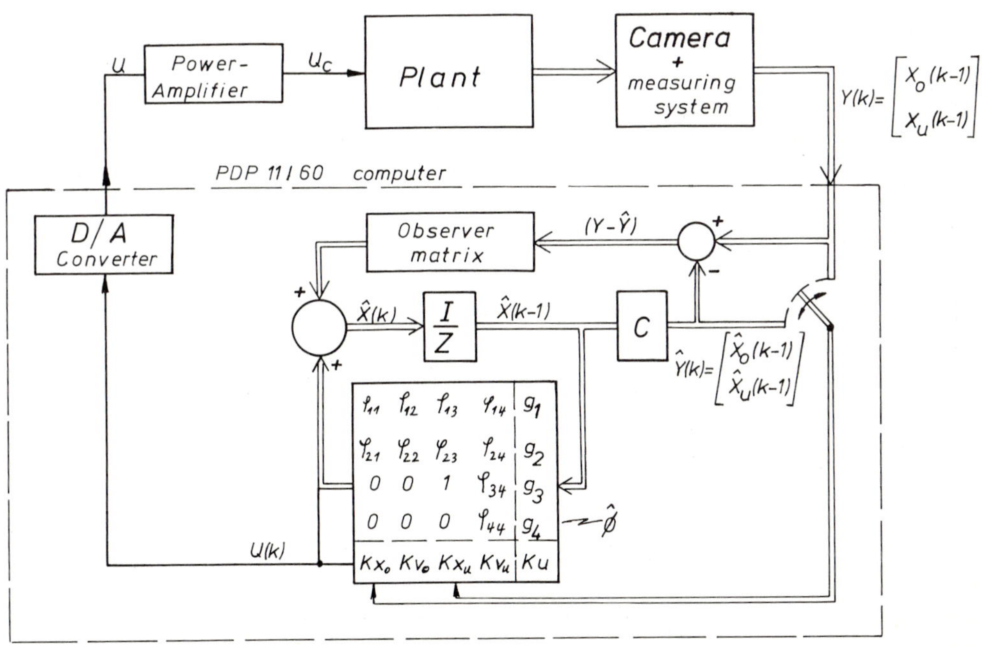

Figure 5. Control loop with time delay and augmented state observer

$$\underline{\hat{x}}^T = [\hat{x}_o, \hat{V}_o, \hat{x}_u, \hat{V}_u, \hat{u}] \tag{14}$$

and the feedback control at time kT is given by

$$u(k) = \underline{K}^T \underline{\hat{x}}(k-1) \tag{15}$$

with

$\underline{K}^T = [K_{xo}, K_{Vo}, K_{xu}, K_{Vu}, K_{\hat{u}}]$ the controller gains yet to be determined. Several state and reduced order observers have been investigated [5]. Figure 5 shows one possible implementation for full state reconstruction. Going from 4th to 5th order to cope with the time delay resulted in a considerable improvement in closed loop performance. The reduced order observers require less computing time and have, therefore, been selected as standard solution for the microprocessor realizations.

3.2 Closed loop dynamics

The gains (eq. (15)) have been determined to yield closed loop poles at $\zeta = 1/\sqrt{2}$ and $\omega_{n1,2} = 3\sqrt{2}$ (double poles) or $\omega_{n1} = 3\sqrt{2}$ and $\omega_{n2} = 6\sqrt{2}\ s^{-1}$. The observer poles were positioned at 12, respectively 24 on the negative real axis. A typical gain vector is (first case above for 1 m rod length)

$$\begin{aligned} K_{xo} &= -170\ V/m \ ; & K_{xu} &= 159\ V/m \\ K_{Vo} &= -44.4\ V/\tfrac{m}{s} \ ; & K_{Vu} &= 24.4\ V/\tfrac{m}{s} \end{aligned} \quad ; \quad K_{\hat{u}} = 0.104 \tag{16}$$

This means that the controller has transfer function zeros at

$$-Z_1 = K_{xo}/K_{Vo} = 3.83 \text{ and } -Z_2 = K_{xu}/K_{Vu} = 6.52\ s^{-1}, \tag{17}$$

the first of which approximately cancels one pole of the inverted pendulum. Equilibrium of the system ($V_o = V_u = 0$) is given by the condition

$$K_{xo}\, x_o + K_{xu}\, x_u = 0 \tag{18}$$

for the cart on an horizontal plane. This allows the determination of the equilibrium cart position for a camera, the lines of which are not adjusted to the horizontal plane (roll angle around the visual axis).

It also indicates, that no calibration of the system to the vertical is required: a misalignment results in a changed equilibrium position. This makes the handling rather simple.

4. Performance analysis of the closed loop system

Stabilization of the real system is not that nice as linear system theory indicates. There is a small dry friction opposing the motion of the cart, so that it comes to rest despite the fact that a small control voltage is still applied. This results in a limit cycling of the order of magnitude of one decimeter at a frequency of 0.3 to 0.5 Hz, which, of course, is driven by the unstable inverted pendulum. Aside from this, closed loop performance may be well explained by linear systems theory. Several aspects will be discussed below.

4.1 Cart track orientation

Figure 6 shows the range of operation within which acceptable performance was achieved with the nominal parameter setting: it is up to $\pm 45^{\circ}$ off the normal to the line of sight of the camera. The limit cycling deteriorates to higher amplitudes before the system goes unstable eventually.

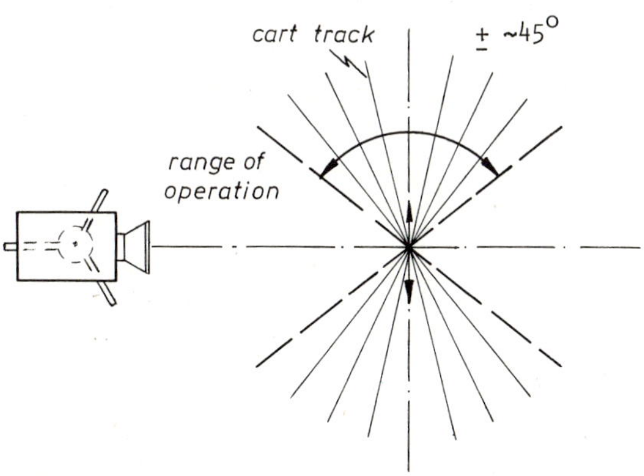

Figure 6. Variation of cart track orientation

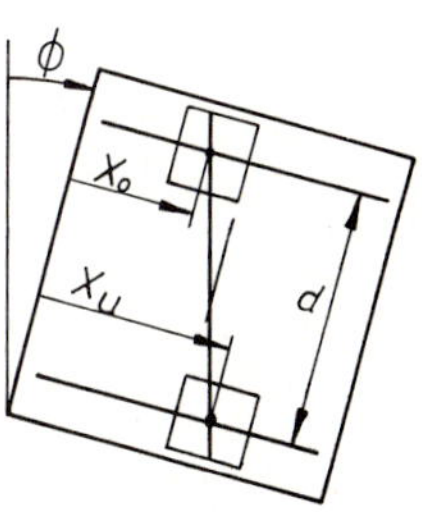

Figure 7. Camera-roll-geometry

4.2 Camera orientation

Since the measured quantities are coded in camera fixed coordinates, motions of the camera lead to different mappings of objects in the scene. The computer, however, commands the interesting object (rod and cart) according to its control law to a state where finally the control output is around 0. The control output is linked to the measured quantities by the dynamic observer and the state feedback law.

4.2.1 Azimuth angle

If the roll angle is 0 and the cart track horizontal, equilibrium is achieved for $x_o = x_u$. This position is specified at the center of the picture. The controller tries to keep the system at this state. If the camera is turned around the vertical axis, this results in a motion of the cart such that it settles down again where the mapping of the pendulum is at the center of the imaging plane in the camera. This is to say, that the cart can be moved back and forth by turning the camera.

4.2.2 Roll angle

A little more difficult to understand is the result of a roll angle of the camera around its optical axis. At first glance one would suppose that if x_o in the camera is not correctly above x_u (above is defined by the earth gravity vector), the system should go unstable because of contradicting informations in the image and the real world. This would happen if there were two independent controls and x_o and x_u would be prescribed independently in camera coordinates. With only one control, only one quantity can be prescribed at will for $\dot{\underline{x}} = 0$; this one quantity may, however, be a weighted sum of several states. In the present case, eq. (18) yields the solution: The cart position $x_c = x_u$ will stabilize at

$$x_u = - x_o K_{xo}/K_{xu} . \qquad (19)$$

Let d be the distance between the upper and lower window (in pixel units), then from fig. 7 it can be seen that

$$x_o = x_u - d \tan\Phi .$$

This into (19) and resolved for x_u yields a relationship between the camera roll angle Φ, the rel. window position d and the state controller gains K_{xo} and K_{xu}

$$x_{u\Phi} = \frac{d \tan\Phi}{1 + K_{xu}/K_{xo}} \text{ , in corresponding pixel units.} \tag{20}$$

It may be noted from fig. 7, that when the cart moves, the vertical window position relative to the hinge line will move resulting in a wrong model implemented in the computer. This will deteriorate closed loop performance. Roll angles up to 15° have been successfully tried; the rolling has to be gradual in order not to induce large transient motions of the unstable basic system.

4.3 Camera zoom

Changing the focal length of the lens results in general in a vertical repositioning of the windows on the rod and in a change of resolution. The first effect would require an adaptation of the observer model (eq. (5) instead of (6)) and corresponding changes of coefficients, while the latter influences the control loop gains directly.

Depending on rod length and prescribed closed loop dynamics, the nominal design parameters (kept constant) resulted in stable operation for a zoom change of up to 50%.

4.4 Hilly terrain

The system is able to drive through hilly terrain without getting wrong measurement results as is the case with a potentiometer mounted on the cart and measuring the pendulum angle relative to the cart body. The camera has a notion of the vertical independent of the cart pitch angle (which is a function of the terrain). Inclined terrain has, however, an influence on the equilibrium point, since the down-hill gravity component has to be countered by a motor torque corresponding to an input-voltage $\neq 0$; this results in a position shift of the cart.

4.5 Higher level intelligence

Beside the effect of measurement redundancy mentioned in section 2.1, the vision system lends itself to a flexible use for adapting the controller parameters and checking the operating conditions.

4.5.1 Positioning for the start

The initial condition for the inverted pendulum is $\pm$ 11° from the normal to the cart body (by hardware design). In order to pick up the pendulum, there is a rather brusque initial motion, by which the cart tries to get x_u underneath x_o. Since this has to be achieved at $V = 0$, the cart overshoots x_o to decelerate the linear motion which the pendulum has picked up in the meantime and to move the system back to the origin (see fig. 8). This initial motion takes a space of about 30 to 50 cm. During this motion, the system must remain in the visible range. Therefore, a check is done as to whether this space is available in the direction of the initial inclination of the pendulum. If not, the cart is driven into a position (by a slow constant motion), from which the start can be safely performed.

4.5.2 Choosing the right dynamics as a function of rod length

There is a set of rods available with different lengths to deal with different dynamics (see table in section 1.1). The set is such that no

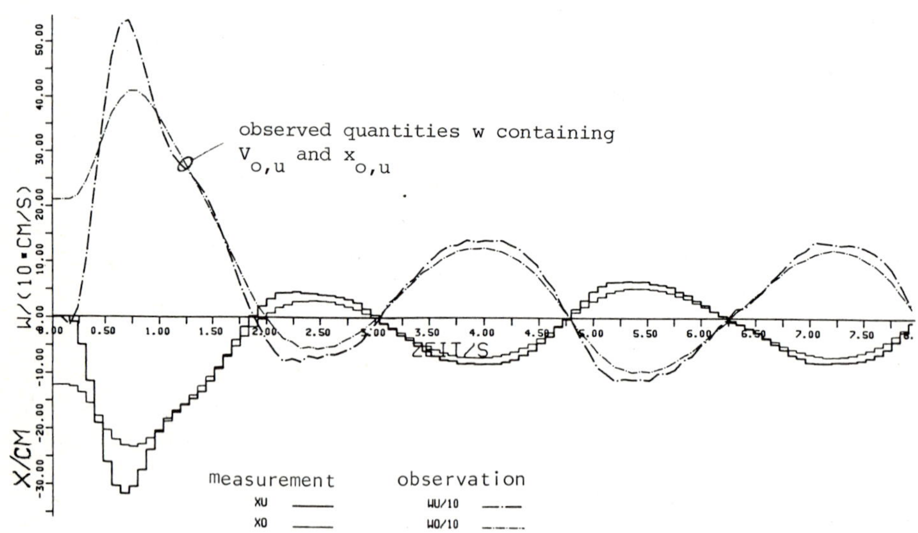

Figure 8. Dynamic behaviour of closed loop system starting from rest with $\varphi_o = 11^o$, $\omega_o = x_{c_o} = V_o = 0$

single controller yields a satisfactory performance for all rod lengths. With conventional instrumentation, the operator has to load the set of controller coefficients into the program corresponding to the actual rod length. With the visual system the length of the actual rod can be determined by inspection in the initialization phase. This is done by moving one window from the lower to the upper endpoint. Then, the initialization program loads that set of coefficients into the real time program that most closely corresponds to the measured rod length.

5. Conclusions

The study revealed that for motion control of systems with a corner frequency of up to about 1 Hz by a conventional TV-system the following points in the measurement signal processing and controller design should be taken care of: position correction for moving objects as a function of picture line element; adaption of the observer algorithm to the time delay, even if this is below 100 ms.

The concept of data flow reduction by assigning small windows with independent data processing capabilities to special features and have them track those independently, with the option to redirect them by command of the higher level computer that uses information from several such windows, has proven to be a valid approach. For more demanding vision systems the data processing and data exchange capability of the presently used microcomputers should be increased by orders of magnitude (which seems to be within range in the near future).

The closed loop performance of the vision system stabilizing an unstable inverted pendulum showed to be remarkably insensitive to parameter changes and to exterior disturbances. The reaction time of a human is too large to allow him to stabilize the system by motor voltage control, even for the relatively slow 2 m rod.

Literature

[1] Haas, G.: Meßwertgewinnung durch Echtzeitauswertung von Bildfolgen. to be published as doctoral thesis, HSBw Munich, 1982

[2] Graefe, V.: A Pre-Processor for the Real-Time Interpretation of Dynamic Scenes. (this volume)

[3] Luenberger, D.G.: Observers for multivariable systems. IEEE Trans. AC, Vol. AC-11 (1966), pp 190-197

[4] Isermann, R.: Digitale Regelungssysteme. Springer Verlag, Berlin, Heidelberg, New York, 1977

[5] Meissner, H.G.: Steuerung dynamischer Systeme aufgrund bildhafter Informationen. to be published as doctoral thesis, HSBw Munich, 1982

REAL-TIME PROCESSING OF RASTERSCAN IMAGES

by

S Grinaker

Norwegian Defence Research Establishment
N-2007 Kjeller, Norway

ABSTRACT

Robotics, industrial automation, automatic inspection and automatic reconnaissance and surveillance include the processing of image sequences. The datarate can be very high in dynamic imaging systems, and often real-time processing requires the use of special hardware. To reduce cost and system size the need for storage should be restricted. As a consequence the amount of data has to be reduced considerably during the scanning of the image. Processing "in the scan" also speeds up the analysis, since total memory access time becomes neglectable.

Some principles are given which allow simultaneous processing and scanning of a TV-like rasterscan image. The video signal flows through special hardware processors which work in parallel at video rate in a parallel-pipeline configuration. Only some postprocessing and the final decisions are left until the completion of the imaging. In most systems a microprocessor can do the final processing during the video flyback time between consecutive frames. This solution will only work for local processors, but relying on knowledge learned from preceding images, the principles can also be extended to most global processes. In a non-static scene this will introduce errors. However, if the image frequency is high compared to the dynamics of the scene, this error is neglectable.

The principles are visualized by examples. The implementation of image segmentation algorithms of various complexities are discussed, and a brief discussion of a complete image analyser is contained. This system allows real time segmentation and recognition of every image in sequences from not too complex out-door scenes.

NATO ASI Series, Vol. F2
Image Sequence Processing and Dynamic Scene Analysis
Edited by T.S. Huang

REAL-TIME PROCESSING OF RASTERSCAN IMAGES

S Grinaker
Norwegian Defence Research Establishment
N-2007 Kjeller, Norway

1 INTRODUCTION

Robotics, industrial automation, automatic inspection and automatic reconnaissance and surveillance are among the topics which may include the processing of image sequences. Very high data rates are typical in such applications, and usually real-time analysis requires parallel processing and the use of special hardware. Even without demands for real-time processing a general purpose computer will not be able to do the processing in a time proper for interactive processing, due to the tremendous amount of data involved. Therefore, even for research and development of algorithms, the use of special computers will be preferred.

A variety of parallel image processing machines are described in the literature (1,2). However, these are developed mainly for research or for processing of single frames. In more recent proposals the machines are usually bus-organized using large image memory systems. These machines are much faster and, though they may be rather expensive, more cost-effective for image processing than traditional sequential computers. Still they may be programmable and fairly flexible. Nevertheless, for real-time processing these machines still may be too slow or they may become too expensive due to the high degree of parallelism needed. This is certainly true for high speed image sequence analysis, and in particular for evaluation and implementation of such systems.

Systems for image analysis, including sequence analysis, are very often composed of similar subsystems, and even different applications may result in similar solutions. In particular, subsystems for segmentation and feature extraction or primitive detection are commonly used. Usually these processes are the most time consuming in the analysis of images due to the large amount of data involved (preprocessing should also be treated in the way described in this paper). Therefore, image sequence analysis may be more effectively performed using even more specialized machines than the "common" array processors and the bus-organized frame store based machines.

NATO ASI Series, Vol. F2
Image Sequence Processing and Dynamic Scene Analysis
Edited by T.S. Huang

This paper will focus on the different categories of parallel processor configurations and how they can be used to obtain real-time processing of image sequences.

2 PARALLEL PROCESSING

Processing speed-up is usually obtained in three ways (figure 1):

1) Identical processors work simultaneously in the same mode on separate data blocks

2) Processors work simultaneously with separate tasks on the same or separate data blocks

3) Pipelining. The data flow through a line of processors which work with separate tasks on the same data. When a processor A has finished the processing of one data block, this data block is shifted to the next processor B, and the processor A receives a new data block for processing.

The type 1 parallelism results basically in a speed-up proportional to the number of processors, and the cost-effectiveness of this configuration is nearly independent of the degree of parallelism.

The type 2 parallelism and the pipeline (type 3) may use specialized processors and their cost-effectiveness may improve with the number of special processors.

3 IMAGE SEQUENCE PROCESSOR SCHEME

The most common camera systems, like TV-cameras, scan the field of view in a rasterscan fashion producing a very high data rate. The special image processing machines are usually based on fast image memories and are effective for processing single frames. Nevertheless, very few of these computers can do advanced image processing at video rate. Advanced macro-operations can be done in parallel at very high speed, but a sequence of operations has to make use of the image memory system. Therefore required memory size will increase with the complexity of the over-all algorithm, i e with number of independent parallel processing sequences and number of processes in these sequences. Further-

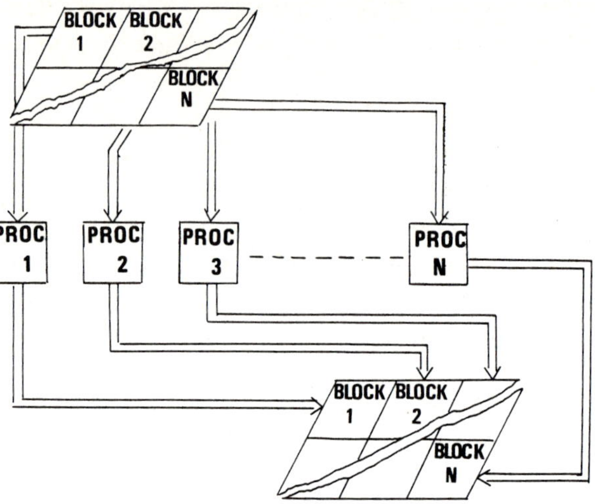

a) TYPE 1 PARALLELISM
Simultaneous and identical processing of separate data blocks.

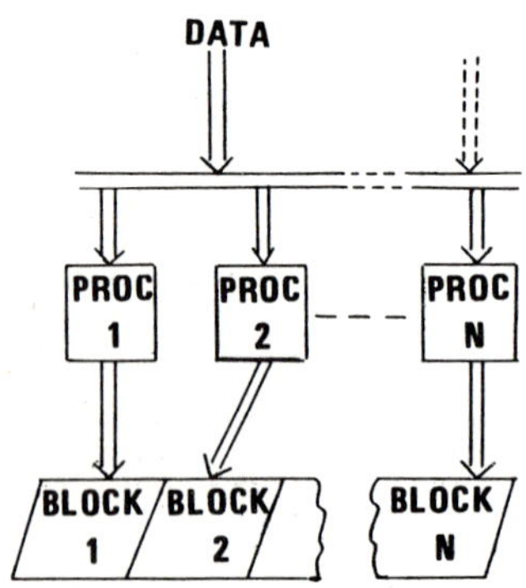

b) TYPE 2 PARALLELISM
Various tasks processed simultaneously on the same or separate data.

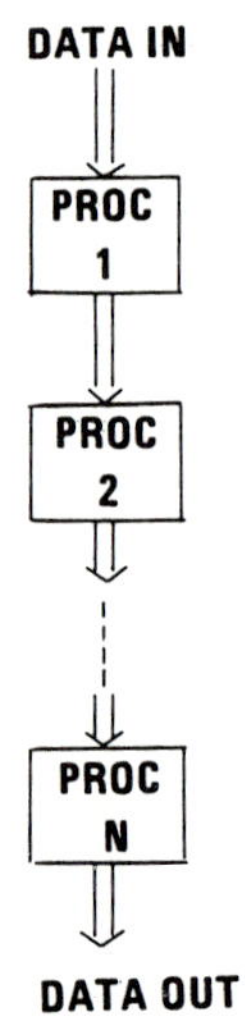

c) TYPE 3 PARALLELISM
Pipelining

Figure 1

PARALLEL PROCESSING CONFIGURATIONS

more, in a memory-organized system one will have problems sharing the databus between the different processors and complex interconnection networks are needed.

The obvious way to overcome these problems is by pipelining, i.e. a cascade of sequential processes is implemented by concatenating the corresponding processors in a pipeline. The data stream is shifted through the pipes at video rate, and therefore the total processing runs at video speed. The shifting and sampling are synchronized at a typical frequency in the range of 4-12 MHz.

Embedded in this principle is that each process is implemented in a separate processor. A process is, in this context, the unit of data handling executed between two data shifts. Sets of such microprocesses will produce macroprocesses like "gradient calculation" and "connected component labeling". Macroprocesses may be combined to subsystems like "enhancement" and "segmentation".

All processors have to be synchronized, but not all have to run at full video frequency (for instance, it is necessary for the column counter, but not for the line counter). Therefore, to not overkill the problem, some processors will work at lower speed, and the parallel-pipeline will be the general configuration for image sequence processors (figure 2). It is inherent in this principle that the need for storage is very limited as long as only local image processes are concerned.

Processing "in the scan" (while scanning) by pipelining is particularly well suited for rasterscan imagery, since the processing can go simultaneously with the imaging, without wasting time waiting for storing data.

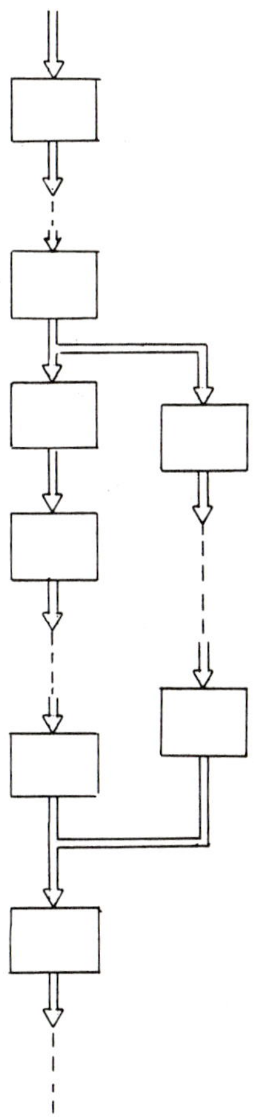

Figure 2
PARALLEL-PIPELINE
CONFIGURATION

The macroprocess in figure 3, producing the city-block gradient

$$G(i,j) = |G_X| + |G_Y|$$

$$G_X = 2 \cdot f(i+2,j) + f(i+1,j) - f(i-1,j) - 2 \cdot f(i-2,j)$$

$$G_Y = 2 \cdot f(i,j+2) + f(i,j+1) - f(i,j-1) - 2 \cdot f(i,j-2)$$

is implemented by using all three principles shown in figure 1. For example, the summing processes (Σ) and the sign removers (ABS) are two type 1 parallel processes, the Σ - ABS connections constitute two parallel pipes (type 3), and the upper part of figure 3 (the data shifting) and the lower part (the computation) represent the type 2 parallel processes.

4 GLOBAL IMAGE PROCESSING

The pipelined image processing configuration is restricted to local operations. Thus the benefits of increased speed and decreased cost are gained at the expense of freedom in choosing kind of process. Of course, by buffering complete images in the pipeline, we also can include global processing units, but this solution defeats the purpose of pipelining. However, the objective being processing of image sequences, the machine can transfer information from one frame to the next. For instance, histogram based thresholding can be performed by computing the threshold from the histogram of the preceding frame. Collecting the histogram and performing the thresholding are type 2 parallel processes. The actual computation of the threshold value is done during the video flyback time.

Of course, relying on information from preceding frames may introduce errors in a non-static scene. A rapid change in the scene illumination may, for instance, require a significant change in the threshold value. However, because of the inherent delay a new global threshold value can not be computed until the completion of the frame, and the use of a bad threshold will "blind" the system (very much like the blinding of the human eye when going from sunshine into a dark room or vice versa). However, if the image frequency is high compared to the dynamics of the scene, this transient error will be of little harm.

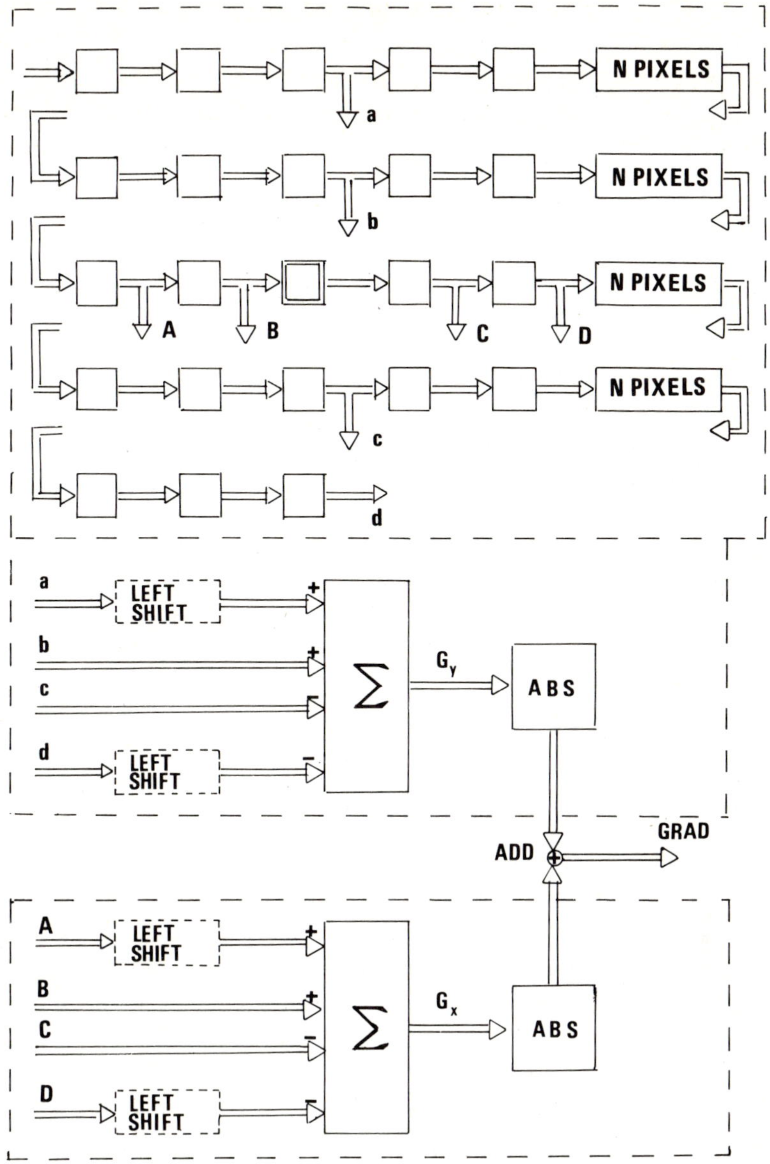

Figure 3

IN-THE-SCAN IMPLEMENTATION OF THE SECOND ORDER LMS APPROXIMATION OF THE CITY-BLOCK GRADIENT

In general, global processes can be decomposed into a cascade of local processes collecting global information and mathematical computations on the global data. Therefore all global processes can be implemented as a pipeline in a multipass solution, i.e. by collecting information from a sequence of consecutive frames. However, the delay may be serious if too many passes are needed.

5 BINARY IMAGE ANALYSIS "IN THE SCAN"

A parallel image processor (PIP) for analysis of binary images has been constructed using the parallel-pipeline structure (see figure 4). Binary video signals flow through an array of latches and shift registers in a rasterscan mode as for the gradient operator shown in figure 3. Local binary video information (LVI) is tapped from the latches synchronously with the shifting of the binary video input. Information from this "logic cell" is used for connected component analysis to produce object labels from a component counter. The LVI is then used in parallel pipes to extract features for all labeled objects.

Local operators will not know for sure whether two regions are connected or not until they either merge or are completed. Therefore, one of the Feature Extraction Modules is an object merge table. True features may be object area, perimeter, elongation, contour features, and other geometrical features including the object's position. By gating the binary video and the output from the ADC, i.e. semi-thresholding, also spectral features may be computed. The Feature Extraction Modules are interchangeable plug-in units and should be selected according to the application. All features are computed during the scanning of the image, and their values are stored in local feature memories. During the video flyback time between two frames these values are transferred to a conventional computer for object classification. Also the final analysis, for instance computation of the steering commands to the camera in an object tracking application, is done by this computer.

6 IMAGE SEGMENTATION "IN THE SCAN"

Among the image analysis subsystems the segmentation unit has to cope with the highest data rate. Therefore, from a time saving point of view segmentation "in the scan", i.e. simultaneously with the imaging is very attractive, making it possible to process image sequences in

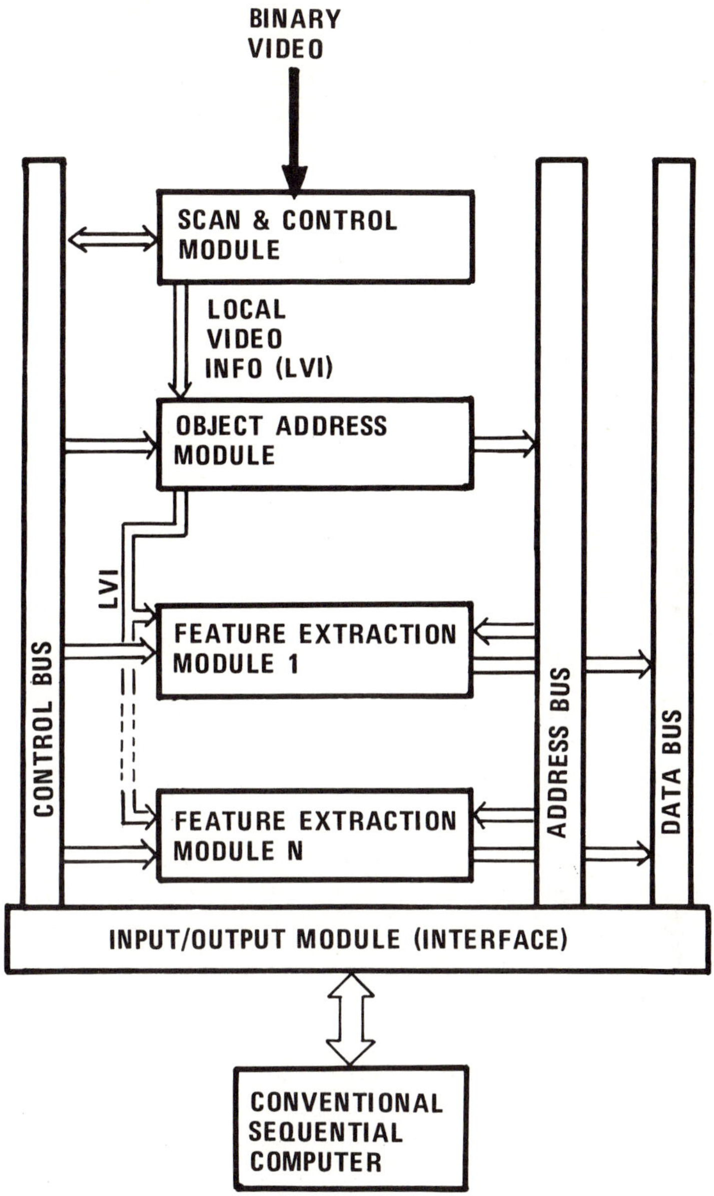

Figure 4

PARALLEL IMAGE PROCESSOR (PIP) FOR REAL-TIME ANALYSIS OF BINARY VIDEO IMAGES real-time.

Of course, simple thresholding using a fixed threshold value (may be given interactively), which can make sense in some industrial applications, is straightforward. Similarly, segmentation using the superslice method is easily implemented by parallel processors. Histogram based thresholding (or slicing) can be implemented in a two-pass solution.

We will now look at the implementation of a high performance local adaptive thresholding described in (3). This is the Norwegian segmentation used in a project run as a cooperative research effort by AC/243 (Panel III) RSG.9 under the NATO Defence Research Group (DRG). The title of the project was "Discrimination and Classification of Operating Military Targets in Natural Scenes from Thermal Imagery" (4).

A gradient picture is computed and binarized using a global threshold. Non-zero pixels in the binary image are clustered using distance measures. A rectangular region around each cluster is defined, and a local gray-level thresholding is applied to each corresponding region in the gray-level image to extract potential objects from the local background.

Without degradation in performance, the segmentation has been adjusted for "in the scan" implementation. A logic filtering is used as a kind of clustering, and the local thresholding procedure is slightly changed. A four-pass implementation of this modified solution is shown in figure 5.

The gradient operator is the one given in figure 3. The gradient threshold value is an optimal threshold computed from a small percentile of the gradient histogram. The computation of this threshold, carried out between two consecutive frames, may need a dedicated microprocessor and/or special hardware. The logic filtering is a cascade of two similar filters implemented in 5x5 pixel cells. Each connected component is then encompassed by a virtual frame defining a window for further processing. The segmentation of the windows are performed in Local Segmentation Processors (LSPs), the number of LSPs giving the upper limit for how many windows that can be segmented simultaneously. The local threshold value is given by a computed percentile of the gray-level histogram over local edge points (defined as non-zero pixels in

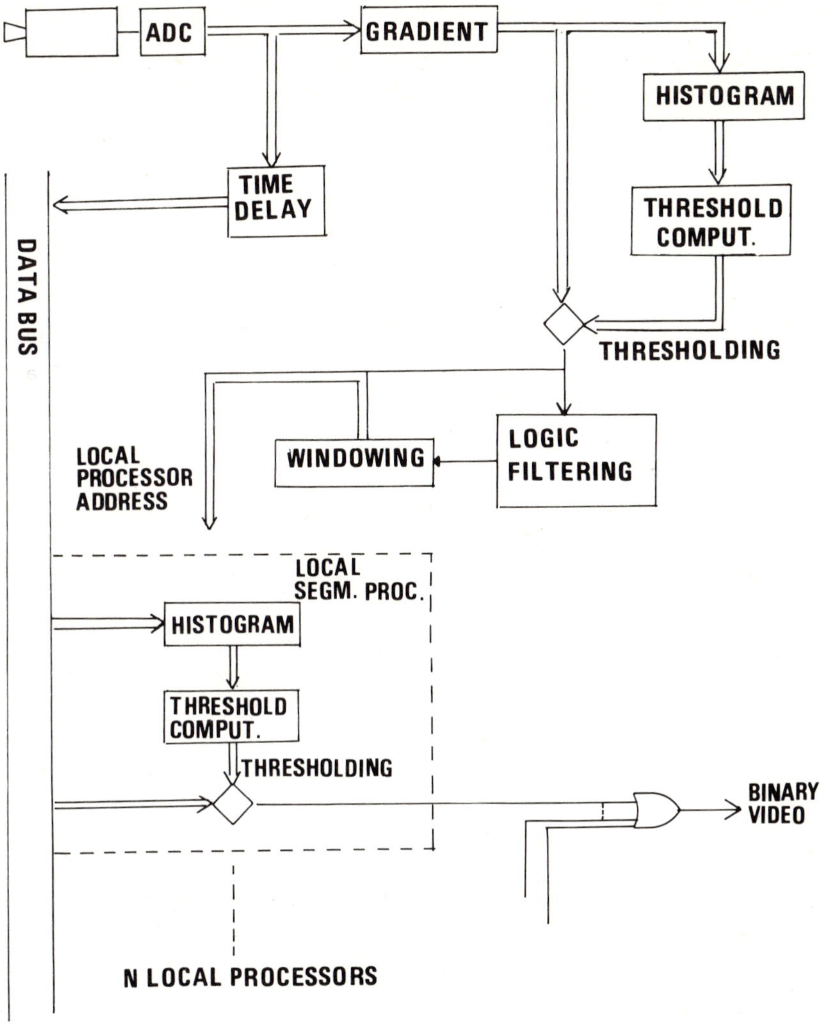

Figure 5

A FOUR-PASS IMPLEMENTATION OF THE MODIFICATION OF THE NORWEGIAN SEGMENTOR USED IN (4) the thresholded gradient picture). The binary video signal representing the union of all local regions, is obtained by a logical-OR of the output from each LSP.

7 FLEXIBILITY AND SYSTEM SOLUTIONS

It is recognized that programmable computers are more flexible for research and development than specialized hardware solutions, and programmability of image processing machines is important. Nevertheless, specialized machines are developed for image processing, to some extent at the expense of programmability. This evolution is a result of advances in electronics, permitting the development of small and cheap hardwired processors for complex operations. Since it seems fruitful to change to specialized image processors for processing of single frames, it is even more reasonable, and it may also be necessary, to do so for image sequence processing. Furthermore, since image processing is fairly modular by nature (preprocessing, image segmentation, feature extraction/ primitive extraction), and since most standard processes can be categorized into only a few families, a rather limited set of processors may suffice for a variety of problems. Several of these processors may contain a microprocessor (and will therefore have increased flexibility), and most of them will contain parameters which can be changed by the host computer. Therefore also specialized parallel-pipeline machines will, to some extent, be programmable. The introduction of ever faster microprocessors will further increase the flexibility of this kind of machines.

The tremendous datarate, causing the problems in image sequence processing, has been reduced before the image analysis (classification, structural analysis, temporal context analysis, etc), and common serial computers can be used for this higher level processing. For instance, by concatination of the segmentation unit in figure 5 and the binary image analyser in figure 4, information is extracted in real-time for discrimination and classification of military targets. However, for the actual detection and classification - and also for target tracking - the information has to be further analysed. This analysis can be done by a serial computer in parallel with the information extraction of the next frame performed by the special computer.

8 CONCLUSION

The parallel-pipeline solution permitting image processing "in the scan" significantly reduces the demand for storage. This is very important for saving time (by reducing wasted time for memory access), but also for reducing hardware volume and cost. However, to be able to utilize global information, multi-pass processing is necessary. Without storing the image this cannot be done. However, if the image frequency is high compared to the dynamics of the scene, we can rely on knowledge learned from preceding frames, and an approximate solution can be found in real-time.

"In the scan" processing is related to sequences of rasterscan images, for which the technique has great advantages, in particular by permitting the processing to start when the image is still scanned. Also, the modularity in image processing makes it reasonable to pipeline very specialized machines (image enhancement machines, segmentation machines, feature extraction machines, etc) for processing image sequences. Though the hardware implementation described in this paper decreases flexibility, the parallel-pipeline configuration may, for the time being, be the most cost-effective way to solve several real-time problems. Furthermore, the introduction of ever faster microprocessors, making them attractive as building elements in this kind of special image sequence processors, will very soon increase the programmability to the level of to-day's special image processing machines.

References

1) Fu, K S - "Special Computer Architectures for Pattern Recognition and Image Processing - An Overview", Proc. Nat'l Comp Conf (1978), pp 1003-1013.

2) - Real-Time/Parallel Computing (Image Analysis). Ed: M Onoe, K Preston jr, A Rosenfeld, Plenum press, N Y (1978).

3) Østevold, E; S Grinaker - Detection and Classification of Military Targets from Infrared Imagery, FFI/RAPPORT-80/4006, NDRE, Kjeller, Norway (1980).

4) Sevigny, L
G Hvedstrup-Jensen
M Bohner
A M Navarro
E Østevold
S Grinaker
J Dehne
- "Discrimination and Classification of Operating Military Targets in Natural Scenes from Thermal Imagery", NATO Report AC/243(Panel III) D/191, AC243(Panel III)RSG.9 D/4.

3D KALMAN FILTERING OF IMAGE SEQUENCES

D. Cano - M. Bénard

Centre de Télédétection et d'Analyse des Milieux Naturels (C.T.A.M.N.)
Ecole Nationale Supérieure des Mines de Paris
Rue Claude Daunesse
Sophia-Antipolis, 06565 Valbonne Cédex (France)

and

Laboratoire Image
Ecole Nationale Supérieure des Télécommunications
46, rue Barrault
75634 Paris Cédex 13

ABSTRACT

The authors are currently engaged in processing time-series of satellite imagery. Processing includes noise elimination and in this context, a 3D Kalman filtering has been developped and is presented here.

The first step is the definition of a class of two and three-parameter Markov discrete processes. The linear filtering of such stochastic processes reduces to a one-parameter vectorial Markov process recursive filtering, described by Kalman's equations. The 3D filter is then broken down into a two-dimensional spatial filter and a one-dimensional time filter. Some more stationnarity hypothesis allows a very simplified algorithm. The CPU time required is about 4 minutes on a middle range computer, for a 512 x 512 pixels picture.

The results are exposed on one-image sequence, which is assumed to verify the Markovian assumptions.

NATO ASI Series, Vol. F2
Image Sequence Processing and Dynamic Scene Analysis
Edited by T.S. Huang

Introduction

In the past years, the 2D Markovian representation of images has been investigated ([2], [3], [4], [5]) and one of the most important application of this model, the 2D-Kalman filtering, is now operationnal in many systems.

When considering the accuracy of the Markovian assumption, one can see that perhaps the application of such an hypothesis should be better in time varying processes than in space varying processes.

With this point of view, we propose a 3D Kalman filtering procedure for image sequences, where the third dimension, time, takes first place.

After a short review on 2D models (Part 1), we introduce the 3D model (Part 2). In the end of Part 2 and in Part 3, we work out a simplification of the calculation in a completely diagonal form, in order to implement the algorithm. The results are exposed on some elements of an image sequence, which was filtered in about 30 s CPU time per frame.

In the conclusion, we give some ideas about applications' future.

Part 1 : 2D Markov discrete processes 2D Kalman filtering

All the random variables will be elements of an Hilbert space $L^2(\Omega, a, P)$, where (Ω, a, P) is a probabilized space. For a centered random process $X = \{X_{mn}, (m,n) \in \mathbb{N}^2\}$ with values on $\mathbb{R}^p$, H^X_{mn} will be the smaller Hilbert sub-space including the set $\{X^k_{ij}$ where k = 1, ... p and $i \leqslant m$, $j \leqslant n\}$.
(X/H) will describe the projection of the vector X on the sub-space H.

1. DEFINITIONS

1.1. Definition 1

A process $X = \{X_{mn}$ where $(m, n) \in \mathbb{N}^2\}$ is called wide-sense Markov, if and only if :

$$\left(X_{mn} / H^X_{\infty q}\right) = \left(X_{mn} / X_{mq}\right) \qquad q \leqslant n$$

$$\left(X_{mn} / H^X_{p\infty}\right) = \left(X_{mn} / X_{pn}\right) \qquad p \leqslant m$$

1.2. Definition 2

$F[(m,n);(p,q)]$ matrix function : $\mathbb{N}^2 \times \mathbb{N}^2 \longrightarrow L(\mathbb{R}^k, \mathbb{R}^k)$ with $p \leqslant m$ and $q \leqslant n$ is a transition matrix, iff :

$$F[(m,n);(m,n)] = Id$$

$$F[(m,n);(i,j)] = F[(m,n);(p,q)] \cdot F[(p,q);(i,j)]$$

with $i \leqslant p \leqslant m$ and $j \leqslant q \leqslant n$.

2. STATE-NOISE PROCESS

2.1. Definition 3

Let $\mathcal{B} = \{B_{mn} ; (m,n) \in \mathbb{N}^2\}$

where : $B_{mn} = X_{mn} - F[(m,n);(m-1,n)]\, X_{m-1n}$

$$- F\left[(m,n);(m,n-1)\right] X_{m n-1}$$
$$+ F\left[(m,n);(m-1,n-1)\right] X_{m-1 n-1}$$

is the orthogonal state-noise process associated with X.

2.2. We have the following important relations between X and B :

$$X_{mn} = \sum_{i=0}^{m} \sum_{j=0}^{n} F\left[(m,n);(i,j)\right] B_{ij}$$
$$H^{X}_{mn} = H^{B}_{mn}$$

These relations are a sufficient condition for X to be a wide-sense Markov process.

The vector process $\left\{(X_{mn}, X_{mn-1}, \cdots, X_{mo})^{T} ; m \in \mathbb{N}\right\}$, where X is a wide-sense Markov process, is a wide-sense Markov process too:

$$\begin{bmatrix} X_{mn} \\ X_{mn-1} \\ \vdots \\ X_{mo} \end{bmatrix} = \begin{bmatrix} F[(m,n);(m-1,n)] & 0 & \cdots & 0 \\ 0 & F[(m,n-1);(m-1,n-1)] & 0 \cdots & 0 \\ \vdots & & \ddots & \vdots \\ 0 & & & F[(m,0);(m-1,0)] \end{bmatrix} \times \begin{bmatrix} X_{m-1n} \\ X_{m-1n-1} \\ \vdots \\ X_{m-1o} \end{bmatrix}$$

$$+ \begin{bmatrix} \sum_{0}^{n} F\left[(m,n);(m,j)\right] B_{mj} \\ \sum_{0}^{n-1} F\left[(m,n-1);(m,j)\right] B_{mj} \\ \vdots \\ \sum_{0} F\left[(m,0);(m,0)\right] B_{mo} \end{bmatrix}$$

We have the same considerations with $\left\{(X_{mn}, \cdots X_{1n}, X_{on})^{T} ; n \in \mathbb{N}\right\}$

3. OBSERVATION PROCESS

$Y = \left\{ Y_{mn} ; (m,n) \in \mathbb{N}^{2} \right\}$ where $Y_{mn} = H_{mn} X_{mn} + W_{mn}$ and X a wide-sense

Markov process given by (Def. 1) is the observation process corresponding to the state X. H_{mn} is a non-random matrix, and W_{mn} an orthogonal and B-orthogonal process called the observation noise.

4. 2D FILTERING

We aim to obtain the recursive estimation $\hat{X}_{mn}$ of the stage X_{mn} with some "past" observations : $\left\{ Y_{pq} \; ; \; p \leqslant m, \; q \leqslant n \right\}$

Mr Korezlioglu (Ref. [1]) considered 3 different approaches :

- Horizontal recurrence $\hat{X}_{mn} \longrightarrow \hat{X}_{m+1n}$ for n constant
- Vertical recurrence $\hat{X}_{mn} \longrightarrow \hat{X}_{mn+1}$ for m constant
- Diagonal recurrence $\hat{X}_{mn} \longrightarrow \hat{X}_{m+1n+1}$

He proved that these three approaches only depend on some vertical and horizontal innovation processes and thus the 1D Kalman equations can describe the 2D Kalman filtering. We will now use these properties in order to give the filtering algorithm.

Only the horizontal filtering will be considered ; the vertical one is similar.

Let A_m be the n-vector $(A_{mn}, A_{mn-1}, \dots A_{m0})^t$ for any process A_{mn} (n is supposed to be constant in the horizontal filtering).

We have : $X_m = F_m \cdot X_{m-1} + B_m$

where :

$$F_m = \begin{bmatrix} F\left[(m,n);(m-1,n)\right] & 0 & 0 \dots & 0 \\ 0 & F\left[(m,n-1),(m-1,n-1)\right] & 0 \dots & 0 \\ \vdots & & \ddots & \\ 0 & & & F\left[(m,0);(m-1,0)\right] \end{bmatrix}$$

and $$Y_{mn} = H_m \, X_m + W_m$$

Kalman filtering algorithm gives the estimation $\hat{X}_{m+1}$ of X_{m+1} :

$$\hat{X}_{m+1} = \left[\left(X_{m+1,n} / H^{Y}_{m+1,n} \right) ; \left(X_{m+1,n-1} / H^{Y}_{m+1n} \right) ; \cdots ; \left(X_{m+1,0} / H^{Y}_{m+1,n} \right) \right]^{T}$$

4.1. Recursive equations system of the 2D filter :

If $\quad Q_m = E\left[B_m . B_m^T \right]$ (State-noise covariance)

$R_m = E\left[W_m . W_m^T \right]$ (Observation-noise covariance)

$$\begin{cases} K_m = P_{m/m-1} . H_m . \left[H_m . P_{m/m-1} . H_m^T + R_m \right]^{-1} \\ \hat{X}_m = F_m . \hat{X}_{m-1} + K_m . \left[Y_m - H_m . F_m . \hat{X}_{m-1} \right] \\ P_m = \left(Id - K_m . H_m \right) . P_{m/m-1} \\ P_{m+1/m} = F_{m+1} . P_m . F_{m+1}^T + Q_{m+1} \end{cases}$$

with boundary conditions : $\hat{X}_{-1} = 0 \quad ; \quad P_{0/-1} = E\left(X_0 . X_0^T \right)$

for a given m, $\hat{X}_m$ gives the estimations $\left(X_{mi} / H^Y_{mn} \right)$ for i = 0, ..., n.

4.2. Remark 1

If $Q_0 = E\left(B_0 . B_0^T \right)$ where $B_0 = \left(B_{0n}, B_{0n-1}, \ldots, B_{00} \right)^T$

$$P_{0/-1} = G_n . Q_0 . G_n^T \quad ; \quad G_n = \begin{bmatrix} F[(0,n);(0,n)] & F[(0,n);(0,n-1)] & \cdots & F[(0,0)(0,0)] \\ 0 & F[(0,n-1);(0,n-1)] & \cdots & F[(0,n-1)(0,0)] \\ 0 & 0 & \ddots & \vdots \\ \vdots & \vdots & & \\ 0 & 0 & & F[(0,0);(0,0)] \end{bmatrix}$$

4.3. Remark 2

It is possible to ameliorate the estimations when "smoothing" the result of the precedent procedure :

$$\begin{cases} \left(X_m / H^Y_{Mn} \right) = \left(X_m / H^Y_{mn} \right) + L_m \left[\left(X_{m+1} / H^Y_{Mn} \right) - F_{m+1} \left(X_m / H^Y_{mn} \right) \right] \\ L_m = P_m . F_m^T . P_{m+1/m}^T \end{cases}$$

After such a "smoothing" procedure, we have the estimations $\left(X_{mn} / H^Y_{Mn} \right)$ where M is the x-size of the picture.

Part 2 : 3D Markov discrete processes 3D Kalman filtering

For a random 3D process $X = \{X_{mnt} ; (m,n,t) \in \mathbb{N}^3\}$ with values on $\mathbb{R}^p$, will be the smaller Hilbert sub-space including the set :

$$\{X^k_{ij\tau} ; k=1,\dots,p ; i=1,\dots,m ; j=1,\dots,n ; \tau=1,\dots t\} \quad , H^X_{mnt}$$

1. DEFINITIONS

1.1. Definition 4

The process $X = \{X_{mnt} ; (m,n,t) \in \mathbb{N}^3\}$ is said to be wide-sense Markov, iff :

$$\begin{cases} (X_{mnt} / H^X_{p\infty\infty}) = (X_{mnt} / X_{pnt}) & p \leq m \\ (X_{mnt} / H^X_{\infty q \infty}) = (X_{mnt} / X_{mqt}) & q \leq n \\ (X_{mnt} / H^X_{\infty\infty t}) = (X_{mnt} / X_{mn\tau}) & \tau \leq t \end{cases}$$

We define F transition function as :

$$(X_{mnt} / X_{pq\tau}) = F[(m,n,t) ; (p,q,\tau)] \cdot X_{pq\tau} .$$

2. STATE-NOISE PROCESS

From the point of view of a recursive filtering of 3D Markov processes, we must introduce the "past" of such a process. With the definition of the past, we will obtain the filtering direction, thus have the possibility to derive an innovation process, which will represent the difference between predicted and observed information at (m, n, t).

2.1. Definition 5

The past of the point X_{mnt} will be the set :

$$\left\{ X_{m-1nt}, X_{mn-1t}, X_{mnt-1}, X_{m-1n-1t}, X_{m-1nt-1}, X_{mn-1t-1}, X_{m-1n-1t-1} \right)$$

The state-noise process of X will be :

$$\mathcal{B}_{mn} = X_{mnt} - \left(X_{mnt} / H^{x}_{m-1nt}\right) - \left(X_{mnt} / H^{x}_{mn-1t}\right) - \left(X_{mnt} / H^{x}_{mnt-1}\right) + \left(X_{mnt} / H^{x}_{m-1n-1t}\right) + \left(X_{mnt} / H^{x}_{m-1nt-1}\right) + \left(X_{mnt} / H^{x}_{mn-1t-1}\right) - \left(X_{mnt} / H^{x}_{m-1n-1t-1}\right)$$

2.2. Proposition 1

B is an orthogonal process, i.e. $E\left[\mathcal{B}_{mnt} \cdot \mathcal{B}^{T}_{pq\tau}\right] = 0$ for $(m, n, t) > (p, q, \tau)$

2.3. Proposition 2

$$X_{mnt} = \sum_{p=0}^{m} \sum_{q=0}^{n} \sum_{\tau=0}^{t} F\left[(m,n,t) ; (p,q,\tau)\right] \cdot \mathcal{B}_{pq\tau}$$

The first proposition is easily proved when considering projections as $\left(\mathcal{B}_{mnt} / H^{x}_{m-1\infty\infty}\right)$, which are zero with application of definition of B. The second demonstration is a recurrence when using the "past-future" direction purposed above.

3. OBSERVATION PROCESS

3.1. Definition 6

$$Y = \left\{ Y_{mnt} ; (m,n,t) \in \mathbb{N}^3 \right\} \quad \text{where} \quad Y_{mnt} = H_{mnt} X_{mnt} + W_{mnt}$$

and H_{mnt} non-random cubic matric,

W_{mnt} orthogonal and B-orthogonal observation noise-process

is the observation process associated with the state X.

3.2. The purpose of the 3D filtering

The purpose of the 3D filtering is now to provide a recursive estimation $\hat{X}_{mnt}$ of X_{mnt} with respect to the observations :

$$\{ Y_{pq\tau} \ ; \ p \leq m \ , \ q \leq n \ ; \tau \leq t \}$$

Notation : $$H_{mnt} = H^{Y}_{mnt} \quad ; \quad \hat{X}_{mnt} = (X_{mnt} / H_{mnt})$$

The innovation process V will represent the difference between the new estimation $\hat{X}_{mnt}$ and the different estimations computed in the previous procedures.

$$V = (X_{mnt}/H_{mnt}) - (X_{mnt}/H_{m-1nt}) - (X_{mnt}/H_{mn-1t}) - (X_{mnt}/H_{mnt-1}) \\ + (X_{mnt}/H_{m-1n-1t}) + (X_{mnt}/H_{m-1nt-1}) + (X_{mnt}/H_{mn-1t-1}) - (X_{mnt}/H_{m-1n-1t-1})$$

If $H^{t}_{pq} = H_{pqt} - H_{pqt-1}$ is the time-innovation space in (p, q), we have :

$$V = (X_{mnt}/H^{t}_{mn}) - (X_{mnt}/H^{t}_{m-1n}) - (X_{mnt}/H^{t}_{mn-1}) + (X_{mnt}/H^{t}_{m-1n-1})$$

which is a 2D innovation process in the time innovation space.

The Markov properties of X allow us to purpose H^{t}_{mn}, H^{t}_{m-1n}, H^{t}_{mn-1} and $H^{t}_{m-1\,n-1}$ as the four sub-spaces, which represent exactly at a given t the new information with respect to the above defined "past". The 3D filter will consequently break down into a 2D (m, n) filter and a 1D t-filter.

Remark : Ohter similar relations are available with permutations of m, n, t.

4. DECOMPOSITION OF THE INNOVATION PROCESS V

We worked-out a decomposition of V on the different innovation spaces and reduced all the 3D and 2D expressions (see the definition of V) to linear combinations of 1D expressions, i.e. with the generator systems of $(H_{mnt} - H_{m-1nt})$, $(H_{mnt} - H_{mn-1t})$ and $(H_{mnt} - H_{mnt-1})$.

Thereafter, this result enables us to consider 3D and 2D filterint of 3D processes only with monodimensional expressions and algorithms.

5. 3D FILTERING

We will consider an image sequence of T pictures [M x N].

a) Filtering the first image (t = 0) with a 2D filtering procedure (Part 1). After this procedure, we have the estimations $\left(X_{mno} / H_{MNO}\right)$.

b) With the result of the above theory, we can have 3D filtering in a 1D pattern :

If
$$\begin{cases} X_{nt} = \left(X_{mnt}, X_{m-1nt}, \ldots, X_{ont}\right)^T \\ X_t = \left[\left(X_{nt}\right)^T, \left(X_{n-1t}\right)^T, X_{n-2t}^T, \ldots, \left(X_{ot}\right)^T\right] \end{cases}$$

X_t is the 1D vector describing the picture at t.

We have $X_t = F(t) \cdot X_{t-1} + B_t$ which can be filtered as in the 2D theory.
A t-1D filter gives $\hat{X}_t$, i.e. the set :

$$\left\{ \left(X_{mnt} / H_{MNt}\right) \; ; \; m = 1, \ldots, M \; ; \; n = 1, \ldots, N \right\}$$

A smoothing procedure gives : $\left\{\left(X_{mnt} / H_{MNT}\right)\right]$

Part 3 : 3D filtering of an image sequence

1. HYPOTHESIS FOR A GLOBAL SIMPLIFICATION

The tremendous calculation of gain matrix and estimations in the 2D and 3D filters does not enable a computer application without CPU times in hours, if not in days ... On the other hand, some more hypothesis on the picture properties are necessary to figure the model in a physical context.

1.1. Stationnarity

This hypothesis is provided by the intrinsic aspect of an image in a first approximation without any other information about this image :

Thus, $F\left[(m,n);(p,q)\right] = F(m-p,n-q)$ (for 2D filter)

$F\left[(m,n,t);(p,q,\tau)\right] = F(m-p,n-q,t-\tau)$ (for 3D filter)

1.2. Noise

The variance of the noise will be considered as constant :

i.e.
$$\begin{cases} E\left[\mathcal{B}_{mn}\cdot\mathcal{B}_{pq}^T\right] = q\,\delta(m-p,n-q)\cdot I_d \\ E\left[W_{mn}\cdot W_{pq}^T\right] = r\,\delta(m-p,n-q)\cdot I_d . \end{cases}$$

1.3. Observation matrix

H_m will also be considered as contrast $H_m = h \cdot I_d$, due to the *a priori* intrinsic relationship between observation and state.

2. 2D KALMAN FILTERING OF THE FIRST IMAGE

With the previous hypothesis, the state equation for 2D processes is :

$$X_m = F_{10}\,X_{m-1} + \mathcal{B}_m$$

The recursive algorithm begins with the boundary conditions :

$$\hat{X}_{-1} = 0 \qquad ; \qquad P_{0/-1} = E\left[X_0 . X_0^T\right]$$

$$P_{0/-1} = E\left[B_0 . B_0^T\right] = q . A . A^T ,$$

i.e. :

$$A = \begin{bmatrix} 1 & F_{01} & \cdots & F_{01}^{n} \\ 0 & 1 & & F_{01}^{n-1} \\ \vdots & & \ddots & \vdots \\ 0 & & & 1 \end{bmatrix}$$

which is positive and symetric.

A diagonalization of $P_{0/-1}$ gives $P_{0/-1} = T . P . T^{-1}$ where P is a diagonal matrix.

We will now prove that diagonalization in the same basis is possible for the whole algorithm :

$$\begin{cases} Q_m = E\left(B_m . B_m^T\right) = P_{0/-1} \\ R_m = r . Id \end{cases}$$

If

$$\begin{cases} K'_m = T^{-1} . K_m . T \qquad ; \qquad P'_m = T^{-1} . P_m . T \\ P'_{m+1/m} = T^{-1} . P_{m+1/m} . T \\ L'_m = T^{-1} . L_m . T \end{cases}$$

and if we put all observations Y_m in the base of the eigenvectors of $P_{0/-1}$ $Y'_m = T^{-1} . Y_m . T$, we have an algorithm only with calculation with diagonal matrix :

$$\begin{cases} K'_m = P'_{m/m-1} . h \left(h^2 . P'_{m/m-1} + r\, Id\right)^{-1} \\ \hat{X}'_m = F_{10}\, \hat{X}'_{m-1} + K'_m . \left(Y'_m - h . F_{10} . \hat{X}'_{m-1}\right) \\ P'_m = \left(-Id + K'_m . h\right) . P'_{m/m-1} \\ P'_{m+1/m} = -F_{10}^2 . P'_m + P \end{cases}$$

The smoothing procedure also reduces to :

$$L'_m = -P'_m \quad F_{10} . P'_{m+1/m}$$

$$\hat{\hat{X}}'_m = \hat{X}'_m + L'_m \left(\hat{X}'_{m+1} - F_{10}\, X'_m\right)$$

The final result supposes a passage in the previous generator system :

$$\hat{X}_m = T \cdot \hat{X}'_m \cdot T^{-1}$$

This final calculation needs the storage of T and T^{-1} in a buffer.

3. 3D KALMAN FILTERING OF A SEQUENCE

As mentionned above, the first step is the 2D filtering of the first image. Then, a vector reorganisation of all images is necessary to process the time filtering.

With the hypothesis also used in 2D filtering, we have for a N x N image the N^2 vector :

$$X_t = (X_{NNt}, X_{N-1Nt}, \dots X_{0Nt}, X_{N-1Nt}, \dots X_{0N-1t}, \dots, X_{00t})^T$$

But :
$$X_{mnt} = F[(m,n,t), (m,n,t-1)]\, X_{mnt-1} + \sum_{i=0}^{m} \sum_{j=0}^{n} \mathcal{B}_{ijt}$$
$$= F_{001}\, X_{mnt-1} + f(\mathcal{B})$$

Then : $X_t = F_{001}\, X_{t-1} + \mathbb{B}_t$

where $\mathbb{B}_t$ is a N^2 noise-vector.

The filtering procedure will be the same as the previous one in 2D filtering, but with :

$$P_{0/-1} = E\left(\mathbb{B}_t \cdot \mathbb{B}_t^T\right) = q\, A \cdot A^T \qquad A : N^2 \times N^2$$

where A is computed considering :

$$E\left[\mathcal{B}_{mnt} \cdot \mathcal{B}^T_{pq\tau}\right] = \begin{cases} 0 & \text{if } (m<p \text{ or } n<p \text{ or } \tau<t \\ F_{001}^{t-\tau} \cdot F_{010}^{n-q} \cdot F_{100}^{m-p} & \end{cases}$$

Due to the dimension (N^4) of A, the size of the image filtered is limited to a little square, which is shifted in the image. But the CPU time is very limited, due to the whole diagonal calculation procedure (see Experiments and Results).

4. EXPERIMENTS AND RESULTS

In order to test the 3D filter properties, we took an image sequence and put some noise on it (see Fig. 1 a and b).

We filtered then the 90 pictures sequence with the 3D Kalman filter implemented on a HP 21 MX with an Array-Processor AP 120 (8 K-words). The CPU time was 35 s per frame. On Fig. 1 to 4, we can see the results on some pictures of the sequence.

It is possible to see that the filter is working with expected results on slow moving sequences (less than 7 pixels between 2 frames). The typical applications are usual films.

A fast moving sequence (more than 7 pixels between 2 frames) becomes non Markovian at all.

Conclusions

We proved that it is possible to implement a stationnary 3D Kalman filtering on a mini-computer with good results on Markov sequences.

The next step is to study the applications on fast moving sequences.

The motion compensation could be very effective to put away the drawback of the lose of the Markovian time properties, when considering fast moving sequences.

Acknowledgments

We want to thank Mr Korezlioglu (ENST) for his support in the first part of the study, and Mr M. Maître (ENST) and Mr J.M. Monget (CTAMN) for the support they gave us in the implementation of the algorithms.

References

[1] H. Korezlioglu, 1978
Filtrage linéaire récursif des processus à 2 indices discrets, Markoviens au sens large. *Rapport ENST* - D - 78004, Fév. 1978.

[2] S. Attasi, 1975
Modélisation et traitement des suites à 2 indices. *Thèse d'Etat*, Université Paris-VI.

[3] Huang, 1975
Topics in applied physics, Vol. 42, 2D Digital Signal Processing. *Springer Verlag, Berlin.*

[4] H. Maître, 1980
Elements of Picture Processing. *Cours ENST*, 1980.

[5] J.W. Woods, 1981
Kalman filtering in two dimensions. Further results. *IEEE - ASSP*, Vol. 29, n° 2, April 1981, pp. 188-197.

Figures 1, 2, 3, 4 :

We considered an image sequence of 90 pictures.

On Figure 1a, we can see the negative of one of the images. We put some additive noise on all the images.

Figure 1b gives the result (disturbance in grey levels and shapes) of the noise input in image 1a.

We thus obtained a noisy sequence (Figures 3a, 4a, 5a).

We filtered then all the 90 images of the noisy sequence with a 3D Kalman filtering described in the paper, and obtained the filtered sequence (Figures 3b, 4b, 5b).

Figure 2 gives the negative of the filtered picture 1, in order to test some properties of the filter.

Fig. 1a

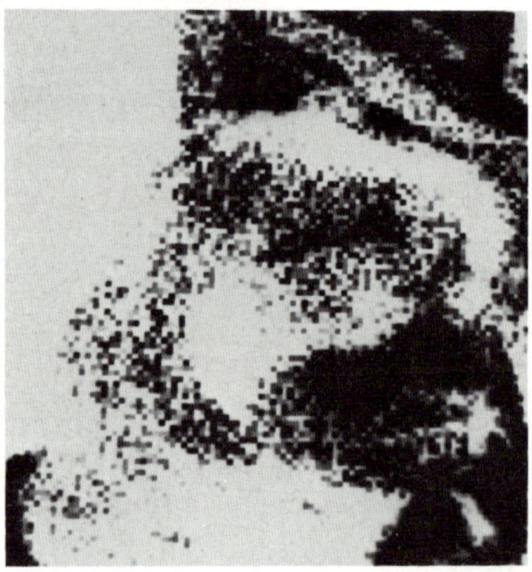

Fig. 1b

Fig. 2

a - Fig. 3 - b

a - Fig. 4 - b

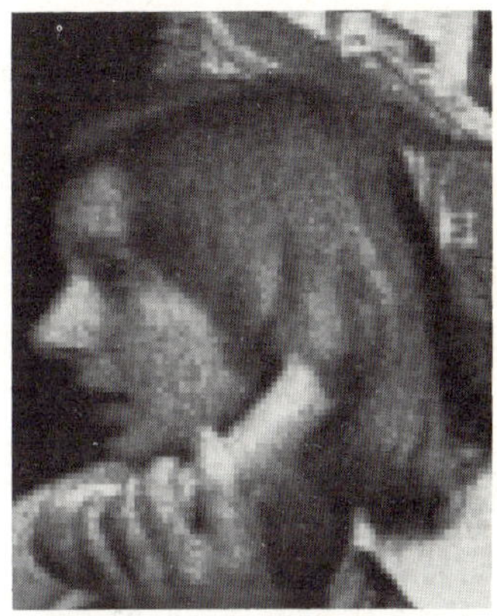

a - Fig. 5 - b

ATMOSPHERIC DISTURBANCES TRACKING IN SATELLITE IMAGES

P. BOUTHEMY, A. BENVENISTE
I.R.I.S.A.
Campus de Beaulieu - Avenue du Gal Leclerc
35042 RENNES Cédex
FRANCE

ABSTRACT.

In this paper, we propose an extraction method of some features, which describe dynamic properties of atmospheric disturbances. We identify a disturbance in a satellite image by means of a spirals model, knowing that the luminance model is constant levels plus noise. Thus, we define a class-conditional log-likelihood functional, what leads to a least-squares procedure. We compute the estimation of the shape change and motion parameters with a gradient algorithm. Such a processing could be an aid in interactive treatments and in the cloud archives development.

KEYWORDS.

Satellite images, clouds, modeling, spiral, motion estimation, gradient algorithm, feature extraction, tracking.

INTRODUCTION.

The satellite images give the opportunity of improving the interpretation and prediction of atmospheric processes. As for us, we want to pay attention to atmospheric disturbances above temperate areas. We propose tools for an estimation of some shape and motion parameters, to describe dynamic properties of these disturbances. The motion analysis presented here is proceeded with a global modeling, and is correlated with pattern recognition field. Such a processing provides a way to extract, from high-resolution satellite images, relevant parameters for storage purposes. It could also be an aid in an interactive treatment.

I. MODELING.

I.1. Description of an atmospheric disturbance.

In a satellite picture (see fig. 1), an atmospheric disturbance looks like a group of clouds whose geometric shape is noised but nevertheless characteristic : that-is-to say two or three areas (including the background) of different mean-intensity, whose edges approximately follow spirals of same centre and different phase. Hence an identification method can be considered.

I.2. Spirals model.

We resort to the following spirals model : the boundaries between the N definite areas are logarithmic spirals S_i, (i=1,N).

Knowing that the point (x_o,y_o) -in image coordinates- is the spirals centre, the angle V is the invariable angle between the radius vector and the tangent at each point of the curve, the angle θ_i is the phase ; the expression :

$$\rho(\theta) = \exp[(\theta-\theta_i) * \text{cotg } V] \qquad (1)$$

is the equation of X_i in polar coordinates, with (x_o,y_o) as origin of the coordinate system.

S_i is transformed in S_{i+1} by the rotation, whose centre is (x_o,y_o) and angle is $\theta_{i+1}-\theta_i$.

A mean-gray level Z_i is assigned to each area A_i whose boundaries are the spirals S_i and S_{i+1}.

I.3. Tracking interpretation.

The position of the centre, the values of the phases θ_i and the angle V, the mean-intensities Z_i are the parameters which will be tracked from image to image.

The variation of (x_o,y_o) and the phase θ_i, from the previous image to the current one, provides the motion estimation of the disturbance, respectively the translation motion of the vortex-ring and the rotation one of the areas boundaries. The variation of V reveals the shape change of the spirals. The greater V gets, the faster the spiral whorl is.

From a meteorological point of view, other interpretations of this tracking may be done. For instance, if the satellite image is acquired in a thermal infrared band, the mean-intensity Z_i informs about clouds temperature. The width of the cloudy area shows the disturbance activity ; its value can be reckoned from the difference between the phases of the spirals-boundaries.

It would be interesting to define a dynamic model for a better tracking of the parameters evolution. At the moment, we have not treated enough data to know whether it is possible or not.

With regard to the cloud archives problem, the features extraction technique, which is proposed here, is an efficient data compression method. Besides, each resulting parameter provides a clear information on lateral displacement, rotational motion or shape changes of an atmospheric disturbance, and in the same time on its location and its size.

I.4. Identification criterion.

Our approach is based on the use of a class-conditional log-likelihood functional and the search of its maximum. This is described in appendix. We define the image model as : constant levels plus noise.

Every homogeneous above-mentioned area A_i is partitioned into a finite number of disjoint elementary pavements which are spotted in polar coordinates. On each pavement P enclosed in the area A_i, the gray level Z_p is given by :

$$Z_p = \int_p Z_i + dN^i$$

dN^i can be considered as the stochastic differential of a two-dimensional spatial white noise $N^i\ (o,\sigma_i^2)$, which is zero-mean Gaussian and whose variance σ_i^2 depends on the region A_i.

If $\mu(P)$ designates the area of P, we get : $\int_p dN^i = N(o,\sigma_i^2\ \mu(P)) = N_p^i$.

Thus, we obtain the following expression for Z_p :

$$Z_p = Z_i\mu(P) + N_P^i \qquad (2)$$

The noises N_p^i are independant.

Under those assumptions, we evaluate the joint probability density function. In fact, the statistically optimum maximum log-likehood identifier leads to a least-squares procedure as identification criterion (see appendix).

The equation (1) of the spiral S_i can also be written as :

$$\theta = \theta_i + \frac{\ln \rho}{\cot V}$$

The observation window is a crown centred in (x_o, y_o). Therefore, the point (ρ, θ) of the image belongs to the region A_i, if :

$$\theta_i + \ln \rho/\cot V \leqslant \theta \leqslant \theta_{i+1} + \frac{\ln \rho}{\cot V}$$

$$\varepsilon \leqslant \rho \leqslant R$$

Then, we must minimize the following function :

$$F[x_o, y_o; \ldots \theta_i \ldots; V; \ldots Z_i \ldots] = \int_\varepsilon^R \rho d\rho \sum_{i=1}^{N} \int_{\theta_i + \frac{\ln \rho}{\cot V}}^{\theta_{i+1} + \frac{\ln \rho}{\cot V}} \frac{[Z(\rho, \theta) - Z_i]^2}{2\sigma_i^2} d\theta \qquad (3)$$

II. ALGORITHM.

II.1. Choice of the gradient algorithm.

The function F described in (3) is somewhat "odd". In particular, it is not convex and the variables are of very different kind. But, studies on synthetic data have shown, that it has a significant global minimum.

Therefore, we use a gradient algorithm to estimate the parameters which minimize F. Besides, since it is not possible to compute the Hessian of F, that prevents us from resorting to Newton methods.

The iterative formula of the gradient algorithm is given by :

$$U^{n+1} = U^n - \Gamma . \text{grad} F \qquad (4)$$

where U is the (2N+3)-dimensional variables vector and Γ denotes the diagonal matrix of the gains or step sizes.

$$U = (x_o, y_o, \ldots, \theta_i, \ldots, V, \ldots, Z_i) \qquad (i=1,N)$$

Hence, we obtain the 2N+3 following expressions :

$$
\begin{aligned}
X_o^{n+1} &= X_o^n - \gamma_{x_o} * \frac{\partial F}{\partial X_o} \\
Y_o^{n+1} &= Y_o^n - \gamma_{y_o} * \frac{\partial F}{\partial Y_o} \\
\theta_i^{n+1} &= \theta_i^n - \gamma_\theta * \frac{\partial F}{\partial \theta_i}, \quad i=1,\dots,N \\
V^{n+1} &= V^n - \gamma_V * \frac{\partial F}{\partial (\tan V)} \\
Z_i^{n+1} &= Z_i^n - \gamma_Z * \frac{\partial F}{\partial Z_i}, \quad i=1,\dots,N
\end{aligned}
\tag{5}
$$

The expressions of all the partial derivatives are given in appendix 2.

Since the function F is linear in tanV, we take the derivative with respect to tanV instead of the angle V. This choice does not affect the convergence for the reason that the directions of the two derivatives make an acute angle.

Furthermore, the form of the derivatives with respect to the Z_i supplies an exact analytical expression for the corresponding minima. At each iteration, we can solve $\frac{\partial F}{\partial Z_i}(Z_i^n) = 0$.

II.2. Further improvements.

The gains values depend on the iteration degree n. They are supplied by the following expression :

$$\gamma_j^n = \gamma_j^\infty * prec_{U_j} * (1+\frac{itc}{n}), \quad j=1,\dots,2N+3 \tag{6}$$

where U_j denotes the j-th component of U and $prec_{U_j}$ the whished accuracy on U_j.

This choice conciliates the convergence rapidity expressed through itc and the results precision expressed by $prec_{U_j}$. The value of γ_j^∞ is predetermined constant according to the component U_j.

Moreover, in order to remove the interference oscillations and the local minima, we introduce a smoothing of the variables U_j with a triangular filter.

II.3. Estimation of the variances.

The choice of the variances is bound by some constraints, since a bad estimation can prevent the algorithm from converging. What matters is the ratios between the variances.

We have chosen the following estimation method. During a start-up phase of fifteen or twenty iterations, the variances ratios are kept to one. The areas are roughly well-defined after this phase. Then on each truncated region A_i, we compute the variance of the gray levels distribution, normalized by its area. Thus, we obtain fitted and significant variances ratios.

III. RESULTS.

The gradient algorithm has been tested on both synthetic and real data.

The presented method can be used both to identify a disturbance in an image with an initialization by the operator and to estimate the disturbance motion from image to image. In the last case, the initialization on the current image is the result of the algorithm on the previous one ; on the first image of the sequence, it can be done in an interactive way. It must be noticed that the iterations frequency for the centre tracking is smaller than the one of the other parameters.

The table hereafter shows the results of the algorithm to identify synthetic data created with a Gaussian noise, on the assumption that the areas number N is equal to 2.

	Original values	Estimation results
Phases θ_1	25° - 26°*	25°.30
θ_2	175° - 176°	175°.23
Angle V	-50°	-49°90
Mean-intensities Z_1	170	171.5
Z_2	140	141.2

* We give two numbers since the sampling step value $\Delta\theta$ is one degree.

The results accuracy can be noted. Moreover, the convergence is rather fast, even with a remote initialization.

We have treated real data acquired from the geosynchronous satellite GOES in October 1980 (Fig. 1). At first, we rectify the distortion due to the parallax. In Fig. 2, we have a visualization of the restored data, within a discus whose radius is equal to 1500 km.

Some results are presented in Fig. 3 and Fig. 4. In the first one, the curves express the way the algorithm converges apart from the centre tracking. In the second one, the found spiral model of the disturbance has been drawn. It must be compared to Fig.2.

Now, we treat an images sequence acquired from the satellite METEOSAT. As its duration is fifteen hours, we should thus obtain a significant motion estimation. In order to process these noisier data a corrective term has been added to the iterative formula (4) ; this term tends to zero in the convergence area, but plays a leading part in case of divergence. Moreover, the mean-intensities Z_1 and Z_2 are kept constant to the upper and lower values of the data histogram. The intensities of the points in the discus centered in (X_o^n, Y_o^n) and defined, inside the greater square, at each iteration of the centre tracking, are filtered in a two-modes way. Results on the first image of the sequence (Fig. 5) are shown in Fig. 6, 7, 8.

REFERENCES.

[1] Rochard G., Brard A. : Nephanalyse Cinématique. Notes techniques de l'EERM Lannion. Avril 78. Avril 80

[2] Sitbon P., Derroir A., Chaume D. : Détermination des champs de vents à partir de photographies de satellites géostationnaires. Rapport du LMD. 1975.

[3] Zick C. : Method and Results of an Analysis of Comma Cloud Developments by means of vorticity fields from upper tropospheric satellite wind data. To be published in "Meteorologische Rundschau".

[4] Haass U., Brubaker T. : Estimation of Cloud Motion From Satellite Pictures. ICASSP. Denver. April 80.

[5] Nagel H.H. : Analysis Techniques for Image Sequences.

[6] Ljung L. : Identification des Systèmes. Ecole d'Eté d'Analyse Numérique. CEA-IRIA-EDF. 1978.

[7] Hestenes M.R. : Conjugate Direction Methods in Optimization. IFIP Conference on Optimization Techniques. Würsburg. Sept. 77.

APPENDIX 1.

Statistically optimum maximum log-likehood identifier.

On each pavement P enclosed in the area A_i, the intensity is given by :

$$Z_p = Z_i\ \mu(P) + N_p^i \tag{A1}$$

where $\mu(P)$ is the area of P and N_p^i a zero-mean Gaussian noise of variance $\sigma_i^2\mu(P)$.

Under these assumptions, the distribution function of the random variable Z_p can be evaluated as :

$$DF(Z) = p(Z_p \geqslant Z) = p(N_p^i \geqslant Z - Z_i\mu(P))$$

Hence : $DF(Z) = \dfrac{1}{\sqrt{2\pi\sigma_i^2\mu(P)}} \int_{Z-Z_i\mu(P)}^{\infty} \exp[-\tau^2/2\sigma_i^2\mu(P)]d\tau$

That becomes with the variable change $u = \tau + Z_i\ \mu(P)$

$$DF(Z) = \frac{1}{\sqrt{2\pi\sigma_i^2\mu(P)}} \int_Z^{\infty} \exp[-[u-Z_i\mu(P)]^2/2\sigma_i^2\mu(P)]du \tag{A2}$$

We then find from (A2) the density function of Z_p :

$$f_p = \frac{1}{\sqrt{2\pi\sigma_i^2\pi(p)}} \exp[-[Z-Z_i\mu(P)]^2/2\sigma_i^2\mu(P)] \tag{A3}$$

We partition the window, which is a crown, in a finite number K of disjoint pavements P_k, $(k=1,\dots,K)$. Thus, we obtain a finite number K of observed samples Z_k, whose functions f_{P_k} are independant and given by (A3). We infer from that the expression of the joint probability function :

$$f(Z_1,\dots,Z_k,\dots,Z_K) = \prod_{k=1}^{K} f_{P_k}(Z_k)$$

The observations within the window are denoted :

$$W = \{Z_k\ ;\ k=1,\dots,K\}$$

We suppose that W is known to be associated with a class labeled H_m of N above-mentioned areas A_i. More specifically, H_m represents the (2N+3)-dimensional vector $(x_o, y_o, \theta_1,\dots,\theta_N, V, Z_1,\dots,Z_N)$.

With assuming the m'th hypothesis acting, the likelihood functional is defined according to :

$$V_m\{W\} = f(Z_1,\dots,Z_k,\dots,Z_K/H_m) \tag{A5}$$

The criterion is to maximize the class-conditional likelihood functional in (A5), and to find the index m_o which verifies :

$$V_{m_o}\{W\} = \max_m V_m\{W\}$$

Consider the logarithmic expressions, we obtain the log-likelihood functional :

$$LV_m\{W\} = \ln f(Z_1,\ldots,Z_k,\ldots,Z_K/H_m)$$

It follows from (A4) that :

$$LV_m\{W\} = \sum_{k=1}^{K} \ln[f_{P_k}(Z_k/H_m)]$$

or

$$LV_m = \sum_{i=1}^{N} \sum_{P_k \subset A_i} \ln[f_{P_k}(Z_k/H_m)] \tag{A6}$$

Using (A3), we can expand (A6) :

$$LV_m = C - \sum_{i=1}^{N} \sum_{P_k \subset A_i/H_m} \frac{[Z_k - Z_i\mu(P_k)]^2}{2\sigma_i^2\mu(P_k)} \tag{A7}$$

with the constant $C = -\frac{1}{2}\sum_{k=1}^{K} \ln[2\pi\sigma_i^2\mu(P_k)]$.

Knowing that Z_k is an integral value on P_k, Z_k can be evaluated as $Z_k = z(\rho_k,\theta_k)\mu(P_k)$ with (ρ_k,θ_k) belonging to P_k and (x_o,y_o) origin of the coordinates system.

From (A7), we can conclude that the identification criterion becomes the minimization of the expression :

$$E_m = \sum_{i=1}^{N} \sum_{P_k \subset A_i/H_m} \frac{(z-Z_i)^2}{2\sigma_i^2}\mu(P_k) \tag{A8}$$

Notice that the area $\mu(P_k)$ of the pavement P_k can be approximated by : $\mu(P_k) \sim \rho_k.\Delta\rho.\Delta\theta$ if $\Delta\rho$ and $\Delta\theta$ are the sampling steps. The expression (A8) represents a discrete version of the integral function given by (3) in the main text.

APPENDIX 2.

Expressions of the partial derivatives of the function F given by (3).

$$F = \frac{1}{\pi R_1^2} \int_{\varepsilon}^{R} \rho d\rho \sum_{i=1}^{N} \int_{\theta_i + \ln\rho/\cot V}^{\theta_{i+1} + \ln\rho/\cot V} \frac{[Z(\rho,\theta)-Z_i]^2}{2\sigma_i^2} d\theta$$

with $R_1^2 = R^2 - \varepsilon^2$.

We denote by Z_{f_i} the gray levels of the points, whose polar coordinates, with (x_o, y_o) as origin of the coordinates system, check :

$$\theta = \theta_i + \ln\rho/\cot V$$

We obtain the following expressions :

$$\frac{\partial F}{\partial x_o} = \frac{-1}{\pi R_1^2} \int_{\varepsilon}^{R} \sum_{i=1}^{N} \left[\frac{(z_{f_i} - z_{i-1})^2}{2\sigma_{i-1}^2} - \frac{(z_{f_i} - z_i)^2}{2\sigma_i^2} \right] * \frac{\sin(\theta_i + \frac{\ln\rho}{\cot V} + V)}{\cos V} d\rho$$

$$\frac{\partial F}{\partial y_o} = \frac{1}{\pi R_1^2} \int_{\varepsilon}^{R} \sum_{i=1}^{N} \left[\frac{(z_{f_i} - z_{i-1})^2}{2\sigma_{i-1}^2} - \frac{(z_{f_i} - z_i)^2}{2\sigma_i^2} \right] * \frac{\cos(\theta_i + \frac{\ln\rho}{\cot V} + V)}{\cos V} d\rho$$

$$\frac{\partial F}{\partial \theta_i} = \frac{1}{\pi R_i^2} \int_{\varepsilon}^{R} \left[\frac{(z_{f_i} - z_{i-1})^2}{2\sigma_{i-1}^2} - \frac{(z_{f_i} - z_i)^2}{2\sigma_i^2} \right] * \rho \, d\rho \qquad (i=1,\ldots,N)$$

$$\frac{\partial F}{\partial (\tan V)} = \frac{1}{\pi R_1^2} \int_{\varepsilon}^{R} \sum_{i=1}^{N} \left[\frac{(z_{f_i} - z_{i-1})^2}{2\sigma_{i-1}^2} - \frac{(z_{f_i} - z_i)^2}{2\sigma_i^2} \right] * \ln\rho * \rho \, d\rho$$

$$\frac{\partial F}{\partial Z_i} = \frac{Z_i * (\theta_{i+1} - \theta_i)}{2\pi\sigma_i^2} - \frac{1}{\pi R_1^2} \frac{1}{\sigma_i^2} \int_{\varepsilon}^{R} \rho \, d\rho \int_{\theta_i + \ln\rho/\cot V}^{\theta_{i+1} + \ln\rho/\cot V} Z \, d\theta \qquad (i=1,\ldots,N)$$

with the notation conventions : Z_o identical to Z_N

θ_{N+1} identical to θ_1

σ_o identical to σ_N

APPENDIX 3.

Figures .

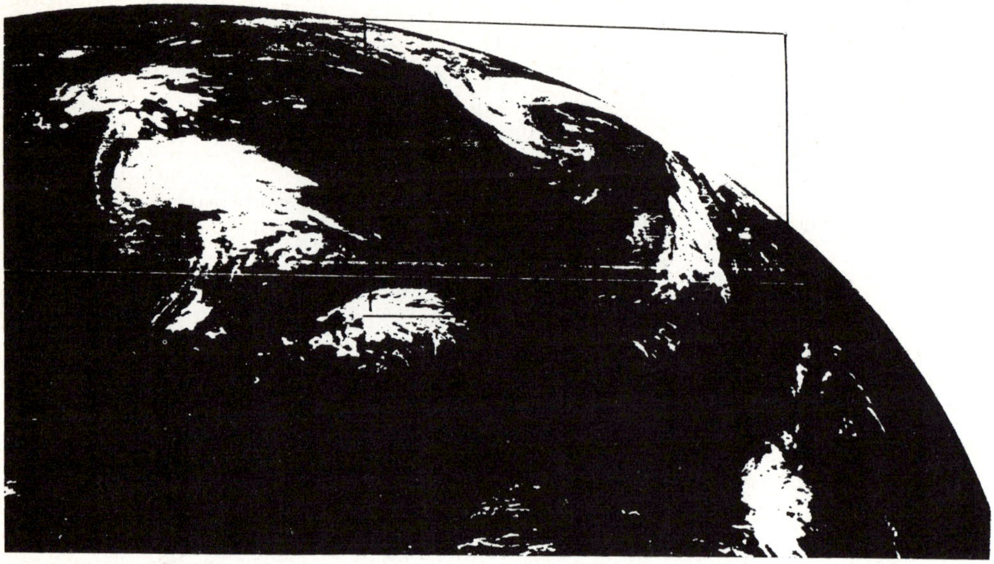

Fig.1. *GOES satellite image 25/10/80*
(thermal IR band)
The disturbance is enclosed

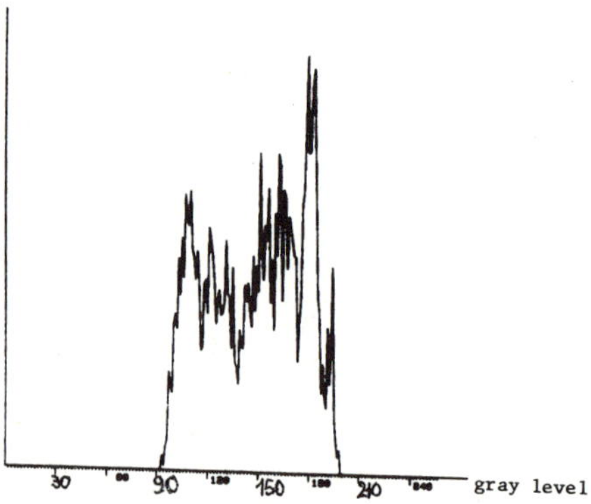

Fig.2. *Two-levels visualisation of the restored data (threshold=155)*

Fig.2bis. *Gray levels histogram for the restored data*

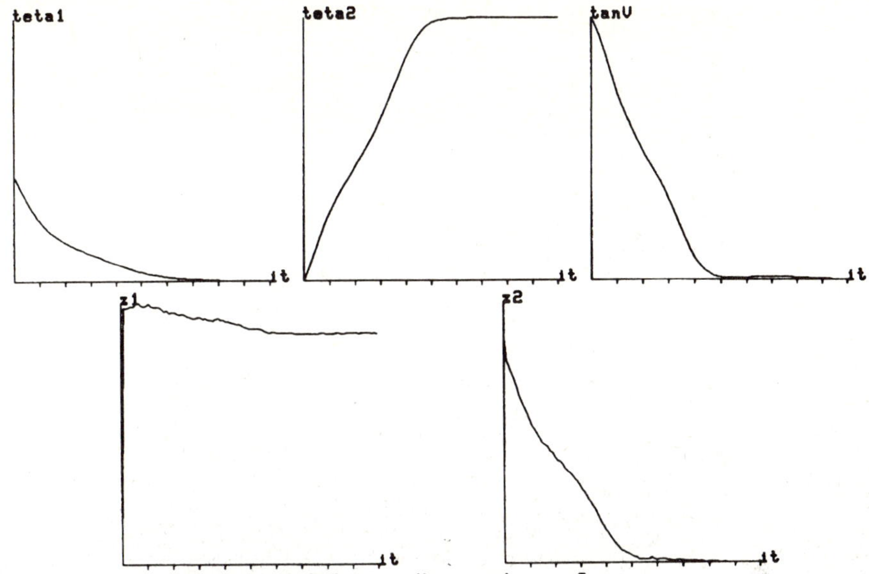

	teta1	teta2	V	z1	z2
init. value	28.00	180.00	-35.00	166.00	138.00
convg. value	19.85	200.48	-39.02	171.88	132.20

iterations: 100

Fig.3. Convergence curves (data are these of Fig.2)

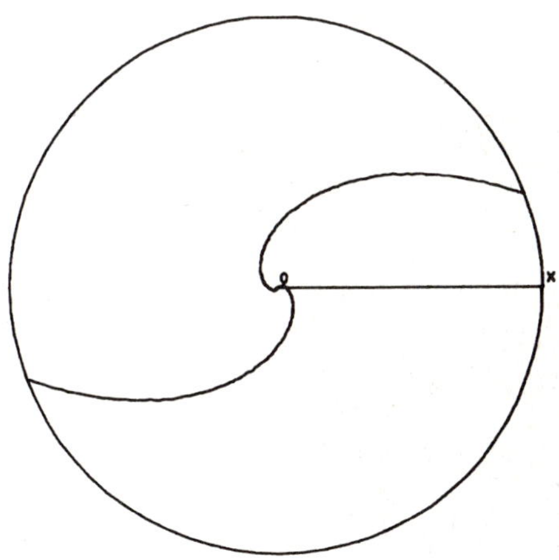

Fig.4. Found spirals model for the data of Fig.2
The parameters are the abovementioned convergence values (N is equal to 2)

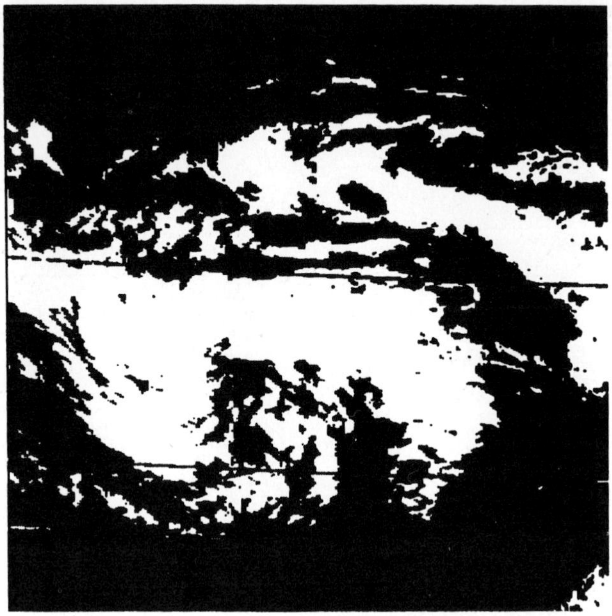

Fig.5. Two-levels visualization of the restored data from METEOSAT IR-image 16/01/79 (side length :2500km ;500-500 pels)

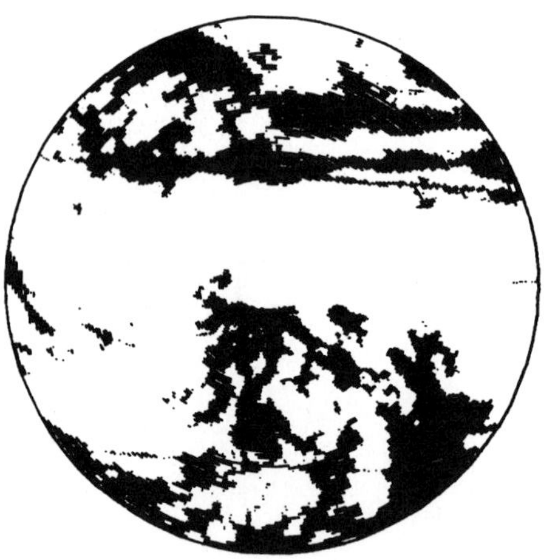

Fig.6. Final discus window of the tracking (radius=750km)

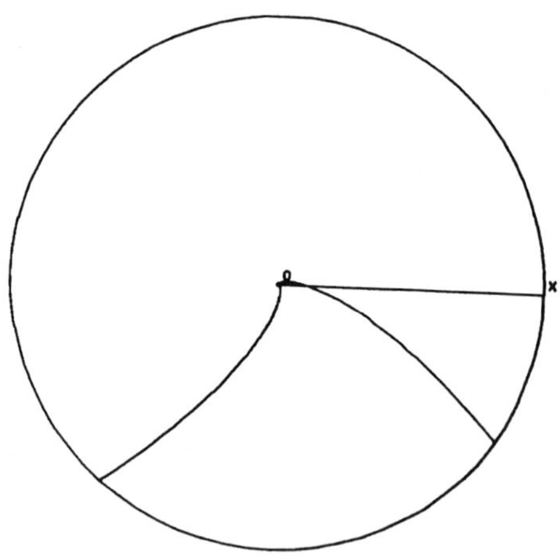

Fig.7. Found spirals model

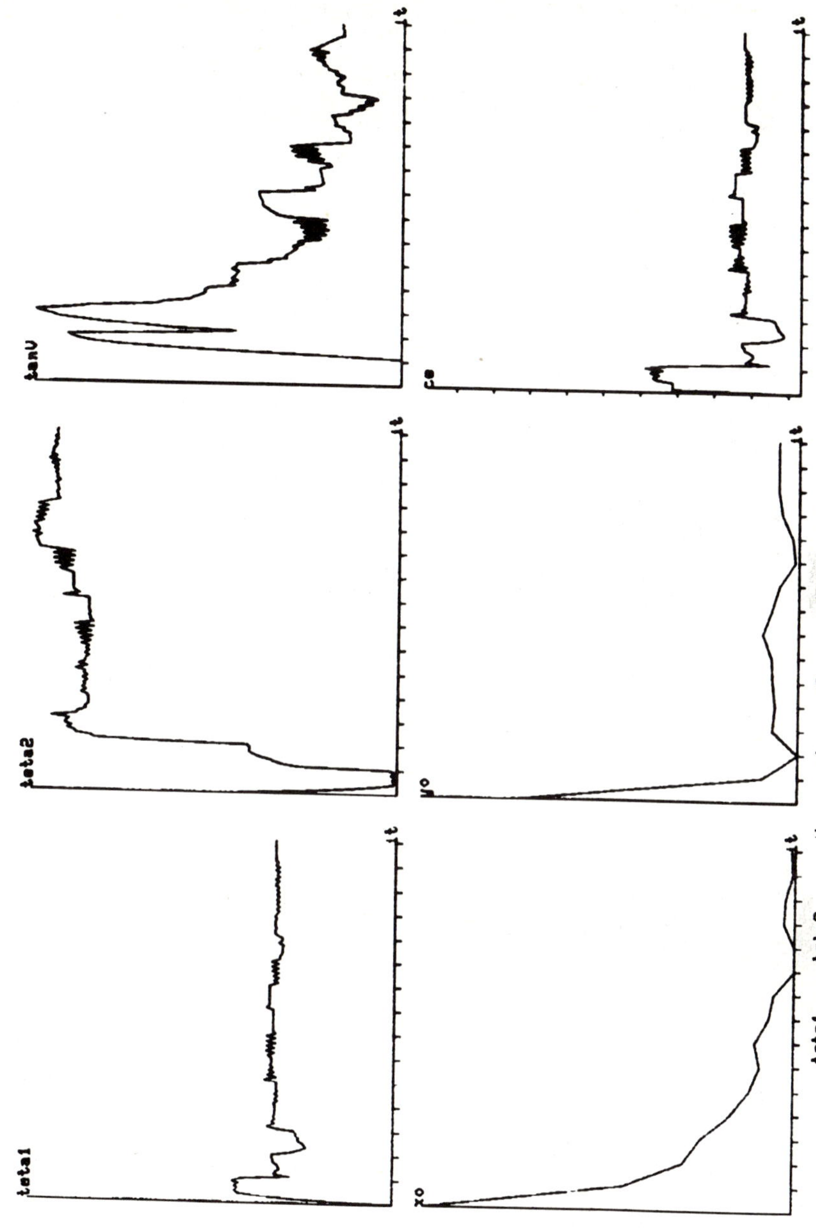

Fig.8. *Convergence curves before the triangular filtering for the data of Fig.5*

ASPECTS OF DYNAMIC SCENE ANALYSIS IN METEOROLOGY.

Guenter Warnecke

Institute for Geophysical Sciences,

Free University of Berlin, Germany

Thielallee 50, D-1000 Berlin 33

In the beginning I like to express some personal impressions gained during the course of this interesting meeting and which mainly urged me to address this contribution to the audience. I am thankful for this opportunity.

Before I came to this Advanced Study Institute, I knew from the program that I would come somewhat as an outsider, although being actively engaged in the analysis of motion scenes; but I did not expect to be so much of an outsider as I felt during the first week of this course. I learned, for example, by how many intelligent ways people are trying to press more and more of those, mostly lousy television programs parallel through one transmission line or how much brain power is being spent just to identify automobiles in TV picture sequences or track military objects. During next week, medicine seems to serve as the only represented field of interest of the applied sciences, Mr. Bouthemy's contributory paper this morning being the only exception. These facts appear to me not so much as a flaw of this meeting but rather a symptom of a more general situation.

During recent years, new technologies in imagery from video presentation of microscopic views of living sub-cell structures to globally viewing the restless atmosphere and oceans from space platforms opened new worlds of

NATO ASI Series, Vol. F2
Image Sequence Processing and Dynamic Scene Analysis
Edited by T.S. Huang

vision and observation to many different disciplines of science. But in most cases, due to lacking experience in the new domain, neither full awareness of the large scientific information content of motion scenes and, partially, of its qualitative newness as a decent scientific tool could be noticed nor proper interpretation or evaluation schemes were available, when these new kinds of information started pouring down in form of huge data streams. Therefore, from a scientific applier's point of view it seems, at present, more urgent to bridge the obvious gaps between existing high-powered technology and its adequate exploitation for scientific applications rather than developing these information streams into a flash flood; and it occurs more reasonable and desirable to intensify its exploitation for the applied sciences rather than primarily for industrial or military purposes.

In the following, I like to illustrate the mentioned situation by referring to the problem of analysis of motion scenes from meteorological satellite images. The main purpose of this contribution is supposed to be an animation of your interest towards dynamic scene analysis of meteorological and oceanographic image sequences. Problems will certainly be hard to solve, but also challenging and awarding. A good reason to start soon is - besides the scientific and practical needs - the availability of tremendous amounts of beautiful data in computer-compatible form.

In order to characterize some fundamental aspects of the problem, a brief historical consideration should be helpful. Instruments developed by Galileo GALILEI and his disciples long ago founded the basis for quantitative physical observations of the atmosphere. But since the Renaissance, observations were mainly presented and interpreted as one-dimensional until the early nineteenth century when H. W. BRANDES, following suggestions by ALEXANDER VON HUMBOLDT, introduced the two-dimensional synopsis of data in form of isopleths in geographic maps, thus initiating the development of modern, "synoptic" meteorology. With the beginning of the twentieth century, kites, balloons, aeroplanes, and - a little later - radiosondes opened the third spatial dimension of instrumental atmospheric observation. In the sixties, artificial earth satellites, in conjunction with the introduction of modern remote sensing techniques, helped to extend three-dimensional observations contiguously over the entire globe. A new dimension was, however, opened when, in 1967, geostationary satellites provided 'high-frequent', i. e. half-hourly picture sequences, and T. FUJITA succeeded in producing the first examples of time-lapse motion pictures of these image sequences. By this step, a four-dimensional visualization of atmospheric and ocean surface processes had been established , an unprecedented means of observation been introduced.

The newness of this means of observation results from a number of different factors:

(a) the perspective, i. e. from space, and the area of view from geostationary altitude, i. e. viewing the full disc of the earth and within this large area all the time-varying cloud (or water vapor) structures from the kilometer range up to global dimensions;

(b) the simultaneous comprehensive perception of the superposition of motions and processes of various atmospheric scales, i. e. conventional observation networks usually do not resolve the weather-intensive sub-synoptic scales, rather the synoptic scale generally serves as the 'work scale' for weather analysis and forecasting, and most of the knowledge has been accumulated from experience in this scale - mainly due to lacking sufficiently dense observations (in space and time) in the sub-synoptic scales; thus the visualization of the real atmospheric behaviour in its actual simultaneous superposition of all scales is absolutely unprecedented, deserving particular appreciation and especially the development of new appropriate analysis techniques;

(c) the tremendous amount of information contained in full-disc motion scenes with its various differential motions and structural patterns continuously changing in space and time, and - in addition - the information accumulation by monitoring the earth day by day with these motion scenes from half-hourly picture sequences; i.e. although a large amount of information and knowledge can be gained qualitatively from these scenes by visual inspection and interpretation, the large amount of data and the desire and necessity for real-time use (which means in this context: minutes to hours) requires extensively automatized quantitative information extraction schemes;

(d) the non-conventional kind of information of these motion scenes, i.e. conventionally dynamic and thermodynamic quantities like wind speed and direction, air pressure, temperature, relative humidity, etc. are measured and thus govern the analysis and interpretation schemes in meteorology, whereas these motion scenes exhibit the spatially and temporally changing patterns resulting from the formation, growing, structural modification, displacement, and dissipation of clouds and cloud systems (or water vapor structures), which are on the other hand produced by a complex of interrelated dynamic and thermodynamic processes.

During the last 15 years, the development of quantitative analysis techniques for satellite picture sequences has, of course, been concentrated on the extraction of conventional parameters. Thus a variety of methods, from cinematographic to interactive or even fully automatic procedures for wind estimation from cloud displacements, have been introduced and highly advanced into refined operational analysis schemes. Recently, H. H. NAGEL gave a brief survey on the applied principles. The derived wind data certainly are very valuable, and for some regions of the earth, mainly the oceanic areas, they meanwhile represent a primary data source. However, the main point I am driving at is the following: Experience from visual inspection of motion scenes from satellite picture sequences evidently shows that there is much more information contained in those motion scenes than the vector field of cloud displacements, because there is more exhibited than horizontal motion. As mentioned above, the motion scenes show the temporal variations of patterns formed by the formation, growing, structural modification, displacement and dissipation of clouds and cloud systems. Thus horizontal displacement, although providing an important and useful parameter of atmospheric dynamics, is only one element of a set of valuable information pieces which the human eye is able to comprehensively perceive. It is all this information beyond cloud displacement which is, at present, largely wasted, due to the lack of effective quantitative elaboration techniques. Present visual analysis methods are rather tentative, of qualitative nature, and purely descriptive, but they brought at least evidence of the large information potential of these motion scenes, particularly when by application of various cinematographic experiences - like e.g. a careful tuning of the time lapse ratio to the time-scale of the observed atmospheric motion systems or the proper choice of motion manipulation (animation mode) -

the visual perception of the characteristics of the observed motion fields and their concurrent pattern changes (see WARNECKE and ZICK, 1981) is being optimized. One should, however, recognize that only a negligible portion of this information potential can ever visually be exploited, compared to the tremendous amount of continuously incoming data. Thus the development of automated, at least interactive motion analysis methods is dearly needed, and all you experts in dynamic scene analysis are cordially invited for help!

Finally I like to point out that I took meteorology as an example. With slight modifications, namely most of the statements are also true for other geophysical disciplines, since time sequences of satellite images are available. In oceanography, for example, motion scenes of the spatio-temporal variations of ocean currents as reflected in the structure and changes of sea surface temperature patterns (NOAA-NESS) or of the annual course of polar and sub-polar sea ice formation and ice transports (GLOERSEN) have been produced as examples of promising and exciting possibilities also waiting for further exploitation.

References

NAGEL, H. H., 1981: "Image Sequence Analysis: What Can We Learn From Application?" In: T. S. Huang, Ed. :Image Sequence Analysis. Springer-Verlag Berlin Heidelberg New York. pp. 19-228

WARNECKE, G. and C. ZICK, 1981: "The use of cinematographic methods for the presentation of atmospheric motions as revealed by remote sensing techniques from satellites." In:A. P. Cracknell:Remote Sensing in Meteorology, Oceanography and Hydrology. Ellis Horwood Ltd. Chichester. pp. 452-473

NOAA-NESS, Applications Group: Film: " Oceanic and related atmospheric phenomena as viewed from environmental satellites". NOAA-NATIONAL ENVIRONMENTAL SATELLITE SERVICE, Camp Springs, Md.

GLOERSEN, P.W.: Film: "Antarctica Ice Sheet Variations." NASA, GODDARD SPACE FLIGHT CENTER, Greenbelt, Md.

PART IV

BIOMEDICAL APPLICATIONS

PROCESSING AND ANALYSIS OF RADIOGRAPHIC IMAGE SEQUENCES

by K.H. Höhne and M. Böhm

Institut für Mathematik und Datenverarbeitung in der Medizin
Universitäts-Krankenhaus Eppendorf and
Deutsches Elektronen-Synchrotron DESY, Hamburg, F. R. G.

Abstract

Image sequence processing for the quantification of dynamic processes as well as for the enhancement of morphological structures from radiographic image sequences is a rapidly growing field. This paper tries to give an overview of the various approaches and their problems.

The motion processes visible in the sequences, the problems of imaging methods and their application are then described: (1) Image sequence restoration and enhancement, which is successfully applied for vascular imaging and visualisation of heart motion, (2) quantification of blood dynamics by densitometric methods and parametric imaging, (3) quantification of heart dynamics by interactive and automatic methods as well as parametric imaging and finally (4) attempts at diagnostic classification. It is concluded that methods which end up with images are already of practical importance, whereas methods of automated interpretation still lie in the domain of basic research.

NATO ASI Series, Vol. F2
Image Sequence Processing and Dynamic Scene Analysis
Edited by T.S. Huang

1. Introduction

The registration of temporal changes in X-ray images for the assessment of functions of the human body has been performed for a long time. In the conventional way this has been done by qualitative inspection of the images. Although the need of a quantitative analysis was felt very early, no real advance was made in this direction until the application of digital methods in the beginning of the seventies. With the development of powerful memory moduls and processors, which are quite inexpensive, promising advances have been made in the past five years. The objectives of the developed methods are twofold:

- Quantification of blood flow and heart motion.

- Enhancement of low-exposure X-ray image sequences, allowing lower patient irradiation without loss of diagnostic information.

The purpose of this paper is to give an introduction to this field and to outline the principles of various approaches and their associated problems. As new results are not as frequent as overview papers, overlaps with other papers of the authors were unavoidable (BÖHM 81[1]).

The overview is certainly not complete because there exists a large number of publications spread out over journals and conference proceedings reaching from radiology, cardiology, and physiology through medical informatics, image processing, pattern recognition, artificial intelligence and system design to X-ray physics.

The processing of X-ray image sequences has to be seen in the larger context of other application areas in which image sequences are processed and analysed. NAGEL 79[3] gives an excellent overview of the whole area with an extensive bibliography.

This paper concentrates on the methods of image sequence restoration and enhancement, and the quantification of the dynamics of blood circulation and heart wall motion. Unless otherwise stated, the examples are from our own work.

2. Principles and Problems

2.1. The Motion Processes

All structures in the human body are subject to temporal change. Only a small subset of those structures and their changes can be observed by radiological methods. Although the human visual system has outstanding capabilities to recognize complex *spatial patterns*, the recognition of complex *motion patterns* as they occur in X-ray image sequences is often very difficult and a quantification is impossible.

There exists therefore a demand to derive a description of the dynamic processes which are *quantitative* and compact enough to allow a human interpretation. Such descriptions depend on the observed objects, the type of motion they perform and the specific question at hand. From an image processing point of view the "objects" in this field and their motion can be divided into several classes:

1. *Rigid objects* can change only their position and orientation in space. There is not much interest in radiology to determine the motion of rigid objects, like bones, automatically.

2. A very important class of moving objects are *elastic objects* which perform an expansion and contraction like the heart chambers or lungs. The function of these organs depends on *variations in size and shape*.

3. Another important class is *liquid motion* like the *blood flow in vessels*. This is a complicated dynamic process because the flow is not continuous due to the heart beat and the vessels are not rigid tubes.

4. Blood flow in vessels which are so small that they are not visible, is here considered to belong to another class, because only spatially differentiated *changes in density* or concentration can be observed. Information about the direction of flow is not obtainable.

This classification is based upon the *functional dependency* within the elements of the moving object. For rigid objects there exists such a dependency between all elements of the moving object. Therefore the motion can be described in a very compact form. The motion of the center of gravity and the rotation of the object is such a description.

For the motion of the heart wall, for example, there exists, in general, only a dependency on some neighbourhood, making a much less compact description necessary. This difference is also important for the algorithms which will detect the motion.

In addition to the differences of the objects, different levels of description can be distinguished for the motion. We can consider:

- The description in quite *general motion terms* like the time of the beginning and end of the motion.

- The *calculation of parameters* like the blood flow in a vessel or the ejection fraction of a heart chamber.

- The *interpretation of the description in diagnostic terms.* An example is the classification of local heart wall motion into classes like diskinesis, akinesis etc.

The situation is much more complex when the observed motion is a mixture of different motions. This makes a *decomposition* necessary. Two important cases are:

- The superposition of changes due to the imaging technique. Examples are the superposition of contrast medium flow in different depths of the organ or the superposition of heart motion and diaphragma motion.

- The superposition of different motions. This can be due to patient motion or breathing. Another example is the superposition of blood motion and heart motion in coronary arteries.

2.2. Problems of Imaging Methods

The imaging systems which are used to observe dynamic processes in the human body cause limitations on the obtainable information. The measured values often represent a mixture of physical effects, most of which are not those one is primarily interested in. Stationary processes such as blood flow can be made recognizable only by an artificial change - the injection of contrast medium - which can itself have an influence on the process which is to be observed.

An ideal imaging technique in radiology would deliver a property of the examined organ (e.g. X-ray absorption) as a function of space and time. With computed tomography (see PFEILER 81[4]) an image representing a slice of the human body can be reconstructed from data taken from many directions. To observe fast dynamic phenomena it would in general be necessary to reconstruct many slices in parallel at a high rate. There exists a prototype system for heart studies (WOOD 79[5]) which allow the acquisition of such sequences of 3-dimensional images at a rate of 60 images per second. But the necessary X-ray equipment (28-X-ray sources), the data rates of 3.2 x 10^9 bits/sec and the produced amount of data (60,000 cross sections for a heart study of 4 sec) seems to be prohibitive for a routine widespread application of such machines in the near future. Significantly simpler tomographic imaging methods for heart studies, which could be important in the future, are based on "gating" techniques. In these the cross sections are recorded during many heart cycles.

But today, in the X-ray field especially when observing fast non periodic motions, one is still restructed to sequences of projection radiographs. To make the blood flow in vessels or a heart chamber visible, a contrast medium has to be injected. To achieve a sufficient contrast, this injection is done via a catheter which has to be guided through arteries to the organ which is to be examined. Recently, however, due to the application of image processing methods, intravenous injection also has lead to fairly good images.

2.3. Problems in Modelling

In some areas of medical diagnostics such as the determination of blood dynamics we have processes which cannot be measured or even recognized without image processing methods. This means that in many cases image processing methods can do more than reproduce abilities already processed by humans. This however raises the problem of how to choose the model of motion description such that new hitherto unknown effects are not lost. This choice is quite difficult, because there is a lack of reliable physiological models which are based on long term clinical experience.

In other words, the application of models which are based on unsure assumptions may result in the fact that interesting effects are overlooked, the existence of which was naturally unknown during model construction. On the other hand the amounts of data which have to be analyzed are to high and their interesting properties can not be recognized by visual inspection that it is often impossible to achieve a description of the dynamics which is compact enough to be interpretable without the application of simplified models. In addition, hemodynamics in different organs has different characteristics so that one has to consider a large anatomical variability, as well. Therefore, one often has a trade off between improving the quality of the model and enlarging the application area.

3. Processing Methods for Image Sequences

Despite the above described problems and limitations various successful methods for extracting quantitative and compact information from radiographic image sequences have been developed. For their application the image sequences are digitized into image matrices with a resolution of typically 512 by 512 elements at a rate up to 60 frames/second. Depending on the problem and the available hardware up to 129 images are taken (see fig. 1). Generally the grey scal resolution is 8 bits. Nearly all applications

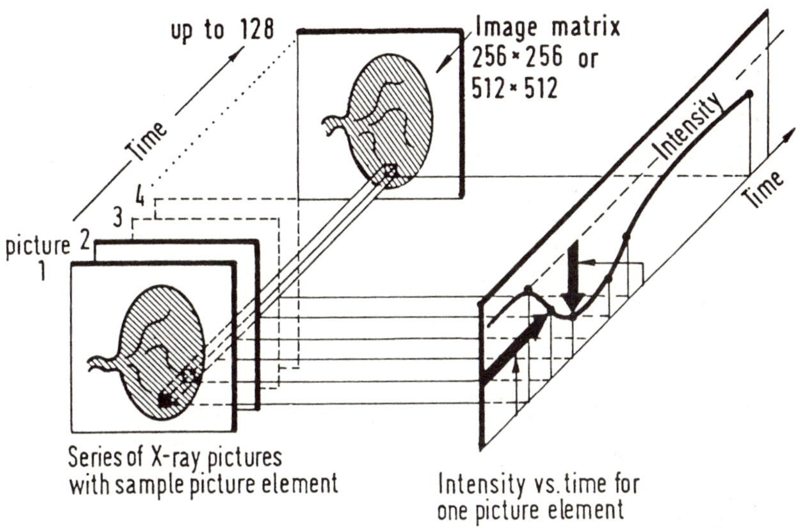

Fig. 1. Typical sizes of digital X-ray image sequences and a typical intensity vs. time curve for one picture element

process data from a video source. In the following sections, the methods applied for the restoration and enhancement of the image sequences as well as for the analysis of blood dynamics and the description of heart wall motion are outlined. When considering the sequence as a cube of intensity values I(x,y,t), the different methods may be classified according to their degree of data compression as shown in fig. 2.

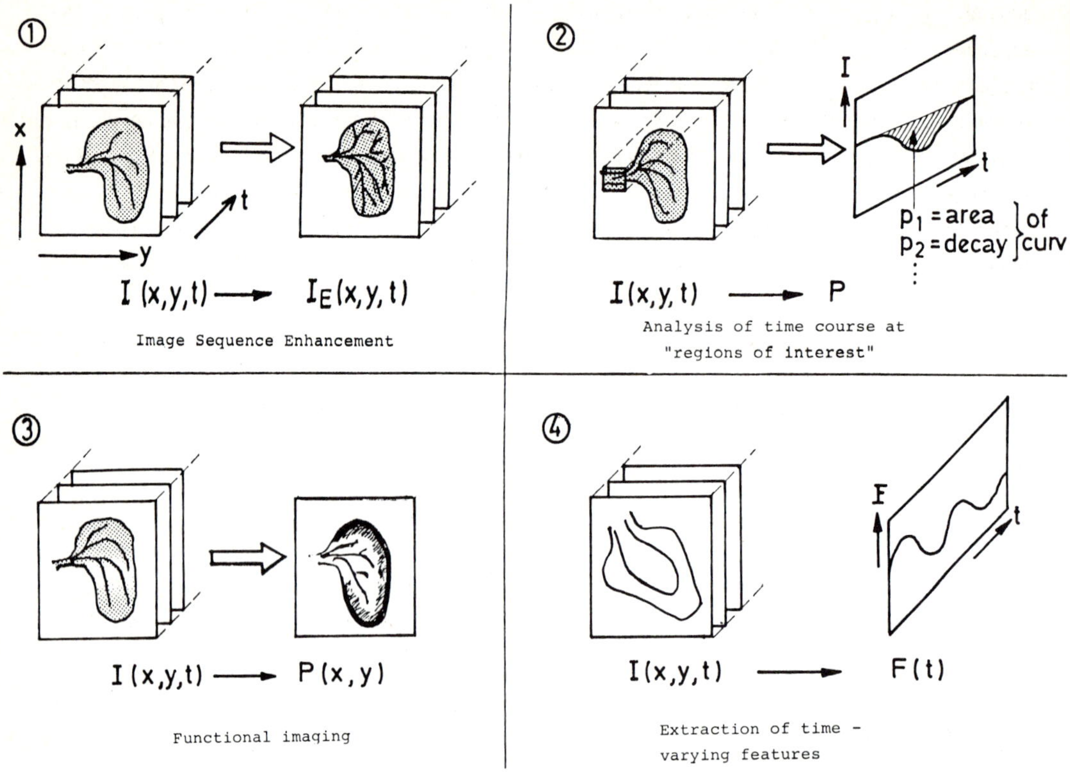

Fig. 2. A selection of methods for the processing of radiographic image sequences.

3.1. Image Sequence Restoration and Enhancement

Many methods have been described to improve the quality of images and image sequences which have deteriorated by whatever mechanism. As X-ray images have inherently good quality, in practical applications only two simple but most effective methods are applied - background subtraction and filtering in the time domain. Examples from vascular imaging and heart motion studies will illustrate this.

3.1.1. Vascular Imaging

Non digital techniques for the enhancement of vessels in angiograms by background subtraction have been known radiology for about 40 years. Many more advantages can be achieved by *digital* subtraction of non opacified background structures. Averaging several images taken before contrast medium injection improves the estimate of the background. To obtain a proper quantifikation of the observed contrast medium amount independent of local variations of the background the exponential X-ray absorption characteristics have to be considered. In most practical cases it is sufficiently correct to make the subtraction logarithmically. X-ray image sequences generally have good quality but when very small opacifications have to be detected filtering techniques are required. For the restoration of single images there exists a number of well known filtering techniques. But spatial filtering is in this case less suitable because one has to deal with abrupt intensity changes and small details. To obtain a significant noise reduction without blurring the images one needs time consuming sophisticated filtering techniques which include a segmentation process. This problem does not occure when applying 1-dimensional filtering in the time domain based on the assumption that the opacification does not change abruptly with time. This filtering is performed independently for the intensity versus time curve (fig. 1) of each picture element.

Recently such enhancement procedures have been applied with very much success in intravenous angiography. After the injection of the contrast medium into a vein, which is much less dangerous to the patients than catheterisation, the arteries were viewed. The contrast is in general much too low compared to the noise to allow a visual inspection of the recorded sequence. As described above digital processing can help to solve this problem.

To obtain images showing the morphology it is sufficient to average several images before and after the intravenous injection of the contrast medium. After subtraction of the background image the resulting opacification is amplified.

These techniques can be implemented in real time avoiding the necessity of expensive large storages for image sequences. Just now commercial equipment from several suppliers has become available. Fig. 3a shows an image of the carotid arteries before processing. Fig. 3b shows the same view, after background subtraction. Fig. 3c shows the reprojection of the enhanced vessels into the background image.

The achievable image quality can of course not compete with conventional angiography, however the more important limitation is the nonselectivity of the technique resulting often in superpositions of veins, arteries and organs. Therefore intravenous angiography will not make selective angiography obsolete but could get an important noninvasive method for screening purposes and monitoring, particularly after surgery.

Sucessful applications of intravenous angiography have been reported in the field of carotid arteries, the aorta and its primary branches and in extremity studies (CRUMMY et al. 80[6], CHRISTENSON et al. 80[7]). The image quality depends strongly on the proper

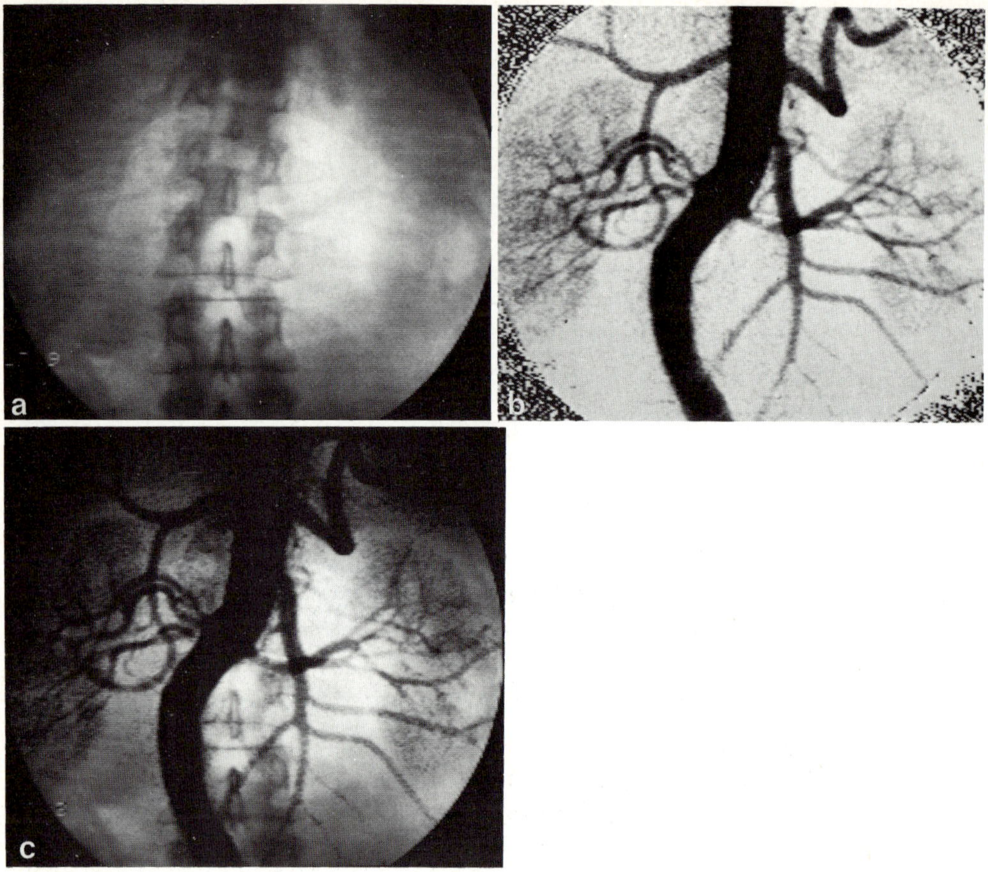

Fig. 3. Intravenous angiography:
(a) one frame of the video image sequence of human kidneys
(b) subtracted image
(c) subtracted image reprojected into the smoothed original

selection of the time intervals in which the images to be subtracted are taken. OBERMÜLLER et al.[8] try to automate this selection by looking at typical patterns in the time course of the opacification. KRUGER et al.[9] apply recursive filters in the time domain for the enhancement of vascular structures. Improvements can be expected not only from digital processing methods but also from better X-ray equipment or more advanced X-ray imaging techniques like K-edge filtering (KRUGER et al. 77[10]). A collection of papers reviewing the state of the art intravenous angiography is found in MISTRETTA and CRUMMY 81[11].

3.1.2. Heart Studies

Integration and subtraction as described above helps to visualize slowly varying structures without any superimposed motions. In intravenous angiocardiography this assumption does not hold. When respiration can be suspended BRENNECKE et al. 77[12] have obtained good results with ECG-synchronized integration of several heart cycles.

To enhance rapid changes due to heart motion and to reduce the slower changes caused by respiration and patient motion KRUGER et al. 79[13] applied a method called time intervall - consisting of typically up to 4 images - is subtracted from the image integrated over the following time interval. BRENNECKE et al. 78[14] achieved similar results by a temporal high pass filtering.

Better results - but at the expense of higher computational complexity - can be achieved by respiration synchronized subtraction (BRENNECKE et al. 78[14]). Here the selection of a background image is performed by spatial cross correlation over a limited region. The simpler time interval differencing can be performed in real time and is implemented in the forthcoming commercial equipment for intravenous angiography.

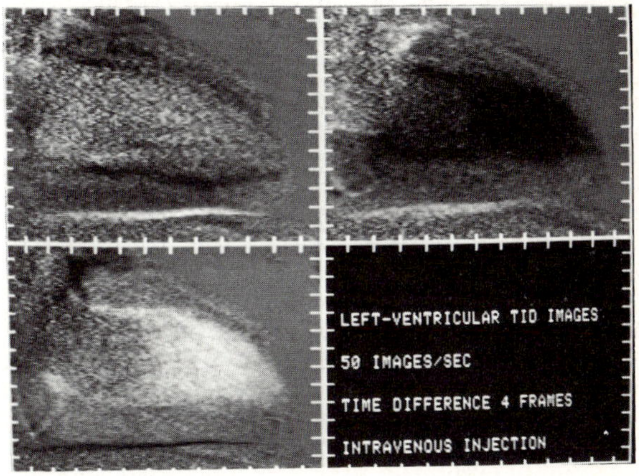

Fig. 4. Time interval differencing (TID) image of the left ventricle

Fig. 4 shows an example of a time interval differencing image. A comparison of the performance of the techniques mentioned here can be found in BRENNECKE et al. 81[15].

3.2. Quantification of Blood Dynamics

Restoration and enhancement procedures deliver images, which still are interpreted by the radiologist in a qualitative way. In a further step quantitative information may be obtained.

3.2.1. Densitometry

Video- and cinedensitometry are the oldest techniques to obtain quantitative measurements from X-ray image sequences. One or more windows called "Regions of Interest (ROI)" are outlined interactively on a representative image and the intensity versus time curve averaged of each window is measured. This can be done with analog equipment.

The most important application is the measurement of blood velocity and blood flow in vessels. The accuracy of absolute flow measurements depends strongly on the correct estimation of the vessel diameter.

Extensive literature on densitometry can be found in HEINTZEN 71[17], HEUCK 73[18] and HEINTZEN and BÜRSCH 78[19]. A comprehensive treatment of the subject is a sequence of papers by VANSELOW, HEUCK and DEININGER 75[20]. A comparison of different methods for blood flow measurement is described by SONNE and HÖHNE 78[21]. For large vessels near the body surface densitometric methods have lost their importance because noninvasive Ultrasound-Doppler-Methods can be applied (PURCELOT et al. 78[22]).

Densitometric methods are used also in ophthalmoscopy. Here a fluorescent indicator is injected intravenously and the fluorescence - induced by strong illumination - during the passage through the very small vessels of the eye fundus is observed (Mc CORMICK at all. 77[23]). Densitometry has also been used for the local analysis of heart wall motion (SILVERMAN 78[24]). Densitometry can provide important quantitative information but except in physiological research it has not found a widespread use in clinical routine.

3.2.2. Parametric Images

With the technique of densitometry only local measurements can be obtained. The blood flow of an organ, however, consists of many spatially differentiated processes. So densitometry is not suited to evaluate the blood dynamics in the whole organ or to detect small areas of reduced perfusion. Due to the spatial differentiation of the processes a large amount of parameters have to be estimated and transmitted to the physician. Because the "transmission channel" with the highest capacity is the visual system of man a pictorial presentation of the results would be the best solution.

In our research group we have therefore developed a method which we have called computer angiography. It produces a compact but spatially differentiated description of blood perfusion in the form of parametric images (HÖHNE et al. 78[25], 78[26], 80[27,28]).

The presentation of motion parameters with parametric images - in the medical context also often called functional images - is known from nuclear medicine. In computer angiography the parametric images are generated by the following procedure. The propagation of a contrast medium through the vessels of the organ which has to be examined is digitized (256 by 256 picture elements, up to 128 images up to 50 images/sec). For the subsequent steps the variation of the intensity with time at each position of the image matrix is treated independently (see fig. 1). Each intensity vs. time curve is filtered to reduce noise and solitary points are removed. Five to twenty images taken before contrast medium injection are used for background subtraction. The resulting curve is then described by a parameter vector (see fig. 5).

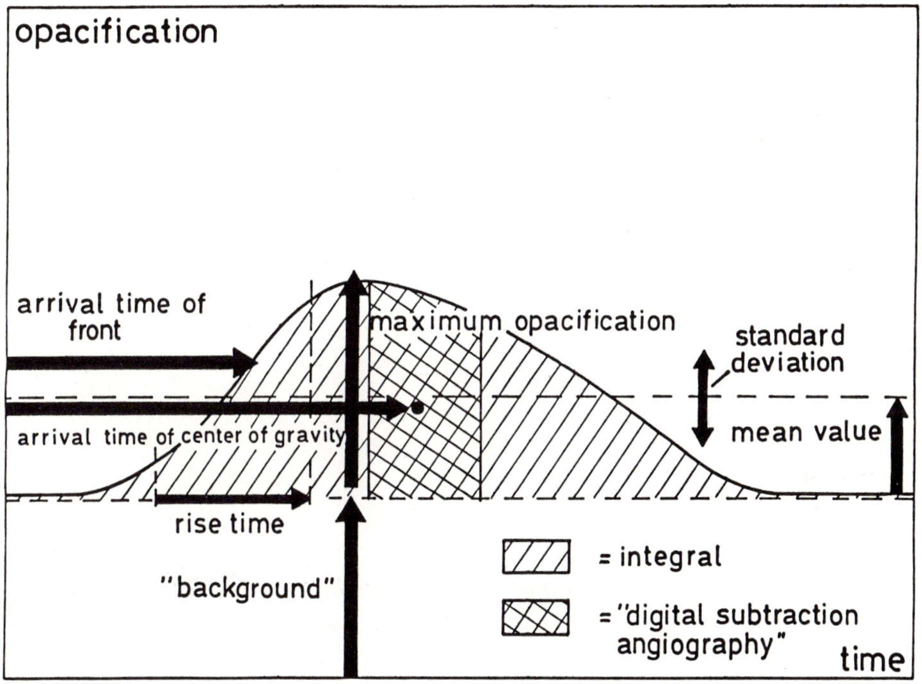

Fig. 5. Typical parameters extracted from the time course of intensity

The parameters can be divided into two groups: Those related to the dynamics, such as time of contrast medium arrival, rise time, rate of decay, and those related to

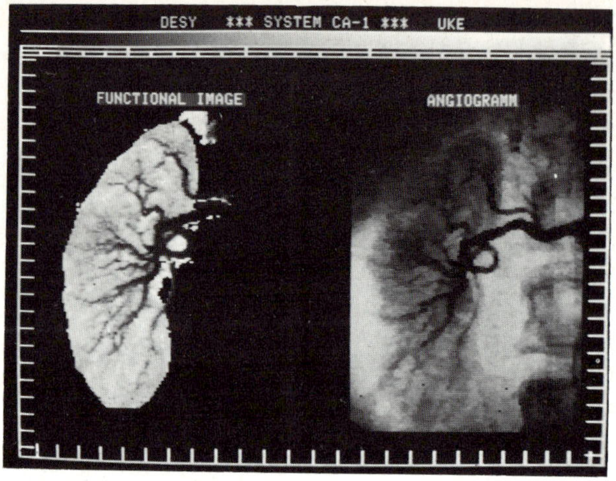

Fig. 6. Comparison of a functional image (left) with one frame of an angiographic sequence (right

blood perfusion, such as maximum or integral of opacification. The appropriate set of parameters depends on the medical problem. Parameters which are more directly associated to physiological processes would be desirable but are often difficult to define because of the lack of reliable models. Nevertheless when presenting the resulting parameters as images, we obtain static images containing one point of view of the dynamic behaviour of the organ.

Fig. 6 shows a functional image showing the arrival time (time of 50 % opacification) of the contrast medium in a kidney together with one frame of the original sequence. One recognizes that as in any subtraction technique the background of non interesting effects disappears. When presenting the image in a colour code a quantitative analysis is possible. The differences in hemodynamics, not visible in the conventional angiogram, become obvious at once.

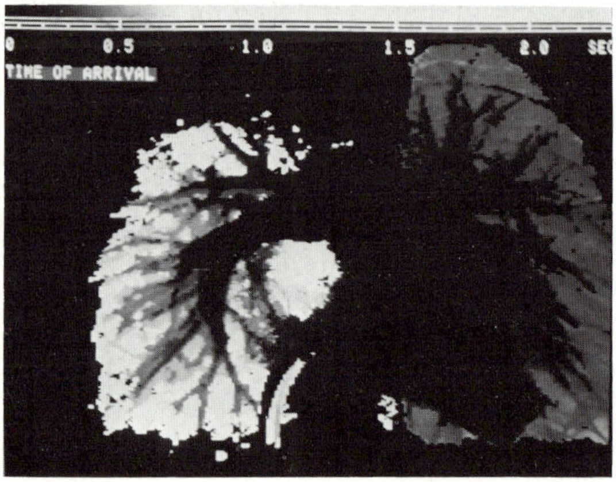

Fig. 7. Arrival time image of the lungs of a child. The left lung (right side) shows a delayed blood flow.

The application of computer angiography to measure blood dynamics in the kidney is described in HÖHNE et al. 78[29]. A promising example is the control of the function of transplanted kidneys (HAGEMANN 81[30]). Nevertheless a striking problem remains: There are often no physiological models which are detailed enough to explain the highly differentiated information delivered by the parametric images. So the main field is the spatially differentiated measurement of *relative* effects (before/after surgery, before/after therapy, different regions of the same organ). One application is in the diagnostics of congenital heart diseases. As an example from this field fig. 7 shows an arrival time image of the lungs of a baby.

An effect which was not expected is the fact that parametric images describing primarily the blood dynamics can also enhance morphological features. Fig. 8 shows the parametric image of a brain, showing the arrival time of the center of gravity of the intensity vs. time curve as parameter. Automatically the arteries show up darker because they are opacified earlier and the veins show up brighter than the parenchyma. Furthermore more details show up because the center of gravity is computed from the *integral* of the intensity vs. time curve, which leads to a noise reduction.

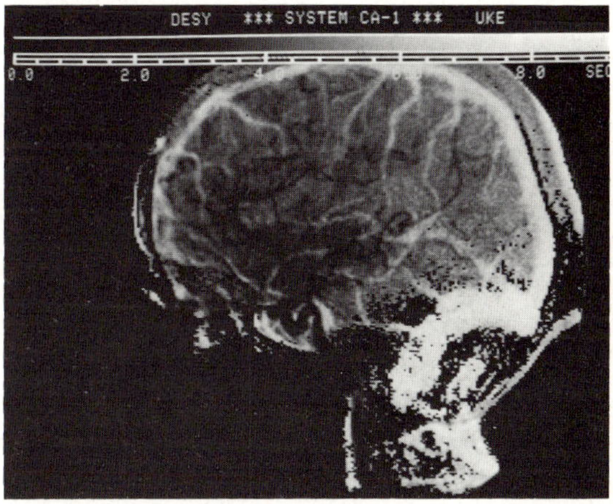

Fig. 8. Parametric image of a brain showing the arrival time of the center of gravity of the intensity vs. time curve as a parameter

Despite the advantages of computer angiography there remain several problems. First there are limitations caused by the imaging method which are mainly due to the fact that only projection images are taken. Other problems result from the technique of angiography itself. The mixture of contrast medium and blood can be inhomogeneous. The injection has a mechanical influence on the blood velocity and can induce vessel reactions. Organ motion caused by patient motion or respiration produces distored images. Background motion, for example the motion of bowels, can produce artefacts. These problems can in principle be solved by image processing methods. Harder to solve are problems concerning the diagnostic interpretation. Since most of these processes could not be measured earlier at all, often there exists no experience whether an observed deviation in blood dynamics has diagnostic relevance or not.

Parametric images can also be generated in a similar way from sequences of computed tomograms (OBERMÖLLER et al. 81[32]). The clinical value however presently seems to be limited by the low image acquisition speed.

3.3. Quantification of Heart Dynamics

The derivation of a quantitative description boundary undergoes some postprocessing consisting of gap filling, removal of redundant points and loops and a smoothing of the contour.

Despite all these efforts it seems that the methods used for the automatic determination of shape parameters are still not stable enough to be applied routinely in broad clinical work. The main reason is obviously the high physiological variability in the imaging devices.

Parametric Image: The goal of this approach is to present the characteristics of the heart wall motion in parametric images rather than to calculate shape parameters (HÖHne et al. 79[43], BRENNECKE et al. 79[49], BÖHM et al. 80 [50]). The generation of parametric images from the intensity versus time curves at each picture element has advantages for the presentation of spatially differentiated motion characteristics.

As an example fig. 9 shows two frames at extreme phases of a heart cycle from a digitized image sequence of a left ventricular angiogram together with the intensity versus time curves (ITC) at different positions of the image. One observes sharp maxima for the raster point near the systolic end position of the ventricle since the contrast-medium - filled ventricular cavity disappears from this position only for a short time. The opposite is true for the raster point near the diastolic

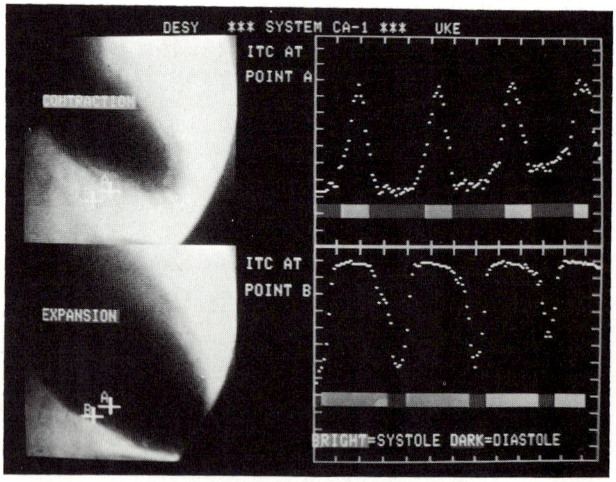

Fig. 9. Intensity variations caused by heart motion at different locations

end position since the ventricle covers this only for a short time.

In general the shape of this curve represents other effects such as opacification changes or superimposed diaphragma motion in addition to heart motion. A decomposition is therefore necessary to extract the compartment representing the ventricle motion (BÖHM et al. 80[47]).

After this decomposition the cleaned curves are then described by a feature vector. Important features include the time at which the ventricle passes during contraction and the time at which the ventricle passes during expansion. From these features a set of parametric images is generated. Fig. 9 shows as an example a parametric image representing the ratio between expansion speed and contraction speed of the heart wall. The regular pattern is typical for a normal heart. It should be mentioned that a significant translational motion of the heart - as caused by patient motion - results in a typically distorted motion pattern. It seems that there is not yet enough clinical experience to answer the question of clinical applicability of parametric imaging.

3.3.1. Diagnostic Classification

Not much work has been reported on an automatic diagnostic classification of extracted motion patterns. An interesting approach based on techniques of artificial intelligence is described by TSOTSOS et al. 81[51]. During open heart surgery tantalum markers can be implanted into the myocardium. This causes most of the previously described problems to disappear: The detection of these markers is quite easy, no catheterization is necessary and much more information can be drawn from the sequence because of the existence of "landmarks". An example is the shortening of the myocardium which can be estimated from the motion of the markers relative to each other.

This method can be applied only for post-surgical examinations but allows for an explicit modelling of the left ventricular activity. In contrast to the techniques described above which concentrate on preprocessing, extraction and description of motion, TSOTSOS et al. concentrate on the classification of already extracted motion into diagnostic classes like normal, akinetic, hypokinetic, diskinetic and paradoxical. The necessary medical definitions are contained in a "knowledge base". The result is a hierarchy of reports describing and classifying the observed motion on several levels from the motion of a single marker up to the motion of the whole ventricle.

4. Implementation Aspects

4.1. Hardware

The processing of X-ray image sequences rises problems because of the high data rates and the large amounts of data resulting from the high spatial and temporal resolution. Two main approaches are possible to deal with this problem - real time processing and processing of stored data.

A pure real-time processing of multi-temporal images has the advantage of high speed and since no temporary storing of large data volumes is necessary, the system is less expensive. On the other hand, pure real-time processing has the disadvantage that more sophisticated algorithms cannot be applied, and the refinement of algorithms is much more difficult if they are implemented in hardware at too early stage of development.

The other approach is to digitize the whole sequence first and perform the processing afterwards on the stored data. This can be done by a general purpose computer or with additional dedicated hardware processors. To achieve efficiency it is necessary to construct fast and large memories which allow the digitization and storage of the whole sequence in real time.

The real time approach has been chosen, for example, by a group of MISTRETTA and coworkers[13,52] to implement a system for the enhancement of intravenous angiocardiographic sequences. Also some commercial systems which have become available for this task are based on real time processing. This is a suitable approach when one restricts oneself to simple algorithms such as background subtraction or time interval differencing (see chapter 3.0).

In our group we decided on real time acquisition and display based on an 8 M byte RAM image sequence store. Computations however are done not in real time but from the stored data (NICOLAE and HÖHNE 78[54]). A somewhat different structure has been proposed by BRENNECKE et al. 79[43]. Here fast dedicated functional units are under development to allow fast processing (BRENNECKE et al. 79[49]). Other systems for processing image sequences are described by GILBERT et al. 76[53] and READ et al. 77[55]. Commercial man--factures recently have announced image sequence processing systems.

4.2. Software

Without appropriate system hardware one cannot start to work with image sequences in practice. But this is only one half of the problem. Due to the difficulties, especially in the past, with system hardware, the other half - the system software - has been somewhat neglected. What is meant here with system software is not the image processing algorithms but the necessary environment in which they are formulated, tested and refined and in which they finally operate.

The main requirements which have to be met by the system software are

- the support of *interactive* communication,
- the support of a problem oriented dialog via a problem oriented *language*,
- high *efficiency* of execution,
- easy *integrability* of special image processing hardware.

Two dialog systems covering this range have been developed in our research group (BÖHM et al. 79[56], PFEIFFER 78, 81[57,58]). Other attempts are described in BOLC et al. 81[59], PRESTON 80[60].

5. Conclusion

We conclude from our overview that real advances for the radiologist can presently be achieved by the methods yielding results in picturial form, as e. g. in subtraction imaging, and parametric imaging. This is certainly due to the fact that these methods provide an adequate data *presentation*, but the *interpretation* is still left to the radiologist. Concerning the much more difficult problem of automatic interpretation advances have been limited; at least in part by the imperfect state of formalization of our medical knowledge.

References

(1) BÖHM, M., and HÖHNE, K.H., The Processing and Analysis of X-Ray Image Sequences, in (2)
(2) HÖHNE, K.H., (ed.) Digital Image Processing in Medicine, Lecture Notes in Medical Informatics 15, Springer, Berlin-Heidelberg, New York, 1981
(3) NAGEL, H.H.: "Analysis of Image Sequences: What can we learn from Applications?", in Springer Series on Information Science, Springer, 1981
(4) PFEILER, M.: CT-Techniques in Medical Imaging, in Medical Informatics 15, Springer, Berlin-Heidelberg, New York, 1981
(5) WOOD, E.H., KINSEY, J.H., ROBB, R.A., GILBERT, B.K., HARRIS, L.D., and RITMAN, E.L.: Application of High Temporal Resolution Computerized Tomography to Physiology and Medicine, in Image Reconstruction from Projections, Implementation and Applications (G.T. Herman ed.), Topics in Applied Physics 32, Springer, New York, 1979
(6) CRUMMY, A.B., et al.: "Computerized Fluoroscopy: Digital Subtraction for Intravenous Angiography and Arteriography", AJR 135 (1980) 1131 - 1140
(7) CHRISTENSON, P.C., OVITT, T.W., FISHER, H.D., FROST, M.M., NUDELMAN, S., and ROEHRIG, H.: "Intravenous Angiography Using Digital Video Subtraction: Intravenous Cervicocerebrovascular Angiography", AJR 135 (1980) 1145 - 1152
(8) OBERMÖLLER, U., BÖHM, M., HÖHNE, K.H.: Restauration von Gefäßdarstellungen in der intravnösen Angiographie, Proc. DAGM-Symposium, Hamburg 1981, Informatik Fachberichte 49, Springer, Berlin-Heidelberg, New York, 1981
(9) KRUGER, R.A.: Time Domain Filtering Using Computerized Fluoroscopy - Intravenous Angriography Applications, Proc. SPIE 314 (1981) 319
(10) KRUGER, R.A., et al.: "Digital K-Edge Subtraction Radiography", Radiology 129 (1977) 243 - 245
(11) MISTRETTA, C.A., and CRUMMY, A.B., (eds.): "Intravenous Video Arteriography", Proc. Conf. Intravenous Video Arteriography, Madison, Wisconsin, USA, 1981
(12) BRENNECKE, R., BROWN, T.K., BUERSCH, J., and HEINTZEN, P.H.: "Computerized Video-Image Preprocessing with Applications to Cardio-Angriographic Roentgen-Image-Series", in Digital Image Processing (H.H. Nagel ed.), Informatik-Fachberichte 8 (1977) 244 - 262
(13) KRUGER, R.A., et al.: "Computerized Fluoroscopy in Real Time for non Invasive Visualization of the Cardovascular System-Preliminary Studies", Radiology 130 (1979) 49 - 57
(14) BRENNECKE, R., HAHNE, H.J., MOLDENHAUER, K., BUERSCH, J.H., and HEINTZEN, P.H.: "Improved Digital Real-Time Processing and Storage Techniques with Applications to Intravenous Contrast Angiography", Proc. Computers in Cardiology, Long Beach, 1979, IEEE Publ. 191 - 194
(15) BRENNECKE, R., BUERSCH, J.H., BOGNER, H.G., and HEINTZEN, P.H.: "Digital Intravenous Imaging Techniques in Pediatric Cardiology", in Proc. Conf. Intravenous Video Arteriography, Madison, Wisconsin, USA, 1981
(16) HÖHNE, K.H., NICOLAE, G., PFEIFFER, G., DIX, W.R., EBENRITTER, W., NOVAK, D., BOEHM, M., SONNE, B., BUECHLER, E.: "An Interactive System for Clinical Application of Angiodensitometry", in Digital Image Processing (H.H. Nagel ed.), Informatik Fachberichte 8 (1977) 232 - 243
(17) HEINTZEN, P.H., (ed.): "Roentgen-, Cine- and Videodensitometry", G. Thieme, Stuttgart, 1971
(18) HEUCK. F.S., (ed.): "Densitometrie in der Radiologie", G. Thieme, Stuttgart, 1973
(19) HEINTZEN, P.H., and BUERSCH, J.H., (ed.): "Roentgen-Video-Techniques", G. Thieme, Stuttgart, 1978
(20) VANSELOW, K., HEUCK, K., and DEINIGER, H.K.: "Neue Grundlagen und Theorien zur Verbesserung der Angio-Cine-Densitometrie, Fortschr. Roentgenstr. 122: 453, Fortschr. Roentgenstr. 123: 268, Fortschr. Roentgenstr. 123: 358, Fortschr. Roentgenstr. 122: 468, 1975
(21) SONNE, B., and HOEHNE, K.H.: "Vergleichende Untersuchungen verschiedener Meßmethoden zur Blutflußbestimmung aus digitalen Angiogrammen", Biomed. Techn. 23 (1978) 208 - 215

(22) POURCELOT, L., BESSE, D., PEJOT, C., and PLANIOL, T.: "Blood-Flow-Display by Doppler Technique", Proc. of FIOSIGMA 78, Paris, 1978

(23) McCORMICK, B.H., READ, J.S., BOROVEC, R.T., and AMENDOLA, R.C.: "Image Processing in Television Ophthalmoscopy", in Digital Processing of Biomedical Images (K. Preston and M. Onoe eds.), Plenum Press, New York, 1976

(24) SILVERMAN, N.R.: "Videodensitometry and Video-Kymography for Evaluation of Myocardial Contractibility", in Biomed. Techn. 23 (1978) 208 - 215

(25) HOEHNE, K.H., BOEHM, M., ERBE, W., NICOLAE, G.C., PFEIFFER, G., and SONNE, B.: "Computer Angiography: A New Tool for X-Ray Functional Diagnostics", Med. Progr. Technol. 6 (1978) 23 - 28

(26) HOEHNE, K.H., BOEHM, M., NICOLAE, G.C.: "Encoding and Analysis of X-Ray Image Series", Proc. of the 4th IJCPR, Kyoto, 1978, 743 - 746

(27) HÖHNE, K.H., BOEHM, M., and NICOLAE, G.C.: "The Processing of X-Ray Image Sequences", in Advances in Digital Image Processing (P. Stucki ed.) Plenum Press, New York, 1980, 147

(28) HOEHNE, K.H., BOEHM, M., OBERMÖLLER, U.: "X-Ray Functional Imaging - An Evaluation of the Properties of Different Parameters", Proc. Conference on Digital Radiography, Stanford 1981, SPIE, 314 (1981)

(29) HOEHNE, K.H., BOEHM, M., ERBE, W., NICOLAE, G.C., PFEIFFER, G., SONNE, B., and BUECHLER, E.: "Die Messung und differenzierte bildliche Darstellung der Nierendurchblutung mit der Computerangiographie", Fortschr. Roentgenstr. 129 (1978) 667 - 672

(30) HAGEMANN, I., GRABBE, E., SONNE, B., BOEHM, M., and HUPE, W.: "Computerangiographie bei transplantierten Nieren", Fortschr. Roentgenstr. 134 (1981)

(31) BOEHM, M., NICOLAE, G.C., PFEIFFER, G., and HOEHNE, K.H.: "Darstellung digitaler Bilder in der Radiologie", Nedizinische Dokumentation und Statistik 14, Springer, Heidelberg, 1979

(32) OBERMÖLLER, U., BOEHM, M., HOEHNE, K.H.: "Functional Images from Serial Computer Tomograms", Technical Note, Med. Progr. Technol. 8 (1981) 99 - 101

(33) WOOD, E.H.: "Evolution of Instrumentation and Techniques for the Study of Cardiovascular Dynamics from the Thirties to 1980", Ann. of Biomed., Eng. 6 (1978) 240 - 309

(34) TASTO, M., FELGENDREHER, M., SPIESSBERGER, W., and SPILLER, P.: "Comparison of Manual versus Computer Determination of Left Ventricular Boundaries from X-Ray Cineangiograms", in G. Thieme, Stuttgart, 1978

(35) CHOW, C.K., and KANEKO, T.: "Computer Calculation of Left Ventricular Volumes from a Cineangiogram", Proc. Quantitative Imaging in the Biomedical Sciences", SPIE, 1971

(36) CHOW, C.K., and KANEKO, T.: "Automatic Boundary Detection of the Left Ventricle from Cineangiograms", Comp. and Biomed. Res. 6 (1972) 388 - 410

(37) CHOW, C.K., HILAL, S.K., and NIEBUHR, K.E.: "X-Ray Image Subtraction by Digital Means", IBM J. Res. Develop. 17 (1973) 206 - 218

(38) SCHELBERT, H.R., KREUZER, H., DITTRICH, J., REITSMA, H., and SPILLER, P.: "Videometrische Ventrikelflächenbestimmung mit haltaufomatischer Korrektur des Bildhintergrundes", Res. exp. Med. 158 (1972) 66 - 74

(39) LEDLEY, R.S.: "Some Clinical Applications of Pattern Recognition", Proc. of the 1st Conf. on Pattern Recognition (1973) 89 - 112

(40) TRENHOLM, B.G., et al.: "Automated Ventricular Volume Calculation from Single Plane Images", Diagnostic Radiology (1974) 299-3o4.

(41) SMALLING, R.W. et al.:"Digital Boundary Detection, Volumetric and Wall Motion Analysis of Left Ventricular Cine Angiograms",Comp.Biol.Med.6 (1976) 78-85.

(42) SLAGER, C.J., REIBER, S.H.C., SCHUUBIERS, J.C.H., and MEESTER, G.T.: "Automated Detection of Left Ventricular Contour, Concept and Application", in (17)

(43) TASTO, M.: "Guided Boundary Detection for Left Ventricular Volume Measurement", Proc. 1st Int. Joint Conf. on Pattern Recognition, 1973

(44) TASTO, M.: "Motion Extraction for Left Ventricular Volume Measurment", IEEE Trans. on Biomed. Eng. 21 (1974) 207 - 213

(45) SPIESSGERBER, W., and TASTO, M.: "The Automatic Measurement of Medical X-Ray Photographs", Philips techn. Review 35, 6 (1975) 170 - 180

(46) EIHO, S., et al.: "Automatic Processing of Cineangiographic Images of Left Ventricle", Proc. of the 4th IJCPR Kyoto, 1978, 740 - 742
(47) SPIESSBERGER, W., and TASTO, M.: "Processing of Medical Image Sequences", Springer Series on Information Science, Springer, 1981
(48) HOEHNE, K.H., BOEHM, M., NICOLAE, G.C., PFEIFFER, G., SONNE, B.: "The Analysis and Visualization of Blood Dynamics and Motion of Organs by Computer Angiography", Proc. Conference on Computer Aided Analysis of Radiological Images, Newport Beach, 1979
(49) BRENNECKE, R., HAHNE, H.J., MOLDENHAUER, K., BUERSCH, J.H., and HEINTZEN, P.H.: "A Special Purpose Processor for Digital Angiocardiography, Design and Applications", Proc. Computers in Cardiology, Long Beach, 1979 IEEE Conf. Publ. 1979
(50) BOEHM, M., OBERMOELLER, U., and HOEHNE, K.H.: "Determination of Heart Dynamics from X-Ray and Ultrasound Image Sequences", Proc. of the 5th International Joint Conference on Pattern Recognition, Miami 1980, IEEE Conf. Publ. 1980
(51) TSOTSOS, J.K., COVVEY, H.D., MYLOPOULOS, J., and McLaughlin, P., ALVER: "A System for the Evaluation of LV Performance from Myocardial Marker Implants", Proc. Conf. Worls Assoc. Med. Inf. 81, Strassbourg 1981
(52) MISTRETTA, C.A., et al.: "Computerized Fluoroscopy Techniques for Non-Invasive Cardiovascular Imaging", Proc. SPIE. 152 (1978) 65 - 71
(53) GILBERT, B.K., STORMA, M.T., JAMES, C.E., HOBROCK, L.W., YANG, E.S., BALLARD, K. C., WOOD, E.H.: "Real Time Hardware System for Digital Processing of Wide Band V ideo Images", IEEE Trans. Comput. 25 (1976) 1089 - 1100
(54) NICOLAE, G.C., and HOEHNE, K.H.: "Multiprocessor System for the Real-Time Digital Processing of Video-Image Series", Elektron. Rechenanl. 21 (1979) 171 - 183
(55) READ, J.S.et al.: "The Television Ophthalmoscope Image Processor. Proceedings of the IEEE Workshop on Picture Data Description and Management, Chicago, 1977, 64-67.
(56) BOEHM, M., NICOLAE, G.C., and HOEHNE, K.H.: "Advanced Hardware and Software Tools for Analysis of Multitemporal Images in a Clinical Environment", Lecture Notes in Medical Informatics 5 (1979), Springer, Berlin-Heidelberg-NewYork, 445
(57) PFEIFFER, G., and HOEHNE, K.H., :"Improvement of Programming Efficiency in Medical Image Processing by a Dialogue Language", Proc. Medical Informatics Europe 78, Lecture Notes in Medical Informatics 1 (1978) 203 - 215
(58) PFEIFFER, G.: "Entwurf und Implementierung eines Dialogsystems zur Erzeugung interaktiver Bildverarbeitungssysteme in der Medizin", Ph. D. Thesis, University of Hamburg
(59) BOLC. L., KULPA, Z., (eds.): "Digital Image Processing Systems", Lecture Notes in Computer Science (1981) 109
(60) PRESTON, K.: "Image Manipulative Languages - A Preliminary Survey", in Pattern Recognition in Practice (E.S. Gelsema and L.N. Kanal eds.), North Holland, 1980, 5 - 20
(61) HUANG, T.S., (ed.): "Image Sequence Analysis", Springer Series on Information Science, Springer, 1981
(62) BOEHM, M., OBERMOELEER, U., PFEIFFER, G., HOEHNE, K.H.: "Image Management in the System CA-1", Proc. SPIE 318 (1981) 161

IMAGE SEQUENCE PROCESSING AND PATTERN RECOGNITION OF BIO-MEDICAL PICTURES.

C. GAUDEAU WITH :

G. ROUTIER, D. DZEIDZINA, C. LONGEVIALLE*

Laboratoire de BIO-INFORMATIQUE
2 bis, Bd Tonnelé
37032 TOURS CEDEX. FRANCE

* D.R.E.T. 26, Bd Victor Hugo 75996 PARIS ARMEES

Already a certain number of systems have been studied and realized in France and abroad for biomedical picture sequences processing.Specialized work has been done on the cineangiocardiography which is concerned principally by the volume rebuilding of the left ventricle and its variations. Extension of this work has been done on the gastric X-ray pictures for the pattern recognition of the stomach and for the reconstitutions of the biomechanical characteristics of the stomach, especially its contractility. In order to perform these analysis, two types of equipment have been developped :

- a semi-automatic system: a video planimeter including video camera, a counter allowing evaluation of differents sequence surfaces for volume estimation (this system, with reduced cost, allows rapid computations of volume changes)
- an automatic system S.A.P.H.I.R. (including an optical scanner with photo-cells and an interactive microprocessor system). This system allows more complex computations : automatic contour detections of cardiac and gastric contours 3 D volume reconstruction, medial axis of cardiac and gastric organs, global and partial cardiac ejections fractions, contractility index. These methods give a precise quantification of organs cinetic and bring new biomechanics parameters as maps of tissue stress repartition.

NATO ASI Series, Vol. F2
Image Sequence Processing and Dynamic Scene Analysis
Edited by T.S. Huang

INTRODUCTION

Among the great number of time sequence data, necessary to caracterize the state of a physiological dynamic system as the heart or the stomach very few are accessible with functionnal exploration methods. In spite of these difficulties, this study imposed to use the accessible data as cineangiographies (cineangiocardiographies and cinegastrographies) to develop methods of time sequence analysis to understand the internal state of the contractile system that represents the myocard or the gastric muscles. Before that, a certain number of pertinent parameters, useful tools for diagnosis, are generally evaluated : (GAASCH 1972)

- contour detection of the organs at different moments of the sequence (left ventricle or gastric contours)
- 3 D rebuilding of the organ at different moments of the sequence
- superposition sequence contours
- global and local contractility indices. (RACKLEY 1964)

But in order that these last parameters should be evaluated precisely, the contour determination is a phase which must be carefully done and comparaison between the contours of each sequence picture must be done - in an automatic way -.

I. DATA ACQUISITION

I.1 ANALOG ACQUISITION

I.1.1. Cineangiocardiographic data

Generally face and profile pictures of the heart are recorded on a 35mm film with a rate of approximately 60 frames/sec. A contrast product is introduced in the left ventricle by a catether.

If each cineangiocardiographic system permit face and profile records, most of the clinicians prefere to record and analyse the pictures taken in one position, in an oblic anterior right position - the longitudinal heart axis appear in this greater projection -.

I.1.2. Cinegastrographic data (BOIRON and COLL. 1978,1979,1980,1981)

All the subjets are in the following conditions for the X-ray investigation : they drink a 250 ml volume of a homogeneous suspension of barium sulfid and stood while drinking. Each subject is put in a prone left posterior oblic position. The subject is in apnea for 30 sec. During the gastric evacuation, the suitable

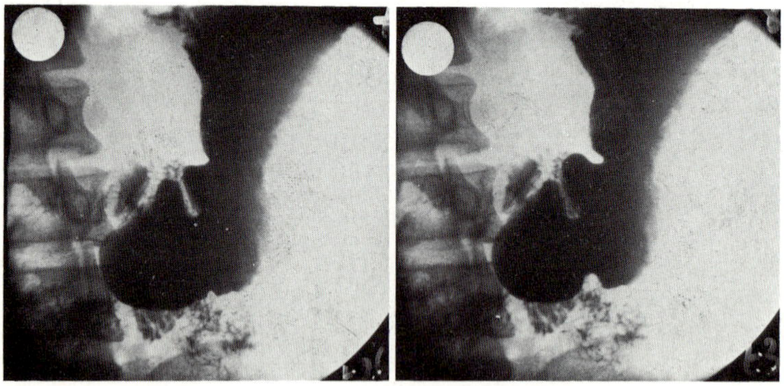

Figure 1 . Two successive gastric X-ray pictures taken every 4 seconds ; changes in contractility can be observed.

moment to start the series of pictures is determined according to the visualisation display. The image intensific fluorography is used to take a picture every 4 seconds during a 30 seconds time span. This system gives us X-ray pictures of 70 X 70 mm format and record the time (fig. 1).

I.2. DIGITAL ACQUISITION

I.2.1. Optical scanners and A/D conversion

For different medical applications several optical scanners have been developed in C.N.R.S. and the Laboratoire de Bio-Informatique in order to fill this need. The first optical scanner was built with a flying spot tube and photomultiplier. The vertical and horizontal explorations of the document are independent and can vary up to 1000 digital characters. The exploration window varies from 4 mm to 50 mm. As high as 32 levels of grey can be achieved, which is largely sufficient for cineangiographic exploration. Precision depends principally on the sampling rate and on the number of grey levels but great precision leads to a rapid saturation of computer memories be-

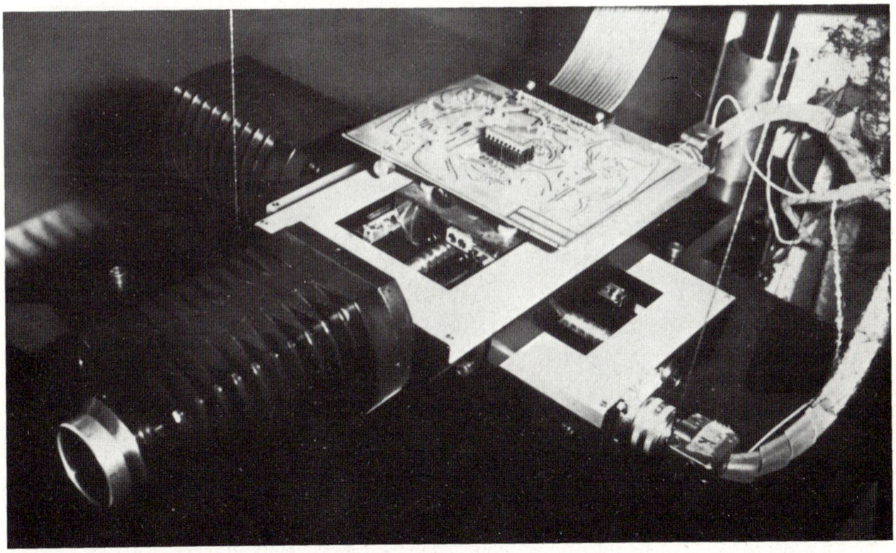

Figure 2 : Bar of 512 photo diodes RETICON with two shifting MOS registers and two step by step motors for X and Y translation.

cause the need to analyze a great number of pictures.

The flying spot , although of great speed and great precision, has several disadvantages. Some geometric distortion occurs and reproductibility is not always satisfactory. For these reasons we have decided to build an optical scanner using photocells.

This system consists of an optical device with photosensitive elements. A uniform source of 6000 lux illuminates a translucent surface on which is laid X-ray picture. The lens has a focus of 25 mm. A bar of 512 photodiodes RETICON (fig. 2) builds up the light sensitive element integrated with two shifting MOS registers of 256 bits on the same support. In the focus plan, the photodiodes of matrix executes a sweeping by successive commutation of each diode on the X-axis.

This X-axis of the photodiodes bar is obtained by mechanic translation of the matrix using a fast heps step by step motor (fig. 3).

Figure 3 : Optical scanner with variable supports for photodiodes matrix and objectif device.

The step of the motor has exactly the same dimension as each of the diodes (25 microns) The number of photocells fixed to the bar (512) and the associated optic determine the maximum resolution for a X-ray picture with a single crossing. The resolution is 1,7 pixel/mm. But the horizontal permits several crossings and the resolution can increase to 20 pixels/mm. With the DA converter, the picture transparence at each point of the analog document is restored by the grey level of each pixel. From 15 to 30 significative levels can be obtained for digital visualisation. The picture is transcripted on a listing by a succession of symbols (letters, numbers, punctuation), each symbol corresponds to a different grey level.

A conversational system (S.A.P.H.I.R.) is built in order to let an operator's choice of processing sub routines to be run by the conversational system which is composed principally with a microcomputer system, Intellec 80 microprocessor which will permit execution of the management programs and the picture-parametrisation programs.

I.2.2. Videoplanimeter

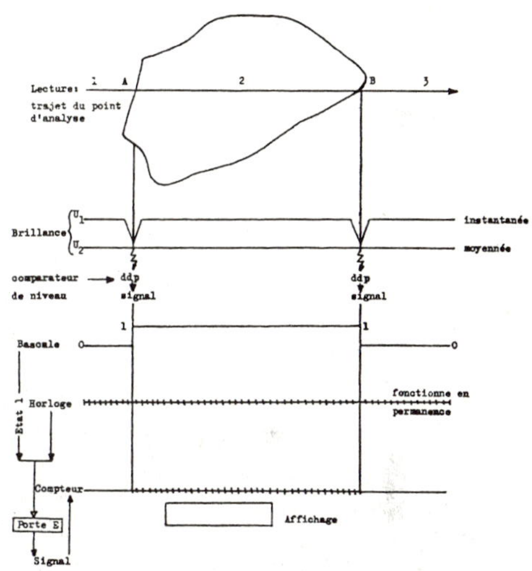

FIGURE 3.A

DIAGRAM OF THE VIDEOPLANIMETER. A COUNTER ADDS UP THE TOPS OF THE CLOCK BETWEEN TWO POINTS OF THE CONTOUR.

The videoplanimeter is a device essentially made for the calculation of relatives surfaces values from which it is possible to estimate the volume of the left ventricle, using the surface-length method either in monoplane or biplane.

A microprocessor is incorporated in order to compute the global ejection fraction. The selected X-ray picture is read by th TV camera, after any necessary contrast reinforcements and is projected on the control screen which permits the detection of measurement errors. The value of the surface included in the detected contour is automatically read and can be introduced in the microprocessor for the volume evaluation.

I.2.3. The conversational system (fig. 4, 5)

It is built in order to let an operator's choice of processing sub programs to be run by the conversational system which is composed with the following elements.

a) Microcomputer system - Intellec 80 microprocessor will permit execution of the management programs and the picture-parametrisation programs.
Hardware organization : the intellec 80 C.P.U. is realized on an I.S.I. chip (Intel's 8080). The display visualizes th e different phases of the operations. It is composed of six buses which connect the principal components of the microcomputer : C.P.U., memory, registers, imput and output devices.
Software organization : the monitor is the user's principle interface to the INTELLEC 80. It is preloaded on PROM'S. It controls the loading and the execution of other systems and user-programs. With this system, the operator manipulates the contents of memory, reads and produces paper tapes, reads or initializes PROM'S. Software is performed by assemblar language.

b) Memory - The R.A.M. (random access memory) is a fast memory of 32 Koctets. In this memory is stored the digitalized data from the D/A converter of the O.D.A.

CONVERSATIONAL SYSTEM BLOCK-DIAGRAM

FIGURE 4

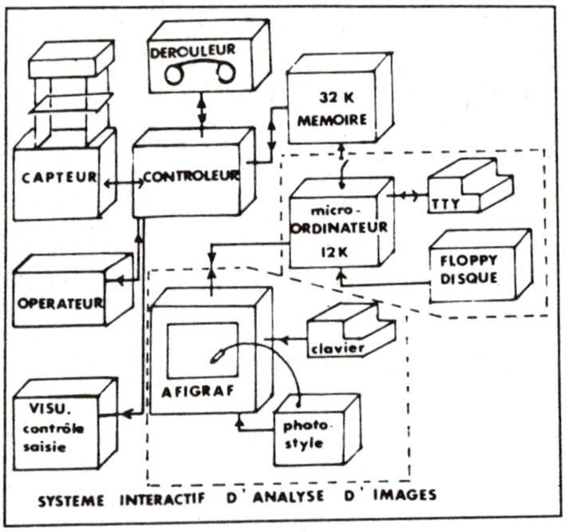

c) The display AFIGRAF system allows to observe, in real time, the graphic data issued from O.D.A. and memory extension. This step constitutes the final result of the computation or the intermediate results (grey levels histogram, grey level compression, ...). With this AFIGRAF,the operator can correct the processing development with a light pen . The operation control can be done in an analog or digital mode.

d) A magnetic tape recorder can be interfaced with large computing systems for processing larger analysis programs. It has its own programs and increases the processing capability of the general system.

e) A.S.R. 33 Teletype permits introduction of the programs in the microprocessor with a clavier or a punched paper tape. It permits a conversational exchange with the microprocessor and a control of the other output devices (fast printer, AFIGRAF screen).

Figure 5 . Conversational system from left to right : AFIGRAF display, microcomputer floppy disk, A.S.R. 33 teletype and magnetic tape recorder.

f) The floppy disk unit ADC behaves as a 240 Koctets mass memory which permits the storage of processing subprograms and input data at the end of the computation or in an intermediate phase.

I.2.4. Charge coupled device

This technology appears in 70's in BELL LABORATORIES.Less complex and less expensive than MOS it permits more intergration.A CCD application is solid state image sensing.The photosensitive elements and the scanning circuitry can be integrated on one silicon chip,leading to a small, very rugged ans low voltage pick up device.

a) Principle ; figure 6 -The device receives a light,creating charge carriers in the silicon.These carriers are integrated during a certain time and subsequently read out.

FIGURE 6

8 by 8 bits image sensor cell and transfer electrodes

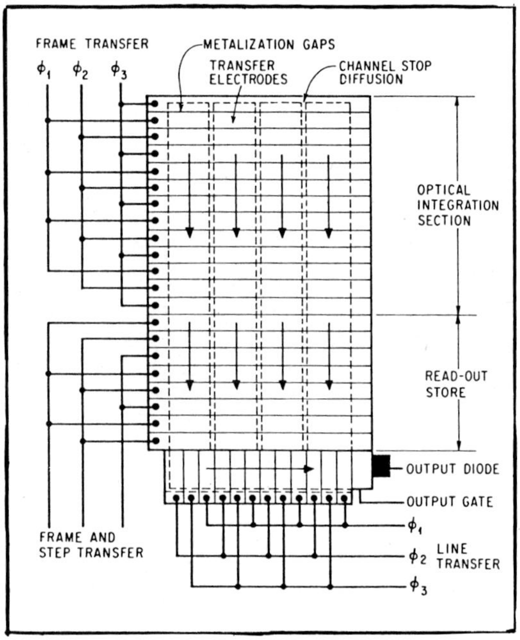

The transfert mode creates a condition that moves the charge to the second adjacent electrode.This movement is accomplished by applying a still more negative voltage to second electrode,during a still deeper potential well under that electrode.Now the holes being stored under electrode 1, attracted to the deeper well transfert to the new electrode.

b) Some numbers - Charge coupled device can operate at more than 10 megahertz data rates and will consume only 5 microwatts per bit, to compare with MOS or other silicongate (5 MHz, 100 microwatts).

The size is also an interesting advantage : with CCD a cell size is 1,5 to 3 square mils ,(other devices:6 to 12 square mils)

CCD bit packing density,at greater than a million bit per square inch.

Now it is possible to build a 1024 x 1024 matrix or more.

c) Disadvantages - In a matrix or in a bar several cells are defective. The characteristics are lower than mean or out..

The laboratory need to detect the false and the cost increase.

II. PREPROCESSING OPERATORS

Preprocessing operators are transformations which filter the undesirable features and / or emphase the informative features. Preprocessing is a selective loss of the information of the picture. One of the first problems consists of optimizing this selective loss. We can consider three kinds of preprocessing operators:

- integration operators
- differentiation operators
- recursive operators

Noise diminution is accomplished by integration operators and information emphasis by differentiation operators. Optimization is an equilibrium between these two tendencies. We must distinguish also:

- context free (or invariant) operators
- context dependant operators which can be:
  - . global context dependant
  - . local context dependant

If $M(i,j)$ is the digitalized picture and $Q[a,b;t(M(i,j))]$ anindex function defined over a rectangle such that

$$-n \leqslant a \leqslant +n \qquad \{n,m\} \in N$$
$$-m \leqslant b \leqslant +m$$

anoutput image R (i,j) is formed by discrete convolution of the input function according to the relation:

$$R(i,j) = \sum_{a=-n}^{a=+n} \sum_{b=-m}^{b=+m} Q[a,b;t(M(i,j))]\, M(a,b;i,j) \qquad (\text{II}.1)$$

more generally we have for a context dependant operator: $Q[k,l;M(i+k,j+l)]$

the relation:

$$R(i,j) = \sum_{k=-n}^{k=+n} \sum_{l=-m}^{l=+m} Q[k,l;M(i+k,j+l)]\, M(i+k,j+l) \qquad (\text{II}.2)$$

The index function can be represented by a rectangular mask $(2n+1 \times 2m+1)$ of which the values are function of the local context of the picture.

An image may be subject to

- electrical sensor noise
- photographic grain noise
- channel error

which usually appears as discret isolated pixel variations that are not spatially correlated.For these kind of noise;

⇒ context free operators (integration operators) are generally applied.

II.1 CONTEXT FREE OPERATOR

II.1.1 Integration operators

This noise has generally a higher spatial frequency spectrum than the normal image components because of its spatial decorrelatedness.Hence simple low-pass spatial filtering can be effective for noise smoothing.

The following mask has been tested:

$$
\begin{matrix} 1 & 1 & 1 \\ 1 & 2 & 1 \\ 1 & 1 & 1 \end{matrix}
\qquad
\begin{matrix} 1 & 1 & 1 & 1 & 1 \\ 1 & 1 & 1 & 1 & 1 \\ 1 & 1 & 2 & 1 & 1 \\ 1 & 1 & 1 & 1 & 1 \\ 1 & 1 & 1 & 1 & 1 \end{matrix}
\qquad
\begin{matrix} 1 & 1 & 1 & 1 & 1 \\ 1 & 2 & 2 & 2 & 1 \\ 1 & 2 & 3 & 2 & 1 \\ 1 & 2 & 2 & 2 & 1 \\ 1 & 1 & 1 & 1 & 1 \end{matrix}
\qquad
\begin{matrix} 0 & 0 & 1 & 2 & 1 & 0 & 0 \\ 0 & 1 & 2 & 3 & 2 & 1 & 0 \\ 1 & 2 & 4 & 6 & 4 & 2 & 1 \\ 2 & 3 & 6 & 9 & 6 & 3 & 2 \\ 1 & 2 & 4 & 6 & 4 & 2 & 1 \\ 0 & 1 & 2 & 3 & 2 & 1 & 0 \\ 0 & 0 & 1 & 2 & 1 & 0 & 0 \end{matrix}
$$

I   Ib   Ia   II

with the relation:

$$
R(i,j) = \frac{\sum_{k=-n}^{k=+n} \sum_{l=-m}^{l=+m} Q(k,l)\, M(i+k, j+l)}{\sum_{k=-n}^{k=+n} \sum_{l=-m}^{l=+m} Q(k,l)} \qquad (II.3)
$$

The use of great dimension masks make the digital picture more homogeneous and attenuates the fast transitions of grey levels (contours and often informative features).

Different authors have compared the advantages and the drawback of different symetricshapes (HUANG, SPEAKE, and MERCEREAU ...)

Precautions must be taken in the application of the integration operators to cineangiocardiographies:

As shown Slager and Coll, (1982) small irregularities at the left ventricular endocardial border (endocardial landmarks) can be detected in the contrast cineangiogram with an automated endocardial outlining system.

These irregularities reflects the motion pattern of the endocardial wall and as we have to detect ventricular motion, elasticity and contraction irregularities on the ventricular contour, smoothing methods can eliminate them.

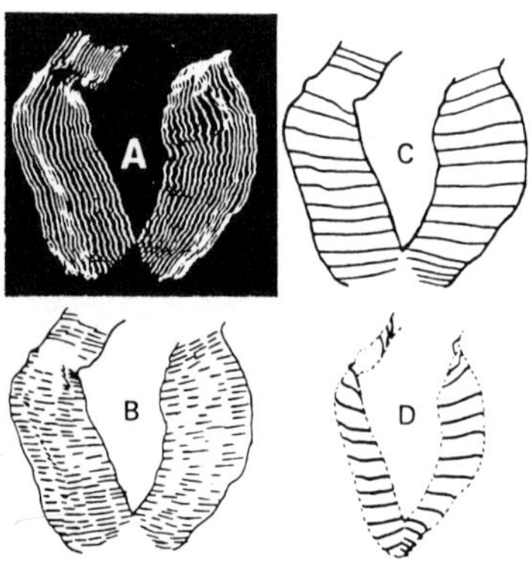

FROM HOOGHOUDT (1982)
GOOD CONTRACTILITY
OF THE MYOCARD

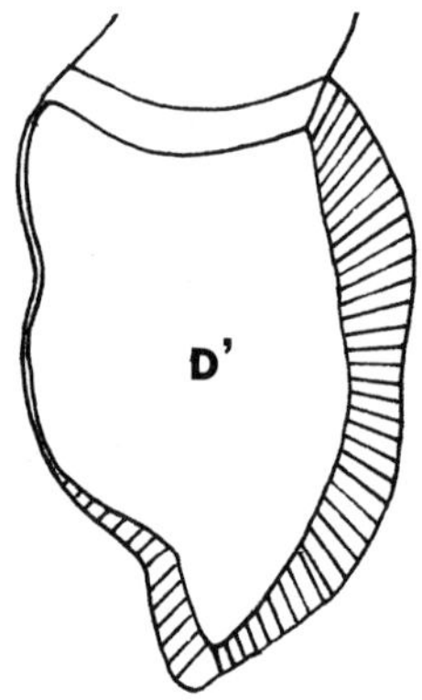

ANEVRISM (SEPT...)

POOR CONTRACTILITY
OF THE MYOCARD

II.3 Small details in the left ventricular contours corresponds with anatomical landmarks and can be used to reconstruct local endocardial motion; smoothing methods can elimate them.

In pathological cases like asyneresis, anatomical landmarks become almost parallels in akinetic areas of the myocard.

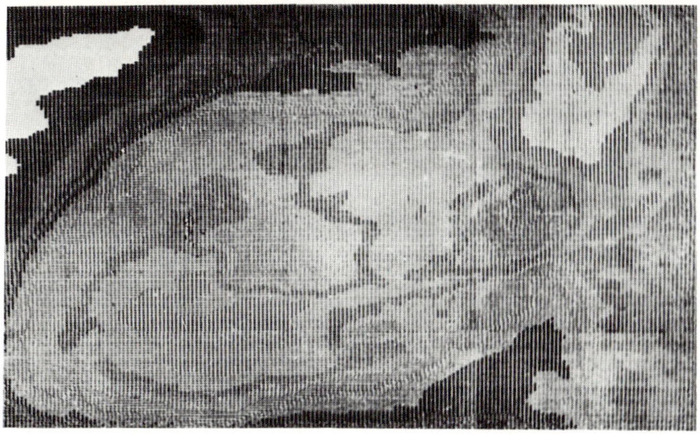

FIGURE II.3A
Digitalized cineangiocardiographie on which is apply the filter II. The contrast product coming from the catheter is apparent on the picture but not apparent to the naked eye.

II.1.2 Differentiation operators

Differentiation operators have been widely applied to biomedical image processing for edge enhancement (generally without regards to edge direction).If integration operators have been more useful for cineangiographies, differentiation operators have been more efficient for cinegastrographies.

The operators generally applied are:

- the gradient operator:

$$\nabla I(a,b) = \frac{\partial I}{\partial a}\vec{i_1} + \frac{\partial I}{\partial b}\vec{i_2} \qquad (\text{II}.4)$$

- the laplacien operator:

$$\nabla^2 I(a,b) = \frac{\partial^2 I}{\partial a^2} + \frac{\partial^2 I}{\partial b^2} \qquad (\text{II}.5)$$

Both the magnitude and the orientation have been successfully used for detection of lung tumors by NADLER (1976).

Approximation of these operators have been done and the following types have been applied to cinegastrographies.

(01)

	−1	
−1	0	+1
	+1	

(012)

		−2		
		−1		
−2	−1	0	+1	+2
		+1		
		+2		

For instance for the filter (01) we have the relatioship :

$$G(j,k) = \left\{ \left[M(j,k+1) - M(j,k-1)\right]^2 + \left[M(j+1,k) - M(j-1,k)\right]^2 \right\}^{1/2} \qquad (\text{II}.6)$$

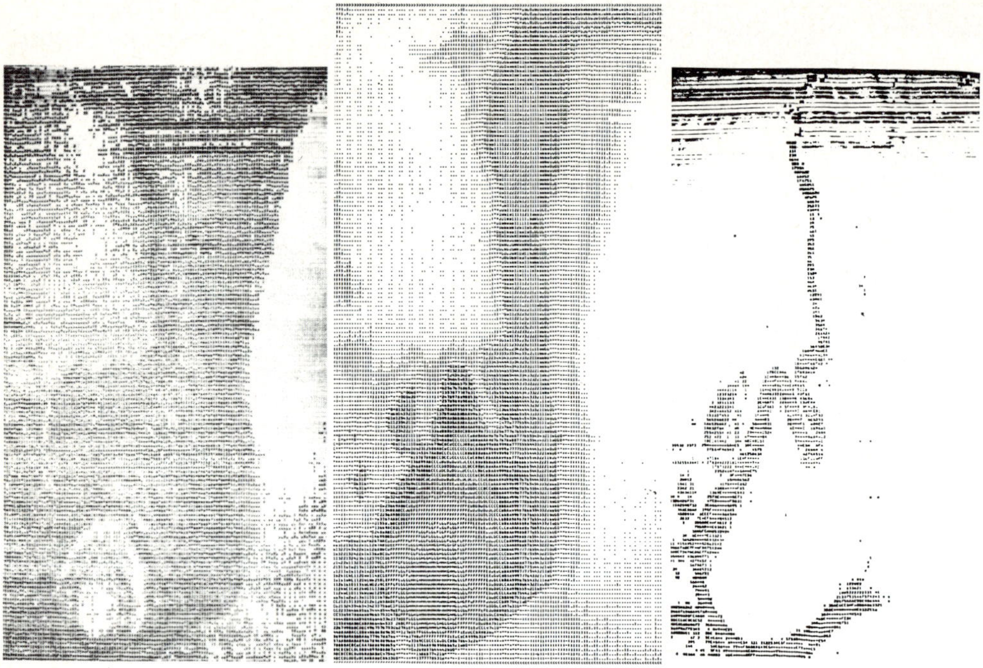

FIGURE II.3.B

ORIGINAL DIGIGITALIZED CINEGASTROGRAPHY	TRANSFORMED PICTURE WITH OPERATOR (01)	TRANSFORMED PICTURE WITH OPERATOR (012)

II.1.3 Recursive operators

Recursive fitering is based on a recursion relationship between input and output images.The one dimension recursive relation is defined as:

M (i) : input sequence $i = \{I,N\}$

R (1) : output sequence $1 = \{I,M\}$

a (j) , b(k) weighting constants

$$R(1) = \sum_{j=1}^{J} a(j)\, M(i - j + 1) - \sum_{k=2}^{K} b(k)\, R(1 - k + 1) \qquad (II.7)$$

The 1 element in the output sequence is dependant not only on the present input and the J-1 past input sample but also on the K-1 previously computed elements of the output sequence.

The 2 - D recursive filtering recursion relation is defined by:

$$R(1_1, 1_2) = \sum_{j_1=1}^{J1} \sum_{j_2=1}^{J2} A(j_1, j_2)\, M(1_1 - j_1 + 1, 1_2 - j_2 + 1)$$

$$- \sum_{k_1=1}^{K1} \sum_{k_2=2}^{K2} B(k_1, k_2)\, R(1_1 - k_1 + 1)\; 1_2 - k_2 + 1) \qquad (\text{II}.8)$$

Recursive filters 3 x 3 and 5 x 5 have been applied advantageously to cineangiocardiographies, (filters : I, Ia, Ib, II) .

II.2 CONTEX OPERATEUR : CONTOUR FOLLOWER AUTOMATA

The automatic contour detection is an ambitious phase of the automatisation. This phase, indeed simulates the human visual perception, a domain which requires a great level of integration in the human brain.(NADLER 1976)

The contour of the left ventricular chamber is generally not always well defined.However from a mathematical approach based upon the systems theory, the contour of the chamber has been detected with a contour following automata.

This automata functions according to the principle of structural recognition,that is to say, by using a reference grammar.

II.2.1 PRINCIPLE

The pertinent information of a picture,as for instance the left ventricle, appears as a structure (a form) which relys on non-specific elements.

This assemblage must be described as an adapted formalism.

The theory of languages brings tools from which the object is to describe the forms (sentences) the recognition of which can be done by verifying the belonging of an unknow sentence to the language engend d by one of the reference grammars.

II.2.2 DESCRIPTION OF THE CONTOUR FOLLOWER AUTOMATA

An automata is generally described by a quintuple:

$$(I , J , S , \lambda , \delta)$$

where: I is a set of inputs $(i_0;i_1 \dots i_T)$

J is a set of outputs $(j_0 ;j_1 \dots j_T)$

S is a set of states $(S_0 ;S_1 \dots S_T)$

λ is the next state function : $I \times S \Rightarrow S$

δ is the output function : $I \times S \Rightarrow J$

Where

S_0 : initial state : $S_1 = \lambda (i_0 , S_0)$

S_T : terminal state: $S_T = \lambda (i_{T-1} , S_{T-1})$

J_T : terminal output: $J_T = \delta (i_T , S_T)$

The state of the automata is characterized by the pixel position (x,y).

The next state function depends on the type of sweeping, done on the picture.The inputs are characterized by the value of the grey level of each pixel.

The automata is initialized from S_0 (pixel at the border of the picture)

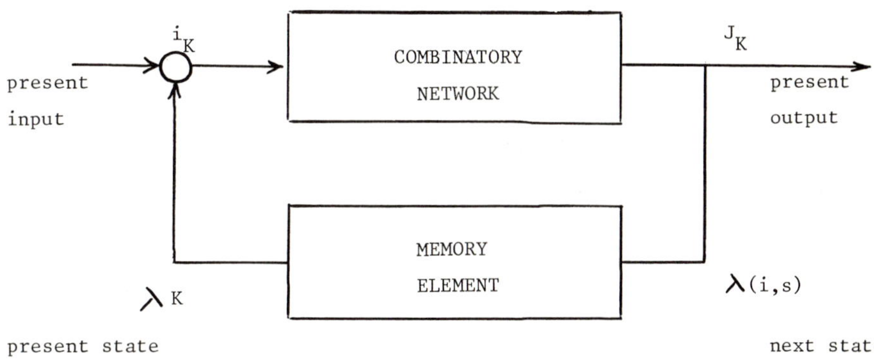

(The terminal state S_T can be also a pixel at another border of the picture)

II.2.3 NEXT STATE FUNCTION

The next state function is related to the sweeping of the automata. Three modes must be distinguish.

a)Canonic sweeping

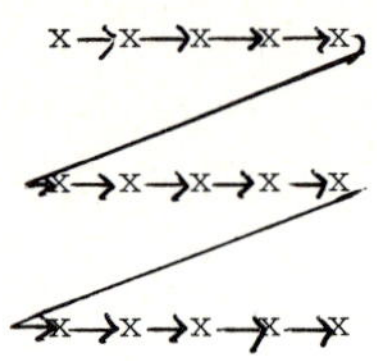

(state independant from the input or context free operator)

b)Peano sweeping (QUINQUETON, 1981)

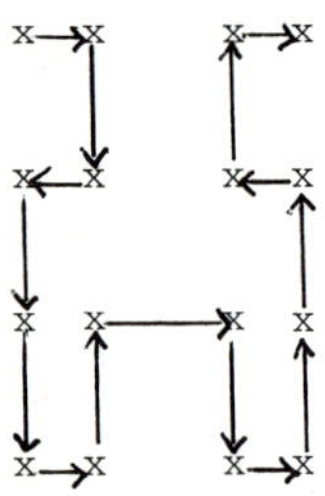

(state also independant from the input) this sweeping is often adopted to generate "image vectors" for pattern recognition purposes.

c) Band sweeping

The automata scans the picture when it supposes the contours presence. The next state function depend of the input (SLAGER 1978)

Principle: it use a scan window which is an interval between two grey level following the contour.

The center of the window is on the n + 1 line and it has the same abscissa that the previous contour pixel.

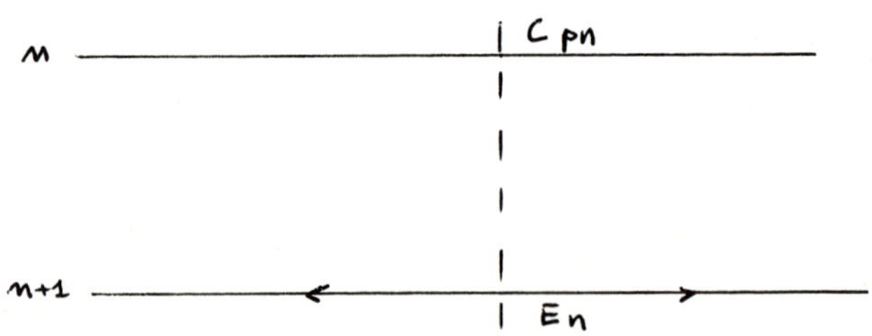

En : scan window

CPn: previous contour pixel

This center is a good approximation of CP n + 1.If this point isn't on the contour the automata scans the other points of the scan window.

This sweeping is an interesting process for unsharp image or discontinuity.

d) Contour following sweeping

With this sweeping,point are on the contour when they have a particular property for exemple: grey level

This process permit to follow exactly the contour and the next point is selected only if it has all properties (figure II.4)

II.2.4 Next state function choice

For the left ventricle,the contour following automata is appropriated because the band sweeping has the disadvantage to give several points on a same line with supplementary choice.

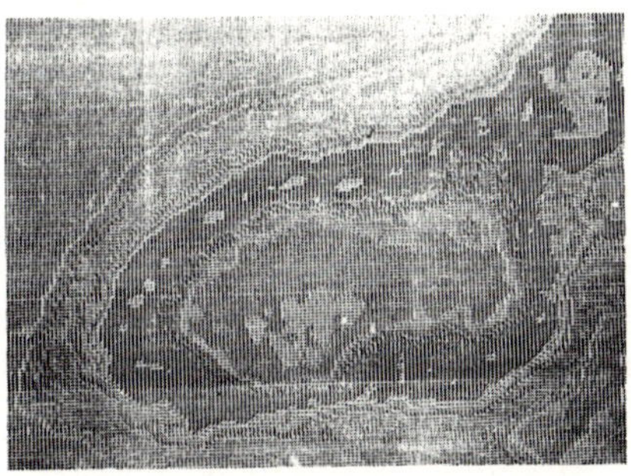

IMAGE (FILTER II)

CONTOUR

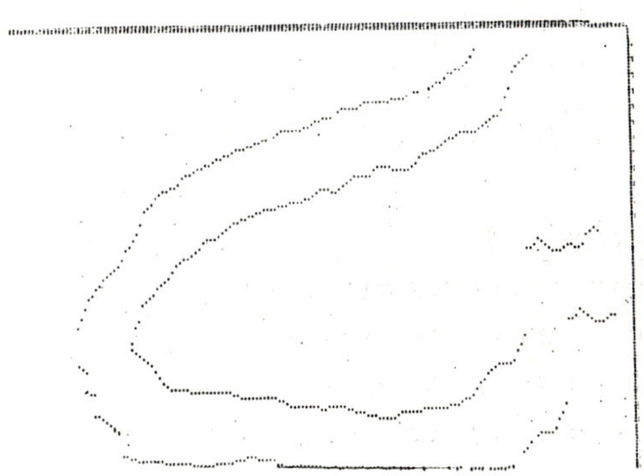

FIGURE II.4

AUTOMATIC DETERMINATION OF THE CONTOURS OF THE LEFT VENTRICULAR CHAMBER AND OF THE MYOCARD.

II.2.3 HISTOGRAMM AND COMPUTATION

The figure (II.5) shows the histogramm of the grey level's distribution.The most points have the same grey level.

They form the interior and exterior of the image.

The other points are less numerous.They form the contours.

The histogramm permits to choose the level for the inner and exterior coats and the first point of the contour.

After the automata compute different functions to determine the next point of the contour:8-1 directions are possible.

The figure II.6 shows an exemple of image.9 points are represented.

The automata makes the means of three points and search for the maximum and the minimum to choose the new point .

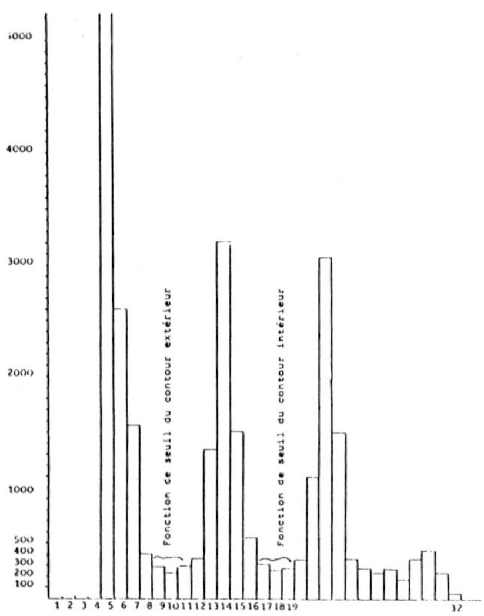

FIGURE II.5

HISTOGRAM OF THE GREY LEVEL DISTRIBUTION.MINIMA PERMIT TO SELECT THE REFERENCES LEVELS FOR THE AUTOMATA

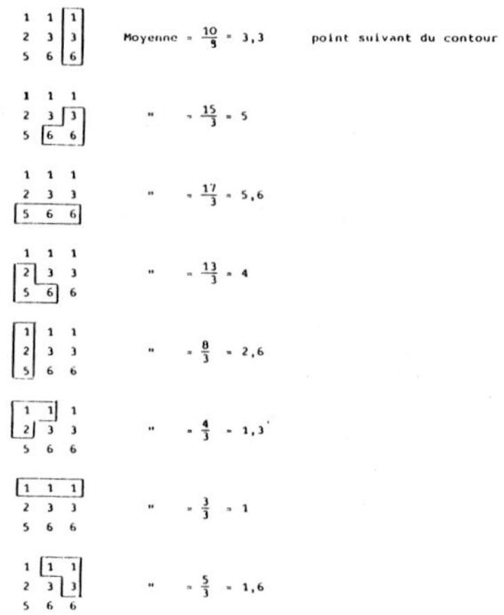

FIGURE II.6

THIS FIGURE DESCRIBE THE AUTOMATA PROCEDURE FOR A LEVEL INTERVAL (3,4) AT THE NEXT STEP THE AUTOMATA WILL MOVE TOWARD THE RIGHT.

III. PARAMETER EXTRACTION

Characteristic parameters extraction is an important phase of the pattern recognition process.

It is also an important phase of the image superposition in the image sequence progressing.Characteristic points and axes will serve as common and intrinsic reference frame for the different objects of interest in the picture.

As much as possible these characteristic points and axes are related to the pattern of the object of interest,here the left ventricle or the stomach.

III.1 CHARACTERISTIC POINTS EXTRACTION

III.1.1 Left ventricle contour

For the determination of the left ventricular volume and for the superposition process of the contour of the left ventricular chamber,it is necessary to close the chamber at the aortic valve level.

This operation is presently performed in an automatic way.

For the automatic closure of the left ventricle at the level of the aortic valves, the algorithm used, follows the contour of the left ventricle in the trigonometric sense and computes at each pixel of this contour the "deviation angle".This angle is formed by the contour pixelswhich preceed and follow it.

The contour turns are thus quantified and form a curve with respect to the followed perimeter.The curve portions below or above a certain level permit us to determine the position of the apex and then the position of points of closure at the level of the aortic valves. (fig III.1 - III.2) (RACKLEY 1964)

Gravity center

Another characteristic point necessary to compute for each contour is the gravity center of the left ventricle outline.

It represents approximately the center of contraction (or dilatation) toward which the different parts of the myocard approximately converges or diverges.

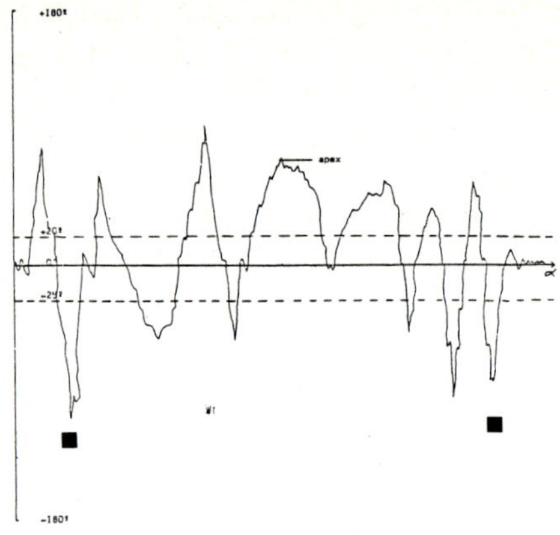

Figure III.1

DEVIATION ANGLE FUNCTION FOR THE DETERMINATION OF THE APEX AND POINTS OF CLOSURE OF THE LEFT VENTRICLE OF THE LEVEL OF THE AORTIC VALVES.

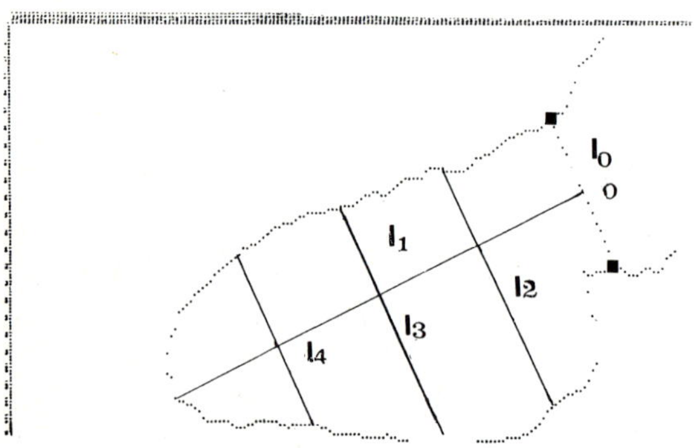

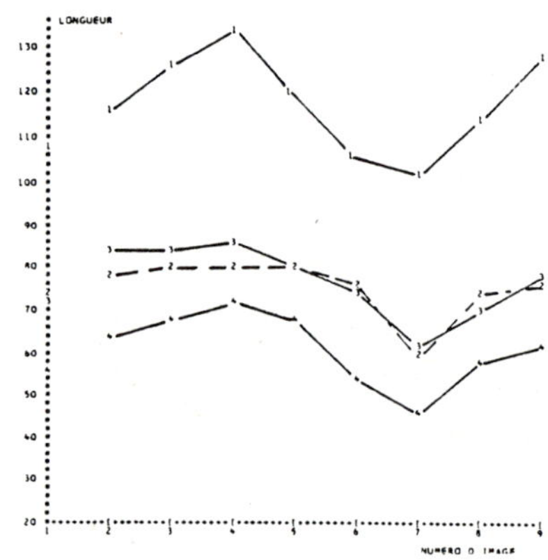

III.2

AUTOMATIC CLOSURE OF THE LEFT VENTRICLE AT THE AORTIC VALVE LEVEL AND COMPUTATION OF THE AXES OF THE LEFT VENTRICULAR CHAMBER.

III.3

VARIATION OF THE AXES LENGHT, L1 L2, L3, L4 DURING THE CARDIAC CYCLE.

III.1.2 Gastric contour

Two orifices define the extremities of the stomach : the cardia and the pylore, but these regions are difficult to define precisely from the X-rays pictures; gastric folds affect the cardial region and the pylore is generally not well contrasted. Also, it has been necessary to take, as a reference point for the superposition of the pictures, a characteristic point on the superior part of the small curvature (BOIRON, 1981) which can be easily detected anatomically.

III.2 CHARACTERISTIC AXES EXTRACTION

Two ways can be followed in the determination of the pattern characteristic axes X,Y of the organ (left ventricle or stomach):

- at each time the axes must be computed and the superposition will be done from each reference frame X (t),Y (t).
- for all the serie of cineangiogram or cinegastrogram a reference intrinsic frame is computed: $\bar{\underline{X}}$, $\bar{\underline{Y}}$.

Generally the first method is used.

III.2.1 Characteristic axes of the left ventricle chamber

Two types of axes have been determined:rectilinear and curvilinear axes, and programs have been developed to calculate the differents axes: (RACKLEY 1964)

- longitudinal axis (Lo) which joint the middle of the aortic valves line to the gravity center (or to the apex).A transvesal axis L3 can be determined perpendicularly to Lo at the level of the gravity center.The variation of Lo during the systole cycle give interesting informations on the circonferential fibers stress.

- inertial axes of the left ventricle chamber by the computation of the eigenvectors of the second order moment matrix.This method take better in account the irregularities of the contour and give a more precise value of the left ventricle volume.But the method is less reliable when the apex is not in the prolongment of the longitudinal axis L1. (FIGURE III.2)

- to avoid this inconvenience we determine curvilinear axis computed by slices perpendicular to L1.The curvilinear axis is formed by the gravity center of each slice.

The transverse axes L2 and L4 parallels to L3 can be considered as parallel to Lo and such that the distance from O to the intersection point with L1 is equal to L1/4.

The intersections of L2 and L4 with L1 are at $\frac{1}{4}$ L1 and $\frac{3}{4}$ x L1.

The computation of these axes are performed on each of the cineangiocardiographic pictures during a diastole and a systole (FIGURE III.3) (GAUDEAU 1972)

III.2.2 Characteristic axes of the stomach

III.2.2.1 Skeleton :

Let F a plane closed and connexe figure limited by a contour $C : C \subset F$

Let $\bar{F}$ the complement of F.

We can define the exterior of F : $\overset{H}{F}$

$\overset{H}{F} = \bar{F} \cup C$ (LAFORGE)

and the partition of F

$\overset{H}{F} = \overset{H}{F}_1 \cup \overset{H}{F}_2$

Let M a point of F, $\exists$ P $\in C$ such that

$d = \inf \{ d(M,P) \} \;;\; P \in \overset{H}{F}$

Let centered in M a disk Dr with a varible radius r; d is also the greatest radius such that Dr $\subset$ F:

$d = \text{Sup} (r \mid Dr \subset F)$

If Dd is greatest disk centered in M and include in F, $\exists$ Mi such that Dd comprised two points P_1, P_2 in commun with $\overset{H}{F}$ $(P_1, P_2) \in C$

$\Longrightarrow$ Mi $\in$ (SKELETON)

III.2.2.2 Medial axis

If we constraint to P_1, and P_2 from the disk to be such that:

$< P_1 \in \overset{H}{F}_1 \wedge P_2 \in \overset{H}{F}_2 >$

Than $(P_1, P_2) \in$ (medial axis)

(or geometrical locus of points Mi)

The median axis is a function of the partition of $\overset{H}{F}$ in $\overset{H}{F}_1$ and $\overset{H}{F}_2$

For gastrix contours, this partition is defined from the middle of the fundus and the pyloric point.

The digitalized picture defines a finite field and let (CS) the curve which divides $\overset{H}{F}$ in $\overset{H}{F}_1$ and $\overset{H}{F}_2$. A and B are the points where (CS) met (C).

The point of radius MB belong to the median axis and not necessary to the Skeleton.

The point S for which $[P'_1 , P'_2] \in \overset{H}{F}$, belong to the skeleton but not to the median axis.

III.2.3 Software method

A variety of methods has been used.

On figure III ,we have partitioned the exterior of the gastric form:

$$\overleftrightarrow{F} = \overleftrightarrow{F}_1 \cup \overleftrightarrow{F}_2 \; ; \; C_1 = \overleftrightarrow{F}_1 \cup C \; ; \; C_2 = \overleftrightarrow{F}_2 \cup C.$$

Median axis is drawn through gastric form, form the cardial point (A) to the pyloric point (B), step by step ,while the disk (Dr) is moving down .

The step moving is conducted by a mean-axis.To the set of points (P_1,P_2) is given an index (i) that represents the direction from the cardial point (A) to the pyloric point (B):

$$P_{1\,(i)} \in C_1 \, , \; P_{2\,(i)} \in C_2 \; ;$$

$$i = \{1,N\} ; \; (1 = \text{cardial-point} \, , N = \text{pyloric-point}).$$

The two points are put in correspondence (Rc) if DM is the diameter of the disk (Dr) :

$$P_{1\,(i)} \mathcal{R} \; P_{2\,(i)} \; ; \; d\left\{P_{1\,(i)} \, , \, P_{2\,(i)}\right\} = DM.$$

The median-axis is defined by:

$$M_{(i)} = m \left\{ P_{1\,(i)} \, , \, P_{2\,(i)} \right\} .$$

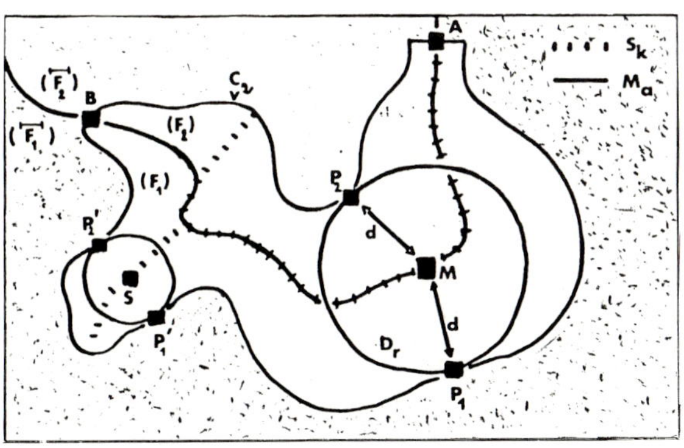

FROM BOIRON,COJAN,1980

FIGURE III.4

DETERMINATION OF PATTERN SKELETON AND MEDIAN AXIS

IV. SUPERPOSITION AND SEGMENTATION

The purpose of the superposition of organ contour procedure is to evaluate overall and local wall motion of the organ after a frame by frame analysis of the angiogram or X rays picture.

IV.1 LEFT VENTRICULAR CONTOUR SUPERPOSITION

The principal purposes of this superposition are to:

- visualize during the cardiac cycle the different type of local motion disturbance or asynergy :(HOOGHOUDT)
  - . akinesis or total lack of motion of a part of the left ventricular wall
  - . asyneresis or diminished motion of a part of the wall
  - . diskinesis or systolic paradoxical motion
  - . asynchrony or a disturbed temporal pattern of contraction.

- quantify these different motion disturbances by the display of left ventricular segmental wall displacement (A.SLAGER, HOOGHOUDT, 1982) and the computation of the partial ejection fraction defined by:

$$F_i = \frac{V_{ei} - V_{ni}}{V_{ei}} \qquad (\text{IV.1})$$

where F_i = partial ejection fraction for the region i,of the left ventricle volume

V_{ei} = tele-diastolic volume of the region i

V_{ni} = tele-systolic volume of the region i.

One of the greatest difficulty for the determination of F_i is the determination of a common reference frame for each frame.During the ejection, the ventricular chamber rotate along its longitudinal axis and translate.Three methods for the determination for a reference frame are used and a new point coordinate,X,Y',Z' of the contour can be obtained from the original coordinates (X,Y,Z)by the relation:

$$\begin{bmatrix} X' \\ Y' \\ Z' \end{bmatrix} = \begin{bmatrix} A \end{bmatrix} \begin{bmatrix} X \\ Y \\ Z \end{bmatrix} + \begin{bmatrix} \Delta X \\ \Delta Y \\ \Delta Z \end{bmatrix} \qquad (IV.2)$$

A = Rotation matrix

IV.1.1 Superposition without translation,with rotation of the longitudinal axis

For clinical application one must avoid to superpose,during the cardiac cycle,the gravity center of each different frame (for instance,on the gravity center of the end-diastolic contour).

This correction during the contour superposition procedure will decrease local akinesis that we try to put in obviousness.This method "normalize" the ventricular motion.But it can be useful to align the apices of the different frames to correct the anti-clockwise rotation occuring during the ventricular ejection.

IV.1.2 Superposition without correction : $A = I, \Delta X = \Delta Y = \Delta Z = 0$

With this method, the contours are superposed without correction such that the cardiolog see them on the screen.This method is used by RASTIER (1977), ISEMBERT (1977), SLAGER (1978), HOOUGHOODT (1982).

If from the biomechanic point of view this method is not satisfying it permit to put in obviousness the motion disturbance.

IV.1.3 Superposition with translation $A = I$

In this translation, each middle point of the aortic valve segment are juxtaposed.The aortic orifice constitute a"fixed point " from which the left ventricle which is a deformable structure is attached.

This representation is performed during the diastole and systole,it reveal concentric motion of the ventricular wall.But in pathological cases,the rotation of the apex provoke local expansions or retractations.But this method is sensible to automatic determination errors of the aortic valve segment.GAUDEAU 1978.

VI.2 SEGMENTATION OF THE LEFT VENTRICULAR CHAMBER

Three methods are developped.

VI.2.1 Cylindric method (HOOGHOODT, 1982)or longitudinal Axis method.

The end diastolic lumen is divided from the level of the mitral valve edge to the apex perpenticularly to the longitudinal axis into 10 slices of equal height ,each slice being divided into two halves by the left ventricular axis of symmetry.The volumes V_{ei} and V_{ni} are computed by the formula.

$$V_{ei} = MR^2_{ei} \cdot \frac{L_{ei}}{20} \; ; \; V_{ni} = MR^2_{ni} \cdot \frac{L_{ni}}{20} \qquad (IV.3)$$

Where R_i = radius of a particular slice

L_i = left ventricular long axis lenght

A correction of this method has been proposed by GAUDEAU 1978 because the geometric pattern of the volume can be far from the oneof the left ventricular chamber itself.Thus it is necessary.

VI.2.2 Gravity center method or sphere sector method : FIGURE IV.2

Three lines are determined respectively from the center of gravity to the apex and to the right and left base of the aortic cups.

The silhouette of the ventricle is separated into two zone by the CG-apex line;9 radii , converting to the CG with equal angulation, aincle each of the two zone into 10 sectors.For each frame and for each sector the segmental wall motion normalized for the end-diastolic radius is defined by averaging the distances between the different points on the LV contour of the studied sector and the gravity center , this quantifies in time the contraction of the 20 sectors.

For the sector volume evaluation in a first approximation we consider sphere sector (GAUDEAU 1978).

IV.2.3 SLAGER method

This method based on studies of the left ventricular endocardial wall motion derived from anatomical landmark trajectories (SLAGER 1979)

The plane figure of the left ventricle is divided by 20 straight lines computed following the description of the landmark trajectories.

This segmentation represent a middle between the two previous methods.

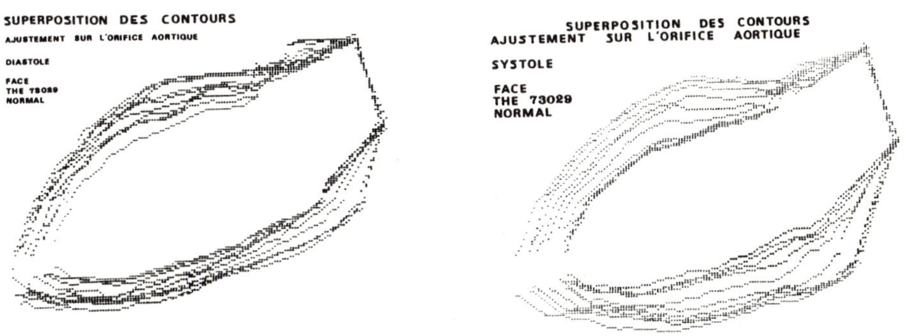

IV.1.SUPERPOSITION WITH TRANSLATION TO THE MIDDLE POINT OF THE AORTIC VALVES.

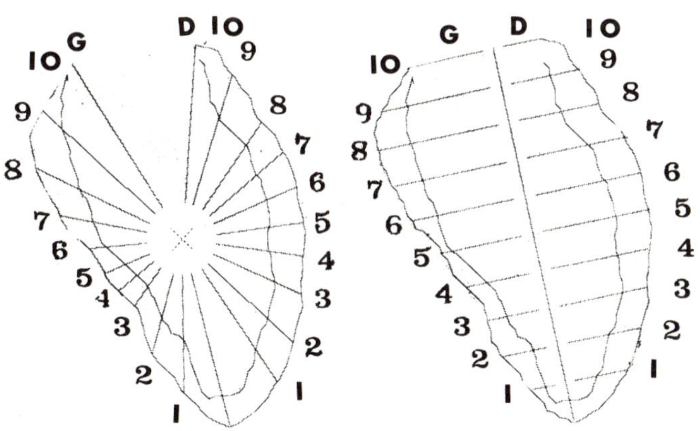

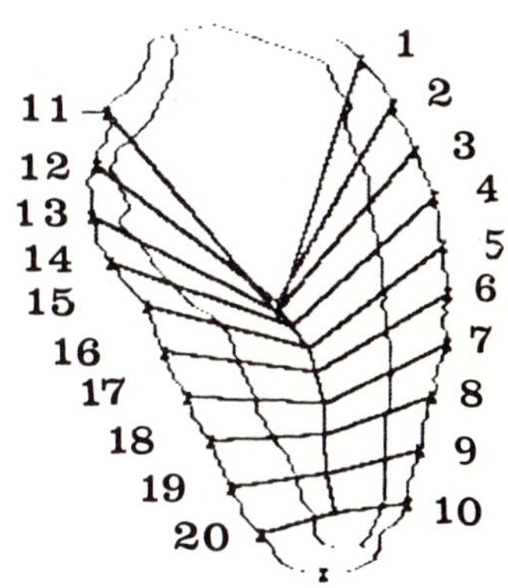

IV.2.SEGMENTATION USING THE GRAVITY CENTER METHOD AND THE CYLINDRICAL METHOD OR LONGITUDINAL AXIS METHOD (FROM AZANCOT AND COLL.)

IV.3.SEGMENTATION USING SLAGER METHOD.

IV.3 GASTRIC SUPERPOSITION AND SEGMENTATION (BOIRON 1981, COJAN 1980)

The principal purpose of the gastric image superposition are:

- to study the motricity of the longitudinal fibers of the stomach through the variation of the length of the median axis.
- to study the speed of the contraction wave from the corpus to the pylore.

For the second purpose from the median axis the half anamorphic form is computed (figure IV.4).

The cardial-point (A) being the middle of the cardial section and the pyloric point (B) being the middle of the pyloric section.

Every section-length is indexed (such as the median-axis-point) and represents thus a local element of a peristaltism data structure.

Let EP (i,j) be the section length supported by the jth median-axis-point of the ith gastric contour ((i) represents temporate scale.

The half anamorphism gastric form is easily obtained from this data-structure. The modulation of the peristaltism of one gastric form is thus, well shown.

The superposition of the straigtern median **axis need reference** point which has been defined by BOIRON (1977)(figure IV.4)

DOCUMENT BOIRON COJAN, FIGURE IV.5 SUPERPOSITION ALL 2 SECONDS OF THE STOMACH CONTOURS WITH REGARD TO MEDIAN AXIS. ↓

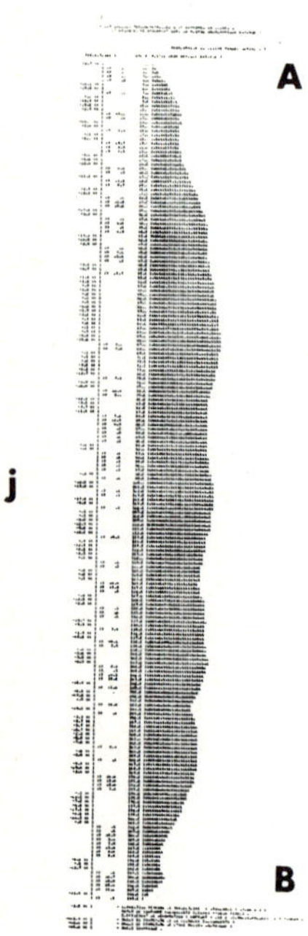

EP(i,j)

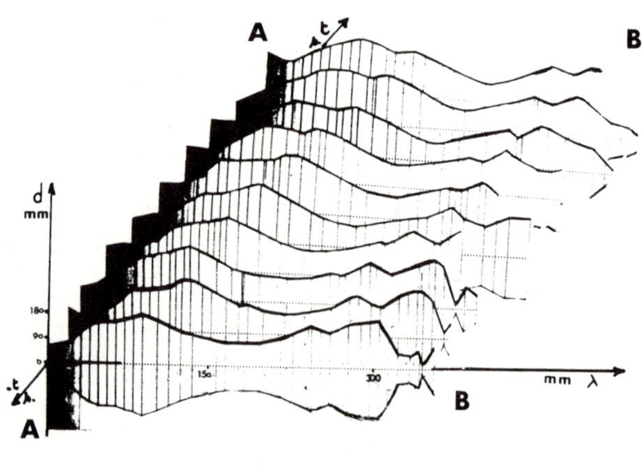

← FIGURE IV.4 . ONE HALF ANAPORPHISM GASTRIC FORM CORRESPONDING TO THE MEDIAN AXIS OF THE GASTRIC FORM.

V. STRUCTURAL ANALYSIS AND EVALUATION OF MYOCARDIC STRESS

The myocardial stress is a biomechanic parameter which is not used frequently in clinical practice.

But,these evaluations of the myocardial performance promise certainly a great development, having been used for stress computation in other domains of research (aeronautics,civil engineering ...)

V.1 EVALUATION OF MYOCARDIAL STRESS

In order to study the complex entanglement of the myocardic fibers,the Rechform program applies the finite elements method to compute the stress distribution on the myocardic wall.

This computation needs simultaneously the record of the cineangiographic pictures and the records of the pressures often not taken simultaneously; in this case it is necessary to make some adjustements.

For intraventricular pressures, it is necessary to have a precise method for defining the forces acting in several directions and the relative position of the elementary surfaces on which these forces act.

On the other hand, the hypothesis based upon the homogeneity and the isotropy of the overall myocardic wall is inconsistent with the observation of the complex structure of the myocard.

V.2 MODEL OF THE COMPOSITE MYOCARD (fig V.1 - V.2)

The myocard can better be considered as a " composite material" with an orthotropic structure.Several anatomical work on the myocard have shown that the orientation of the myocardial fibers change with a continually way from epicard to endocard.

The fibers are disposed in a helicoidal structure from which Voukydis (1972) has proposed a mathematical description.

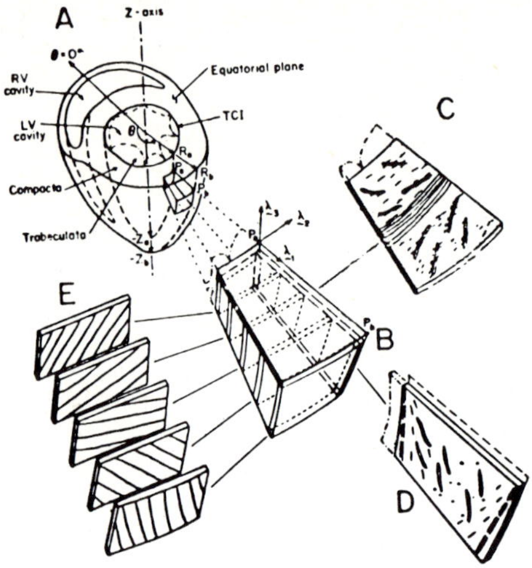

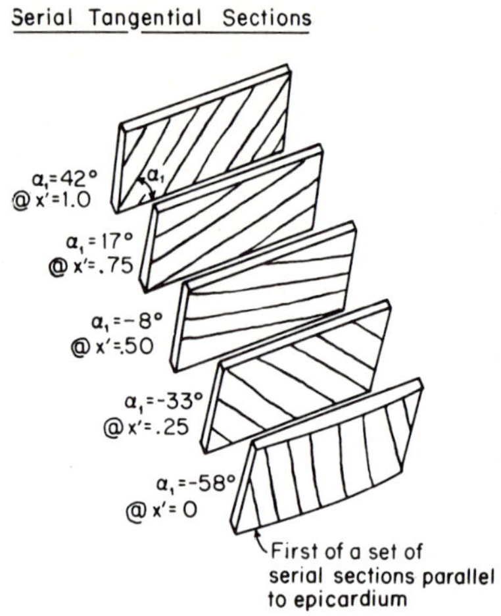

V.1

MODEL OF THE MYOCARD CONSIDERED AS A "COMPOSITE MATERIAL"WITH AN ORTHOTROPIC SRTUCTURE.

V.2

ORIENTATION OF THE MYOCARDIAL FIBERS CHANGE CONTINUALLY FROM EPICARD TO ENDOCARD.

If we consider in the myocard a thin plate to be parallele to the epicardic wall, we can suppose that in this volume all the muscular fibers go in the same direction.

We can consider this element as an orthotropic plate and apply to it, and to all the other, the Hook Law.

However,each of these plates do not slip freely with respect to each other.For this reason we must compute the myocard fiber stress in considering the myocard as an inhomogeneous material but locally homogeneous.

Thus,Voukydis has computed the myocardial fiber stress in 55 rings of 30 plates ,the fiber stress being constant for each plate ring and the myocard struture being considered one of revolution.

V.3 FINITE ELEMENT METHOD

The finite element method is a method which replaces the continum by an assembly of different areas or elements related each to the other by a finite number of elements. (NADLER 1976)

Thus the knowledge of the strain and the stress- strain relationship for each of these elements permits the deduction of the properties of the assembled structure and the evaluation of the stress.

The figure (V3) gives a representation of the cutting out of triangular elements from the left ventricle wall. (GAUDEAU 1975)

The figure (V4) shows theresults obtained and the different values of the stress in each of the modes previously defined.It shows that in the endocardic and equatorial area the stress for a defined pressure is greater than in the apex area for this case.

Discussion:

If the computation of the local myocardial stress is a progress with regard to the precise calculation of the global stress in which the left ventricular volume is considered as an extensible ellipsoid of inertia (representation far from the real pattern in pathological case) on the other hand we must take care of the following problems:

- the computation of the stress need an estimation of the elasticity modulus which is not constant during the diastole and systole.
- it need a good estimation of the left ventricle contours.The superposition of the contours can permit to eliminate now physiological contours.
- the orientation of the vector pressure must change during the systole and the diastole.

They can be considered as perpendicular to the endicardial wall.

When there is no blood going in or out the left ventricle.

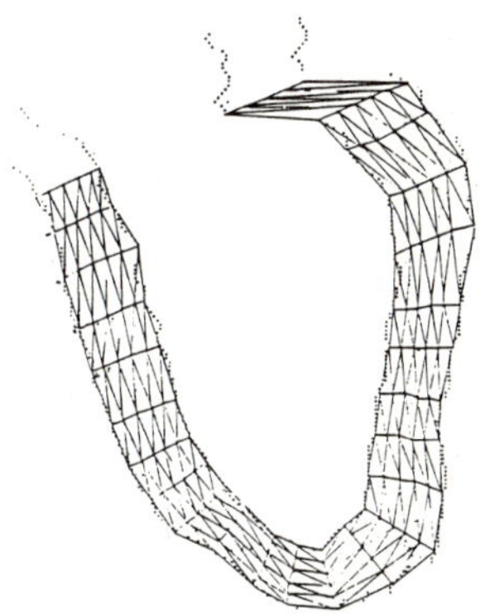

FIGURE V.3

MYOCARD SEGMENTATION IN TRIANGULAR FINITE ELEMENTS FOR THE DETERMINATION OF THE LEFT VENTRICULAR STRESS.

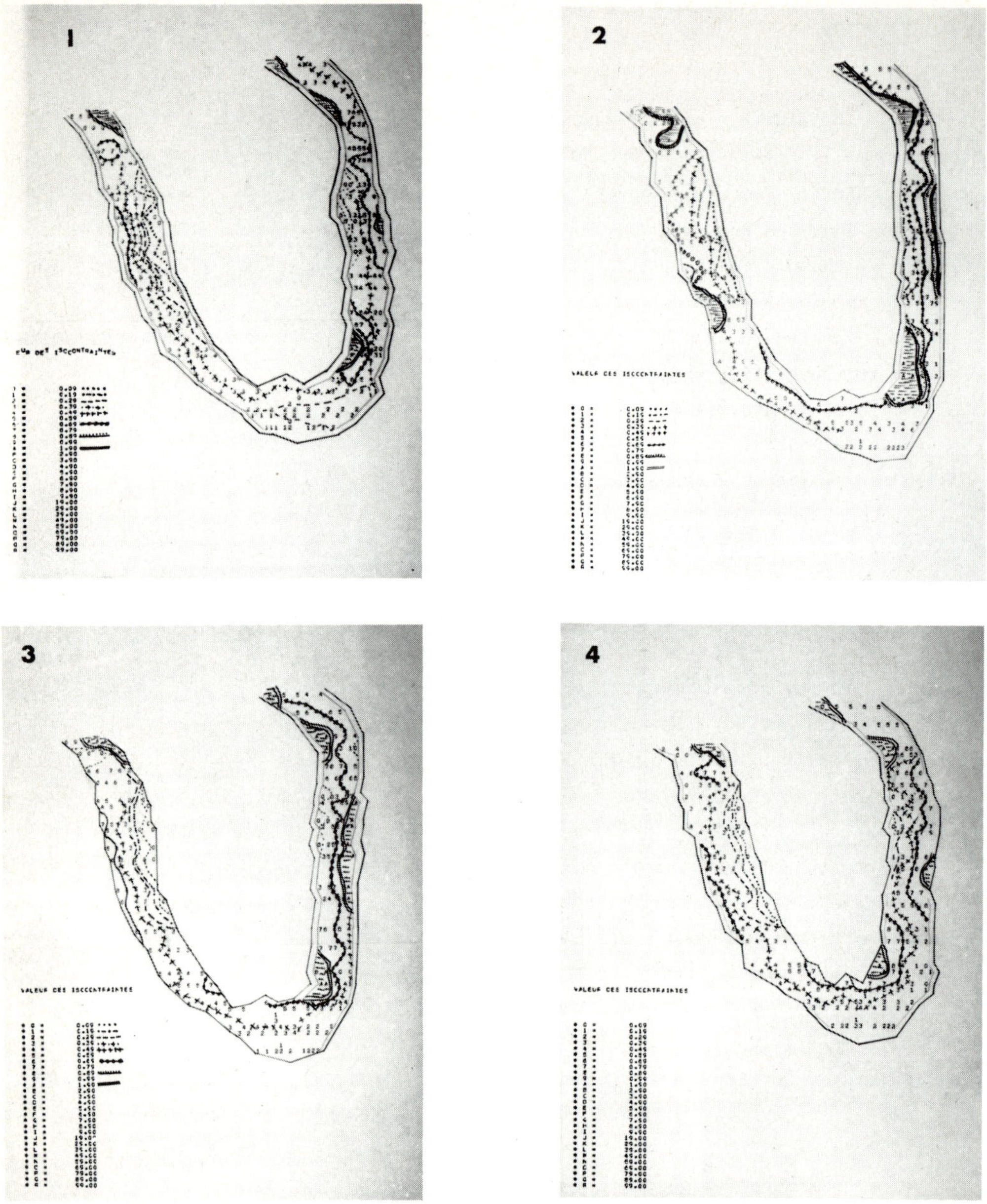

V.4.VARIATION OF ISO-STRESS CURVES DURING THE CARDIAC CYCLE:

WITH THIS MODEL CIRCONFERENTIAL ENDOCARDIC PARTS SUPPORT HIGHT STRESS BECAUSE WE HAVE NOT TAKEN IN ACCOUNT THE ORIENTATION OF THE PRESSURE VECTORS WHICH ARE NOT PERPENDICULAR IN THIS AREA TO THE ENDOCARDIAC WALL DUE TO THE EJECTION.

CONCLUSION

In this study we have successively considered:

- the informational means utilized in the analysis biomedical image sequence.
- the problems set by the contours.
- the determination of the different pattern characteristics parameters.

Its clinical interest is basic since it is necessary to compare it with other data such as the evolution of intracavitary pressures. (BOIRON 1978)

These methods permit to consider a functional exploration of the internal dynamic organs such as the heart and stomach more detailed and more selective than previously performed.

BIBLIOGRAPHY

I. - BOIRON M, GAUDEAU C, LONGEVIALLE C, BOUABDALLAH M, KURC C et THOUVENOT J :
Analyse automatique des radiographies du système digestif par capteur optique.
Path. Bio., 25 n°4 ,215-224 1977

- BOIRON M,GAUDEAU C, THOUVENOT J:
Etude dynamique et paramétrisation des déformations gastriques.Biomecanics C.R Séminaire Franco-Bulgare, VARNA,volume 7,45-50 Edit. Acad Sci.Bulgare Sofia 1978.

- BOIRON M,COJAN P,GAUDEAU C, ROULEAU PH, THOUVENOT J:
Analyse automatique d'ampliphotographies.Etude chez l'homme de la contractilité gastrique et de sa propagation.
BIOMED 80.Second Mediterranean Conference on Medical and Biological Engineering. 15-19 Sept.1980 MARSEILLE.

- BOIRON M,GAUDEAU C, ROULEAU P,THOUVENOT J:
Etude cinéradiologique quantitative de la motricité antrale.
Reprod. Nutr. Dévelop., 1980, 20 (4B) 1181-1186.

- BOIRON M,ROULEAU P,GAUDEAU C,THOUVENOT J:
Sur la motricité longitudinale de l'estomac chez l'homme.
Reprod. Nutr. Dévelop., 1981 ,21,(5B),775.

II - COJAN P,BOIRON M,GAUDEAU C,ROULEAU P et THOUVENOT J:
Etude automatique de la propagation des ondes de contraction gastrique à partir d'ampliphotographies gastriques.
C.R Séminaire sur l'analyse et traitements des images biomedicales, Bois-des Hâtes, 30-31 Mai 1980.

- COJAN P,GAUDEAU C, BOIRON M, ROULEAU P et THOUVENOT J:
Détermination automatique de contours radiologiques:application à la caractérisation de contours d'organes cavitaires opacifiés.
idem.

III - CRETTEZ J.P, GAUDEAU C, BOIRON M, BOUABDALLAH M, LONGEVIALLE C et THOUVENOT J:
Traitement automatique des images appliquées à l'étude de la motricité digestive.
Journées d'Inform.Méd., TOULOUSE, 13-17 Mars,1,259-278.

IV -DZEIDZINA D, GAUDEAU C, LONGEVIALLE C, MORAND P et THOUVENOT J: Contribution des moyens informatiques à l'étude des propriétés elastiques et contractiles du ventricule gauche. C.R Séminaire Franco-BULGARE, VARNA , 7-12 Sept;1978,51-56

V -GAASCH W.H, BATTLE W.E, OBELER A.A,BANAS J.S, LEVINEH J: Left ventricular stress and compliance in man,Circulation, Vol. XLV,April 1972,746-762.

VI -GAUDEAU C, REYNAL M: Utilisation des capteurs optiques pour le traitement des plans et graphiques,Europautomation,n°22,JANVIER 1972,25-27.

-GAUDEAU C, GIBRAT J.C : La saisie automatique des informations biométéorologiques par capteur optique,La Météorologie,Vol.28, 1973,9-18.

-GAUDEAU C, GAZEAU J.C, HUGENHOLTZ P,PIGNON D: Surveillance du système cardiovasculaire au moyen du vectocardiogramme. Simulation par méthode de régression séquentielle optimisée,C.R du Symp.d'Infor. TOULOUSE MARS 1973.

-GAUDEAU C: Reconnaissance des états d'un système physiologique et de ses processus aléatoires.Identification par des modèles stochastiques.Thèse de Biol. Hum. TOURS 1975.

-GAUDEAU C, LONGEVIALLE C, MALET C, BOUABDALLAH M, KURC C, MORAND P, ET THOUVENOT J: Automatic analysis of cineangiocardiography,Computers in Cardiology, 1975 Edit.I.E.E.E. 1976

-GAUDEAU C, DZEIDZINA D, LONGEVIALLE C, MORAND P,THOUVENOT J ET LAFARGE M: Analyse automatique des cineangiocardiographies : Méthodologie. Biosigna 1978,1,Comm. B1.6, 56-65.

-GAUDEAU C, DE LA SOUCHERE J.C, DZEIDZINA D, BOIRON M, THOUVENOT J: Détection des contours par des automates dans la reconnaissance des formes C.R Séminaire sur l'analyse et traitements des images Biomédicales, Bois-des Hâtes, 30-31 Mai 1980.

-GAUDEAU C, BOIRON M, LONGEVIALLE C, COJAN P, ROULEAU P, THOUVENOT J: Automatic analysis of gastric deformation from X-ray films. MEDINFO 80,TOKYO 29 Sept./4 Oct. 1980,part 1 , 190-195.

-GAUDEAU C, BOIRON M, THOUVENOT J: Modélisation des arborescences vasculaires viscérales.Séminaire Interdisciplinaire sur la modélisation des systèmes vivants complexes et des interactions avec leur environnement,19-20 Juin 1981.Bois-des-Hâtes TOURS

VII -HUANG T.S:
Image Sequence Analysis Vol. 5,Springer Verlag Series in Information Sciences 480 p. 1981.

VIII -HOOGHOUDT T.E.H:
Computer Analysis of left Ventricular Cineangiograms. THES. Retterdam University Avril 1982.

IX -MORELLE O, GAUDEAU C, RIVALLIER J et THOUVENOT J:
Apport du traitement d'image radiologiques à l'étude du transit digestif: cinétique quantitative comparée d'une suspension barytée avec ou sans glucides chez le rat éveillé.
Ann.biol. Anim.Bioch.Biophys.1979, 19 (3B),911-914.

X -NADLER N :
Traitement et classification d'image par ordinateur.Reconnaissance de formes, Cast-Lyon , Février 1976.

XI -QUINQUETON J:
Le concept de dimension intrinsèque en reconnaissance des formes. Thèse Université Paris VI, 1981.

XII -RACKLEY C.E, DODGE H.T, COBLE Y.D, HAY R.E:
A method for determining left ventricular mass in man.Circulation Vol. XXIX, May 1964, 666-671.

XIII -RIVAILLIER J, LONGEVIALLE C, GAUDEAU C, et THOUVENOT J:
Système d'acquisition et de traitement d'images en mode conversationnel. Journées d'Inform. Méd. TOULOUSE, 13-17 Mars,1 , 295-310.

XIV -ROUTIER G, GAUDEAU C, BRULE J, WARUSFEL:
Evolution des capteurs optiques, perspectives d'avenir.Journées sur le traitement des images Bio-Méd.TOURS 1982.

XV -SLAGER C.J, REIBER J.H.C, SCHUURBIERS J.C.H, MEESTER G.T:
Contouromat: a bardwired left ventricular angioprocessing system I. Design and application.Comput.Biomed.Res.11: 491, 1978.

XVI -VOUKYDIS P.C:
Fiber stress profiles in the left ventricle of the heart during diastole. Effects of distension and hypertrophy,Bull.Math.Biophysics 34,1972,378-392.

A RULE-BASED SYSTEM FOR CHARACTERIZING BLOOD CELL MOTION

Martin D. Levine[1], Peter B. Noble[2], Youssry M. Youssef[1]

1 Computer Vision and Graphics Laboratory
Department of Electrical Engineering

2 Faculty of Dentistry

McGill University

Montreal, Quebec, Canada

KEY WORDS: DYNAMIC IMAGES, WHITE BLOOD CELLS, IMAGE UNDERSTANDING, RULE-BASED SYSTEMS.

ABSTRACT

The main function of a blood cell's surface is to receive information from the environment. Recently, experiments have indicated that the cell membrane plays a vital role in the life, development, and regulation of cells. However, there is no existing method to quantify the observable changes in membrane shape that occur in cell locomotion. The main goal of this research is to develop an image interpretation system capable of analyzing the structural changes in the morphology of cells from a sequence of pictures. This paper describes a scene analysis system for understanding cell motion. The system has been successfully used to analyze the pseudopod kinetics of the cell and relate it to the overall motion patterns.

1. INTRODUCTION

1.1 Biomedical Problem

Cell movement is a fundamental process of some importance to aspects of cell biology as diverse as migration of cells in embryological development and to host defense mechanisms. Advances have been made recently in the characterization of locomotory paths taken by cells in vitro and how these are affected by various substances. The internal mechanisms for cell locomotion are also reasonably well understood. However, progress has been much slower as to how the cell monitors external substances in order that the internal mechanisms involved in locomotion might be regulated. This interaction between external factors and cell internal processes has to occur at or within the cell membrane.

The thirty-eighth Symposium of the Society for Developmental Biology, held in Vancouver B.C. in June of 1979, was mainly devoted to summarizing the current status of knowledge about cell surface (plasma membrane). At the conclusion of the symposium, Wessells stated [Wessells, 79]:

> "It has become increasingly evident that the cell surface plays a truly pivotal role in the life, development, and regulation of cells. On one hand, the surface functions in the transmission of information from the environment to the cell, and here I mean not only molecular signals, but also mechanical forces stemming from adhesions and junctions that affect the cytoskeleton and

NATO ASI Series, Vol. F2
Image Sequence Processing and Dynamic Scene Analysis
Edited by T.S. Huang

so intracellular activations. The surface is also, in a real sense, an expression of the cell's genetic information and developmental state. Embryologists and developmental biologists must pay increasing heed to the cell surface and to its changing properties."

Development in multicellular eukaryotes [1] must depend on mechanisms that extend beyond the usual notions inherent in our concepts of sequential gene activation. For example, development of an embryo requires that cells know where they are and where they should be. There must be mechanisms that regulate this social behaviour of cells, and more than intuition informs us that the cell membrane is involved both as the donor and receptor of such social signals [Branton, 80].

The important activities of the cell membrane have now been documented in terms of transport function, receptor function, and mechanical function. Study of receptor function and mechanical function have led to the realization that components of the cell surface are mobile in the plane of the membrane, but that this lateral mobility is subject to regulation [Singer and Nicolson, 1972]; Nicolson, 1976a]. Such regulated, lateral mobility has been the basis for many hypotheses on the molecular mechanisms mediating cell recognition and growth control [Edelman, 1976; Nicolson, 1976a,b]. In particular, it is often suggested that the interaction of extracellular ligands[2], with their cell surface receptors, alters the distribution of transmembrane elements that can bind to mobility-related proteins such as actin or tubulin at the cytoplasmic surface of the plasma membrane [Edelman, 1976; Bourguignon and Singer, 1977].

Although many observations give credence to such hypotheses, it is only recently that a direct chemical demonstration of the binding between membrane and cytoplasmic components has been demonstrated. If transmembrane elements can interact with components at the cytoplasmic surface of the membrane, one would like to know the precise nature of the binding sites, the affinities and specificities of the interaction, and the manner in which these affinities are regulated. Some progress has been made in this direction with the polymorphonuclear leucocytes (PMN) in that binding studies with the chemotactic agent, N. formylmethionineleucinephenylalanine[3], have been carried out [Williams et al., 77]. The locomotory organ of these cells is the pseudopod and changes in pseudopod activity are the first morphological events visible as the cells respond to chemotactic agents. Similarly, lymphocytes also show pseudopod activity during locomotion [Lewis and Webster, 21], and many studies have been done on membrane bound ligand interactions in this cell type. Little is known of the pseudopod kinetics of these cells during locomotion.

1.2 Objectives

As indicated in the previous section, pseudopod formation is an important property of locomoting cells. Yet presently, there is no existing method for quantifying the observable changes in the membrane shape that occur during locomotion. Consequently, it is difficult to study the interaction at the membrane level, between the cell internal processes and the external factors, which modify cell locomotion. Therefore, what is required is a system to analyze a sequence of images of a moving cell to provide a quantification and descritpion (numerically and symbolically) of the dynamic changes in membrane shape. For example, this study would prove useful in assessing whether leucytes sensed a chemotactic gradient by a "temporal" or "spatial" mechanism [Zigmond et al., 81; Gerisch and Keller, 81]. Such a study would be of relevance to the understanding of the role of the cell membrane in the mechanisms which regulate the social behaviour of the cell, i.e. directed cell migration.

The main goal of this research is to develop an image interpretation system capable of analyzing the structural changes in the shape of a moving cell from a

---

1 Organisms made up of cells with nuclei bounded by nuclear envelopes.

2 A molecule that will bind to a complementary site on a given structure.

3 Sometimes abbreviated as: f Met-Leu-Phe

sequence of pictures using automatic techniques of digital image processing. To do this, the system would have to be able to: recognize the various image patterns, segment and interpret the desired object, and detect the significant (global) changes in the location, shape, and structure of the moving object.

Using this system to analyze the dynamics of blood cell motion, biologists can obtain the quantifications and descriptions of the data necessary to understand or answer questions pertaining to the cell behaviour. Figure (1) illustrates certain concepts for analyzing the observable changes in the membrane shape,in order to understand its role in regulating and modifying the cell locomotion. This figure refers to a moving cell under observation, and a biologist might be interested in the answers to three basic questions pertaining to the cell's dynamic behaviour. First, how can one recognize a subpart developed on or by the membrane as a "pseudopod"? And,if it is a pseudopod, is it stationary, growing, or contracting? Second, is a pseudopod "dominant" (the dominant pseudopod is one which contributes to the cell locomotion) or not? If so, what is its degree of domination? Finally, of all of

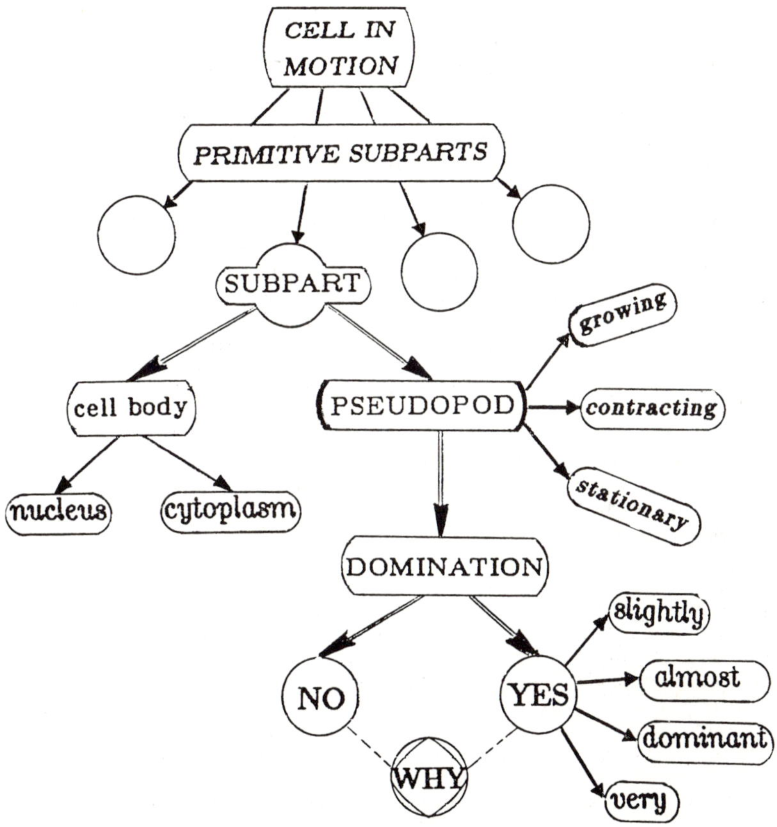

Figure(1) Characterization of the observable changes in the cell membrane of a moving cell, in order to understand the nature and distribution of "receptors" on or within the membrane. The properties of these receptors might explain the interaction between external factors and cell internal process, and how these mechanisms regulate the social behaviour of the cell.

this information were at hand, the third question would pertain to the interpretation of this data. For example, why is a specific pseudopod dominant and another not?

Our system has successfully provided all the quantification, description, and characterization information whereby the first two cited basic questions can be answered. Analysis of the data obtained may provide the answer to the third question.

1.3 Computer Analysis

In the automatic digital image processing of cell images, most work has dealt with static pictures of blood smears (or frozen cells) for the purpose of classification or counting. Previous research applied to the study of moving cells has concentrated on tracking cell paths only, rather than studying cell interaction characteristics [Levine et al., 80]. Furthermore, there is no existing system which concerns itself with the analysis of the structural changes in the cell shape. Consequently, there is no method extant which quantifies and characterizes the observable changes in the cell membrane occurring during locomotion. Indeed, a terminology for these descriptions is not even available, and an objective of this research was to develop one in conjunction with biologists.

Even if we extend our consideration to the general problem of processing dynamic images, this field has been largely restricted to motion detection, recognition, and tracking [Nagel, 79]. Thus, two main issues have been ignored: (a) shape and structural changes of a non-rigid moving object, and (b) motion understanding and description. In our current work, we are dealing with non-rigid objects which change their shape and structure randomly. We assume that the objects are essentially two-dimensional and ignore the effects of the property changes which actually do vary in three-space. In order to describe the motion, it is not enough to merely determine the incremental movements or changes that occur between consecutive images [Tsotsos et al., 80]. What is required is a system which abstracts a description of the global motion characteristics from the complete set of static and incremental data. Development of such a system represents the approach being taken in our present research in image sequence analysis.

An understanding system, either as a computer vision system for static scene interpretation, or visual motion description from a sequence of images, requires the explicit use of knowledge about the objects. The system should utilize diverse knowledge sources at multiple levels of description, and be supported by an efficient control structure mechanism. The progress towards the development of knowledge-based systems for visual motion understanding is slow, and the work which has been done is very limited. The most relevant research to that reported here is by Badler [Badler, 75], Yachida et al. [Yachida et al., 78], and Tsotsos [Tsotsos et al., 80]. A detailed discussion of these is given in [Youssef, 1982].

The main input for any image sequence analysis system is a series of two-dimensional digital images representing the variation in a specific scene along a third-dimension[1]. The function of the system is to generate a description of the consistent characteristics and behaviour of the moving object(s) recorded within the sequence. A theoretical model for a general dynamic scene analysis and motion understanding system has been developed [Levine et al., 82]. It involves three basic entities (Figure 2):

(a) Static (constant) data: The data which remains unchanged during the course of analysis. They contain constraint knowledge pertaining to the class of scenes and motion under analysis, as well as the control information describing the pertinent computational processes.

(b) A collection of analysis processes, each of which is assigned to a particular task.

(c) Dynamic data: The data which are continually changing as a result of the functioning of the different analysis processes.

----

1 Laboratory methods and facilities, including blood cell preparation, are discussed in [Youssef, 1982].

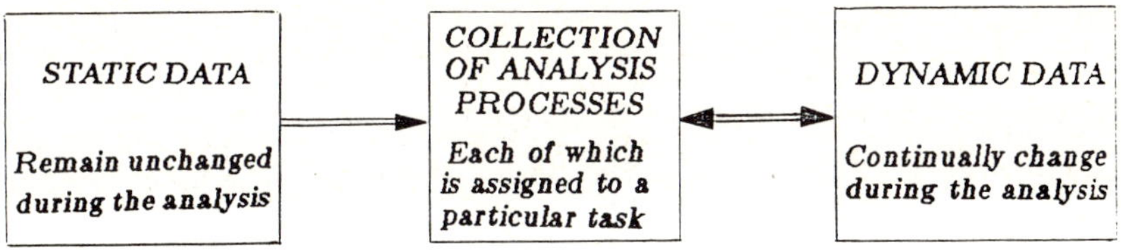

Figure(2) *Basic structure units of a dynamic scene analysis system.*

The analysis is specified in terms of condition-action rules, and these will be presented in later sections of this paper. Readers are referred to [Levine et al., 82] for a detailed discussion of the system knowledge representation and control.

The different computational processes are designed to execute through a hierarchial structure consisting of three basic levels: static, incremental, and global (see Figure 3).

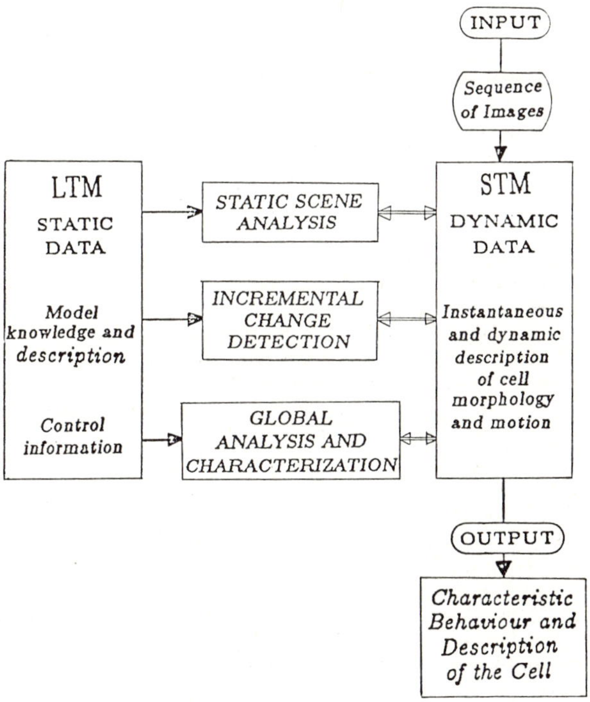

Figure(3) *Main processing stages and data structure of a motion understanding system*

(a) Static Scene Analysis: This step is similar to a conventional image processing system in that the input is a single digital image, and the output is a description and interpretation of the scene. However, in the analysis of dynamic images, the information extracted from the previous frames of the same sequence may also be used as knowledge to assist in the analysis of the current frame. The main objective of this stage is to identify the desired moving object, segment it, and describe it in each frame of the sequence.

(b) Incremental Change Detection: This is an intermediate step between the static and the global. The main objective is to detect and describe the incremental changes in shape, structure, and motion of the moving object.

(c) Global Analysis: This is the highest level in the hierarchy of the dynamic scene analysis system. The goal is to analyze the static and incremental data in order to detect and describe the global observable changes within the sequence of frames. In this way, the characterization of the consistent dynamic behaviour of the cell may be obtained.

The following section describes processes for global locomotion analysis and description. The quantification and symbolic description of the observable changes in the cell shape are given in Section 3. Global structural analysis is discussed in Section 4. Section 5 is concerned with the integration of the global aspects pertaining to locomotion, shape, and structure, in order to understand and characterize the dynamic behaviour of the cell. Also in this section, we discuss one of the basic questions related to understanding the role the cell membrane plays in the mechanisms which regulate its social behaviour. That is, the domination of a pseudopod in the global locomotion of the cell. Finally, in Section 6, the research is summarized and discussed. An evaluation of the computer analysis and characterization is given based on a comparison to those obtained by a biologist.

2. LOCOMOTION ANALYSIS

Motion analysis and description of a moving object can be achieved in two steps. The first is concerned with motion detection and tracking of an object in the image sequence. The second step involves the analysis of the extracted data in order to generate a summary of the locomotion behaviour. In the present system, the processes associated with the first step are responsible for locating the cell in each image of the sequence, and detecting the incremental change in its location between sequential frames. The objective of global motion analysis and description is to analyze the multitude of static and incremental data to characterize the cell's locomotion. To accomplish this, it is necessary to isolate the significant global movements from the irrelevant and noisy ones. From this data, the cell path is constructed in a sequence of vectors, as shown in Figure (4). Each motion vector is described by a set of motion properties such as: time, distance, direction of motion, velocity, and acceleration. These data are then utilized to quantify and describe the cell locomotion behaviour and the effectiveness of an external factor on it. This analysis is described below.

2.1 Chemotaxis Analysis

Chemotaxis is one of the most important descriptors of cell locomotion. It is the directional response of a motile cell to the influence of an external factor, i.e., chemical substances produced by cells, such as bacteria and tumours. The importance of chemotaxis lies in the fact that it results in the efficient localization of invading agents. This section is concerned with the quantification and description of the relationship between cell locomotion and the influence of such an external factor. The objective of this chemotaxis analysis is accomplished in three steps:

(a) Compute the global directional movement of the cell.

(b) Quantify the cell response to the influence of an external factor, thereby determining the effectiveness of the influence.

(c) Characterize and describe the global behaviour of the cell locomotion.

A typical example of the quantification of the global cell locomotion is given in Description (1).

DESCRIPTION (1)

GLOBAL LOCOMOTION ANALYSIS AND DESCRIPTION

INTRODUCTION

The following is a global characterization and discription of the locomotion analysis of a NEUTROPHILE cell. The cell motion was recorded in real time on 16mm cine film at rate of TWO frames per second. The cell was under the influence of BACTERIA which is located in the SOUTH-WESTERLY direction of the original location of the cell. The total observation time was 225 seconds (450 frames). The following is a description of the cell locomotion between frame number 200 and 450 (125.0 seconds)

CELL PATH and MOTION ANALYSIS

The space domain of the cell motion is divided into EIGHT equal directions(states). First, the cell path was obtained by sampling the displacement between frames in increments of approximatly 2.0 microns. Then sequences with the same incremental direction were merged into one to produce the final cell path which consists of 34 steps. The description of the time, distance, speed, direction, and acceleration of the cell at each step is as follows :

TIME	DISTANCE	SPEED	DIRECTION	ACCELERATION	
VERY SHORT	VERY SHORT	VERY SLOW	NORTH-EASTERLY	NONE	
VERY SHORT	SHORT	VERY FAST	EASTERLY	VERY FAST	POSITIVE
VERY LONG	VERY LONG	AVERAGE	WESTERLY	AVERAGE	NEGATIVE
SHORT	VERY SHORT	VERY SLOW	NORTH-EASTERLY	VERY FAST	NEGATIVE
SHORT	MEDIUM	VERY FAST	SOUTH-WESTERLY	VERY FAST	POSITIVE
SHORT	VERY SHORT	VERY SLOW	NORTH-EASTERLY	VERY FAST	NEGATIVE
MEDIUM	SHORT	AVERAGE	SOUTH-WESTERLY	VERY FAST	POSITIVE
VERY LONG	MEDIUM	SLOW	SOUTH	NONE	
LONG	MEDIUM	AVERAGE	SOUTH-WESTERLY	SLOW	POSITIVE
SHORT	MEDIUM	VERY FAST	EASTERLY	VERY FAST	POSITIVE
MEDIUM	LONG	FAST	SOUTH-WESTERLY	VERY FAST	NEGATIVE
SHORT	MEDIUM	VERY FAST	EASTERLY	VERY FAST	POSITIVE
SHORT	MEDIUM	VERY FAST	NORTH-WESTERLY	VERY FAST	NEGATIVE
MEDIUM	VERY LONG	VERY FAST	SOUTH	AVERAGE	NEGATIVE
MEDIUM	MEDIUM	AVERAGE	SOUTH-WESTERLY	AVERAGE	NEGATIVE
LONG	MEDIUM	SLOW	SOUTH	AVERAGE	NEGATIVE
MEDIUM	MEDIUM	AVERAGE	SOUTH-EASTERLY	AVERAGE	POSITIVE
MEDIUM	MEDIUM	AVERAGE	SOUTH-WESTERLY	SLOW	POSITIVE
SHORT	MEDIUM	VERY FAST	SOUTH-EASTERLY	VERY FAST	POSITIVE
MEDIUM	MEDIUM	FAST	SOUTH-WESTERLY	VERY FAST	NEGATIVE

LONG	MEDIUM	SLOW	SOUTH	AVERAGE	NEGATIVE
MEDIUM	MEDIUM	AVERAGE	SOUTH-WESTERLY	AVERAGE	POSITIVE
MEDIUM	MEDIUM	AVERAGE	SOUTH	NONE	
SHORT	MEDIUM	FAST	SOUTH-WESTERLY	VERY FAST	POSITIVE
VERY LONG	VERY LONG	AVERAGE	SOUTH	AVERAGE	NEGATIVE
MEDIUM	LONG	VERY FAST	WESTERLY	VERY FAST	POSITIVE
MEDIUM	MEDIUM	AVERAGE	SOUTH-EASTERLY	FAST	NEGATIVE
VERY LONG	MEDIUM	SLOW	SOUTH	AVERAGE	NEGATIVE
LONG	MEDIUM	AVERAGE	SOUTH-EASTERLY	SLOW	POSITIVE
VERY SHORT	VERY SHORT	VERY SLOW	NORTH-EASTERLY	VERY FAST	NEGATIVE
SHORT	MEDIUM	FAST	EASTERLY	VERY FAST	POSITIVE
MEDIUM	VERY SHORT	VERY SLOW	NORTH-EASTERLY	VERY FAST	NEGATIVE
SHORT	SHORT	AVERAGE	EASTERLY	VERY FAST	POSITIVE
MEDIUM	MEDIUM	AVERAGE	SOUTH-EASTERLY	AVERAGE	NEGATIVE

CHEMOTAXIS ANALYSIS :

Chemotaxis is the response of a motile cell to the directional influence of a chemical substance or any external factor (BACTERIA in this film). The following is a summary of the directional movements of the cell under analysis when compared to typical random motion of a similar cell:

DIRECTION	TOTAL DISPLACEMENT
EASTERLY	AVERAGE
NORTH-EASTERLY	NONE
NORTH	NONE
NORTH-WESTERLY	SHORT
WESTERLY	AVERAGE
SOUTH-WESTERLY	VERY LONG
SOUTH	VERY LONG
SOUTH-EASTERLY	AVERAGE

CONCLUSION :

The resultant directional locomotion is in a SOUTH direction, which is ALMOST THE SAME direction in which the BACTERIA is located. This motion represents THREE-FIFTHS of the total displacement of the cell.

From the above analysis we may conclude that :

THE CELL HAS AN AVERAGE POSITIVE CHEMOTAXIS MOTION

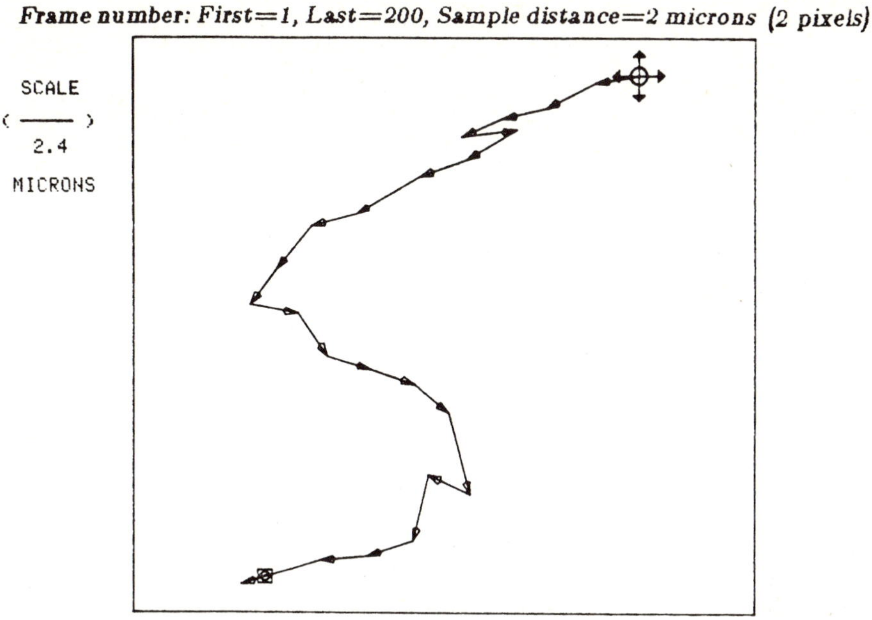

Figure(4) Distance sampling of the cell path.

2.1.1 Directional Movement of the Cell

The objective of this analysis is to determine whether the cell motion has a tendency towards a specific direction or if it is random. The input consists of the incremental displacement of the cell between sequential frames. The output is a quantification of the cell movement in the main directions of the plane.

Let us define the incremental displacement and direction of motion between two sequential frames (i, i+1) as L_i and D_i, where:

$$L_i = [dX_i^2 + dY_i^2]^{1/2} \tag{1}$$

$$D_i = \mathrm{Tan}^{-1}(dY/dX) \tag{2}$$

and $$dX_i = X_{i+1} - X_i,\ dY_i = Y_{i+1} - Y_i$$

We divide the plane of cell motion into nd equal directions $S_1, S_2, \ldots, S_j, \ldots, S_{nd}$. The extent of each direction is equal to 360/nd. The incremental displacement is considered to be in direction S_i if the displacement makes an angle D_i with the X axis that is between S_{j-1} and S_j. The vector that represents the total locomotion in a specific direction (TDR) can be computed as the resultant of all the cell displacements in that direction. Figures (5a) and (5b) illustrate the total displacement of the cell, where the space has been divided into four and eight equal directions, respectively. The vector (RR), which represents the global locomotion of the cell, is the resultant of all the vectors defining the cell locomotion in the different directions. Thus:

$$RR = \sum_{i=1}^{nd} RD_i, \tag{3}$$

Frame number: First=1, Last=250 4 state analysis

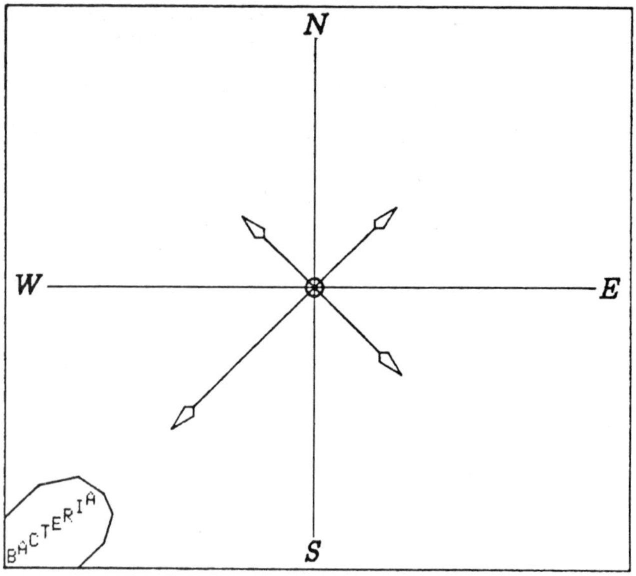

Figure(5a) Vector sum of locomotion in each of four directions.

Frame number: First=1, Last=250 8 state analysis

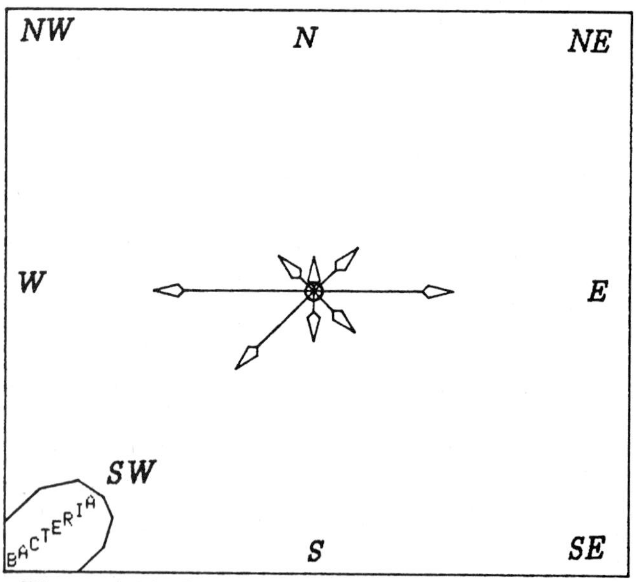

Figure(5b) Vector sum of locomotion in each of eight directions.

where "sum" in this equation indicates a vector summation.

The directional tendency of the cell is the amount of movement that it exhibits in a specific direction. In the case of random locomotion, the probability of motion in the different directions is approximately equal [Levine et al., 80]. Consequently, the resultant vector RR of the cell locomotion is approximately zero. In the case of chemotactic locomotion, the cell exhibits an inclination to move in a specific direction.

The movement in the different directions, as well as the global locomotion, can be described as NONE, VERY SHORT, SHORT, AVERAGE, LONG, and VERY LONG[1]. This qualification is based on a comparison with typical random motion. Thus, the Total Translation Distance, TTD, indicates the translation from the initial to the final point in the temporal sequence. It can be computed as:

$$TTD = [(X_b - X_a)^2 + (Y_b - Y_a)^2]^{1/2}, \quad (4)$$

where (X_a, Y_a) and (X_b, Y_b) are coordinates of the cell in the first and last frames. Then let

$$TDR_{av} = TTD/nd \quad (5)$$

Locomotion in a specific direction can be normalized as follows:

$$TDRN_k = TDR_k/TDR_{av}. \quad (6)$$

Using the stored representational rules, the normalized values in equation (6) can be assigned to the appropriate symbolic qualifiers. For example, we may describe the cell motion as being AVERAGE if the resultant vector is about equal to TDR.

2.1.2 Chemotaxis Quantification

The chemotactic motion of the cell can be measured as the ratio of its movement in the direction of influence to the total movements in the space domain. Thus, let the total locomotion of the cell be represented by the vector RR = {LRR, DRR}, where LRR and DRR are the length and direction of the vector RR, respectively. Then the percentage chemotactic motion PCM can be computed as:

$$PCM = LRR/TLD \,.\, 100\ \%, \quad (7)$$

where TLD is total locomotion distance in the space domain. It is given by

$$TLD = \underset{i=1}{\overset{m-1}{SUM}} LS_{i,i+1}, \quad (8)$$

where m is the number of path points, and $LS_{i,i+1}$ is distance between two sequential points. Thus, using the value of PCM, the cell motion can be characterized as random or chemotactic by the following rule:

RULE:

IF PCM . LE . E_1 ==then==> RANDOM

IF PCM . GT . E_1 AND . LE . E_2 ==then==> ALMOST RANDOM

IF PCM . GT . E_2 AND . LE . E_3 ==then==> PARTIALLY RANDOM

IF PCM . GT . E_3 AND . LE . E_4 ==then==> ALMOST CHEMOTACTIC

IF PCM . GT . E_4 ==then==> CHEMOTACTIC

Note that the thresholds E_1, E_2, E_3, and E_4 are equal to 0.1, 0.3, 0.7, and 0.9, respectively [Youssef, 82].

---

1 Readers are referred to [Youssef, 82; Levine et al., 82] for a detailed discussion of the symbolic representations presented in this paper.

Chemotaxis can be described as being positive or negative depending on whether the cell moves towards or against the direction of influence. To compute and describe this behaviour, we consider the cell as having started its motion from point O (X_o, Y_o), the origin of the two dimensional plane, with the external factor concentrated at a point E (X_e, Y_e). The line OE which connects the original position of the cell (O) to the center of the external factor (E) represents the direction of the influence, as shown in Figure (6). In this figure, the line SOS, which passes through the point O and is normal to the line OE, represents the border line which divides the plane into two regions.

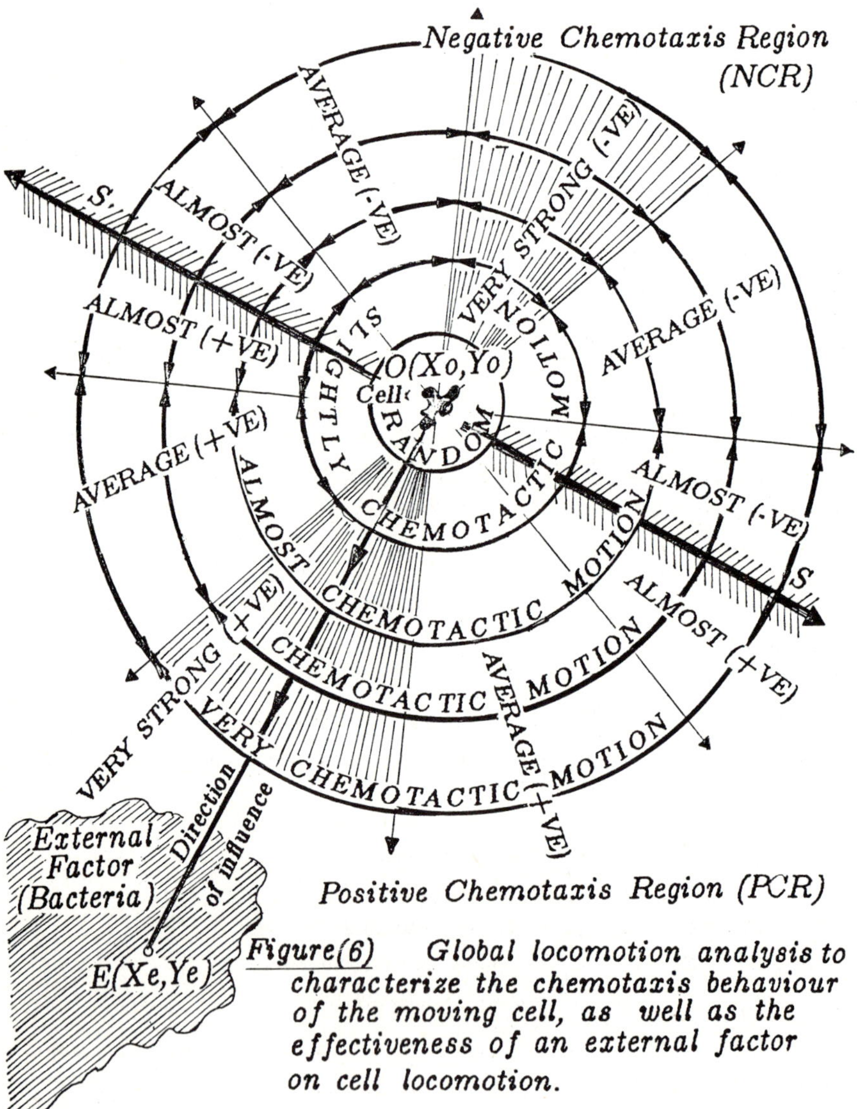

Figure(6) *Global locomotion analysis to characterize the chemotaxis behaviour of the moving cell, as well as the effectiveness of an external factor on cell locomotion.*

One is PCR or "positive chemotaxis region" (the region where the external factor is located), and the other is NCR or "negative chemotaxis region". Thus, if the vector that represents the global locomotion is in the region PCR, this indicates positive chemotaxis. If it is in region NCR, negative chemotaxis is implied. This can be quantified as follows:

RULE:

If DRR . GE . E_1 AND .LT. E_2 ==then==> NEGATIVE CHEMOTAXIS

IF DRR . GE . E_2 AND .LT. E_1 ==then==> POSITIVE CHEMOTAXIS

where DRR is the angle of the global locomotion with the X axis. E_1 and E_2 are the angles of the border line between the positive and negative chemotaxis regions.

2.1.3 Quantifying the Effectiveness of Influence

The effectiveness of an external factor (EF) on the locomotion of a moving cell can be defined as its ability to attract the cell in the direction of its influence. This force is a function of two factors: (a) the percentage of the chemotactic movements (PCM), and (b) the angle between the direction of motion and influence (DRI). Thus,

$$EF = f(PCM, DRI). \qquad (9)$$

PCM is defined in the preceding section. The angle DRI can be computed as:

$$DRI = |DR - DI|, \qquad (10)$$

where DR and DI are the directions of the global locomotion and the influence, respectively. DRI can be utilized as an indicator of whether or not the cell will ultimately reach the location of the external factor. To determine this, the value of DRI can be normalized (0-1) as:

$$DRI_n = (90 - DRI)/90. \qquad (11)$$

Thus, $DRI_n = -1$ if the global locomotion is opposite to the direction of influence (purely negative chemotaxis), and $DRI_n = +1$ if the locomotion is exactly in the direction of influence (purely positive chemotaxis). Using simple representational rules, the value DRI_n can then be used to describe the direction of global locomotion, as compared to the direction of influence. Typical descriptions are QUITE OPPOSITE, ALMOST OPPOSITE, PERPENDICULAR, ALMOST THE SAME, THE SAME.

Finally, the effectiveness EF of an influence can be computed as the geometric mean (GM) of PCM and DRI_n. Thus

$$EF = (PCM \cdot DRI_n)^{1/2} \cdot 100\ \%. \qquad (12)$$

Note that EF lies between plus and minus 100%.

We observe that EF summarizes the chemotactic movement and the effectiveness of the influence. The sign indicates positive or negative chemotaxis, and the amplitude represents the strength. This can be utilized by representational rules to describe the global behaviour of the cell locomotion as follows:

RULE:

IF EF . LT . 0 ==then==> NEGATIVE CHEMOTAXIS

IF EF . EQ . 0 ==then==> RANDOM MOTION

IF EF . GT . 0 ==then==> POSITIVE CHEMOTAXIS

IF $|EF|$. LE . E_1 ==then==> SLIGHT

IF $|EF|$. GT . E_1 AND . LE . E_2 ==then==> ALMOST

IF $|EF|$. GT . E_2 AND . LE . E_3 ==then==> AVERAGE

IF $|EF|$. GT . E_3 AND . LE . E_4 ==then==> STRONG

IF $|EF|$. GT . E_4 ==then==> VERY STRONG

Description (1) is a typical example of the global locomotion characterization and description generated by the system. Figure (7) shows different characterizations of the same cell under the influence of bacteria placed in different positions of the field.

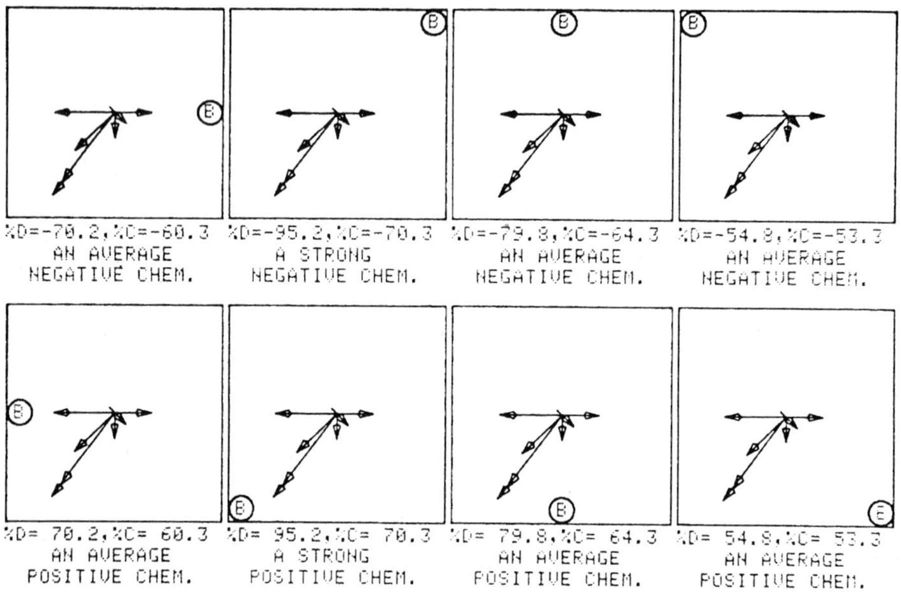

Figure(7) *Global quantification and characterization of moving cell locomotion.*

The symbol B represents the location of the external influence.

%D refers to the symbol DRI, and %C to PCM (see equation (7)).

3. SHAPE ANALYSIS AND DESCRIPTION

3.1 Introduction

The perception of shape plays a prominent role in both human and computer vision. It may be achieved through two stages of processing: shape analysis and shape description. In shape analysis, a digitized image of an object is transformed into a scalar vector whose elements are measurements of some of the shape properties. The second task of shape analysis is to transform the image of an object into a graph whose properties express the object's shape and structure. Shape description represents the higher level process of shape perception by computer. In this case the scalar vector or graph, the result of shape analysis, is analyzed using syntactic analysis [Pavlidis, 76]. A symbolic description is generated which contains all the relevant information pertaining to the shape of the object.

Most of the work which has been done in dynamic scene analysis has been restricted to the study of rigid body motion. Sometimes consideration is given to change in shape due to altering viewing conditions. For example, a car moving in front of a fixed camera [Nagel, 78b], fish swimming in a vat [Yachida et al., 78], and left ventricular motion [Tsotsos et al., 80]. However, in each of these cases the dynamic change is restricted by a fixed object model. Our study differs from previous work in that we examine the motion of non-rigid moving objects. Also, the changes in cell membrane shape are due to what appears to be randomly varying physical properties which are not predictable.

The approach taken to shape analysis and description involves the following stages: (a) segmentation and boundary tracking, (b) polygonal approximation, (c) polygonal decomposition and labeled graph representation, and (d) rule-based syntactic analysis and shape description techniques. The last stage is introduced to quantify and symbolically describe the shape properties. In addition to the usual difficulties of describing an arbitrary shape in a static image, we face the following problems:

(a) Estimating the incremental change in the shape and structure of a non-rigid moving object.

(b) Detecting and characterizing the global shape changes in the morphology of a non-rigid moving object over a period of time.

(c) Presenting all the above descriptions in a meaningful form to the user.

The problem of detecting and characterizing the global changes in the shape will comprise the remainder of this section. Figure (8) shows the main processes and data structures used. The input to this stage of the system consists of the static and incremental description of the shape properties in and between each frame. An example of the input data is shown in Figure (9). It consists of the static values of one of the shape properties (circularity) in a sequence of 150 frames. The output of the analysis is a summary describing the global changes in cell shape and their characterization. A typical example of this characterization is given in Description (2). The methodologies and processes which are used to generate these descriptions are discussed below.

3.2 Detecting Global Shape Changes

The technique for detecting and describing shape changes in this research consists of three basic steps:

(a) Global change detection: to detect relevant changes in shape from the static and incremental data.

(b) Symbolical qualification: to transform the global data into symbolic qualifiers.

(c) Characterization and description: to analyze the symbolic qualification in order to summarize the characteristic behaviour of the changes in the cell shape.

The objective of the first step is to reduce the static and incremental data to those representing information relevant to the global changes. This can be achieved by detecting the times (frames) in the temporal sequence where a significant change

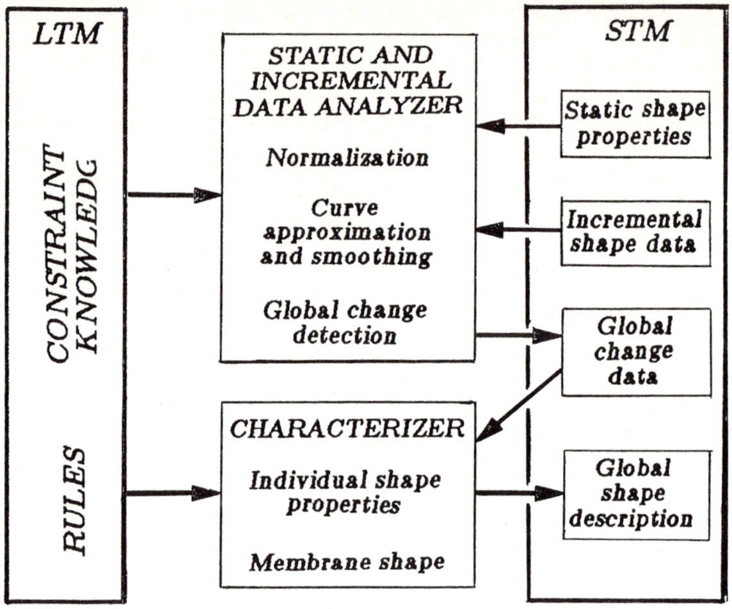

Figure(8) Processes and data structure for global shape analysis.

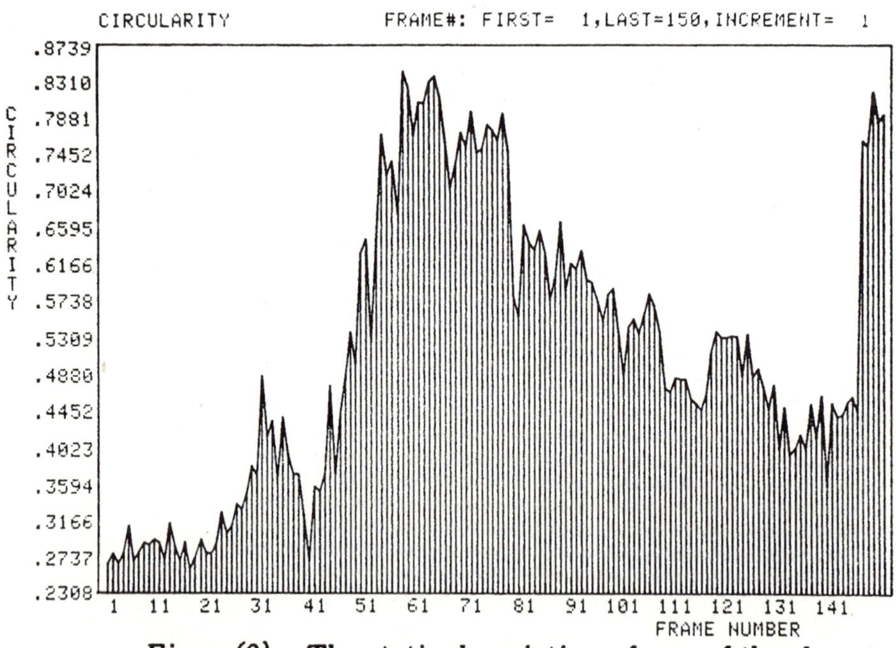

Figure(9) The static description of one of the shape properties (circularity) used as input data for global shape analysis.

DESCRIPTION <2>

GLOBAL SHAPE ANALYSIS
AND DESCRIPTION

INTRODUCTION

A summary of the global changes in the cell shape characteristics are described below. The description is given in two sections. First, the characterization of each of the main cell shape properties, and then, the global changes in the cell membrane shape are described. The description given is for the period including frame number(1) to frame number(150), a duration of 75.0 seconds.

SHAPE PROPERTIES

AVERAGE BENDING ENERGY

PERIOD NUMBER	FRAME NUMBER	TIME	DESCRIPTION
1	1 --> 2	VERY SHORT	SMOOTH
2	3 --> 18	SHORT	ALMOST SMOOTH
3	19 --> 20	VERY SHORT	JAGGED
4	21 --> 30	SHORT	ALMOST SMOOTH
5	31 --> 35	VERY SHORT	JAGGED
6	36 --> 37	VERY SHORT	ALMOST SMOOTH
7	38 --> 44	VERY SHORT	SMOOTH
8	45 --> 78	MEDIUM	ALMOST SMOOTH
9	79 --> 85	VERY SHORT	JAGGED
10	86 --> 88	VERY SHORT	VERY JAGGY
11	89 -->103	SHORT	JAGGED
12	104 -->127	MEDIUM	ALMOST SMOOTH
13	128 -->138	SHORT	JAGGED
14	139 -->150	SHORT	ALMOST SMOOTH

CIRCULARITY

PERIOD NUMBER	FRAME NUMBER	TIME	DESCRIPTION
1	1 --> 25	MEDIUM	NOT CIRCULAR
2	26 --> 43	SHORT	SLIGHTLY CIRCULAR
3	44 --> 53	SHORT	ALMOST CIRCULAR
4	54 --> 57	VERY SHORT	CIRCULAR
5	58 --> 78	SHORT	VERY CIRCULAR
6	79 -->131	LONG	ALMOST CIRCULAR
7	132 -->137	VERY SHORT	SLIGHTLY CIRCULAR
8	138 -->145	SHORT	ALMOST CIRCULAR
9	146 -->149	VERY SHORT	CIRCULAR
10	150 -->150	VERY SHORT	VERY CIRCULAR

ANGLE REGULARITY

PERIOD NUMBER	FRAME NUMBER	TIME	DESCRIPTION
1	1 --> 13	SHORT	ALMOST REGULAR
2	14 --> 21	SHORT	IRREGULAR
3	22 --> 23	VERY SHORT	ALMOST REGULAR
4	24 --> 28	VERY SHORT	IRREGULAR
5	29 --> 32	VERY SHORT	ALMOST REGULAR
6	33 --> 36	VERY SHORT	IRREGULAR
7	37 --> 46	SHORT	VERY IRREGULAR
8	47 --> 52	VERY SHORT	IRREGULAR
9	53 --> 57	VERY SHORT	ALMOST REGULAR
10	58 --> 64	VERY SHORT	VERY REGULAR
11	65 --> 67	VERY SHORT	ALMOST REGULAR
12	68 --> 70	VERY SHORT	REGULAR
13	71 --> 77	VERY SHORT	VERY REGULAR
14	78 -->109	MEDIUM	ALMOST REGULAR
15	110 -->116	VERY SHORT	IRREGULAR
16	117 -->147	MEDIUM	ALMOST REGULAR
17	148 -->150	VERY SHORT	REGULAR

MEMBRANE SHAPE DESCRIPTION

The following is a summary of the global shape of the cell membrane based on the GEOMETRIC MEAN of the following properties: AVERAGE BENDING ENERGY, CIRCULARITY, and ANGLE REGULARITY.

PERIOD NUMBER	FRAME NUMBER	TIME	DESCRIPTION
1	1 --> 18	SHORT	COMPLEX
2	19 --> 20	VERY SHORT	ALMOST SIMPLE
3	21 --> 27	VERY SHORT	COMPLEX
4	28 --> 36	SHORT	ALMOST SIMPLE
5	37 --> 44	SHORT	VERY COMPLEX
6	45 --> 49	VERY SHORT	COMPLEX
7	50 --> 57	SHORT	ALMOST SIMPLE
8	58 --> 60	VERY SHORT	SIMPLE
9	61 --> 64	VERY SHORT	VERY SIMPLE
10	65 --> 70	VERY SHORT	ALMOST SIMPLE
11	71 --> 77	VERY SHORT	SIMPLE
12	78 --> 86	SHORT	ALMOST SIMPLE
13	87 --> 88	VERY SHORT	SIMPLE
14	89 -->113	MEDIUM	ALMOST SIMPLE
15	114 -->116	VERY SHORT	COMPLEX
16	117 -->145	MEDIUM	ALMOST SIMPLE
17	146 -->150	VERY SHORT	SIMPLE

has occurred, called "key frames" [Burtynk and Wein, 78]. There are two definitions for such points. The first involves an alteration of the dynamic behaviour; for example, from stationary (dP/dt = 0) to increasing (dP/dt > 0) or decreasing (dP.dt < 0). The second refers to a time at which a significant change has occurred in the qualification (description) level. Thus $Q_i(P_k)$ ---> $Q_j(P_k)$, where $Q_i(P)$ and $Q_j(P)$ are two different levels of qualification for the same property (P_k). For example, from SIMPLE to COMPLEX, or from SMOOTH to JAGGED.

The technique developed for cell analysis is designed to detect and provide descriptions for both types of global changes. In this section, the detection of the changes in the dynamic behaviour will be discussed. The following section describes a methodology for detecting and summarizing the significant changes in the qualification level.

In order to maintain a uniform approach within the analysis rules, the static and incremental data must first be normalized. In this operation, the values of a specific property in a sequence of frames $P_1, P_2, \ldots, P_n$ are normalized to range between zero and one to give $p_1, p_2, \ldots, p_n$, as follows:

$$p_i = (P_i - P_{min}) / (P_{max} - P_{min}), \tag{13}$$

where p_i is the normalized factor of P_i, and

$$P_{min} = \text{Min } (P_1, P_2, \ldots, P_n) \tag{14}$$

$$P_{max} = \text{Max } (P_1, P_2, \ldots, P_n) \tag{15}$$

Note that this is not an absolute form of normalization since the data within the sequence are employed to provide P_{min} and P_{max}. Thus the amplitude of the normalized curve at any point (frame) represents the static value of the property at that frame (see Figure (9)). The variations between neighbouring points in the curve represent the incremental changes of the property between sequential frames. Curve smoothing and approximation can be used to detect the key frames.

The curve approximation algorithm is due to Ramer [Ramer, 72]. The procedure consists of the following steps:

(1) Connect the first and the last point on the curve by a straight line AB.

(2) Find the point C on the curve at the maximum distance L_m from the line AB. If L_m is greater than the approximation threshold E, then split the curve at point C.

(3) For each new segment, repeat steps (1) and (2).

Figure (10) shows the curve approximation of the data in Figure (9).

The result of curve approximation consists of a set of vertices {KF} representing the key frames:

$$\{KF\} = \{KF_1, KF_2, \ldots, KF_j, \ldots KF_m\}, \tag{16}$$

where m is the number of key frames. Each key frame is defined by time (frame number) and the property value at that time, as follows:

$$KF_j(P_i) = [t_j, P_i(t_j)]. \tag{17}$$

$P_i(t_j)$ is the value of the property (i) at time (t_j).

A property has a constant dynamic behaviour (increasing, decreasing, or stationary) between each two sequential key frames. This behaviour can be described from the information included in the key frames. At this level, the following global data can be computed for each period between two sequential key frames:

(i) Time of Change

$$TKF_{i,i+1} = [t_{i+1} - t_i + t_o]\, t_o \quad \text{seconds}, \tag{18}$$

where t_i, t_{i+1} are the times of key frames i and i+1, respectively, and t_o is the time interval between two sequential frames.

(ii) Amount of Change

$$CKF_{i,i+1} = P_{i+1} - P_i, \tag{19}$$

where P_i, P_{i+1} are the property values at key frames $KF_{(i)}$ and $KF_{(i+1)}$, respectively.

(iii) Rate of Change

$$RKF_{i,i+1} = \tan^{-1}[CKF_{i,i+1} / TKF_{i,i+1}]. \tag{20}$$

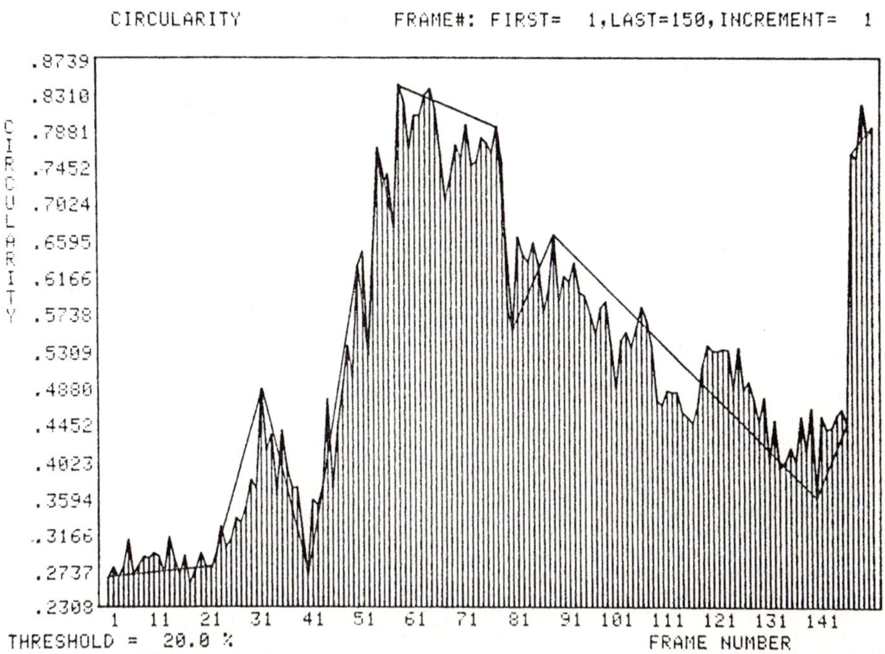

Figure(10) ***Curve fitting of the static data shown in figure(9). An iterative splitting algorithm is used to compute a piecewise linear approximation to the curve.***

The values of TKF, CKF, and RKF can be qualified in order to summarize the dynamic changes in the property under analysis, as follows:

TIME OF CHANGE	TYPE OF CHANGE	AMOUNT OF CHANGE	RATE OF CHANGE
TKF	CKF(0/+/-)	|CKF|	RKF
VERY SHORT	STATIONARY	NEGLIGIBLE	VERY SLOW
SHORT	INCREASING	SLIGHT	SLOW
AVERAGE	DECREASING	AVERAGE	AVERAGE
LONG		CONSIDERABLE	FAST
VERY LONG		SIGNIFICANT	VERY FAST

The above analysis is used to describe the global changes in the dynamic behaviour of a specific property. For example, the generated description of the ELONGATION property could be:

"For a SHORT time, the ELONGATION was INCREASING at a VERY SLOW rate, causing a SLIGHT INCREASE in the ELONGATION. Then for a LONG time it was STATIONARY. This was followed by a VERY FAST DECREASE in a VERY SHORT time."

3.3 Qualification of Property Level Changes

In order to summarize the global changes in the qualification levels of a specific property, two methods can be used: (a) applying the representational rules directly to the static and incremental symbolic qualifications; (b) applying the representational rules to the key frame data. Method (a) is recommended only if the generated description from curve approximation is not required. Otherwise, method (b) is obviously faster and will result in a more precise description. This is because, in the curve approximation, most of the irrelevant and noisy data has already been removed. Thus, the data associated with the key frames represent the input for the analysis rules.

The objective is to split and merge periods between key frames to generate a summary of the quantification level changes. The output of the curve approximation consists of a number of periods, each period being bounded by two key frames. If the symbolic qualifier of the property in these two key frames is different, we split the period bounded by them into a number of periods, such that the property in each period has the same qualifier. A key frame is initiated for each new period and the key frame set is updated. This can be achieved by applying the following rule:

RULE:

IF $Q(P_k, KF_i)$ .NE. $Q(P_k, KF_j)$

==then==> 1) SPLIT T_{ij} INTO $T_{i1}, T_{12}, T_{23}, \ldots, T_{mj}$

2) KF_{u+m} <--- $KF_{(u)}$, u = j,n

3) INITIATE $KF_{i+1}, KF_{i+2}, \ldots, KF_{(i+m)}$

where $Q(P_k, KF_i)$ and $Q(P_k, KF_j)$ are the property qualifiers in key frames $KF_{(i)}$ and $KF_{(j)}$, respectively. For example, suppose that T_{ij} is the period bounded by key frames KF(i) and KF(j), respectively, where the descriptor is different at those key frames. The interval T_{ij} is then split into subintervals, possibly $T_{i1}, T_{12}, T_{23}, T_{3j}$. The three key frames $KF_{i+1}, KF_{i+2}, KF_{i+3}$ should be initiated and inserted in the key frame set:

$$KF_j \text{-->} KF_{j+3},\ KF_{j+1} \text{-->} KF_{j+4}, \ldots, KF_n \text{-->} KF_{n+3}.$$

Figure (11a) shows two sequential periods resulting from the curve approximation, and Figure (11b) shows the result of splitting them into a number of periods.

The splitting operation might yield sequential periods having the same qualifier, as shown in Figure (11b), or it may produce undesirable short periods. The latter are

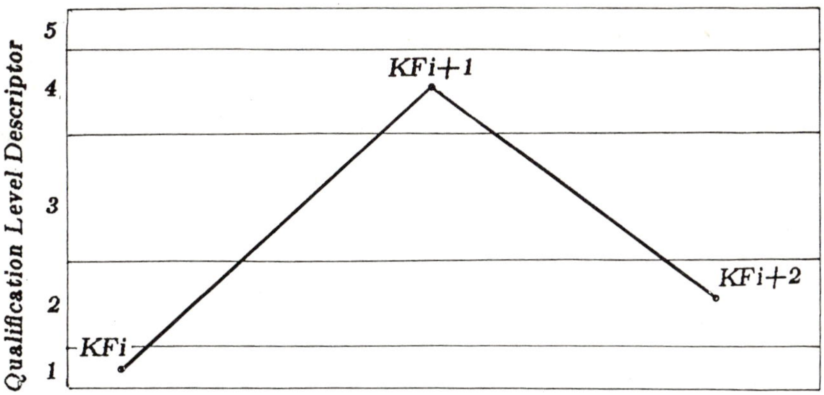

Figure(11a) *Two periods bounded by three sequential key frames, resulting from the curve approximation of the static data. In each period, the dynamic behaviour of the characteristic is constant. Between KF(i) and KF(i+1) the characteristic is increasing. During the next period it is decreasing.*

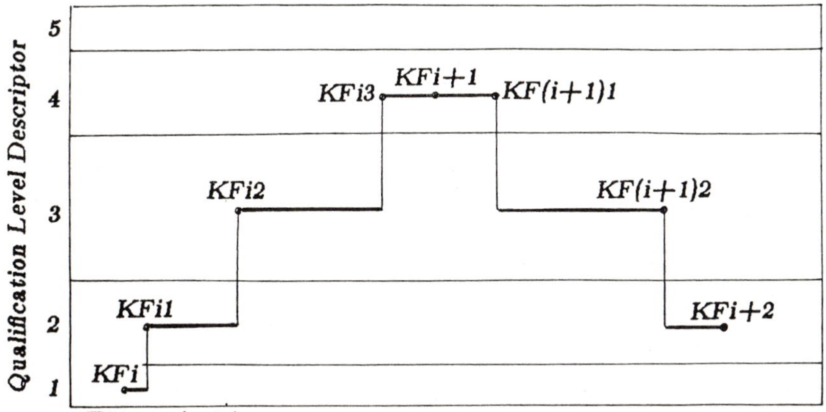

Figure(11b) **Splitting the above periods into a sequence of periods, where in each period the property has the same qualification descriptor. By further analysis, these periods may be merged using high level representational rules.**

Figure(11) Splitting periods in order to generate a descriptive summary of the global changes in a specific property. (a) input, (b) output.

usually caused by noise or changes due to experimental conditions. These periods can be merged according to other rules. For example, if a specific property has the same qualifier for a sequence of periods, then these periods can be merged. The property during the merged period can be described by the same descriptor, and the key frames updated. This merging strategy may be modeled by the following rule:

RULE:

IF $Q(P_k, T_{i,i+1})$. EQ . $Q(P_k, T_{i+1,i+2})$. EQ ... EQ .
$Q(P_k, T_{j-1,j})$. NE . $Q(P_k, T_{j,j+1})$

==then==> 1) MERGE:

$T_{i,i+1} + T_{i+1,i+2} + \ldots + T_{j-1,j}$ -INTO-> T_{ij}

2) $Q(P_k, T_{ij})$ <--- $Q(P_k, T_{i,i+1})$; | Assign a qualifier to the merged period.

3) $KF_{u-j+i+1}$ <--- KF_u, u=j,n | Updating

4) DELETE KF_u, u=[n-(j-i-1),n

5) n <--- n-(j-i-1) | key frames

where n is the number of key frames. Note that T_{ij} is the merged period that includes the periods $\{T_{i,i+1}, T_{i+1,i+2},\ldots,T_{j-1,j}\}$, and the key frame KF_j is updated to KF_{i+1}, KF_{j+1} to KF_{i+2}, ... and so on.

The above is an example of a simple rule, based only on the dynamic data of the different properties. The second step in creating global periods, is concerned with distinguishing the significant changes from the irrelevent ones. In this case, additional rules would utilize the dynamic description of the different properties, in conjunction with logical knowledge constraints. For example, an inference process could eliminate the very short events. Thus, suppose a specific property P_i is measured through three sequential periods T_{01}, T_{12}, and T_{23}. If the time T_{12} is very short relative to both t_{01} and T_{23} ($T_{01}, T_{23} >> T_{12}$), then we might eliminate T_{12} by one of two actions: first, by merging all three periods T_{01}, T_{12}, T_{23} into one (T_{03}), or second, by merging T_{12} to either T_{01} or T_{23}. Thus, if the cell was described in three sequential periods as SMALL, VERY SMALL, and SMALL, then a merging of these periods would result in small as the description. This may be achieved by the following rule:

RULE:

IF (T_{01} . GTT . T_{12} . LTT . T_{23}) . AND .
$Q(P_i, T_{01})$. EQ . $Q(P_i, T_{23})$. AND .
$Q(P_i, T_{01})$. NE . $Q(P_i, T_{12})$,

=then=> 1) MERGE THE THREE PERIODS INTO T_{03}

2) UPDATE THE KEY FRAMES

where $T_{03} = T_{01} + T_{12} + T_{23}$, and $Q(P_i, T_{13}) = Q(P_i, T_{01}) = Q(P_i, T_{23})$. Figure(12a) illustrates the function of this rule. In this figure, the values $Q(P_i, T_{13})$, $Q(P_i, T_{13})$, $Q(P_i, T_{01})$, and $Q(P_i, T_{23})$ are presented as Q_1, Q_2 and Q_3, respectively. The details of the merging action, steps 1 and 2 in the above rule, one similar to that described above.

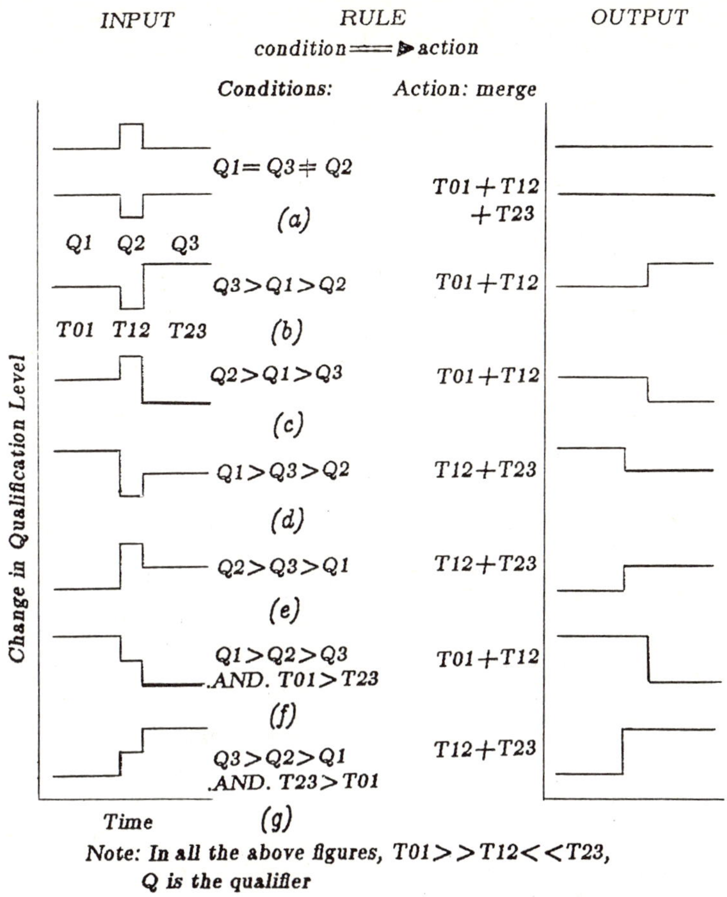

Figure(12) Using condition==>action rules to eliminate irrelevant and noisy changes.

A second action, in which T_{12} may be merged to either T_{01} or T_{23}, is based on logical inference. For example, if the maximum extent of the cell was described in three sequential periods as being SHORT, VERY SHORT, MEDIUM, then one may logically deduce that it has changed from SHORT to MEDIUM, if the description of the very short period (T_{12}) is the same as that of either the preceding or following period. Thus, if the description of the VERY SHORT period is closer to the SHORT period rather than to the MEDIUM period, period T_{12} may be merged to T_{01}. This logical deduction may be accomplished by the following rule:

RULE:

IF (T_{01}) . GTT. T_{12}, . AND. T_{12} . LTT. $T_{23})$

.AND. (Q_1 . GT . Q_2 . AND . Q_2 . LT. Q_3 . AND Q_1 . LT . Q_3),

.OR. (Q_1 . LT . Q_2 . AND . Q_2 . GT. Q_3 . AND Q_1 . GT . Q_3),

=then⇛ 1) MERGE PERIODS T_{01} AND T_{12} INTO T_{02},

2) UPDATE KEY FRAMES

where $T_{02} = T_{01} + T_{12}$, and $Q(P_i, T_{02}) = Q(P_i, T_{01})$.

Figure (12b) illustrates the function of this rule. Other examples of similar rules are shown in Figures (12c-12g).

The final step in this analysis is to characterize the data and describe them symbolically. For example, the time for each period T_{ij} may be normalized and described symbolically as VERY SHORT, SHORT, MEDIUM, LONG, or VERY LONG. The properties used to describe the cell shape must be translation, rotation, and size independent. From our experimental work, we have found that the most efficient properties are [Youssef, 82]:

(i) Circularity

(ii) Average Bending Energy

(iii) Angle Regularity

(iv) Elongation

(v) Number of Concave Angles

(vi) Number of Subparts

These are referred to as information nonpreserving [Pavlidis, 76].

The normalized static data for each property are analyzed according to the approach discussed above. That is, splitting, merging, and finally generating summaries describing their global changes. Typical examples of these summaries are given in Description (2). Examples of cell shapes that are described on the basis of different qualifiers are given in Figures (13a-13b).

3.4 Membrane Shape: Global Change Description

The task of describing the shape of the cell membrane is similar to that of describing the silhouette of an arbitrary shape. From the previous analysis of the individual shape properties, one can see that each of them describes only a specific characteristic of shape. But the actual change in object shape does not necessarily match the equivalent change in a particular property. In other words, we may find that many different shapes have the same value for a specific property. In fact, there is no known single shape property that gives a unique value for each different shape.

In order to illustrate the above observation, we use as an example, one of the earliest and most established shape descriptors, that is, circularity. Figure (14) shows twenty different shapes designed such that all of them have the same area and perimeter. From this figure, one can see that a wide variety of different shapes can have the same property value. Therefore, in order to generate different descriptions for different shapes, we must use more than one property. The perfect description is the one that fits only one shape.

Our approach to this problem is based on forming a mathematical expression which can be computed from the individual shape properties. This expression represents the degree of membrane shape complexity.

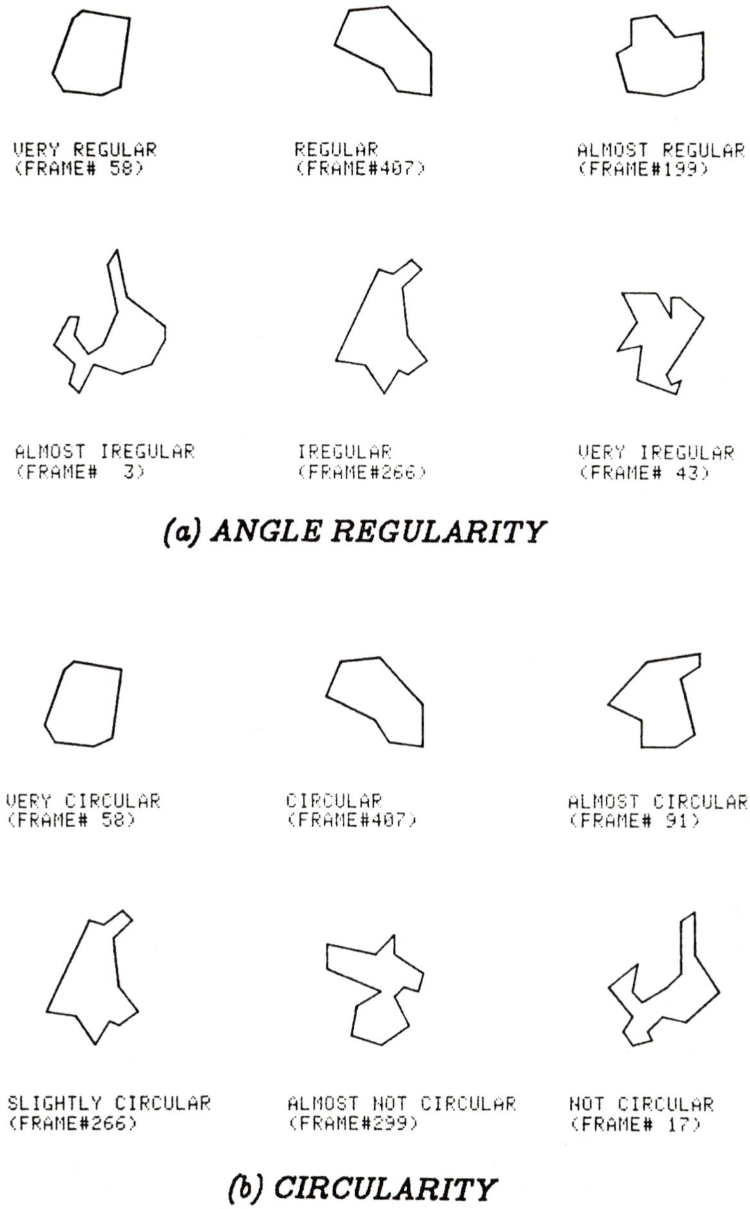

Figure(13) Examples of cell shape property characterization and description. (a) Angle regularity (b) Circularity.

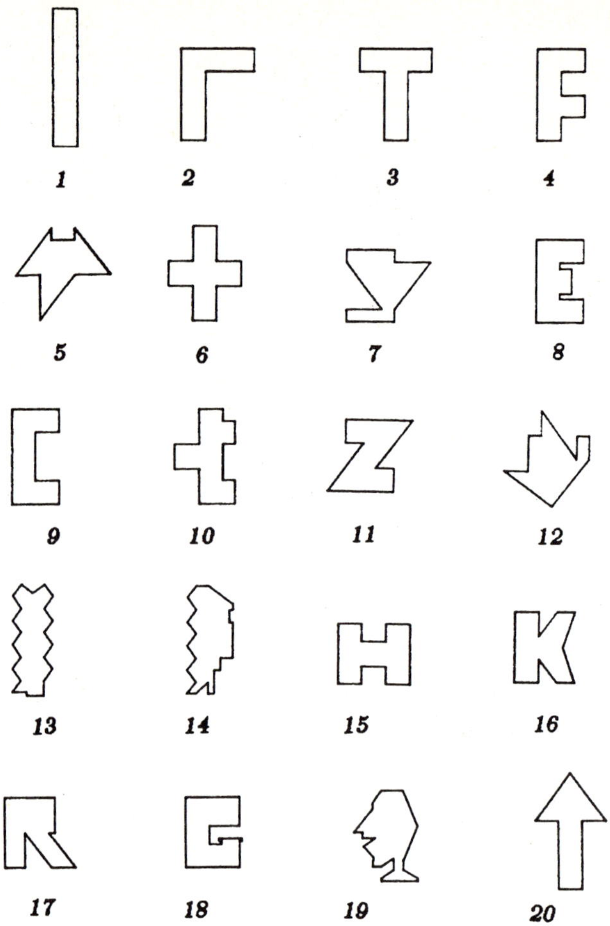

Figure(14) **Different shapes having the same area and perimeter, and hence the same circularity.**

In order to define a measure for an arbitrary shape, we choose the minimum set of properties based on two criteria: (a) maximum discrimination between the different shapes, and (b) descriptive power related to shape complexity. To satisfy these criteria, we have used the shapes given in Figure (14) as a training set. We computed the value of each shape property for every shape in the training set. Then, we selected the properties that: (a) produced the same value for a minimum number of shapes, and (b) sorted the complexity of the shapes in a manner similar to that of human sorting. In this experiment, we did not include circularity, because the shapes were originally designed to give the same value for circularity.

The following shape properties were included in the membrane shape measurement: circularity, average bending energy, and angle regularity. Figure (15) is a sort of the shapes according to complexity, using each individual property. To combine the three selected properties into one expression, we have repeated the above experiment, using the geometric means of the properties. The result is shown in Figure (16).

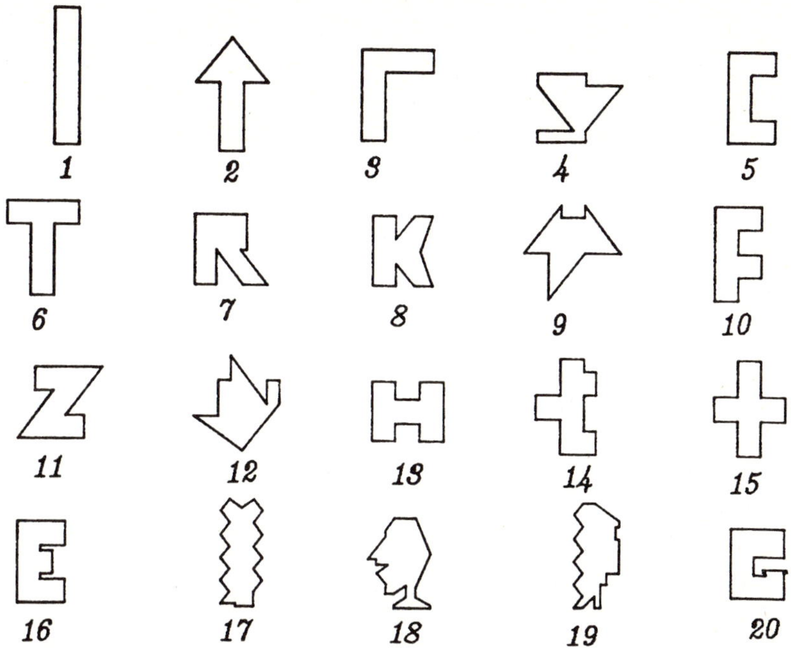

Figure(15a) *Different shapes sorted according to their complexity using average bending energy.*

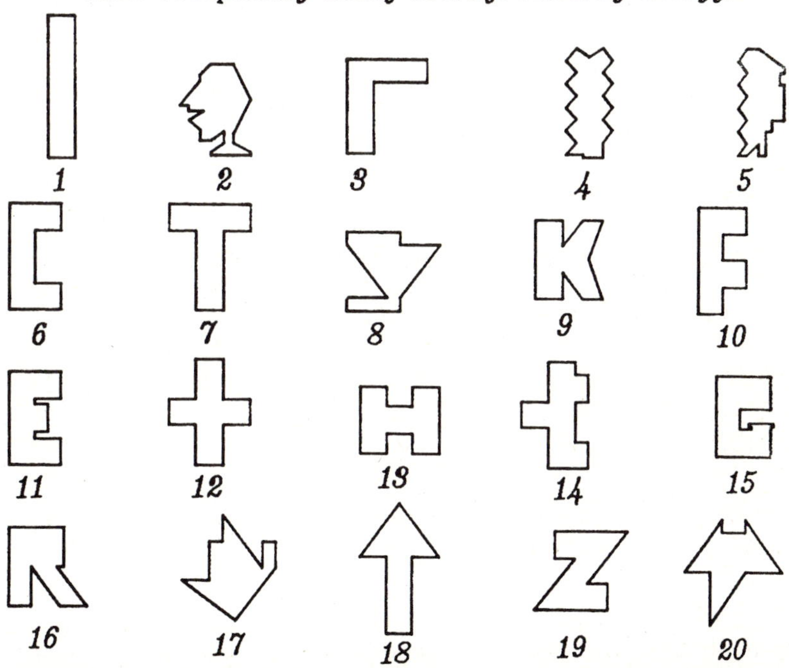

Figure(15b) *Different shapes sorted according to their complexity using angle regularity.*

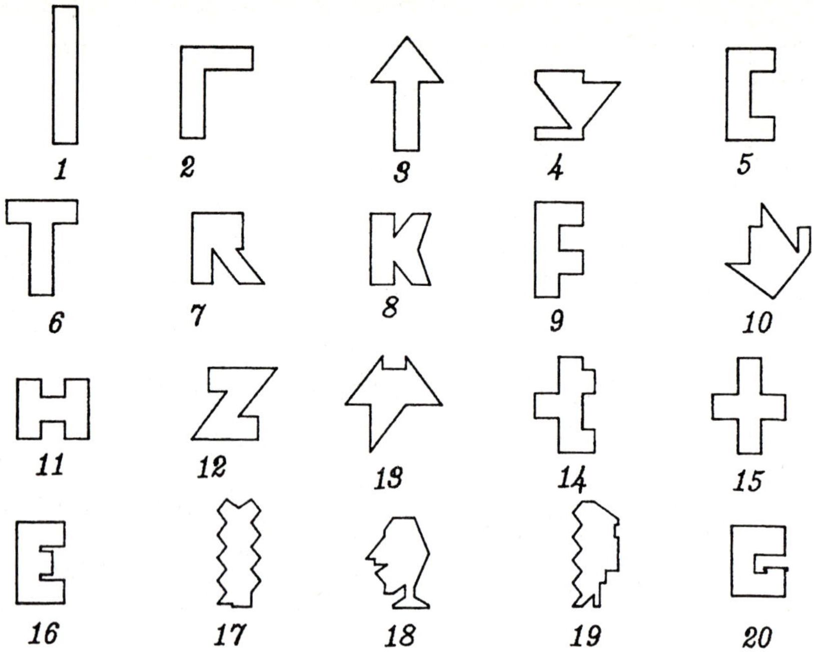

(b) Sorting using geometric mean.

Figure(16) Different shapes sorted according to their complexity using the geometric mean of the average bending energy and angle regularity.

The geometric mean of the selected properties shown in Figure (17) can be computed from static data. This shape measure can be analyzed in a similar fashion to the individual shape properties in order to generate a summary of the global changes in the cell membrane. As we observed above, the process includes detecting key frames and assigning descriptors to both the dynamic trends and the actual values of membrane shape. The different symbolic qualifiers that are used to describe the membrane shape are:

VERY SIMPLE, SIMPLE, ALMOST SIMPLE, ALMOST COMPLEX, COMPLEX, VERY COMPLEX. Figure (18) illustrates examples of cell shapes that are characterized by the different descriptors. A typical example of the generated summary of the global membrane shape changes is indicated in Description (2).

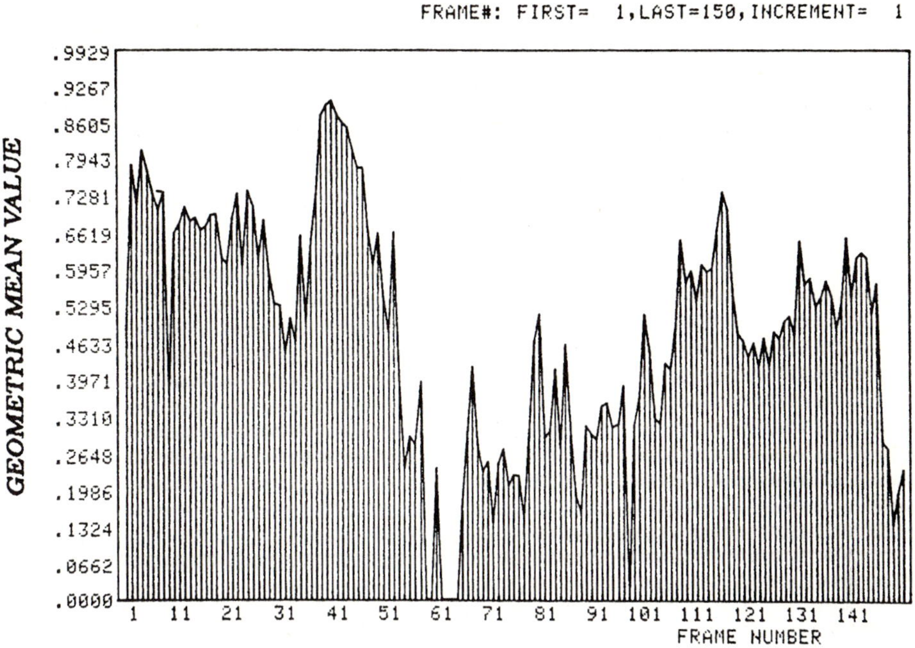

Figure(17) ***Geometric mean of circularity, average bending energy, and angle regularity computed for cell shapes in a sequence of 150 frames.***

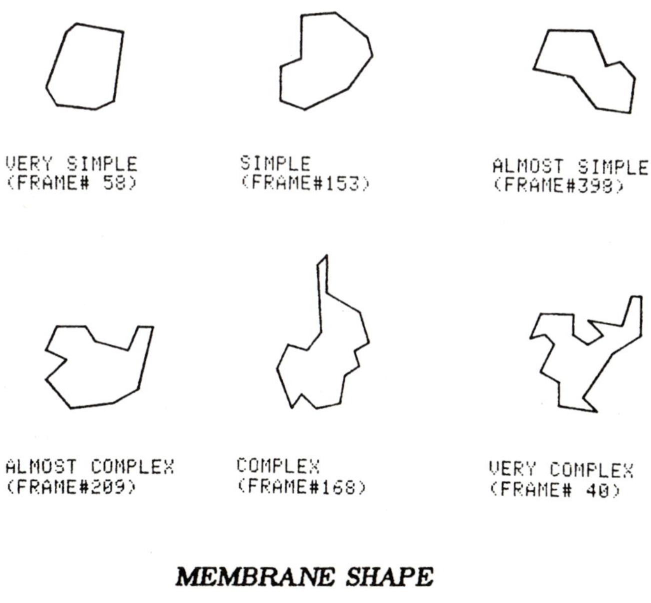

Figure(18) **Examples of cell shape characterization.**

4. STRUCTURAL ANALYSIS

4.1 Matching

The primary objective of analyzing the structural changes in a moving cell is to characterize and describe the dynamic behaviour of the different pseudopods that are formed during cell locomotion. Also, we need to study the relationships between the changes in the shape of subparts, their movement, and the global locomotion of the cell. The approach taken is based on the same philosophy as syntactic pattern recognition techniques. These analyze patterns by a parsing process of hierarchical decomposition. Such techniques appear to be quite amenable as a strategy for cell shape description, including the global structural and membrane shape changes.

Quantification and characterization of the structural changes of a moving cell can be accomplished in three stages of analysis: static, incremental, and global. In static analysis, the cell is decomposed into its primitive subparts. The latter can be represented as a labeled graph that conveys the topological properties of the cell. An example using the sequence in Figure (19a) is shown in Figures (19b and 19c). The incremental analysis is responsible for detecting and quantifying the structural changes between two sequential frames.

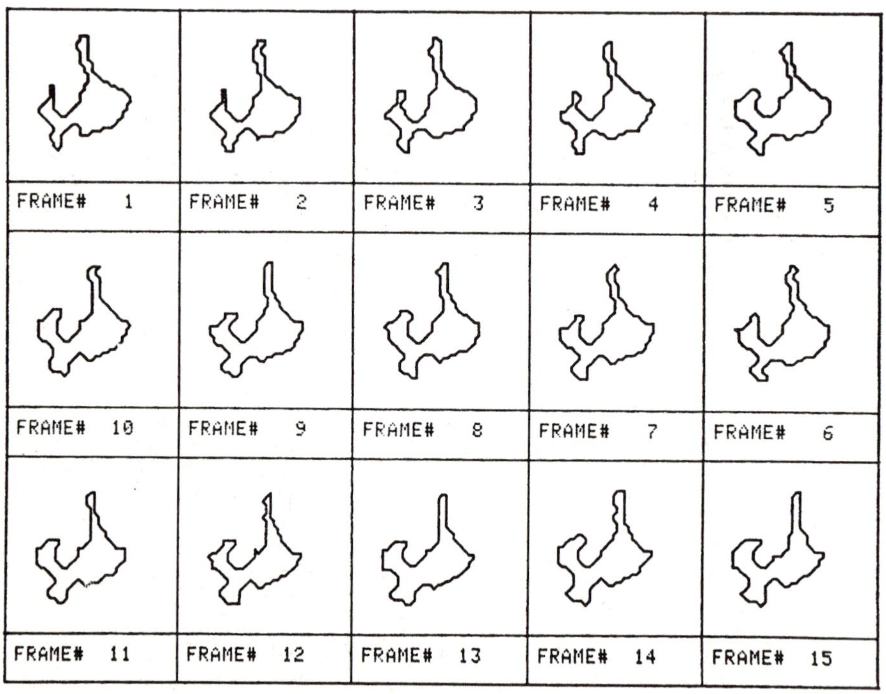

Figure(19a) Eight connected boundaries of a sequence of cells.

To accomplish this, the labeled graphs in both frames are matched in order to detect the correspondence between the different subparts. This is achieved as follows:

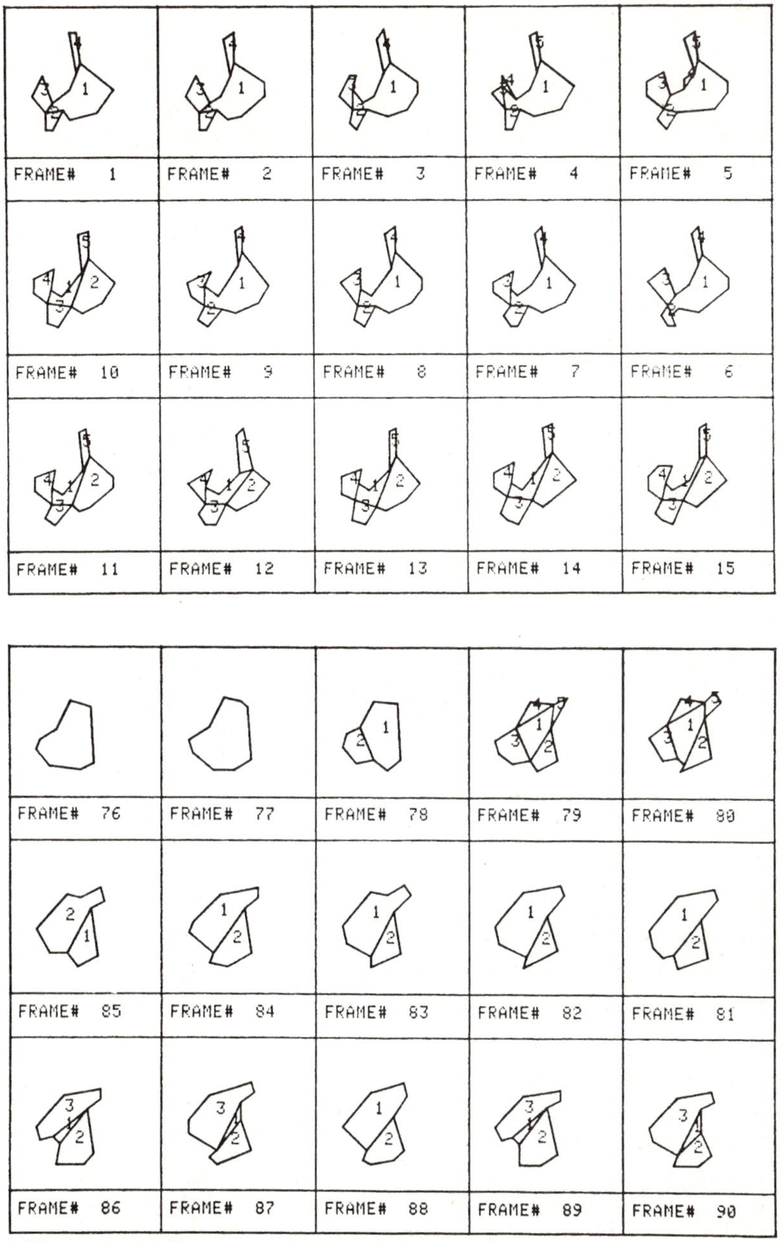

Figure(19b) Decomposition of the cell in a sequence of frames. Note that frames 76 and 77 have simple shapes which cannot be decomposed.

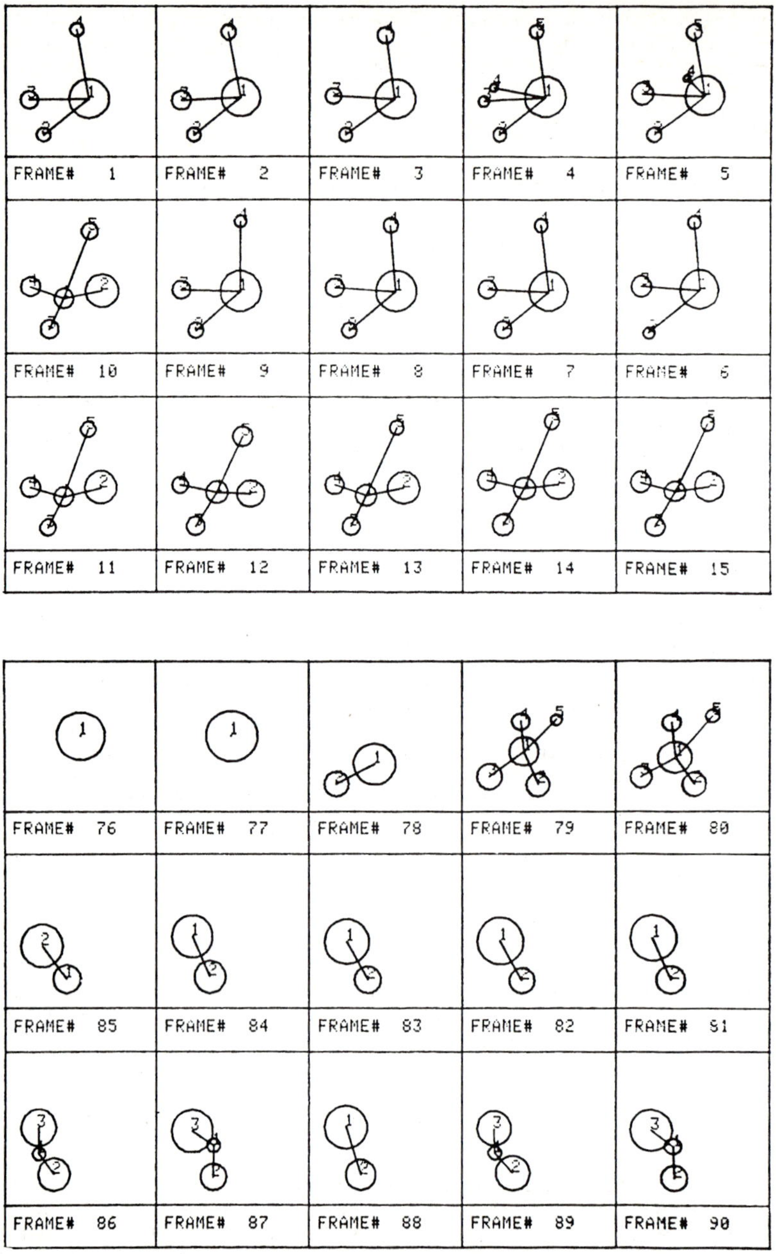

Figure(19c) Graphs representing the geometrical structure of the cell.

(1) Consider the subparts of the first frame as models (prototypes). For each subpart i in the second frame, find the subpart j in the first frame which yields the best match (M_{max}). M_{max} is given by

$$M_{max} = \max (M_{i1}, M_{i2}, \ldots, M_{ij}, \ldots, M_{in}), \tag{21}$$

M_{ij} is the match between subpart i (second frame) and subpart j (first frame), and is defined by

$$M_{ij} = [\ \sum_{k=1}^{k=m} \left| \frac{P_{k(i)} - P_{k(j)}}{P_{k(i)} + P_{k(j)}} \right| \cdot W_k] \ / \ m. \tag{22}$$

Here m is the number of subpart properties, and $P_{k(i)}$, $P_{k(j)}$ are the values of k for subparts i and j, respectively. The subpart properties which are used in this matching process are: area, perimeter, circularity, regularity, average bending energy, centroid coodinates, base-line, connective-line, and orientation. We may note that the first six properties pertain to shape, whereas the rest specify the relative position of a subpart in the cell structure.

(2) Repeat step number (1), considering the subparts of the second frame as the models of the subparts in the first frame.

(3) Subparts i and j correspond if $M_{ij} = M_{ji}$. Thus, two subparts that have been matched in both steps (1) and (2) are said to correspond.

Then, the total structural match (SM) of the cell between two sequential frames is computed as a percentage value which depends on the number of corresponding subparts and their degree of match. Thus, SM will equal 100% if (a) the cell has the same number of subparts in both frames, (b) there is a one-to-one match between all subparts, and (c) there is no change in any of the subpart properties. The structural match of the cell may equal 0% if there are no corresponding subparts between the two frames. The value of SM is computed as:

$$SM = (M_1 + M_2 + \ldots + M_k)/m, \tag{23}$$

where M_1, $M_2 \ldots, M_k$ is the percentage match of the corresponding subparts (see equation (22), k is the number of the corresponding subparts, and m the number of subparts in the second frame. This quantified measure of the cell structure match is described symbolically on a scale beginning with VERY DIFFERENT and ending with VERY SIMILAR.

4.2 Analysis

The objective of the structural analysis is to generate a summary of the significant structural changes, from which questions pertaining to the subpart and/or pseudopod characteristics can be answered. To achieve this, we first detect the frames where false structure has occurred. This is accomplished by an analysis of the incremental structural changes. Use is made of representational rules that incorporate both dynamic data and constraint knowledge pertaining to the cell under consideration. Most cases of false structure are corrected by feedback to the lower level decomposition process. Then, observable changes in the properties of each subpart are analyzed and described. The final step in this analysis is to generate a description of the significant changes in the structure of the cell and its primitive subparts.

Two main objectives are cited: (a) To analyze the static structure data and the incremental structural matching data in order to detect and correct (if possible) false structure resulting from the low level decomposition. (b) To find the sequence of frames in which individual subparts have appeared. These two issues are discussed below.

Errors in the low level decomposition can produce false structures. This is mainly caused by artifactual shape changes due to noise, image segmentation, and/or the inherent three-dimensionality of the cell. These artifactual situations cannot be easily detected by the static analysis stage. However, at the incremental analysis stage, false structures can be interpreted as ambiguous situations. These can be detected through a low value of the cell structural match between two sequential frames. On the other hand, if a false decomposition were to appear in several sequential frames, it obviously could not be detected from the incremental matching.

Our approach to detecting the cases of false structure uses representational rules that utilize both the dynamic data and constraint knowledge pertaining to the structure of the cell. For example, if the structural match $SM_{i-1,i}$ between two sequential frames (i-1) and (i) is less than a specific threshold value E_m, this indicates either a false decomposition at frame (i) or a significant structural change. This ambiguous situation can be addressed by the incremental analysis at frames (i) and (i-1), by using the following rule:

RULE

IF $SM_{i-1,i}$. LT. E_m

==then==> SIGNIFICANT STRUCTURAL CHANGE

. OR. FALSE DECOMPOSITION

However, this can be further clarified at the global level by the following rule:

RULE:

IF $SM_{i-1,i}$. AND. $SM_{i,i+1}$. LT. E_m

==then==> FALSE DECOMPOSITION AT FRAME (i)

where $SM_{i-1,i}$, $SM_{i,i+1}$ are the structural matches between frames (i-1), (i) and (i), (i+1), respectively, and the threshold value E_m specifies the acceptable structural match SM.

These two rules are based on the cell structural data which are derived from frames (i-1), (i), (i+1), and the constraint knowledge pertaining to the cell dynamic behaviour. For example, a pseudopod or subpart of the cell membrane cannot be formed and deformed in three sequential frames (1.5 seconds in this particular application).

Figure (20) shows examples of sequential frames where false decompositions have been detected. In the figure, the structural matching of the cell in the false frames with the previous and the successive ones is indicated. The bound E_m (normally about 0.8) defines the distinction between false and appropriate decomposition[1].

When a false structure is detected, more information about the cell structure and its match with the previous and successive ones is extracted. For example, information such as OVERDECOMPOSITION or UNDERDECOMPOSITION can be computed and utilized by the low level decomposition processes to produce a correction. Examples of the rules that accomplish this objective are:

----

[1] See [Youssef, 82] for method of computing E_m.

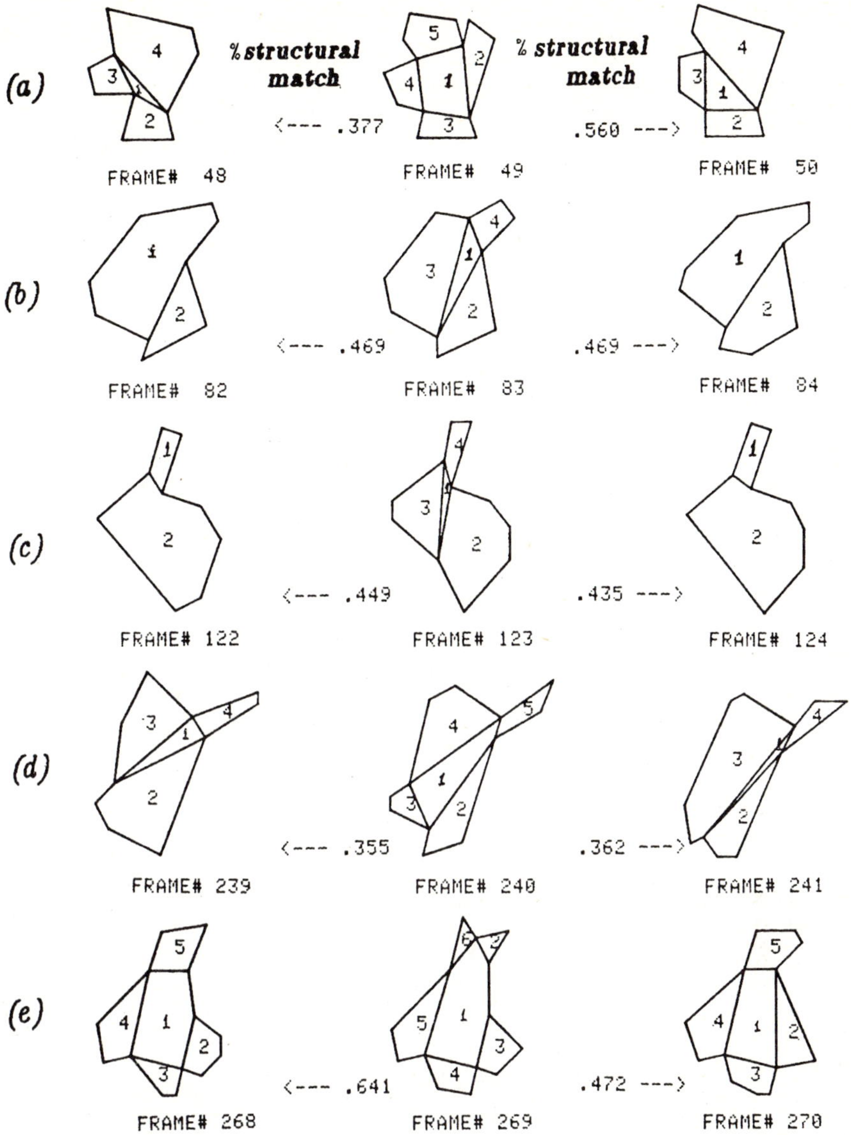

Figure(20) *Examples of frames having false structure (decomposition) detected at the global structural analysis stage.*

Note that in each example the middle frames, where the false structure appears, is associated with a low value of structural match compared with both the previous and successive frames.

RULE:

IF FALSE DECOMPOSITION AT FRAME (i)

$.AND.\ NSP_{i-1}\ .EQ.\ NSP_{i+1}\ .LT.\ NSP_i$

==then==> OVERDECOMPOSITION AT FRAME (i)

RULE:

IF FALSE DECOMPOSITION AT FRAME (i)

$.AND.\ NSP_{i-1}\ .EQ.\ NSP_{i+1}\ .GT.\ NSP_i$

==then==> UNDERDECOMPOSITION AT FRAME (i)

where NSP_{i-1}, NSP_i, NSP_{i+1} are the number of subparts at frames i-1, i, and i+1, respectively. For example, in Figure (20) an overdecomposition is detected in each of the middle frames.

Using this type of feedback data, specific rules are then activated in order to modify the result of the original low level analysis. For example, in the case of the overdecomposition, the following rules will be activated:

RULE:

IF OVERDECOMPOSITION AT FRAME (i)

==then==> 1) MERGE SP_j and SP_k $j=1,2,\ldots,NSP_{i-1}$, $k=j+1, j+2,\ldots,NSP_i$

2) FOR EACH MERGE, MEASURE $SM_{i-1,i}$ and $SM_{i,i+1}$

3) SELECT THE MERGE THAT RESULTS IN MAX. OF $[SM_{i-1,i}, SM_{i,i+1}]$.

Figure (21) shows the result of activating this rule on the cases shown in Figure (20). The corrected structure and the resulting modified structural matches are shown in Figure (22). We may note in Figure (22) that a false structural change has occurred between frames 240 and 241; however, a significant change in structure has occurred between frames 239 and 240. Using the above rule, most occurrences of false structure can indeed be corrected. Nevertheless, in some situations it may fail, especially if the false decomposition is due to noise or artificial changes. In this case, the frame is simply deleted:

RULE

IF $SM_{i-1,i+1}\ .GT.\ SM_{i-1,i}\ .OR.\ SM_{i,i+1}$

==then==> DELETE FRAME (i) FROM THE SEQUENCE

where $SM_{i-1,i+1}$ is the structural match between frames i-1 and i+1.

The result of activating the above set of rules can be one of the following:

(a) The structure of frame (i) is corrected and the matching is improved; this is the most common case.

(b) The false decomposition of frame (i) cannot be corrected, but frames (i-1) and (i+1) exhibit an acceptable structural match. In this case, frame (i) is considered as noisy data and deleted from the sequence.

(c) The structure of frame (i) cannot be corrected, and the decomposition in frames (i-1) and (i+1) do not match. This case is not very common. However, if it does occur, it will cause noise which can be removed by a global analysis similar to that described in Section 3.2.

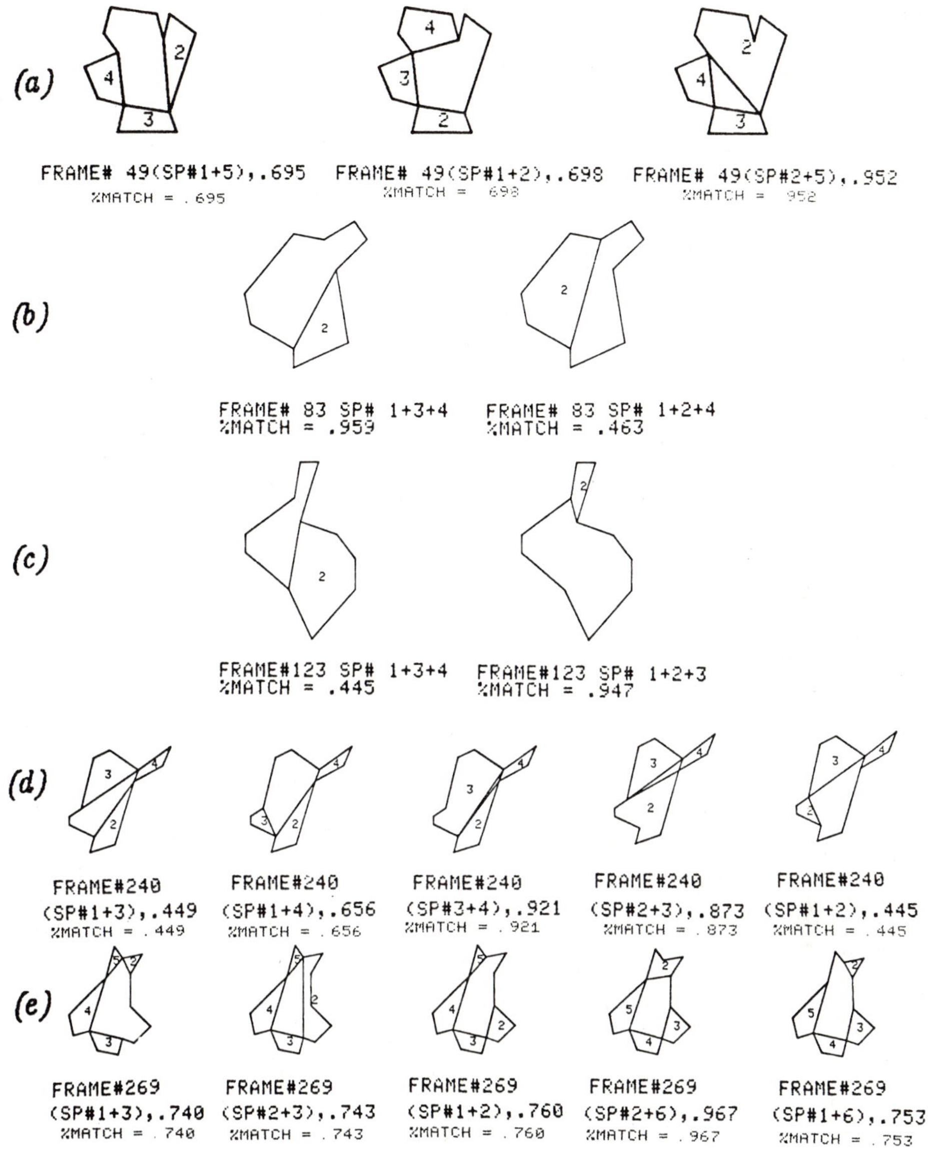

***Figure(21)** Possible variations in the decompositions of the cells having false structure in Figure(20).*

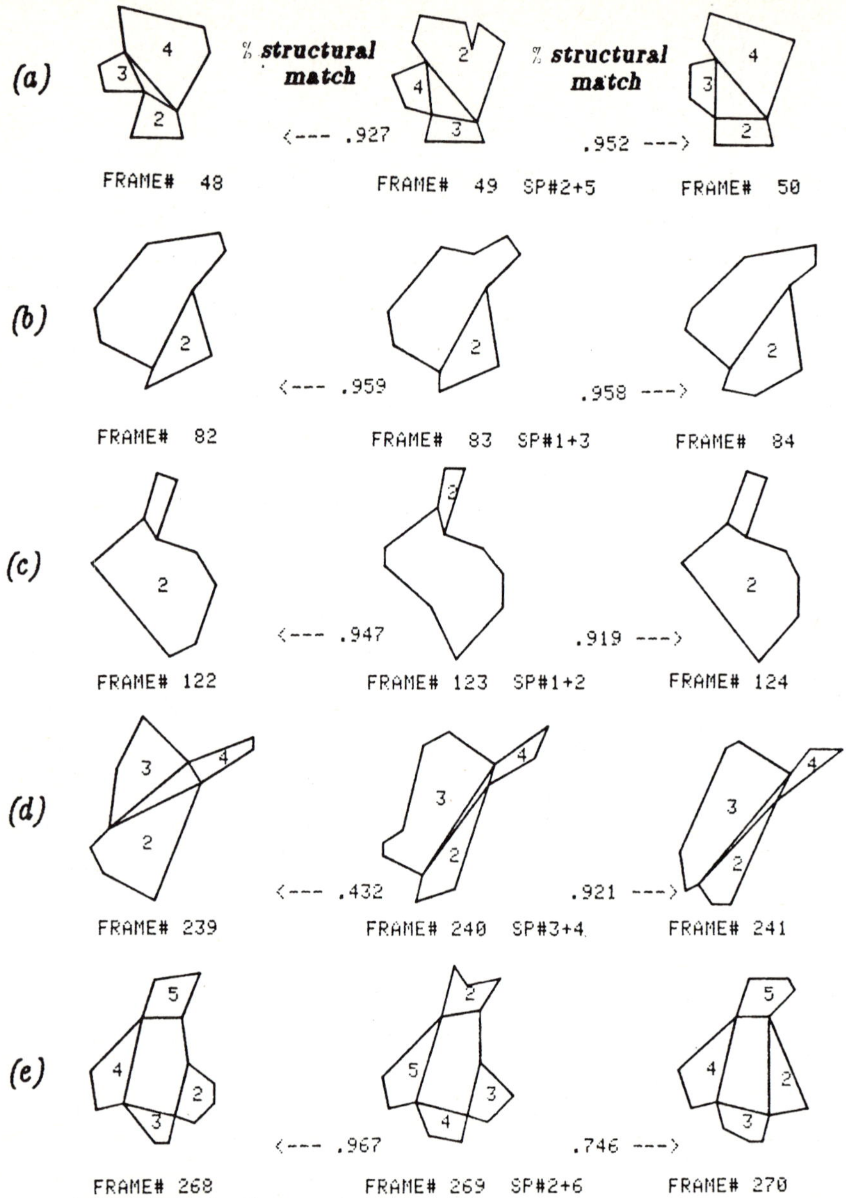

Figure(22) Correcting the false structure of the examples shown in Figure(20).

Note that in case (d), the structural match has only been corrected between frames 240 and 241. There still remains a significant structural discrepancy between frames 239 and 240. Thus, subpart 3 in frame 239 has merged with the cell body, and in frame 240, the new subpart 2 appears.

The output of the structural analysis is a set of key frames representing the times at which a subpart has appeared or disappeared. In addition, we are interested in characterizing the dynamic behaviour of the individual subparts. Therefore, each one is extracted and its properties analyzed. Then a subpart can be classified as being (a) irrelevant change or noise, (b) part of or the cell body, or a (c) pseudopod.

We are specifically interested in those subparts which may be good candidates for pseudopods. This distinction is based on criteria which embody the expert's knowledge. For example, the subparts that appeared for MEDIUM, LONG, or VERY LONG period are selected. Also, those which started after the cell had a VERY SIMPLE shape (one part) are likely candidates. In fact, these are the criteria which are currently in use. This is because, at this stage of research, our knowledge about the pseudopod characteristics is limited. With further experimentation more sophisticated rules could be developed to improve the selection of the pseudopod candidates.

4.3 Pseudopod Characterization

The main goal of characterizing the individual subparts of the cell is to determine whether or not a specific subpart is a pseudopod. In order to accomplish this, each subpart is treated individually as a moving object. The properties pertaining to their location, shape, and topological structure are analyzed using similar techniques to those for the entire cell.

In order to characterize a subpart as a pseudopod, we use representational rules that utilize the dynamic data associated with it, as well as constraint knowledge pertaining to pseudopods in general. The function of these rules is to compare the description of specific subpart properties to the pseudopod constraint knowledge. For example, a pseudopod has an area that should not exceed half of the cell area, has elongated shape, concave corners, short baseline, and its main axis is almost normal to the cell body. However, these properties change gradually through the growing and contracting of the pseudopod. For example, a pseudopod starts with a relatively long base line, very small area, and main axis parallel to the cell body. Thus, we cannot employ the properties of a specific subpart in a static frame to decide whether it is a pseudopod or not. We have used the normalized values of the baseline and relative area to characterize a subpart as a pseudopod or part of the cell body, by using the following rule:

RULE:

```
IF            BASE LINE (SHORT . OR. VERY SHORT)
              .AND. AREA (SMALL. OR. VERY SMALL)
==then==>     SUBPART  <---PSEUDOPOD
IF            BASE LINE (LONG .OR. VERY LONG)
              .AND. AREA (LARGE .OR. VERY LARGE)
==then==>     SUBPART <---CELL BODY
```

In the analysis, a subpart is characterized by a percentage value PCB indicating the confidence of its interpretation as a pseudopod or cell body. This value is computed as:

$$PRA_{j,i} = PA_{j,i} / A_i \tag{24}$$

$$PRB_{j,i} = BL_{j,i} / PP_{j,i} \tag{25}$$

where A_i is the cell area in frame (i), and $PA_{j,i}$), $BL_{j,i}$, $PP_{j,i}$, $PRA_{j,i}$ and $PRB_{j,i}$ are the area, baseline, perimeter, relative area and relative baseline of pseudopod (j) in frame (i), respectively. Using the values of PRA and PRB can be computed as:

$$PCB_i = [PRA_i \cdot (1-PRB_i)]^{1/2} \tag{26}$$

In addition, we are interested in questions pertaining to the pseudopod kinetics. For example, we would like to know whether a pseudopod is growing, contracting, or stationary.

Suppose a specific subpart is characterized as a pseudopod during the period T_{ij} (t_i and t_j are the initial and final frame numbers of the pseudopod existence). Then the different properties which describe the changes in location, shape and structural relationships during this period are analyzed to generate a description of its dynamic behaviour.

At this stage of study, expert knowledge about growing, contracting, or stationary pseudopods is limited to a definition based on the change in its size. However, in many cases, although a pseudopod may be stationary, the size of the entire cell may change, which consequently alters the pseudopod size. The effect of this can be avoided by using, as a feature, the change in the relative area of the pseudopod compared to the entire cell. Using this parameter and the change in the absolute area of the pseudopod, the change in the pseudopod size can be quantified as a percentage value (PCA), such that:

$$PCA = \frac{PA_i - PA_{i-1}}{A_{i-1} - A_i} \tag{27}$$

where PA_{i-1}, PA_i and A_{i-1} are the areas of the pseudopod and cell in frames (i-1) and (i), respectively. This value is used by the following rule in order to characterize a pseudopod as GROWING, CONTRACTING, or STATIONARY:

RULE:

IF PCA . LT. E_1 ==then==> CONTRACTING

IF PCA . GE. E_1 . AND .LT. E_2 ==then==> STATIONARY

IF PCA . GE. E_2 ==then==> GROWING

Using the resulting descriptions generated by the analysis rules discussed in this section, a summary of the pseudopod characterization is obtained.

5. DYNAMIC BEHAVIOUR DESCRIPTION

The observable changes in the three basic aspects (locomotion, shape, and structure) of the dynamics of a white blood cell have been studied indivudually. However, in most cases, the change in any of these aspects is related to change in the others. For example, the shape and structure of the cell are functions of each other. Therefore, in order to understand its dynamic behaviour it is not enough to study the changes in each aspect individually. We wish to determine the relationship between the changes in the membrane shape, pseudopod characteristics, and the global locomotion of the cell. In this section, we will discuss the behaviour of a moving cell in order to answer questions pertaining to the domination of the different pseudopods in locomoting the cell.

A pseudopod can be defined as being dominant if it leads the motion. We have employed two measures to compute the domination. The first is based on two parameters, and the second on three. The first approach includes the following steps:

(a) Represent the dynamic motion of a pseudopod by a vector PV. This vector is the resultant of two vectors; the first representing the locomotion of the centroid of the pseudopod, and the second, the direction of growth or contraction. The latter can be computed from the increase or decrease in the pseudopod elongation in the direction of the main axis of the pseudopod.

(b) Represent the total cell locomotion during the period of pseudopod existence by the vector CV.

(c) Initiate the vectors PV and CV from a point t_i, where t_i is time (initial frame number) where the pseudopod commenced to appear.

(d) From the analysis of the vectors PV and CV, we can characterize the pseudopod domination. This analysis is similar to that described in Section 2.1 for the characterization of the chemotaxis behaviour.

In this case, the direction of the vector PV is the direction of influence, and the vector CV is used as the vector sum of the cell locomotion. Then, by using equation (12), the domination of a pseudopod in locomoting the cell is quantified and characterized as NOT DOMINANT, SLIGHTLY DOMINANT, ALMOST DOMINANT, DOMINANT, or VERY DOMINANT.

In the second approach, an additional factor can be considered in the pseudopod characterization. This parameter is the relative location of the pseudopod with respect to the entire cell. It is embodied by the vector PSV, which represents the connective line between the centroid of the pseudopod and that of the entire cell. The three vectors PV, PSV, and CV are shown in Figure (23). The domination DOM of a pseudopod is a function of the three vectors:

$$DOM = f\ (PV, PSV, CV). \tag{28}$$

Thus, given the cell locomotion, we assume that specific patterns in the pseudopod dynamics result in cell movement. Analyzing these vectors, a pseudopod can be characterized as VERY DOMINANT, DOMINANT, ALMOST DOMINANT, SLIGHTLY DOMINANT, or NOT DOMINANT. These symbolic qualifications are used to discriminate between the different classes of pseudopod domination. However, the terminology is not established and may be changed in the future when the characteristic behaviour of pseudopods is better understood.

A final step pertaining to pseudopod characterization is to study the global rate of formation and deformation of the different pseudopods, and the effectiveness of each in dominating the cell locomotion. We may examine the cell locomotion path and the vectors representing the dynamics of the different pseudopods in one vector diagram, as shown in Figures (24a) and (24b). The vectors representing the pseudopods branch from the cell path at the points where each pseudopod started to appear. From this figure, one can see the rate of pseudopod formation and the effectiveness of each in dominating the cell path. The cell locomotion in this particular figure was under the influence of bacteria located in the bottom left corner of the plane (South-West). Cell locomotion was characterized as POSITIVE CHEMOTAXIS. One can see from Figure (24) that the dynamic behaviour of the cell is due to the effect of the pseudopod dynamics. Also, we note that the pseudopods guide the cell towards the bacteria. Thus, if we consider the cell starting at the origin of the plane, in order to reach the bacteria, it should move in the South-West direction. As we can see from Figure (24), whenever the cell started to lose this direction, a pseudopod formed and guided the cell correctly.

The accuracy of the results obtained by this system were tested and evaluated on the basis of a comparison to those obtained by a biologist. This is discussed in the next section.

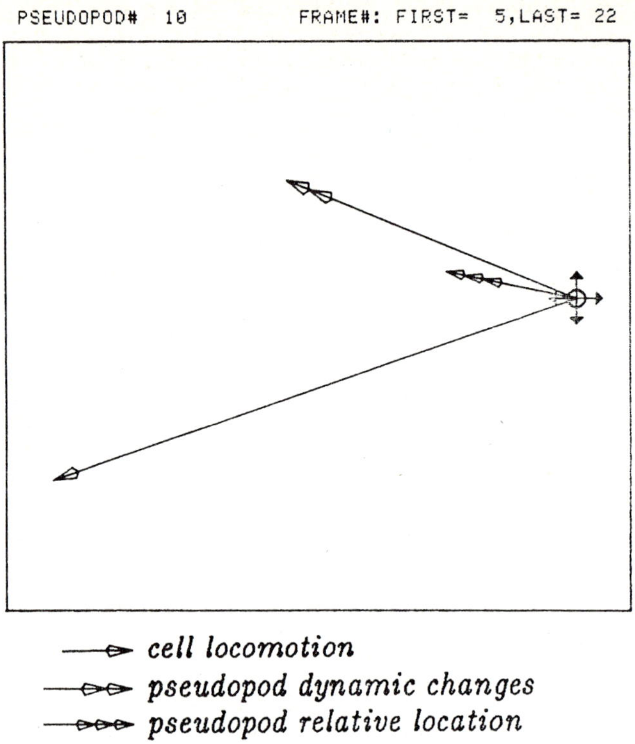

Figure(23) ***A vector diagram representing the dynamic changes of a pseudopod, its relative location, and the total locomotion of the cell.***

6. CONCLUSIONS

The system has been tested by quantifying and characterizing the dynamic behaviour of a polymorphonuclear leucocyte (PMN), as well as its pseudopod kinetics, in both chemotactic and neutral conditions. In the chemotactic case, the cell motion was recorded under the influence of a small sample of 'E.coli' (bacteria). With respect to the neutral condition, the cell motion was recorded without the presence of external chemotactic factors. The same experiments conducted for chemotactic locomotion were used to analyze the cell dynamic behaviour under the neutral conditions.

Examples of the result of these experiments are shown in Figure (25) and (26). In order to understand the difference between the cell dynamics under chemotactic and neutral conditions, we present in these figures the cell path using the same sampling parameters for both chemotactic and neutral conditions. The same number of frames (60) is shown in each case. From these figures, we can observe the following: (a) The cell path under chemotactic conditions is straighter and more directed than the one under the neutral conditions (see Figure 25). (b) The vector representing the total locomotion of the cell in the different directions of the plane is ALMOST SIMILAR for the neutral conditions. This situation is described by the system as RANDOM LOCOMOTION. In the chemotactic case, the vector is LONGER in the direction of the bacteria (see Figure 26). The vector sum of the cell locomotion is almost zero in the neutral case, whereas in the chemotactic condition it has a value in the direction of the bacteria. This is the situation even though a relatively short path is examined in the example shown in these figures. The amplitude and direction of the motion vector are utilized to quantify and describe the chemotaxis behaviour (see Section 2.1).

The quantification and characterization of the changes in location, shape, and structure of a moving cell that are obtained by this system have not been computed before, either visually (biologist) or automatically (computer). In order to test the accuracy of the resulting data, the output information at each stage of analysis was presented to a physiologist (P.N.). Most of the computer results agreed with his a priori knowledge about cell and pseudopod dynamic behaviour. However, some of these data, especially those pertaining to specific membrane shape and structural changes, cannot be evaluated at this time because they have not been reported before, and are not yet well understood.

The same sequence of frames that was analyzed by the system has also been examined visually by a biologist. We have compared the two sets of pseudopod descriptions, and a percentage error was estimated based on the differences in the characterization. The results indicate an error of 18% in the case where only the cell locomotion and pseudopod dynamic changes were used as characterization parameters. This error is reduced to 10% by using the relative location of the pseudopod as a third parameter.

From this comparison, we claim that the system has successfully characterized the kinetics of the pseudopods that contribute to the cell locomotion. Using the technique developed, it is now possible to study why, in random motion, one pseudopod becomes dominant and to compare these pseudopod characteristics with pseudopods during chemotactic locomotion. The approach, being applicable to other leucocyte types, i.e. lymphocytes, will enable one to study pseudopod kinetics in positive and negative chemotaxis.

It is observed that we have investigated the dynamic behaviour of a moving cell from a sequence of two-dimensional images. Most of the false segmentations, decompositions, and artifactual changes are due to the effect of the three dimensionality of the cell motion. Fortunately, in the present application, the three-dimensional motion of the cells is restricted to some extent, because these cells (PMN) require a substratum on which they flatten before commencing locomotion. However, in future studies, it might be necessary to employ rules that consider the three-dimensional changes in motion, shape, and structure of the moving object.

The structural changes in the cell morphology that occur during locomotion have not been previously reported in the literature. Furthermore, in spite of the importance and great interest in understanding the role that the cell membrane plays in the locomotion, there is no existing method for quantifying and analyzing the observable changes in the membrane shape. In this research, we have developed an image interpretation system capable of quantifying, analyzing, and describing the structural changes in the morphology of a moving cell.

The system has successfully provided all the quantification, description, and characterization information whereby basic questions pertaining to the cell's dynamic behaviour can be answered. This study might provide clues to the nature and distribution of "receptors" on or within the membrane which would be a vital link in the interaction between the external factors and cell internal processes. Also, it might lead to the understanding of the roles the cell membrane plays in the mechanisms which regulate the social behaviour of the cell.

REFERENCES

Badler, N., Temporal Scene Analysis: Conceptual Description of Object Movements, Dep. Computer Sci. University of Toronto, Rep. T.R. No. 80, 1975.

Branton, D., Molecular Interaction Governing Plasma Membrane, in "The Cell Surface: Mediator of Development Process", Subtelny, S. and Wessells, N.K., (eds.), Academic Press, 1980, pp. 3-8.

Bourguignon, L.Y., and Singer, S.J., Proc. Nat. Acad. Sci. U.S.A. 74, 5031-5035, 1977.

Burtynk, N. and Wein, M., Interactive Skeleton Technique for Enhancing Motion Dynamics in Key-Frames Animation, Communication ACM Vol. 19, pp. 504-509, 1978.

Edelman, G., Science 192, 218-226, 1976.

Gerisch, G. and Keller, H.U., Chemotactic Reorientation of Granulocyte Stimulated with Micropipettes Containing f Met-Leu-Phe, J. Cell Sci. 52, 1-10, 1981.

Levine, M.D., Youssef, Y.M., Noble, P.B., Boyarsky, A., The Quantification of Blood Cell Motion by a Method of Automatic Digital Picture Processing, IEEE, Trans. on Pattern Analysis and Machine Intelligence, Vol. PAMI-2, No. 4, Nov. 1980, pp. 444-450.

Levine, M.D., Youssef, Y.M., Noble, P.B., Understanding Blood Cell Motion, Computer Graphics and Image Processing, to appear.

Lewis, W.H., and Webster, L.T., Migration of Lymphocytes in Plasma Cultures of Human Lymph Nodes, J. Exp. Med. 33, pp. 261-269, 1921.

Nagel, H.H., Formation of an Object Concept by Analysis of Systematic Time Variation in the Optically Perceptible Environment, Computer Graphics and Image Processing, Vol. 7, 1978, pp. 149-194.

Nagel, H.H., Image Sequence Analysis: What Can We Learn from Applications? Report No. IfI-HH-M-79/80, Univ. Hamburg, Sep. 1979. To appear in "Image Sequence Analysis", T.S. Huang, (ed.), Springer Verlag, Berlin-Heidelberg, New York, 1981.

Nicolson, G.L., Biochem. Acta 457, 57-108, 1976.

Nicolson, G.L., Biochem. Acta 458, 1-72, 1976.

Pavlidis, T., "Structural Pattern Recognition", Springer-Verlag, New York, 1977.

Ramer, U., An Iterative Procedure for Polygonal Approximation of Plane Curves, Computer Graphics and Image Processing, Vol. 1, 1972, pp. 244-256.

Singer, S.J., and Nicolson, G.L., Science 175, 720-731, 1972.

Tsotsos, J.K., Mylopoulos, J.H., Covey, H.D., and Zucker, S.W., A Framework for Visual Motion Understanding, IEEE Trans. on Pattern Analysis and Machine Intelligence, Vol. PAMI-2, No. 6, November 1980, pp. 563-573.

Wessells, N.K., and Subtelny, S., The Cell Surface: Mediator of Development Process, Proceedings of the 38th Symposium of the Society for Developmental Biology, June 25-27, 1979, Vancouver, B.C., Canada.

Williams, L.T., Snyderman, R., Pike, M.C., Lefkowikz, R.J., Specific Receptor Sites for Chemotactic Peptides on Human Polymorphonuclear Leukocytes, Proc. Nat. Acad. Sci. U.S.A. 74, 1204-1208, 1977.

Yachida, M., Asada, M., and Tsuji, S., Automatic Motion Analysis System of Moving Object from the Records of Natural Processes, Proceedings of the Fourth International Joint Conference on Pattern Recognition, Nov. 7-10, 1978, Kyoto, Japan, pp. 726-730.

Zigmond, S.H., Levitsky, H.I., Kreel, B.J., Cell Polarity: An Examination of its Behavioural Expression and its Consequences for Polymorphonuclear Leukocyte Chemotaxis, J. of Cell Biology, Vol. 89, June 1981, pp. 585-592.

Youssef, Y.M., Quantification and Characterization of the Motion and Shape of a Moving Cell, Ph.D. Thesis, Elect. Eng. Dept., McGill Univ., May 1982.

ACKNOWLEDGEMENTS

The research of M.D. Levine was supported in part by the Medical Research Council of Canada under Grant No. MA-3236, and the Province of Quebec under Grant No. EQ-633. P.B. Noble was supported in part by a grant from the National Cancer Institute, Canada. Y.M. Youssef was supported by the Ministry of Education, Egypt.

THREE DIMENSIONAL IMAGING FROM COMPUTED TOMOGRAMS

H. OSWALD
Technical University
and
GRAZ Research Center

Institute for Image Processing and
Computer Graphics

Wastiangasse 6
A-8010 GRAZ AUSTRIA

INTRODUCTION

Image sequence processing is a very large field, with very different techniques. In common three different types of image sequences can be separated,namely the temporal, the spectral and the spatial image sequences. An excellent overview of the processing and analysis techniques for all kinds of image sequences is given by NAGEL 79 (1). This paper concentrates on the methods of imaging from spatial image sequences. Nevertheless our techniques can also be applied on other image sequences.

Spatial image sequences represent a three-dimensional (3-D) space. This space is built up by volume elements (voxels for short) and can be considered as a 3-D array. With the advent of a number of three-dimensional imaging technologies, three-dimensional digital images have become quite common in many scientific fields. A very large and popular source of such three dimensional data arrays is the area of computed tomography.

By using computed tomography, it is possible to measure the spatial distribution of the x-ray absorption coefficient within the human body (2). Usually such measurements are obtained by consecutively scanning parallel sections through the human body. A single cross-section or slice is then represented by a square matrix containing thousands of picture elements or pixels. These slices are usually viewed as monochrome or sometimes pseudocolored pictures. Compact structures such as organs or bones appear in different sections as regions of nearly constant gray values. The shape of these structures or objects generally varies from one cross section to another. To have an impression of the true appearance of an object, the observer has to synthesize mentally a picture of the three-dimensional shape. Even for the trained physician this could be very difficult and requires considerable imagination.

NATO ASI Series, Vol. F2
Image Sequence Processing and Dynamic Scene Analysis
Edited by T.S. Huang

In this article we discuss algorithms for defining, detecting and displaying the surface of objects in such 3-D arrays. Our software is designed for implementation on a minicomputer PDP-11/34 and is written in FORTRAN in a portable fashion. For displaying our results we used a DeAnza IP 5000 system connected to our PDP 11/34.

CURRENT TECHNIQUES

Several attempts have been made to design mathematical tools for the perspective and realistic display of three-dimensional objects inside the human body. Essentially there are two approaches to this problem. The difference stems from the different methods of representing objects in the three-dimensional image. In the first approach a set of contours is used to define the object regions in a sequence of slices. The boundary surface of the object is produced either by tiling contours in succesive slices with triangular patches (4) or by fitting surface elements to the contours by the method of lofting using cardinal splines (3).

In the second approach a set of volume elements represents an object in the three dimensional image (5). A boundary surface of the object consists of a set of faces of voxels. These voxels are determined by tracking through the connected set of faces comprising the boundary surface (6).

If we compare the two approaches there are also great differences in the resulting shaded-surface display. The first approach yields objects with a very smoothed and realistic appearence. However, this method has proved to be rather time-consuming. Furthermore the method cannot handle objects with arbitraryly complex shapes. But such complex objects are commonly encountered in medical applications.

The second method is not limited by the complexity of the shape of an object. The restriction in this method is in shading the object's surface, because a voxel has only six faces. Therefore a surface can only consist of three distinct directions for the shading model. These three directions are to the three possibly visible faces of a voxel. A more detailed comparison of the two methods is published by HERMAN (5).

Our approach is closely related to the second, but with the essential difference of interpreting the cubic voxels as single points in a discrete space. A surface of an object in this case is formed by a set of points sampled in some appropriate sense. We also can handle objects with very complex shapes. The surface shading model can generate more or less smoothed displays, only by changing the chosen neighbourhood of a surface element.

THE OBJECT SPACE MODEL

The object space or universe we are concerned with is a discrete 3-dimensional space or scene for short. It consists of cubic voxels in the size of the discretisation unit. These voxels are the basic units of our universe. Each of these voxels can have the value 0 or 1. The object space in our definition is therefore a binary scene with empty (value 0) or full (value 1) voxels.

The universe itself is a subspace of R^3. Together with the universe a rectangular coordinate system is determined. The origin of the universe is the point of intersection of the coordinate axis. In this universe to each voxel a triple (i,j,k) of integers can be assigned.

Definition: U is a universe or object space if

(i) $U \subset R^3$

(ii) $U = \{V(i,j,k) \mid V(i,j,k)=\text{cubic and } (1 \leq i \leq I) \text{ and } (1 \leq j \leq J) \text{ and } (1 \leq k \leq K)\}$

$i,j,k,I,J,K \in \mathbb{N}$

Now we have to establish the relation between a CT-scene and our universe.

Both are digitized 3-dimensional scenes. The differences are in the sizes of the voxels and in the range of values assigned to the voxels. In general a CT-scene has voxels in the size of 1.0 mm x 1.0 mm x 5.0 mm to 1.0 mm x 1.0 mm x 13.0 mm. The values for a voxel are in the range of -1024 to 1024 and usually are called Hounsfield Units.

To deal with these differences, we have to to some scene processing, the 3D analog to image processing. Thus we get a binary scene or universe out of a CT-scene. In our scene processing we use two operations: interpolation and segmentation.

Interpolation produces a scene in which the voxel size is different from that in the original scene. The densities assigned to the new voxels are estimates based on the values of the neighbouring voxels in the original. In our example the original has 1.0 mm x 1.0 mm x 5.0 mm voxels in a 160 x 160 x 10 scene. To get cubic voxels, we have to interpolate linear subslices between each two slices.

After interpolation the CT-scene has the same discretisation unit as the universe we defined.

Segmentation, as we use it, produces a binary scene, the second property of our universe. The advantage for segmentation of CT-scene is the well know range of density values of the various organs. Therefore one can use a priori information to segment the scene (9). For our segmentation (7), (8) we use a two step process. The first step is a two level threshold and the second is a region growing process. The result of scene processing is an universe with the chosen object.

OBJECTS AND THEIR SURFACES

After having defined the object space, we now indicate an approach to defining surfaces of objects.

For a human observer there is no difficulty to identify objects in three dimensional scenes. It is therefore surprising that it is far from trivial to give a precise mathematical definition of an object and its surface (5). In the large area of computer graphics, to which these problems belong, VOELCKER and REQUICHA (10),(11), classified the existing techniques to the following six categories:

(1) Primitive Instancing - Families of objects are defined parametrically; shape type and a limited set of parameter values specify an object.

(2) Spatial Enumeration - An object is represented by a list of the cubic spatial cells which it occupies

(3) Cell Decomposition - A generalized form of spatial enumeration in which the disjoint cells are not necessarily cubic or even uniform.

(4) Constructive Solid Geometry (CSG)- Objects are represented as collections of primitive solids (cuboids, cylinders, etc.) They are connected via the Boolean operations.

(5) Sweep Representation - A solid is defined as the volume swept by a 2-D or 3-D shape as it is translated along a curve.

(6) Boundary Representation - Objects are represented by their enclosing surfaces (planes, quadric surface, patches, etc.).

Our approach to define an object in the binary scene uses a technique in category two. Thereby an object is a set of voxels V(i,j,k) with the value 1. The specific point in our approach is that we interprete a single cubic spatial cell, a voxel, as a single point in the universe, disregarding its finite surface. By this assumption an object is formed by an O(3)-adjacent set of voxels and has always to be of the same order as the universe. In our case the universe is of order three and the objects therefore have to occupy volume. A plane or line (2-D objects) with one voxels thickness cannot exist in our universe.

If we project our object to an image plane, only the shape of the object will appear, without any structure on the surface. For a shaded display we use the surface and its direction.

Definiton: A surface S of an Object O is a set of points V with the following properties:

$$\text{(i)}\quad S = \{V(i,j,k)\,|\,V(i,j,k) = 1 \text{ and } |V(i-1,j,k)-V(i+1,j,k)| + |V(i,j-1,k)-V(i,j+1,k)| + |V(i,j,k-1)-V(i,j,k+1)| \geq 1\}$$

$$\text{(ii)}\quad \|S\| \geq 3 \quad \text{(numbers of connected points)}$$

The surface of an object consists of points which are border points of the objects. A border point has at least one 1-adjacent object point or at least one 1-adjacent non-object point. To avoid singularities of the surface, the number of connected surface points has to be not less than three. With these properties we have given an exact definition of a surface of an object in the universe.

In short, let us make some remarks of the discretisation errors. In most application areas the discrete 3-dimensional scene is only an approximation to the underlying domain of interest. Usually, we assume that there exists a function f(x,y,z) of three real variables and the value assigned to a voxel V(i,j,k) is an estimate of the average value of f(x,y,z). In computed tomography f(x,y,z) is the x-ray attenuation coefficent of the human body at location (x,y,z). The object of interest, an organ or a part of the skeleton will occupy some voxels completely, will miss some others and will partially intersect a third set of voxels. In the process of discretization some information regarding this true object is irrevocably lost. Nevertheless the recovery of its surface in the discrete 3-D scene should be the best approximation to the true one. All processes and techniques - scene processing and shaded displaying have to be understood before this background.

SHADED DISPLAY OF SURFACES

The realistic display of an object involves the calculation of shading parameters for the surface. There exists a large quantity of possibilities for a shading mode, but computer costs increase rapidly with the comlexity of the model (12). We use a shading model according to the Lambertian Law, which consists of diffuse illumination and diffuse reflection. In our opinion this model is satisfactory in medical applications for the display of organs of a CT-scene. In the following we describe the principles of our model.

The light energy Ep emanating from a surface point P can be considered as the sum of the diffuse illumination Epd and the diffuse reflection Eps:

$$\begin{aligned} Ep &= Epd + Eps \\ Epd &= Rp * Id \\ Eps &= (Rp * \cos\varphi) * Ips \end{aligned}$$

In these equations Rp is the reflectance coefficient at P, in the range of 0 to 1, Id is the diffuse illumination falling onto the entire scene and Ips is the energy arriving from a light source. The angle φ is the angle between the surface normal Np and a ray to the light source.(see Figure 1.)

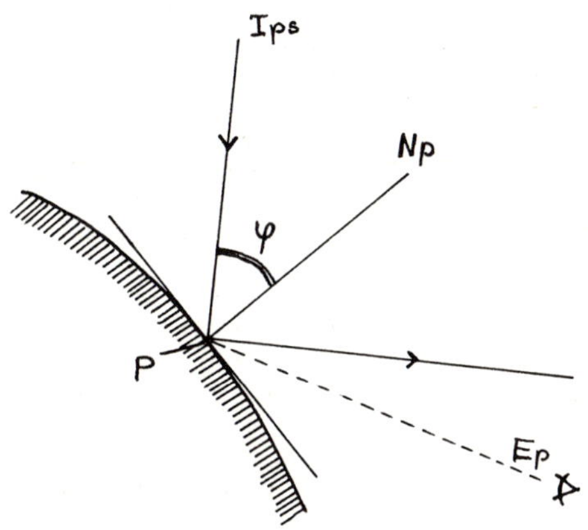

Fig.1.The shading of a point P

Given the shading model, we see that the normal vector on a surface point is used. But the direction of the surface in a point depends on the point's neighbours. A first order normal vector is implicitly defined in our surface definition in the form of

$$Np = \begin{bmatrix} ni \\ nj \\ nk \end{bmatrix} = \begin{bmatrix} V(i-1,j,k)-V(i+1,j,k) \\ V(i,j-1,k)-V(i,j+1,k) \\ V(i,j,k-1)-V(i,j,k+1) \end{bmatrix}$$

This definition of the normal vector generates a sample out of 26 distinct directions of a surface element. As we can show in our illustrations, this discretisation of an object surface is satisfactory for our problem. For a finer discretisation with a more smoothed surface, one may define the normal vector in a greater neighborhood than the first order neighbourhood. In the second order neighbourhood we get 98 distinct directions of surface elements. This problem is the 3-D analogue to chain-coding in 2-D space (13).

To a surface point P of our object O, we assigne now a six-tuple (i,j,k,ni,nj,nk) of integers. This six-tuple consists of the coordinates of the surface element and the normal vector in it. We calculate these values slice by slice from our binary scene and get a linear list of surface elements EL, sorted by increasing coordinates. The fact that coordinates increase is important for our hidden surface algorithm for reasons to be explained below.

PERSPECTIVE PROJECTION

In order to create a realistic view of an object, we implemented a perspective projection. This projection corresponds to natural view and camera imaging. In general, there is no restriction on projecting techniques to our object model. Sometimes one can find an advantage in orthographic projection, for example, if he wants to measure some distances in the object.

If we remember that the universe in our definition is a discrete three dimensional space and that the image space onto which we are projecting is a discrete, but two dimensional space, too, the interpretation of our objects as a set of points reduces the problem of projecting surfaces to the problem of projecting points. However,we have to guarantee that the projection of voxels, which are O(3)-adjacent in the universe, yields O(2)-adjacent pixels in the image space. To do this we project the diameter of the sphere enveloping the object space onto the image space. In more detail, we project those two orthogonal diameters, which are parallel to the image plane and one of them has to be parallel to the (i,j)-plane of the object space. The length of these projected diameters has to be chosen in such a way that it is as large as possible and still yields a connected projection. By this we calculate the size of the image plane. It is very important that the image plane has square formed discretisation units to provide a deformation of the object.(see Figure 2.)

By observing these considerations, the projection of our object, sampled by points, will give a correct view of this object. If one uses another projection as the perspective projection the same considerations have to be observed.

In the opinion of the author the great advantage of our projection technique is the low cost of computer time. Especially for medical applications, this is a very important feature. This opinion is confirmed by other authors (5), (14).

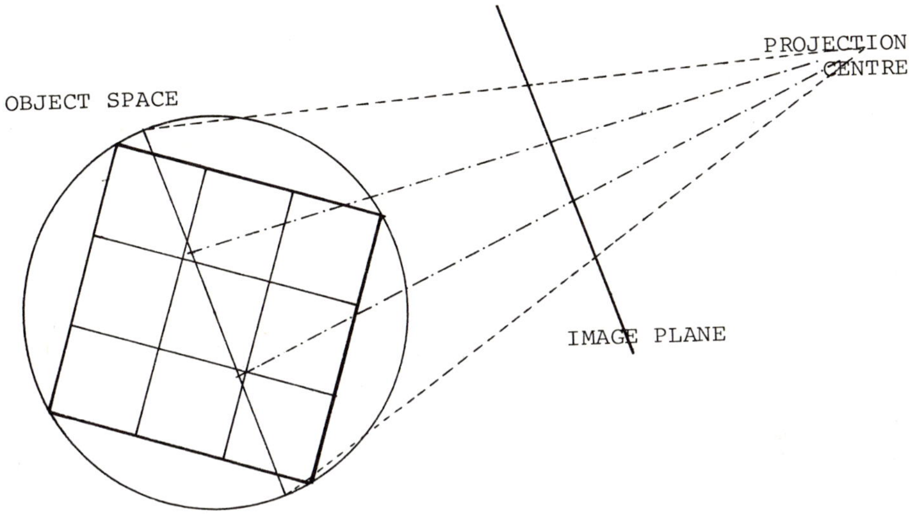

Fig.2. The 2-D to 1-D perspective projection and the calculation of the image space discretisation.

HIDDEN SURFACE ALGORITHM

The fast calculation of the hidden surface points and their removal is essential for the medical application of our display system. There exist some techniques for this hidden surface problem. One of the simplest algorithm is the depth-buffer or z-buffer algorithm (12). In this algorithm two arrays for the intensity and for the depth or distance to the observer are used. For each surface point we have to calculate the shading and the distance. The great disadvantage of this technique is the use of two arrays as an image of 500 x 500 pixels requires 250.000 storage locations for one array. Our consideration was the following:

if the observer is in the fist octant of our coordinate system in the universe, a projection of the surface elements in the way of increasing coordinates will cover those elements which are hidden. Instead of calculating the depth of each surface point, we simply sort the surface elements. The sorting of the surface elements need not be changed, if the observer moves his position in an octant.

But the great advantage of our hidden surface algorithm is that it needs only one array for the display. (see Fig. 3). In our hardware configuration we use the image memory of the display system. In this way, the algorithm can be implemented on a minicomputer with a small display system.

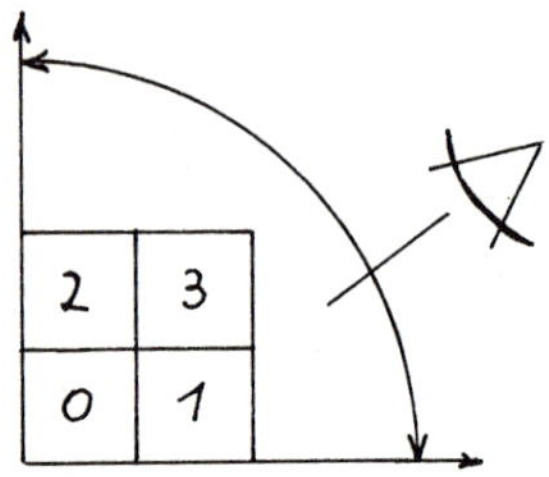

Fig.3. 2-D hidden line sequence. For a viewer located between 0 degree and 90 degree, a display sequence of 0,1,2,3, will generate a 2-D hidden line view.

EXAMPLES

For illustration of our results we have chosen a cranial CT-Scene. It consists of 10 slices, each slice with a thickness of 5 mm. Our aim was to display the objects skull and lesion (see Fig. 5). The first step of our scene processing was the interpolation of subslices to get a CT-Scene of cubic voxels. In the second step we segmented the scene. The resulting binary scene consists of 160 x 160 x 46 elements and the detected surface of our object has about 8000 elements. The tree-dimensional displays show the skull and the lesion in shaded perspective views from different directions. (see Fig. 6 and Fig. 7).

DISCUSSION AND OUTLOOK

We have described problems and solutions associated with the detection and display of surfaces of complex objects in a discrete three-dimensional space. We have tested the methods in a specivic application area: organ surfaces from computed tomograms. There exists some interest among physicians in the tree-dimensional organ surface display, especially for instruction and for planning a surgery.

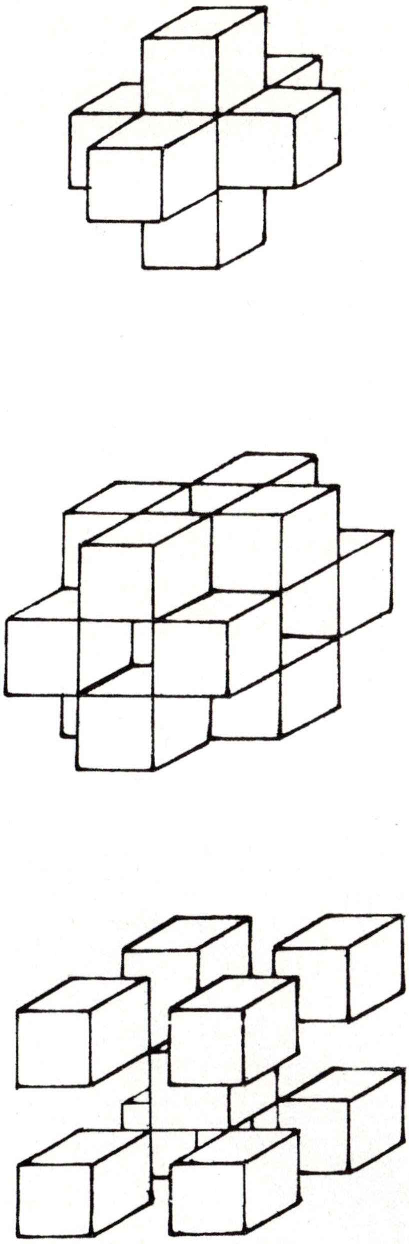

a)

b)

c)

Figure 4. The geometric appearance of the three possible adjacencies in R^3:

a) 1-adjacent,

b) 2-adjacent,

c) 3-adjacent.

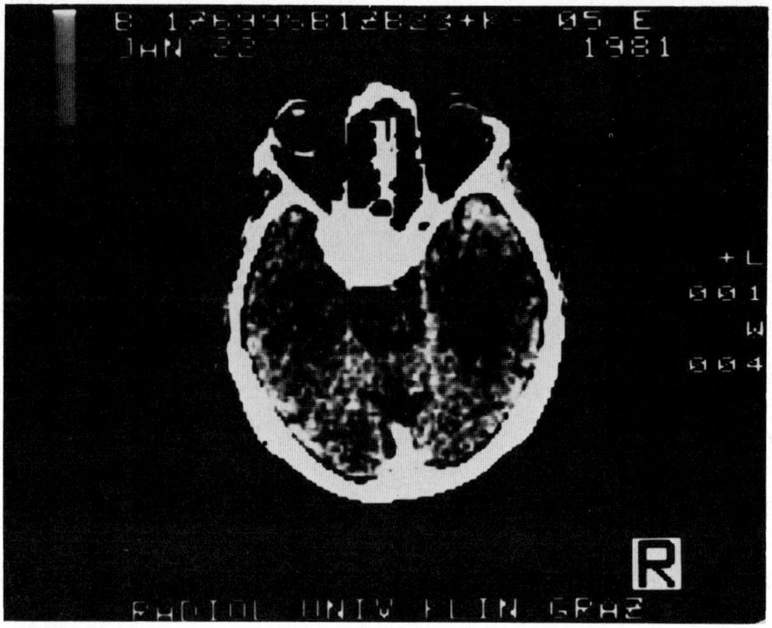

Figure 5. Slice 30 out of the 46 1.0 mm thick slices of a head scann. This slice is produced bz an EMI CT1010 scanner.

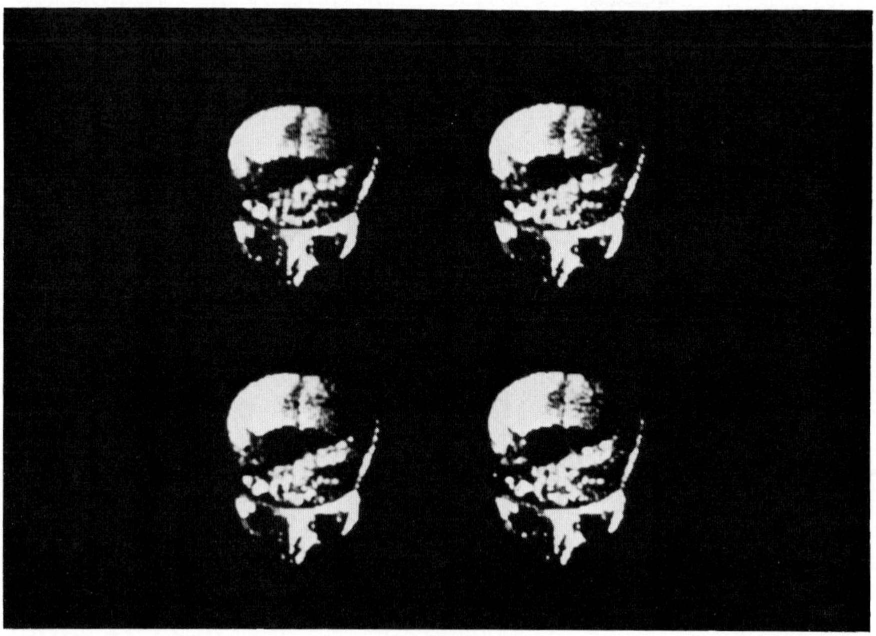

Figure 6. Four views of the detected surface of the skull and lesion as displayed by our program.

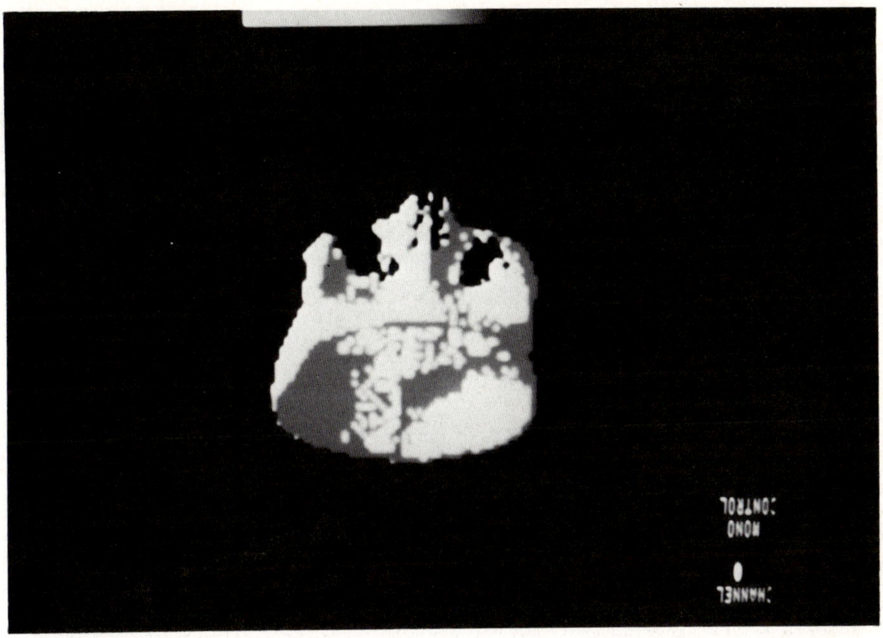

Figure 7. A single view of the detected surface of the skull and lesion. This picture is zoomed with the factor 2 by linear interpolation.

Nevertheless, the medical usefulness of this object representation is not as yet proven or generally accepted.

In the next step we want to implement a new efficient data storage structure for the surface elements, namely the octree representation (15). There are several advantages of this data structure. One of them is that all objects are kept spatially pre-sorted at all times. Thus the hidden-surface display algorithm itself requires no sorting. The trees representing the objects to be displayed are simply traversed in a specific order, depending on the view direction.

ACKNOWLEDGEMENTS

The research reported in this paper is supported by the Fonds zur Foerderung der wissenschafltichen Forschung. We are gratefurl to Mr. Gell for his many interesting comments and helpful discussions and for supplying the CT-data for our example. Thanks also to Mr. Ranzinger for his comments on an early draft.

APPENDIX

Adjacenty and Convectivity in n-Dimensional Discrete Space (16)

A point p in a n-dimensional space R^n is an n-tuple of integers (pl,..,pn). Two points p and p' are n-adjacent if and only if

$$\text{(i)} \quad 0 \leq |pi - pi'| \leq 1 \qquad i=1,...,n$$

$$\text{(ii)} \quad \sum_{i=1}^{n} | pi - pi' | = n$$

Two points are O(n)-adjacent if they are k-adjacent for all $k \leq n$. The geometric situation in R3 is shown in Fig.4.

REFERENCES

(1) M.H. Nagel, Analysis of Image Sequences: What can we have from Applications?, in HUANG. T.S.(ed): Image Sequence Analysis, Springer, 1981

(2) R.A. Brooks and G. Di Chiro, Priciples of Computer Assisted Tomograhpy (CAT) in

Radiographic and Radio-isotopic Imaging, Phys. Med. Bid. 21, (1976), 689-732

(3) A. Sunguroff and D. Greenberg, Computer Generated Images for Medical Applications, SIGGRAPH '78, Proceedings, 196 - 202

(4) H. Fuchs, Z.M. Kedem and S.P. Uselton, Optimal Surface Reconstruction from Planar Contour, Comm. ACM (1977), 693-702

(5) G.T.Herman, Three-Dimensional Imaging from Tomograms, Digital Image Processing in Medicine, Proceedings, Hamburg, October 1981, Springer

(6) J.K. Upuda, Interactive Segmentation and Boundary Surface Formation of 3-D Digital Images, Computer Graphics and Image Processing, 213-235, Vol 18, 1982

(7) H. Oswald, Merkmalserkennung in Dreidimensionalen Szenen der Computer Tomographie, 219-233, OEAGM-Tagung 1981, OECG-Proceedings

(8) H. Oswald, Segmentieren von Dreidimensionalen Computertomographischen Szenen und raeumliche Darstellung der Ergebnisse, DIBAG-Publication Nr.3, 1981, FZG, Graz, AUSTRIA

(9) W. Kropatsch, Segmentation of Digital Images Using A Priori Information about the Expected Image Content, Interval Report, 1982, FZG, GRAZ, AUSTRIA

(10) A. Requicha and H. Voelcker, Geometric Modeling of Mechanical Parts and Machining Processes, COMPCONTROL 1979, Sopron, Hungary.

(11) A. Requicha, Representations for Rigid Solids: Theory, Methods, and Systems, comput. Surveys 12, No.4, December 1980

(12) W.H.Newman and R.F.Sproull, Principles of Interactive Computer Graphics, International Student Edition, 1979.

(13) H. Freeman and J.A. Saghri,Comperative Analysis of Line-Drawing Modeling Schemes, Computer Graphics and Image Processing, 203-223, Vol 12, 1980.

(14) M.S.Stiehl, Automatische Verarbeitung and Analyse von kranialen Computer-Tomogrammen, Berlin 1980

(15) D.Meagher, Geometric Modeling Using Octree Encoding,

Computer Graphics and Image Processing, 129-147, Vo. 19, 1982

(16) J.K. Upuda et al., Boundary Detection in Multidimensions, IEEE Trans. on Pattern Anal. and Machine Anal. and Machine Itell., PAMI-4, Nol, 1982

MODEL BASED ANALYSIS OF SCINTIGRAPHIC IMAGE SEQUENCES OF THE HUMAN HEART

H. Bunke, G. Sagerer, H. Niemann
Lehrstuhl für Informatik 5 (Mustererkennung)
Universität Erlangen-Nürnberg
Martensstr. 3
D-8520 Erlangen
F. R. Germany

Abstract: A system for the automatic analysis of scintigraphic image sequences of the human heart is described in this paper. The aim of the system is the automatic detection of motion abnormalities. The system has a modular structure consisting of four components, namely 'Methods', 'Model', 'Instances', and 'Control'. The Module 'Methods' comprises several routines for image preprocessing and segmentation. The knowledge needed for analyzing an image sequence is contained in the module 'Model'. The model is represented by means of a semantic network. Instances of model concepts are kept separately from the model. Each instance has the same structure as its corresponding concept. The 'Control' module has two principal concerns. First, it provides some means for communication between user and system. Secondly, it controls the analysis of an image sequence based on the model creating instances of model concepts.

1. Introduction

A system for the automatic analysis of scintigraphic image sequences of the human heart, which is presently under development, is described in this paper. Since they are noninvasive, scintigraphic methods play an important role in medicine today. A radionuclide is injected into the patients bloodstream and a scintillation camera is used for acquiring an image sequence. In our system, we are concerned with gated cardiac blood pool studies which show the interior space of the ventricles /Fig.1/. The visual data are provided as follows: After equal distribution of the radionuclide Tc-99m in the blood pool, the ECG is used for triggering the scintillation camera. The time intervall between two R-waves of the ECG is divided into n segments ($12 \leq n \leq 64$) of the same length and one image is taken in each such segment. In order to get a sufficient number of gamma particles per image, the image matrices of about 500 cycles are synchronously added. This procedure leads to a representative cycle of the beating heart.

Image sequences obtained in this way are used in medicine to asses the motility of the heart. The main applications are in diagnosis and in periodical supervision of patients after heart surgeries. By means of such image sequences, one is able to detect location and gravity of abnormalities of the heart, especially of the left ventricle, which is the most important part.

NATO ASI Series, Vol. F2
Image Sequence Processing and Dynamic Scene Analysis
Edited by T.S. Huang

The aim of our development is a system for both quantitative and qualitative automatic analysis of scintigraphic image sequences. The system is designed model driven. Systems described in literature for left ventricle wall motion analysis are designed for X-ray images. There exist qualitative approaches /1/ based on artificial intelligence methods as well as quantitative approaches as discussed in /2/. In our approach, we want to combine quantitative procedures, special procedures developed for gated cardiac blood pool studies like /3/, and methods of computer vision systems to give a description of the dynamic behaviour of the heart, especially the left ventricle.

2. System Overview

The overall system design /Fig.2/ is based on the concept proposed in /4/. The solid lines in Fig.2 represent control and data flow, while the dashed lines stand for links between data structures. The system is modularily built up consisting of the components 'Methods', 'Model', 'Instances', and 'Control'. In this section a short overview of these four components and their tasks is given.

The module 'Methods' comprises several submodules for image processing and extraction of basic model components from an image sequence. There are different classes of methods. Besides methods for smoothing, contrast enhencement, and contour extraction, there are special methods developed for left ventricular wall motion analysis, like ventricle segmentation and volume curve calculation.

The module 'Model' contains knowledge about the structural properties of the patterns - in our case, knowledge about the structure of the human heart as it occurs on the images and about the motion in one heart cycle. The representation formalism has its roots in semantic network theory /5/. The basic structures which built up the net are named concepts and relations. The slots of each concept are distributed into several groups, which describe the structural properties and refer to parts and associates, like specializations and similarities. Furthermore, the model contains prognosis about location and motion properties of different objects. The knowledge itself is divided into several hierarchies. A detailed description is given in section 4.

An other module of the system contains the 'Instances' of various concepts of the model. Each instance has the same structure as the corresponding concept and is hierarchically built up from data received from the image sequence to be analyzed.

The module 'Control' has two principle concerns. First, it guides the interaction between user and system. Secondly, it controls the analysis of an image sequence based on the knowledge stored in the model creating actual instances of model concepts and relations from the actual image sequence. Using model and control as in our system is similar to other approaches to computer vision systems known from literature, e.g. /6/.

A detailed description of these four components is given in the next sections.

3. Methods

3.1 General Remarks

All image processing routines for enhancement and segmentation with the goal of extracting components which are primitive to the model are combined in the system module 'Methods'. An important model primitive is the left ventricle contour on each image of the sequence. Since we are interested in detecting motion abnormalities not only globally, i.e. for the left ventricle as whole, but also on a local basis, the left ventricle is split up into sectors. These sectors are primitive to the model, too. Therefore, image enhencement and segmentation routines are directed to extracting the left ventricle and splitting it into sectors.

Input to the system is a sequence of images obtained by a scintillation camera operating under a GAMMA 11 - system on a PDP 11 computer. The images are taken under a left anterior projection of 45 degrees. An image sequence shows one representative cycle of the human heart. Presently, the spatial resolution is 64 x 64 picture points with 8 bit intensity resolution per pixel; one heart cycle is represented by a sequence of 12 up to 32 frames. For the extraction, the following sequence of operations is carried out: smoothing, thresholding, left ventricle extraction, left ventricle smoothing, splitting up the left ventricle into sectors.

3.2 Image Smoothing

After image acquisition, local preprocessing operations are carried out for both improving the subjective image quality and enhancing the histogramm as further aid to finding a threshold for separating the heart from the background. These operations are median smoothing and extremum sharpening as defined in /7/. For median smoothing, the intensity at pixel p is replaced by the median in a local neighborhood, while for extremum sharpening it is substituted by either the minimum or maximum intensity in a local neighborhood, dependent whichever of two extremum intensities is closer to the value of p. Experiments have shown that a window size of 7 x 7 in both median smoothing and extremum sharpening yields best results. A sequence of

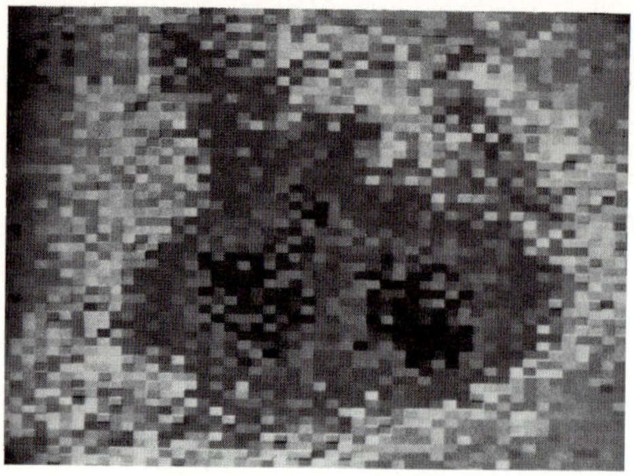

Fig. 1a.
Original image: 45 left anterior projection showing the enddiastolic heart (the first image in the sequence)

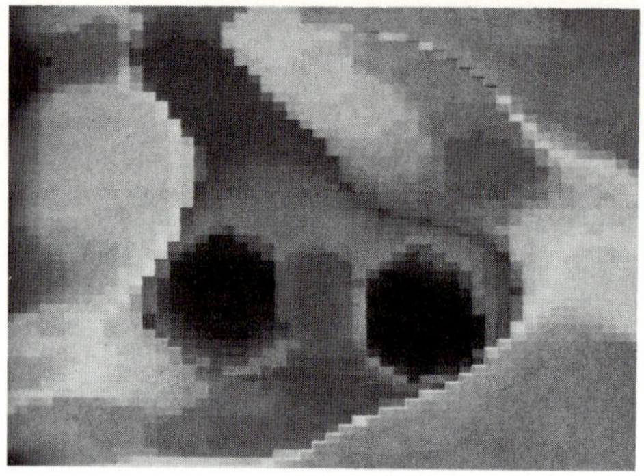

Fig. 1b.
The image in Fig.1a after filtering: 7x7 median smoothing, followed by 7x7 extremum sharpening, followed by 7x7 median smoothing

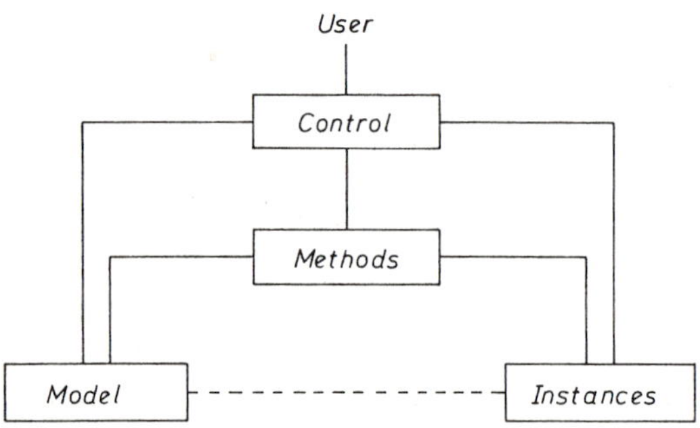

Fig. 2. System Overview

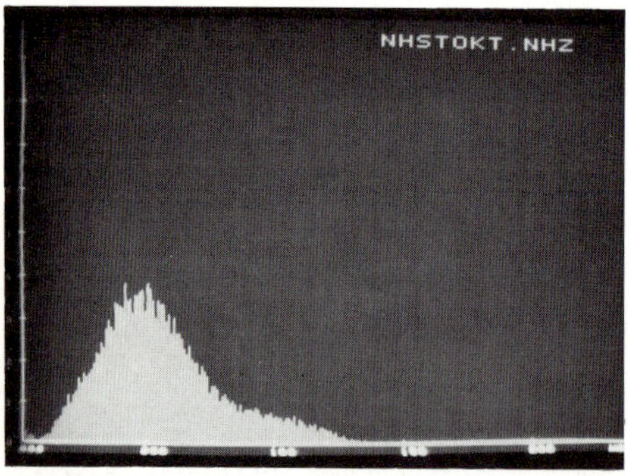

Fig. 3a.
Intensity histogram corresponding to Fig. 1a

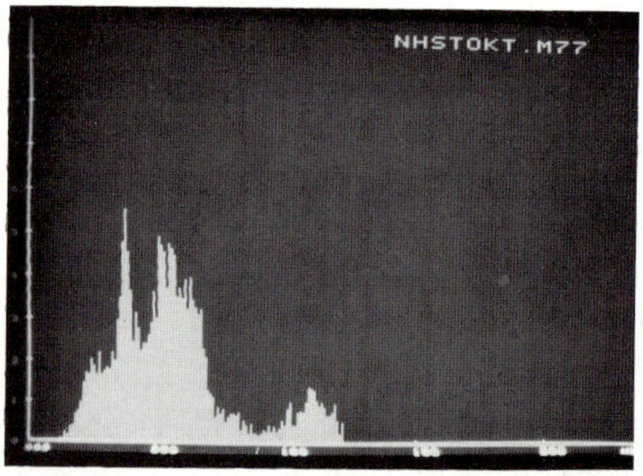

Fig. 3b.
Intensity histogram corresponding to Fig. 1b

first median smoothing, followed by extremum sharpening, followed by a second median smoothing has turned out to perform best on our images. An image and its filtered version is shown in Fig.1. The corresponding histograms are shown in Fig.3. A fast implementation of median smoothing according to /8/ has resulted in a speed up of ten times (900 s vs. 9000 s for 7 x 7 median smoothing of an image sequence of 12 frames on a PDP 11/34).

3.3 Thresholding

After filtering, thresholding is performed in order to separate the heart from the background. Our approach to thresholding is based on the assumption that both the heart and the background shows a normal intensity distribution. As a consequence, the point of intersection between those two distributions is chosen as threshold. This idea is due to /9/. In contrast with /9/, however, where local thresholds in image subwindows are computed, we determine one global threshold from the histogram of a whole image sequence. This saves considerable computing time and has worked sufficiently well for our application so far. After histogram smoothing, the deepest valley is taken as initial point of separation and both the left and the right part of the histogram is approximated by a normal distribution. A hill climbing method is then applied for finding a leastsquare fit of the sum of the two normal distributions to the histogram.

For an example, look at Fig.3b. The threshold obtained by the method sketched above is right in the valley between the two modes. This threshold is suitable although there is a significant notch in the left (background) mode of the histogram.

3.4 Locating the Left Ventricle

By means of thresholding, the heart is separated from the background. Afterwards, the goal is to separate the left ventricle from the rest of the hearts. For this purpose, we track the septum, which appears on our images as a valley between the left and the right ventricle. In detail, the process of separating the left ventricle consists of three steps:

1) Find a suitable starting point.
2) Track the septum valley in both directions north-northeast and south-southeast until the outer contour of the heart is reached.
3) Refine the result by means of backward tracking.

A starting point is found by convolving the image with the mask shown in Fig.4. This mask has been designed to match the intensity function of the septum. Only the heart area is taken into account for the convolution. Starting point is the point with maximal response to the mask in Fig.4. Since the septum separating the left from the right ventricle is slightly bended while the mask in Fig.4 is adapted to a straight vertical valley, a point in the middle, i.e. between the southernmost and northernmost point of the septum, is likeley to be choosen as starting point.

The scheme for tracking the septum in direction north-northeast is shown in Fig.5a. When point z is arrived, we move in the next step either to point x or y dependent which one has a lower intensity. If both intensities are equal, point y is chosen. Fig.5b illustrates tracking in direction south-southeast. Tracking in either direction stops when a contour point of the heart has been reached. In order to improve the robustness of locating the septum with respect to the starting point, backward tracking along the septum is performed. Backward tracking starts at the northern and southern point of intersection between septum contour and outer heart contour, respectively. The tracking scheme is analogous to Fig.5a and b. Backward tracking stops when both directions south-southwest and north-northeast meet each other. The result of locating the left ventricle when the above procedure is applied after thresholding is shown in Fig.6.

3.5 Smooting of the Left Ventricle

Due to thresholding and septum tracking as described above, the contour of the left ventricle is usually ragged. Since we are interested in analyzing left ventricular motion in terms of the area of the left ventricle and its sectors, protruding points as well as gaps in the contour need to be smoothed out. For this purpose, we apply expand and shrink operations as proposed in /10/. When expanding an object, we change a background point p into an object point, i.e. we change its gray level from 0 into 1, if the distance between p and the object is equal or less than a threshold d. Conversely, shrinking an object is done by changing an object point q into a background point if the distance between q and the background is less than a threshold d'. In our application, we work with thresholds d=d'=1. A greater value is not feasible due to the rather coarse spatial resolution of 64 x 64 pixels. The overall sequence of operations is first EXPAND followed by SHRINK for gap filling; afterwards, we apply SHRINK followed by EXPAND for removing protruding points.

3.6 Segmentation of the Left Ventricle

As last operation for model primitive extraction, the left ventricle is split up

$$\begin{matrix} 2 & 1 & -6 & 1 & 2 \\ 2 & 1 & -6 & 1 & 2 \\ 2 & 1 & -6 & 1 & 2 \\ 2 & 1 & -6 & 1 & 2 \\ 2 & 1 & -6 & 1 & 2 \end{matrix}$$

Fig. 4:
Convolution mask for finding a starting point for septum tracking

a) $\begin{matrix} x & y \\ z & \end{matrix}$ b) $\begin{matrix} z & \\ x & y \end{matrix}$

Fig. 5:
a) Tracking scheme for direction north-northeast
b) Tracking scheme for direction south-southeast

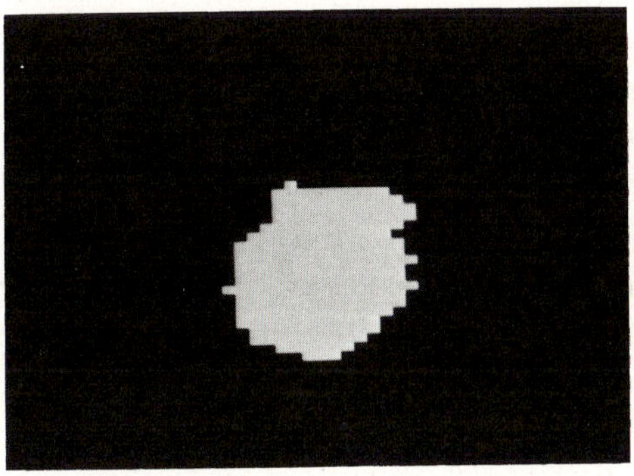

Fig. 6.
A binary image showing the left ventricle

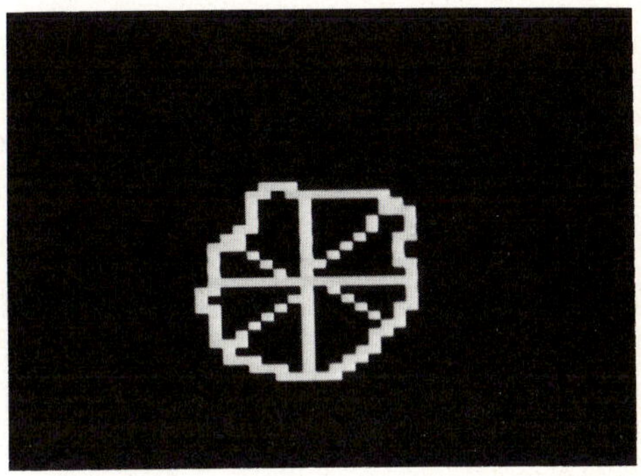

Fig. 7.
Result of splitting the left ventricle into 8 sectors

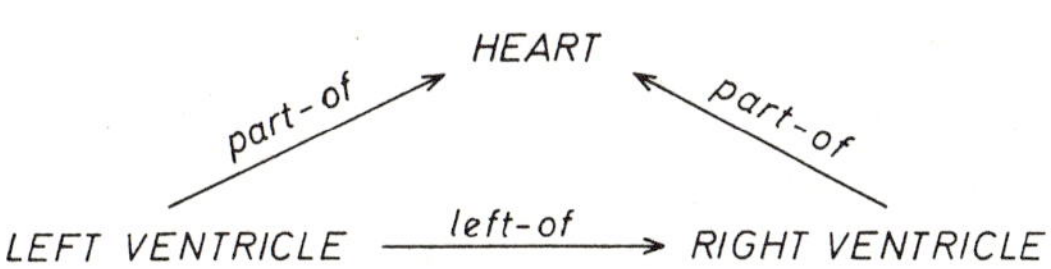

Fig. 8. A simple network

into equal angular sectors. The center of gravity of the left ventricle on the end-diastolic image, i.e. the first image in the sequence, is taken as reference point for each image in the sequence. This is in accordance to /2/, for example. Since we are experimenting with different model parameters, our present implementation allows a flexible number of sectors. The result of splitting the left ventricle in Fig.6 into 8 Sectors is shown in Fig.7.

4. Model and Instances

4.1 Organization of the Model

The representation formalism for the model has its roots in semantic network theory /5/. Semantic networks are labeled directed graphs where the nodes represent descriptions of objects and the links relationships between these objects. The graph in Fig.8 is a representation of the fact that the left and the right ventricle are adjacent and that both are parts of the heart. There are two different types of vertices in this example, namely 'left-of' and 'part-of'. While the first of these two relationships is a user defined one, the relation 'part-of' is a standard relation built in each network. This means 'part-of' is a standard component in the structure representing a node. This basic structure is also called 'concept'. The second basic structure of our representation formalism is called 'relation', for example 'left-of'. Both structures consist not only of structural but also of procedural information. This approach is similar to the procedural semantics for semantic networks described in /11/.

According to our task, there are concepts for heart diseases to be detected, for organs and their parts occuring on the images, and for motion the heart and its parts can have. Each concept has an arbitrary but fixed number of slots, which are distributed in several groups. Each slot has an associated type that refers to another concept, to a relation, or to a simple type like an integer value. In the following, the syntax of a concept and the interpretation of its different slots is described.

A concept is composed of the following slots, respectively groups of slots.
Metainformation: This slot contains the name of the concept and an arbitrary text.
Instances: In order to simplify the management of the actual instances of a concept during the interpretation an image sequence, there exists a pointer in each concept to the first element of a linked list of instances of the concept.
Generalizations: Here, concepts are refered to of which the defined one is a specialization.
Specializations: This is the inverse relation of generalization. The specialization

of a concept consists of all the semantic components (see below) which the general concept has and eventually additional ones. The types of those slots can be specializations of the types refered in the more general concept, i.e. by restrictions, a new concept is built up. For example, a detected object can be specialized to a left ventricle if it satisfies restrictions concerning form, location, and size. Both slots generalizations and specializations are organized as linked lists.

Argument-of: Relations with the defined concept as argument are linked by a list and refered in this slot.

Similarities: These links connect the concept to others which are similar in the following sense: if the instantiation of a concept fails because of unsatisfied restrictions, this slot refers to concepts which are possibly the correct ones. Besides structural information, also procedural information how to transfer single entities between similar concepts is provided in this slot.

Semantic-components: This group contains the semantic definition of a concept. It is divided into three subgroups each consisting of an arbitrary but fixed number of named slots. Each such slot has an associated type, and, besides, procedural information to select methods which can be helpful for calculations. Furthermore, each slot refers to a default instance of the corresponding type. The three groups are:

Parts: These slots refer to those concepts which are parts of the defined concept, i.e. are prerequisities for creating an instance. The members of this group are independent from each other. For example, parts of the left ventricle are the sectors shown in Fig.7.

Attributes: These slots represent the 'value' of the defined concept, e.g. the contour of an object, the characterization of a motion and so on. They depend in general on the parts or on other attributes of the concept.

Structures: Relations, i.e. conditions on parts or attributes are collected in this group. They form the structural description of a concept dependent on the parts and attributes. In a concept for the heart, the relation 'left-of' with left and right ventricle as arguments is such a structure, for example.

Part-of: This linked list represents the inverse relation to the semantic components.

The structure relation is built up like the structure concept. It has a slot 'meta-information', and, analogously to the slot 'instances', a slot named propositions. Besides, there are named pointers to those concepts which are arguments, and to a procedure which tests whether the arguments satisfy the relation.

4.2 Implementation of the Model

The syntax of the representation formalism as described above has been implemented as a macro library. This library has been written using the macro processor described in /12/ which allows text replacement with parameters. The host language of this macro processor is RATFOR /12/, a preprocessor for FORTRAN.

For obtaining the internal model representation of the network on our computer, the following steps are performed. First, the network written down according to a special syntax is transformed into a RATFOR program by means of the macro library. Second, a FORTRAN program is generated by the RATFOR preprocessor. Finally, running the FORTRAN program produces the desired internal model representation.

4.3 Knowledge Represented in the Model

The different concepts in the model are ordered along the two main hierarchies specialization and part-of. The knowledge necessary to analyze image sequences of the human heart can be divided into three main components:
1) Concepts for objects and their parts occuring on the images.
2) Concepts for motion description of an object by means of its appearence on one or more images in a sequence. With respect to our problem domain, it is sufficient to look at motion as change in area.
3) Concepts for the interpretation of the form of an object and its motional behaviour from a medical point of view.

The hierarchy 'part-of' is mainly built up by lines, objects, collection of an object through all images, motion, and diagnosis. Specializations of the different levels of this hierarchy lead to anatomical interpretation and to different judgements. Fig.9 shows a representation of these two hierarchies.

4.4 The Concept Left Ventricle

As an example for a concept in the model, the left ventricle (LV) is described in this section. This concept belongs to the group of anatomical objects, see Fig.9. Therefore, LV is a special object. It is a part of the collection of left ventricles in the whole image sequence and a part of the heart.Possible judgements from the medical point of view are the conclusions that the observed LV is too great in volume or that it has normal properties. As mentioned above, the LV is argument together with the right ventricle of the relation LEFT-OF. 'Parts' inside the group 'semantic components' are the different sectors (cf. section 3.6). These sectors must satisfy the 'structure' by means of relation ADJACENT. Attributes like location, contour, and area are common to all objects and therefore defined in the ge-

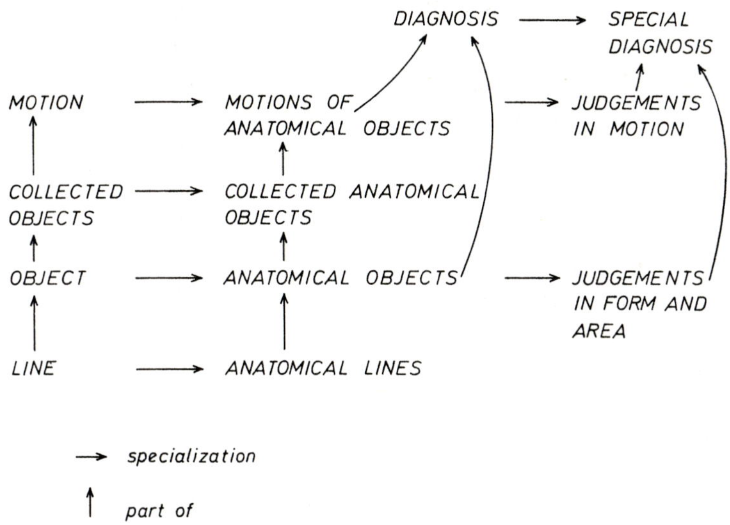

Fig. 9. Main hierarchies in the model

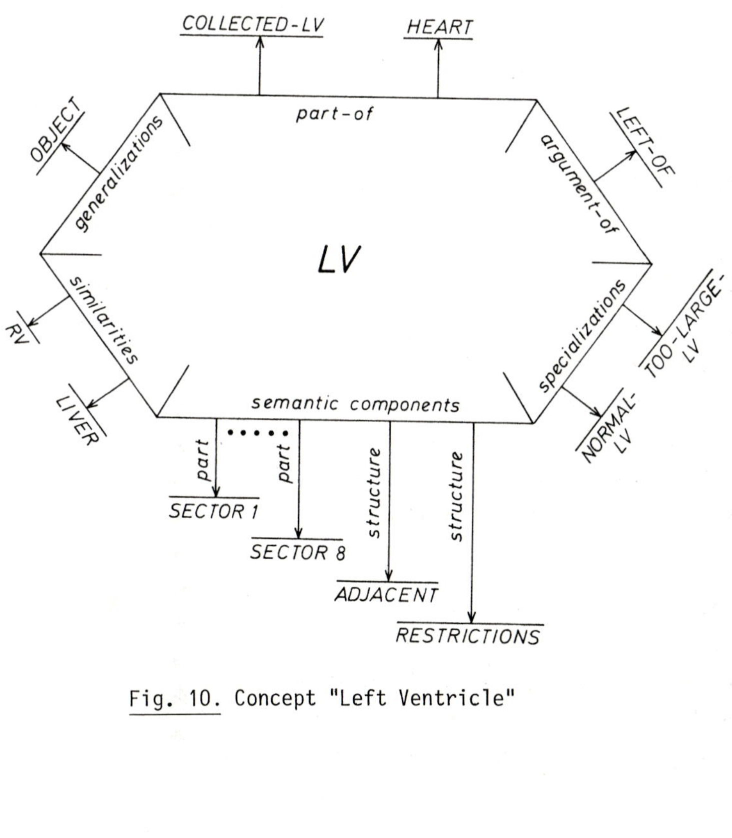

Fig. 10. Concept "Left Ventricle"

neral concept OBJECT. In the concept LV, these attributes are restricted by a 'structure', named RESTRICTIONS. This relation makes the LV distinguishable from other objects similar to it in form and area, like the right ventricle (RV) and the liver. Fig.10 shows the concept LV with its outgoing links.

5. Control

In this section, an outline of the control module, which is presently under development, is given. This module has two principal concerns. First, it guides the interactions between user, i.e. physician, and system. Secondly, it controls the analysis of an image sequence based on the knowledge stored in the model.

5.1 General Concerns

The possible interactions between the user and the system are controlled through a menue. A list of goals is displayed to the user who may choose one of these goals. Examples of goals are diseases like 'HYPOKINESIS' or 'INFERO-APICAL DISKINESIS'. Generally speaking, each goal selectable from the menue is the name of a concept which is part of the model. After the goal has been choosen by the user, the systems task is to verify, resp. falsify that concept. I.e. the task is to create instances of the goal concept. When such instances have been created, we are interested, additionally, in creating instances of specializations of that concept. Consider the concept 'DIAGNOSIS', for example. When an instance of that concept has been created, we are additionally interested which special diagnosis, e.g. HYPOKINESIS (which means restricted motility), or AKINESIS (which means no motility) can be concluded from the image sequence. The process of instantiating specializations iterates several times since there are again specializations of concepts like HYPOKINESIS, e.g. INFERO-APICAL HYPOKINESIS (hypokinesis in the infero-apical part of the heart). In this paper the concept name DIAGNOSIS is used in a rather informal way by utilizing only information obtainable from the image sequence to be analyzed for instantiating that concept. This is in contrast with a diagnosis given by a physician where additional knowledge, like the patients personal history, is utilized.

The model in our system is built up on the basis of the two relations 'part-of' and 'specialization'. Since they are independent from each other, we have two orthogonal views of the network, the first involving only the relation 'part-of', and the second based on the relation 'specialization' only. As we have seen in the last paragraph, exploiting the network is done in a top-down fashion with respect to the relation 'specialization', i.e. after a concept has been instantiated, its specializations are candidates for instantiation. Afterwards, specializations of the specializations are to be instantiated, and so on. Similarily, we proceed top-down for evaluating the 'part-of' relations in the network. Top-down processing is feasible here since the goal concept choosen by the user acts as a constraint restricting the search to that part of the network which is relevant with respect to the goal concept. Focusing on a part of the network is similar to the sketchmap approach introduced in /13/.

5.2 Outline of the Algorithm

Generally speaking, the task of the control is to generate instances of those concepts which belong to the relevant part of the network defined by the goal concept. Our method for doing this is basically heuristic search according to /14/ where the search space consists of (concept,instance)-pairs. In other words, the control algorithm is concerned with assigning instances to concepts in such a way that a global cost function is minimized.

In addition to model and instances, we need two data structures, namely SEARCH-SPACE and EXPANDED-MODEL. The SEARCH-SPACE is for storing the complete history of alternatives in creating instances for concepts. In the EXPANDED-MODEL, we keep record what parts of the model have been exploited at a certain time. This means that each state, i.e. node, in the SEARCH-SPACE has associated an EXPANDED-MODEL showing that part of the network actually evaluated so far. In expanding the model, we start with the goal concept and successively augment the EXPANDED-MODEL by concepts which are 'part-of' or 'specialization' of the concept focused on. Our approach to control is a simplified version of the strategy followed in /6/, where both top-down and bottom-up exploitation of the network is provided.

An outline of the control algorithm is given in Fig.11. Following are comments on that algorithm. Initialization of the SEARCH-SPACE is done by creating its root node. The initial EXPANDED-MODEL consists of the goal concept selected by the user. Choosing an ACTUAL-CONCEPT from the EXPANDED-MODEL means selecting a bottom node which has not yet been focused on. If an instance of that ACTUAL-CONCEPT already exists, another ACTUAL-CONCEPT is choosen next. Having selected an ACTUAL-CONCEPT which is not yet instantiated, three cases are to be distinguished.

In case 1, the ACTUAL-CONCEPT is primitive. Methods described in section 3 are called for creating instances of that primitive concept. In general, there exist several alternative instances for one concept. Therefore, after linking all instances to the model, the SEARCH-SPACE is to be expanded by creating one successor of the cur-

```
begin
   initialize SEARCH-SPACE;
   initialize EXPANDED MODEL;
   while (not all concepts in EXPANDED-MODEL
         have been focused on)
   do choose ACTUAL-CONCEPT from EXPANDED-MODEL;
      if instances of ACTUAL-CONCEPT don't exist
      then if ACTUAL-CONCEPT is primitive
           then call methods for creating instances
                of primitive concepts;
                link instances to model;
                expand SEARCH-SPACE:
                pick up best state in SEARCH-SPACE;
           if ACTUAL-CONCEPT isn't primitive and all
              parts of ACTUAL-CONCEPT are instantiated
           then create instances of ACTUAL-CONCEPT;
                link instances to model;
                expand SEARCH-SPACE;
                pick up best state in SEARCH SPACE;
           if ACTUAL-CONCEPT isn't primitive and not
              all parts of ACTUAL-CONCEPT are instantiated
           then expand model;
   od;
end
```

Fig. 11.
Outline of the control algorithm

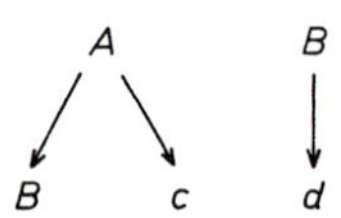

Fig. 12.
Two concepts and their parts

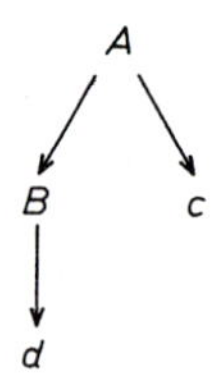

Fig. 13.
The EXPANDED MODEL after concept B has been expanded

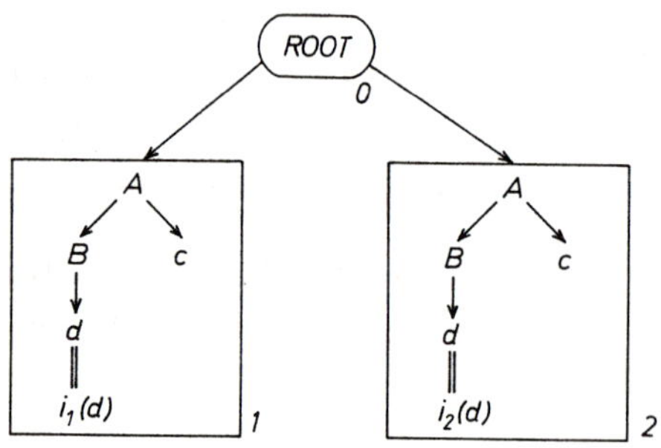

Fig. 14.
SEARCH-SPACE after instantiation of concept d

rent state for each alternative instance. An example of a concept possibly having several alternative instances is 'ENDSYSTOLE' (moment of maximal contraction) which may vary dependent on the particular sequence.

In case 2, instances of the ACTUAL-CONCEPT can be created since all its parts are instantiated. The following operations are analogous to case 1.

In case 3, finally, the ACTUAL-CONCEPT cannot be instantiated since some of its parts are uninstantiated. These parts are to be instantiated before an instance of the ACTUAL-CONCEPT can be created.

A very simple example is given to illustrate the algorithm in Fig.11. In Fig.12 two simple concepts are defined. Capital letters refer to nonprimitive concepts, while small letters denote primitive concepts. The arrows stand for the relation 'part-of'. No specializations are considered in this example. Let A be goal concept. The EXPANDED-MODEL after expansion of concept B is shown in Fig.13.Let i1(d) and i2(d) denote concurrent instances of concept d. Fig.14 shows the SEARCH-SPACE after instantiation of primitive concept d. Assuming $i_1(d)$ being the better alternative, n successors of state 1 will be created, one for each concurrent instance of the next concept a.s.o.

5.3 Cost of Search

According to /14/, the SEARCH-SPACE state with lowest cost is to be picked up for further expansion. Equivalently, we use the notion of certanity factor (cf) in our approach. High costs correspond to low cf and vice versa. The cf of a SEARCH-SPACE state is built up hierarchally on the basis of primitive concepts.

On the primitive concept level, a cf is used for assessing the confidence of an instance of a primitive concept. For that reason, we allow conditions, i.e. structures on primitive concepts, respectively their attributes. A condition may have a truth value, varying between 0 and 1. The conditions as well as their truth value, i.e. the cf, are part of the concept. A similar approach has been proposed in /6/.

Let nonterminal concept A consist of parts $B_1 \ldots\ldots B_n$ and let $cf(B_i)$ denote the cf of concept B_i. Let furthermore $COND_1 \ldots COND_m$ be conditions imposed on A, resp, its attributes. Then cf of concept A is given by

$$cf(A) = f(cf(B_1), \ldots, cf(B_n), cf(COND_1), \ldots, cf(COND_m))$$

where f is some function and $cf(COND_i)$ is the truth value of $COND_i$, $1 \leq i \leq m$, $0 \leq cf(COND_i) \leq 1$. Functions f being implemented in our system are standard functions known from fuzzy set theory /15/, e.g. MINIMUM, MAXIMUM, or SOFT-DISJUNCTION as well as user defined functions.

Finally, for assessing the cost of a state S in the SEARCH-SPACE, we take an optimistic assessment of the cf corresponding to the goal concept in state S. 'Optimistic' means that we assume a cf equal to 1 for all primitive concepts not yet instantiated. Under this assumption we can calculate the cf of the goal concept. Look, for example, at Fig.14. There exist cfs for both instances of d. Assuming a cf equal to 1 for concept c, we can calculate the cf of concept A, which is the goal concept, in both state 1 and state 2. The cf of the goal node A in each state is taken for representing the certanity of the whole state.

Acknowledgement

The work reported in this paper is sponsored by the German Research Foundation (Deutsche Forschungsgemeinschaft DFG). The authors gratefully acknowledge the help of Prof. F. Wolf and Dr. H. Feistel (Institut und Poliklinik für Nuklearmedizin, Universität Erlangen) who provided data and contributed to this work with many hours of fruitful discussion.

Literature

[1] J.K. Tsotsos, H.D. Convey, S. Zucker: A Framework For Visual Motion Understanding, in IEEE Trans. PAMI, Vol.2, 1980, 563-573
[2] Protocol of the 2nd session of the working group 'Cardiovascular Medicine' of the German Association for Nuclear Medicine, Frankfurt/Main, 1981
[3] H. Geffers, W.E. Adam, F. Bitter, H. Sigel, H. Kampmann: Data Processing and Functional Imaging in Radionuclide Ventriculographie, in Information Processing in Medical Imaging. Proc. 5th Int. Conf., Nashville, 1978, 322-331
[4] H. Niemann: Pattern Analysis, Springer-Verlag 1981
[5] N.V. Findler (ed.): Associative Networks. New York, 1979
[6] A.R. Hanson, M.E. Riseman: A Computer System for Interpreting Scenes, in Hanson, Riseman (ed.): Computer Vision Systems, New York, 1978, 303-333
[7] J.M. Lester, J.F. Brenner, W.D. Sellers: Local Transforms for Biomedical Image Analysis, Computer Graphics and Image Processing 13, 1980, 17-30
[8] G.J. Yang, T.S. Huang: Median Filters and Their Applications to Image Processing. Technical Report TR-EE 80-1. Purdue University, 1980
[9] C.K.Chow, T. Kaneko: Automatic Boundary Detection of the Left Ventricle from Cineangiograms. Computers and Biomedical Research 5, 1972, 388-410
[10] A. Rosenfeld, J.L. Pfaltz: Distance Functions on Digital Pictures, Vol.1, 1968, 38-61
[11] H. Levesque, J. Mylopoulos: A Procedural Semantics for Semantic Networks, in /5/
[12] B.W. Kernighan, P.J. Plaugher: Software Tools, Reading, Massachussets, 1976
[13] D.H. Ballard, C.M. Brown, J.A. Feldman: An Approach to Knowledge-Directed Image Analysis, in /6/, 271-281
[14] J.N. Nilsson: Principles of Artificial Intelligence, Palo Alto, California, 1980
[15] L.A. Zadeh, K.S. Fu, K. Tanaka, M. Shimura (ed.): Fuzzy Sets and Their Application to Cognitive and Decision Processes. Academic Press, 1975

SUBJECT INDEX

Adaptation, human visual system 473-474
Adaptive coding 125, 237
Adaptive filter 447
Adaptive hybrid picture coding (AHPC) 237
Adaptive quantizer 337
Adjacenty, n-D discrete space 722
Affine mapping 15
Aggregation technique 3, 6, 7-8, 11-12, 27, 343
Air to ground video signal 275, 425
Akinesis 605, 617, 650, 736
Albedo 450
Analog image data acquisition 626
Angiocardiography 4
Angiography 609
Angle Regularity 688
Apparent depth 56
Area Feature 401
Articulated objects 67
Artificial intelligence 617
Artificial limb 4
Asynchrony 650
Asyneresis 650
Atmospheric disturbances 580
Atmospheric scales 596
Attribute 11, 733
Autocorrelation function 107-108
Autocovariance 104
Automation 294-295, 640
Autoregressive model 238
Average bending energy 688

Behavioral studies 4, 5
Binary image analysis 556
Binary Scene 712
Binocular disparity 11, 12
Biomedical images 624
Blob 337, 341, 472, 479
Block coding 319
Blockworld 11, 19
Blood dynamics 602, 611-616
Blood flow 604
Blood cell analysis 479-481
Blood cell motion 663-707
Blood circulation 4
Body surface potential map 4
Bottom-up analysis 396
Boundary surface tracking 711
Broadcast TV 197, 293

Calibrating stereo camera 19, 65
Camera movement (see also sensor motion) 135, 136
Cardiac blood pool 725
Cardiac organs 624
Cardinal splines 711
Carotid arteries 609
CCIR 203
Cell tracking 4
Center of gravity 645, 652
Centered difference 108-110
Central limit theorem 239
Central projection (see also perspective projection) 366
Centroid 47
Change analysis 26-27
Change area 26
Change detection 24-26, 665, 668, 677
Characteristic axis 647,648
Charge coupled device (CCD) 632-633
Chemotaxis 688, 705
Chrominance signals 158, 165, 184, 200, 207
Cineangiocardiography 636, 639
Cineangiography 625
Cinegastrography 626, 637-638
Cinemicrophotography 4
Circularity 688

Closed loop system 532, 542

Cloud displacement 4, 5, 598

Cluster analysis 12, 13, 339, 340-347, 433, 558

Clutter 504

CMOS technology 211

CO (Containing the origin) 366

Component coding 156, 184, 196, 208

Composite coding 156, 157-165, 184 196, 208

Composite prediction 196

Composite signal 161

Computed tomography (CT) 605, 710-722

Computer angiography 612

Computer graphic method 10

Computer-internal representation 2

Concave angle 688

Condition-action rule 667

Conditional replenishment 3, 75, 76, 101, 137, 337

Connected component analysis 67

Connected component labeling 553

Context dependent operator 634

Context free operator 634

Continuity hypothesis 11

Contour
- coding 319, 340
- detection (see also edge detection) 473-474, 625
- extraction 314, 726
- following (tracing) 319, 349, 640-644, 677
- image 318
- piecewise approximation 459

Contrast enhancement 726

Contrast medium 605, 608, 613

Control 725, 726, 736-740

Control point (see also Critical point; Interest point; Landmark; Prominent point; Token) 459

Convectivity, n-D discrete space 722

Convergence 169, 226

Convex hull 448, 454

Corner 7, 15, 17-18, 42

Corpus geniculatim laterale (CGL) 472, 476

Correspondence (see also Line Correspondence; Point Correspondence) 42, 45, 53, 61, 236, 407

Cover type mapping 4

Cranial CT scene 718

Critical point (see also Control point; Interest point; Landmark; Prominent point; Token) 457, 458

Crop yield 4, 5

Cross correlation 7, 11, 16, 49-53, 107, 215, 307, 493, 611

Curvilinear boundary 43-47

Data Structure 5, 27

Deaf 314

Densitometric method 602, 612

Diagnostic classification 602, 605, 617

Difference pictures 104

Differencing 47-49, 270

Differential geometry 7

Differential pulse-coded modulation (DPCM) 141, 158, 200, 201-213, 279, 284, 307
- implementation 211-212

Differentiation operator 637

Diffuse illumination 451, 715

Digital broadcasting system 213

Digital controller 540

Digital image data acquisition 197, 627

Digital recording 213

Digital television for studios, CCIR recommendation 203, 208

Digital video sequence store (DVS) 197

Discretization errors 714

Diskinesis 605, 617, 650, 736

Disparity 11, 12, 457

Displaced frame difference (DFD) 77, 118, 166, 169, 174, 223, 224, 236, 242, 308

Displaced prediction 177

Displaced previous frame 182

Displaced redisual difference (DRD) 242

Displacement compensation 76

Displacement estimation 12-18, 161, 249
adaptive gradient approach 235
autocorrelation function 107-108, 116-117, 220
background motion 431-434
convergence of algorithm 169, 226, 232, 236, 244
crosscorrelation method 215, 224-230, 279, 427
differential method 104-124, 215-232
edge based 148-152, 292-300
first-order approximations 12-15, 105, 219-224
Fourier method 215
hardware implementation 111-112
higher-order function approximations 15-18
Kalman filtering approach 235
maximum likelihood estimation 14
minimization approach 15, 19
multi-frame problem 19
noise effects 185-192
objects on moving background 424-436
rate of convergence of algorithm 169
recursive 77-87, 116-119, 166-182, 223-231, 235-245, 252, 297
second order approximation 16-17
transform-domain techniques 85-87, 236

Displacement prediction 428

Displacement vector field (see also Optical flow) 3, 5, 11, 12-18, 165, 266, 402
decomposition 22-24

Dissimilarity grading 26

DMA Controller 525

Dry friction 537

Dynamic behavior description 704-706

Dynamic model 533, 540, 582, 666

Dynamic mosaic model 137

Dynamic phenomenon 2

Dynamic scene 438, 519

ECG (EKG) 725

ECL technology 211

Edge 180, 293, 396
detection (see also Contour detection; Gradient; Kirsch operator; Laplacean; Mexican hat operator; Pseudo-Laplacean; Robinson operator; Sobel operator; Zero crossing) 24, 294, 314, 483-492
enhancement 637
following (see also Contour following) 294-295, 319
orientation estimation 141-145
parameters 126, 130
spatiotemporal 148, 294
statistical properties 130-133
straight 133
vector 322

Elastic object 604, 677

Elongation 688

Enhancement 553, 602, 608

Entropy 165

Error probability 309

Event 6, 10

Exponential-area transform 505

Extremum sharpening 727

Eye movement 133-135

Eye-tracking capability 135-136

Feature-based segmentation 42-47

Feature operator (see also Interest operator) 395, 401

Feature vector 617

FIFO buffer 525

Figure-ground problem 55

Finite element method 657

Focus of expansion (FOE) 21-22, 24

Four-dimensional visualization 596

Fourier amplitude 351

Fourier phase 351

Fourier spectra of TV pictures 127-129, 350-362, 437

Fourier transform 502

Frame averaging 609

Frame difference 104, 113, 308, 338

Frequency response 437, 442, 473, 483

Functional approximation 12

Functional dependency 604

Functional image 614

Gain compensation 82-85

Gastric organs 624

Gate array 211-212

Gating technique 606

Gaussian curvature 459, 460

Gaussian distribution 239

Geometric description 6, 8-9

Geosynchronous satellite 502, 585, 596

Global analysis 668, 694

GOES 585

Gradient 401, 429, 553, 558

Gradient algorithm 583

Grammar 640

Graph 11

Grayvalue corner (see also Corner) 458

Hardware 618

Heart dynamics 602, 611, 616-617, 725-740

Heart wall movement 5, 603, 616, 650, 726

Hemodynamics 614

Hidden surface algorithm 10, 717, 718

Hierarchical match 11

Histogram of interframe Fourier amplitude differences 350, 351-355

History 6, 10

Homing system 498

Homogeneous coordinates 368

Hough transform 8, 12, 13, 505, 509

Hueckel operator 489

Huffman code 340

Human body model 8

Human visual system 471, 483

Hypokinetic 617, 736

Illumination variation 76, 449

Image model 265

Incremental analysis 694

Infrared (IR) imagery 500

Innovation 571

Instance 732

Integration operator 635

Interest operator (see also feature operator) 7, 396, 414, 457

Interest point (see also Control point; Critical point; Landmark; Prominent point; Token) 418

Interframe coding 3, 75, 264, 275, 307, 337, 350

Interframe difference 3, 26

Interframe match (see also matching) 12, 18, 395, 457

Interpolation 169, 264, 270, 275, 277, 712

Intraframe coding 200-213, 275, 307, 337

Inverted pendulum 526, 532

Isodata 345

Jamming 275

Jitter 505

Jointed object 8

Just visible contrast 474

Kalman filter
- three-dimensional 563, 569-579
- two-dimensional 565-568

K-edge filtering 610

Kidney 614

Kirsch operator 401

Knowledge-based system 666

Knowledge representation 734

Labeled graph 677

Lambertian surface 451, 715

Landmark (see also Control point; Critical point; Interest point; Prominent point; Token) 617

LANDSAT 504

Laplacean 321, 460

Least-mean-square estimator 77, 80, 454

Left ventricle (LV) 624, 636, 645, 650, 727, 729, 734

Levenberg-Marquardt algorithm 19, 63, 454

Line correspondence (LC) 365-393

Line of sight 22

Linear filtering
- spatial 484
- spatiotemporal 138-139, 437-447, 563
- temporal 609

Linear predictor 202

Lip reading 315

Liquid motion 604

Locomotion analysis, cells 668

Low exposure image 603

Low-level description 6, 7

Luminance signal 158, 200

Lymphocyte 664

Macroprocessor 553

Major axis 47

Markov process 563
- three-dimensional 569-572
- two-dimensional 565-568

Mask (window) 402, 439, 636

Masking function 204-207, 213

Matched filter 512

Matching (see also Interframe match) 11, 45, 236, 395-422, 694-697

Maximally flat 159

Maximum likelihood estimate 25, 582, 587

Mean square prediction error 202, 238

Medial axis 648

Median clipping 476

Median smoothing 727

Medical image sequence 10

Metabolism 4, 5

Meta information 732

Meteorological modeling 4

Meteorological satellite images 595, 598

Meteorology 594

METEOSAT 585

Mexiacn hot operator 7, 459

Microcomputer 519, 533, 629, 630

Microtone slice 5

Minimal mapping 12

Minimum mean square error (MMSE) 202, 249

Minimum spanning tree 12

Minor axis 47

Model-based analysis 725

Model-based interpretation 25

Modulation transfer function (MTF) 484

Morphological structure 602, 615, 663, 677, 707

Mosaic 5, 12, 25, 128, 137

Motion adaptive extrapolation 275, 278, 281

Motion adaptive interpolation 275, 277

Motion-compensated interframe coding 13, 87-101, 157-196, 312

Motion compensation (see movement compensation)

Motion constraint 65

Motion description 666

Motion detection (see movement detection)

Motion pattern 604, 617

Motion understanding 666

Movement compensated prediction 184, 185

Movement compensation 3, 75-101, 113-114, 120-122, 157, 161, 196, 264, 576

Movement detector 137, 425, 668

Moving area 255

Moving background 424

Moving target indication (MTI) 509

Multicellular eukaryote 664

Myocardial stress 655

Natural language 9-10

NCO (not containing the origin) 366

Nerve cell 5

Newly exposed background 180, 242

Newton search method 229

Noise reduction (see also smoothing) 563, 634

Nonlinear temporal filtering 185

Nonrecursive filter 437, 442, 443

Nonstationarity 125, 127-137, 239

Normalized mean squared error (NMSE) 339, 347

NTSC color video signal 156
- chrominance signals (I and Q) 158, 165, 184
- component coding 156, 184, 196
- composite coding 156, 157-165, 184, 196
- colour subcarrier 159
- demodulation process 184
- luminance signal (Y) 158

Object detection (see Target detection)

Object movement 135

Object tracking (see Target tracking)

Observation model 249, 566, 570

Occluding edge element 14

Occlusion 10, 24, 53, 55-60, 67-69, 264, 401, 414, 496

Octree 722

Ophthalmoscopy 612

Optical flow (see also Displacement vector field) 3, 19, 22-23, 366

Orbit determination 250

Organ 5

Orthographic projection 9, 18, 43

PAL color video signal 200
- chrominance signals (R-Y, B-Y) 200, 207
- component coding 208
- composite coding 208
- cross color distortions 208
- DPCM 202-213
- luminance component (Y) 200

Panning 26, 136

Parallel processing 522, 532, 551

Parameter extraction 645-649

Parametric imaging 602, 612-617

Pass-region of filter 442

Path model 249

PCM hierarchy, European 201

Peano sweeping 642

Perspective projection (see also Central projection) 18, 19, 366, 716

Photogrammetry 19

Photometric stereo 448, 450

Picture element (pel) 76, 237

Picture processor 524

Piecewise linear curve approximation 682-683

Pipelining 551

Pixel-based segmentation 47-53

Planar rotation (see also Rotation, two-dimensional 13

Planning 319

Point correspondence (PC) 366, 372, 376

Point feature 401

Point spread function (PSF) 504

Polar equation 18

Polygonal approximation 677

Polygonal decomposition 677

Polygonal figure 42-43, 49

Polyhedron 9, 448

Polymorphonuclear leucocytes (PMN) 664, 706

Portatel 317

Post-processing 616

Prediction (see also Linear predictor; Previous frame prediction) 174, 307

Predictive coding 157, 201, 276, 337

Preprocessing 519, 634-644

Previous frame prediction 185

Primitive region 7

Principal curvature 458, 461

Projection (see also Orthographic projection; Perspective projection) 8, 716

Prominent point (see also Control point; Critical point; Interest point; Landmark; Token) 448, 458

Pseudocolor 710

Pseudoinverse 105

Pseudo-Laplacean 318, 321

Pseudopod 663, 694, 703

Pulse-code modulation (PCM) 200

Quantization 204, 240, 337

Radiographic image sequence 602
Radionuclid imagery 5
Random dot pattern 8
Rate of convergence 169, 226
Real-time implementation 519, 549, 618
Real-world scene 27
Recursive filter 437, 442, 443, 563, 638
Recursive operator 638
Recursive temporal filter 166
Reflectance function, analytic 448, 450
Reflectance map 448, 450
Region growing 49, 268, 713
Region merging 27
Region of interest (ROI) 612
Region of support of filter 445
Registration (see also Matching) 5, 603
Relation 733
Relational structure 11
Relaxation 11, 12
Restoration 602, 608
Rigid object 604
Robinson operator 489
Robotics 5
Root-mean-square prediction error (RMSPE) 165, 185
Rotation, two-dimensional (see also Planar rotation) 249, 256, 307, 434
Rule-based syntactic analysis 677
Rule-based system 663
Runlength coding 319, 340

Sampling 203
  spatiotemporal 139-141
Satellite images (see also Geosynchronous satellite; GOES; LANDSAT; METEOSAT) 580
Scale change 13, 498
Scaling 249, 256, 498
Scene analysis 264, 594, 663, 668, 712
Scene-domain constraint 55-56
Schema 6
Scintigraphic image sequence 725
Search 739
Segmentation (see also Edge detection; Region growing) 8, 41-54, 177, 273, 402, 433, 447, 471-481, 553, 556, 650, 652, 654, 665, 677, 712 726, 730
Semantic network 726, 732
Semantic noise 58
Semantically oriented approach 4
Sensed correspondence 369
Sensor baseline 13
Sensor motion 11, 136, 425
Shaded-surface display 711, 714
Shape 45, 581, 616, 666, 677, 688
Shape analysis, cells 677-693
Sign language 314, 315-316
Singular value decomposition (SVD) 390
Slot 732
Smoothing (see also Noise reduction) 726, 727-729
Sobel operator 318, 489
Software 619
Solid modeling 9
Space trajectory 3
Spatial image sequence 5, 710
Spatial pattern 604
Spatial-temporal interpolation 264
Spatiotemporal filtering 138-139, 437-447, 563
Spectral analysis 508
Spirals model, cloud 580, 581
Split and merge 684
State feedback controller 540
State transition 2
State vector 239, 250
Static analysis 694

Steepest descent algorithm 78, 238, 243
Stereopsis 9, 11
Stomach 624, 645, 648
Straight edges 42
Structural match (SM) 697
Structure analysis 694-704
Subjective contour 56
Sub-pixel target 502
Subtraction 608, 611
Superslice 558
Surface 714
Synoptic meteorology 596
Syntactic pattern recognition 694

Tantalum marker 617
Target detection 493, 498
Target tracking (see also Tracking) 493
Task-specific knowledge 520
Taylor expansion 15, 219, 244
Temporal filtering
  linear 609
  nonlinear 185
  recursive 166
Temporal-spatial gradient 53
Texture 475
Three-dimensional motion and structure estimation 18-24, 61-67, 365-393
  fixed-axis assumption 65-67
  Gaussian sphere (unit sphere) 366
  general rigid motion 18-24, 279-387
  nonlinear equation approach 18-19
  photometric stereo 448-455
  point correspondence (PC) 366, 372, 376
  pure rotation 21, 369-372
  straight line correspondence (LC) 365-393
  three-dimensional correspondence 378-387
  three-frame case 387-392
  uniqueness 19, 453
Three-dimensional (3-D) object description 3
Three-dimensional (3-D) rigid point configuration 18-24
Thresholding 554, 558, 729
Time lapse ratio 598
Token (see also Control point; Critical point; Interest point; Landmark; Prominent point) 45
Tomography (see also Computed tomography) 5
Top-down analysis 396
Track acquisition 502
Tracking 4, 493-500, 528, 536, 580, 668
  atmospheric disturbances 580-593
  cell 4
  target 493
Traffic monitoring 4, 10, 12
Transform coding 201, 349, 350-362
Transient response 446
Two-dimensionam motion estimation 249-263, 582

Unit sample response 438
Unitary transform 239
Unstable plant 532

Vascular imaging 602, 608-610
Velocity-matched filter 442
Ventricle (see also Left ventricle) 725
Video bus 525
Video cassette recorder 213
Video conferencing 213, 293, 317
Video telephone 213
Virtual slice 5
Visual cortex 472
Volume description 9, 62
Volume element (Voxel) 710

Weather prediction 4
Wind estimation 598
Wire-frame model 8

X-ray absorption coefficient 710

Zero crossing 323, 459, 460, 461, 473
483, 489

Zooming 26, 136, 307, 498

Z-transform 438